Flood Risk and Social Justice

So birgt denn, was wir am wenigsten vermuten,
das Wesende der Technik den möglichen Aufgang des Rettenden in sich

(Thus the coming to presence of technology harbours in itself what we least suspect,
the possible arising of the saving power)

Martin Heidegger, Die Technik und die Kehre, p. 32
(The Question Concerning Technology, p. 32)

Flood Risk and Social Justice

From Quantitative to Qualitative Flood Risk Assessment and Mitigation

---

Zoran Vojinović and Michael B. Abbott

IWA Publishing
London • New York

Published by **IWA Publishing**
Alliance House
12 Caxton Street
London SW1H 0QS, UK
Telephone: +44 (0)20 7654 5500
Fax: +44 (0)20 7654 5555
Email: publications@iwap.co.uk
Web: www.iwapublishing.com

First published 2012

British Library Cataloguing in Publication Data
A CIP catalogue record for this book is available from the British Library

ISBN 9781843393870 (Hardback)
ISBN 9781780400822 (eBook)

MIX
Paper from responsible sources
FSC® C007785

Printed by Bell & Bain Ltd., Glasgow

Contents

Chapter 22

Chapter 23

Foreword by J. Philip O'Kane

This foreword in praise of this book is addressed to staff and students in Schools of Engineering, especially those concerned with flooding, and attempts to answer three questions for the prospective reader:

Why should the reader take up the book at all?
Why should the reader engage with it now?
In what way should the book be read?

The answers are followed by a brief comment on *hydroinformatics – the pursuit of social justice by stakeholders and the concept of system.*

Why should the reader take up this book at all?

The engineering professions throughout the English speaking world[1] have set criteria for the accreditation of educational programmes in Schools of Engineering that prepare women and men for recognition as chartered engineers. In the case of Engineers Ireland[2] the criteria form a matrix of generic *programme outcomes* and *areas* that is applied as a template to all branches of engineering, including therefore that branch of water engineering which deals with floods and social justice – the subject matter of this book.

To receive accreditation, graduates of programmes must be able to demonstrate, in addition to the classical academic content of engineering, *"an understanding of the need for high ethical standards in the practice of engineering, including the responsibilities of the engineering profession towards people and the environment, the ability to work effectively as an individual, in teams and in multi-disciplinary settings together with the capacity to undertake lifelong learning, the ability to communicate effectively with the engineering community and with society at large"*. Furthermore, this understanding must also be demonstrated in application to six programme areas: *Sciences and Mathematics, Discipline-specific Technology, Information and Communications Technology, Design and Development, Engineering Practice, Social and Business Context*. We will call a graduate of such a programme a *new model engineer*.

[1]http://www.washingtonaccord.org/Washington-Accord/

[2]http://www.engineersireland.ie/services/programme-accreditaton/how-it-works/

Three quarters of the resultant matrix of criteria are new. We may interpret the new criteria as a *grand challenge*. They ask us to enrich the engineering curriculum with the human sciences so as to open the hearts and minds of new model engineers to honourable and successful practice in the cultures of the world. This educational task cannot be delegated in its entirety to University departments of Philosophy and Sociology, Government and Law, Social and Economic Science, Business and Commerce, etc. It requires, instead, a dialogue between engineering as an academic discipline and the human sciences. Academic engineers in the English-speaking world are not well prepared for this task, but only they in their world can initiate it. This unique book is a contribution to that dialogue, and is the fruit of years of thought and contact with the practice of engineering in relation to flooding and social justice. Consequently, its most important audience in the long term are those engineers in Universities who wish to rise to the grand challenge and to bring their model engineers with them, and ultimately all their academic colleagues as well.

Why should the reader engage with this book now?

This book will help raise the quality of both engineering-degree programmes and CPD courses concerned with flooding, by helping to ensure such programmes and courses meet the new accreditation criteria requested by the engineering profession; and be seen to do so. There are many books that address the technical issues of flooding, but there are far fewer books that address the social and ethical aspects of flood risk. The book fills this gap in the current literature by broadening the focus of traditional flood management practice. It offers, in one place, an overview on issues that range from the social and ethical, to the scientific and practical.

In what way should the book be read?

The present book is concerned with the socio-technology of flooding and its amelioration. It is therefore in some parts cultural, in others philosophical and theoretical, and in some other parts again simply technical and practical. It draws upon concepts beyond those of traditional engineering which will strike some readers as unfamiliar and challenging – but so too are mathematics and the physical sciences, which engineering students have always taken in their stride. The philosophical elements of the book are the biggest challenge for English speaking engineers and engineering students. Those readers who want to master the text should expand their library in the obvious way and when necessary seek the assistance of a tutor from the nearest University department of philosophy.

Hydroinformatics – the pursuit of social justice and the concept of system

At the time of writing this book there is still a common perception that hydroinformatics is exclusively concerned with computational modelling and decision support. Although hydroinformatics first emerged when Abbott (1991)[3] identified a synergetic relation between numerical modeling, data collection, and processing, at the end of the 1980s, it now has much broader objectives. Since its inception, it has

[3]Abbott M. B. (1991). "Hydroinformatics: Information technology and the aquatic environment", Avebury Technical, Aldershot, UK, and Brookfield, USA.

incorporated the social dimension of water infrastructure by attending to those social and technical processes that cannot be separated from each other. Therefore,

> "hydroinformatics integrates knowledges from the social and technical domains to create so-called conjunctive knowledges, that are concerned with an understanding of how technical interventions have social consequences and how the resulting social changes in turn generate new technical developments."

The study and practice of *sociotechnology* have now become essential aspects of hydroinformatics.

A new series of books on urban hydroinformatics was launched earlier this year with the publication of a volume addressing urban water systems[4]. The present book is the second in the series and is concerned with *flood risk and social justice* and draws on the substantial practical experience of the authors in dealing with flood related problems around the world.

The socio-technical nature of water problems arises from the multiplicity of stakeholders and their diverse interests and attitudes. The authors are convinced of the limitations of older technocratic modes of thinking, and justify their position by quoting from philosophers such as Locke, Hume and Heidegger, and from theologians such as Kierkegaard and Barth. These limitations have excluded stakeholders from participation in the development of their water infrastructure. A new paradigm is presented for the inclusion of stakeholders as active participants in guiding the technical aspects of water engineering projects. The role of the hydroinformatician in this process is to

> "create socio-technical environments in which the transmutations necessary to provide states of social justice can be catalysed through the creation of appropriate socio-technologies".

The paradigm provides a completely transparent decision-making process involving all stakeholders that is directed towards realisations of social justice. This whole, and indeed holistic, practice is defined in this book as a transcendental praxis.

As hydroinformatics cuts across the practice of all water business, water engineering and water science it is neither a branch of water engineering, nor a branch of informatics. It rejects the concept of system as its organising principle, and consequently the centrality of systems analysis as well. However, the word *system* occurs roughly once on every page of some parts of the book. Is this a contradiction?

The roots of the classical Greek word *system* are the prefix *sym/syn* meaning 'together', and the verb *histanai* 'to cause to stand'.

The energised and synchronised machine, composed of many parts, is the archetypal system. When all its parts mesh properly together, the machine functions correctly. It stands ready to hand, as designed, as intended, as a useful working system.

In contrast, an object is not a system when two or more of its parts stand, not together, but in contradiction to each other; the object is then called a *complex*. Such is the case when water and stakeholders are brought together in developing and managing their hydraulic world. The contradictions among stakeholders feed conflict over their water, its associated engineering hardware, and aquatic life. It is a world lacking in system, to a greater or lesser extent, and it is abhorrent to human values to impose *system* on stakeholders by force, that is to treat them as if they were parts of a machine.

In an ideal world the first task for each stakeholder of good will is to understand how past actions and data are implicated, through strings of actions, decisions, judgments, positions, attitudes, and beliefs grounded on facts derived from the data - the Abbott programme of sociological research into past decisions. The second task is a collective task of identifying the contradictions between all the implication strings. The third task for

[4]Price R. K. and Vojinović Z. (2010). "Urban Hydroinformatics: Data, Models and Decision Support for Integrated Urban Water Management", IWA Publishing.

each stakeholder of good will is to participate in a reflective process to resolve the contradictions (pacé Hegel), moving one-step-at-a-time from beliefs grounded on facts derived from data, to attitudes, to positions, to judgments, to decisions, and finally to agreement on particular actions – the Abbott programme of decision-making.

Informatic tools on the web may facilitate this process at every stage, in a new kind of participative democracy, for stakeholders of good will, who seek social justice through transparency, the bedrock of sustainable development, and transcend their former selves in the process. In this book the stakeholders are placed at the centre of engineering design for sustainability, and never in a subordinate position as parts of an optimized simulated *water resource system*. I commend this most interesting book to people of good will who are concerned with the risk of flooding, and with the education of new model engineers, and say to them: Read on!

J. Philip O'Kane
Emeritus Professor
University College Cork, Cork, Ireland.
Honorary Fellow
UNESCO-IHE, Delft, The Netherlands.
September 2011

Foreword by Jean A. Cunge

This is a kind of fundamental book in the sense that it is concerned with the basic problems of today's society, using urban flood mitigation as a starting point and an example: it places engineering, engineers, and, more generally speaking, an entire "materialistic and technocratic mindset", within a more general framework and environment of activities. This book is sometimes demanding to read and requires a particular effort on the part of those who are not engineers but are deeply involved in these matters (let us call them, in the manner of the authors, "stakeholders"). Indeed, the book states that we are here in a *sociotechnological* environment where "Sociotechnology is concerned with the study of processes in which social aspects cannot be separated in thought from technical aspects: the social and the technical are so related to one another in the process that the presence of the one makes no sense without the presence of the other". Most of us, belonging to the engineering world, are not accustomed to linking such ideas to our everyday occupations.

The book is really composed of two parts. The second part is a description of urban flooding and flood mitigation problems seen from an engineering and technological point of view: from climate and hydrology to structures, through observations and data collecting and real-time operations, including numerical simulation models. This is an excellent source for engineers or engineering postgraduates who want to learn about urban flood problems. Even for those who are more advanced, it will certainly be useful as a reminder. Because it is encyclopedic, however, one cannot expect everyone to become an urban flood hydraulics specialist after the reading. And, incidentally, this second part is easy to read and understand for one coming from a civil engineering field. ***But***, as pointed out in the book, nowadays there are other stakeholders than engineers (e.g., decision makers) and they should read this part of the book to be able to discuss with engineers. And, since these readers on average have no civil or hydraulic engineering background, they certainly may have a hard time to go through this part. But if they do, it will also certainly help them in their activities and, in the first place, in understanding engineers and their projected solutions.

The first part of the book, to the contrary, may be demanding to read for engineers. This part concerns the notions of social justice, ethics and their evolution in Europe – easily read by political scientists, philosophers and historians but virtually *terra incognita* for engineers (with some exceptions, of course).

It is this first part that will be dealt with in what follows because it is vital that as many engineers as possible should get interested in the problems that are raised here. Consider the list of stakeholders involved in the flood mitigation sector of a great urban area! There will be "various government representatives (from national, regional and local levels), water utilities, disaster prevention and mitigation officers, planners, all those at risk of flooding, technical experts, private sector interests, environmental agencies, NGOs, and, most important of all, those members of the general public who really care, and indeed in many cases care passionately about the causes at stake." Note that these stakeholders are also, at least in democratic societies, decision makers: who, among engineers, have not experienced the necessity of changing an original design or even abandoning a project under the pressure of some of such groups? Obviously the times are gone when technical city services (i.e., engineers) decided alone what is good or bad. Engineers theoretically were always *servants of the social requirements* (and of available budgets ….) but until some 60 years ago they were the ones who decided what social requirements were. With time there may came an obvious deviation in some engineers' minds: "we are the only ones who know how to do things and we do them according to our professional ethics such as efficiency, beauty as we see it, etc." And, then comes a further step: "since we are the only ones who know, it is our calling to tell the society what should be done".

These times are gone and it is a necessity for engineers to be accepted as rightful members of such stakeholder-decision-making groups. To this end we must understand the decision-makers' problems and provide them with useful inputs. Since the stakeholders are only too often isolated from one another in thinking and purpose (if not downright hostile to one another!) there is a place for engineers to help them reach a consensus. But, again, to be able to do so, they must know the stakeholders' ways and problems. And here the reading of such a book can certainly help them. Especially the first part, even if an engineer's first reaction may be 'that's philosophy, it is not my department'.

Note that the above basic points do not exclusively concern urban flooding; to the contrary, the latter is only an example since this problematic concerns all the domains of engineering and most of the sciences. But urban flooding is an excellent example for several reasons. Indeed, most of the Planet's population will in the near future congregate around enormous metropolises that are at risk of suffering inundation. An important part of this population is and will still be poor and deprived of a position as stakeholders, while politicians and stakeholders organisations and lobbies are of significant importance to all decisions. Now, when the flood mitigation is concerned, the question arises: who and what should one protect if there is no means to protect everybody? The districts where the losses would cost more (usually where the well-off population lives) or the districts where the worst-off are living? This 'simple' question shows why urban flood mitigation could be taken as an example, and why the understanding of the problem involved can help towards a projection of the questions and approaches to other domains.

It is immediately obvious from the above that entering such a domain raises questions of ethics and social justice and, consequently, of politics. We are not on the technical engineering ground any more. And for most of us, as engineers, these are problems of personal attitudes, 'soft sciences', 'small talk', etc., hence not worthy of being associated with projects of dams, buildings or dykes. And yet … let us come back to the book. The European history of the last thousand years shows that the technological developments were supportive of the political power mutation and vice versa. As the authors write: "There then occurred in certain places a synergy between the exercise of political power and the development of industry through the application of an increasingly science-based technology, and it was this synergy that provided the motor that drove Europe to its 'pinnacle' of world domination." This remark is essential to understand why we, as Europeans, have some moral obligations to our past and, hence, to our present. And, of course to our future: if we break the link between the social and technological developments, we also let go of our moral obligations to the future. Not only to the European future politically speaking, since domination is gone because of globalisation, but to European civilization and its influence on human progress as we understand them: democracy, freedom, etc. And the links we are writing about are founded on the notions of ethics and social justice.

These two points, ethics and social justice, are preeminent in the first part of the book. These two philosophical categories go together and we cannot skip the problems that arise here, especially since the title of the book includes one of the terms. What is social justice in general? Each religion and every political or economic theory provides its own definition and, in order to achieve social justice, also introduces its own ethics. The Authors propose "… to explain some of the fundamental aspects of social justice theory and practice, and for their relation to the subject of the present book we tap into the works of Aristotle, Plato, Hobbes, Locke, Hume, Kant, Schopenhauer, Kierkegaard, Heidegger, Godwin, Proudhon, Barth and other philosophers and theologians who have made their invaluable contributions to a history of thought that is now of inestimable value to us."

The readers will find the references and basic description of each approach in the book but what comes out as a conclusion is that *there is no way to define in general what social justice is*. Obviously social justice as generally accepted varies with the evolution of societies. What was considered as social justice in ancient Greece and Rome based on slavery is not acceptable today. In general the concept of social justice cannot be dissociated from historical context, political ideas and overall ethical principles. For the purposes of the present book, social justice is that which provides all its citizens with the possibility to transcend the limits of their earlier selves, a definition that is prepared by references to several sources.

But we are talking about the specific domain of urban flood mitigation – what does social justice mean in this context? The Authors list four principles: "*Maximising welfare; Respecting individual rights and equality*; *Fairness* (public resources should be used fairly, e.g., favouring the most vulnerable and worst-off); *Promoting virtue* (encourage sharing risks)". Is it a utopia, even in Western democracy? It is important to strive towards utopias and that is why it is important to read and think over such works as the present book. But the Authors also describe the means through which such objectives can be reached and even give an example of the successful enterprise. In a nutshell: there are technological means available that allow all stakeholders to be informed (about policies, projects and their implementation) and to intervene in order to arrive at a consensus among at least the majority of stakeholders. These means are supplied by hydroinformatics technologies and by other technologies (ICT) such as those of the Internet and mobile telephony. Then, as the Authors proceed, the sociotechnological mutation in society, linking these new technologies and political structures, leads towards one or the other kind of social justice in our projects. The example of an application described in the book is the link between Denmark and Sweden – an example that should certainly be studied by all civil engineers.

We do not intend to say more: the readers of the book should develop their own opinions on subjects that, somehow, form the very *raison d'être* of our profession. The book is no text-book in a sense that whoever expects to find in it recipes for anything and everything will be disappointed. Many things in it, at least as far as first part is concerned, are controversial, subject to discussion or possibly true in the future while being nearly utopian today. Thus it is obvious that what was done in the Denmark-Sweden connection is possible only in long-existing democratic and free societies (and probably next to impossible in many European countries, not to speak about other continents). Internet, mobile phones and social networks are not in themselves the engine that modifies societies: those means can be censured and controlled at least to some extent, by political power as it is clearly shown in China or Syria and even in Europe. When the Authors write: "The ultimate aim of the urban flood risk mitigation practice is to strive … to a level that is just and acceptable in the society", the reader may ask who defines what is "just and acceptable"? And, no doubt many things will provoke eye-brow rising from many readers. But precisely, all this should be taken as an encouragement to all those who have survived reading this preface to continue and read the book!

Jean A. Cunge
Hon. Member IAHR, M. ASCE
Euro-Aquae Erasmus Mundus EU Project, Nice, France
Former Development and Marketing Director of
Laboratoire d'Hydraulique de France

Preface

"If you want things to stay as they are, things will have to change"
Giuseppe Tomasi di Lampedusa (1896–1957)

Why should anyone write, yet alone read, another book again on flood risk, and why should this now be connected to social justice? The answer has already been given very clearly above, by Giuseppe Tomasi di Lampedusa:

"If you want things to stay as they are, things will have to change"

This message is now been sounded so loudly in so many languages and within so many contexts that it is beginning to be heard more widely, as its imperative tone is experienced ever more directly 'on their own skins' by an ever larger part of the world's population. This awakening call is especially clearly heard in those parts of society that like to associate themselves with 'The First World', even as the process of 'changing things' is confronted with the greatest difficulties in what this 'First World' likes to call 'The Third World'. Alongside this again, we are confronted with what is commonly called 'The Change in the Balance of World Economic Power', with Asia generally and China in particular overtaking the self-styled 'First World' at ever higher levels of science and technology with ever more applications — and exports. These 'sea changes' themselves are leading to the growth of cities, where already we have a position where fifty percent of the world's populations are already living in cities and there are prognoses for eighty percent by 2030. However the consequences of this immigration to the cities are those of increasing risks of new kinds, and the most dangerous of these are the increases in frequencies and magnitudes of flooding.

The thesis of this book is not only that *things will have to change*, but that *they will have to change fundamentally*, if society is even so much as to maintain its present way of living in its 'First World', while for much of the rest of the world the prospects remain dismal unless the paradigms that have the capacity to make these fundamental changes can be transferred to that part of the world too. Since these paradigms are not themselves of a social nature, and neither are they of a technical nature, they cannot be realised within the current practices of flood mitigation, and especially so in cities. It will be argued here that the only paradigms that have the possibility to resolve the difficulties that nearly all societies face in respect to urban flooding, as indeed in respect to most other risks, *are sociotechnical in nature*, understanding that *every sociotechnology is a conjunctive knowledge*, that is as something that joins two or more other knowledges together without being subsumed in any of these knowledges that it brings together. By doing this it may, and in fact in most cases must be another kind of knowledge that *transcends* the knowledges that it joins together by introducing knowledges and practices that cannot normally be found and cannot even be applied in the individual knowledges that are so brought together. It introduces new and essentially holistic approaches to the work of those who care to follow its own very special ways, such as are introduced in this work.

Fortunately for Hydroinformatics, this more holistic understanding has now penetrated quite widely in other fields and a considerable literature has accumulated on this subject, some parts of which are introduced as we proceed through this book. There is throughout a deep understanding that only such an approach can save our own societies – and bring our fellow men, women and children in the so-called 'Third World' into a more commodious way of living at the same time.

The book is in four parts, two of which start out primarily from the social sides and explain how these are connected, and absorbed into one part of the overarching sociotechnical context, changing qualitatively, and so *transmuting*, during the process, as this is realised within *hydroinformatics environments* (i.e., by the use of modelling and measuring equipment with its surveillance, control and data acquisition systems and radio-linked, on-line and so real-time operating systems, GIS maps, computer animations, multimedia technologies, World-Wide-Web, mobile telephony and, currently as work in progress, such devices as the IPod and the IPad: See Abbott and Vojinović 2011, as referenced in Part II of this book). The third and fourth parts concentrate upon the technological aspects even while intimating how these can come to be employed by the new kinds of arrangements that we are introducing, and so be introduced into another kind of social environment. It concludes with some case studies that demonstrate how the technical aspects can be merged with the social requirements to give sound project conclusions, even though these cases were not directly connected to their stakeholders.

A special emphasis is placed throughout the book upon the role of active stakeholder participation, which is sociotechnical through and through, as exemplified as a paradigm case by the Denmark-Sweden road and rail connection over the Øresund. The understanding obtained from this movement is then transposed to the so-called 'Third World' with its own basic problems, so as to round-off that part.

If we should try to introduce what we are doing in our work as hydroinformaticians in this field we should say that our principal task is to construct hydroinformatics' environments, as web-based systems that deliver content to those stakeholders who care so much about the issues at stake that they are prepared to sacrifice their time and their energy to participate in the decision-making process – and then to such effect that they can come to lead and control this process. The hydroinformatics environment is then composed of data streams from instruments and other means of observation that are processed into GIS maps, screen displays and multimedia animations that have a heightened capacity to project into the minds of the active stakeholders the realities of events on the ground. This environment has then demonstrated, as in the Øresund project, that through these and some other means it becomes possible to cause qualitative changes, or transmutations, in the minds of the active stakeholders, so that their imaginations are mobilised and activated, albeit initially for the most part on an individual basis. The processes of interaction that then follow between the so-activated stakeholders leads to a further qualitative change, or transmutation, *from the individual active imaginations to their creative imaginations*, whereby they can influence the course of the project development. This process can of course become iterative, as changes are proposed to accommodate the findings of the active stakeholders as a whole and this is linked to the process of transcendence in that deeper understanding which is the conditio sine qua non of social justice. Indeed, with this approach we not only catalyse the processes of transcendence for ourselves, but also for our fellow creatures and their environments, rescuing them from the mindless and soulless destruction that is currently being increasingly imposed upon them. Since this has been the task of humanity from the beginning of time, we see this as a task that is even more than commensurable with the overcoming of the destruction that is being visited upon our own kind.

The present book is a second book in the series on Urban Hydroinformatics and it is our response to the rising significance of floods and flood related disasters world-wide, an initiative to promote a hydroinformatics viewpoint of how to approach the problem of flood risk mitigation in a socially just way. The book is written to support a wide range of audiences and it seeks to improve the dialogue between researchers and practitioners from different disciplines (including post-graduate engineering and social science students, academics, planners, environmental advocacy groups and environmental law professionals) who have a strong interest in a new kind of social justice work that can act as a continuous counter-force to the various mechanisms that unceasingly give rise to injustices. As such, the book also supports UNESCO-IHE's regular/on-line module on Urban Flood Management and Disaster Risk Mitigation. It contains a range of aspects from the social and ethical, to the scientific and technical. Due to some overlapping aspects of the subject matter, certain parts of the text presented in Part III are common with the *first book*[1] published in the Urban Hydroinformatics book series. This was done to maintain the consistency between the two books and to provide the necessary background and theory of modelling to those readers who may not necessarily follow the same courses as those for which the first book was written.

In addition to the book, there is also a website (www.urbanhydroinformatics.com) where we intend to provide access to relevant information and to maintain a conversation on the subject of urban hydroinformatics. In particular, we welcome critically constructive comments on the material in this book. Furthermore, there is a web-site of the second author (www.knowledge-engineering.org) which also contains material that relates to this book.

We are aware that we have only touched on the basic aspects of the subject – there is so much more that can be written. Therefore we have had to be selective about the material we include. In doing so, we have explored some profound issues and some of our ideas may appear as incomplete, and indeed in fact they may never be complete. They represent our current thinking of a new paradigm where the concepts of social justice and its ethics are used as a bridge between a European science and higher levels of consciousness. We trust that this book will do much to raise the awareness of social justice, and improve the debate between different disciplines on what it is so that we have a better platform for communicating ideas and concepts in the future. We hope to have the opportunity to correct any errors or serious omissions in future editions of this book. We sincerely hope that the book will challenge and encourage students and practitioners all over the world to advance their knowledge and practice of dealing with flood related problems and hydroinformatics in general.

Zoran Vojinović
Michael B. Abbott

---

[1]Price R. K. and Vojinović Z. (2011). Urban Hydroinformatics: Data, Models and Decision Support for Integrated Urban Water Management, IWA Publishing.

Acknowledgements

We are grateful to colleagues around the world for their contributions to this book. We express our appreciation of those who have allowed us to include specific case studies for which they were responsible: Vanessa Cançado, Prof dr. Nilo Nascimento, and Sylvie Barraud (Flood Vulnerability Assessment, Belo Horizonte, Brazil), dr. Sutat Weesakul, P. Supataratan, and dr. Ole Mark (Flood Control Centre, Bangkok, Thailand), David McConnnell (Flood Forecasting – From Rainfall to Emergency Response, Australia), and Ahmad Fikri Abdullah (The SMART Tunnel, Kuala Lumpur, Malaysia). We are also grateful to dr. Neelke Doorn from the Department of Philosophy of Delft University of Technology who has made contribution to the Section 7.4 (An Overview of Ethical Theories), to Prof dr. Sabine Roeser from the Department of Philosophy of Delft University of Technology for inspiring conversation and suggestions and to Prof dr. Slobodan Djordjević from the University of Exeter for valuable comments and inputs into some parts of the text. We would also like to thank our colleagues at UNESCO-IHE and elsewhere who have supported this initiative: Prof dr. Andras Szollosi-Nagy, Jan Luijendijk, Prof dr. Roland K. Price, Prof dr. Damir Brdjanovic, Jetze Heune, dr. Ole Mark, dr. Dusan Sakulski, Subasing Wijesiri, John Tetteroo and Hans Emeis; then present and former PhD students: Wilmer Barretto, Solomon Seyoum, Arlex Sanchez Torres, Ahmad Fikri Abdullah, Vorawit Meesuk, Yared Abayneh Abebe and the many UNESCO-IHE Masters students who have stimulated us by their enthusiasm for the subject and their many questions and insights which have caused us to look afresh at the issues involved. We are especially grateful to Prof dr. Philip O'Kane and Prof dr. Jean A. Cunge for valuable comments and support. We are indebted to Maggie Smith and the production team at IWA for helping us to publish this book. Finally, we could not have even begun to contemplate writing without the encouragement of Jana and Louise: thank you for your patience and loving support.

Zoran Vojinović
The Hague
Michael B. Abbott
Brussels
October, 2011

Introduction

At the time of writing this book the world is being overwhelmed by increasingly disastrous manifestations of the disturbances and disruptions of Nature's processes. Record disasters such as earthquakes, hurricanes (or tornados), tsunamis, widespread flooding, droughts, extremes of snowfalls and temperature, all of which are associated with devastating losses and suffering, occur almost daily. However, the majority of these events, although commonly referred to as *natural disasters*, are not in fact the result of nature related processes alone. They are to an ever increasing extent directly attributable to the actions of human beings. These actions are the result of human thinking and decisions, which in turn are based on understandings, attitudes, needs and values. Unfortunately, in many cases these are guided by greed, ideology of money, the desire for power and self-centredness. Thus the roots of many of our problems nowadays lie in the human mind as much as in anything else.

This book is concerned with floods and flood-related disasters and it argues that a part of the solution is not in discovering any new knowledge as the basic wisdom already exists. What is needed is to move beyond the materialistic and technocratic (or piecemeal) mindset and to add other dimensions to our approaches of flood risk mitigation that can give us the will to achieve viable states of *social justice*. This view raises some important questions. Why should professionals who are dealings with floods and flood related disasters be interested in *social justice*? How is *social justice* related to flood risk mitigation and why should we be thinking of introducing it into our current practice? There are potentially many answers that can be given to these questions and this book is an attempt to do so. In short, one of the simplest answers would be: because of the complex and multiple nature of flood risk! Whether the challenge is with the protection of a part of the community or the catchment area as a whole, whatever risk mitigation measures are taken they will never be able to bring equal benefits to all members of the society, and even if they do so for the present generation they may not do the same for future generations. Consequently, on one hand, there will be members of the society who will benefit from such measures in one way or the other, and on the other hand, there will be other members of the society who will be more burdened by the same measures. Since such inequalities, or *injustices*, tend to arise throughout our lives due to events which are sometimes in our control and sometime beyond our control, there is an urgent need to engage in a new kind of social justice work that can act as a continuous counter-force to the various mechanisms that unceasingly give rise to injustices. It is through this process that we arrive at what is for most of us a new definition of social justice, even as this was pronounced already by Blaise Pascal in the seventeenth century and has been adumbrated over the last two decades especially, by the Catholic Church, and now developed further within the present context of this book.

The current 'flood management' practice has undergone a significant transformation over the last thirty years. Inspired by the realisation that the risk can hardly ever be completely eliminated the traditional 'flood defence' culture has been replaced with the culture of learning how to live with flood risk and how to be better prepared to respond to its negative impacts. However, despite our deep respect for modern science, most of the current research and practical efforts appear to be the same as those that have become so characteristic of modern science generally, namely those of *ordering, numbering, counting and computing*. Science has set humanity on a path to address flooding by using its logic, its science, its technology and its observations, while ignoring the deep emotions and the inner senses of social justice as means for uncovering the *reality* of aspects which concern the results of *flooding as a collection of phenomena*.

The problem then presents itself of how to address the needs and concerns of all members of a society in a socially just and ethical manner. This book is accordingly about *socially just* flood risk mitigation practices in which the process of engaging in social justice is viewed partly as a continuous and partly as a discontinuous transcendental process which starts from social needs and concerns and proceeds in initially indefinable feedback cycles where the acceptable social, ethical, technical and environmental norms and standards continuously change, leading to transcendences in levels of understanding and *consciousness*. In this process, the question of how to judge whether a risk is acceptable or not requires much deeper social and philosophical considerations rather than simple calculations of probabilities multiplied by unwanted outcomes and their balancing between costs and benefits. Therefore, the material presented in this book is in some parts philosophical and holistically theoretical and in some other parts it is purely technical and practical. The aim of the book is not to provide readers with any kind of preset theoretical and practical apparatus for dealing with any and every problem that is associated with floods and mitigation of floods, but to provide them with the means to develop their own specialised apparatus for each occasion as it presents itself to them.

The notion of *social justice* which emerges as a central concept of this book is to enforce the sociotechnical point of view within a flood risk mitigation practice in which technical parts cannot be separated from social parts and in which principles of ethics and social justice play a central part. This could be expressed as follows:

> *What kind of societies do we need to develop in order that our flood mitigation measures can meet the challenges posed by an ever continuing urbanisation, population growth and climate change and how are these measures to develop in order to support such societies?*

This is then to say that, just to the extent that if such a 'problem' can be posed at all, we must already pose it as *an essentially sociotechnical problem* which concerns not only technical but also social and ethical issues. This means that we need to develop a truly sustainable mode of human *consciousness* for which the knowledge that has been gained through moral philosophy and other related fields is applied to seek answers of what is correct and what is true. In developing human *consciousness* researchers from some other disciplines have looked to quantum physics, some to information theory, some to neuropsychology, and others to spiritual teachings. In the field of our endeavour, we could go and say that what is nowadays needed for successful flood risk mitigation is in fact an integration of our scientific understanding of the physical world with the wisdom that is held in the world's philosophical (or, rather, spiritual) traditions. Just to introduce what is surely the most significant of these sources and inspirations for the present work is the resurrection of the way in which science itself is experienced, such as was introduced by Johann Wolfgang von Goethe, who was not only one of the supreme literary geniuses of his time, and acclaimed as the greatest in German history, but also a very capable and thorough biologist, theoretical physicist and polymath. For our present purposes we shall follow Henri Bortoft's work which, after being published in French and in a German translation, appeared in an English translation in 1996 and attained its fourth reprinting in 2010. As will be explained in Part II, this work, drawing also upon some of its predecessors dating from 1950 onwards, makes possible an understanding of previously incomprehensible processes, such as those in which a higher level of consciousness across a variegated group of active stakeholders *transmutes* into a unified state of conscientiousness over the group as a whole. It thus provides a postmodern theory of science which, following Bruno Latour (1991: see References to Part I), most commonly resorts to a pre-modern theory, and so in the first place to the one or the other form of alchemy: thence the word 'transmutes' instead of the word 'transforms'.

In current practice there is a clear evidence of a lack of interaction between social aspects and engineering and this appears to be a major hindrance for solving some of the greatest problems associated with floods and flood-related disasters. This can be largely explained by engineering education, where for example, many engineers have not even read Locke, Hume, Kant, Kierkegaard, let alone Heidegger, even as they have a special kind of moral responsibility to understand the social needs of our time – and to develop the technical measures accordingly.

This problem has at this time of writing to be reposed, and indeed radically reposed, as a consequence of the results of the current flood mitigation practice which has been dominated by the development and application of almost exclusively *technocratic* and *materialistic* approaches which consider floods as natural events that can be overcome and put under control by the application of engineering measures alone. There are many such examples of dealing with various water related projects whereby large amounts of human and financial resources have been thrown into technical developments without regard to their related social interactions. Only by thinking within the realms of the sociotechnical and the authentic ethical can we hope to attain to some measure of completeness in our analyses and thus obtain not only cost-effective, but often even more importantly, socially and ecologically acceptable results.

It is now widely recognised that, to achieve better practices for flood risk mitigation, traditional measures also need to be supported by a number of non-traditional elements of a kind that, although they have been realised to a certain extent in some applications, have so far been insufficiently analysed and have consequently been even less understood within the present context. Thus, the flood risk mitigation framework is nowadays concerned not so much with protecting from floods but rather with understanding flood risk, preparing for flood risk and living with flood risk. This implies that risk assessment and gauging how the risk is perceived and accepted and how people will react to risk are some of the crucial aspects of the successful flood risk mitigation practice. Therefore, the traditional risk assessment process which is based on calculations of probabilities and applications of cost-benefit analyses must now be combined with moral judgements and social considerations in order to come up with what the right strategies are. This gives rise to the following questions. What are ethically legitimate considerations in judging the acceptability of risk? Does the scientific analysis provide an adequate method to come up with a decision or should we introduce an additional point of view which cannot be addressed within the spirit of modern science, and indeed within the ambit of European science altogether?

Another feature of this book is correspondingly that it attempts to address the sociotechnical dimensions of flooding and flood risk not only in the context of 'developed' but also so-called 'developing' countries. In this much greater world, where the sizes of its cities are increasing at a much faster rate than in the 'developed' world, whereby the problems of urban flooding are also rapidly becoming so much greater and their solutions correspondingly so much more urgent, the problems that have to be solved are that much greater correspondingly. This book accordingly discusses these and some other related aspects and it suggests that the present 'flood risk management' framework should adopt the social justice approach and pay particular attention on *what is the right thing to do* within a much wider context rather than simply to maximise the benefits and minimise the costs.

The book strives to broaden the traditional technocratic view of flood defence culture and to bring other necessary aspects together with technical aspects for those involved (or those who wish to be involved) in flood risk mitigation in order to become empowered with the necessary knowledge to engage in processes that can lead to social justice. In the course of the book we shall draw upon concepts from outside of engineering that may come across to some readers as unfamiliar and challenging. This may have the effect that the book is demanding in some parts and may require the reader to further explore some of the ideas that we reference from other sources. In order to explain some of the fundamental aspects of social justice theory and practice and their relation to the subject of the present book we tap into the works of Aristotle, Plato, Hobbes, Locke, Hume, Kant, Schopenhauer, Kierkegaard, Husserl, Heidegger, Godwin, Proudhon, Barth and other philosophers and theologians who have made their invaluable contributions to a stream of thought that is now of inestimable value to us.

The book is divided into four parts covering four different aspects of flood risk mitigation, even as it addresses different types of floods that may affect urban communities. Part I opens with a discussion of different root causes and it brings together as many as possible of the various threads of the causes of flood risk. We make a distinction between *four levels* of such causes: natural, technical, social and those that are related specifically to the human mind. We then go on to a discussion of *the nature of urban flood risk*. Already here, the notion of social justice and its relevance to flood risk mitigation practice is introduced. The notion of social justice is introduced in the context of what has commonly been referred to as 'flood risk management', but then in a context in which management itself is placed directly in the hands of the most active stakeholders rather than being in the hands of what have so far passed as "professional managers".

Part II reflects *social and ethical considerations* and it starts with the review of different aspects that relate to social justice theory and practice with the intention of characterising what we mean by social justice and how it can be related to flood risk mitigation practice. We draw particular attention to the essential limitations of modern science and the *technocratic* way of thinking which engenders and imposes the limits upon our seeing the larger context of the problem with which we are concerned in our endeavours. A new approach (or a new paradigm) that aims towards a continuous process of *transcendence* in the flood risk mitigation practice while being more responsive to social justice concerns and needs is formulated and described. We also provide a discussion of the key ethical theories and concepts that are needed for effective engagements in social justice work. We further identify strategies for achieving such changes. The means of enabling active stakeholder participation and creation within this process are also discussed. This part of the book also discusses the role of hydroinformatics within the flood-related social justice work and it outlines the associated concepts that are needed to enable such a process. It ends with a first paradigm case of a stakeholder participation process which can be regarded as a socially just process, while contrasting this with what we observe in the so-called "Third World".

Part III is concerned with *scientific and technical considerations* and it starts with discussion of different types of floods that are directly related to urban flood risk. We describe different types of collection systems and their individual elements. We discuss the purpose and objectives of data acquisition and processing and describe the role of modelling technologies which are invaluable means for dealing with floods and flood-related disasters. With instantiated models and specialist GIS flood mapping techniques it is possible to explore the generation of floods and evaluate the effects of different measures in response to any extreme event. Such means enable the development of a more comprehensive approach to estimating the risks involved and addressing the uncertainties that may arise due to urbanisation and climate change by simulating a number of alternative scenarios. Information obtained from numerical modelling work can be than presented in a set of related plans that can enable stakeholders to better respond to risk and to build more resilient societies. We also discuss the theory and application of models and describe different modelling approaches such as one-dimensional (1D), two-dimensional (2D) and coupled 1D/2D approaches together with their associated numerical solutions, schemes and practical issues.

Part IV deals with *practicalities of the flood risk assessment and mitigation planning* process and it discusses aspects such as risk identification, risk perception, qualitative and quantitative risk assessment, and estimation of hazards and vulnerabilities from numerical model results as well as from remotely-sensed data sets. An approach based on the holistic perspective which takes into account not only technical but also social and ethical considerations is presented and discussed. We also provide an overview of some of the typical structural and non-structural flood mitigation measures that can be applied to different kinds of floods and flood-related problems. We also discuss the production of different plans: integrated water resources management plans, river basin management plans, catchment flood management plans, coastal management plans, disaster management plans and asset management plans, while understanding that leadership is for the most part in the hands of the active stakeholders. We describe both theoretical and practical aspects of how to prepare for disaster before it occurs, how to respond during disasters (e.g., emergency evacuation), and how to support, and rebuild a society after a disaster has occurred. We connect to most of the previously discussed aspects for formulation of strategies for flood disaster prevention, mitigation, response and recovery. The focus in this part of the book is not only on technical measures but also on what can or should individuals and social groups do to reduce their own vulnerability when situated within suitable *hydroinformatics environments*, and how external actors can best work with these developments at the individual or group level in order to build more resilient societies. This part of the book ends with a discussion of different concepts concerning asset management theory and practice. It also includes four international case studies where some of the practical aspects of flood risk assessment and mitigation are described and illustrated.

Part I

The Nature of Urban Flood Risk

Chapter 1

Urban areas and flooding

1.1 OVERVIEW OF PART I

The predominance of almost exclusively *technocratic* and piecemeal approaches has led to the development of less effective and less efficient means of responding to floods and flood-related disasters. In current practice there is a clear evidence of a lack of interaction between social aspects and engineering and this appears to be a major hindrance for solving some of the greatest problems associated with floods and flood-related disasters. There are many such examples of dealing with various flood-related projects whereby large amounts of human and financial resources have been thrown into technical developments without regard to their related social interactions. It is now obvious that flood risk mitigation processes require multiple approaches and disciplines and the aim of this part of the book is to bring into a synergic relation a range of aspects concerning both social and technical issues that can provide states of *social justice*.

In the following sections we illustrate the nature of flooding in the case of a 2007 Jakarta flood event and we bring other related aspects from different places in the world. We introduce a new holistic view of looking at root causes of urban flood risk. We touch upon the evolution of flood mitigation practice and current state of risk assessment work. We also introduce social justice as an important dimension within the flood risk mitigation process. At the same time, we observe that further examples can be seen on our television programmes on an almost everyday basis, even if there is rarely any explanation of the causes and effects in these presentations.

1.2 URBAN AREAS AND FLOODING – *WHAT CAN WE CONCLUDE SO FAR?*

Urban areas are usually defined as geographic locations where secondary (industrial, manufacturing) or tertiary (service) sectors dominate over the primary (agriculture, forestry, mining, etc.) sectors, which display high population density or size, constitute an administrative region and where all land uses and activities lying within such a metropolitan district are referred to as "urban". In urban areas, society is popularly perceived as being in control of a physical environment where natural hazards such as floods can be managed to a larger extent by the means of technical (or engineering) measures.

Floods are commonly defined as an overwhelming flow of water onto land that is normally dry and which under certain circumstances can cause unprecedented losses and devastation. These can result from a wide range of events and processes. These factors can be diverse (natural, human or technology related) as well as site-specific. Amongst others they can include heavy rain, storm surges, tsunamis, overtopping river banks and raised groundwater levels. A review of natural catastrophes between 1950 and 2000 and economical losses can be sourced from Munich Re (2002).

The threats from floods and other natural disasters were initially described as "those elements of the physical environment harmful to man and caused by forces extraneous to him" (Barton and Kates 1964). Such a perception has led to the preference for physical rather than social treatment of floods with urban societies being allowed to live under the shadow of increasing risk. Certainly, there is much to be gained from physical engineering-based interventions, but without an appropriate inclusion of social and ethical aspects into the analysis our interventions into the physical environment are in danger of ignoring differences in the influence of individual and social characteristics on the susceptibility or response to urban flood risk. It is the objective of this part of the book to explore not only the technical and natural but also the closely interacting social dimensions of at-risk elements and examine the ways in which such interactions shape the vulnerability to floods that take place in urban areas in developed as well as developing countries. Perhaps, the following example of Jakarta 2007 flood illustrates best the fundamental characteristics of urban flood risk and highlights the greater susceptibility to losses of those living in the cities.

On 2nd February 2007 there was a major flood event in the city of Jakarta, the capital of Indonesia, which affected not only the central area but also several other areas around the city, such as West Java and Banten. This flood is considered as one of the worst floods that took place in Jakarta in the last three centuries, including the 1996 and 2002 flood events. On that day in 2007, over 70,000 homes were flooded, resulting in the displacement of some 200,000 people: see Figure 1.1. This event has caused almost US$1 billion in tangible damages. Approximately 190,000 people have fallen ill due to flood-related illnesses and the officially reported death toll was close to 70. This was one of the greatest flood disasters that had ever occurred in Jakarta. What stands out is that the residents of an officially illegal neighbourhood – a squatter (or slum) area on the flood plain – suffered an unprecedented impact. This flood event illustrates some of the fundamental characteristics of urban flood risk and vulnerability which forms an important theme for discussion in this chapter. It certainly reminds us that urban areas are not immune to the forces of nature, but more importantly, it indicates that some community groups are more exposed to threats from flood disasters (i.e., they are more *vulnerable*) than others as they continue living and working on floodplains as well as other hazardous areas.

Figure 1.1 February 2007 Jakarta flood (Photograph: Hendra Murtidjaja)

The same flood event also proves that flood-related disasters are not only the result of natural events. They are also the product of social, economic, historical, political and even cultural issues. People live in different socio-economic situations which often dictate their choice of regions and places for living. Some of these regions and places are often affected by floods, and to such an extent that hardly anybody would ever expect that they could be considered as habitable. However, due to political, economic and some other reasons these areas are not only partially occupied but they are in many cases even encouraged for full scale developments. This can only aggravate the situation and cause even more severe impacts of floods. In the case of Jakarta, the slum areas represent home to more than half of its inhabitants. The informal sector comprised a majority section of the inhabitants involved in plantations with tofu and in temporary industrial units, while many earn their livelihood working in metal shops or numerous small-scale factories and vending food. Battling through heavy traffic congestions, smutty rivers, recurring flooding, air reeking of diesel odours and living with unfenced rail lines, the residents of areas such as Ciliwung and Galur continue to endure, work and raise their children.

Another important aspect to note in the Jakarta 2007 flood event is the perception of flood risk. Siringoringo *et al.* (2008) observed that the perception of flood impact on life and daily activities is significantly different amongst Jakarta's inhabitants and that the level of knowledge and understanding of flood risk in a given area is directly related to people's decisions to either adjust their living to such a risk or simply to ignore it.

In addition to the above observations we can also note that in some other parts of the world the gender, race and ethnicity issues represent yet another aspect which in certain conditions may contribute to disastrous outcomes. All these observations suggest that the development of effective flood mitigation measures requires not only sound engineering knowledge but also a much deeper understanding of social and ethical aspects, while any ignorance, either intentional or unintentional, of such aspects is likely to create not only ineffective solutions but also conditions for ever increasing risks and greater disasters.

What can also be observed from the same example of the Jakarta 2007 flood event is that the process of flood risk reduction and the associated social and institutional responses to such events represent an immense challenge, and then not only to planners, engineers and scientists but to urban societies as a whole. The resources that are available at the city level are often limited, and the task of setting up priorities, so that even making the right choices so as to achieve some measure of "*distributive justice*" in distribution of risks and allocation of flood protection funds has been a difficult, and in many cases unachievable, task. This is mainly due to the technocratic and piecemeal way of thinking which has been dominated by engineering problem-solving and misplaced applications of cost-benefit analyses.

The same Jakarta 2007 flood event also demonstrates that dealing with floods *is a sociotechnical endeavour through and through*: that very few flood-related problems are purely technical or purely social, but they are for much the greater part sociotechnical. In dealing with flood-related problems, sociotechnology constitutes a particular species of *conjunctive knowledge*, which is one that joins together two (or more) different and otherwise separate knowledges (i.e., social, technical, economic, environmental, etc.), thereby providing a new field of knowledge with qualitatively different properties than those of its constituent parts. Therefore, the study and practice of *sociotechnology* for the purposes of dealing with *flood risk mitigation* problems is essential if one wishes to understand how technical measures have social consequences and how the resulting social changes react back again on the technical developments, and so on, again and again, in what has the potential to become a sequence of *positive feedback* cycles.

The debate that arose in the aftermath of the Jakarta 2007 event not only posed a number of social and technical questions but it also posed hard questions of morality and justice. How are flood protection standards generally defined and by whom? Do they reflect the needs and concerns of all community members? Which parts of the community are more protected and which parts are less protected and why? What kinds of standards can be regarded as ethically and socially legitimate? Who should be involved in setting such standards? If we cannot provide an equal level of protection to all residents in the city, how can we decide who should be protected first? What is the fair and just distribution of risks? How should the community assess and perceive risk and how should the victims be treated and compensated when disaster strikes? What are socially-just considerations in judging the acceptability of residual risk (i.e., a risk that remains after a particular flood protection measure has been implemented)? Is the application of cost-benefit analysis to identify protection measures the best approach to reaching a decision or do we need to introduce additional considerations that cannot easily be incorporated into the current framework? In order to answer these and other such questions, we have to first explore and understand the meaning of *social justice* that is applicable in our field of endeavour and then to seek for the right principles that need to be incorporated into our practice. In fact, by raising these questions we have already begun to provide answers! Our intention here is not to try to provide a universal answer to these questions but to present the readers with the theoretical and practical backgrounds that may assist them in posing such questions themselves and so come up with their own answers and judgements. Therefore, the book aims at rethinking our work practices and developing of skills and knowledges based upon *social justice* considerations that are needed to take practical actions for change within the current practice.

The book introduces the notion of social justice in the context of what has popularly been referred to as *"flood risk management"*, but then in a context in which management itself is placed directly in the hands of the most active stakeholders rather than being in the hands of what have so far passed as "professional managers". Furthermore, if we recall that the meaning of management has to do with *establishing and providing order* it then becomes doubtful whether it is appropriate to use the word "management" in a combination with the term "flood risk" at all, since neither a person nor a group can ever be in a position to bring an order into something that is intangible and which contains a complex mix of social, technical and natural phenomena. Hence, we prefer the term "flood risk mitigation" over the term "flood risk management".

1.3 FURTHER OBSERVATIONS

As illustrated by the 2007 Jakarta flood event, the overall consequences of such events highlight the position of flooding as the most deadly and destructive of all natural disasters. These more general observations must however be broken down into their particular instances, as the measures that are being put into place to reduce the impact of flooding are now changing at an historically unprecedented rate in some countries even as so little attention is currently being given to these measures and its successes elsewhere. There is in the current English-language literature an almost total concentration on a few "western" nations whose investments in infrastructure generally are really quite small and piecemeal, amounting to only a few percentage points of the world total, even as urban flooding in the much greater regions of the so-called third (or developing) world is given only a cursory "headline" treatment with only a negligible attention being given to the true magnitude of these disasters and only a miniscule amount of assistance being proffered. At the same time, in other parts of the world again, among the so-called "emerging economies" which are often far better organised, this investment has long been some orders of magnitude greater and appears to have been in certain instances substantially more successful, even as it has received only the most perfunctory and superficial attention in the western media, with such attention as has been given being concentrated on a relatively few specific and particularly tragic failings. To take the most dramatic and recent examples at this time of writing, the averting of the floods following upon landslides that could have had much more disastrous consequences than actually occurred during the

Great Sichuan Earthquake of 2008, as well as the fact that not one of the 139 major dams in the region failed, indicated the capability of engineers in that nation. Similarly, the speedy and effective responses to the flooding that struck southern China in 2010 have graphically illustrated the magnitude of the resources that could be brought to bear in reducing the loss of life and property in reasonably well-organised, well-equipped and efficient operations. In both instances these actions were led on the spot by the President of China, himself a well-qualified hydraulic engineer, and with the authority, for example, to deploy army helicopters to deliver powered assault boats with their crews to otherwise isolated villages and individual families, as well as to organise effective emergency food supplies and temporary shelter. In this case there was a very adequate television coverage of events which even showed heavy earth- and rock-moving equipment being parachuted off rear-facing ramps, presumably from Russian heavy-lifting aircraft, into otherwise inaccessible areas.

One of the first priorities in this book is to take some distance from the current largely- Eurocentric view of the sociotechnical dimensions of flood risk and its mitigation. Alongside the now currently ongoing lament in the west about the necessity of "drawing the contours of a new world order" in economic terms, there is a lack of understanding of the developments of cities and other urban areas, within the context of this new world order. It should be enough to understand that China alone now accounts for about half of all the world's investment in infrastructure to realise the importance of comprehending something of the driving forces that are at work within that newly emerging economy. Alongside this, it should be understood that the industrial development of China remains the principal motor behind the development of the ASEAN region generally, even as its influence also filters further again into African and Latin American environments, catalysing further new developments in these regions also. Corresponding to this sea change in economic developments that must ultimately alleviate, even if not obviate, flood risk, we shall first concentrate in this section of the book upon the way in which socioeconomic development in China is currently proceeding in the direction of mitigating the effects of flooding.

If however we first try to seek out just one issue that appears to be common to all urban areas and especially to problems of losses of life and property due to flooding, it must be that of inequalities, and indeed of gross inequalities, in family incomes. Thus, even in the self-styled "first world", this is becoming an increasing problem as a consequence of a currently ongoing financial-economic crisis. Thus, to quote from an article of Edwin Heathcote that appeared in the Financial Times (October 30, 2010, p. 7):

> London has always been something else; scrappy, incoherent, mixed and flexible. But that may be about to change. In its recent spending review, Britain's Conservative-led coalition government announced changes to housing benefits that may soon force some 82,000 low income families to move out of the city centre. So severe are the coming cuts that Boris Johnson, London's Conservative Mayor, caused a political firestorm this week taking on his own party as he warned of "Kosovo-style social cleansing", as poorer residents were forced to flee to the suburbs.
>
> … But such benefit cuts are not the only incidences of austerity measures now creating new tensions in our urban landscapes. Throughout the developed world worries are being expressed that shrinking state services risk turning diverse centres into deadening ghettos of wealth. Worse, the same trend may cripple the capacity of global cities to do what they do best: change, adapt and mould themselves to each new age.

This example indicates how a major part of a population, and specifically the poorest part, is induced to move from a more commodious and safer area to less attractive environments, and specifically to less safe areas, and not infrequently one component of this change is that from an area that is relatively safe from flooding, and thus so much more desirable and correspondingly more expensive, to one that is less safe from flooding, and correspondingly less expensive. As we shall subsequently come to emphasise, *flood risk is to a large extent a function of poverty*, and this is one of the important themes that we shall follow when we proceed further into the world of *social justice* within the context of urban flooding. It is then indicative that in the so-called First World, the incidence of poverty in the United States, currently defined as an annual income for a family of four of $22,314, or $11,139 for a single person, has increased to its highest levels since records began fifty years ago – and rising from 31.6 million persons in 2000 to 46.2 million in 2010 (Matt Kennard and Shannon Bond, Financial Times, pp. 1 and 4, September 14th, 2011), This number overtook the numbers of persons described as "affluent", that is with gross incomes, before tax, exceeding $100,000 a year, such as would be considered for a large part as middle class in mainland Europe. The consequence of this tendency could be intimated by the social divisions that became so evident during the various flooding events of New Orleans in 2005 – and its economic aftermath.

Very similarly, at least superficially, but then proceeding at the other end of the world, in China as our primary paradigm case in this section of the book, we may refer at some length to a very self-critical article of Fan Gang that appeared in the *People's Daily*, the official journal of the Chinese Communist party (October 30, 2010, pp. 11 and 17, with italics added) entitled *China's war on inequality*:

> A major new target in the "Five-Year Plan for Economic and Social Development" that China just unveiled is to boost the growth rate for household (disposable) income so that it equals the growth rate of the country's GDP. The reason is simple: over the past 10 years or so, China's household income grew more slowly than GDP, making it a smaller and smaller proportion of total national income.
>
> Many important structural problems have resulted from this trend. Lagging household income has held back private consumption, even though the economy has the capacity to produce more consumer goods. It has also driven up corporate savings, because firms' earnings are growing faster than household income (and, for that matter, faster than overall GDP). That, in turn, may cause higher investment or asset bubbles, as businesses seek to reinvest their savings somewhere. And lagging household income clearly contributes to China's trade surplus, because low domestic consumption tends to keep exports higher than imports.
>
> But there are even more problems related to China's disproportionately small household income, particularly growing income disparity. Indeed, not all of China's "households" have benefited alike from rapid GDP growth. Some social groups, such as skilled workers, engineers, and financial-sector employees, have seen their wages rise strongly. Urbanites – people with formal registration as urban residents – have recorded income gains as well, owing to their coverage by the government-run education system and social safety net. And, as corporate profits have grown, those who share the capital gains in one way or another have also seen their incomes increase faster than the national average.
>
> Those with less education, however, such as migrant workers and farmers, have fared much worse. The former earn an annual salary (including fringe benefits!) totalling $2,000 [that is, in the Purchasing-Power Parity (PPP) terms appropriate to that income level, some $12,000]; the later may earn only half that. They comprise, in roughly equal parts, the low-income workers who account for up to 65-70% of the total workforce. Their average income has grown, but more slowly than the 8–10% annual GDP growth rate of the past 20 years.
>
> The target set by the new five-year plan is thus also a policy manifesto to battle these social disparities, which are now a burning issue for the country. But why did these disparities arise in the first place?
>
> Thirty years ago, 80% of China's labour force was composed of farmers. But, while that figure is down now to about 30%, rural education has continued to suffer from inadequate funding and human capital relative to urban, industrialising regions. That has caused rising inequality between the urban privileged and the rural poor. Geographical differences and government social policies that have been more favourable to urban and industrial sectors may be also be to blame.
>
> But the most fundamental problem is that development does not take place overnight. It takes a long time to bring education and better-paid jobs to everyone. During this long period, some get rich first, which means that others lag behind.

Thirty years of rapid growth brought only half of China's farmers (about 200 million people) into the industrial or service sectors, where their wages doubled. But at least another 150 million famers are still coming into the labour market, competing for higher-paying jobs.

The old generation of farmers may stay put, but younger generations will continue to leave the land, creating a seemingly infinite supply of labour, which suppresses wages in all industries and services for lower-educated workers. As a result, the income of almost 70% of the labour force cannot increase as fast as their labour productivity, and average household income thus cannot grow as fast as the economy as a whole. Given China's massive stock of "surplus labour," this trend will not reverse itself soon. Indeed, the situation may worsen for another decade or two before it improves.

This is why the government wants to intervene. In recent years, the central government has increased its spending on rural compulsory education and poverty relief, and local governments have changed regulations to increase the minimum wage by 20-30% in all 30 provinces.

In the next five years, more efforts in that direction are planned. The social-security system is to be extended to cover all workers and farmers. Fiscal and taxation reform is to be accelerated in order to transfer more income from the corporate sector to households and public budgets for social programs, including low-rent housing for the urban poor and more and better services for new immigrants from rural areas.

But none of this will solve the problem. *The best a government plan can do is to prevent the worst.* Government subsidies may (temporarily) redress income disparity in a developed country where farmers account for only 2% of the labour force, or in a country in which low-income groups count for only 10% of the total population. But in China, where farmers make up 30–35% of the labour force and 70% of the population falls into low-income categories, government can play only a marginal role.

In such circumstances, the old adage that the best social-welfare program is economic growth is all the more applicable. All the lessons from developing countries from the 1950's through the 1980's, and recent lessons from some advanced developed countries, tell us that over-burdening social programs with too great a focus on redistribution may cause large fiscal deficits, debt crises, hyper-inflation, or financial meltdown, with the end result being not a narrowing of disparity but an enlargement.

China must keep these lessons in mind over the next 20–40 years. Indeed, continued growth and job creation is the only real solution for lifting hundreds of millions of Chinese out of poverty permanently.

(*It should be explained that Fan Gang is Professor of Economics at both Beijing University and the Chinese Academy of Social Sciences, Director of China's National Economic Research Institute, Secretary-General of the China Reform Foundation, and a former member of the Monetary Policy Committee of the People's Bank of China.*)

Even in the case of China, the problem of flooding often arises from the arrival of migrants from the countryside who are seeking better job opportunities and higher incomes than they can expect to receive if they stay put, as they tend to live in those parts of cities and towns that are cheaper because they are more liable to flooding. This tendency is of course much more pronounced again in nations such as Bangladesh, India and Pakistan as these are even more exposed to risks from rising water levels in their rivers and their oceans, even as they are less well prepared to face the challenges involved. Our particular interest in China in this place is due to the efforts that are being made there to confront these problems and to mitigate their more serious consequences. The nature of these efforts thus claims a suitable introduction to the present work.

At the same time as the understandably pessimistic view of Professor Fan Gang cannot be ignored, it should be added in all fairness that the situation in central and western China, where poverty has long been most marked, started to change with a totally unprecedented – and almost completely unexpected – rapidity in 2009 and that this change has accelerated further through 2010, causing the Web-Site of the

Financial Times, *FT China Confidential*, that is dedicated to following Chinese developments, to exclaim (with the original italics) that:

> ... Rural China barely merits a mention in most of the unitary visions of the country's future ... That is 721m people unaccounted for. There are good reasons for this. The economy of the countryside is difficult to understand in official statistics, which are patchy at best. Investment analysts rarely travel to rural areas because the general assumption is that they are economically inert, places where people merely eke out a living.
>
> To say this view is mistaken is an understatement. *It is a colossal error*. Building on our rural research last year (CC Apr 30 2009 Big Call and CC Sept 3 2009 All Sections), our network of grassroots researchers has conducted interviews with nearly 300 people in rural China, covering Anhui, Zhejiang, Jiangxi, Hubei, Shaanxi, Shandong, Chongqing, Shanxi, Guangdong, Sichuan and Jiangsu provinces. The story they have uncovered is one of rural take-off. Agricultural land is being traded at a faster rate, farm size and efficiency are increasing, investment in machinery is climbing, basic infrastructure is proliferating, a property construction boom is underway, remittances back to villages from migrant workers is surging, financial services are spreading and a new cohort of – we estimate – 250m-300m rural consumers is starting to flex its muscles (see Consumer China, Big Call, Capital Intensive China and Financial China).

The reasons for this extraordinary change that now looks sustainable – *The FT China Confidential* web site speaks of "tens of millions" of new enterprises! – were quickly identified even as they had previously been largely ignored. The origin was identified as the new (2008) law on the long-term leasing of land – understanding that land cannot be privately owned in China – whereby the newly-minted leases were quickly used as collateral for loans from the secondary banking sector, and the proceeds used to form mostly small and medium-sized companies capable of acquiring the means to increase productivity substantially. This development has also led to a building boom using local resources for the most part, and all of this is currently being further stimulated by extensive rural road building developments by regional governments as initiated by the Chinese central government's 2008 financial stimulus.

This development has gained momentum also due to extensive failures of crops in many other parts of the world, with corresponding dramatic increases in food prices internationally. This serendipity (at least for the Chinese farmers!) has made possible a rapid increase in food prices in China, as in most other parts of the world. Thus with *FT China Confidential* of 4 November 2010:

> Rising food prices are the main impetus behind the China's Consumer Price Inflation's (CPI's) gain in October by around 4% year on year, or possibly slightly higher. In September, the CPI rose to 3.6% year on year, a 23-month high. According to National Bureau of Statistic (NBS), food inflation grew 8% in September year on year. The Ministry of Commerce, however, reported a more than 10% jump in its edible agricultural product index in the same period – suggesting NBS figures may be understating the true level of inflationary pressures
>
> The wholesale price index for agricultural products, published by the Ministry of Agriculture, rose 2.6 points to 176.8 in the first five days of November, an increase of 1.5%. The index, based on prices of more than 400 products, jumped 13.6% in October from a year earlier, the largest increase in two years. We think higher food prices are likely to persist throughout November as producers and distributors hoard supplies. In Heilongjiang, home to China's biggest paddy fields, farmers have sold less than a third of their produce by the end of this week, according to local agricultural officials. The ratio was more than 60% in the same period of last year, officials said.

Less than a year after this phenomenon was identified, however, another, albeit complementary, dimension of this transmutation had become visible, as industrial capacity has migrated ever more rapidly from the already developed East to the cities, towns and villages in the centre and west of the country, even as the coastal areas, in particular, have been forced to move up-market to higher-technology products and services. Of course this has been happening for some years now, albeit little noticed, but as the *FT China Confidential* described the industrial component in mid-2010:

> (This) process is shifting into high gear as the costs of labour, land, utilities and other key inputs in the coastal "workshops of the world" climb significantly. The accelerating shift of manufacturing hubs to towns and cities across inland China has several profound implications for the wider economy.

Following this same source we can outline some of the principal features of this kind of development (again from *FTChinaConfidential*, 29.07.10, but with italics added):

> The migration of manufacturing capacity from coastal provinces to the hinterland is gaining a new momentum that may last for five years or more, transforming the structure of the economy in the process. As an extreme example, Chongqing, which is also now home to a number of appliance plants, reported a seven-fold year on year jump in television production in the first half of this year. The migration is also propelling the urbanisation of central and western areas, which is probably unfolding more quickly than the official estimate of 1% annually, or 13m people moving from the countryside to cities each year. We think the true pace of urbanisation in China (CC Dec 10 2009 Big Call) might, as a rough estimate, be around 3%, or 39m people, moving to urban environments each year. The shift of manufacturing capacity inland is likely to boost industrial output in central provinces – Hubei, Hunan, Jiangxi and Anhui – to grow by between 5 and 10 percentage points faster than in eastern provinces .... Retail sales in central China are also likely to keep growing at around 5 percentage points faster than in eastern provinces ....
>
> *A qualitatively new phase of the capacity shift has started.* Whereas in the past, the migration of capacity tended to involve the relocation of individual factories, supply chains are now shifting too. The implication of this is that, over time, central and western China may start to catch up with the east coast in terms of manufacturing technology and quality.
>
> Beijing's latest plans to spur development in the far northwest is also likely to become a focal point of inland development. By 2015, Xinjiang region is planned to have six new airports, bringing the total in the region to 22, and to increase the number of international flights. The rail network in the region is set to be increased from 3,599 km currently to 12,000 km by 2020 at a cost of Rmb 310bn. Another Rmb 120bn to Rmb 140bn is to be spent on improving roads, the government announced in July.

In much of same vein of these and other previously noted observations, we shall now proceed with discussion of the roots of urban flood risk in order to introduce some of the issues that need to be raised within the context of social justice, albeit a concept that is of the essence of this book.

Chapter 2

Tracing the roots of urban flood risk

2.1 INTRODUCING THE HOLISTIC POINT OF VIEW

As illustrated with the case of a 2007 Jakarta flood event, the roots of urban flood risk can be traced and analysed from different perspectives. The natural and technical (or engineering–drainage, infrastructure-related) roots are for the most part obvious, and in many cases these provide us with the dominant causes of floods and flood-related disasters. However, there are also other roots that need to be taken into account, and by approaching the problem from the most immediately observed and technology-related processes alone we find that our search for so-called *sustainable* and *adaptable* measures cannot yield satisfactory results. Therefore, the important thing is to introduce and maintain *an holistic perspective* throughout the entire flood risk mitigation process. This is one which will enable us to look at risk as a whole, as this emerges through its parts as these become connected to one another in their own rights so as to provide a *phenomenon* (i.e., through an analysis of its root causes and their interactions, as a phenomenon in the exact Husserlian sense: Husserl 1900, 1901/1913) rather than as the whole which emerges through an 'integration-by-superposition of parts' which is the case in the conventional way of working.

The conventional way of working places the whole as secondary to the parts, because it implies that the whole comes after the parts. Since the evolution of our practice has become stuck in this conventional paradigm, which can only pose incomplete questions, and correspondingly provide incomplete answers, it is not surprising that many of our solutions nowadays are limited in their reach and their effect. What is essentially needed is a *new paradigm* that can bring a truly holistic perspective into our view and which can provide us with the ability to recognise what is part of a whole – and what is not. Such a paradigm has to focus on the wholeness of risk and search for meanings and essences of its experience through *phenomenological enquiry*[1] rather than to act within the sole consideration of its parts through

[1]Phenomenological enquiry is a method of enquiry based on the premise that reality consists of objects, situations, experiences and events (i.e., phenomena) as they are perceived or understood in human consciousness and not of anything independent of human consciousness. Phenomenology (from Greek: phainómenon "that which appears"; and lógos "study") is a broad philosophical movement emphasising the study of *conscious* experience (it is often divided into three types: hermeneutical, existential and transcendental). It was founded in the early years of the 20th century by Edmund Husserl, German philosopher and mathematician and the founder of the 20th century philosophical school of phenomenology. Phenomenology, in Husserl's conception, is primarily

measurements, computations and explanations. In order to aid this process and to visualise the complex composition of different root causes (i.e., 'the parts') of flood risk (i.e., a whole risk, or the *holistic risk*) and to introduce the holistic point of view more generally, we shall apply a simplified representation of four distinctive levels of root causes: natural, technical, social and those that are related to the human mind, Figure 2.1. Even though it might be possible to describe most of the root causes with three levels (natural, technical and social) we shall expand our discussions by adding another critical level of the whole – one that is usually ignored or forgotten – namely a level that combines causes that may arise from the very nature of the human mind. A truly holistic paradigm must also include this level in its approach, or otherwise it will be no different to any other piecemeal way of thinking.

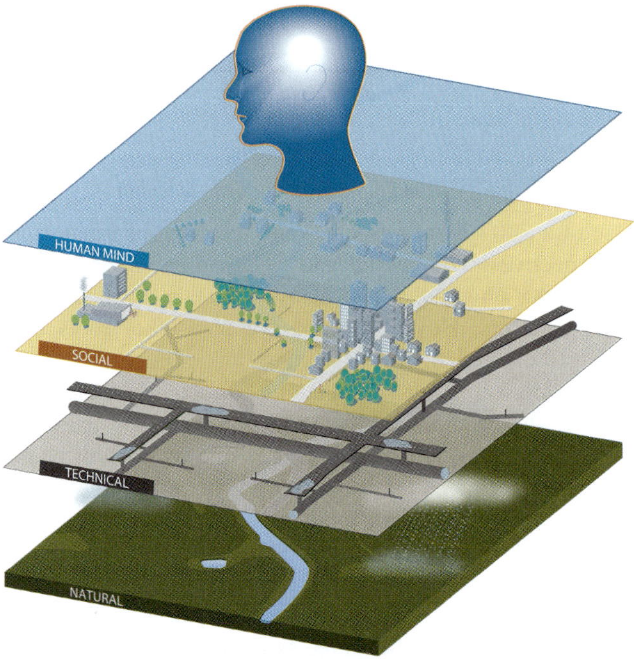

Figure 2.1 Root causes of flood risk: the whole and the parts

concerned with the systematic reflection on and analysis of the structures of *consciousness*, and the phenomena that appear in acts of *consciousness*. It studies the structure of various types of experience ranging from perception, feelings, thought, memory, imagination, emotion, desire, and volition to bodily awareness, embodied action, and social activity, including linguistic activity. It seeks to reflect on the conditions of possibility of experience and cognition, and it suspends metaphysical assumption about the existence of a mind-independent world (i.e., Cartesian Dualism) in order to investigate the phenomenon critically. Thus, *conscious experience* is the starting point of phenomenology and it includes perception, imagination, thought, emotion, desire, feeling and other similar forms of inner experience. What makes an experience conscious is a certain awareness one has of the experience while living through or experiencing it. This form of *inner awareness* has been a topic of considerable debate, centuries after the issue arose with Descartes' sense of consciousness (conscience, co-knowledge) and Locke's notion of self-consciousness. In this vein, we may say phenomenology is the study of consciousness – that is, conscious experience of various types – as experienced from the first-person (or subjective) point of view. For an overview of phenomenology, see for example Moran (2000), Farber (1943) and Spiegelberg (1982), and for discussion of Husserl, see for example Zahavi (2003).

In the holistic approach, the whole (i.e., that constitutes the risk), as this depends upon the parts (i.e., the root causes) emerges simultaneously with the accumulation of the parts, not because it is the sum of the parts, but because the parts depend upon the significance of the emergence of the whole. Therefore, the recognition of a part is possible only through the emergence of the whole. Such a view of the part is particularly important because it shows us the way to the whole. The whole is not to be encountered by stepping back to take an overview. It is to be encountered by stepping right into the parts and then to move into the whole as we pass though the parts in our mind and its productions. Despite the complex and multiple nature of risk, this holistic approach does not lead to fragmentation and multiplicity. Instead, it leads to the perception of different parts within the whole, whereby the unique aspects of a particular root cause are seen holistically within the context of other root causes. In Husserl's famous words, we must proceed in this way "to the things themselves"[2].

Typical examples of natural root causes are heavy rainfall, increasing groundwater levels, river-related issues, tidal surges, and their combinations thereof. The technical (or technology related) root causes are associated with flood mitigation practices, inefficient drainage systems (minor and major) and flood defence structures. The social root causes refer to issues that reflect the way that the society is organised and how it functions in relation to flood risk. The root causes that may arise from the human mind are associated with wrong choices and decisions. For this purpose, we shall make a distinction between two groups of reasons: unintentional and intentional. Examples of unintentional reasons for making wrong choices and decisions are: incompetence, inability to distinguish *right* from *wrong*, unawareness, lack of holistic thinking, fear and so on. Examples of intentional reasons for making wrong choices and decisions are: selfishness, ignorance, greed, wrong values, ideology of money, the desire for power, corruption, self-centredness and other such failures of human character, as these are leading to the destruction of the Earth's ecosystems at a rate that has not occurred, so far as we can judge, since the destruction of the dinosaurs.

The essence of presenting different root causes in the form of levels, as illustrated in Figure 2.1, is to encounter the whole (i.e., the risk) and to better understand the nature of its parts (i.e., its root causes) and their mutual interactions and interdependencies. It is obvious that the root causes contained in the bottom and top levels will influence the two intermediate levels as they are acting as limiting as well as enabling factors concerning how the society reacts to its intimations of flood risk. For instance, if decision makers (or stakeholders) are not acting in the favour of those *primary values* (i.e., human values and those of nature) and the needs of the community which is at risk and if they are selfishly favouring their personal interests instead, then those actions that arise through such decisions are likely to create not only ineffective solutions but also conditions for ever-increasing risks and ever-greater disasters. Similarly, if the natural root causes are not sufficiently understood and carefully considered, then measures that will be implemented may not be necessarily aligned with natural processes, and as such they can also cause escalations of risks and disasters. Therefore, an approach of encountering flood risk through different levels aims to facilitate the holistic view and to provide better insights into the formulations of strategies and actions that need to be undertaken at different levels in order to meet the challenges posed by an ever ongoing urbanisation, population growth and climate change.

Perceptive modes of consciousness play critical roles in the holistic way of encountering phenomena. There is now a growing body of evidence to support the view that there are two major modes of human

[2]This statement is often interpreted as a criticism of scientism, and as an appeal for a disclosure of a more original relation to the world than the one manifested in scientific rationality. It is an appeal for a return to the perceptual world that is prior to and a precondition for any scientific conceptualisation and analysis (see e.g., Zahavi 2003).

consciousness (see e.g., Bortoft 2010). Pursuing engineering education we specialise in only one of these two modes: the *analytical mode of consciousness*. The analytical mode of consciousness is dominated by a quantitative thinking (i.e., a form of thinking which is rooted in modern science) which attempts to calculate the phenomena with the intention of controlling it in a precise way. Attempting to encounter risk from the holistic perspective, which is promoted throughout this book, requires us to look at risk from the *perceptive mode of consciousness* (i.e., intuitions, feelings and other forms of subjective experience) and not only from the mere *analytical mode of consciousness* (i.e., a mode which breaks the whole into parts and approaches each one of these parts in isolation). As the word *consciousness* can be used in so many different ways, the way we shall use it here is in reference to the capacity for inner experience. The perceptive mode of consciousness looks into the whole from an intuitive perception and the nature of the senses which we already carry in ourselves, but which can be further stimulated and enhanced by engaging in promoting *social justice* and its corresponding *ethics*. It is that mode of perceptive consciousness that has the right capacity to transcend the mundane and average workings of the human mind, taking it into higher levels of consciousness, and hence into *the higher states of an ever more scrupulous conscientiousness* (i.e., the states of inner wisdom, concern, respect, responsibility and care). The perceptive mode of consciousness is a way of seeing and experiencing the phenomenon with the sensuous-intuitive mind, with the sensorial–imaginative mind, and as such it cannot be understood by the analytical and intellectual mind alone. As Wittgenstein[3] so memorably observed, "understanding means that we see it in action". In our case this refers to our thinking in terms of situations, both physical and emotional, and thereby bringing the danger into coincidence with the potential flood victims. This can be only experienced on its own terms. Therefore, the holistic mode of consciousness, even as it is complementary to the analytical one, transcends it. It is this that leads us to a *sociotechnology of flood risk mitigation* that is ultimately based upon a social justice that applies universally, such as is treated later in this work. This approach depends upon modern science and its attendant technology even as in this case it correspondingly also transcends it, and thereby it leads to an entirely other way of treating these human, all to human, disasters than those which are currently employed. We shall later introduce the notion that:

"a fascination with technology is a fascination with truth",

but then in such a way that this fascination should never degenerate into a one-sided 'idolatry of technology', such as would threaten the most basic spiritual values of *benevolence, benignity and compassion*, those timeless values which are our only certainties in this life.

It is important to emphasise that our discussion here does not imply that we want to disregard modern science. On the contrary, the purpose of superimposing perceptive modes of consciousness (i.e., subjective experience) is rather complementary and it is to give the phenomena its intensive depth so we can have a more comprehensive and complete encounter with it. Such a form of experience is perceived through the mind when the mind functions as an organ of perception and not only as a medium of

[3]Ludwig Josef Johann Wittgenstein was an Austrian philosopher who held the professorship in philosophy at the University of Cambridge and played a central role in 20th-century analytic philosophy. His work continues to influence current philosophical thought in topics as diverse as logic and language, perception and intention, ethics and religion, aesthetics and culture. He is the author of one of the most influential works by the 20th-century: Philosophical Investigations (Philosophische Untersuchungen). In it, Wittgenstein discusses numerous problems and puzzles in the fields of semantics, logic, philosophy of mathematics, philosophy of psychology, philosophy of action, and the philosophy of mind.

logical thought. A truly holistic thinking is a vision much needed today in view of the limitations of modern science which have now become so evident. Such thinking does not view modern science as an autonomous activity which stands outside of history, culture, human and social contexts, pursuing knowledge on its own. In fact, it recognises that there could be two different kinds of perspectives of a phenomenon as they reveal a phenomenon in different ways: one that describes the casual order in a phenomenon and the other that describes the wholeness (as illustrated with example of reversible cube given in Figure 4.2, Chapter 4, Part II). If we consider the Jakarta 2007 flood event, we can note that the same event can be experienced in one way by those who were directly affected by this event and in another way by those who read about it as it was reported in the newspapers. The same event can be experienced in yet another way by those who were involved in the planning and implementation of flood protection measures. In this respect, qualitative *phenomenological* methods offer the necessary means to encounter the variety of different experiences, thoughts, reasons and perceptions and to take them into consideration for decision making. Correspondingly, the role of *hydroinformatics environments* becomes an important element in this process as it can assist in initiating and stimulating cognitive processes of flood risk perception (discussion concerning *hydroinformatics environments* will be taken up in the subsequent chapters of Part II and again with some illustrations in Chapter 20, Part IV).

In what follows we provide further discussions of each of the four levels of root causes illustrated in Figure 2.1.

2.2 TRACING THE NATURAL ROOTS OF URBAN FLOOD RISK

As mentioned earlier, typical examples of natural processes are heavy rainfall, increasing groundwater levels, river-related issues, tidal surges, and their combinations thereof. In this section we shall provide a brief overview of these processes, while a more detailed discussion will be taken up in Part III of this book.

The amount of runoff that can be produced by a rainfall event over an urban area depends on the rainfall characteristics (type, pattern, intensity, duration, spatial variability, etc.) and the surface characteristics (slope, permeability, extent of impermeable surface, surface depression storage, type of surface vegetation, degree of moisture already present on the land surface, efficiency of a natural or man-made drainage system to convey the flood water, etc.). Surface runoff occurs when the infiltration of water from rainfall or melting snow is limited due to the paved nature of the surface or the underlying or exposed soil is fully saturated and the excess water flows over the surface.

High groundwater levels lead to water rising to the ground surface, usually following prolonged periods of excessive rainfall. Groundwater flooding differs from surface water flooding by its persistence, with a typical duration that is measured in weeks rather than hours and days and has a tendency to occur throughout the winter, often extending into spring and sometimes into the early summer.

Similar to groundwater processes, fluvial and estuarial processes can also lead to flooding if there are prolonged or intense rainfall events. If significant rainfall occurs after a wet period the soil could become saturated and the resulting runoff volumes and rates could also achieve significant proportions such as could lead to an overtopping of flood defences and flooding.

Coastal processes can also be a cause of flooding in situations where high tides and extreme weather conditions have the potential to affect urban areas along the coastal zone. The water pushed towards the land area is often referred to as a "storm surge" and this process is caused by a combination of high tides (which are of course controlled by the movements of the sun and the moon) and waves. Such conditions operate across a range of space and time scales and they can be predicted with varying degrees of certainty. The storm tides and associated waves can rise quickly to overtop coastal barriers, threatening coastal dwellers, disrupting road or rail traffic, and flooding and eroding valuable coastal real estate and productive farm land. Furthermore, erosion of the coastline can reduce the capacity of a beach to restore itself, leaving it increasingly vulnerable to future storms.

Tsunamis represent yet another cause of coastal floods which can cause some of the most devastating disruptions. They are triggered by undersea earthquakes that cause massive changes to the ocean floor. Tsunami can also be caused by seismic events, landslides, or even asteroid impacts. When an undersea earthquake or other major disturbance causes a section of the ocean floor to suddenly rise or sink, the mass of water above the affected area also rises or sinks. This unexpected movement of the water creates a series of powerful waves which can be devastating for urban areas along the coastal zone. In a recent event following a major 8.9 offshore quake that occurred on 11th March 2011 at 2:46 pm local time a massive wave of water, as high as 38 metres in some parts, reached more than ten kilometres inland of Japan's northern Pacific coast and swamped entire towns. The Japanese National Police Agency has confirmed close to 15,000 deaths, about 5,000 people were injured and almost 11,000 people were missing, as well as over 125,000 buildings damaged or destroyed. According to the US Geological Survey and Japanese seismologists, this quake was the largest to hit Japan, the fifth strongest tremor worldwide since 1900 and the seventh strongest in history.

In addition to all this, climate change has become one of the defining challenges for those concerned with floods and flood related disasters. The reports show unprecedented increases in currently measured global temperatures. The duration and timing of seasons are changing. The frequency and severity of floods and cyclones, accompanied by rising sea levels, are increasing. Hence, we can add that some consequences of climate change also contribute towards urban vulnerabilities and pose a threat of an increasing flood risk. Even though climate change may be seen as an act of a nature, with many earlier but unmeasured precedents, it is now becoming evident that such phenomena may be also attributed to human-induced greenhouse gas (GHG) emissions. Increased levels of heat trapped in the atmosphere have set-off a process that is modifying weather patterns, which in turn affect temperatures, sea levels and storm frequencies. As such, this is likely to have an impact on cities and other urban areas, especially those in coastal zones. Therefore, in order to address increasingly frequent and extreme climate change events, considerable efforts should be made to develop the appropriate strategies for proactively reducing activities that involve GHG emissions.

2.3 TRACING THE TECHNICAL ROOTS OF URBAN FLOOD RISK

In addition to the natural root causes, technical root causes can also be identified as another dominant cause of floods and flood-related disasters. Such issues are often associated with poor drainage systems (e.g., lack of the system's capacity, poor design, bad construction practice, lack of operation and maintenance, etc.), inadequate flood defence structures and local obstructions across overland flowpaths.

Engineering and its product, technology, undoubtedly play an important role in the process of flood risk mitigation. *Technology, for the immediate purpose of this part of the book, shall be defined as a system of knowledge, skills, expertise, and organisation used to produce and utilise goods and services that are needed to minimise flood risk and satisfy social demands.* An obvious example of such products is an urban drainage system which is primarily composed of pipes, channels, culverts, manholes, catchpits, outfalls, pumping stations, bifurcations, treatment and other ancillary structures, detention/retention/infiltration basins, watercourses, dams, etc. Such systems are principally designed to control or convey stormwater to the main receptors, such as the main river channel or the nearest large body of water (e.g., lake, coastal waters, groundwater, etc.). Since these systems are man-made creations intended to collect and transport the stormwater runoff they are regarded as structural stormwater management (or flood mitigation) technologies. In many cities around the world drainage systems are not in a good shape due to a variety of reasons. For example, in some cases poor design and construction practices can be a primary cause for a system's malfunctioning, whereas in some other cases the main cause of flooding can be the lack of sufficient cleaning and maintenance activities. The flooding in slum areas occurs many times a year because there are very few drains, most of the ground is highly compacted and pathways between dwellings become streams after heavy rain.

For those cities that are situated on low-lying areas in the middle or low reaches of rivers the flooding may occur due to the low capacities of the river channel, dykes, levees, dam failure or the inability of the local drainage system to discharge the stormwater due to the elevated water level at the outlet. Thus cities located along river and estuarine reaches may have to bear the combined impacts or fluvial (as well as coastal) and local (pluvial or groundwater related) floods.

The ultimate aim of the appropriate flood risk mitigation practice is to strive for the right combination of technical and non-technical measures that can reduce flood risk from all kinds of floods to the level that is socially just and ecologically sustainable. A further and more detailed discussion of the technical as well as the natural causes of urban flood risk is given in Part III of this book.

2.4 TRACING THE SOCIAL ROOTS OF URBAN FLOOD RISK

The following discussion is focused towards those issues that reflect the way that the society is organised and how it functions in relation to flood risk.

In geographic terms, many societies are situated at locations where large river systems meet the ocean and as such they are subject to meteorological and hydrological hazards. For most of the cities in developing countries this is a colonial legacy as these sites provided access into the colonised territories for the colonial powers. These cities have grown considerably over time despite the efforts to shift the capitals into other territorial locations. Apart from other threats, these cities also remain at risk from storm surges, hurricanes, typhoons and tsunamis. Many examples of such cities can be found in Africa, Asia, Latin America and the Caribbean.

With the continuous process of urbanisation, it is now becoming obvious that urban areas are increasingly becoming sites of environmental risks for their residents (see e.g., IFRC/RC 1999). The process of urbanisation has been viewed as a reflexive one, which in social theory is regarded as a circular relationship between cause and effect, or in other words, the one that contributes towards its own increasing risk through different scale of degradation (local, regional and global: see Hardoy *et al.* 2001). From the urban flood mitigation point of view, it can be observed that there is a nonlinear relationship between the ever changing social, technical and natural processes that are acting in an urban area. Such relationship has been described as *coevolutionary* (see Pelling 2003). This phenomena can be illustrated by the fact that flood mitigation interventions have social consequences and the resulting social changes react back on further improvement of flood mitigation practices by creating ongoing positive feedback cycles. It is inevitable that an urbanisation process brings with it a transition of a catchment from an initially natural one to an urban one which has modified the environment through deforestation and slope instability, increases in the frequency and severity of flood events, contamination of surrounding waterways and lowering of groundwater tables, commonly followed by land subsidence. These changes have a direct impact on the risk to urban communities and it is often the case that the urban poor takes the most of the negative impacts of such changes.

The United Nations reports state that more than half of the world's population lives in cities, with about one billion people living in urban slums. There are now 7 billion people in the world and this figure is expected to surge dramatically to over 9 billion by the year 2050, with Asia and Africa leading the growth. It is anticipated that by 2030 81% of the world's urban population will live in developing countries. The largest migration pressures will come in the 18 megacities (i.e., the cities with populations in excess of 10 million) situated in coastal areas. About 75% of people residing in low-lying areas are in Asia, and the most vulnerable are, again, the poor. Slums (a.k.a., 'informal' urban areas, squatters, shanty towns, etc.) are growing at the same rate as urban growth. It is expected that most of the growth will occur in developing countries, particularly in regions which are already under water stress, and which lack access to safe drinking water and inadequate flood protection and sanitation facilities. These people live in difficult economic situations in places which are often severely affected by floods and other kinds of disasters and as such they represent the most vulnerable group of urban populations.

It is the case for many cities that the low- and middle-income residents are excluded from the formal housing market because of their economic poverty and inappropriate financial legislation that makes it very difficult for them to obtain loans to build or buy dwellings in formal settlements. Such exclusion *forces* people to occupy urban floodplains and other unwanted land. For example, in Manila, informal settlements at risk of coastal flooding make up 35 percent of urban population. In Calcutta, 66 percent of the population lives in informal settlements and as such is exposed to regular flooding and cyclones (Wisner *et al.* 2004). Furthermore, poverty and social exclusion are often related to the lack of education, which again makes these communities more susceptible to floods and flood-related disasters. Inability to access critical communication facilities and/or to understand the messages released by disaster management authorities during flood events can deepen the effects of flood disasters yet further.

It is also argued that the dense pattern of urban living can contribute to generation of risk and vulnerability to natural disasters. Hewitt (1997) argues that high population densities can lead to increases in urban vulnerability through increasing concentrations of energy and energy transportation routes and their proximity to residential areas. Secondly, living in crowded conditions can increase risks though traffic congestion which can also constrain the relief efforts. It is further argued that cities are complex systems with many interactions between different elements where some elements are less predictable than others. This means that during and after disastrous events some groups of an urban population can become the targets for exploitation and repression.

It is also the case that the lack of expertise to connect urban planning practice with drainage infrastructure management is yet another aspect that can contribute towards floods and flood-related disasters. The lack of appropriate institutional, financial and legal arrangements has also been regarded as one of the causes for flood related disasters. A piecemeal approach to operation and management of different urban water phases is one such example (for illustration of urban water phases see Figure 12.14, Chapter 12, Part III). The rapid supply of housing to meet rising demand without appropriate building codes and regulations can further increase the immediate impacts of floods. In this regard, unintentional or even intentional, inappropriate planning and legislation activities can increase the vulnerability of urban areas by dramatic proportions. Unintentional forms of inappropriate planning can be found in societies with an absence of relevant expertise and poor or nonexistent communication lines between drainage authorities, town planners and other water/environment-related organisations, whereas, intentional form can be found in cases of corruption and conflicting political interests. There are many cases where urban planning decisions are driven by greed, ideology of money, the desire for power, self-centredness or other such failures of human character. We can say that they are related to selfish and ignorant minds. Thus the roots of some of our flood-related problems lie in the human mind as much as in technology, or economy.

Also, for some societies, race, class, ethnicity, immigration status, gender and other forms of inequalities represent significant issues that can have detrimental effects during floods. In these societies, for people that are in the targeted categories, such issues will restrict their capacity to anticipate, cope with, resist and recover from floods and flood-related disasters. In Dhaka, Bangladesh, the urban poor are mainly rural migrants whose lack of access to secure housing and livelihoods is compounded by the absence of support (Rashid 2000). Hurricane Katrina and the disaster that unfolded provided an example of race and class inequalities in the United States where such differences have demonstrated the serious disadvantages to people in targeted racial categories during and immediately after this disastrous event (Alterman 2005). Gendered social science rests on the social fact of gender as primary organising principle of societies and the conviction that gender must be addressed if we are to include all people living in risky environments. In the UNDP report (Wiest *et al.* 1994) it is stated that girls are vulnerable to sexual abuse and exploitation following disasters, and displaced girls especially. For example, CARE International (2002) report states that in Cambodia women and girls were found to be at risk of rape and sexual abuse as they were forced to migrate to find work after floods as a debt repayment strategy. Following the 2004 Indian Ocean tsunami, there were numerous media accounts of violence against women and sexual exploitation of girls. All of these issues underline the importance of local cultural and socio-economic factors in the flood risk assessment process and the dangers inherent in any search for overly simple generalisation about different aspects of vulnerability.

The crucial point about understanding what creates flood risk is that it is not only a result of natural events and insufficient engineering (or technical) measures alone. It is the combination of different aspects in which the natural and sociotechnical aspects cannot be separated from each other. Therefore, the useful contribution to the urban flood risk mitigation practice is to depart from the well established piecemeal approaches that only privilege technocratic and materialistic measures and to move towards sociotechnical approaches that are driven by legitimate ethical and social considerations defined by stakeholders who can reflect all the dimensions of urban flood risk in order to build more socially acceptable and long-term solutions. This is not to undermine the contribution of flood protection made by engineering works or building codes, but rather to suggest that the flood risk mitigation practice that omits a social and ethical component will be severely restricted in its scope and its reach.

2.5 TRACING THE ROOTS OF URBAN FLOOD RISK RELATED TO THE HUMAN MIND

So far we have discussed the threats facing urban areas in terms of natural, technical and social issues and concerns and we have explained how their impacts can contribute to the formation of urban flood risk. All these issues and concerns are very real and very urgent and as such they certainly should receive our fullest attention. However, if we consider that our current way of thinking is very much dominated by the values of an ever more materialistic world with its *ideology* (or idolatry) of money, with greediness, self-centred and ignorant attitudes, then many of the previously mentioned issues may appear only as symptoms of a much deeper underlying problem. In more exact terms *this problem is an epistemological*[4] *one.*

So the questions that we wish to address at this point need to be placed in a much broader perspective: What is at a deeper root of so many of our water-related and environmental problems? Will we be able to solve these problems in a truly (i.e., socially and ecologically) sustainable terms? The answer to these questions certainly depends on the conditions under which we can hope to be successful in contributing to

[4]Epistemology is the branch of philosophy concerned with the nature, origin and scope (limitations) of knowledge. The field is sometimes referred to as the theory of knowledge. It addresses the questions of *What is knowledge? How is knowledge acquired? How do we know what we know?* Epistemology (i.e., how we know things) is combined with ontology (i.e., what things exist) to form the branch of philosophy which is referred to as *metaphysics*. Besides epistemology, the common literature defines the other three branches of philosophy: ethics (which attempts to understand and prescribe conduct), politics (which attempts to describe how we should interact with one another), and aesthetics (which discusses questions of beauty and taste).

a sustainable and just environment. This is to say that the many crises which we face in our societies nowadays are the result of decisions that we have made in relation to our values. They may have been made by individuals or groups, without awareness of their impact or in ignorance or out of selfish motives. Whatever the way or the reason, we can say that they are the result of our *thinking, beliefs* and values.

Values are commonly regarded as aspects of people's identities that reflect what they deem to be desirable, important, and worthy of striving for in their lives: see for example, Rokeach (1973), Schwartz (1992) and Crompton and Kasser (2009). There is a substantial amount of empirical and theoretical work available which demonstrates that values have important ramifications for people's attitudes and behaviours: see for example Feather (1992) and Crompton and Kasser (2009). This is because values are understood to be higher-order cognitions that influence human attitudes. We shall now discuss how those values that are based on *ideology of money* can lead to the formation of various kinds of problems, including the formation of urban flood risk: see also Chapter 20, Section 20.4.10.

Our starting point in this venture is the observation that the threats and acts of wars, the greenhouse effect, the destruction of the rain forests, soil erosion, pollution, the energy crisis, the economic crisis, the food crisis, the water crisis, the housing crisis, the sanitation crisis, the AIDS crisis, and all the many other crises that humanity faces nowadays are symptoms of a deeper *ideological crisis*, which now attains to the dimensions of a *spiritual crisis*, which has been driven by our beliefs and values towards our so-called 'prosperity'. Unfortunately, prosperity is all too often interpreted as wealth and measured exclusively in terms of money, whereas 'to prosper' actually means 'to succeed and to thrive', and money is simply one of the primary constructs that our society has invented to facilitate this process. What is terribly missing in such a notion of prosperity is the true essence of the prosperity of a society, namely the success and the wellbeing of its citizens as a whole, as this remains dependant on so-called 'soft' issues which are difficult if not impossible to measure in monetary terms, such as happiness, peace and health. When speaking of societies, and particularly those in the 'Anglo-Saxon' world, our thinking is caught up in a welter of traditional economic indicators, such as 'growth in GDP' and other measures of economic performance that are expressed exclusively in monetary terms. What does it mean and what does it really matter if Germany and Japan have had a lower GDP growth than the USA and the UK for almost a decade, when the real standard of living of most of their people has been increasing faster than in the US and the UK? We then argue further that to idolise money in this way as the exclusive measure, not only of economic, but also of social performance, and to follow such figures slavishly correspondingly, is to miss the point that the real purpose of measures of social performance should be those of indicating the prosperity of a society, which should be equivalent to the wellbeing of the society's citizens as a whole, despite differences in incomes. That the richest 400 families in the USA have the same wealth and incomes as the 150,000,000 less fortunate citizens is just such a monetary measure, albeit in this case indicating something of the gross inequalities that increasingly prevail in that nation at this time of writing – and in this case the differences in wellbeing are manifestly flagrant.

So far as the so-called 'Third world' is concerned, these differences are often even more flagrant again, as they then surpass in their inequalities in wellbeing any measure of money estimation, as vividly described, for example, by the Arandhati Roy, the much prized Indian novelist and political activist.

Since GDP, GNP and other similar measures are based on aspirations for financial success they are closely linked to *outer* and *materialistic* values and as such they fall far short in reflecting *self-transcendent values of benevolence* (being honest, ethical, helpful and loyal) and *universalism* (caring about environment, a peaceful world and social justice): see also Crompton and Kasser (2009). Furthermore, a variety of studies have shown that such measures are more likely to create negative rather than positive attitudes (i.e., people are less likely to work with each other and support social and ecological well-being): see for example Saunders and Munro (2000), Schultz *et al.* (2005), Good (2007) and Crompton and Kasser (2009). In response to this several alternative indicators have been proposed. These include proposals to directly measure citizens' psychological well-being, the Kingdom of Bhutan's gross national happiness measure, Redefining Progress' Genuine Progress Indicator (Talberth *et al.* 2006) and the New Economics Foundation's Happy Planet Index (NEF 2006). Such indicators include measures of social cohesion and trust, of life satisfaction and feelings of vitality, of volunteering and caring for others, and of environmental health. Crompton and Kasser (2009) have shown that if such alternative indicators are adopted and reported, decision makers and citizens would receive a fundamentally different set of messages which in turn could enhance social and ecological wellbeing, respect and care for others, social cohesion and sustainability.

Despite these observations, however, we still continue to see much of the business and, especially, the financial media, bombarding us more and more with "financial information". To put it in simple words, we live in a world where just one measure, the value of money, dominates so many other values. Correspondingly, it is then explained to us that this great construction and all the others around it in, for example, The City of London, are engaged in *making money* – and then by legal means and not by counterfeiting! – and even a banknote is produced to show such dilettantes as ourselves what money *is* – we observe that what is shown to us is only a piece of coloured paper bearing a thread and embossed patch of metal, such as would be all but valueless in most parts of the world without the immediate presence of a money-changer. One of the biggest dangers is a mindset that is obsessed with money is that it leads to a particularly destructive, and even self-destructive, *ideology of money* or *idolatry of money*, which has been a main cause for so many inequalities between people in so much of the world. Being driven by such a destructive ideology it is not surprising that our motivations, which are associated with those deformities of human character such as greed, ideology of money, desire for power and self-centredness, are leading to increasingly destructive actions and outcomes in many areas of our lives. Since, as again we shall describe later in this book, the long-established Meinongian object-value theory (Findlay, J.N., Mainong's Theory of Objects and Values, Oxford University Press, 1963 // Gregg Revivals, UK, 1995) demonstrates *that neither numbers nor money **ex**ist in our external world, but can only **sub**sist in our internal or **sub**jective worlds.* Certainly the numerals **ex**ist and bank notes, share certificates, etc. **ex**ist, but, when viewed epistemologically, these only become numbers and money so long as we believe that they represent numbers and money.

The *ideology of money*, as this descends further to the *idolatry of money*, seduces with false promises, such as those of higher money incomes, of better 'life styles', and in brief with promises of success, fame and fortune – promises which it has no intention whatsoever of keeping. It correspondingly causes us to place higher values on a personal, egocentric, materialistic world than on anything else. This is evident in many professions and in many cultures. Obviously, the roots of such a problem lie deep within our minds, at the very heart of our functioning: as our minds condition and decide the choices that we make, we are lead to use our creativity and our knowledge of technology in ways which may benefit us, but only in the short term – and even this is questionable.

What drives us in our daily lives is not so much associated with our physical needs. It is much more associated with our ideological, or even psychological, needs. Led by the greater dominance of materialistic values we face an ongoing need to satisfy our egos, which we experience collectively as our *outer self* even though this is totally different from *the inner, or authentic, Self* that we shall define later in this book. We may think that we can become happy through accumulating material wealth and we may worry about our outer appearances, but the only result is that we are spending a considerable amount of time and energy on protecting, maintaining and developing this *outer self* and with this our *egos*. Unfortunately, despite the tremendous advances our culture has made in material development, we have made little, if any, progress in developing our *inner self*, and so of allowing *the true, the authentic,*

Self to manifest itself. What we need nowadays, apart from the basic commodities necessary for our individual well-being, is not more things in our outer worlds, but awakenings of our true inner natures (i.e., our *authentic consciousness and, inseparately from this, our conscientiousness*). This is not any special insight of the authors of this book or any particular one else's own teaching, but it is very much a universal teaching that is embodied in most philosophy and almost all serious theology. It is now almost two thousand years since the Greek philosopher Epicitus explained the reasons why we may be happy or unhappy. He wrote that happiness has nothing to do with what we have or what we do not have; it is a matter of how we perceive things. The Vedic Masters, the Buddha, Christ and Mohammed, among many others, have put forward very similar ideas and for the most part long before Epicitus, and so have many other teachers and philosophers over the ages. Thus the roots of many of our problems nowadays lie in the states of the human mind. A part of the solution is not in discovering any new knowledge because *the basic wisdom* already exists, even though it has been less and less practised over the last millennium. What we need to do now is to incorporate this timeless wisdom within the field of our endeavour in a way that can give us the will to achieve the states of social justice to which we aspire. It has been the great achievement of a *Notabile* of authors, such as Rudolf Steiner, Ernst Cassirer and Hans-Georg Gadamer, that they have prepared the way to this deeper understanding that Henri Bortoft has so brilliantly summarised and as has already been very briefly introduced above. We shall return to an outline of the nature of this deeper understanding in Part II of this work.

With the development of technology we have amplified our ability to change the world around us to serve most of our purposes, and in ways far beyond the capacities of any other creature. However, as soon as the greedy and ever more hungry ego takes dominance over the needs of our society and of our fellow creatures then the use of any higher technology starts heading for trouble. This was expressed already in Gregory Bateson's *Steps to an Ecology of Mind: Collected Essays in Anthropology, Psychiatry, Evolution and Epistemology*, and his most relevant aphorism that "all experience is subjective", the consequences of which we shall introduce with Bortoft's own contributions, also in the second Part. For the moment we may just take one quotation from Bateson's work (Bateson 1972):

> The myth of power, is of course, a very powerful myth; and probably most people in this world more or less believe in it … But it is still epistemological lunacy and leads inevitably to all sorts of disasters … If we continue to operate in terms of a Cartesian dualism of mind versus matter, we shall probably also come to see the world in terms of God versus man; élite versus people; chosen race versus others; nation versus nation and man versus environment. It is doubtful whether a species having both an advanced technology and this strange way of looking at the world can endure …
>
> The whole of our thinking about who we are and who other people are has got to be restructured. This is not funny, and I do not know how long we have to do it in. If we continue to operate on the premises that were fashionable during the Pre-Cybernetic era, and which were especially underlined during the Industrial Revolution, which seemed to validate the Darwinian unit of survival, we may have twenty or thirty years before the logical *reductio ad absurdum* of our old positions destroys us. Nobody knows how long we have, under the present system, before some disaster strikes us, more serious than the destruction of any group of nations. The most important task today is, perhaps, to learn to think in the new way.

That is to say that a great deal of the problems that confront us nowadays come as a result of the materialistic and piecemeal (and, we can add, technocratic) way of thinking. We are not likely to solve these problems in a sustainable way with the mindset that has brought us into this situation. Hence, we have to change the mindset and its corresponding values! But that is again the principle objective of the Part II of the present book as it comes to terms with the paradigm of decision making by the active stakeholders and the role of hydroinformatics in realising this.

Following closely upon different causes of floods and flood related disasters and aiming to create the means to minimise their negative impacts we say that we engage in a sociotechnical development. In doing this, we can subsequently realise the distinction between that which drives the ever ongoing processes of nature and that which drives the processes of human creativity. As we follow these in several manifestations and in depth, we shall see that this difference is reduced and even in some cases disappears when we extend our technology so as to comprehend its human interventions, making of it a *sociotechnology*. We shall then see that we cannot really understand the creative aspects, the *poietics*[5], of technology without integrating them into the parallel *poietics* of those social, including institutional, changes that catalyse technology and that are in their turn catalysed by technology again, a process that occurs once more, so to say, 'in its own proper time'. The following section correspondingly introduces the nature and role of sociotechnology within the context of flood risk mitigation.

2.6 SOCIOTECHNOLOGY AS A MEANS FOR FLOOD RISK MITIGATION

Whether operating a stormwater collection network, or developing an early warning system, it is clear that there are social and technical aspects to almost all the activities involved. True as this may be, we traditionally separate the social aspects from the technical aspects. The social domain refers to the design and implementation of social and institutional participations, policy analyses (including economic analyses), recommendations, legislation, regulatory and legal frameworks, etc. The technical domain is associated with engineered tools and systems for efficient transformations of the material world to minimise flood risk. More succinctly, we can define the social domain as the realm of human-to-human relations and the technical as human-to-artefact relations.

Within our modern society, what we shall subsequently identify a *modern science* as instituted in the *Condemnations of 7th March 1277*, has become the epitome of human rationality as the only legitimate form of knowledge, and both the domains of the social and the technical claim a reliance on modern rational-scientific knowledge as then instituted. The scientific method of argumentation, experimental validation, and logical proof in revealing laws of nature as scientific theories has enjoyed a spectacular success and as a result its methods have been extended to almost all domains of human thought and action, and particularly to the understanding of physics of floods and our dealings with such phenomena. In this paradigm, all other knowledge that is not susceptible to modern scientific argumentation, validation and proof, is regarded as something inferior and is marginalised, regardless of its power to maintain social bonds and structures. This then has the effect of categorising all other knowledge coming from customs, rituals, authorities, religion, tradition, spirituality, etc … as 'underdeveloped', 'savage', or 'primitive', as pre-scientific residues that require science to 'enlighten', 'develop', 'civilise' and 'educate' them.

With this construction, science is supposed to be objective and value-neutral or value-free (i.e., free from the influence of any values that are not purely epistemic), but this approach obscures the enormous difficulty of making the transition from establishing scientific truth to prescribing actions that are just and ethically legitimate.

Technology, still often naively regarded as 'applied science', then also becomes independent of any social context, supposedly suitable for solving any problem as and by itself. One of the many consequences is that negative uses of technology are not blamed on technology, but only on those who

[5] *Poietic* is a term used in semiotics, the study of signs, to describe perceptive and productive levels, processes, and analyses of symbolic forms. The meaning of poietic (which comes from Greek: ποιητικός "creative") refers to productive or formative and æsthesic (from αἴσθησις "sense") refers to receptive or perceptive processes. We shall return to these terms in the subsequent chapters.

misapply it. A classic example of this is the argument that 'guns don't kill people; people kill people' whereby the technology itself is divorced from the social environment within which it has developed and continues to dominate. Hence a 'modern' technology, like any modern science, is seen as autonomous and neutral. This is referred to as *technological determinism*. Problems spawned from technology will, in this view, be solved by technology alone.

As we move from a modern condition to a post-modern condition of society, that is, from a society of knowers to a society of consumers of knowledge, so we see that our current stage of development no longer allows the continuation of this compartmentalised approach. In the new post-modern condition of society, social arrangements are sustained with technological support (particular examples being the various developments based upon the real-time communication, information and control systems, Internet and mobile telephony) and their corresponding technological developments are often strongly conditioned by our social arrangements. The two aspects are inseparable and must be recognised as such. As mentioned earlier, the social and technical have never been separated in reality, but they have been separated conceptually. This conceptual separation may well have been useful in that it allowed for the accelerated development of these subjects, both as sciences and as technologies. However, this rapid development has now had such far-reaching social and environmental impacts (including both past and currently-perceived future impacts) that the implications can only be managed by admitting the 'hybrid' or sociotechnical nature of so many of our contemporary problems.

As introduced earlier, sociotechnology is a *conjunctive knowledge* in the sense that it joins together the social and technical domains, thereby creating a new field of knowledge quite different from the fields that it unites. Thus, we repeat, sociotechnology is not then the sum of the social and the technical but the ability to combine them in the widest sense, and specifically when solving problems. This apparently simple concept provides a new perspective on how we approach flood risk problems.

This *sociotechnical approach* applies to all flood-related problems irrespective of scale, although larger scales bring more 'hybridisation'. The awareness of the sociotechnical nature of urban flooding, as it was demonstrated earlier on the case of Jakarta 2007 flood event, brings with it the awareness that there are increased numbers of stakeholders, a correspondingly increased diversity in interests and attitudes and sub-divisions of the stakeholder groups, even down to the level of individual citizens. Recognising this and understanding the sociotechnical nature of problems leads to a greater integration of stakeholders and, within the context of flood risk mitigation, this in turn can be enabled using various kinds of technologies.

To illustrate this with examples, the effectiveness of an early warning system depends strongly upon a community's level of understanding of the hazard type and acceptance of the warning issued to them. Equally important is the credibility of the authority that issues the warnings, whose reliability in alerting the public is measured according to its track record in predicting an authentic warning. Furthermore, the effectiveness of such technology is also judged on whether the warning enables appropriate and timely decision making by disaster mitigation authorities and those who are at risk. This example is quite often used for the promotion of public and stakeholder involvement, but such an involvement needs to be at a much deeper level than even this example suggests. As another example we look to the selection and construction of flood defence structures which necessitate not only agreement with stakeholders on a particular structure (or a set of structures) and the limits of the impacts that such construction works could bring, but also their active involvement and empowerment to stop the work at any point if the pre-agreed limits are succeeded. So in this example such projects would need technological developments, namely advanced off-line and on-line monitoring, remote sensing and earth observation, numerical modelling, GIS, multimedia and the Internet, together with such social developments as stakeholder participation and empowerment, through providing ethically legitimate information. The combination of these technological and social environments provides an essentially sociotechnical solution to one of the greatest challenges involved in flood risk mitigation.

Both of these examples draw on the importance of active stakeholder participation in the decision-making process when looking at the problems and solutions from a sociotechnical point of view.

In the next section we shall describe the evolution of flood risk mitigation practice and discuss some of the latest concepts and definitions that are used in the current literature.

2.7 THE EVOLUTION OF FLOOD MITIGATION PRACTICE

Flood mitigation practice has evolved from the need to mitigate and minimise negative impacts of floods. Given the scale of impacts of flooding on the community and the economy, it is not surprising that there have long been attempts to mitigate its effects. Most of the early efforts were focused on preventing floods by constructing various kinds of structures (e.g., levees, floodwalls, dams, embankments, storage basins, diversions, etc.) according to a certain level of protection (a.k.a., *safety standard* or *standard of protection*) across the region even though it was unclear what such protection may mean to the surrounding environment and how it's effects and limitations can be measured. A weighing-up of costs and benefits was used as a primary means for selecting different flood protection standards and measures. We can recall that the origin of such methods can be traced back to the time of Jeremy Bentham (1748 – 1832) who developed the maximum utility concept ('the greatest pleasure for the greatest number', or in other words, 'maximising benefits and minimising costs' – the doctrine of utilitarianism) and which is still very much present in the current flood risk mitigation practice. As will be further exemplified later, it is an appealing but very much a simplistic way of assessing the benefits of different measures (including flood protection standards) even as other aspects, such as ethical consideration of impacts on society and the ecosystem, are completely left unattended. Another drawback of such an approach is that in some areas the costs of ensuring a uniform level of protection could considerably exceed the benefits, whereas in some other areas they could be grossly insufficient to prevent losses. Indeed, for rural flood mitigation this may have been less of a problem as the related impacts were not always negative (e.g., spread of sediment containing beneficial nutrients to topsoil) but for urban areas such an approach is inadequate due to the fact that the consequences are much more adverse and severe.

In terms of the design practice, engineers have come up with what has been commonly referred to as a "design event" (or a return-period event). Since then, various approaches for estimating the nature of this design event have been developed. Some of these approaches include using a historical worst-case scenario that happened, or could plausibly have happened, which is often referred to as *storm transposition*. Other approaches attempt to maximise theoretically the meteorological factors that could occur in an area leading to the worst possible storm, such as that producing the worst possible flood. These are known as the Probable Maximum Storm and Probable Maximum Flood, respectively. There are also approaches which use a probability-based (or frequency-based) analysis wherein systematic records and historical information on past flooding events are used to develop a relation of probability-of-occurrence versus magnitude.

The frequency (or probability) based approaches have been dominating the flood management practice for quite some time. Typically, the peak flood discharge and corresponding water level are established for various frequencies of occurrence or return periods of events such as once in 10 years (1:10), once in 20 years (1:20), once in 50 years (1:50), once in 100 years (1:100) and so on. Not long ago it was a common practice for flood mitigation projects to use the return period to describe the probability of a flood event of a given magnitude. However, it is now generally accepted that this concept can be misleading to the general public and the preferred terminology is to use the percentage probability of a flood event of a given magnitude (i.e., an event that is equalled or exceeded in any year). For example, 1:100 year flood has 1% of annual probability and 22% chance of occurring at least once in a 25-year period, and a 53% chance of occurring at least once in a 75-years period.

More recently, it is becoming more popular to adopt the concept of *acceptable risk* rather than adopting preset levels of protection associated with a specific probability of occurrence (e.g., the 100-year flood). This concept has emerged with the realisation that floods are a phenomenon which cannot be completely eliminated nor brought under total control and as such it has raised the importance of residual risks and ways of dealing with them. Therefore, the current paradigm of flood mitigation has included the economic analysis of costs and benefits for different flood protection measures under the umbrella of risk assessment with a consideration of potential damages that might be expected in case of their failure. Hence the term 'flood risk mitigation'. Nowadays, flood risk mitigation practice is very much based on an integrated rather than fragmented approach and it treats floods as part of the whole water cycle: it integrates land and water management by adopting a mix of strategies, it promotes cooperation between different organisations and it ensures a participatory approach (see e.g., WMO 2004, 2008 and 2009). However, there are limitations to this recent approach as it is still very much limited to the economic analysis of costs and benefits, whereby it overlooks other considerations that are morally important aspects of risk. We argue in this book that there is a need to further develop the present approach by incorporating principles of social justice in which the ethics of risk would play an important part in the entire risk assessment process.

Within the recent approaches to flood risk mitigation, the procedures for the evaluation of different structural flood protection measures have been dominated by estimations of flood hazards and flood damages. Physically-based computational modelling has been invaluable for this purpose. Initially, computed flood depths were used as a main indicator for identification of different levels of flood hazards and, more recently, other variables such as velocities and duration of flood were also incorporated into the analyses. Furthermore, it can be observed that many researchers and practitioners adopt the more general flood risk management framework which defines floods in terms of state variables that characterise their sources (e.g., rainfall, wind, waves, etc.), pathways (e.g., spilling manholes, overtopping river banks, flood protection structures, overland flow-paths, etc.), receptors (e.g., property, people, environment, infrastructure, etc.), and consequences (DETR *et al.* 2000; ICE 2001; Fleming 2001). Within this concept, sources and pathways are related to the flood hazard, receptors are related to the resilience and consequences are related to the magnitude of the harm caused by the flood event, Figure 2.2.

In general terms, flood risk has been defined as the probability of the negative effects of a flood (e.g., the flood hazard posed by a certain rainfall event) multiplied by the cost of damages (Sayers *et al.* 2002). In a traditional cost-benefit type of analysis (i.e., CBA analysis), for evaluation of different measures (or alternatives), an optimal measure is defined by summing all costs multiplied by their probabilities for each measure, and choosing the least expensive measure. In a risk-based approach, a flood probability distribution is calculated together with an estimation of potential risks (which is very often reflected by estimation of damages) from different flood levels. Normally, a discount rate is applied to future costs to give a net present value for evaluating different protection levels. With this approach, typically, insurance companies, governmental agencies and some stakeholders are asked to determine an acceptable risk based on different preferences regarding potential loss of life, infrastructure investment, or economic loss. The main benefit of the risk-based approach is seen as its ability to allow choices based on a comparison of expected outcomes and costs of different solution measures (Sayers *et al.* 2002; Hall *et al.* 2003; Vis *et al.* 2003). Although this approach has dominating the practice, in essence it is still very much the same utilitarian concept which follows the maximum utility concept. Again, such an approach is limited in a sense that it considers that the right thing to do is whatever maximises benefits and minimises costs. It is very much a quantitative type of analysis which cares only about the sum of benefits and costs and it overlooks other more qualitative considerations that are morally important (and *socially just*) aspects of risk. For example, the benefits of constructing a nuclear plant are immense and they can exceed the costs which might be unforeseeable at the time of the risk assessment process. However, in the event of a disaster, as was the case with Fukushima Daiichi Nuclear Plant on 13th March 2011 following the tsunami event, the consequences can be far greater and indeed devastating. Therefore, considerations based on social justice and ethical principles should take an important part, if not the central part, in the risk assessment process and in deciding whether to accept the risk or not.

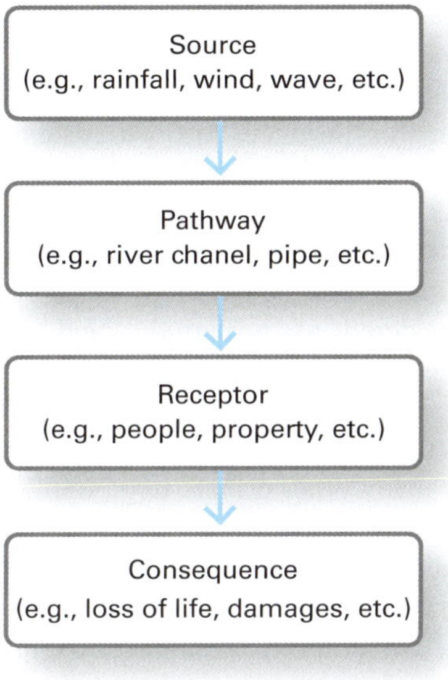

Figure 2.2 Source-Pathway-Receptor-Consequence model (Source: ICE 2001)

Furthermore, the current approach tries to bring 'rationality' to complex social aspects by translating them into monetary terms, statistic, indices or scorecards and comparing these[6]. Unfortunately, such translations are often meaningless, as placing a monetary (or index) value on social aspects (including the possibility for loss of human life) is certainly morally obtuse and naive. These are some of the main pitfalls of the present practice and the aim of this book is to correct this view by broadening, or even shifting, the focus to what is the *socially just* (or morally legitimate) thing to do.

[6]In common practice, a statistic is a regarded as a relatively unprocessed measure. An indicator is the transformation of the statistic into a meaningful measure over time and across the city. For example, a statistic can be a number of flooded houses in a given area. An indicator could be a rate of impact of a flood event on different income groups in the same area. An index would be then the weighted or unweighted combination of two or more indicators to summarise an issue of interest.

2.8 AN OVERVIEW OF CURRENT DISASTER RISK ASSESSMENT PRACTICE

The application of risk assessment methods has been more dominant in the chemical, aeronautics and nuclear industries than in the area of flood management. One of the reasons may be attributed to the fact that dealing with floods concerns characteristics of a much larger and complex picture rather than characteristics of an individual component, such as may be more suitable in other application areas. In their seminal paper, Kaplan and Garrick (1981) have introduced some of the early risk related concepts which are nowadays agreed upon in risk theory. Since then, many researchers and practitioners have considered risk assessment as a systematic approach to organising and analysing scientific knowledge and information for potentially hazardous activities and events (see e.g., National Research Council 1994). Unfortunately, such a view of the risk assessment process leaves a hole in the treatment of risk whereby ethical and other social values are left unattained.

In dealings with floods, it has been a common practice to start with identification of the risk, the development of strategies to reduce that risk, and to end up with the creation of policies and programmes to put these strategies into effect. In this instance, the risk assessment process is considered to be a fundamental activity geared to the evaluation of measures for reducing, but not necessarily eliminating, the overall risk, as in many cases risk cannot be entirely eliminated. Moreover, it is important to realise that the nature of flood risk itself is rather dynamic since the risk may arise from events in and out of the urban area and their character and magnitude can change significantly over time. Therefore, the risk assessment process should be considered as a periodic and an iterative process in which any phenomenon that may cause a change to the state of the flood risk is commonly referred to as a *driver*. Such consideration yields the conclusion that some of the drivers can be under the control of the flood management authorities (e.g., construction of structural flood protection measures, undertaking of maintenance activities, etc.), whereas many other drivers (e.g., rainfall, occupation of the floodplain area, etc.) are outside of their control (Hall *et al.* 2003).

Most of the terms and concepts used nowadays in the flood management practice have found their roots in the more general literature on 'disaster risk management'. A variety of definitions that relate to aspects such as disaster risk, hazard, vulnerability, resilience, coping capacity, resistance and exposure have been already offered in the disaster management literature (see, e.g., UN/ISDR (2002), Pelling (2003) and Schneiderbauer and Erlich (2004)). The term *disaster risk* has been referred to the existence of the possibility that the natural phenomena can cause harm to those elements that are exposed or susceptible to being affected. *Hazard* has been defined as the potential to cause harm to humans and objects. *Vulnerability* is usually defined as the state of being prone to or susceptible to harm. The term *resilience* is referred to as the capacity to adjust to threats and to mitigate or avoid them. *Coping capacity* has been defined as a combination of all strengths and resources available within a community or organisation that can reduce the level of risk, or the effect of disaster. The term *exposure* is used to reflect the number of people or other elements at risk that can be affected by a particular event. Therefore, the risk assessment process concerns assessment of different components of risk and it is usually expressed as:

$$R = H \times E \times V \times R \times CC \tag{2.1}$$

where H denotes hazard, E denotes exposure, V denotes vulnerability, R denotes resilience and CC denotes coping capacity.

Such conventional expression of risk where risk appears numerically as the "unity in multiplicity" between different levels of its root causes (or parts) is what we refer to as analytical way of thinking and it is the result of analytical mode of consciousness (see also Bortoft 2010). This view of risk comes from the perspective of modern science and it reveals risk in a way that is not necessarily holistic. As we shall observe, a lot of confusion has risen through attempting to capture human dimension and to understand unity (i.e., a *holistic risk*) in the wrong mode of consciousness. This now has to be approached with the *sensuous-intuitive mind* which comes from the perceptive modes of consciousness and which plays a critical role in the holistic way of encountering phenomena. As we realise that almost any phenomena can manifest itself in more than one way, we do not want to argue that one way is more fundamental than another. Hence, we could potentially have two views which will act in a complementary rather than in a contradictory way.

Vulnerability is nowadays seen as a crucial feature needed for the estimation of the potential consequences of different hazards and the identification of appropriate mitigation measures. The concept of vulnerability has evolved from the social sciences and as such it was introduced into the area of disaster risk management in the 1970s (Schneiderbauer and Erlich (2004)). One of the early works concerning vulnerability comes from Hewitt and Burton's Hazardousness of Place (Hewitt and Burton 1971) and James Lewis' work on place-based vulnerability (Lewis 1979). There are several schools of thought which may view vulnerability in their own ways, but essentially they are all the product of analytical way of thinking. At one end of spectrum there are those who believe that there is a double structure of vulnerability (Bohle 2001), Figure 2.3.

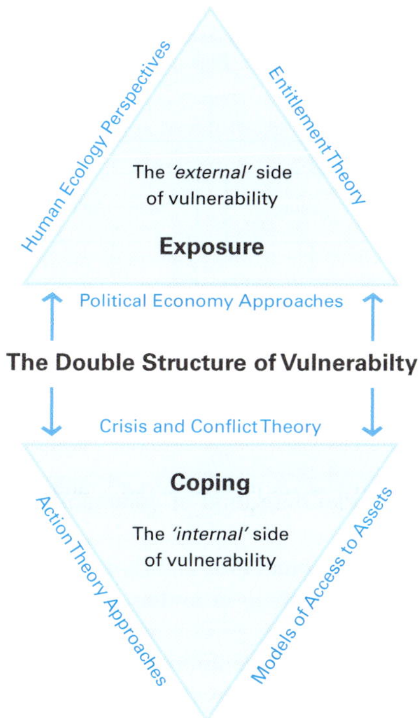

Figure 2.3 Bohle's Vernerability framework. Source: Birkmann (2006)

Some other authors focus on the social context of hazards by addressing aspects of social vulnerability (see e.g., Blaikie *et al.* 1994; Watts and Bohle 1993; Wisner *et al.* 2004), Figure 2.4.

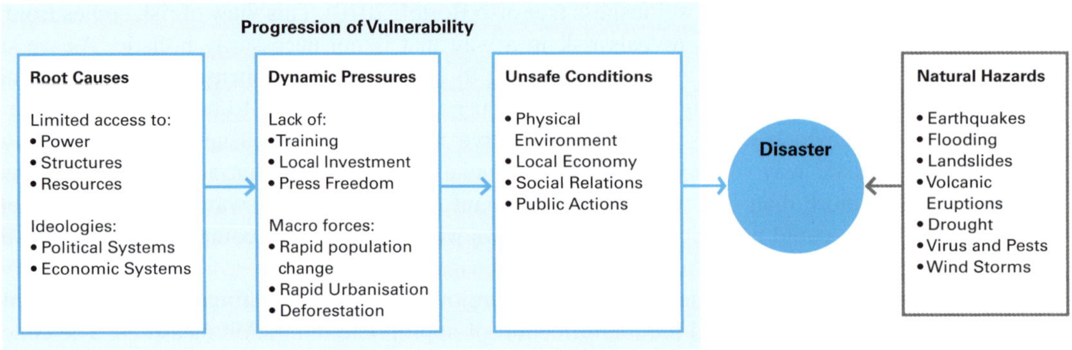

Figure 2.4 Pressure and Release (PAR) model. Source: Blaikie *et al.* (1994)

Others define vulnerability as a hazard of place which encompasses biophysical risks as well as social response and action (see e.g., Cutter 1996; Weichselgartner 2001). Davidson (1997) and Bollin *et al.* (2003) separate vulnerability from coping capacities and exposure, Figure 2.5.

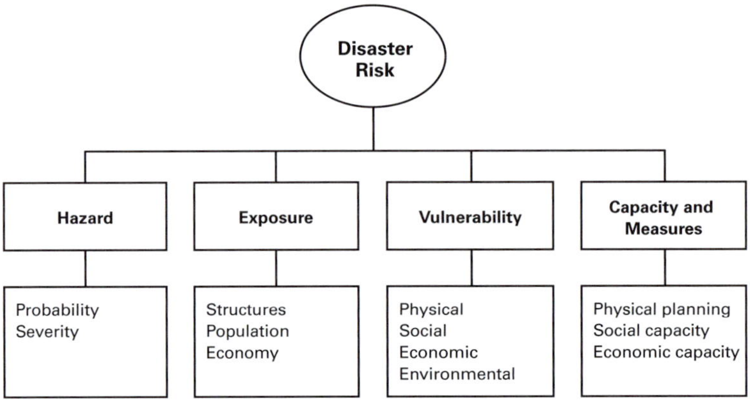

Figure 2.5 Davidson's conceptual framework. Source: Davidson (1997); Bollin *et al.* (2003)

Turner *et al.* (2003) maintains a broader definition of vulnerability by combining exposure, sensitivity and response capacity, Figure 2.6.

Bogardi and Birkmann (2004) and Cardona (1999/2001) have defined the so-called BBC framework which places vulnerability within a feedback loop system and links it to a sustainable development, Figure 2.7.

Even though it has been a long time since vulnerability was introduced, not only is there still a difficulty in defining its meaning precisely, but due to the prevalent analytical way of thinking there is still a lot of confusion in finding the right means for measuring and assessing its different components. Because the concept of vulnerability is a cross-cutting one and often ill-defined, it is then difficult, if not impossible, to define such analytical method that can take into account all important aspects. As a consequence, there are more than thirty different definitions of vulnerability and their assessment methodologies (see, e.g., Thywissen 2006).

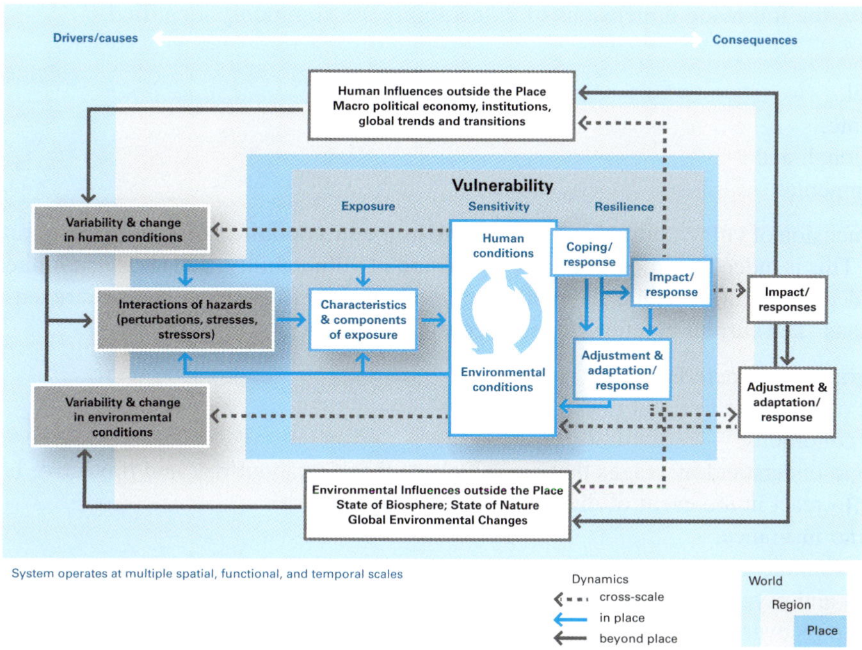

Figure 2.6 Turner *et al.*'s vulnerability framework. Source: Turner *et al.* (2003)

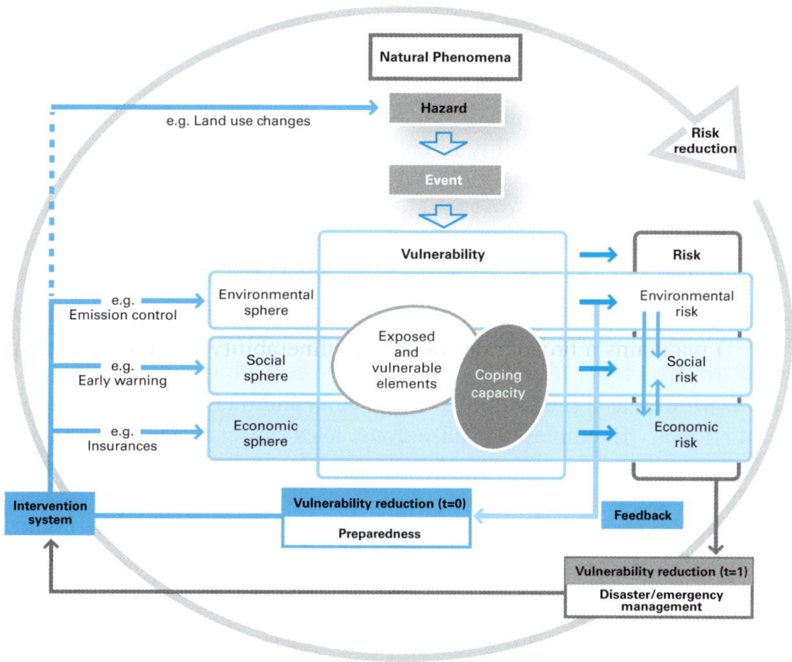

Figure 2.7 BBC Vernerability framework. Source: Bogardi and Birkmann (2004) and Cardona (1999/2001).

In summary, the following dimensions of vulnerability are commonly identified:

- Social,
- Physical,
- Economic,
- Institutional, and
- Environmental.

The social dimension of vulnerability has been defined as the susceptibility of a society to a particular hazard (e.g., floods). This is indeed the most complex dimension of vulnerability, and to date, a limited success has been achieved in capturing social vulnerability as most of the efforts have been restricted to the use of analytical means. The current practice has been confined to the following aspects:

- Demographic characteristics,
- Degree of exposure to flood risks,
- Living conditions,
- Ability to understand messages that come from authorities about risk and protective behaviour,
- Ability to react in the event of disaster,
- Access to insurance,
- Socio-economic conditions,
- Physical and psychological well-being,
- Ability to recover from flood impacts,
- Access to health care and other support.

The physical dimension of vulnerability usually refers to the susceptibility of the built environment (e.g., buildings, structures, etc.) to cope with floods and flood-related disasters and it encompasses characteristics of objects to sustain or resist to potential physical (i.e., structural) damage.

The economic dimension of vulnerability typically takes a broader context as it refers to the ability of different economic sectors (e.g., wholesale and retail, transportation, services, finance, insurance, tourism, construction, manufacturing, agriculture, mining, etc.) to resist and recover from negative impacts of floods.

The institutional dimension of vulnerability refers to the ability of key institutions to provide adequate support, protection and response to those affected by floods as well as to resist and recover their functioning in case of such events. These key institutions may include the following: hospitals and nursing homes, fire and rescue services, police, utilities, communications, transportation, schools, shelters, government offices, etc. The lack of appropriate actions by such institutions may lead to chaotic circumstances in times of floods, and therefore, the existence of adequate institutional and organisational arrangements is certainly an important factor to achieve more stable and supportive circumstances.

The environmental dimension of vulnerability refers to the condition of ecosystems to resist and recover from flood events. It is viewed as a property of such systems and their ability to absorb and recover from negative impacts of floods. Examples of hazardous sites are sewage treatment plants, pesticide manufacturing plants, places where pesticides or chemicals are stored, nuclear facilities, underground storage tanks, oil facilities, etc.

Indeed, assessing different aspects of vulnerability is a great challenge and there is still a long way to go from concepts and experiments to the full practical use of anticipative vulnerability evaluations. Thus (Alexander 2000, p. 103):

> *"The many kinds of vulnerability in the urban environment can indeed only be understood with sophisticated information measurement systems. But is to be hoped that the human element does not get lost in the race to acquire technology …."* (Alexander 2000, p. 103).

What follows from this statement is that by focusing on technical and engineering issues there is always a danger of overlooking or downplaying the social aspects in a rush to provide policy-makers with clear recommendations – even at the expense of appropriateness. We would go even further and argue that limitations of our analytical views to account for different dimensions of vulnerability, and particularly human dimension, are obvious signs that the existing paradigm has encountered an anomaly – an observation that cannot be neither explained nor encapsulated by the current scientific worldview. What is needed is a more radical change, perhaps a *paradigm shift* in which *human dimension* and *the new holistic model of reality, which approaches risk as a phenomenon through qualitative phenomenological enquiry*, would represent an important addition to our risk assessment practice. This is not a matter of discovering any new knowledge as the basic wisdom already exists. We need to change our views and values and move beyond the materialistic and technocratic mindset in which our culture has become stuck. By adding this wisdom we will be able to develop *holistic* and *socially acceptable* (or socially just) solutions for present and future generations. Certainly, to put this in practice not only in the context of 'developed' but also so-called 'developing' countries is a challenging work as it requires combination of knowledges from different domains. However, it is important to realise that working on *social justice* issues is a matter of learning to read qualities and it is part of a long struggle that has come before and a longer struggle that will continue through generations. In what follows we shall discuss some of the complexities and limitations facing less developed countries in their efforts towards socially just flood mitigation practice.

2.9 SOME NOTABLE COMPLEXITIES OF LESS DEVELOPED COUNTRIES AND THEIR LIMITATIONS IN MITIGATING RISK FROM FLOOD DISASTERS

Less developed or so-called developing countries are often defined as nations with a lower level of material well being. An overview of countries with relation to different income groups is given in Figure 2.8 UN reports indicate that 75% of total world's population can be classified as poor (i.e., they earn less than 20% of the world's income) and the majority are in Asia, Africa and Latin America including the Caribbean.

As reported by the UN, nearly all of the urban growth projected in the 21st century will occur in developing countries. In the 1950s only New York and Tokyo were large enough to qualify as megacities (i.e., cities with more than 10 million residents), but population experts project that six cities – São Paulo, Mexico City, Mumbai (formerly Bombay), Delhi, Dhaka, and Tokyo – will have exceed the 20 million mark by 2015, with New York, Jakarta, and Lagos close behind: see Figure 2.9.

According to UN reports, increasing population growth and rapid urbanisation are alarming signals that it is incumbent on the world community to address some of the most essential urban infrastructural aspects in the developing world in order to avoid catastrophic consequences (Figure 2.10).

The significant difference in population growth figures between developed and developing countries (Figure 2.10) can be largely explained by the difference in educational levels and career prospects. Typically, higher education levels are linked to lower population growth rates. Also, many developing countries have large indigenous populations which have a very early marriage culture and lower educational levels.

In terms of the urbanisation process, human societies were originally predominantly rural-based. The urbanisation process has advanced dramatically since 1950 and recent urban population figures indicate an increase from 2.9 billion in 2000 to 5 billion in 2030 (UNPD 2002). In terms of the provision of water and wastewater/drainage services in developing countries, the megacities have gained the greatest attention. They certainly do pose enormous challenge to researchers and practitioners but overall they account for a relatively small percentage of the global population. It is observed that medium to small sized urban centres of developing countries are experiencing much of the urban growth and their trend is likely to continue or even accelerate.

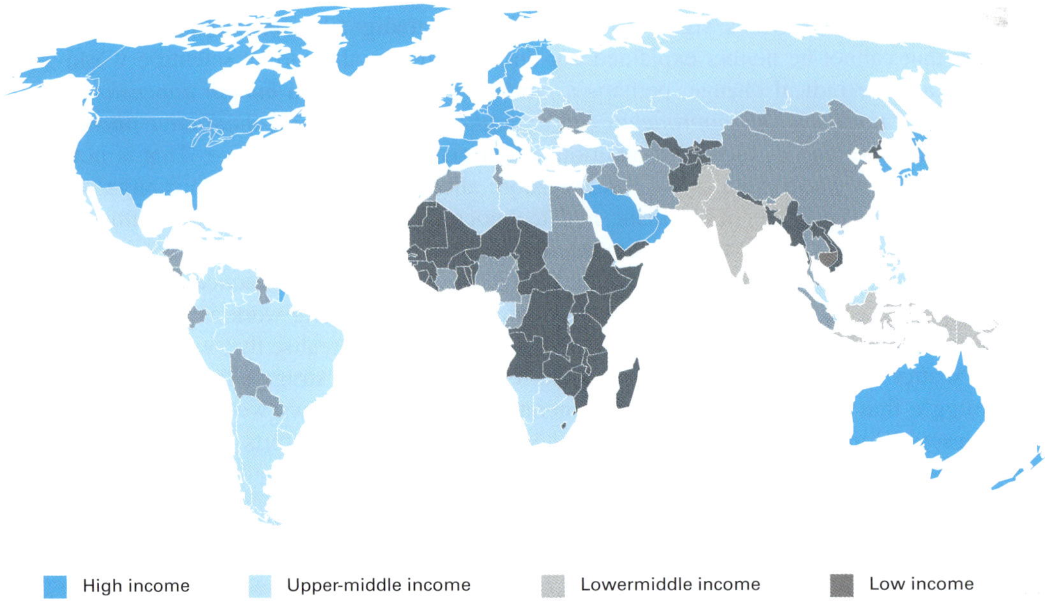

Figure 2.8 World Bank income groups in exchange-rate terms. (Accessed on 12th October 2011 from: http://en.wikipedia.org/wiki/Developing_country)

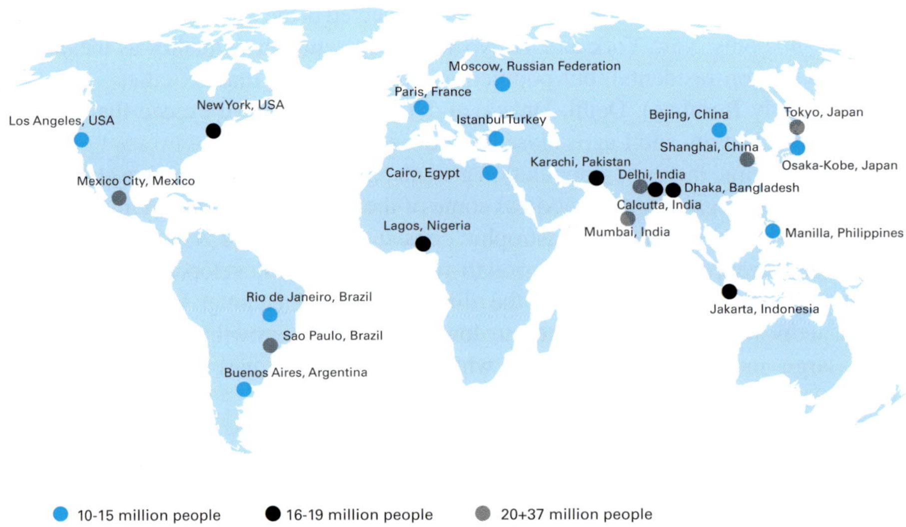

Figure 2.9 Cities with more than 10 million people by the year 2015 (Accessed on 12th October 2011 from: http://www.learner.org/courses/envsci/unit/text.php?unit=5&secNum=6#worlds_largest_cities)

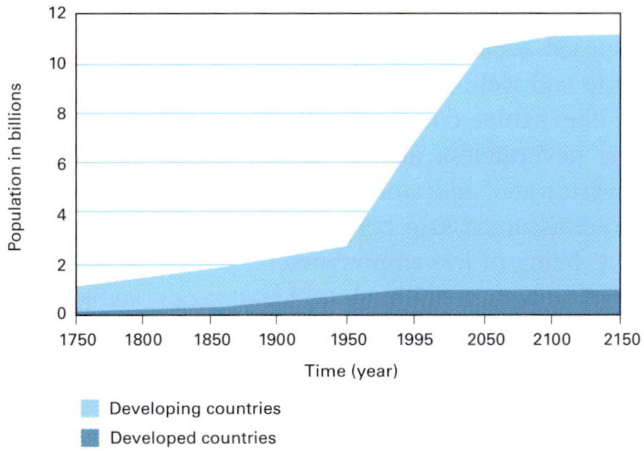

Figure 2.10 The worldwide trend of population growth (The Economist, 1998)

Over the past decades, the economies of developing countries have not performed well. This situation has made its impact on possibilities to allocate adequate resources to tackle some of the basic flood related issues. Furthermore, the lack of proper planning and management practices, insufficient or even nonexistent financial resources and the poor condition of existing infrastructure have only aggravated the situation. A more formal proactive approach to the management of drainage systems and flood defence structures is hardly existent and such assets are typically utilised until their failure. Rapid urbanisation and population growth have caused the limited amounts of water available to be transformed from the agricultural sector to the urban water sector. In return, farmers who have suddenly lost their livelihood from agricultural work are now forced to migrate to cities and look to their survival in the new environment. The poorly managed urbanisation process has also resulted in extensive air, water, land and noise pollution as well as in the rise of flood water levels. Furthermore, rapid urbanisation is one of the key contributors towards the high levels of criminal activities.

The presence of informal or squatter settlements (i.e., slum areas or shanty towns) in the cities of developing countries brings a special dimension into the perspective when dealing with provision of adequate water/wastewater/drainage services. The United Nations Human Settlements Programme (UNHABITAT) defines these areas as those that lack basic services (water, sanitation and electricity), have substandard housing, experience overcrowding, are unhealthy and situated in hazardous locations, with insecure tenure and social exclusion. As reported by the UN, almost 1 billion people, most of them in developing countries, live in informal settlements, and this figure is expected to double over the next 30 years. In some cities, these areas may account for up to 80% of urban population and they are always highly congested, leaving virtually no space for construction of proper facilities. As more people from rural areas migrate to cities, the densities of such settlements are continuing to increase. The location of such settlements is either in the centre of a city near to employment opportunities, or in the peri-urban area where residents are more isolated.

In addition to the above, there are also a number of institutional and financial issues that are limiting their capacity for urban flood mitigation. As the related projects require considerable financial investments there have to be adequate economic and fiscal arrangements put in place. In particular, there should be facilities for tax collection in order to pay for the capital investment in constructing drainage systems and protection facilities, and to maintain them in an efficient and effective manner. Whereas it is comparatively easy to justify financial charges for the amount of potable water delivered, it is considerably more difficult to arrange corresponding charges for stormwater and wastewater collected, or for the treatment of such wastewater before being disposed back into the environment. This is because everyone is aware of the value of freshwater to their life and wellbeing, even if they are also aware that such water is in one sense freely available as rainfall: the actual collection, storage, treatment, distribution and availability of potable water on demand is nevertheless an important service directly benefiting the individual. In contrast, the collection of wastewater and stormwater and preferably the treatment of the wastewater component are services that are assumed as a benefit more to the community than to the individual, and therefore they are perceived as being of less importance.

Technology transfer and cooperation tailored to local needs and capacity building are important aspects for sustainable development of urban areas. One of the positive elements in situation of developing countries is that they are in a position to benefit from the lessons learned by developed countries and they can avoid repeating the mistakes of past technologies. They do not have to follow the same trajectory of technological progress that has taken place in the developed world. They can rather make a shortcut to the newest, most productive and environmentally sound technologies available. This however requires not only effective technology transfer and cooperation but also a detailed knowledge of social and cultural issues which can in many cases play a crucial role for successful implementation of these technologies.

In the view of the above, there is an exponentially growing need for properly trained professionals who would be able to bring both social and technical aspects together in a way that can serve the needs of communities and engage in a socially just flood mitigation practice. Such engagement requires a broader and truly holistic perspective and one of critical issues is the awakening of true *consciousness* and empowerment of people to meaningfully participate in the decision making process.

2.10 URBAN POOR AND DISASTER RISK

Irrespective of the geographic location, city or a country, all urban poor share a number of common issues which affect their day to day life. Some of these key issues are: lack of financial resources and employment, inadequate housing, exposure to natural disasters, environmental hazards and health risks, poor infrastructure and services, and inequality which is closely linked to problems of exclusion.

In terms of the employment issues, the urban poor face the challenges of possessing only low skills, low wages, lack of insurances, engagements in the informal sector and poor working conditions. Since they are heavily dependent on the cash economy, with very little savings, they are particularly vulnerable to shocks. The World Bank reports suggest that in Dhaka, Bangladesh, unemployment rates for the poorest male workers are about 10 percent, twice that of the wealthiest (5 percent), Baker (2008). For women, about 25 percent of the poor are unemployed compared to 12 percent of the non-poor (World Bank 2007). Youth unemployment is a major problem in many cities, and increasingly linked to growing social problems and can create urban unrest. Average youth unemployment rates were highest in the Middle East and North Africa Region (25.6 percent) and Sub-Saharan Africa (21 percent), and lowest in East Asia (7 percent) for 2003 (ILO 2004).

Child labour exists in many countries, and it is at its highest in Sub-Saharan Africa. The latest ILO estimates for several African countries show that more than 26 percent of children aged 5–14 were economically active in 2004. Child labour is directly related to low levels of school enrolment and this in turn affects children's opportunities later in life.

A significant portion of the urban poor lives in slums. Estimates by UN-HABITAT (2006) show that around one third of the urban population in developing countries (around one billion) lives in slums. In Africa, the proportion of urban residents living in slums is an astounding 72 percent.

The poor often end up on insecure public or private land as it is their only option. Such situations create conditions whereby the urban poor are at a constant risk of natural disasters and remain unable to change this situation.

The issue of accessing infrastructure and having reliable services represents a particular problem for the urban poor. For example, water distribution services may be available only for a few hours a day. They are often left to seek alternative sources which may be of a lower quality and unreliable.

The urban poor are often located at sites vulnerable to floods and landslides, earthquakes and fires, infrastructure is weak or lacking, and housing is substandard and prone to collapse. Recovering from disasters is a particularly difficult task for them as they do not have resources or adequate safety nets, and public policies often prioritise rebuilding in other parts of the city (Fay, Ghesquiere, and Solo 2003). Public health issues and particularly HIV/AIDS rates are very high in urban areas, exceeding 50 percent in some African cities. Indoor air pollution and acute respiratory infections contribute to child mortality in slums, accounting for 18 percent of deaths among children under five (UN-HABITAT 2006). Reports suggest that 12 million children in Sub-Saharan Africa have lost parents to AIDS.

A majority of the urban poor live on sites that are undesirable to others. While some portion of them live in the city in order to be located close to job opportunities and markets, there are also others who choose to live in peri-urban areas where access to job opportunities and markets is much more difficult. Living in peri-urban areas, particularly without adequate access to transport services, can mean exclusion from a range of urban facilities, services, and jobs, exacerbating problems of social exclusion. Such situations can have socio-economic, environmental and political repercussions such as: exclusion from equitable access to economic/financial, social, human and natural resource assets; exclusion from access to basic services, and social exclusion restricting people from participating on fair terms in local and national social life (see World Bank 2006a and World Bank 2006b). Different forms of exclusions can be observed in many instances and they range from educational inequality across schools to limited access to jobs.

The above mentioned issues are alarming and they suggest that without *social justice* for the poor in the larger society our efforts aimed at improving the overall flood-related problems will be limited in their scope and success. Social justice for the urban poor involves in the first place the recognition of their official existence and tenure rights. They should be enabled to take advantage of economic opportunities through access to education and skills. A provision of infrastructure and services and the control of pollution are necessary to address their health problems. Also, strong commitment from the government, good local governance, empowerment, together with equal recognition and participation in a dialogue with all stakeholders are prerequisites for their engagement in the social justice process.

Chapter 3
The nature of risk

3.1 DIFFERENT CONTEXTS OF RISK

As stated earlier, many researchers and practitioners have placed their efforts towards the development of various indicators and indices to assess different aspects of risk for different elements at risk. However, very little progress has been made in an adequate incorporation of social dimensions into the analysis. Since the impacts of floods and flood-related disasters on societies are much deeper as they are associated with considerable losses (e.g., the loss of family members, destruction of property, temporary or permanent loss of jobs, disability) and unpleasant feelings (e.g., psychological traumas) it is then in the first place necessary to develop an understanding of the human dimension of vulnerability in order to find answers to the question of how to incorporate it into the analysis. As pointed out earlier, within the decision process itself we cannot determine what is good for society simply by aggregating preferences: we should also assess the risk from an ethical point of view and judge what are the good reasons for or against options for a certain action. Furthermore, our obligations to future generations (which can be referred to as issues concerning *intergenerational justice*) have also been inadequately represented within the current practice and the ultimate task of a socially-just flood risk mitigation practice is to mend these defects.

Since risk and uncertainty have become fundamental aspects of almost any decision making process, such aspects need to be carefully understood and addressed prior to embarking on any meaningful assessment work. The current literature is rather inconsistent in defining these terms and it is interesting to note that the notion of risk has probably suffered more than most of the notions used in the water sector of our economies from a lack of critical analysis and understanding. Obviously the word 'risk' itself can be used in many different contexts with many different meanings, of which just a few will be considered here. Correspondingly the notion of risk has become an easy victim to any and every kind of ideology, where we may borrow from the work of Jean-Pierre Dupuy (2002) the beautiful definition of Louis Dumont: "L'idéologie est le système d'idées et des valeurs qui gouverne l'imaginaire d'une société donnée" // "The ideology is the system of ideas and values that govern the imagination of a given society". As introduced earlier, the most widespread of such ideologies is that which characterises modern science as that which is dominated by an obsession with the orderable, the numerable, the countable and the computable (Heidegger 1963; Abbott 1991). So much is then done to reduce the representation of risk to the one or the other mathematical formalism and the estimation of risk to the one or the other system of mathematical manipulation. Of course this has its own utility in its own place, but it is really quite circumscribed in its validity and limiting in its range of applications. The purpose here is to indicate that the notion of risk gives rise to many concepts that cannot be accommodated within this ideology by demonstrating that most of these are essentially *qualitative* despite the fact that most of the current literature nowadays promotes mainly the so-called *quantitative risk assessment* approach. Qualitative risk assessment is usually based upon perceptions, opinions, beliefs, feelings, judgments, consent, liability in the absence of consent, public or political consensus or combinations of these, whereas quantitative risk assessment is based on numerical values and the use of calculations and models to express various components of risk while taking into account the probability that the loss will occur. As we shall show later, neither the qualitative nor the quantitative approach can fully express the risk by itself. Whether the challenge is with building the flood protection structure for a community, or the catchment area as a whole, or whether to issue a flood warning or not, both types of approaches should be considered and integrated into the analysis to attain to a safe risk estimation.

Correspondingly, risk assessment can be regarded as a process of determining qualitative and quantitative aspects of risk related to a concrete situation and a recognised threat (also referred to as hazard). Therefore, the main purpose of the flood risk assessment process is to gain a comprehensive understanding of flood risk before identifying those mitigation measures that are likely to reduce such risk. The same process also needs to take into account the knowledge of other risks such as residual risks (i.e., the risk that remains after implementing the preferred flood mitigation measures), technological risk (i.e., the risk associated with the presence of different flood mitigation technologies), legal and other risks and to identify the measures that can deal with them. These issues will be taken up further in Part IV of the present book.

In what proceeds now we shall address different contexts of risk and show that risk is shaped not only by the power of natural forces but also with our own human activities: see Figure 2.1. We also introduce a more appealing mathematical expression of risk by using a variant of Category Theory.

3.2 RISK AS AN EXISTENTIAL CATEGORY

That an intimation of risk is inseparable from the very existence of many creatures is indubitable and mankind is just one of these many creatures. Given the capacities of humans to reflect upon such intimations, however, risk has certain features in the case of mankind, and so *for us*, such as we do not normally associate with our fellow creatures. Thus for us to say that risk is existential is to say that it is so in ways that are intimately related to specifically human, and so human social and cultural patterns of existence. Risk, and especially so in more recent times, is then less and less perceived as emanating from nature, but more and more as arising from our own actions upon nature as a function of our own power over nature. The greatest risks to humans nowadays are perceived as those that arise from human activity, whereby nature is but the instrument whereby these risks become manifest. Risks arise more and more when we ourselves make them possible, or to anglicise a term of Sartre, when we ourselves 'possiblise' them.

The greatest source of risks to ourselves nowadays is comprised of the actions of our own selves. Thus when we speak about 'a risk of flooding' we increasingly mean a risk to ourselves of a situation that we have ourselves created, such as by obstructing a downstream watercourse or by urbanising a flood plain. It is for the most part in this more special sense that we nowadays say that risk is *existential*.

To say that risk is a category is to say something quite else again. Very generally we can speak of a category simply as an element in a scheme of classification, but it has long been more usual to speak of a category, with Kant, as 'a pure or *a priori* conception of the understanding'. It became possible in the second half of the last century however to give this a more mathematical expression in that branch of modern mathematics called Category Theory, and we shall follow this approach to develop a formalisation of risk in a different way from the standard formalisations and applications of standard mathematics. The essential point in the present work is that this is a mathematics of what are usually regarded as non-mathematical objects, so that it can be used to formalise *qualitative* relations, as will now be demonstrated.

In such an approach we regard risk as a function, which we can write as *risk(x)*, which maps a perception of a state of affairs that we signify by x (which can also consist of multiple components and expressed as a vector) into an intimation of risk that we associate with this same signifier x. We then regard the perception of the state of affairs as the *source* of the function *risk(x)*, denoting this source by $\Box risk(x)$ and our intimation of risk that this arouses as the *target* of *risk(x)*, which we write as $risk(x)\Box$. Thus, simply by definition of its terms:

$$risk(x)\colon \Box risk(x) \longrightarrow risk(x)\Box \tag{3.1}$$

In the more general terms of Category Theory we may write this for any category f as:

$$f\colon \Box f \longrightarrow f\Box \tag{3.2}$$

where f is any function that with its *source* $\Box f$ and its *target* $f\Box$ define a category. Thus $\Box f$ and $f\Box$ constitute unary operations on f.

Since an intimation of risk is itself a perception, and in this case of the state of affairs of being at risk, we can also speak of 'the risk of a risk', or more commonly of 'the risk of taking a risk'. Thus, for example, the risk to one's job when taking the risk of not issuing a flood warning may be just such a risk of taking a risk. We then have a compound function that maps the risk that we associate with the signifier of flooding with the risk that we associate with the signifier of losing our job if we misinterpret that risk. If we represent the function that realises the first mapping by f and the function that realises any second mapping by g, we are in the presence in the general case of the composite function fg. Since in this case the $f\Box$ of one function f may act upon the $\Box g$ of another function g to provide this composite function, fg, we introduce further a binary partial operation to accommodate this. Clearly further levels of risk may be incorporated in this way by introducing further functions h, i, etc. Then *Category Theory* is defined by these operations and by the axioms (See Freyd and Scedrov 1990):

$$\begin{aligned}
&fg \text{ is defined if and only if } f\Box = \Box g,\\
&(\Box f)\Box = \Box f \quad \text{and} \quad \Box(f\Box) = f\Box,\\
&(\Box f)f = f \quad \text{and} \quad f(f\Box) = f,\\
&\Box(fg) = \Box[(f(\Box g))] \quad \text{and} \quad (fg)\Box = [((f\Box)g)]\Box,\\
&f(gh) = (fg)h.
\end{aligned} \tag{3.3}$$

It may then be shown that, although two different perceptions of states of affairs may provide one and the same intimation of risk, no two different intimations of risk can be associated with one and the same perception of a state of affairs. One state of affairs can certainly lead to several different intimations of risk of course, but *the perception* of this one state of affairs will vary according to the kind of risk that is intimated. Under these circumstances the whole structure of Category Theory, with all its theorems and their proofs, can be employed to delineate structures of risk. (We should just remark in passing that a major part of the financial 'industry' nowadays likes to talk incessantly about "the risk of not taking a risk"! This can arrive in our case when, for instance, we have to breach a levee or a dyke in order to lower the water level behind this structure and so relieve various stresses behind it, even at the cost of accentuated flooding elsewhere. It can as well be used to "take the top off a flood wave" but then at the cost of flooding other areas again. This strategy is described as "sacrificial flooding").

3.3 THE REFLEXIVITY OF RISK

As we live in a world of uncertainty and risk, we face situations where trade-offs among alternatives are continuously donc on a daily basis. We try to identify and assess the severity of the risks and such reasoning aims to help us to decide what we ought to do. However, we do not only view the risk from a logical (or analytical reason) point of view, but we also use our intuitive senses which come from experiential modes and feelings. Analytical reasoning is logical and behavior mediated by conscious appraisal of events, whereas experiential (or sensuous-intuitive) perception of risk is typically associated with feelings, intuitions, emotions and affect. The social and psychological literature on risk uses notions such as 'risk perception', 'risk as feelings' and 'social amplification of risk' to relate to public risk perception.

Risk has the rather special property that it depends upon the way in which it is perceived and thereby intimated. A certain level of risk may be acceptable for one individual but it may not be acceptable for another. When a risk is correctly identified then often some measures can be taken to limit its anticipated deleterious consequences. On the other hand, if it is incorrectly perceived or intimated then the risk may well be augmented as compared even with the situation when it is not perceived, and thus not intimated, at all. Thus, generally:

$$risk = risk\left(intimation\ of\ risk\right) \tag{3.4}$$

Thus many measures for reducing risk devolve upon means to identify the correctness or otherwise of our perception of it.

Since the intimation of risk is that intimated to a person, risk is a function of the person to whom the risk is intimated and so in the first place it becomes a function of the risk-relevant knowledge of that person and the capacity of that person to take measures that will reduce the disembursements inherent in the state of affairs that provides the corresponding $\Box risk(x)$. The provision of risk-relevant knowledge when combined with the capacity to take measures to reduce disembursements provides one of the most cost-effective ways of reducing the untoward consequences of making risky decisions. Thus:

$$risk\,(x) = risk\left(presence\ of\ risk\text{-}relevant\ knowledge\ about\ x\right) \tag{3.5}$$

Such risk-relevant knowledge is usually most reliably acquired through experience, so that:

$$risk\,(x) = risk\left(available\ risk\text{-}relevant\ experience\ of\ \ x\right) \tag{3.6}$$

Of course this risk-relevant experience does not necessarily have to be acquired by the person subject to the intimation of risk, but can be acquired from another person who has such experience when this is translated into knowledge and communicated. This other person may of course in turn draw upon the experience and the associated ratiocinations of other persons again. Such experience-based knowledge is then most effectively communicated by *advice* given by the person who is the more experienced and knowledgeable to the person who intimates the risk inherent in the state of affairs *x*. This corresponds to a flow of knowledge from a more knowledgeable person to a less knowledgeable person and so presupposes the existence of a *knowledge gradient* between these persons: these and some other aspects concerning the theory of knowledge are discussed in Part II of this book. This in turn presupposes that the knowledge that flows due to this gradient provides means for the person intimating the risk to take action to reduce this risk, so that the knowledge concerned is *actionable*.

In post-modernist studies of societies, the more knowledgeable person who provides the knowledge is described as a *knower* and the person who receives and employs the knowledge without necessarily retaining it personally is described as *a consumer of knowledge*. As already introduced earlier, a society in which an emphasis on knowing is transforming into an emphasis on the consumption of knowledge is then said to be *moving from a modern to a post-modern condition of society*. Within any such context, the purpose of giving and receiving advice is to increase the *competence* of the person or persons when reacting to their perception of risk. Thus, further,

$$risk = risk\left(competence\ of\ the\ person\ perceiving\ and\ reacting\ to\ this\ risk\right) \tag{3.7}$$

We must add that the moral status of particular instances of risk-taking will also depend upon the motives and the character of the risk-taker. Thus:

$$risk = risk\left(motives\ and\ moral\ character\ of\ the\ person\ taking\ decision\ about\ risk\right) \tag{3.8}$$

3.4 RISK AS A DYNAMIC PHENOMENON

Risk is very much a dynamic phenomenon in a sense that it can change its existence, shape and form throughout the space and time. It is rather invisible (or intangible) and its consequences can be felt only when they take place. For example, we can move flood effects, and therefore flood risk, from one point to another by implementing various kinds of flood mitigation measures. The construction of levees may reduce the flooding on one parcel of land, but increase flooding on upstream or downstream properties. There is also a temporal dimension of risk as it can change within and outside the rainfall event period. Correspondingly, its impacts should not only be considered during the actual time of floods but also in relation to the time taken to re-establish community life. Some parts of urban community may be found to be more resilient and able to respond at the time of the event but in terms of the subsequent consequences they may feel them for a very long period of time, whereas some other parts of the community may experience a completely opposite situation.

As introduced in Section 2.7, flooding is a phenomenon which cannot be completely eliminated nor brought under total control and as such it raises the importance of residual and other risks and ways of dealing with them. In this respect, the following forms (or categories) of risk can be identified:

- *Primary risk* is referred to a risk in the absence of mitigation (or protection) measures;
- *Residual risk* is a risk that remains after implementation of flood protection measures (e.g., risk associated with events which are greater than the design standard event),
- *Technological risk* (i.e., a risk due to negative side-effects of flood mitigation technologies).

Certainly, in addition to the above forms (or categories) of risk there could be also other risks such as legal, institutional, organisational, political, etc. and they all need to be integrated into the solutions to attain to sustainable flood risk mitigation: see Figure 3.1. It is also important to note that risk can change from one form (or category) to another in the view of the changes that can take the place in one or more of the levels of root causes (e.g., primary⟶residual⟶technological⟶other): see also Figure 2.1.

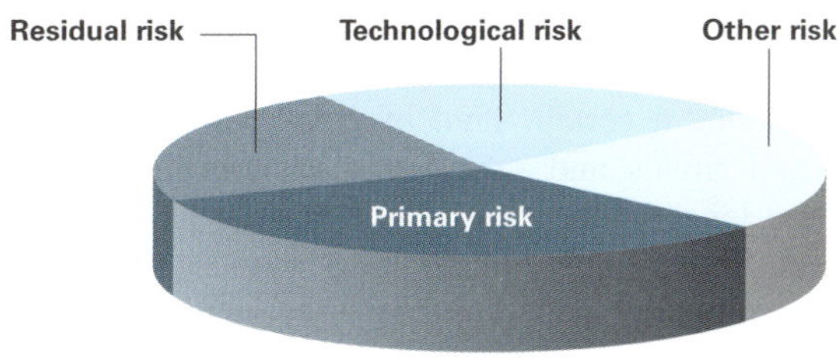

Figure 3.1 Illustration of different risk categories

3.5 RISK AND SCALE OF COMMUNITIES

The risk can be also viewed as a function of the scale of the communities that are set at risk, as the size of a community affects the capacity (financial, physical and human) to respond to a flood event. From experience to date, we can observe that the greatest losses from flood events are found in mega-cities with population in excess of 10 million. This can be explained by the fact that smaller cities are less likely to receive large-scale of densely populated areas as well as informal settlements and they also are less likely to attract large-scale industrial complexes which can contribute to the magnitude as well as frequency of hazards. Therefore, high population growth rate and poverty can affect the city's adaptive potential to resist and recover from flood events.

3.6 RISK AND POVERTY

The urban environment has been identified as a large source of risk for much of the urban population, most especially the urban poor who live in more vulnerable physical and human environments. In flood events it is almost always the case that the losses incurred by rich people are significantly less when compared to the losses brought to the poorer population. The wealthier population can afford better housing conditions with safer locations and they can maintain more comprehensive insurance covers. Those houses that are built with flood proof material will withstand different levels of floods and as such they will minimise the frequency of such events for the rich, even if they are living in a flood prone area. Therefore, poverty is an important aspect of flood risk and it cannot be overlooked.

3.7 ON RACE, CLASS, ETHNICITY AND GENDER

It can be also observed that in many places around the world the race, class, ethnicity and gender of urban population play important factors that can contribute to the formation and magnitude of risk. The best examples of such a nature could be observed with the Hurricane Katrina and the disaster that subsequently unfolded in the city of New Orleans. Tens of thousands of mostly low-income African Americans were left for a large part on their own as the city flooded from breached levees on Lake Pontchartrain. As described by Alterman (2005), this event called attention to issues of race and class and it showed that such social differences can be the core conditions that shape the risk.

3.8 REPETITION AND FAILURE

Repetition is of the essence of experience: experience is gained only to the extent that a particular state of affairs repeats itself so often that we can learn something of its behaviour - and of course it is precisely this that we learn that we call 'experience'. However, in existential philosophy, as also in religious ritual, repetition becomes an *æsthetic* category. This is to say that the experience of repeating something has the power to induce a certain kind of pleasure, extending further to a certain kind of satisfaction. All ceremonies and rituals and sacraments provide illustrations of this power, providing 'experience' at the most exalted levels. The reading of the same 'fairy story' to one's children, over and over again, provides another and more homely example.

Repetition then provides a certain kind of *bonding*, such as between the members of a congregation and their church or between the parents and their children in these examples. Repetition in the sense that it is used in the present connection refers to a bonding between certain states of affairs and persons who are repeatedly witnesses to these states of affairs, whether by direct experience or through drawing upon the experience of others.

In existential thinking, the concept of repetition is closely related to that of *failure*. Risk is always in the one or the other sense a risk of failure, so that a failure itself is a risk that has become an actuality. Failure in this context is again a category, but it is one that transcends the æsthetic when it is combined with repetition, for when we fail but then 'pick ourselves up again' so as to try again, that is, to repeat, so we acquire another kind of experience, as *one that no longer extends, but transcends our previous experience*. Thus, with Wahl (1959, p. 74):

> Jaspers' interpretation of the idea of *repetition* is different from Heidegger's and a little closer to Kierkegaardian repetition: our failures, he believes, are nothing but the affirmation of transcendence. The fact that we fail is an indication that there is some greater thing, a sphere higher than existence. We become aware of this sphere in what Jaspers calls 'boundary situations'.
>
> Thus failure makes it possible for us to mount from existence towards transcendence, and for this reason, in so far as it conducts us to repetition, it is the supreme 'cipher', the supreme symbol of transcendence ....
>
> It is thus that the two ideas of failure and repetition are not so opposed as they seem at first, since in both Heidegger and Jaspers it is failure that brings about the possibility of repetition, it is failure that leads us to triumph and to authenticity.
>
> It is this transcendence however that leads us to study risk in a very different way to that usually followed, for (ibid. p. 67):

It is this transcendence however that leads us to study risk in a very different way to that usually followed, for (ibid. p. 67):

> The ideas of passion and uncertainty are interrelated for Kierkegaard; it is the unsure thing that solicits our passion. We are not impassioned by certainties, but by things involving a risk. The subjective thinker is not in possession of any universable truth in the rational sense of the word; rather, he always has the sense of being in danger, of being on the high seas amid a storm. It is this uncertainty as regards what he believes that heightens and intensifies his subjectivity. He is on a stormy, bottomless sea. In the sphere of existence, in the sphere of subjectivity, there are no proofs, no demonstrations. When he has come to know this risk the individual is thoroughly transformed; there are no results in the sense that science gives us results, but the individual is *internally* transformed and, in this sense, everything is transformed.

We then have to do here with an exponentiation of competence whereby the risk of an enterprise can be greatly reduced again. This notion is of course fundamental to the understanding of entrepreneurship: see Abbott (2004).

3.9 FOR AN ENLIGHTENED THEORY OF CATASTROPHE: WHEN THE IMPOSSIBLE IS CERTAIN

This is a translation of the title of a book of Jean-Pierre Dupuy, *Pour un Catatrophisme Éclairé: Quand l'Impossible est Certain*. It has to do with a situation where an event, corresponding to a state of affairs, is regarded as impossible until it actually occurs, at which point it is not only perceived as being possible but as having been certain. Major economic depressions, such as the so-called 'Great Crash' of late 1929 that initiated the 'Great Depression' that swept through most of the world during the 1930s, provides a paradigm case. Currently a default on the sovereign debt of the USA is widely regarded as impossible despite the USA's astronomical levels of endebtment, but if this default should occur it would surely be regarded once again as being not only possible but certain. Dupuy's first example was taken from the reflections of Henri Bergson on the outbreak of the First World War on the fourth of August 1914: "The war thus appeared to Bergson 'at one and the same time as probable and impossible', a complex and contradictory idea which persisted right up until the fatal date". This thus corresponded to a state of *aporia*, as a sustained perplexity of the kind that is studied extensively in the literature on post-modernism. Its most superficial but familiar paradigm case is provided by the white forms of two identical faces seen in profile that are facing one another against a black background, but which can then just as well be perceived as a black vase seen against a white background.

States of aporia are often described as 'states of denial', whereby a person or persons persist in regarding a risk as nonexistent, so that its actualisation is regarded as impossible, even as when it does occur it is regarded as certain, to the great detriment of those in denial. In this case we have to do with a 'secondary risk' which is the risk of serious and perhaps terminal damage to the reputations of the persons in denial. We can however better refer to Dupuy for a thorough discussion of this dimension of 'risk' which in such cases uses the term 'risk' itself at its own peril.

3.10 THE INTRODUCTION OF AUTOCHTHONIC KNOWLEDGES

The role of failure and its overcoming as a means of transcendence is of vital importance in the design and implementation of mass-customised advice-serving systems (Abbott 2001; Abbott and Jonoski 2001). These are systems for the provision of soundly-based advice at a very low cost to farmers, aquaculturalists, health care providers and others working in the aquatic environment, enabling these to increase their productions substantially in economically and environmentally sustainable ways. As such, they are primarily directed to users in the so-called 'third world'. Such systems must then make provisions for the refusal of the advice that is given while then following this up by eliciting autochthonic knowledge*s* (written always in the plural) as means for augmenting and supplementing the modern-scientifically-based knowledge with which these systems are instantiated. Simply by virtue of the physical survival of the end-users of the proffered knowledge, these and their forbearers must have transcended their failures and so accumulated knowledges that are not otherwise accessible and these knowledges must then be merged with the existing knowledge base in order to maintain the integrity and reliability of the system.

Such systems can also be used within so-called 'first-world' contexts and specifically in cases where strategies of *sacrificial flooding* are introduced (e.g., disaster management and risk sharing strategies). We have just defined sacrificial flooding as a process where certain low-lying lands along a watercourse are flooded deliberately in order to reduce the damage due to flooding of more valuable lands riparian upon this watercourse, usually further downstream. In this case the autochthonic knowledges of persons dwelling along the watercourse must be accessed in order that these persons, who are the most directly effected by the sacrificial flooding strategy, should be represented through their knowledge during the decision-making process. The accessing of these autochthonic knowledges in such cases can also be enabled through the use of an intranet whereby the concerned parties can intervene in the decision-making process on-line: see also Torkilsen and Dynesen (2001) and Price and Vojinović (2011).

3.11 THE EXISTENTIAL NATURE OF INSTITUTIONAL RISK

Several practitioners have emphasised that the greatest risks in the water sector have their origins in institutional arrangements in the water sector, including such features as the *politicisation of science* and the *pseudo-scientification of politics* (e.g., Cunge and Erlich 1999; Cunge 2003; Abbott 2004 and Abbott *et al.* 2005). It is probably most pertinent here however to introduce two very specific phenomena that introduce some of the greatest risks in this area, leaving the reader to fill in the examples. The first of these is called by the name of *publicness*, which in the work of Heidegger is associated with distancing everyone concerned from the essential problems, reducing everything to the level of the most average understanding and bringing everything and everybody down to the lowest possible common denominator. In all such cases (Heidegger 1927, p. 127):

> The distantiality, averageness, and levelling down, as ways of Being of the 'they', constitute that we call 'publicness'. Publicness proximally controls every way in which the world and Dasein [that is 'our being there' or, more colloquially, our 'being with it'] get interpreted, and it is always right ... because it is insensitive to every difference of level and of genuineness, and never gets to 'the heart of the matter'. By publicness everything gets obscured, and what has been covered up gets passed off as something familiar and accessible to everyone.

The basic concept that follows from this, which goes by its German name, there being no appropriate English equivalent, is called *Gerede*. We can only translate it into English as 'gossip, chatter and idle talk', which has the effect of trivialising what is really something quite sinister and potentially highly dangerous. Returning to Heidegger (ibid, pp. 212, 213):

> In the language which is spoken when one expresses oneself there lies an average intelligibility; and in accordance with this intelligibility the discourse that is communicated can be understood to a considerable extent, even if the hearer does not bring himself into such a kind of Being towards what the discourse is about to have a primordial understanding of it. We do not so much understand the entities that are talked about; we already are listening only to what is said-in-the-talk as such. What is said-in-the-talk gets understood; but what the talk is about is understood only approximately and superficially.
>
> ... What is said-in-the-talk as such, spreads in wider circles and takes on an authoritative character. Things are so because one says so. Idle talk is constituted by just such gossiping and passing the word along – a process by which its initial lack of grounds to stand on [*Bodenständigkeit*] becomes aggravated to complete groundlessness [*Bodenlosigkeit*].
>
> The groundlessness of idle talk is no obstacle to its becoming public; instead it encourages this. Idle talk is the possibility of understanding everything without previously making the thing one's own. If this were done, idle talk would founder; and it already guards against such a danger. Idle talk is something that anyone can rake up; it not only releases one from the task of genuinely understanding, but develops an undifferentiated kind of intelligibility ...
>
> ... The fact that something has been said groundlessly, and then gets passed along in further retelling, amounts to perverting the act of disclosing [*Erschliessen*] into an act of closing-off [*Verschliessen*]. For what is said is always understood proximally as 'saying' something – that is, an uncovering something. Thus, by its very nature, idle talk is a closing-off, since to go back to the ground of what is talked about is something which it *leaves undone.*
>
> ... The dominance of the public way in which things have been interpreted has already been decisive even for the possibility of having a mood – that is, for the way in which Dasein lets the world 'matter' to it. The 'they' prescribes one's state-of-mind, and determines what and how one 'sees'.

All of this will be familiar enough from the interminable 'management meetings' which contribute no improvement whatsoever in management but only increase existing muddles, the column-kilometres of print in newspapers which never arrive at any fast conclusion but leave everything in confusion, and the endless waffling of 'commentators', 'analysts' and so-called 'panels of experts' that provide the regular diet of television programmes and which leave one more disorientated at the end than one was at the beginning. But at the same time of course one must not underestimate the great success of all this in 'forming public opinion', manipulating the minds of this 'public' to acquiesce in the one or the other political, economic, business or whatever other objective.

The risks incurred in this way should be obvious to everyone.

3.12 INTRODUCING THE NOTION OF SOCIAL JUSTICE IN FLOOD RISK MITIGATION

As described previously, most discussions of urban flood risk are nowadays developed in broadly consequentialist (or utilitarian) terms, focusing on the quantitative outcomes of risk as such. The major defect of the current approach is that it excludes information that is highly decision-relevant (i.e., it cannot differentiate between justice and injustice). The alternative account of risk is an approach based on social justice principles which considers the risk not only from quantitative point of view but also from qualitative point of view which in the first place refers to ethics and social needs. The field of ethics (also known as moral philosophy, or a branch of philosophy that addresses questions about morality) becomes an important aspect in a socially just risk assessment process as it prevents us from being little more than blind users of statistical methods and numerical models. Instead of uncritically calculating risk in terms of probabilities multiplied by unwanted outcomes and aggregating preferences we should also use ethical and social considerations to answer what are good reasons for or against options for action. Ethical approaches to risk assessment and evaluation involve systematising, defending, and recommending concepts of right and wrong behaviour. Philosophers usually split ethical theories into three general subject areas: metaethics, normative ethics, and applied ethics. Metaethics explores where our ethical principles come from, and what they mean. Are they merely social inventions? Do they involve more than expressions of our individual emotions? Metaethical answers to these questions focus on the issues of universal truths, the will of God, the role of reason in ethical judgments, and the meaning of ethical terms themselves. Normative ethics takes on a more practical task, which is to arrive at moral standards that regulate right and wrong conduct. This may involve articulating the good habits that we should acquire, the duties that we should follow, or the consequences of our behaviour on others. Finally, applied ethics involves examining specific controversial issues, such as abortion, infanticide, animal rights, environmental concerns, homosexuality, capital punishment, torture or nuclear war.

The practical implementation of these concepts within the area of flood risk mitigation clearly necessitates a complete rethink of current work practices and a shift of focus. We will never be able to achieve sustainable solutions of our flood-related problems for present and future generations if we continue to focus only on external materialistic circumstances. By adding the inner human and ethical dimension to our many other approaches to sustainability we may achieve a sustainable environment and we will be developing a truly sustainable mode of *consciousness*. We define *consciousness* in this context as the capacity to experience justice and to differentiate it from injustice. Thus the focus of our efforts should not be only on weighing-up costs and benefits but also on differentiating between justice and injustice, and on human values and the values of nature as much as on technology, or economy.

Since modern science is only concerned with value-free or value-neutral (i.e., objective) knowledge, the knowledge and skills required for engaging in social justice work obviously need to be sought outside of engineering subjects. This is to say that if we are to engage in a socially just practice, where social aspects are essentially qualitative, besides our technical knowledge we also need the knowledge that comes from moral philosophy and social sciences. What is needed at the end is a species of conjunctive knowledge, which is one that joins together two or more different and otherwise separate knowledges, thereby providing a new field of knowledge with qualitatively different properties than those of its constituent parts.

Engaging in a socially just practice necessitates at the first place the kind of critical thinking that can augment traditional engineering education and transform critical thinking into reflective action, or *praxis*. This correspondingly leads us to one of the possible definitions of social justice which can be expressed as a *state of transcendental praxis*, where the word *transcendental* refers to the process of arriving at a higher level of consciousness, and the word *praxis* refers to the sociotechnical development process which is rooted in the society as the basis for theorising, and always returning to the practical with the view to social transformation. Praxis has meaning in political, educational, and spiritual realms and as such it has been discussed in the writings of Plato, Aristotle, St. Augustine, Immanuel Kant, Søren Kierkegaard, Karl Marx, Martin Heidegger, Hannah Arendt, among many others.

Another observation that we should make is that any work concerning social justice is only as strong as those who are involved in it. Therefore, the role of authentic leadership in the social justice work appears as an important one and it must be realised at various levels and contexts: within industry, organisations, teams and within our projects. Its purpose is to facilitate transformation to a just practice by catalysing the work of individuals and organisations towards continuous transcendence. A true leadership requires understanding of the significance of a reflective practice, a culture that is based on strong human values and those of nature, and which comes from authentic social needs and concerns. We shall return to this later.

Moreover, the awareness of the sociotechnical nature of problems brings with it the awareness that there are increased numbers of stakeholders, a correspondingly increased diversity in interests and attitudes and sub-divisions within the stakeholder groups, even down to the level of individual citizens. Their involvement is a fundamental prerequisite for social justice work. Stakeholder participation is a process that requires active engagement and flows of relevant and authentic knowledge that can lead towards an awakening of our own true values that are in turn rooted in our inner nature, in our true Self. Knowledge is a distinctive entity that deeply characterises human relationships and the semantic sphere of human groups. When we speak about stakeholder participation we speak about flows of knowledge and understanding. Within our postmodern society we often refer to an 'information revolution', an 'information society' or even a 'knowledge society'. We should now extend this notion to an 'understanding society' in order to bring to light critical aspects of the issue that we might otherwise have overlooked: see Abbott *et al.* (2005).

In our efforts to provide sustainable solutions for present and future generations we need a theory of social justice that takes the human values and those of nature as primary values and which takes into consideration various social justice principles. Examples of such principles are:

- Every person has the right to be protected from negative impacts of floods (this is also known as the restricted liberty principle);
- All flood mitigation activities must be given permission by those who are at risk through their informed consent (this is also known as the principle of reciprocity);

These and some other social justice principles together with transparent stakeholder-driven decision-making environments in which fairness, maximising welfare, respecting individual rights, equality and virtue are continuously sought so as to form the basis for socially just flood risk mitigation practices. Such practice cannot be addressed by probabilistic cost (or risk) benefit analyses, as it requires much deeper philosophical and social considerations. Therefore, finding the right answer poses a formidable challenge as it requires the development and application of advanced technologies to enable such environments. A deeper discussion of these and some other related aspects is given in Part II of this book.

3.13 FROM A HYDROINFORMATICS OF THE QUANTITIES TO A HYDROINFORMATICS OF THE QUALITIES

Hydroinformatics was born when numerical modelling and data collection and processing came into a synergic relation at the end of the 1980s (Abbott 1991). By that time the field of numerical modelling had expanded its range from one that was restricted to the modelling of flows of water exclusively to a much wider ranging field that combined flows and all that these flows transported with them, which increasingly included living creatures that had, in turn, their own means of motion. Data collection had expanded its range of activities similarly, passing from recordings of water levels and velocity distributions to recordings and samplings of distributions of sediments, chemical substances, and aquatic vegetation and other forms of aquatic, including marine life. This development was driven by engineering practice applied to many construction and management projects.

With the extension of hydroinformatics onto the sociotechnical dimension, as projected already in Abbott (1996), the role of hydroinformatics has turned into a transformation of the corresponding social environment from a reactive one, (in which each stakeholder reacts individually during the course of a purely technical transformation and solely on the basis of the most immediate social consequences for that individual stakeholder) into an interactive one (in which each stakeholder interacts with the other stakeholders within a community, whereby the different stakeholders come to cooperate in the forming and shaping of the transformation as a whole, making of it an essentially sociotechnical transmutation). Therefore, when speaking about flood risk mitigation, with all of its advantages, the primary role of hydroinformatics is in the development and installation of sociotechnical arrangements that can truly enable engagement in social justice work.

We shall conclude at this point that although the resources of hydroinformatics in our endeavour are largely based upon modern scientific (which are here regarded as *quantitative*) assessments, the ways in which these translate through stakeholder participation into experiences of social justice are essentially qualitative. Thus, although we are currently concerned for the most part with a hydroinformatics of the quantities, where the human dimension has largely been overlooked, *the focus now, and increasingly, has to be shifted towards a hydroinformatics of the qualities.* This implies that what is necessary needs to be identified and satisfied so that we can achieve the right balance between quantitative and qualitative considerations and apply them in the flood risk mitigation practice.

As we shall later define social justice as *transcendental praxis*, the notion of praxis is used to describe the relationship between our efforts to develop and implement effective flood mitigation measures on one side and to meet social needs and concerns on the other. This is to say that instead of starting from a theory, then moving to practice to test that theory, praxis refers to the social needs and concerns as the basis for theorising, and then returning to the practical with a view to social transformation. In doing so, the thinking and doing are very much tuned towards a *qualitative* change (or what has been referred here as something much higher again, namely a *transcendence*). With this in mind, it is obvious that our traditional engineering thinking has to change into one where ideas emerge from social (and nature related) concerns and which serve humanity (and the planet). This concept, which supersedes, and therefore *transcends*, the traditional relationship between an engineer (or a hydroinformatician when speaking in the context of hydroinformatics' environments) and technology.

Part II

Adding Social and Ethical Aspects into Flood Risk Mitigation

Chapter 4
The technocratic way of thinking

4.1 OVERVIEW OF PART II

Throughout its history, but especially over the last millennium, the engineering profession, qua profession, has been associated with a narrow technocratic focus that, although fascinated with technology, from its beginning served ever expanding militaristic and other such materialistic interests (see also Riley 2008). It has correspondingly been perceived as a profession, unlike medicine or law, that is relatively unresponsive to broader public concerns, even as it led to the profession of the *Engineer* in the eighteenth century. This is mainly due to the fact that engineers are not educated and trained in how to relate their work to the wider needs and interests of society and how to recognise the imperatives of social justice. Even less again are they trained to identify the origins and natures of untruths and injustices, and how these are brought into being and are fed so as to do their damage within a given environment. For this reason, if for no others, engineers and other technologists who are not forearmed cannot expect to succeed in environments in which the negative forces that theology describes as *nothingness* (which can be also regarded as the primary source of all injustices and other evils) has already taken hold (see, e.g., Barth 1938–1944//1956/1973, to which we shall return later here). It will then be explained how the weapons required to defeat nothingness are the means of telling and indeed inculcating the highest possible level of truth – preferably without destroying one's own self in the process! – and differentiating as clearly as possible between justice and injustice, and it is these means that we have in our turn to inculcate in this place.

The subject that concerns us in this book deals with problems in which engineers have a major impact and therefore have deep ethical and social responsibilities. They can no longer afford to be ignorant or distant from the hard social questions that now arise: they can no longer rely blindly on so-called 'regulative frameworks' which are often left short in providing any kind of moral guidelines on how to deal with risks and how to incorporate meaningful social requirements into functioning frameworks. Some of these hard questions are:

(1) What is an 'acceptable' level of risk?
(2) What is 'a fair and socially just' distribution of risk?
(3) What should be the 'priorities' in our efforts to attain to risk reduction?

These and many other such questions do not concern technical aspects alone. They are also, for example, occupied with what the law should be and about how urban areas and societies should respond and be organised correspondingly. As they devolve upon such concepts as maximising welfare (i.e., improving an often nebulous 'quality of life'), respecting individual rights (i.e., whereby each of us has an established right to be protected from the negative impacts of floods) and promoting virtue (i.e., recognising and taking morally worthy actions) we can say that they are very much questions about *social justice*! To address them, we could take an approach to evaluate risk, which often takes a form of a scientific undertaking (e.g., setting up a number of scenarios and running numerical models to evaluate the impacts of these scenarios), and then using ethical analyses as our primary means, primarily through engaging the attention of the active stakeholders, so as to decide upon 'the right thing to do'. This already raises the problem of recognising what passes as 'ethical' in a given social environment. In doing so, once such a first question is addressed, a second question arises, namely of relating the first problem to problems of selecting a particular measure (or a set of measures) amongst all possible measures (or sets of measures) that can meet the acceptable level of risk. The third question relates to the necessity of being able to define priorities, as most flood defence (or drainage management) authorities lack adequate resources to address all flood-related problems at once. Should we give equal priority to reducing risks to all members of the community regardless of their circumstances, or should we give a higher priority to reducing risks to those members of the community that are found to be the most vulnerable? Certainly, there are no universal answers to such questions as they require 'ethical' considerations of what is the right thing to do in response to a particular situation. Each and every answer should be sought from within the community itself and in the light of the needs and concerns of those at-risk.

In order to be able to address the above-mentioned questions in a way that can enable us to better serve humanity and the world of nature we need to develop the right combination of skills and knowledge. What is then needed is a new paradigm that integrates a variety of social justice considerations into the flood risk mitigation framework. The aim of this part of the book is to bring what is at stake into that framework and to reason our way through the contested terrain of justice and injustice, ethical and unethical, equality and inequality, individual and social rights and 'the common good'. This is certainly a challenging task as it requires a philosophical, and ultimately theological discussion which carries risks of its own and which may often be labelled as 'personal' and 'political'. These risks spring from the fact that philosophy teaches us and unsettles us by confronting us with what we think that we already know, while theology confronts us also with that which we cannot know but in which we must have faith, and it is our task to do this as best as we can. We need to formulate a new approach in contemporary terms, educate a broader audience in its value, and help this mindset to move beyond the materialistic rut in which our culture has now become for the most part stuck. By adding the human (including whatever passes as the ethical) and the more spiritual dimensions to our many other approaches to sustainability, we should be able to achieve sustainable solutions and develop a truly sustainable mode of *awareness* and *care*. As already introduced earlier, this requires a new approach to what we call 'understanding', as first introduced in one particular way in the European civilizations but which is now being totally reformulated in the second half of the twentieth century, even as it has drawn so much on the nineteen-century pioneering work of Johann Wolfgang von Goethe. As intimated earlier here, we shall draw primarily upon Henri Bortoft's (1996/2010) mind-bending book that completely transforms the way in which we go about understanding what we experience with our senses.

This part of the present book starts with a critique of the Western (and apparently 'universal') scientistic and technocratic way of thinking (which has long been dominating the mainstream of engineering culture) and it connects in the first place to the origin and nature of *modern science* which provides the essence of engineering education and its currently associated doctrine. We then continue by reviewing some of the key aspects associated with social justice theory and practice with the intent of characterising what we mean by social justice and how we may incorporate it into the flood risk mitigation practice using the new tools that have now become available.

We shall also consider some of the important skills and knowledges from the social perspective of flood risk mitigation practices that are needed to conduct the stakeholder participation process effectively. These are the *knowledge of ethics*, *authentic leadership* and *understanding of the nature of knowledge*. Leadership represents one of the critical aspects in the social justice process and the purpose of it is to lead the change and to assist in the process of creating a just society by catalysing a new generation of individuals and organisations to provide a renewed social justice practice. The objective of this discussion is then to develop a capacity to recognise and analyse situations where this wider knowledge is required and to evolve the modes of practice that are best suited to each such situation.

4.2 THE ENGINEERING PROFESSION

The typical mindset in the mainstream of engineering culture is dominated with problem-solving as it applies mathematics and physics to achieve what appear to be useful ends. We shall recall that the earliest uses of the word *engineer* date back to the thirteenth century when this term was coined to describe someone who designs and constructs military works. It is derived (c. 1250) from the Latin word *ingenium*, meaning "a clever invention". Such a historical orientation of engineering reveals a distinctive traditional mindset focused almost completely on technological aspects and the practical side of problem solving while largely ignoring human relationships. By thinking in terms of technology exclusively, without taking account of its multiple and recursive interactions with society, we are dealing with an incomplete system, usually with negative consequences for our understanding with which we have long been acquainted. We say in such a case that we are *thinking technocratically*, Figure 4.1(a).

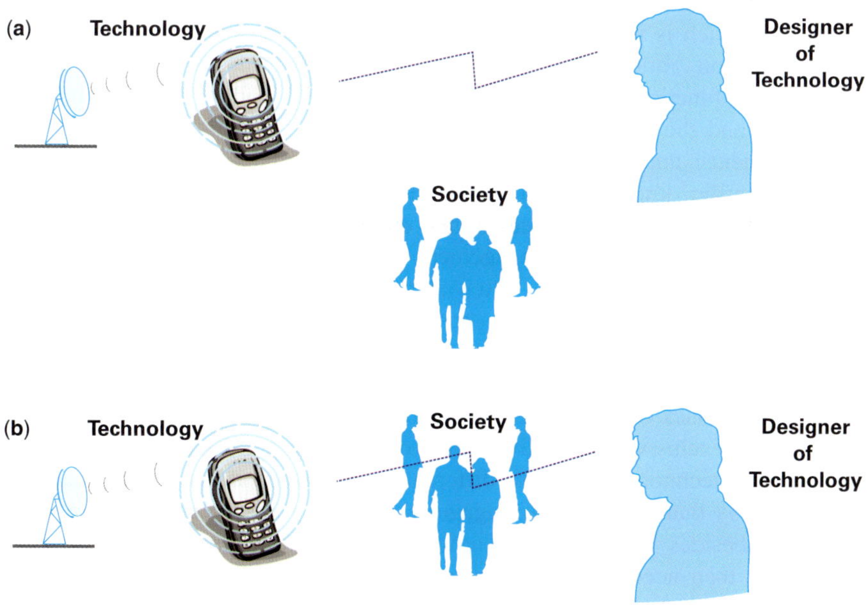

Figure 4.1 Illustration of a technocratic mindset (a) and a sociotechnical mindset (b)

This exclusive focus on the technical-scientific aspects of human society is also known as *positivism.* In *positivism*, the main criterion is *verifiability*, which comes from two different sources: empirical statements, which come from science, and analytic truth, as contained in statements which are true or false by definition. It promotes the view that all true knowledge is scientific and that all things are ultimately measurable. It is closely related to *reductionism*, a notion that promotes the notion that an entity of any kind is reducible to its smaller, component, entities and that the analysis of such smaller component entities can fully explain the workings of the system as a whole.

Other related theories are: *technological determinism*, *naturalism* and *nominalism. Technological determinism* presumes that a society's technology drives the development of its social structures and cultural values. *Naturalism* is the theory which accepts that only natural forces and entities make up the world: the clear implication of naturalism is that, since science alone is competent to examine these things, there is no non-scientific knowledge of the world, or indeed any non-scientific knowledge at all for that matter, and so no entities that science cannot examine and describe. *Nominalism* is the theory which promotes the view that only concrete things exist; abstract entities such as species do not. All of these theories have an immediate appeal to a scientific (and corresponding engineering) mindset which has often progressed by utilising one or more of these theories to sweep away non-scientific (and for the most part *subjective* and *qualitative*) ideas.

The technocratic line of thought is very much associated with the engineering profession where the main focus is on the technology itself and not on how society constructs, accepts and shapes the technology. As the above theories form the basis for engineering education they lead engineers to believe that their work is objective (a.k.a. value-free or value-neutral), and correspondingly we could add that engineers follow the myth of objectivity, as *modern science* forms the basis for such thinking. It is interesting to recall at this point that the motivation of modern science, as stated over 400 years ago by Francis Bacon, one of the inventors of the scientific method, is to obtain the knowledge that can be used to dominate and control nature. All true knowledge, he asserted, derives from observation and experiment, which are the essence of the modern scientific method. René Descartes started with the now famous/notorious statement, "I think, therefore I am," which is often interpreted in a way that his objective was to build a system of knowledge based purely on 'rational' thought. Descartes ended up with a view of the universe as a colossal machine, all the functions of which could be measured. Today, his mechanistic worldview is known as "Cartesian," and the belief that all questions can be reduced to calculation is, of course, a property of modern science and thereby a property of the engineering profession. Such a privilege extended towards scientific knowledge over all other kinds of knowledge seeks for simplistic explanations of complex human and social phenomena and eschews the knowledge outside of engineering and creates limitations to the engagement of the engineer in the promotion of social justice. The antithesis to the Cartesian paradigm is often associated with the writings and teachings of Blaise Pascal and the corresponding world view in this case is described as "Pascalian": we shall come to meet Blaise Pascal later in this and the fourth part of the book.

Furthermore, a mindset that for the most part considers the modern-scientific method as the only way of knowing can often lead to a lack of questioning of certain types of information. By viewing the world from a 'purely scientific' (i.e., exclusively objective) point of view we are disregarding the social forces that demand certain forms of technology or pose different kinds of questions that typical engineers would never raise. This is to say that if we are to engage in a socially-just practice, where social aspects are essentially qualitative, besides our technical knowledge we also require knowledge that comes from other sciences. What is then needed is a species of conjunctive knowledges, which are those that join together two or more different and otherwise separate knowledges, thereby providing a new field of knowledge with qualitatively different properties than those of its constituent parts. In this respect, the study and practice of sociotechnology is essential if one wishes to engage in a socially-just flood risk mitigation practice as well as to understand how technical innovations have social consequences, and how the resulting social changes react back again on the technical developments, and so on and on, again and again, in what become sequences of 'positive feedback' cycles (which in turn provide what is often referred to as *praxis*). It can then be observed that many subjects which are rarely if ever regarded sociotechnically, take on another aspect when they are so considered, and indeed that within a modernistic ideology 'progress' itself is often defined in a specific sociotechnical way.

Before going any further, however, we should spare at least a glance at what is to be understood as 'modern science', from its origins in Europe and thence onwards through its inner development and outwards through its influence on technology, and thence its geographical dissemination.

4.3 THE ORIGINS AND NATURE OF MODERN SCIENCE

The way of thinking about the world that we call by the name of 'modern science' has spread so widely around the world itself and has become so accepted as 'something self-evident' that its limitations, and thence its constraints, have been largely ignored. Thus modern science has come to be seen as something universal, as something that applies irreducibly in all places at all time, while other and much earlier forms of science, such as those embodied in the ancient practices of alchemy[1], have been for the most part consigned to oblivion. Today, it is largely only in certain arcane practices of traditional Indian and Chinese medicine and surgery that some parts of earlier, 'pre-modern' sciences have survived within the 'modern-scientific' context, even as these are in some places reasserting themselves in everyday life at a more spiritual level.

There are of course other kinds of sciences, which for the most part are commonly grouped together as the social sciences, but these are for the mostly so far removed from modern science that they are not regarded as sciences at all when seen from within the confines of this modern science. Some of these 'outliers' have been relabelled as 'postmodern sciences' but in many such cases their claims to scientific propriety continue to be strongly contested (e.g., Christopher Norris' "The Truth about Postmodernism" and Zygmunt Bauman's "Postmodern Ethics").

It is then easily overlooked that what nowadays passes as 'modern science' is essentially a particular form of science that was established at a very specific time, namely the Seventh of March, 1277, and in a very specific place, namely the University of Paris, and indeed was enunciated by a very specific person, namely Étienne Tempier. This ignorance of the origins of what nowadays passes in the West, not just as the first among the sciences, but often as 'the only true science', leads to a situation whereby all other forms of science are ignored, being consigned to the realm of the imaginary, which is indeed where they do to a large extent nowadays belong, and to the world of the inessential, where they do in fact nowadays remain for the most part hidden and ignored. However, as our western societies enter deeper into so-called 'post-modern conditions', that is, as societies composed of 'knowers' transmute into societies of 'consumers of knowledge', as catalysed by Wikipedia, Google and other such 'knowledge providers', so the limitations of this 'modern science' become increasingly apparent. This is because, as

[1]Alchemy refers to a quest for a fabled elixir or *kimia* (Old Persian for "elixir", later arabicised as *alchemy*) capable of turning copper and other base metals to gold and also a quest for something to prevent human beings' bodies from aging. Alchemy is both a philosophy and an ancient practice that seeks to prepare the "elixir of longevity" or philosophers' stone, accomplish the transmutation into gold, and attain ultimate wisdom. Some alchemical sources treat the various substances, equipment and processes in an allegorical sense, as metaphors for a spiritual discipline: see Jung (1944//1953).

we pass into the tasks of realising such long-standing demands as those for 'Social Justice' within our increasingly post-modern societies, so we find ourselves obliged to extend our views of our inner worlds (which can be defined as inseparable from our individual Beings and are contained entirely within our own minds), and with this the perspectives of our outer worlds (experienced as being entirely outside our own selves), and then necessarily in such ways as to proceed beyond the realm of the modern scientific paradigm.

The 'scientistic' process introduced above, which is proceeding with increasing intensity in the West, corresponds to a misappropriation of the languages and methodologies of a modern science which, we must never forget, was at its origins a gift of the Christian Church, even as this Church drew upon numerous Islamic and Judaic precedents which it prized and valued correspondingly – and even as it did not fail to criticise them. The Condemnations of 7th March 1277 were however in the first place condemnations of a pagan science, an extreme Aristotelianism that associated another instrumentalism again, as a kind of proto-science, with for example the motions of each of its planets ruled by its individual deities, so as to provide a *polytheistic cosmology*. Following other, and specifically Islamic and Judaic precedents, the Christian Church then came to understand that there could be but one set of natural laws as preordained by one unique God – and it then became the task of 'natural science' to discover and elucidate this unique set of natural laws (where the word 'natural' itself has the connotation of 'as given at the creation', or, as we shall later re-emphasise with Blaise Pascal, as 'that which we cannot lose') so as to provide a monotheistic cosmology (alongside Duhem and Brenner, e.g., 1936/1997, see also Corbin 1964 and 1974/1986/2006, for the choice that the Western Church made to follow the teachings of Averroes instead of those of Avicenna as it separated from its deeper sources, of Islamic knowledge, thus separating from the *Orient* so as to provide an *Occident*). We may follow the relevance of this by following the words of Seyyed Hossein Nasr, as taken from his *Studies in Comparative Religion* (6, 3), as follows:

> The West which had been in many ways an "Orient" in the *ishraqi* sense of the term and had possessed a traditional civilisation which more than in any other period of its known history resembled the great Oriental civilisations, was now becoming an Occident, not only geographically but also in the *ishraqi* sense of concerning itself with the domain of ratiocination that is cut off from the illumination of the Divine Intellect. The migration of Ibn 'Arabi from Andalusia to Syria and the lack of receptivity in the West to the *ishraqi* doctrines of Suhrawardi are symptoms of this event and in fact symbolic of the parting of ways between two worlds which until now had been treading a similar course. Needless to say, men like Dante and Eckhart were yet to appear but the tendency had already begun in the West in the direction of the "Occident" or *maghrib* of the intellectual world, in the direction of a rationalism which finally, with the razor of Ockham, was to put an end to the life of scholastic theology, at least as the main intellectual force in the West. On the margin of this movement, however, the mystical and illuminative tendencies continued to assert themselves and in this domain there may have been contacts with the teachings of Suhrawardi, although this will not be fully known until all the manuscripts are studied. This influence is most likely in the Oxford school of the thirteenth century and with such figures as Roger Bacon and even Robert Grossteste. The latter has been credited by Crombie as being the founder of the experimental method, so that some may think him to stand at the antipode of illuminative theosophy. But both in Islam and medieval and even Renaissance Europe interest in the observation of nature and experiment often went hand in hand with gnostic and illuminationist tendencies rather than with rationalism, for the mystic and not the rationalist sought after the "vision" of things in their essential reality and tried to remove the mental image which separates the subject from the object. Many *ishraqis* in the East also, foremost among them Qutb al-Din Shirazi, were well-known physicists and observers of nature … Be that as it may, we … know of the interest of the Oxford school in illumination and that Roger Bacon even wore the dress of the *ishraqis* and lectured upon them.

A book review by Tom Cheetham (2010) quoted the following passage of his former teacher, E. Edward Cranz (2006), to an even greater effect:

> There was a general reorientation of categories of thought c. 1,000 AD, say in the generation of Anselm and Abelard. Against the ancient position … in which sensation and intellection lead to conjunction and union with what was sensed as intellected, we find a dichotomy between the mind and what is outside it, between meaning and things. [Rather than the] ancient *extensive* self, a self open to the world around it, we find a move to an *intensive* self, a universe of meanings separated by a dichotomy from the world of things. Finally, against an ancient reason which is primarily a vision of what is, we find a movement towards a reason based on the systematic coherence of what is said or thought. [These phases] are held together in the experience of what we call language.

We shall briefly return to this level of understanding when we define 'Reality' and 'Truth' later here, where we may appear to make use of just this dichotomy.

It is necessary here to understand the influence of the consequent 'deprived' modern science in terms of the change that it brought about in the essential nature of technology, and so ultimately in the engineering professions, as another kind of product of our inner worlds again, albeit a product that can be traced back far beyond the advent of modern science. Since modern science first developed within Europe, however, it was this innovation, as it apparently functioned as a clarification, that came to influence technology in very specific ways, so that technology itself has also to be considered within a specifically European environment, to provide what we might now call a *modern technology*. Correspondingly, in order to understand what it was that marked out this European technology from all other technologies, we have to follow its relation to modern science, as a specifically European innovation.

Now if we define a science conventionally, as a totality established through an interconnection of true propositions, then each science can be regarded as the body of knowledge that is expressed by, or encapsulated in this totality. When we speak about the physics- and biologically-based sciences upon which most of the technology used in the flood risk mitigation work draws for its sustenance, we speak in the spirit of modern science, following the historical studies of Pierre Duhem (initially with his ten-volume *Épitomé du système du monde* from 1913–1917; but now mostly read in its condensed form, as edited as *L'aube du savoir* by Anastasios Brenner 1997) alongside most historians of science (dating from Gilson 1938; Copleston 1953; and Böhner and Gilson 1954, onwards). These conceptual origins of science were however traced back by some of these historians from way beyond the Condemnations, to a distinction in the nature of knowledge that had been made already by Plato some fifteen hundred years earlier again (see, e.g., Duhem-Brenner 1997, p. 24):

> There are three degrees of understanding.
>
> The lowest degree is that of the knowledge of the senses (αἴσθησις); it perceives what is born and what dies, that which endlessly changes and passes away; it grasps nothing permanent, nothing that is forever, indeed nothing that can be called true.
>
> The highest level is that of the pure intelligence (νόησις); the pure intelligence contemplates the eternal and, above all the astral bodies, the sovereign Good.
>
> Through the union of pure intelligence and the knowledge of the senses there arises a sort of mixed, bastard (λογισμός, νόθος) reason which occupies an intermediate position; one species of knowledge that is born of this level is geometry. This knowledge attains to propositions that are precise and permanent; indeed that are true …. It prepares the way to participate in the νόησις, which alone reveals the eternal species.

The ancient Greek science, with astronomical prediction at its head, scarcely progressed beyond Plato's intermediate level, where each and every astral body had its own laws of motion determined by the god with whom it was identified, and which could be emulated by geometrical constructions, such as in the geometry of homocentric spheres and that of excentrics and epicycles. It was in this way that it then corresponded to *a polytheistic cosmology that was anathema to the Roman Church.* This level of science is often called instrumentalism, since it satisfied itself with providing an instrument – a geometrical construction in this case – for computing the trajectory and instantaneous position of each astral body individually without enquiring into any underlying 'cause' that would apply to all such bodies, let alone to all material bodies whatsoever. Plato's highest level of science, however, was concerned with identifying 'laws of the universe' such as would apply to all processes in all places at all times, and this only came to realisation when the theologians of the late middle ages recognised this level as the only one appropriate to their one and only God, so that it corresponded to a *monotheistic cosmology*. This level of science might then be called essentialism. All science that did not satisfy these conditions thus came to be described as 'pre-modern'.

4.3.1 The limitations of modern science

The development of modern science from Buridan's[2] recognition of the law of conservation of momentum – naming this 'impetus' – onwards, has been recorded so extensively elsewhere that no further commentary on it should be needed here. Clearly any science of this kind must always deal with ideal situations, which rarely if ever correspond at all exactly with the real situations treated by technology. Indeed, it was originally only in the case of planetary and other such celestial motions, where frictional resistance and other effects were so insignificant that the laws of mechanics could apply to them with negligible error, that modern science could convincingly demonstrate the generality, and thence the power, of its doctrine. However, the introduction of empirical, but carefully measured, coefficients into these universal laws in such a way that these coefficients could be constructed from combinations of the variables entering the ideal laws, subsequently made it possible to extend these laws to many other and less idealised processes of specific kinds so as to apply also over wide ranges of processes proceeding at all places and at all times, albeit within certain limits, a process that has been greatly advanced over the last four centuries (See Newton's *Principia* for one of the first clear formulations of this methodology). The ethos of the modern-scientifically-based technologies that follow this tradition can then also be described as 'essentialist', while the regressions to devices that disregard many, if not all, of the productions of the physical, and biological, sciences, such as are still being very actively promoted today in many places, can continue to be described as 'instrumentalist' in that they are only concerned to reproduce some observed results, and even then only for particular combinations of processes occurring in specific places at specific times – and often further only under specific anterior conditions. As regressions to an earlier era, preceding that of modern science, they have the same concerns as had the ancient Greek astronomers, of *saving the appearances* (σώζειν τά φαινόμενα) of certain observed phenomena, while disregarding all, or almost all, concern with understanding. For example, it has been one of the disappointments of most of the developments in data-mining technologies that much of the excellent pioneering work that has been done in this area over the last two decades has already been widely misused in this way, and it is at least partly for this reason that we here regard such disciplines as 'economics' as a pseudo-science rather than a science in its own right, as will be explained later here.

[2]Jean Buridan (in Latin, Johannes Buridanus) (ca. 1300 – after 1358) was a French priest who sowed the seeds of the Copernican revolution in Europe. Although he was one of the most famous and influential philosophers of the late Middle Ages, he is today among the least well known. He developed the concept of impetus, the first step toward the modern concept of inertia, and an important development in the history of medieval science.

We repeat, and in places word-for-word here, that the process described above, of saving the appearances by disregarding all, or almost all concern with understanding, which continues to proceed with increasing intensity in the West, corresponds to a misappropriation of the languages and methodologies of a modern science which, we repeat again, was at its origins a gift of the Christian Church, even as this Church drew upon numerous Islamic and Judaic precedents which it prized and valued correspondingly. We have already emphasised that the Condemnations of March 1277 were in the first place a condemnation of a pagan science, an extreme Aristotelianism that associated another instrumentalism again, as a kind of proto-science, with each of its deities, so as to provide their polytheistic cosmology. We repeat that the Christian Church only then came to accept and to understand that there could be but one set of natural laws as preordained by one unique God – and it then became the task of 'natural science' to discover and elucidate this unique set of natural laws so as to provide a monotheistic cosmology (see again Duhem and Brenner, e.g., 1936/1997; as also Corbin 1964 and 1974/1986/2006). If one should have to identify the most characteristic trait that separates modern science from what we shall later identify as *dogmatic science*[3] it is just this: *that modern science abhors the paradox while dogmatic science lives by it.* It was of course inevitable that modern science should evolve its own paradoxes, but it has constantly struggled to expunge them, while dogmatic science continues to glory in them.

There are seven features of modern science and the modern technology that flowed from it that are of particular significance within our present context:

(1) Modern science was in the first place the creation of the Christian Church, and even of the Western Christian Church, and, although severe in its judgements, the Condemnations that established its principles were generous in the acknowledgement that they gave to Islamic and Judaic scholars for their scientific insights, even as it pointed clearly to the ways in which these and their derivatives deviated from Christian doctrine. The Condemnations were originally, in the best senses of the words, as catholic in their tastes as they were severe in their disapprobations.

(2) An instrumentalist science was of little or no value to technology since it was impractical to construct special geometrical systems for each and every individual piece of technical equipment. There was thus little relation between science and technology during the period preceding the era of modern science. Such applications as there were, such as those of aids to timekeeping, and hence to navigation, and to surveying, were themselves essentially of a geometric nature rather than a physical one, and indeed in France a surveyor is still called *un géomètre*. Only through modern science, with its emphasis on universal laws as expressed algebraically, could the physical and biological sciences come to serve technology. Moreover, this kind of development in technology in turn served to provide the scientific apparatus necessary to advance the frontiers of modern science itself still further again. Since, however, modern science came to presence almost exclusively in Europe, and even then in that part of Europe in which the Catholic Church and subsequently its reformed offshoots became dominant, it was just this part of Europe that first developed a major 'competitive advantage' over all other parts of the world through its elaboration of what we may introduce as 'modern technology' – and this advantage still exists, albeit increasingly tenuously, to this day. These origins of modern technology in social developments show us already that technology cannot be understood or

[3]Dogmatic science is generally regarded as the introduction to and foundation of dogmatic theology. Dogmatic theology is that part of theology which treats of the theoretical truths of faith concerning God and His works (*dogmata fidei*), whereas moral theology has for its subject-matter the practical truths of morality (*dogmata morum*).

explained exclusively on its own terms, but must be understood and explained in terms of an interweaving of social and technical threads of development, the one sometimes reinforcing and at other times weakening the other. We then return once more to the capital role of sociotechnology in all such understandings and explanations.

(3) The progress made possible by technologies that were increasingly supported by modern physical science, which together we have called 'modern technology', was further accelerated by the socio-religious changes brought on by the Reformation. In the words of the German protestant theologian and educational reformer Friedrich Schleiermacher, "The antithesis between Protestantism and Catholicism may be provisionally conceived thus: the former makes the individual's relation to the Church dependent on his relation to Christ, while the latter contrariwise makes the individual's relation to Christ dependent on his relation to the Church". The resulting 'freedom' from this ethical determinism of institutionalised religion in the form of the Roman church led, in Protestant Europe, through what the German sociologist and political economist Max Weber called *'The Protestant Work Ethic'*, to new forms and intensities of capital accumulation and thence an ever accelerated industrial development. Weber saw in this the true dawn of industrial capitalism. Alongside its many positive aspects, unfortunately, this also provided the individualistic ethic and, inseparable from this, the industrial potential to prosecute ever greater and ever more destructive wars, so that Europe, and especially its Protestant Western parts, gradually developed new sociotechnical means, primarily those of sea power, to extend these wars to any place on Earth. Correspondingly, the deterioration of authentic Christian values, as expressed in its dogmatic science, proceeded even further and even faster, as manifested in the primarily Protestant Republic of The Netherlands, the leading seafaring nation of its time, and as exemplified by the works of Salomon van Til (1643–1715) and J. Franz Buddeus (1667–1729) whereby, in the words of Barth (Vol. I, Part 2, pp. 288–291): "It shows that at the end of the period which started with Buddeus theology had lost any serious intention of taking itself seriously as theology". It was at this point that the fascination with truth that drove technology became completely out of balance with the Christian virtues of benevolence and benignity, even though many individual divines, authors, anarchists, political reformers and industrialists, among others again, did subsequently try to redress the balance by their own independent actions.

(4) The mutually reinforcing social-religious and economic developments over the period roughly from 1450 to the 1960s had in turn the consequence that one part of Europe, essentially the North-Atlanticist part, came to colonise almost the whole world, for the most part directly, and for the remaining part indirectly, building new cities as they went, and then mostly along the coasts from where they exported their ill-gotten gains. Indeed, some parts of the resulting colonial empires, and especially those in South and North America and in Australasia, ultimately became populated predominantly by Europeans and persons of European descent. However, a military and naval rivalry between France and Great Britain led to a coalition of the British colonies in North America achieving independence, leading to the establishing of the United States of America, as a predominately Protestant nation. Beyond this again, as the sociotechnical capacities of the Protestant Western-Europeans and the United States, our "*North Atlanticist Powers*", came to overtake those of their Catholic neighbours, so the former tended to displace the latter, either by direct conquest of their bordering territories, as in the case of the USA, or by aiding and abetting the subversion of the European Catholic powers by their own local colonists, such as occurred in Latin America. Whatever their origins, however, these newly independent colonists then came to play their own part in the continuation of the colonial adventure, and especially through their virtual extermination of most, and in the United States and Australia

almost all, the original, indigenous, populations. The colonial activities of the European powers continued also, and the attendant atrocities were sometimes also on the very largest scale.

(5) The resulting new slave economies, such as that which 'transformed the black gold of slaves into the white gold of cotton', although it led to the accumulation of vast wealth in the North-Atlanticist nations, brought their economies into conflict with the new kinds of industry-based economies of European origin and especially within the United States, ultimately and indirectly leading in this last case to the American Civil War and the gradual elimination of the world's slave economies as major influences in economic affairs as a whole Slavery was found to be incompatible with a modern-industrial economy and, with the slave economies virtually eliminated as major economic competitors, the growing economies became ever more predominantly modern industrial, following the European model. This was also however a model of an increasing social tolerance and, over time, the rise of an institutionalised respect for 'human rights' and 'social rights', such as entitlements to unemployment payments and retirement and disability pension rights, subsequently extended further to the provision of medical, educational and other social services. Outside of direct government legislation, these rights also became anchored in the power of trade unions, cooperative societies and other such institutions. Although some attempts were subsequently made to combine major part-slave and part-industrial economies, and notably in the then Soviet Union between 1930 and 1955, primarily as a means of building an industrial base in Asiatic Russia, and in Germany between 1940 and 1945, following heavy losses from aerial bombardment and from massive casualties on Germany's Eastern front, slavery could not be reconciled with social structures which were symbiotic with a balanced and sustainable industrial development. In the Americas the former slaves have become a part of the population, albeit in many cases a deprived part, and then especially in the cities: the 2005 New Orleans flooding gave evident witness to this deprivation.

(6) The process of 'positive feedback' in weaponry, strategy and tactics from the colonial wars into European conflicts led inexorably to the two great wars and subsequent upheavals that dominated the history of Europe, and eventually came to influence the whole world, in the first half of the twentieth century. The first of these wars and its aftermaths exhausted most of Europe, although, outside of Russia, it did not devastate it, while the second both exhausted and devastated almost the whole of Europe, even as the neutral nations and others, such as the United States (which played scarcely any military part at all in the first such conflict and only a secondary part militarily in the second – see e.g., Liddell Hart 1970, p. 530 – but which obtained all the economic advantages of a neutral country in both conflicts) prospered greatly. By 1945 the United States had some 75 percent of the world's industrial production, while most of Europe was in ruins, with its population reduced for a large part to penury. By the 1950s it started to become clear, despite the continuation by some European powers of colonial wars that attained the largest scales of all times and entailed the suppression of local resistance with unprecedented ferocity, that the European domination of the world was in fact at an end. The continuation of these wars by the United States, with some support from Australia, which unleashed a totally unprecedented paroxysm of destruction, similarly led to such internal tensions that these nations had to admit a de-facto defeat. As the ancient Greeks had long since discovered: "One cannot enslave others without becoming enslaved oneself". Once again, already in the 1950s in the case of Europe, it became clear that one could not suppress and dominate others without damaging the entire social fabric upon which one's own prosperity and development was dependent. The project of European Unity, still only getting under way at that time, had its origin in this experience: in the rejection of all forms of suppression and

domination and with this the rejection of all aspirations towards hegemony. It was finally understood that one's own human and social rights, freedom of expression and rule of law were incompatible with the suppression of these rights in the case of others. The resulting understanding of the necessity of European Unity has been a direct consequence of the many dreadful events and experiences that flowed to a large extent from modern technology and the modern science upon which it so much depended and continues to depend. European Union can then be seen as a turning away from a path that not only brought unprecedented misery to so many of the peoples of this planet outside Europe, but which finally came to visit these miseries upon the peoples of Europe and European origin themselves.

(7) All of these social events are then seen to be closely associated with modern technology as this originated in Europe, even as it had such far reaching and ultimately its mostly negative consequences, initially outside of Europe but finally within Europe also. It is then necessary to understand what kind of technology this was, and indeed still is, that could be and can still be turned to such negative as well as positive purposes, and thence to penetrate to some understanding of the forces that drive technology in the one or the other direction. After all, how is it possible that modern technology, founded as it was upon modern science as a gift of the Church, could have forgotten its compensating Christian virtues to such an extent that it could be so misused as to cause such great sufferings? It is essential for every person who wants to engage in social justice work to recognise and understand this Janus-like face of modern technology and the modern science upon which it draws so much of its strength, and to understand how modern technology must itself be served by Christian or other religious and spiritual values if it is to show its true face of goodness.

4.3.2 The global consequences of modern science

We should already here state clearly that, although we point to an understanding of these religious and specifically Christian influences as essential to an understanding of European industrial history, it is by no means necessary to be of the Christian faith or indeed of any religious faith at all in order to accept this role of Christianity within European industrial history. As observed in Abbott (2004), in its third chapter as partially reproduced later in this book, this history remains most relevant today because it is essential to explaining the discrepancies in industrial and mercantile development between an Atlanticist Western Europe on the one hand and Southern and Eastern Europe on the other, differences which are still with us and with which the enlargements of the European Union have still to contend to this day as it struggles with its financial, economic – and finally existential – crises.

These other non-technical sciences that come into play as we enter ever deeper into a postmodern condition of society and as we try to understand such processes as the *transmutation* (now established here as an alchemical concept!) from a higher level of awareness and capacity to experience (which we have referred to repeatedly as *consciousness*) to a higher level of inner wisdom and caring (which we have referred to repeatedly as *conscientiousness*), and as we connect the honesty and moral qualities of an engineer (or a hydroinformatician) to someone possessing something that must have the capacity to transcend the ethical (as is, or rather was, again essential to the earlier practice of alchemy). In both of these cases we ground our present-day observations mainly in the praxis of psychiatry, and we come to regard psychiatry correspondingly as a science – even though it does not, and indeed cannot, comply with the exigencies of modern science.

These 'deviations' become much more evident when we introduce the dogmatics of the churches as sciences, albeit as sciences that are as far removed in practice from modern science as can be imagined, as will be explained later in this chapter. Indeed we may in retrospect see the formation of modern science by the Western-Christian church which we may nowadays construe as a device for separating one specific kind of science, which eschewed subjectivity, from a totally other science, as *dogmatic science*, which gives itself over almost exclusively to subjective judgements as instruments of *faith*. This dogmatic science is thus at the opposite pole of modern science and we shall come to speak of it later here as it impinges necessarily upon our efforts to arrive at some recognitions of *social justice* in our subsequent studies and their applications within specifically flood risk mitigation within its urban hydroinformatics contexts.

This space between what we usually regard as modern science and our living experiences of nature, as what Henri Bortoft (1997, Fourth edition, 2010), who started his career working with David Bohm on the holism of quantum theory, has described as "a sense of a different kind of dimension in nature", as something that "turns our habitual way of thinking inside out", can be briefly introduced as follows (p. xi):

> Over the past few decades, we have been increasingly aware of the importance of the cultural context within which modern science has developed. The new field of history and philosophy of science has shown us what is referred to now as a historicity of scientific knowledge, the way that cultural-historical factors enter into the very form which scientific knowledge takes. We have, for the most part, given up thinking of science as an autonomous activity which stands outside of any human social context, pursuing its own absolute, contextless way of acquiring pure knowledge. In fact we have begun to recognise that this view of science itself first arose within a particular cultural-historical context, and that it is an expression of a style of thinking which has its own validity but does not have access to "ultimate reality". We can now recognise, for example, that the fact that modern physics is true – which it certainly is – does not mean that it is fundamental. Hence it cannot be a foundation upon which everything else, human beings included, depends. Recognising that the foundations of science are cultural-historical does not affect the truth of science, but it does put a different perspective on the fundamentalist claims made on behalf of science by some of its self-appointed missionaries today. Looked at in the light of new discoveries in the history and philosophy of science, such claims to have found the ultimate basis of reality look like no more than quaint relics of a bygone age.

This is again the point of view taken here, but then not only on modern science but also on the 'modern technology' that has been so influenced, and was even created, by 'modern science' and which we still have to introduce. We may begin this task by following how, in the first place, Hydroinformatics, as a sociotechnology, deals with the notions of reality, truth, space and time, and we must in so doing spare a glance at the foundations of these concepts before discussing and defining their terms in a deeper context, such as we shall do subsequently.

4.3.3 The definability of reality, truth, space and time

Although most of us may think out of hand that we know what a definition is and how it is used, reference to any philosophical encyclopædia will no doubt persuade us that the way in which we define things is neither simple nor straightforward. The question concerning 'right definitions' has puzzled many of the greatest thinkers in every age, while in the last century it emerged clearly as an ideological question. The way in which one 'defines a definition' was seen to mark a division in ways of looking at the world as a whole, and especially at the position of mankind within this world (Abelson 1967). This division has caused an understandable confusion in the minds of those who have been exposed to it. The resulting confusion has in turn had the gravest consequences, both for humanity alone, and most clearly so in the first half of the twentieth century, and for humanity and its natural environment inseparably in the second half of that century – and into the present century again. Hydroinformatics is intensely concerned with the second of these consequences. Thus a discussion on definitions used in a sociotechnical discipline like hydroinformatics is not the relatively simple, technical matter that it might at first appear to be, but carries deeper issues along with it. As will be explained, this question concerning definitions carries with it also the future of this and many other subjects and is thus of vital importance: (see Abbott 2002). Therefore, we shall now discuss the *definitions* of the following terms: *reality*, *truth*, *space* and *time*.

In our present essentialist view, each self has an *inner world*, which it normally experiences as something inseparable from its Being and is contained entirely within itself, and indeed for most persons within the individual mind, and an *outer world* which it normally experienced as being entirely outside the self, and so outside the body and outside the mind. At the same time, of course, the qualities that appear in the inner world – like colour, sound, smell, or whatever – become objects of perception and its inseparable cogitation, to provide an impression of an outer world within the mind. Hence, the inner experience is a fundamental component of *reality*. However, in some occasions we may come across phenomena that reveal doubts in our construction of reality. A classic example is a reversible cube illustrated in Figure 4.2. Depending on our perspective we can form alternate perceptions: a cube can be perceived that it is seen either from one side or from the other. This illusion appears due to our inner experience of cube as an object with depth.

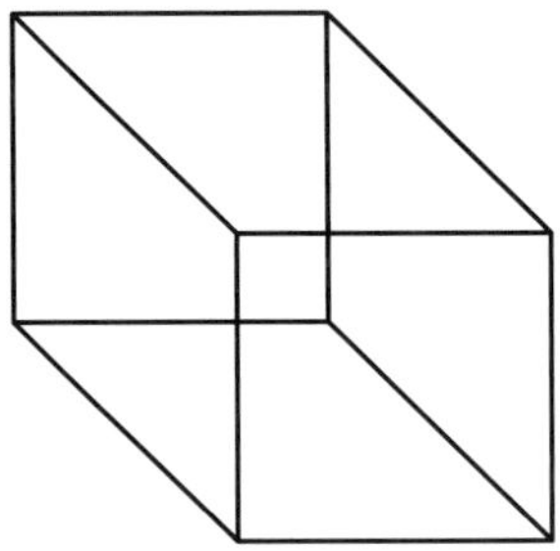

Figure 4.2 Reversible cube

There are many limitations in our interpretation of reality, over and above the fact that no two persons can have an identical sense of reality over all its manifestations. What we see is less than physical reality due to the fact that our eyes are sensitive only to light in a relatively narrow frequency range: see Figure 14.8b (Section 14.3, Chapter 14, Part III). At the lower frequencies we find infrared, microwaves and radio waves. At higher frequencies are ultraviolet, X rays and gamma rays. Our eyes cannot detect any of these frequencies that are outside our range of visible light. The same holds for other qualities that appear in our inner world: sound, smell, taste and so on. We can only detect a fraction of the outer world, whereas some of our fellow creatures are able to detect more and even much more, than we do. Therefore, the reality that we construct in our inner world is simply the human form of reality. A long time ago, Immanuel Kant drew a clear distinction between what he called *phenomenon* (that which appears to us to be in our inner world) and the *noumenon* (that which we suppose to be in the outer world, as a 'reality' that we cannot know).

The outer world includes the material world or others with whom the self may conduct a discourse in their status as subjects, and other, and usually less animate, objects with which the self does not normally conduct a discourse. Then, roughly following Augustine:

Reality is the name that we give to the interface between our inner and our outer worlds and a truth is the intimation of the oneness of these two worlds.

Thus, reality is only defined through the name of a something that may very well not be 'objectified', or 'made objective', or indeed may conceivable not even 'exist' at all. Concomitantly, truth is defined in terms of an intimation that can only be 'objectified' or 'made objective' by well-chosen and illustrative descriptions of such special feelings as pleasure and satisfaction, or displeasure and dissatisfaction, which it evokes within the self, which can again only be described and characterised in terms of its associated objects through the devices of a finely-tuned *rhetoric*. We refer to Augustine (398//e.g., 1961, and specifically sections IV.15; X.23; XI.3–7 and XII.25–32; see also Chisholm, 1966 and 1982, and Findlay, 1961) for the paradigmatic explanation of the processes of objectification in these cases. It is remarkable that the experience of truth so beautifully delineated by Augustine is reaffirmed even in the logical-positivistic writings of William James (e.g., 1976). The relation of this essentialist and experiential definition of truth to the experience of 'mathematical truth' is very well described in popular form by Huntley (1970). We should however observe in passing that the English 'reality' does not correspond in one important respect to the German '*Realität*'. Findlay (1963, 1995, p. 116) expressed the vital difference as follows:

> The notion expressed by the word *Realität* is an extremely important one, but it is impossible to give the same meaning to the English word 'reality'; this could certainly not be applied to something non-existent [as and by itself]. The distinction between the abstract and the concrete comes nearer to Meinong's distinction between the ideal and the real.

Thus when we speak, for example, of such entities as numbers and money and their derivatives as ***non-ex****istent* objects, we in no way discount their reality in the minds of those who are thinking about them, in that they ***subsist*** in their inner worlds even as in such cases these entities might be said to be ***real****ised* in their outer world in such forms as numerals, coins and promissory notes, which in themselves can have (a mundane transitory) existence as an 'imaginary money'. It is within this context that we may introduce what Meinong called 'ideal objects', following Findley (ibid, p. 115) as follows:

> Based on the distinction between existence and subsistence is the distinction between 'real' and 'ideal' objects. An ideal object is no sense of the word a mental object — like any other object it has a nature which is independent of our knowing — but it is the sort of object which can subsist but is incapable of existence. Real objects, on the other hand, are objects whose nature does not preclude them from existing.[4]

---

[4]Findley defended Meinong against Heidegger, but It has to be accepted at this point that Heidegger was developing in philosophy what Barth was developing in theology, so that Heidegger's *Sein und Zeit* was much better suited to an increasingly atheistic European world than was Barth's (one world-war later!) *Kirchliche Dogmatik*. One might even go so far as to describe Heidegger's masterpiece as a 'Godless theology' and his negative remarks concerning, for example, Kierkegaard's grasp of philosophy reinforce this impression. Indeed, it is at this point that we find ourselves, unwillingly it must be confessed, in partial agreement with Findlay when he wrote in the preface to his work on Meinong (p. xi):

> Meinong's thought had a relatively poor and rapidly dwindling reception in the German philosophical world. In the obliteration of practically all philosophical landmarks by the blanketing snowfall of Heidegger – one of those great natural cataclysms to which philosophy seems liable, and to which recent British philosophy offers analogies – the work of Meinong was plunged into neglect.

However, reading other works of Findlay, very much like those of other philosophers in the Anglo-Saxon tradition, shows that Findlay did not grasp the *relevance*, let alone the *significance*, of Heidegger's role in the transliteration of theological studies into modern-philosophical, and even into potentially postmodern-philosophical forms. Moreover, Findlay's treatment of theology in general in his other writings was, to say the least of it, cavalier. Still, it is indicative that the seventeenth-century writings of Johann Heinrich Heidegger are much more cited in Barth's *Church Dogmatics* than are the writings of the Martin Heidegger of the twentieth century!

Reverting now to the observation of the inner division of the self, as is commonly identified with various agencies, and thus agents, we see that there may be any number of realities and truths even within any one self and certainly over a considerable number and range of perceptions of 'the others'. In everyday discussion we speak, using visual metaphors, of a multitude of 'points of view' or 'perspectives', and then essentially upon objects. Then:

An object that is seen from a particular point of view or within a certain perspective is called a *phenomenon.*

The study of the ways in which the inner world of the individual self influences its view of its outer world, and how this in turn influences its own inner world again, is the subject of *phenomenology*. Phenomenology, following Husserl (1900, 1901) is the study of 'the things themselves' as these are received and cogitated in the human mind. We can of course only speak about *a* truth: *the* truth, the Platonic *noumenon*, remains forever closed to us.

We experience reality as extending over space and proceeding in time, while we usually experience truth as coming to presence in space and time. Now so long as we are treating of reality and truth in a purely modern-scientific manner, so that space is totally orderable, perfectly measurable, additive within a dimension and multiplicative over dimensions, and time is also totally orderable, perfectly measurable and additive, we feel no need to enquire further into the definitions of space and time. However, as soon as our endeavour moves out into its sociotechnical dimensions, these idealities of modern science become inadequate and we are obliged to consider our definitions of space and time much more carefully and generally. We then recall from Kant (ibid, p. 42//p. 71) that "Space is … the subjective condition of sensibility under which alone outer intuition is possible for us". This space is then not even a component of content, but 'only' the necessary condition for the operations of content in its dealings with our outer world. How then is this to be objectified for the purpose of defining space? The solution of this problem was long sought, and was largely found, through the introduction of processes of *measurement*: space became definable just to the extent that it could be measured. Entirely consentaneously again, *geometry*, as the first pure science, began with this means of definition through its ideal objectification of space. However, the *arithmetic* theories of measure demonstrate that 'an exact measurement' would necessitate an infinite thermodynamic entropy production, i.e., it would consume an infinite quantity of negentropy or 'free energy' (Kolmogorov and Fomin 1961; Brillouin 1956). An 'exact definition' is thus unrealisable. Apart from this, however, the hydroinformatician has, it seems, little to learn about the definition of space.

The situation regarding time, however, is of a much greater concern to the hydroinformatician, as well as to many others (see, e.g., Wittgenstein 1953, p. 26). For Kant (ibid, p. 46//p. 74, 75) "Time is a necessary representation that underlies all intuitions …. Time is … given *a priori*. In it alone is actuality of appearance possible at all. Appearances may, one and all, vanish: but time (as the condition of their possibility) cannot itself be removed." Thus the device of objectifying time by introducing measures of it directly, in imitation of space, is not possible: a further step in abstraction is required. This step, which appears to have dated back at least to Pharaonic Egypt, was to employ space as a source of metaphors for time. These metaphors were sought, and largely found, in the ever-changing positions of the sun, the moon, the planets and the stars relative to an observer on Earth, whereby measurements of distance could be used to provide measurements of time, as exemplified by the sundial. By these means time could be objectified and thereby defined. This metaphor so impressed the ancient Greeks (who were very sensitive to such conceptual movements) that they could not ascribe it to any human agency, but only to that of a god, who they then called *Cronos*, and who was himself correspondingly the fruit of the union between the goddess of the Earth (*Gæa*) and the god of the Heavens (*Uranus*). Time which is measured and thus defined in this way is accordingly called *chronological time*. We employ the word *clock* generally for any device, artefactual or natural, that realises this correspondence between time and recognisable, and basically measurable, representations in space. This time is clearly something quite other than time as it is experienced in our everyday affairs, which instead provides a wide variety of *experiential*, or *subjective, times*. Thus although the hydroinformatician working in a purely modern-scientific domain needs to be concerned only with chronological time, when working in the sociotechnical domain, which includes the ecological domain and the domain of the natural environment generally, he or she must take account of the times of human and other creatures' experiences, which are for much the greater part experiential, or 'subjective', times. Thus the hydroinformatician working with natural organisms and their habitats can for the most part only use chronological time for reference when reporting results, working otherwise with the experiential times of the organisms and their habitats. For example, the blossoming of certain plants may occur in April in one year and in February in another year, depending on the sequences of weather conditions which they experience. Only when the sequence of experiences is strongly influenced by the movement of the heavenly bodies, as occurs for example in tidal processes, or, more weakly and, over a longer span, with the intensity of solar radiation, do experiential and chronological times come into some kind of correspondence. (See, especially, Cloudsley-Thompson 1978, and Palmer 1978, and, more generally again, the whole work edited by Fraser *et al.* 1978). We may observe in passing that nearly all the arts proceed exclusively in experiential time.

4.3.4 On numbers and money

While dealing with flood related problems, we are much too often in a make-believe world that attempts to treat real-world social, economic and political issues in abstract terms that appear increasingly inadequate as time goes by. In particular, we are almost always confronted to the posing 'optimal cost-benefit analyses' in terms of numbers of monetary units exclusively. What gives cause to surprise, and even of downright amazement for an 'outsider' entering this world, is that such works appear to make no reference at all to Meinongian object-value theory, so that they cannot properly distinguish between ***ex***isting objects, as those ***ex***isting in our ***ex***ternal world, and ***sub***sisting objects, as those ***sub***sisting in the inner worlds of our minds, and all the categories that follow from these. Just as the economists refuse to accept that *numbers* do not ***ex***ist in our outer or ***ex***ternal world – whereas mathematicians and computer scientists know that only *numerals* ***ex***ist – but can 'only' ***sub***sist in our inner and even ***sub***conscious world, and that money similarly cannot ***ex***ist in our outer or ***ex***ternal world, but can only ***sub***sist in our inner world, with only such physical objects as banknotes, share certificates and bank accounts ***ex***isting in our outer world, so that sums of money are 'only' *doubly-non-existent objects*, so the current purveyors of 'social justice' remain similarly wandering as lost souls when they try to objectify Social Justice in any kind of monetary form. In this respect most current talk about 'social justice' resembles 'economics' in that it is *essentially unscientific in the broader sense of this word*: it attempts to remain modern scientific and so to be totally exoteric, suppressing all the esoteric workings of the active – let alone the creative – imagination, and so simply wandering around in a miasma of misunderstanding. It lacks, so to say, any *esoteric dimension*. In a full-page article in the Financial Times (26.08.11, p. 9) entitled 'A realm dismal in its rituals of rigour', John Key gave his personal view, which we share, concluding as follows:

> Economics is not a technique in search of problems, but a set of problems in search of a solution. Such problems are varied and the solutions will inevitably be eclectic. Such pragmatic thinking requires not just deductive logic but an understanding of the processes of belief formation, of anthropology, psychology and organisational behaviour, and meticulous observation of what people, businesses and governments do.

> The belief that [some] models are not just useful tools but are capable of yielding comprehensive and universal descriptions of the world blinded proponents to realities that had been staring them in the face. That blindness made a big contribution to our present crisis, and conditions our continued responses to it. Economists – in government agencies as well as universities – were playing *Grand Theft Auto* while the world around them was falling apart.

Just as talk about economics can be downright dangerous, corresponding to Warren Buffet's description of financial derivatives as "financial weapons of mass destruction" as these became ***triply*** and ***even higher-order non-ex***istents, so also can this kind of talk about social justice be dangerous too. This was exemplified by an article of Catherine Audard – who had translated Rawls' *Social Justice* into French – who observed (*Philosophie magazine*, October 1993, p. 55) how this work came to be read in China, and was actually taken seriously by some Chinese students, contributing in its own way to the misguided protests that culminated in the Tiananmen Square massacre. We shall return however to the 'theories' of Rawls in more detail later in this chapter, mentioning at this stage only its reception outside the USA.

The only European author of any note who had at some stage embraced something of the Rawlsian pseudo-ethic was Jűrgen Habermas, but he soon seems to have passed way beyond that stage. After he had come to some accommodation with Jacques Derrida – for whose work we retain a deep respect – he even came to relativise his nominally atheistic position following a dialogue with Cardinal Joseph Ratzinger (now H. H. Benedict XVI). In 2007, Ignatius Press published an account of this dialogue entitled *The Dialectics of Secularisation*.

However, already in 1999 – and now we quote from Wikipedia – Habermas had stated that:

> For the normative self-understanding of modernity, Christianity has functioned as more than just a precursor or catalyst. Universalistic egalitarianism, from which sprang the ideals of freedom and a collective life in solidarity, the autonomous conduct of life and emancipation, the individual morality of conscience, human rights and democracy, is the direct legacy of the Judaic ethic of justice and the Christian ethic of love. This legacy, substantially unchanged, has been the object of a continual critical reappropriation and reinterpretation. Up to this very day there is no alternative to it. And in light of the current challenges of a post-national constellation, we must draw sustenance now, as in the past, from this substance. Everything else is idle postmodern talk.

This is, of course, essentially the position taken in the present work concerning the numinous substance that is at the foundation of all our thoughts and all our actions associated with *social justice*, including those that have found their inspiration in the anarchist movements, both of the past and of the present, with all their deep spiritual motivations. It is only within this context that *social justice* can make a genuine entrance, but then necessarily as a paradox, and that is of the essence of its subliminal attraction. We shall return later in this book to the essential role of the paradox in any serious discussion of *social justice*, as it is also of all dogmatic sciences – and this leads us into a less combative and indeed more emollient use of language.

4.3.5 The limitations of modern technology

There is, of course, a vital difference between an act of nature, such as the blossoming of a flower, and an act of technology, in that the one proceeds out of itself with no external agency, other perhaps than a change of temperature – which is again an act of nature – whereas the manufacture of a new product necessitates an intervention from outside technology itself, as realised by the craftsman or the artist or whatever other human agent. We shall subsequently have cause to return to this distinction between that which drives the ever ongoing creative processes of nature and that which drives the processes of human creativity.

As we subsequently come to follow these in several manifestations and in depth, we shall see that this difference is reduced and even in some cases disappears when we extend our technology so as to comprehend its human interventions, making of it a *sociotechnology*. We shall then see that we cannot really understand the *poietics*[5] *of technology* without integrating them into the parallel poietics of those social, including institutional, changes that catalyse technology and that are in their turn catalysed by technology again, a process that occurs once more, so to say, 'in its own proper time'.

Thus, echoing Goethe, once we take their interacting social developments into account, the productions of the arts and crafts, like the productions of nature, come to presence 'in their own given time'. As an example that will be considerably amplified later in this work, the development of modern communication technologies, as manifested by the real-time communication and information systems, Internet and the radio-networked mobile telephone, occurred precisely as the societies in which these technologies came to presence moved from a modern condition into a post-modern condition, that is, as their populations transformed from predominantly 'knowers' to predominantly 'consumers of knowledge'.

If however we return to follow, but now with Heidegger (Heidegger 1977), the thinking, and with this the language of the Greeks, we may arrive at a much deeper understanding again. At this level we see technology as something that is *active, revealing, challenging, creative*, or again in a single word: *poietic*. We then have to do with a bringing-forth, which can be either something of nature or of the arts and crafts or of any other modes of creative production. However, any bringing-forth involves *something being revealed that was previously concealed*, and we accordingly call the process through which this occurs *revealing*. But then the Greek word for revealing, *alētheia*, was used also with the connotation of *truth*: to say that "it is revealed to me" was the same as saying that "it is true" and it was only the more legalistic Romans who translated this by the word *veritus*, which then in turn translated most directly into our word, 'truth' Even to this day in fact we may still talk of a truth that has struck us with great force by saying that "it was a revelation", and this usage remains firmly grounded in Biblical and in standard literary usage. Similarly, as we shall shortly exemplify, we can still describe something that we experience in technology as an expression of profound truth as "a revelation".

Arriving at this point and within this higher level of understanding, we can now do no better than to repeat one of Heidegger's most memorable and most quoted passages (ibid, pp. 12 and 13):

> Technology is therefore no mere means. Technology is a way of revealing. If we give heed to this, then another whole realm for the essence of technology will open itself up to us. It is the realm of revealing, i.e., of truth.
>
> This prospect strikes us as strange. Indeed it should do so, should do so as persistently as possible and with so much urgency that we will finally take seriously the simple question of what the name 'technology' means. The word stems from the Greek. *Technikon* meaning that which belongs to *technē*. We must observe two things with respect to the meaning of this word. One is that *technē* is the name not only for the activities and skills of the craftsman, but also for the arts of the mind and the fine arts. *Technē* belongs to bringing-forth, to *poiēsis*, It is something poietic [i.e. creative, formative, productive, active].
>
> The other point that we should observe with regard with *technē* is even more important. From the earliest times until Plato the word *technē* is linked with the word *epistemē*. Both words are names for knowing in the wider

---

[5]*Esthesic* (UK *aesthetic*) and *poietic* are terms used in semiotics, the study of signs, to describe perceptive and productive levels, processes, and analyses of symbolic forms. We shall return to the word *poietic* in greater depth when concluding this chapter.

> sense. They mean to be entirely at home in something, to understand and be expert in it. Such knowing provides an opening up. As an opening up it is a revealing ….
>
> Technology is a mode of revealing. Technology comes to presence in the realm where revealing and unconcealment take place, where *alētheia*, truth, happens.

We may now provide a *definition* of 'technology' in the Heideggerian sense that will be true for every one of its instances, for every one of its denotations and connotations, applying to all possible forms of the Platonic Form of 'technology', and so providing an *authentic technology*:

Authentic technology is the continuation of the Creation by human means as this comes to presence in the realm where revealing and unconcealment takes place and so where alētheia*, truth, happens.*

Here we are introducing the word 'Creation' with an upper-case 'C' as an event that is outside of modern science even as it provides what is commonly called *a creation myth* in the Judaic, Christian and Islamic 'Religions of the Book', to use this primarily Islamic denotation. However, we are in this case following a specifically Islamic Sufi tradition that nowadays has its echoes in the Roman Church and which regards the creation as a continuously ongoing process through which the Divinity becomes manifest, rather than a six/seven-day event that occurred once and for all, 'once upon a time'. In his masterpiece on the creative imagination of Ibn 'Arabi, to which we shall return later, Henry Corbin (1958/2006//1969/1997, p. 200) explained, proceeding to the Plato's highest level, to the level of the *noetic*, that:

> *Creative Being* is the *manifested*, diversified, successive and evanescent forms, which have their substance not in their fictitious autonomy but in the Being that is manifest in them and by them. Thus creation signifies nothing less than the Manifestation of the hidden Divine Being in the form of beings; first in their eternal *hexeity*, then – by virtue of a renewal, a recurrence that has been going on from moment to moment since pre-eternity – in their sensuous forms.

If, correspondingly, we extend our view beyond mankind, so as to encompass the whole world of nature and so recognising the marvels of technology that our fellow creatures here on Earth produce, we arrive at a more general and compact definition again:

Technology is the continuation of the creation by creaturely means.

It may be useful at this stage to exemplify this very compact and concentrated presentation with a few familiar instances of its application. Clearly a product cannot be expected to be successful if it does not respect the true properties of its materials, if it is not made with the truly necessary and sufficient forms and dimensions, if it is not properly suited to its true purpose and if it is not manufactured by those who express the truth about how it can best be manufactured and subsequently used. (These correspond indeed to the four types of 'causes', the *causae*, of antiquity: the *causa materialis*, the *causa formalis*, the *causa finalis* and the *causa efficiens*.) Consequently, the engineer waiting at a nineteenth century railway station may while away his time by 'reading' the construction of the wrought-iron roof structure above him, tracing the lines of force passing through the structure and their associated shearing forces and bending moments under the different loading conditions to which the structure may be subject. In the case of a truly great structure, such as that of a Brunel, the engineer reads this with much the same pleasure, and indeed with much the same deep satisfaction, as he would read a poem of Keats or Shelley. Such a structure *exudes the truth*. It is a revelation! We can trace what is essentially the same aesthetic in the efforts of the conservation movement as applied to industrial artefacts, such as in the painstaking work of restoring steam locomotives and their other rolling stock, together with track and signalling arrangements, or even in the modelling movement, in which a certain aesthetic is recreated in miniature as an alternative source of pleasure and satisfaction, as another kind of revelation again.

Taking the opposite case, when confronted with a product that does not satisfy these conditions of truthfulness, the person with passion for technology experiences a deep revulsion, bordering on disgust, as an ethical as well as an aesthetic reaction. For example, the monotonous reproduction of what is structurally the same and totally anonymous apartment block or the same totally anonymous office building, again and again and again, ad nauseam, almost everywhere on Earth, is deeply repulsive to the creative instincts of the structural engineer who really cares about structural engineering – just as it should be also to the architect who genuinely cares about architecture! As we shall see in more detail later, it is primarily because of these kinds of aesthetic and ethical impulses that social justice work requires *engineers*, understood as those who know and apply technology, to develop technology based upon a variety of authentic social justice considerations. It is then this property of technology, as the place where truth happens, where a certain revelation occurs, which generates the aesthetic fascination, the ethical passion and even the quasi-religious fervour that so often emanates from the engineer. Of course we have every right to be suspicious of such exhilarations, as we shall subsequently discuss later, but there is no denying their presence in the engineer – and indeed in many other feeling persons besides – when confronted with the products of technological excellence. If we may try to concentrate these experiences into a single aphorism, it is then this:

A fascination with authentic technology is a fascination with truth

We have already here ventured into the domain of Creative Being and the concept of The Creator as a spiritual necessity, and we should now spend a moment to introduce this ongoing process of creation, It is then generally understood that the human economy, which most humans like to think of as a 'first economy', is supported by or, figuratively speaking, 'floats upon' the natural economy, which then becomes a 'second economy', where this 'floating' metaphor introduces water, which is that which most binds these two economies together. To the extent that we can descry any structure at all in something as diverse and inhomogeneous as the human economy, so we might characterise it as a *subset of the natural economy*. For the greater part of humanity, however, this all-embracing natural economy is founded upon *another economy again*, which created and continues to support the natural economy and with this the human economy, but then in such a way that it makes its presence known to those humans within the first economy who are charged by this 'other economy' with sustaining and supporting the 'second economy'. This 'other economy', this *Third Economy*, this third support that has served to preserve the stability of the other two economies for as long as human existence can be traced, and continues to sustain some semblance of stability even in the West today, is commonly subsumed within the general category of *the religious* to the extent that it is concerned, under the heading of 'Economy', with *the divine governance of the world*, to use the *Shorter Oxford Dictionary* definition. In the *Greater Webster's Dictionary* it is defined, again under 'Economy', as "*God's plan or system for the governance of the world …; a special divine dispensation suited to the needs of a nation or period*". The term 'economy' itself refers back to the Greek word for the governance of the household, as *husbandry*, implying as it does an attribution of values. Another term that is widely employed is that of *stewardship*. Within the context of the anticipated audience of the present book, however, this association with 'the divine' has to be handled with great care and circumspection.

We are however obliged to add further that the ordering of these economies is changed, and in some cases even reversed, in the Islamic faith as also in Buddhism and some other religions and beliefs besides, so that what we have introduced above as the third economy then becomes the first economy, the world of nature in some cases remains the second economy, and in that case the world of humanity becomes the third economy (see, especially, Ibn 'Arabi [1194+/−] in the 1997 translation – itself taken from an Ottoman Turkish translation – of Shaykh Tosun Bayrak). Thus, with p. 67 of this work:

> Help comes from the Lord who addresses the world saying: O world, I order you, be a lowly servant to the ones who serve Me, and be a tyrannical master to the ones who serve you.

We may repeat once again here that the reason for introducing this, in Western eyes, obscure branch of human knowledge, and then, following a theological and indeed Sufist tradition, as an economy, is that it presents as great a challenge to the West as does anything occurring in the human and natural economies, for this nowadays' Western ignorance of, and even rejection of the very existence of such an economy, is not shared elsewhere, and in the first place in Islam. It has been said of the book that introduced Hydroinformatics (Abbott 1991) to the world that "everyone has read the first part and no one has understood the second part". This was because, although the first part was written in the language of modern science (see also in this context Duhem and Brenner 1997; Gilson 1955/1980/1982; Copleston 1962, 1963), the second part was written in an onto-theological language that few practising engineers and scientists in the West could be expected even at that time to understand – while, unfortunately, even fewer in the West can be expected to understand it today. Thus, with Abbott (1991, p. 106):

> *Hydroinformatics is the means whereby the number myth strengthens the covenant between the Creator and His creature in the realm of the arteries and the veins of the biosphere.*

Here the word 'myth' is to be understood in its authentic meaning, such as with Lévi-Strauss (1958, p. 231, with emphasis added):

> A myth is always related to past events, 'before the Creation', 'during the earliest times', or 'once upon a time'. But the intrinsic value that is attributable to a myth comes from this: that the events that it recalls, appearing to take place at some moment in time, also form *a permanent structure*. It is this that relates itself simultaneously to the past, the present and the future.

Thus a myth is an experience which is reified in our minds whenever we encounter its representation. We can provide a more exact definition, following Jung (1944/1953), when we express it in (pseudo) category-theoretical terms, such as:

$$\text{def}(M) = \text{text}(M) \rightarrow \text{context}(M), \quad M \neq M(s), \quad M \neq M(t)$$

where the text can be any communicable message and the context is the state of mind that this myth induces when its 'message' is 'read', Here s stands for space and t stands for time, so that the myth is the most translatable of all grand narratives while the message that it conveys remains invariant in time, commonly over thousands of years

Such forms of language as the myths are archetypal and recur incessantly. For example, in the words of Carey (2002):

> We are the children of transience: our imaginations cannot grasp eternity. That is why myths – those events which, in the words of the philosopher Sallustius, 'never were, and always are' – are said to have happened 'once upon a time', when in truth they exist beyond all time. The error of [Western] fundamentalism lies in its attempt to crush the symbol, to freeze and flatten the archetypal deeds of myth into the schematic chronology of history books. It is idolatry.

Once we come into the world of the sociotechnical, everything functions in terms of the most potent myths and their symbolisation processes, as illustrated in Abbott (2000) writing on the gender issue in hydroinformatics, as will also be taken up briefly towards the end of Chapter 8. As Donington expressed this matter (1963 p. 21):

> A symbol invariably arises from the heart upward. This is because a symbol is not an intellectual counter at all. Thoughts are intellectual counters; symbols are emotional counters, compounded at once of feeling and intuition. Not only do symbols represent inner realities; they are inner realities.

The great varieties of number myths, dating back at least to Pythagoras but still being produced in our own time, were also treated in the *Hydroinformatics* (ibid, pp. 94–128).

We shall now provide an overview of historical perspectives of social justice and discuss some of the important streams of thought which have made their contributions to the present subject.

Chapter 5

Historical perspectives of social justice

5.1 WHAT IS SOCIAL JUSTICE?

As reported in *The Free Encyclopaedia, Wikipedia*, the term *social justice* was coined by the Jesuit Luigi Taparelli (an Italian Catholic scholar) in 1840 based on the teachings of St. Thomas Aquinas and given further exposure in 1848 by Antonio Rosmini-Serbati (Figure 5.1).

Figure 5.1 Luigi Taparelli D'Azeglio(1793–1862), Source: Consciencia.org, Filosofia e Ciências Humanas (Accessed on 26th February 2008 from: http://www.consciencia.org/bancodeimagens/albums/userpics/thumb_Azeglio.jpg).

Many religious traditions consider it as one of the most important components for their activities. It has commonly referred to the idea of creating a society or institution that is based on the principles of equality and solidarity, that understands and values human rights, and that recognises the dignity of every human being. Despite this common reference, the universal meaning of *social justice* is still very difficult, if not impossible to define in its exact terms. The conception of such a term is contextual and subjective and its precise definition can vary across individuals, subject areas, time and place. It appears that each of us knows what we mean by it but when we want to define it with a precise universal definition it then becomes a difficult task. Therefore, it is not surprising to find that social justice movements rely on their own functional definitions such as can best express their scopes of work. Hence, we shall also try to define what we mean when we use this term within the context of the present book so that we can better define our goals and strategies for working towards these goals.

The traditional literature on social justice has been primarily focused on the following aspects:

- Equal rights and equal dignity;
- Realisation of civic, social and economic rights;
- A vision of a future more inclusive society in which the basic needs of all people are met;
- A more equitable distribution of rights, power and opportunity;
- Empowering marginalised groups to advocate and defend their interests;
- Changing power relationships between citizens, government, business and civil society;
- Changing at source the causes of social inequities;
- Testing innovative approaches and enabling change;
- Bringing stakeholders together to understand their role in an unjust system;
- Transforming the people who are part of the problem to become part of the solution;
- Access to assets and capabilities to reverse poverty and injustice;

By summarising the above mentioned aspects the following key principles can be derived:

- Maximising welfare;
- Fairness;
- Respecting individual rights and equality;
- Promoting virtue;

If we place these principles into the context of flood risk mitigation we could interpret them in the following way:

- Maximising welfare – maximising the benefits of flood mitigation measures according to stakeholder defined values;
- Fairness – public resources should be used fairly (favouring the most vulnerable) and efficiently in terms of resources, spatial scales (national, regional, local) and time (existing and future or inter-generational);
- Respecting individual rights and equality – everyone has the right to be safeguarded from the effects of floods and/or another person's actions in relation to floods;
- Promoting virtue – public policy should encourage sharing risks and working together to protect the most vulnerable members of the community;

It is important here, however, to draw attention to the theological and philosophical roots of these principles in order to understand their relevance and meaning. This is important as the links between the philosophical, theological and spiritual traditions and the justice principles are strong and, in turn, they influence which decision rules are regarded as the most appropriate when distributing resources or burdens (or, in our case, flood risks) across society. Hence, we shall first review some of the important streams of thought which have made their contributions to this subject. In doing so, we will quickly realise that there are many intellectual and activist traditions that have made such contribution and their thorough discussion would easily fill up this book on its own. Therefore, we can only mention some of them and more interested readers are encouraged to seek further information from our references and other sources.

5.2 ANARCHISTIC PRECEDENTS AND PRECURSORS

Anarchist perspectives have made considerable contribution to the social justice literature, policy and practice. Anarchism is defined by George Woodcock in the *Encyclopedia of Philosophy* (1967/1972) as "a social philosophy that rejects authoritarian government and maintains that voluntary institutions are best suited to express man's natural social tendencies". William Godwin was the first known anarchist in Western history, being the first to write about anarchism in a primarily European context, although anarchist movements have a much longer history, stretching back at least to the ancient Greeks (Figure 5.2). Godwin did not in fact use the word 'anarchy' at all in his writings. He was in fact preceded by Winstanley (e.g., 1649) but his anarchism followed another track than that of subsequent exponents of this way of living.

Figure 5.2 William Godwin (1756–1836) Source: Wikipedia (Accessed on 12th October 2011 from: http://en.wikipedia.org/wiki/William_Godwin)

Godwin published a book called, *Political Justice*, in 1793 where he introduced his ideas about anarchism. After his death, Pierre Joseph Proudhon, subsequently became a leading anarchist figure in the Western world. He incorporated a wider meaning to the word anarchism in his book entitled *What is Property?*. Citing Proudhon:

> "anarchism became not only a rejection of established authority but a theory opposing ownership of land and property as well" (Calabrese 96).

Other important anarchists were the Russians Mikhail Bakunin (1814–1876) and Peter Kropotkin (1842–1921). Bakunin had a major influence in the world and introduced anarchism to many people. Kropotkin was one of the many people inspired by Bakunin and he wrote many books on anarchism, including *Fields, Factories and Workshops*, and *The Conquest of Bread*, and greatly aided in the evolution of the theory of anarchism. Anarchistic movements have long flourished in Latin America in response to extreme differences in social environments.

The origin of the word anarchy comes from the Greek, with anarchy, anarchism, and anarchist implying coordination on equal footing, without superiors and subordinates, that is, horizontal organisation and co-operation without coercion, ideally or practically. Anarchistic systems represent a movement towards more human rights, fairness and efficiency, with less rank and income differences. The main principles of anarchism are the negation of authority and all of its oppressive power and juridical laws, and it teaches a tendency towards freedom, equality, solidarity, social justice, free contract, free initiative, atheism, antimilitarism, internationalism, decentralism, autonomy and federalism, self management and '*comunismo libertario*', that is not communism, but libertarian communalism – "from each according to ability, to each according to needs".

In the words of anarchists, social justice refers to:

(a) anarchist law and court systems, compatible with the negation of hierarchy, etc., that is, alternatives to authoritarian juridical laws; and
(b) antimilitarist corps broadly defined, sufficiently strong to keep order and keep up the balances of strength, as well as to stop militarism, intra- and internationally.

5.3 FROM WINSTANLEY TO GODWIN

Gerrard Winstanley's *The New Law of Righteousness* was published in 1649 and William Godwin's *Enquiry Concerning Political Justice* appeared in 1793. Neither work used the word 'anarchism' as an approbatory description, understanding it only in its original Greek sense, to mean 'no government', or 'without a ruler'. Both, however, proposed what have since been recognised as anarchistic social arrangements as the means of overcoming the gross inequalities and injustices of their times. What then can we learn from these great preceptors?

In his *Encyclopaedia of Philosophy* article on anarchism, George Woodcock identified in the work of Winstanley (Figure 5.3) a clear motivation arising from Christian-religious faith, observing that (1967, p. 113):

> Advocating a rational Christianity, Winstanley equated Christ with 'the universal liberty' and declared the universally corrupting nature of authority. He saw 'an equal privilege to share in the blessings of liberty' and detected an intimate link between the institution of property and lack of freedom. In the society he sketched, work would be done in common and the products shared equally through a system of open storehouses, without commerce.

In fact Winstanley organised what we might now call 'co-operatives' with essentially anarchistic self-structuring organisations for the production and equitable distribution of food and other necessities at the time of great poverty and hunger in England that accompanied and succeeded the Civil War.

Figure 5.3 Gerrard Winstanley (1609?–1676) Source: Wikia (Accessed on 12th October 2011 from: http://counterculture.wikia.com/wiki/Gerrard_Winstanley)

It is difficult to connect mentally with Winstanley's work after such a passage of time, for although so much is profoundly suggestive and in some deeper way indicative of later events, it all appears as if "seen in a glass, darkly" when read nowadays. It is for example indicative and remarkable that it opens with the figure of Abraham, who in present-day leadership studies stands as the paradigm case for *irregular indirect leadership* when placed within the context of his own time. Since this form of leadership is the best fitted to anarchistic social arrangements and, correspondingly, was employed by Kierkegaard when developing his concept of 'the teleological suspension of the ethical' which we shall discuss again later here, its primacy in Winstanley's work is highly indicative. Winstanley appears clearly already from the beginning as one who was profoundly serious in his reformed Christianity and who had at the same time thought through its teachings very thoroughly and in considerable depth – a position that naturally placed him at odds with much of the church-teaching of his time. Thus (p. 13):

> Now search the scriptures, you that stand up to be Teachers, that say that I deny the Scriptures, and let them judge me, whether I deny them or not; but one thing you will find to your shame, that those scriptures of the Prophets and Apostles, which you seem to perceive with such love and zealous tenderness, shall cast the first stone at you, to stone you out of your Pulpits; for you do not profess those scripture in love of them, but in your zealous covetousness to uphold your trade.

As we shall illustrate later here, Kierkegaard expressed much the same sentiment some two hundred years later, but then using the power of irony to attain to a much more devastating effect. But Winstanley was formulating 'The New Law of Righteousness', taking it as the very title of his book, and this Law was to be 'God's Law brought back to Earth' as governing all creatures on Earth, while Kierkegaard was aiming more accurately because more narrowly focussed, addressing exclusively *Hiin Enkelte*, the one, the individual, to whom he had the honour to address as "my reader".

Posing the question of what it was that was so opposed to Righteousness, which Winstanley called "the kingdom of the devil" – echoes of Martin Luther! – he himself answered (p. 108):

> … Covetousness, the selfish power ruling in a man. And as this power corrupted the Creation (mankind) so it rules, or hath ruled in every one more or less; but is the curse. And every one that hath lain under the bondage of this selfish power, and is in any measure delivered, he can from that experience declare, what the power of darkness is in whole mankind, as I have showed how secretly and closely this selfish power branches himself forth in every one, till by the right of Reason's Law shining within, he is discovered and cast out.

However, that which men covet in the first instance is that which they called their 'property', even as in Winstanley's work this is not said as such, at least in part because inheritance is that which can only be bequeathed by God in Winstanley's account, so that this which we shall shortly introduce as 'naked' property cannot appear. Still, the thrust of the argument is cogent enough.

We must however quickly move on from Winstanley because during the period between 1649 and 1793 so very much happened in the world of thought that is philosophy, and not least in what had become by 1793 a unified, predominantly English-speaking nation state. Consentaneously, this became the only period in history in which philosophical works in the English language entered decisively into the mainstream of European culture.

What had intervened was, in the first place, the effect on the reading public of Thomas Hobbes' *Leviathan*, published already in 1651, which laid out the principles of an authoritarian state which was acceptable to its subjects so long as it subscribed to and exemplified the maxim that "The safety of the people is the supreme law" and protected them correspondingly – and then in the first place from the depredations of their fellow subjects. He thereby outlined the form of a state that was necessarily attuned to a market economy in which property was enshrined as a social right – and with this came the right of privilege too. At the same time he insisted that, at least in principle, justice should be available to all, regardless of their station in life: that there should be equality before the law of the land. As his editor remarked, "not only was this equality of rights required for the operation of a market society, but at the same time it was the existence of the market that made this new equality possible". In effect, he outlined and defended a form of society based upon individual self interest, as the form that was necessary to avoid his notorious *bellum omnium contra omnes* // the war of all against all, and without which he memorably foresaw "the life of man, solitary, poor, nasty, brutish, and short". At the same time, his editor remarked ironically in his introduction that "Hobbes was born in 1588 and lived to be ninety-one. His longevity was due not least to the care he took for his political life". Clearly Winstanley could have had no truck with Hobbes even when he could have responded to him in later editions of his major work, while Godwin ignored him, referring only to "English writers upon politics in general, from Sydney to Locke to the author of *The Rights of Man*" (that is, to Thomas Paine, this work having been published in 1791). He further drew upon Helvetius and Rousseau, and specifically the latter's *Émile*. He also gratefully acknowledged the contributions of Jonathan Swift and referred frequently to Voltaire. On the other hand, he described Burke's *Vindication of Natural Society* as "a treatise in which the evils of the existing political institutions are displayed with incomparable force of reasoning and lustre of eloquence, while the intention of the author was that these evils were to be considered as trivial". A major influence on Godwin was however also exerted by David Hume (1739 and 1740, and also 1748) even though Godwin took issue with him on some occasions. Although Godwin made no mention of Winstanley, it seems likely that he was acquainted with the ideas that Winstanley had promoted as these lingered on in the traditions and teachings of English Protestant sects.

Thus we can view Godwin, coming so long after Winstanley, as a man of the Enlightenment, with all the predispositions and limitations thereby entailed. In particular, a religious "Righteousness" has been superseded by an agnostic "Political Justice". Woodcock expressed the consequences of this position as follows:

> Godwin differed from most later anarchists in preferring to revolutionary action the gradual and, as it seemed to him, more natural process of discussion between men of good will, by which he hoped truth would eventually triumph through its own power. Godwin … put forward in a developed form the basic anarchistic criticisms of the state, of accumulated property, and the delegation of authority through democratic procedure.

Writing in the 1978 edition of the *Encyclopaedia Britannica*, this same commentator repeats the judgement of Sir Alexander Grey (in his work entitled *The Socialist Tradition*, of 1946, p. 134) that "Godwin sums up, as no one else does, the sum and substance of anarchism, and thus embodies a whole tradition".

From the point of view which will be advanced here however, Godwin tried to achieve through the power of reasoned argument alone that which could not possibly be attained through reasoned argument alone, and it is in this sense, as much as in his 'naïve faith' in a God-given human virtue, benevolence and goodwill, that his work has since been viewed as *utopian*, and thus as an attempt to work out in practice what Sir Thomas More had only sketched in principle, as some kind of ideal, in his celebrated *Utopia* of 1516. Despite this, or perhaps precisely because of it, Godwin's masterpiece was none the less highly *inspirational* and consequently it appealed to those who were similarly inspired, and most notably such internationally renowned poets as Percy Bysshe Shelley and William Wordsworth among other great writers, such as William Hazlitt and Robert Owen. Thus, although Godwin was himself all but forgotten by the time of his death in 1836, the *doctrine* that he had propounded persisted in the *miracles* of poetry, as represented iconically by Shelley's *Prometheus Unbound* and his *Queen Mab*. This is then to say, in accordance with the great French thinker, Blaise Pascal, that these miracles arose from Godwin's doctrine, and not the doctrine from the miracles. Thus:

Those who would expect of others that they shall perform miracles
must first provide them with a doctrine that will enable them to do so

Accordingly also however, turning directly to Pascal:

Il faut juger de la doctrine par les miracles. Il faut juger des miracles par la doctrine.
Tout cela est vrai, mais cela ne se contredit pas

(One must judge the doctrine by the miracles.
One must judge the miracles by the doctrine.
All of this is true and there is no contradiction)

Les miracles et la vérité sont nécessaires, à cause qu'il faut convaincre l'homme entier,
en corps et en âme

(Both the miracles and the truth are necessary to convincing the entire person,
both in body and in soul)

Instinct et raison, marques de deux natures

*(Instinct and reason, the marks of two natures)**

But then also and inevitably (Pascal/Lafuma 1962/2003, p. 161):

Les miracles ne servent pas à convertir mais à condamner

(Miracles do not serve to convert, but to condemn)

And to what is it that those who truly experience the miracle are condemned? We see that they are *condemned to accept*. And to what are they condemned to accept in the present case? They are condemned to accept the tautology that Kierkegaard so beautifully expressed in his aphorism: "And faith is a miracle, otherwise it is not faith". And with this they are condemned to accept that all of this that drove Winstanley, Godwin, Proudhon and Bakunin, just as it drove Shelley, Mozart, Bolzano and Cantor and so many others and which in turn drives our present enterprise, succeeds and can only succeed, following Kierkegaard, "*I Kraft af det Absurde*", which, although it is normally translated as "By virtue of the absurd" can be better translated more literally in our present case as "Through the power of its absurdity".

Something of this understanding penetrated into philosophy, primarily with Kant, who analysed so far as it could be analysed philosophically the experience of *The Sublime*, making of it an *æsthetic-religious category* in his *Critique of Judgement* of 1790. Kant saw this as something that was universally communicable, exercising an innate capacity for the most exalted feelings and inducing an experience of elation at the highest level of moral experience, as the ethos of the nineteenth century. In our present case, we are forever seeking *paradigm cases* that will serve to convey the deepest emotional experiences, including those of terrible and frightful events, combining ethos and pathos, necessarily paradoxically. Such a catastrophe was exemplified as a paradigm case in the 2011 Japanese tsunamis which devastated so much of a Japanese coastal region and led to further deep concerns over nuclear electricity generation.

5.4 FROM PROUDHON TO BAKUNIN

Godwin's subject was a *political justice* that was consentaneous with a *social justice*, and it was said of Pierre-Joseph Proudhon that he "spent his whole life in quest of social justice" (Figure 5.4).

Figure 5.4 Pierre-Joseph Proudhon (1809–1865) Source: Wikipedia (Accessed on 12th October 2011 from: http://en.wikipedia.org/wiki/Pierre-Joseph_Proudhon)

And yet the subject of Proudhon's greatest work was posited at what must at first appear as something subsidiary to this objective, namely: *Qu'est-ce que la propriété? // What is property*? (To which he answered: *La propriété c'est du vol // Property is theft*.) Of course the question concerning property was raised indirectly already by Winstanley and emphasised more directly by Godwin, but in both cases it took a secondary position relative to the concept of divine righteousness in the first case and the expression of divine righteousness in social justice in the second. Of course Proudhon also *cared* passionately about justice, but his primary *concern* was with property: property had, so to speak, 'moved to centre stage'. Clearly something had changed within the social environment through which the spirit of anarchism was progressing – and its motivations were changing correspondingly. What, correspondingly again, can we learn from Proudhon and Bakunin concerning this change in emphasis that appears to be reversing in our own time and in our own case?

Now at one level the reason for this change in emphasis appears to be quite obvious. Thus, from the introduction to the Cambridge edition (1994, p. xxvi):

> Proudhon grew up in what E. M. Forster was to call the 'age of property'. Property was indeed an object of worship among the *bourgeois conquérants* of the early nineteenth century; and jurists like the elder, the Professor, Proudhon were the high priests of this institutional embodiment of the Revolution – "its internal support" as Hyppolyte Taine wrote, "its primary force, and its historical meaning". Although the product of an exercise in what Raymond Troplong (another of Pierre-Joseph's targets) called "dividing the spoils", property became an "inviolable and sacred right"…

> In Proudhon's day the institution of property, which was the most significant legacy of the Revolution, represented a sensitive issue on practical as well as theoretical grounds. First there had been the massive dispossessions following revolutionary confiscations, further embittered by debates over the Law of Indemnity of 1825; and then there was the threat to private fortunes posed by the crucial qualification of the 'absolute' character of private property … which was that in some cases 'public utility' might override private rights … For Proudhon this was simply another form of political confiscation. The confiscations, the irony, and the hypocrisy thus produced by the lawyers, especially in their claims to have separated property from all political considerations, were insupportable; and Proudhon proposed to do something about this situation.

Now although this motivation might appear at first sight as entirely adequate to explain Proudhon's development of the theory and practice of property, we have long ago learnt from Michel Foucault that far-reaching revolutionary changes of this kind are moved by deeper currents than those that are perceived at the surface of such everyday events. Indeed it would be quite normal for the principle exponent of such a revolution to be himself unaware of the presence of these deeper currents that are in fact carrying him on his passage through life. In his *Les mots at les choses*, unfortunately but necessarily translated as *The Order of Things*, Foucault introduced his own way of analysing the nature of such deeper currents as follows (1966, omitted / 1970, p. xi):

> I did not operate … at the level that is usually that of the historian of science – I should say at the two levels that are usually his. For, on the one hand, the history of science traces the progress of discovery, the formulation of problems, and the clash of controversy; it also analyses theories in their internal economy; in short, it describes the processes and products of the scientific consciousness. But, on the other hand, it tries to restore what eluded that consciousness: the influences that affected it, the implicit philosophies that were subjacent to it, the unformulated thematics, the unseen obstacles; it describes the unconscious of science. This unconscious is always the negative side of science – that which resists it, deflects it, or disturbs it. What I would like to do, however, is to reveal a *positive unconscious* of knowledge; a level that eludes the consciousness of the scientist and yet is part of scientific discourse, instead of disputing its validity and seeking to diminish its scientific nature. What was common to the history, the economics, and the grammar of the Classical period was certainly not present to the consciousness of the scientist; or that part of it that was conscious was superficial, limited and almost fanciful…; but, unknown to themselves, the naturalists, economists, and grammarians employed the same rules to define the objects proper to their own study, to form their concepts, to build their theories. It is these rules of formation, which were never formulated in their own right, but are to be found in widely different theories, concepts, and objects of study, that I have tried to reveal, by isolating, as their specific locus, a level that I have called, somewhat arbitrarily perhaps, archaeological.

It is this kind of transformation, which in its essence is a transmutation in motives, that we need to identify here as this is formed by the deeper currents of a more general and increasingly socio*technical* transformation. If we may put a name to it, we may identify it as the extension of the notion of physical property to encompass a wide class of non-physical properties, now conventionally called collectively '*intellectual property*'. In the language of Foucault, we have to distinguish a *discontinuity in the history of thought* as this extended – and indeed continues even today to extend, and then ever more insidiously and widely – the idea of property into territory that this idea never previously occupied. The idea of property passes from tangible objects (or ***ex***istent objects in the sense that they are to be found in our ***ex***ternal world) to intangible objects (or non***ex***istent objects, in that they are, as 'thoughts' in the first place, to be found nowhere in the ***ex***ternal world but only ***sub***limely in the minds of men and women where, once again, they may only be said to ***sub***sist). Of course as intimated earlier, these objects may be *represented* in the *ex*ternal world, such as in patents, contracts, laws and so on in the present case, but this is not, so to say, their place of birth. Such *re-presentations* may serve to transfer some part of the thought world of one person to the thought world of another person, but they remain and can never be more than *re-presentations of things that are in their essence, as presentations in the thought world of the individual person, intangible*. Clearly this change was already inseparable from the development of a modern condition of society that placed a new kind of emphasis upon 'knowers' while both promoting and demoting their various interests correspondingly.

The essential point to be made here is that this process was initiated precisely during the period in which Proudhon thought and wrote, even though he did not write about this process but instead wrote almost entirely about matters that concerned tangible property. The growths of *the patent* and of *the copyright*, events that were contemporary with Proudhon's working life and must have concerned him at least to some extent in his work as a printer, were not at all the immediate subjects of his writings, even as we are hypothesising that the occasions of his work and his life must have made him aware of their presence as at least potential extensions of the idea and the ideology of property, but then only subconsciously, being things that probably appeared to be of a lesser immediate significance at that time. This implies however that when we now read Proudhon within our present context of 'intellectual property', we must do so from another point of view, or from another angle, or within another perspective. To use a term that has been so much popularised by the work of Jacques Derrida, we have to *deconstruct* Proudhon's *Qu'est-ce que la propriété? // What is property*? when we now read it.

If we regard a patent in the common way, as a right granted to a person (including a 'legal person', such as a company) to derive material, and primarily financial, benefit from it, then, although we may trace its origins back to the fifteenth century, we see that it only really established itself legally and institutionally between the last decade of the eighteenth century and the middle of the nineteenth century. After this period the patenting of inventions became commonplace and almost universally accepted. The institution of copyright arrived somewhat later and was first accepted only within the countries of the copyright-holders origin. The first law to extend copyright protection to non-nationals of the protecting state was legislated in France in 1852, but this precedent was only extended to a wider application, initially covering 14 countries, by the *Bern Convention for the Protection of Literary and Artistic Works*, of September 9th, 1886.

Let us now place some of the most immediately relevant parts of Proudhon's masterpiece within this more specialised context. He opened his analysis of the origins and nature of property by its formulation in Roman law – "*jus utendi et abutendi re sua, quatenus juris ratio patitur* – as the right to use and abuse a thing within the limits of the law" and showed how, "in matters of property, use and abuse are necessarily indistinguishable". He showed further how the corresponding sections of the French Constitution of 1793 and the *Code Napoléon* differed in no essential way from this formulation. It follows at once that anyone acquiring a patent upon a piece of software, for example, has just as much right to suppress it as to employ it, just so long as the law of the land is not violated.

Proudhon then made the classical distinction between *property* and *possession*, and how both had come to be subsumed under the one rubric of 'property', albeit as two different forms of property. Thus he made the commonly held distinction between:

> (i) property pure and simple, the right of domain over a thing, or as they say, 'naked possession' and (ii) Possession. 'Possession', says [Alexandra] Duranton, 'is a matter of fact, not of right.' [Charles] Toullier: 'Property is a right, possession is a fact'. The tenant, the farmer, the shareholder, and the usufructuary are possessors; the owner who rents and lends for use and the heir who waits for the death of the usufructuary are proprietors …
>
> This double definition of property – domain and possession – is of the highest importance, and it must be clearly understood in order to comprehend what is to follow.

Thus, within this classical tradition, the knowledge of a person, being as much a possession of that person as are his or her own arms and legs, is that person's property in the second sense. On the other hand, someone who had acquired a slave as his property, as was entirely possible in the USA in particular at the time that Proudhon wrote, could not have had the arms and the legs of the slave as his or her possession since *that was not possible in physical fact*, so that this slave was only property in the first, 'domain' or 'naked' sense. Similarly then, a person who takes out a patent or a copyright on *the representation of this knowledge, as that which only encapsulates the knowledge that is so represented, such as in the form of a code for example*, by registering it in his own name or in the name of his company regardless of who produced this representation and whose knowledge is therewith encapsulated in it, is also only a proprietor in the first, 'domain' or 'naked' sense.

But to continue this summary of the classical position with Proudhon, but then further to introduce for the first time Proudhon's opposition to it, with some italics added (pp. 36, 37):

> From the distinction between possession and property arise two sorts of rights: the right *in* a thing (*jus in re*) is the right by which I may reclaim the property which I have acquired, in whatever hands I find it, and the right *to* a thing (*jus ad rem*), which gives me a claim to become a proprietor. Thus the right of marriage partners over each other's person is a *jus in re*; that of two betrothed is only a *jus ad rem*...
>
> This distinction between the *jus in re* and the *jus ad rem* is the basis of the famous distinction between *possessoire* and *pétitoire* – actual categories of jurisprudence, the whole of which they encompass within their vast boundaries. *Pétitoire* refers to everything relating to property, *possessoire* to that relating to possession. In writing this memoir against property I bring an *action pétitoire* against all society. I prove that those who do not possess today are proprietors by the same title as those who do possess; but instead of inferring from this that property should be shared by all, I demand, *as a measure of general security*, its entire abolition. If I fail to win my case, there is nothing for us proletarians to do but to cut our throats. We can ask nothing more from the justice of nations; for as the *Code of Procedure* [1807] (art. 26) tells us in its energetic style, the plaintiff who has been denied his rights as *pétitoire* is prohibited from acting as *possessoire*. If, on the contrary I win my case, we must begin an *action possessoire* in order to be reinstated in the enjoyment of the wealth of which property deprives us. I hope that we shall not be forced to this extreme, but these two actions cannot be prosecuted at once because, according to the same *Code of Procedure*, "the *possessoire* and the *pétitoire* may never be combined".

(J. A. Cunge, in a private communication, has contributed the French dictionary definition of *pétitoire*: *un action qui a pour objet la reconnaissance, la protection et le libre exercice d'un droit réel immobilier*//an action that has as its objective the recognition, protection and the free exercise of a real right to property. Similarly with *possessoire: relative à la protection judiciaire de la possession (immobilière)*// relative to the judicial protection of the possession of property.)

However, Proudhon's masterpiece extends this understanding again. In the first place it exemplifies what Proudhon made very clear in his great work, that the movement that was transforming everything into property must become exceedingly destructive of the society so afflicted and thence it would ultimately undermine the competitiveness of that society. With remarkable perspicacity, Proudhon explained that (with italics added): "property and society are completely irreconcilable with one another. It is as impossible to associate two proprietors as to join two magnets by their opposite poles. *Either society must perish, or it must destroy property*" (*loc.cit.* p. 43). Reading any business newspaper and listening to any television news broadcast as this is written demonstrates the prescience of this observation. We shall return to this when we consider the contributions of Bakunin in a moment.

Proudhon's work thus leads us to a first understanding of this something else again, namely why it was that it was precisely the area of software production, before all others, that attracted a specifically anarchistic form of working, and thence of governance. To some extent, these reasons have already been adumbrated: the treatment of software as property in the 'naked' sense, through the oppressive power of patents, was becoming increasingly destructive of the creative impulse of those who wrote software and those whose knowledges were encapsulated in this software. This was experienced by these creative persons as an invasion, as an act of aggression carried out by the forces of property and directed against their creative instincts. In effect, this invasion threatened directly their most vital interest: their liberty to employ their knowledge in the most creative ways possible. In this sense the programming community followed in the footsteps of an earlier artistic community, as recorded by George Woodcock in his *Encyclopaedia Britannica* article:

> During the 1890s, especially in France, Anarchism was adopted as a philosophy by avant-garde painters and writers. Gustave Courbet had already been a disciple of Proudhon; among those who in the 1890s accepted an Anarchistic philosophy were Camille Pissarro, George Seurat, Paul Signac, Paul Adam, Octave Mirabeau, Laurent Tailhade, and, at least as a sympathiser, Stéphane Mallarmé.

But with this we can now best continue with Bakunin (1871//1972, p. 261), Figure 5.5:

> I am a fanatical lover of *liberty*. I consider it the only environment in which human intelligence, dignity, and happiness can thrive and develop. I do not mean that formal liberty which is dispensed, measured out, and regulated by the State, for this is a perennial lie and represents nothing but the privilege of the few, based upon the servitude of the remainder … No, I mean the only liberty worthy of the name, the liberty which implies the full development of all the material, intellectual, and moral capacities latent in every one of us; the liberty that knows no other restrictions but those set by our own nature. Consequently there are, properly speaking, no restrictions, since these laws are not imposed upon us by any legislator from outside, alongside, or above ourselves. These laws are subjective, inherent in ourselves; they constitute the very basis of our being.

Figure 5.5 Mikhail Alexandrovich Bakunin (1814–1876) Source: Wikipedia (Accessed on 12th October 2011 from: http://en.wikipedia.org/wiki/Mikhail_Bakunin)

Bakunin directed his ire in the first place against the power of the state, and especially as this was developing in his own time. In 1873 he published his *Anarchy and Statism*, in which he warned against these new-found powers that were to lead to what we now call *the technologies of persuasion*. Thus (ibid, p. 114):

> There was a ... path, taken long ago, in truth, by the Roman Emperor Augustus and his successors, but abandoned after them, rediscovered in recent times by Napoleon III and widened and improved by his pupil, Prince Bismarck. That is the path of state, military, and political despotism, camouflaged and embellished with the broadest, and most innocuous, forms of popular representation.
>
> In 1815, however, that path was completely unknown. At that time no one suspected the truth which has now become obvious even to the most stupid of despots, that so-called constitutional forms, or forms of popular representation, do not impede state, military, political, and financial despotism. Instead they have the effect of legitimising it and giving it a false appearance of popular government, and they can significantly enhance its internal strength and vigour.

From these quotation alone, it will be clear that Bakunin's anarchism was as much an anathema to Marx and Engels, who labelled it *Utopian Socialism*, as it came to be to more religiously inclined intellectuals, for whom it represented a *Naïve Christianity* that took no account of the negative force, which we have introduced earlier as *nothingness* within theology, and *nihilism* within philosophy, that necessarily worked within the human psyche – and which was treated by Søren Kierkegaard, Friedrich Nietzsche, Martin Heidegger (see, especially, 1967), Jean-Paul Sartre, Karl Jaspers and Karl Barth, as the most prominent among many religious writers, philosophers and theologians – as something inseparable from human existence itself. For our present purposes, however, it is only necessary to observe further the *anti-statist* position introduced and argued by Bakunin that nowadays penetrates also into the open-source movement. We repeatedly observe today, just as occurred in earlier times also, that everything that we do that is alive, creative, productive or, in a word, *poietic*, must be done in despite of a large part of the efforts of the state apparatus as well as most other organisations, including many businesses, to crush genuine originality, to subvert socially valuable developments and to derail promising initiatives, even as these public entities make the most specious claims for supporting all the valuable endeavours that they in practice suppress. The consequence is that all practical developments in the field of concern here must be made through what is called the *creative business enterprise*. Then, from Abbott, 2004 (p. 90):

> A creative business is one that is driven in the first instance by an urge to create which is natural to most, if not all, human beings. In order that this business can continue its creative activities in our current societies it must of course be financially viable, but that is rarely the primary reason for its existence. From the point of view of the born leader, businesses are not built in order to make money, but money is made in order to build businesses. Other persons than the leader may not necessarily see it that way, but we are here talking about leadership.

Steve Jobs might serve as a paradigm case, at least within the confines of our present-day societies. It is in this sense relevant that existing creative businesses in the field of hydroinformatics are not themselves property in the most immediate sense that they cannot be bought or sold: no person or group of persons can dispose of them or any of their assets at will. The aim then is always to provide an atmosphere where creativity and personal satisfaction should have full rein, while maintaining very adequate pecuniary benefits. It was anticipated that the creative businesses that came to build around the open-source software provided by the Water Knowledge Initiative[1] would concern themselves with setting up the means to access electronically encapsulated knowledge and supporting human knowledge

[1]The *Water Knowledge Initiative* brought together seven persons at its core. One of its major achievements, albeit so-far little reported in publications, was its success in placing a proprietary computational engine on a server in such a way that it could be accessed and used over the web while using a standard desktop computer.

and understanding simultaneously and interactively using the full orchestra of semiotic devices. These will have the capacity to catalyse such social processes as active stakeholder participation, making new processes of 'maximising social justice' in the sense of Proudhon viable, while being far more effective than current technocratic devices devoted to minimising costs expressed in terms of currency units.

Figure 5.6 Pjotr Aleksejevitsj Kropotkin (1842–1921) Source: Wikipedia (Accessed on 12th October 2011 from: http://nl.wikipedia.org/wiki/Peter_Kropotkin)

5.5 FROM KROPOTKIN UNTIL TODAY

It is remarkable that Eric Raymond, in his book on the open-source initiative in software production, entitled *The Cathedral and the Bazaar* (1999), only referred to anarchism on one page (p. 52 – not p. 51 as given in the index) and then only in relation to a quotation from Kropotkin (Figure 5.6), who was one of the last of the great anarchistic leaders in Europe, living until 1921 and dying in what had by then already become the kernel of the USSR. Raymond, however, could never get beyond a motivation of "their own ego satisfaction and reputation among other hackers" as a driving force of the participants in the open-source movement, a position that is completely untenable within the context employed here. Although Raymond takes up and develops somewhat Kropotkin's "principle of shared understanding" as a basis for "sustained cooperation", he does not at all investigate the deeper roots of Kropotkin's work.

At this time of writing this conflict between authentic, creative technology and inauthentic, destructive technology in our field of the Web and the internet is taking yet another turn again, in the form of a new software bubble that imitates in several ways the forms and natures of the 2000–2001 boom-and-bust in the software market place. In a *Financial Times* (28 July, 2011, p. 15) article entitled "*Patent hunting is the latest game on tech bubble circuit*", Richard Waters wrote:

> A new kind of bubble is inflating in tech and, like all the most alluring bubbles, this one is being pumped up by a potent combination: expansive claims from its promoters, some early deals that appear to "prove" the validity of the high prices being talked of, and the lack of transparency that makes it hard to assess the real value of the assets at issue.
>
> You could call it the Great Patent Bubble of 2011. In the month since an auction of patents from the bankrupt Nortel Networks ended with a shockingly high bid of $4.5 billion, or five times the initial offer, the favourite game of tech circles has been to find the next big chest of buried gold …

> As always in tech bubbles, it is the "pure plays" that have drawn the most interest – in this case, the companies set up mainly to exploit the value of pure IP (Intellectual Property), rather than actually to build things ….
>
> Yet the value of individual patents is almost impossible to assess, given how lax patent issuing practices over the last years have brought about such a jungle of overlapping claims. As we pointed out this year, as many as 250,000 patents could be at stake in a smartphone ….
>
> As the stock market starts to tune in to the value of patents, this exposes an unedifying reality. To keep their shareholders happy, corporate managers are likely to take a more aggressive stance with their IP. Whether shareholders will see any benefits is likely to end up (as) a roll of the dice.

This new 'tech bubble' is potentially even more dangerous than its earlier progenitor in terms of the damage that it has the capacity to do as it subverts the endeavours of creative initiatives such as hydroinformatics' activities in the field of active stakeholder participations and other such sociotechnical initiatives. Our praxis has not so far attracted the attentions of these financial-legal adventurers and their attempts to 'cash in' on the praxis of the kind that we are developing, but through their exercises of patenting in all-encompassing ways they may well subvert our efforts to maintain quality and by so doing subvert our enterprise.

Such multi-billion dollar investments into these potentially empty boxes is in crass contrast with the absence of adequate funding for the SaaS, fifth generation, of modelling (Abbott and Vojinović 2009) which would change public participation out of all recognition. However there are no such sources of funding in these and other cases, with that which has so far been offered being insufficient to launch even an exploratory initiative. This seems to be a very general experience in the cases of authentic technologies, and even more so with authentic sociotechnical initiatives, where this anarchistic license remains unacceptable within current sources of investment. HRH The Prince of Wales has experienced this and written about it on many occasions Thus, already in 2002 he wrote (*Temenos Academy Review 5*, pp. 13, 14):

> It never ceases to amaze me how powerful is the rearguard action fought by those who seem to have developed an underlying hatred for anything that cannot be explained, or 'proved' by Science – even if, in the case of what has come to be known as complementary medicine, it can be demonstrated on many occasions to be beneficial to the recipient. Such is the fury of this rearguard action that it is virtually impossible to attract any funding from official bodies, or even charitable trusts, for the purpose of carrying out research into the efficacy of such complementary therapies. I remember, for instance, when three Government research bodies were carrying out proper research into the organic farming system at Highgrove, that there was a complete refusal of my request to include the homeopathic treatment of our livestock in the research programme. Despite approaching other relevant organisations to carry out the research instead not one of them would agree to do so.

HRH's speech of 2010, also published in the 2011 edition of the *Temenos Academy Review*, was even more dismissive of the industrial powers-that-be in the West in this respect, seeing a resurgence of spiritual life as the only solution to the problems of the West. For ourselves, we consider the application of hydroinformatics to organic farming systems – which we would characterise as *knowledge-intensive systems* – as essential to the sustainability of the first and second economies, and so surely pleasing to our third economy (See again, Abbott 2000, Jonoski 2002, and, more critically, the books of Arundhati Roy, e.g., 1999).

We should in conclusion to this essay on anarchistic forms of cooperation observe that from Kropotkin's time onwards anarchism itself developed most extensively as anarcho-syndicalism, as in Spain until it was destroyed by Fascism using brutal military and police repression from 1936 onwards. Anarcho-syndicalism as such developed primarily in the trades-union movements and in pacifist circles as means of establishing working, sustainable arrangements that would breed other such developments through emulations of their success. Our Water Knowledge Initiative was in this sense an anarchosyndicalist arrangement and it is on the basis of this understanding that the present contribution has been written.

Although rarely identified as such, most civil-rights movements, anti-war movements, genuine pro-democracy movements, human-rights movements, wildlife-conservation movements and others of this kind are anarchosyndicalist in their origins, even though they commonly acquire more formal ‘command structures’ as they develop and, in the course of time, become institutionalised.

Anarchistic arrangements naturally developed outside Europe during the twentieth century, and then in the first place in India under the leadership of Mahatma Gandhi in the manner of a pacifist civil-rights movement. They took other forms again in other Asian countries and in Latin America of course, and are at this time of writing developing rapidly within Islamic cultures, and most obviously in the Middle East.

In the Europe of the nineteenth and twentieth centuries already, however, it became possible to identify the deeper currents that were carrying some parts of society in this general direction, and therewith to provide a conceptual framework and a language with which to describe and so identify the nature and the overall direction of the resulting movements, such as are now increasingly manifesting themselves in our own time at the surface of society. The person who may be said to have initiated this movement – which subsequently came to be called *existentialism* – was the one described by Ludwig Wittgenstein as “by far the most profound thinker of the nineteenth century” and by Martin Heidegger as “the only one of his kind”, namely the Søren Kierkegaard who has already been introduced earlier here. It is impossible in this place even to mention the immense range of insights into transformations occurring in the human psyche and the social movements that arise from these that Kierkegaard provided, so that it is only possible to take up a few basic notions and to adumbrate their consequences in our own time and in relation to our own field of endeavour.

Kierkegaard identified three ‘levels’ at which our thinking proceeds within its subliminal depths, which he described as the *æsthetic level*, the *ethical level* and the *religious level*. In his first great work, *Enten-Eller//Either-Or*, as long since translated into all common languages, the first part of the work was written from an almost exclusively æsthetic perspective, and so one that was situated at the level of the æsthetic, and the second part almost exclusively from an ethical perspective, and so one that was situated at the level of the ethical, while Kierkegaard himself adopting different pseudonyms for these two parts. The spiritual perspective, as this would appear for a psyche situated at the level of the religious, was only introduced in the form of a short but shocking *Ultimatum* that concluded the work. However, Kierkegaard left Copenhagen as this book was still in the press and travelled to Berlin, where he had followed the lectures of Schlegel a few years earlier. There he wrote the first drafts of three major works in three weeks. One of these took up a theme of Schlegel, that the æsthetic constituted a category that was being subverted, moving away from one that attached itself to an eternal beauty to one that attached itself to whatever happened to be *interesting* at the very moment. The present-day obsession in the Western world with ‘personalities’, such as footballers and pop-stars, who are regarded as the ‘interesting people’ *par excellence*, may be seen as a culmination of this movement.

This transformation of the category of *the æsthetic* into its degenerate subcategory of *the interesting* led Kierkegaard to consider something at the other extreme to this movement, namely to *the transformation of an ethical category into a superior religious category*. It was abundantly clear to Kierkegaard, as it had earlier been to Winstanley and other thinkers, that the paradigm case that established this category as a superior one, as one that was more profound than the category of the ethical in Christian faith, was the Judaic, Biblical and Koranic person of Abraham.

We recall how Kierkegaard spoke in this case of *en teleologisk Suspension af det Ethiske // a teleological suspension of the ethical*, but we should now introduce at least the title of the book that described this process in detail. namely *Frygt og Bæven // Fear and Trembling* – simultaneously excusing its scarcely appropriate English translation.

Let us now see how this leads us to a deeper understanding of our own motives by quoting the first paragraphs of the main body of the text of Kierkegaard's masterpiece, where we can already associate Kierkegaard's 'bread' with that which so often provides 'bread' to some even as it takes it from others, at least in the eyes of the anarchists, which is 'property':

> An old proverb pertaining to the outward and visible world says: "only one who works gets bread". Oddly enough, the saying doesn't apply in the world to which it most properly belongs, for the outward world is subject to the law of imperfection: there it happens time and again that one who gets bread is one who does not work, that one who sleeps gets it in greater abundance than one who labours. In the outer world everything belongs to whoever has it, the outer world is subject to the law of indifference, and the genie of the ring obeys whoever wears it, whether it be a Noureddin or an Aladdin, and whoever holds the world's treasures does so however he came by them. It is otherwise in the world of the spirit. Here there prevails an eternal divine order, here it does not rain on the just and the unjust alike, here the sun does not shine on both good and evil, here only one who works gets bread, and only one who knows anguish finds rest, only one who descends to the underworld saves the loved one, only one who draws the knife gets Isaac. He who will not work does not get bread, but will be deluded, as the gods deluded Orpheus with an airy figure in place of the beloved, deluded him because he was tender hearted, not courageous, deluded him because he was a lyre-player, not a man …
>
> Conventional knowledge [*Kendskab*] aims presumptuously to introduce into the world of the spirit that same law of indifference under which the outside world groans. No other work is necessary. But then it does not get bread, it starves to death while everything is transformed into gold. And what else does it know?

Thus it is the presumptuous extension of the concept of 'naked' property into the world of the intellect, so as to make of it 'intellectual property' also in this 'naked' sense, that spreads the disease of indifference among those who produce it, whereby the world in which they labour produces gold for those who succeed in claiming this property as their own while starving emotionally-intellectually those who produce this wealth, dispossessing them correspondingly. The movement that opposes this movement is then one that opposes the spread of indifference into its own realm, and this movement is a movement of the spirit, as a teleological suspension of what has come to pass as the ethical in its own sphere. This movement then proceeds in its essence 'at the level of the religious'. It should thus be no surprise to observe that all the great anarchists whose works have been mentioned above were religiously inclined – and here we must include Tolstoy – and in several cases were devoutly religious.

Of course, a movement such as our own proceeds also at the level of the æsthetic, just as it does again at the level of the ethical, but its most profound layer is that which moves at the level of the religious. But then any statement of this kind is almost certain to be widely misinterpreted in our present-day western world that has so little understanding of these categories, and so some further explanation is required to correct what is almost certain otherwise to create an even deeper and even more widespread misunderstanding. For this purpose we leave to the last the greatest exponent of the anarchist tradition founded upon faith, which is that of the Mahatma, Gandhi (Figure 5.7). We can here simply quote from Wikipedia, as follows:

> Gandhi was born a Hindu and practised Hinduism all his life, deriving most of his principles from Hinduism. As a common Hindu, he believed all religions to be equal, and rejected all efforts to convert him to a different faith. He was an avid theologian and read extensively about all major religions. He had the following to say about Hinduism:
>
>> Hinduism as I know it entirely satisfies my soul, fills my whole being … When doubts haunt me, when disappointments stare me in the face, and when I see not one ray of light on the horizon, I turn to the *Bhagavad Gita*, and find a verse to comfort me; and I immediately begin to smile in the midst of overwhelming sorrow. My life has been full of tragedies and if they have not left any visible and indelible effect on me, I owe it to the teachings of the *Bhagavad Gita* ….

Figure 5.7 Mohandas Karamchand Gandhi (1869–1948) Source: Wikipedia (Accessed on 12th October 2011 from: http://en.wikipedia.org/wiki/Mohandas_Karamchand_Gandhi)

Gandhi believed that at the core of every religion was truth and love (compassion, nonviolence and *The Golden Rule* or *ethic of reciprocity*) that essentially states either of the following:

One should treat others as one would like others to treat oneself (the positive form).
One should not treat others in ways that one would not like to be treated (the negative/prohibitive form, also called the *Silver Rule*).

Gandhi also questioned what he saw as hypocrisy, malpractices, and dogma in all religions, including his own, and he was a tireless advocate for social reform in religion. Some of his comments on various religions are (with italics added in conclusion here):

> Thus if I could not accept Christianity either as a perfect, or the greatest religion, neither was I then convinced of Hinduism being such. Hindu defects were pressingly visible to me. If *untouchability* could be a part of Hinduism, it could but be a rotten part or an excrescence. I could not understand the *raison d'être* of a multitude of sects and castes. What was the meaning of saying that the Vedas were the inspired Word of God? If they were inspired, why not also the Bible and the Koran? As Christian friends were endeavouring to convert me, so were Muslim friends. Abdullah Seth had kept on inducing me to study Islam, and of course he had always something to say regarding its beauty.
>
> As soon as we lose the moral basis, we cease to be religious. There is no such thing as religion over-riding morality. Man, for instance, cannot be untruthful, cruel or incontinent and claim to have God on his side.
> The sayings of Muhammad are a treasure of wisdom, not only for Muslims but for all of mankind.
>
> *I like your Christ, I do not like your Christians. Your Christians are so unlike your Christ.*
>
> *God has no religion.*

Later in his life, when he was asked whether he was a Hindu, he replied, "Yes I am. I am also a Christian, a Muslim, a Buddhist and a Jew."

Gandhi was a self-described *philosophical anarchist*, and his vision of India meant India without an underlying government. He once said that "the ideally nonviolent state would be an ordered anarchy." While political systems are largely hierarchical, with each layer of authority from the individual to the central government having increasing levels of authority over the layer below, Gandhi believed that society should be the exact opposite, where nothing is done without the consent of anyone, down to the individual. His idea was that true *self-rule* in a country means that every person rules his or herself and that there is no state which enforces laws upon the people. This would be achieved over time with nonviolent conflict mediation, as power is divested from layers of hierarchical authorities, ultimately to the individual, which would come to embody the ethic of nonviolence. Rather than a system where rights are enforced by a higher authority, people are self-governed by mutual responsibilities. On returning from South Africa, when Gandhi received a letter asking for his participation in writing a world charter for human rights, he responded saying, "in my experience, it is far more important to have a charter for human duties." A free India for him meant the existence of thousands of self sufficient small communities (an idea possibly from Tolstoy) who rule themselves without hindering others. It did not mean merely transferring a British established administrative structure into Indian hands which he said was just *making Hindustan into Englistan*. He wanted to ultimately dissolve the Congress Party after independence and establish a system of *direct democracy* in India, having no faith in the British-styled parliamentary system.

The passages on non-violence express to perfection the position taken in the 2010–2011 reform movements in Tunisia and Egypt, as well as in the present book, even as these current applications depend upon the immense advances that have occurred in the communications technologies since Gandhi's time.

5.6 MARXIST TRADITIONS

Many of the writings of Marx (Figure 5.8) and Engels are closely related to the subject of social justice even as both were strongly opposed to all forms of anarchism. The notion of praxis which refers to the process by which a theory, lesson, or skill is enacted or practiced, embodied and/or realised can be viewed as a practical framework for engaging in the work of social justice.

Figure 5.8 Karl Marx (1818–1883) Source: Wikipedia (Accessed on 12th October 2011 from: http://en.wikipedia.org/wiki/ Karl_Marx)

The notion of praxis has been a popular topic in the field of philosophy, and Plato, Aristotle, St. Augustine, Immanuel Kant, Søren Kierkegaard, Paulo Freire, Martin Heidegger, Hannah Arendt, and many others, have written about this topic. "Praxis" may also refer to the act of engaging, applying, exercising, realising, or practicing ideas. Therefore, citing Karl Marx:

"Practice without theory is blind. Theory without practice is sterile. Theory becomes a material force as soon as it is absorbed by the masses."

In much the same vein (Witkovsky, M. S., 2011, (editor): *Avant Guarde Art in Everyday Life: Early twentieth-century European modernism*, Art University of Chicago and Yale University Press) observed (p. 13) that:

Addressing an audience of politically committed academics in Paris in 1961, the French theorist Guy Debord made clear his impatience with research unconnected to revolution. "To study every-day life", he began "would be a completely absurd undertaking, unable even to grasp anything of its object, if this study was not explicitly for the purpose of transforming everyday life".

It is intriguing that HRH The Prince of Wales (2010/2011, pp. 19 and 22) has said something very similar, such as is now reflected in the present work, but of course not using the word 'revolution':

Now there are many examples where communities have replaced the short-term impulse with the long-term plan. But part of that strategy – to my mind at least at the heart of it – is the need for a new public- and private-sector partnership which includes NGO and community participation. To work effectively this will require governments to provide policies which support community participation.

From the historical point of view, the notion of praxis describes, among so much else, the social transformation that took place during the industrial revolution and gave the insight into struggles for economic justice and economic equality. With the industrial revolution new technological developments created social conditions that highlighted the need for workers to demand better working conditions. This has also brought forward the need for a synergic relation between the engineering professions and social justice and has further emphasised the importance of the realm of praxis.

Besides Marx, Paulo Friere, the Brazilian educationalist, has left a significant mark on thinking about progressive practice and he defined the corresponding praxis as:

"the action and reflection of people upon their world in order to transform it."

A figurative illustration of the notion of praxis is given in Figure 5.9.

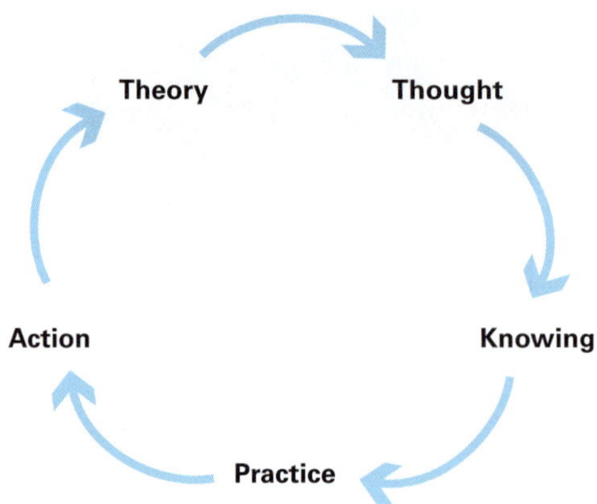

Figure 5.9 Illustration of the idea of "Praxis"

The idea of praxis has influenced many theorists working on the subject of social justice. It is important to emphasise that praxis combines theory and practice in a very profound way. Instead of starting with theory, then moving to practice, praxis starts from social relationships as the basis for theorising and always returns to the practical with a view to social transformation, and indeed to individual transcendences at the level of a whole social group. In this cycle, there is a constant exchange between theory and practice, and between thinking and doing, directed towards changes that are meaningful to the social group concerned. Therefore, praxis is not just any thoughtful act but an action with the power to change individuals and their society and we shall use this notion in order to develop our own definition of social justice within the context of flood risk mitigation.

5.7 RIGHTS TRADITIONS

Social movements are referred to as groups dedicated to realising the goals of a particular movement. As introduced earlier, such groups are often regarded as anarchosyndicalist in their origins, even though they commonly acquire more formal 'command structures' as they develop and, in the course of time, become institutionalised. Examples of this kind are civil-rights movements, anti-war movements, genuine pro-democracy movements, human-rights movements, wildlife and environment conservation movements … and there are many others.

Many social movements draw upon the concept of *a human-rights* framework to realise their goals. *Rights* approaches are concerned with what people (and sometimes other entities such as animals) deserve or are entitled to. They are also known through their developing different approaches to distributing rights equitably. The roots of their designs are often founded in the work of Enlightenment philosophers like Hobbes, Locke and Rousseau.

Thomas Hobbes (1588–1679) | John Locke (1632–1704) | Jean-Jacques Rousseau (1712–1778) | John Rawls (1921–2002)

Figure 5.10 Enlightenment philosophers Source: Wikipedia (Accessed on 12th October 2011 from: http://en.wikipedia.org/wiki/)

In recent time, the theory of the writer John Rawls has been regarded as one of the most influential theories of distributive justice. In '*A Theory of Justice and Political Liberalism*', Rawls outlined his view on justice and proposed that justice is equivalent to 'fairness', being based on two 'principles':

(1) "equality in the assignment of basic rights and duties;" and

(2) "social and economic inequalities … are just only if they result in compensating benefits for everyone, and in particular for the least advantaged members of society."

Some of his main arguments are that charitable giving cannot achieve social justice and there has to be substantial institutional change in order to create greater social and economic equality for all. Rawls proposed the difference principle, which advocates improvement of lives of the worst-off. As there are literally hundreds of books and thousands of articles which refer to Rawls' political thoughts, there are also many again of criticism that point out the difficulties/impossibilities involved in realising such ideas. The critiques of Rawls often raise the question of what would be an 'equal amount of goods and services', given people's differences in needs and preferences, and how could society maintain distributive justice over time?

5.8 ECOLOGICAL JUSTICE

Among the many pages of criticism concerning Rawls' theory, some of these critiques note that his theories of justice do not necessarily account for nonhuman entities. Thus, a different tradition has emerged to address issues of ecological justice which has been associated with a praxis of ecological justice. This whole field has struggled over the question of definitions, and different traditions have emerged to address issues of ecological and environmental justice. The emerging theories are *deep ecology* and *shallow ecology*. Deep ecology distinguishes itself from shallow ecology by valuing nonhuman species and biodiversity and by viewing humans as only one of many species present in the world of nature (see e.g., Lovelock and Margulis 1974). Other related theories are: *social ecology* (Bookchin 1980), *ecofeminism* (Waren 2000; Shiva 1988), *animal rights* (Regan 1983), *environmental racism* (Chavis & Benjamin 1993), *environmental inequality* (in Bullard 2005) and *environmental justice* (Bullard 2000).

Environmental Racism is a term used to describe the disproportionate effects of environmental pollution on racial minorities. The term *Environmental Inequality* has emerged more recently to encompass such aspects as class, gender and immigration status, as well as the inter-connections between these aspects. The field of environmental justice has emerged with the aim of reducing the level of environmental risk in communities. A fundamental question in environmental justice is concerned with environmental equity – whether the spatial distribution of environmental risk is indeed equitable among different racial and socioeconomic groups (see e.g., Nussbaum 1995).

Chapter 6

Characterisations of social justice

6.1 SOCIAL JUSTICE AS AN EXISTENTIELL

Social justice is inseparable from human existence itself and indeed it is a concept that we share with so many of our fellow creatures here on earth that we may suppose that it in fact pre-dates human Being, which is to say that it is pre-ontological. The word "existentiell" is commonly used to translate Heidegger's *existenziell*, as referring, reflexively, to our own understanding of our own modes of existence. (In the words of Heidegger himself [1927, p. 12 // 1962, p. 33]: "Die Frage der Existenz ist immer nur durch das Existieren selbst ins Reine zu bringen. Das hierbei führende Verständnis seiner selbst nennen wir das existenzielle" // "The question of existence never gets straightened-out except through existing itself. The understanding of oneself which leads along this way is called existentiell".) Thus it can only be known either by living through it, and more often than not, by experiencing social injustice. Social justice refers to the one of these modes that has to do with what we perceive as "our just relations" with our fellow beings within our social context and its presence is inseparable from the emotional force with which this perception is experienced. We should observe that this characterisation speaks only of "The understanding of oneself": it does not speak of an understanding of that through which one is actually living, that one is experiencing. Social justice can certainly be known, and social injustice even more so, but it does not necessarily follow that this very personal experience of either of them can itself be understood. We have already here *to understand that there are things that we cannot understand within the currently predominant paradigms of understanding*. Central to the corresponding paradigm shift that is required is a sea of change in the way of *experiencing* the matters at hand more directly and more intimately. Bortoft 2010, pp. 23, 24, introduced this requirements already in his ground-breaking book as follows (with our italics added):

> The experience of authentic wholeness requires a new style of learning largely ignored in our schools and universities today. Typically, modern education is grounded in the intellectual faculty, whose analytical capacity alone is developed, mostly through verbal reasoning. One notes, for example, that science students are not interested in observing phenomena of nature; if asked to, they become easily bored Their observations often bear little resemblance to the phenomenon itself. These students are much happier with textbook descriptions and explanations, a fact readily understandable once one recognises that most educational experience unfolds in this mode of consciousness – the verbal rational mode.
>
> The experience of authentic wholeness is impossible in this mode of consciousness, and a complimentary style of understanding could be usefully developed. This can be done first by learning to work with mental images in a

> way emulating Goethe – i.e. *forming images from sensory experiences*. In turn, this process requires careful observation of the phenomenon. Authentic wholeness means that the whole is in the part; hence careful attention must be given to the parts instead of to general principals. In contrast, an intellectual approach to scientific education begins by seeing the phenomenon as an instance of general principals.
>
> Working with mental images activates a different mode of consciousness which is holistic and intuitive. One area where this style of learning is now used practically is in transpersonal education. Experiments with guided imagination indicate that a frequent result is the extensions of feelings, whereby the student experiences a deeper, more direct contact with the phenomenon imagined, In this way, a more comprehensive and complete encounter with the phenomenon results and aspects of the phenomenon otherwise unnoticed often come to light. In addition, students feel themselves to be more in harmony with the phenomenon, as if they themselves were participating in it. This leads to an attitude towards nature more grounded in concern, respect and responsibility.

This corresponds of course to the alpha and omega of active stakeholder participation, where the imaginations of the individual stakeholders are first mobilised and then activated by the visual images that appear with the utmost sense of reality on their screens, where-after the coalescence of the group develops as it enters into the phase of the creative imagination. This is the sense of the, at first individual and subsequently collective identification of more preferable alternatives to what has been earlier proposed as "the best solution". It is along this path, which as we shall see is recursive, that we approach the aim of social justice.

Correspondingly however, simply because social justice is an existentiell, any definition of it must be immediately circular, and in this case we speak, following Dilthey (e.g., 1976), of hermeneutic circles of words and their meanings that can only feed upon themselves, such as were symbolised by the Greeks by a snake and by the alchemists by a crowned dragon, each eating its own tail (e.g., Jung 1944//1953, p. 103). Of course, every dictionary definition ultimately has this property – a dictionary is always "a book that only speaks about itself" – whether defining a lexeme of the word for the most part by what its sememe is or by what it does (e.g., Klinkenberg 1996; Abbott 2002). In much this same vein, we shall not enter here into the relation between social justice and "the rule of law", only observing that, in many places, and especially in the so-called "third world", "the rule of law" is as often employed to subvert social justice as it is used to promote it (Roy 1999).

Although the emotional moment of social justice must be essentially personal, it can become social though its expression within a social group, such as an association of active stakeholders. In the already introduced and now-classical unified theory of objects and values, as originally established by Meinong between 1880 and 1920 (see, e.g., Meinong [1913], but taken much further since that time) expressions are properly so called because they are capable of expressing something about the thoughts of the expressing (and consequently expressive) agent. Each expresses something that is not itself a belief or a judgement – which are essentially private, subjective and mental – but a state of affairs, which is essentially public, objective and tangible. At the same time, of course, the expression has a meaning which is mental and in that way private. In the words of Simons (1996, p. 173, with italics added): "In the matter of meaning, it is the relation of presentation that wears the trousers, because it gets us from the private and the mental to the public and objective."

Clearly the intimation of justice is then both personal and social. At the same time, although the evidence presented by the hydroinformatician – the presentation – may be, and in most cases should be, scientific, the act of judgement that it engenders is itself pre-scientific. All processes of stakeholder participation have this property, which is to say that they cannot be conducted entirely scientifically in the currently accepted sense, and certainly not exclusively in the spirit of modern science, but they must be conducted at least in part pre-scientifically, taking due account of their pre-predicative and even pre-linguistic foundations.

Thus, although we are currently still concerned for the most part with a hydroinformatics of the quantities, the focus now, and increasingly, has to be shifted towards *a hydroinformatics of the qualities*. Indeed it is a common experience that many of those who engage themselves and their fortunes in such endeavours are possessed of an engagement that often rises up to a passion: they often *care passionately* about the causes that they represent. It is this passion that largely drives the environmental movements and so often engages its proponents in body and soul, and it is to this that those involved in flood risk mitigation (and very much more so in urban hydroinformatics more generally), must respond, supporting it and protecting it whenever and wherever necessary. It is then important to emphasise here that hydroinformatics is no longer only about numbers, even as it continues to be based to a large extent upon the use of numbers, but it is also about emotions, including some of the deepest emotions of mankind and, if only through surrogation, of so many other creatures in the world of nature besides. This naturally implies that we must also proceed here much more cautiously and with more circumspection than has previously been considered necessary.

6.2 A STATE OF SOCIAL JUSTICE AS A TRANSCENDENTAL STATE

As introduced earlier, the theory that underlies this part of the book has many and various precedents, primarily in the fields of philosophy and theology, but it is associated in the first place with one person whose writings themselves transcended the bounds of both philosophy and theology. Just to take the views of two of the most eminent philosophers of the twentieth century, he was described by Wittgenstein as "by far the most profound thinker of the last [nineteenth] century" (see Drury, in Rhees 1981, p. 102) and by Heidegger as "the only one of his kind". This person was Søren Kierkegaard. From the side of theology, we may quote from a work that was published under the imprimatur of H. H. John Paul II in 1998 entitled *Fides et Ratio*, or *Faith and Reason* in English, where we read (on p. 111): "Saint Paul, the Fathers of the Church and, closer to our own time, philosophers such as Pascal and Kierkegaard reproached such presumption [on the part of certain philosophers]" where the presumption to which mention was made there and which was to be reproached is not only still with us, but is continuing to gain ground, whereby the combating of this presumption provides one of the principle motivations for the present book:

> Another threat to be reckoned with is *scientism*. This is the philosophical notion which refuses to admit the validity of forms of knowledge other than those of the positive sciences; and it relegates religious, theological, ethical and æsthetic knowledge to the realm of mere fantasy. In the past, the same idea emerged in positivism and neo-positivism, which considered metaphysical statements to be meaningless. Critical epistemology has discredited such a claim, but now we see it revived in the new guise of scientism, which dismisses values as mere products of the emotions and rejects the notion of being in order to clear the way for pure and simple facticity. Science would thus be poised to dominate all aspects of human life through technological progress. The undeniable triumphs of scientific research and contemporary technology have helped to propagate a scientistic outlook, which now seems boundless, given its inroads into different cultures and the radical changes it has brought.

Our starting point in the applications that are considered here is a work of Kierkegaard called *Repetition* (as translated from the nineteenth-century Danish *Gjentagelse*, translated literally as "taking-again"). Like nearly all of Kierkegaard's published works, it is readily available in most major languages, such as any visit to a major international bookshop, such as the chain of Takashimaya bookshops in Asia, will demonstrate. It is important here, however, to draw attention to the subtitle in this work, namely *Et Forsøg i den Experimenterende Psychologi*, which Hong & Hong translated into English, after *Repetition*, as *A Venture in Experimenting Psychology*. However, in his incisive comparison of this work and the much more recent work of Jacques Lacan, Yves Depelsenaire (2004) translated Kierkegaard's Danish subtitle as *Un essai d'expérience psychologique* and this seems more appropriate within our present context, since the Danish infinitive *at forsøger* most commonly translates as "to try", and thus *à essayer*: we have to do here with a trying, an attempt.

We need examples, and even paradigm cases, and we can then best follow Wittgenstein (Rhees 1982) initially in considering the structure of games, whether physical or mental. All games exhibit this property of repetition, whether they are ball games or card games or word games or whatever other kind of games. Each repetition can lead to success or failure in the short term, but in the longer term all games other than those governed by "pure chance" have the property that each failure, to the extent that the player overcomes the despondency caused by such failure, can function as an incentive to learn, and it is the repetition of this process that leads to many, and indeed in nature to most kinds of learning processes. This is very obvious in the case of many wild animals where the young offspring play games through which they develop their hunting and other survival skills. Here we observe the repetition of failure and the overcoming of failure that leads to a higher level of being on the part of the creatures concerned. In such a situation the very self of the creature concerned comes to *transcend* its earlier self, and in so doing it changes qualitatively. We commonly call the process concerned "maturing" and its transcendent state "adulthood" although it obviously extends further to the development of "alpha" or "lead" individuals. Among pack and herd animals this is a process of both individualisation and socialisation. Since it is so widely experienced in natural as well as within human societies we must again accept that it is pre-ontological, and thus pre-predicative and pre-scientific. Within human societies, however, it acquires a further feature, in that this process of transcendence can attain to a social dimension where it acquires other kinds of attributes again. To take a particularly popular example, a football match may demonstrate skills honed over thousands upon thousands of repetitions and failures and recoveries from failures that lead to such a level of transcendence that the match can create states of euphoria among the spectators as well as among the players themselves. This process is associated with a transcendence of the individual players that is interwoven with, and is thus inseparable from, a transcendence of the group as a whole, as a team, and that which is said to emerge here is called a team spirit. In this case, as in several others, the presence of up to a hundred million television spectators for some events demonstrates the power of this process and the fact that this sense of team spirit can be transmitted way beyond the team itself. The Olympic Games, similarly, can be perceived as a celebration of transcendence within the mind–body relation, and in this case the television audience in some of the events is estimated at some billion of persons world-wide. Certain events can then appear to many of the spectators as nothing less than "miraculous" and it is important to bear this accolade in mind, for it corresponds again to a qualitative transformation, as one that in such a collective case corresponds to what will later be reintroduced more precisely as a mutation. To anticipate our later development of this phenomenon, we adopt from the open-source movement in software production the term 'halo' to describe this kind of social extension. We are at this point clearly already in the domain of deep feelings and strong emotions.

In order to comprehend the possibilities inherent in this mutation it is necessary to recall the Kierkegaardian definition of the authentic *Self* that is employed here and how this is compromised by the human, all too human, attachments of the *ego*.

Both of these are present in every one of the persons engaged in active stakeholder participation and it is necessary to identify the fundamental difference between them, and how, for the great majority of humanity, the authentic Self and the ego must necessarily come into conflict with one another in many situations. We need first to define the Self of an individual and then to see how this is totally different from the ego, whereby we can come to see how the pure self is one that is totally free of all attachments, while the ego is that which is subject to strong attractions to, and even identifications with, its attachments. It is within this context that

Eknath Easwaran, in his popular three-volume commentary on the *Bhagavad Gita*, even went so far as to speak of *the tyranny of the ego* within certain contexts, an expression that goes back at least to the twelfth century in the Christian calendar, harking back to one of the earliest of Ibn ‘Arabi’s works. Easwaran introduced his work as “a practical manual for daily living in any age, in any religious tradition, and my commentary as an attempt to apply its teachings specifically to the problems facing us in modern life”. The second paragraph of the first volume expressed the present-day situation as he experienced this in the USA, where he long taught at Berkeley, as follows:

> Today there is an urgent need for such a manual. We see this reflected in our newspapers, magazines, books, movies and television programs, but most of all in our everyday living. Life has never seemed more futile. In spite of all our technological advances and material prosperity, we have no peace of mind and live in fear and anger in the midst of increasing violence. We are caught in the lurid dream that the pursuit of pleasure will lead to joy, the pursuit of profit will lead to security, and most of us have no other purpose in life than this driving urge to bring about our own private fulfillment even if it is at the expense of other persons, races, or countries.

This is the *reality* that is confronting the hydroinformatician in pursuing the path of active stakeholder participation in the cause of social justice in the broadest sense, as we have defined this *reality* in terms of our inner and outer worlds in the previous section and as we shall repeat this again elsewhere in this book. This situation is particularly prevalent in urban areas, and then most strongly in the USA and the UK in the cases of the Atlanticist nations.

We shall introduce these concepts of Self and ego by first defining ‘the Self’ in its most perfect and so for the great majority of us most idealised form, so as to define this *S*elf with an upper-case ‘S’. In Abbott (2002) definitions are themselves defined, already following Locke (1690/e.g., 1976, p. 208) in terms of ‘functions of the mind’, but we have now increasingly to enquire into whose mind this is. In a world in which nature is being subjected to so much destruction and desolation, and where an ever greater number of humans are being reduced to destitution, this question comes to be posed ever more urgently. As hydroinformatics as a whole, regardless of its different approaches, attempts to reduce and even reverse this tendency to increasing desolation and destitution through its application of advanced information and communication technologies within the worlds of the waters, so it must inevitably pose this question. Hydroinformaticians are themselves proxies for their fellow humans and for the creations of nature, and in this position they must relate their own selves and their own selfhoods to the selves and selfhoods of others through that substance, water, that is common to all life on Earth. In our present case this also proceeds through its ecological associations, in which hydroinformaticians participate in alleviating and restoring the lives of their fellow creatures, and then necessarily and specifically in the sociotechnical direction, in which hydroinformaticians strive to provide the means for their fellow humans (directly) and other creatures (indirectly, through their proxies) to participate in the decisions that so influence their lives, whereby questions concerning the ‘self’ and ‘selfhood’ are constantly and increasingly posed. We may conceive that these ‘other selves’ create objects in their minds and associate these objects through their own content, and these processes cannot be indifferent to us. How are we then to *define* ‘self’ and ‘selfhood’ in this general context? There are obviously very many approaches to this question, and correspondingly a considerable literature has been devoted to it, as introduced again in Abbott (2002). The works closest to our present theme are more associated with the humanities and are often traced back to the work of William James (1890/1902) on the psychology of the self. In particular, James’ characterisation of the ‘self-as-knower’ with both a ‘sense of personal identity’ and a ‘sense of sameness’ remains apposite to our present theme despite its logical-empiricist origins (see also the collection of essays in James 1957). In a very different vein, however, it relates back to the starting point for one of Kierkegaard’s three great explorations of an already rapidly advancing nihilism as this was experienced at the level of the authentic self, as indicated by using an upper case ‘S’, as Self. Kierkegaard’s (1849/1920–31//1989) formula becomes, in one of the better translations into English:

The Self is a relation that relates itself to itself,
or is the relation relating itself to itself in the relation;
the self is not the relation but is the relation relating itself to itself.

Thus, in the notation of Category Theory, the definition of the authentic self, *S*, as denoted by *def*(*S*), is:

$$def(S) = S{:}S \longrightarrow S \quad (4)$$

so that the source, $\Box S$, = *S* and the target, $S\Box$, = *S*.

This is then to say that the function, its source and its target are, all three, one and the same. We are then back with The Trinity; of The Father, The Son and The Holy Spirit, while recalling that man, and subsequently woman, were made in the image of God, The Father, and lived correspondingly in Paradise with no need of attachments. We then have the foundation myth (where we use the word ‘myth’ in its exact meaning, as defined earlier in this book) whereby first Eve and then Adam acquired a taste for attachments by eating of the Tree of Knowledge as conspired by the serpent – as Denmark’s national poet Oehlenschlager observed in his definitive *Aladdin*: “there is a snake in every paradise”! The Father immediately identified their sinfulness through their covering their bodies as he approached them and correspondingly expelled them from Paradise. With their eating of the tree of knowledge they acquired an ego and it is this that has nowadays, at least in the West, all but eliminated the experience of the Self in most philosophical studies, the main interest then being reduced to solipsistic wrangling.

Formulae similar to that of Kierkegaard have taken a prominent place in the works of several later authors (e.g., Sartre 1943//1958). Among these authors, however, it is probably Sartre who has most popularised the concept of (or, more strictly, ‘the thought of’) the *other* as something that is defined by the more mundane, with a lower case ‘s’, *self* in the image of this self, and which in this process simultaneously defines another, and higher, state of the self. This self then relates the other to itself in the same manner as its own self, and thereby it is said to attain to a higher state of *selfhood.* To the extent that a self creates objects within the same value system as itself, so it transcends its view of its world in terms of objects, as items of *primary reflection*, and instead attains, if only by degrees, to states of *secondary reflection*, often popularly called ‘self-realisation’. We then call the resulting self a *subject*, being one who views the others as subjects too. This then applies not only to fellow humans, but to all other creatures. Thus, when Gabriel Marcel said (see Troisfontaines 1968, part 2, p. 8) that ‘to be a subject is neither a fact nor a point of departure, but a conquest and a purpose of life’, he was regarding the subject in just his way. Thus, when the other is experienced in this way, it is said also to have attained to *selfhood* in the mind of the self, and this is then, entirely symmetrically, the necessary condition for the individual experiencing the other, him*self* or her-*self* also to attain to *selfhood.* It is only within this relation that any kind of authentic interaction can take place at all between the self and the other.[1]

This is a long discursion but it contains the essence of what we need before introducing the resources that the flood risk mitigation practice (and so urban hydroinformatics) must deploy and the new means that are becoming available to employ these resources in order to realise states of social justice. Within this context,

[1]These notions are already fundamental to the entire ‘ecological’ and ‘green’ movements, as these function at both their individual and social levels, and so they are fundamental to hydroinformatics. They also motivate aid programmes of most types and descriptions. For more current and mainly psychologistic views on self and selfhood, reference may be made to the extensive bibliography of Hermans (1996).

it is essential to accept that the general concept of "self" and "selfhood" proceeds way beyond humans, clearly extending over the world of nature and even extending further in many cases into the world of artefacts, covering, in effect, *all the things about which we care*. Our attitude towards all such entities must then always be one of respect, and indeed in the "Religions of the Book" our only purpose here on earth is to serve as the stewards, or even only as the servants of the creation, so that we are obliged to treat all of its creatures and other created objects with respect. We are within this perspective only the proxies of the creation. In this same vein, we are ourselves only the temporary custodians of our genes. As Jung constantly emphasised, the self must be clearly differentiated from the *ego*, and indeed "owing to the fundamentally indefinable nature of human personality, the self must remain a borderline concept, expressing a reality to which no limits can be set" (Jung 1944//1953, p. 355). Correspondingly, anyone who wants to know "what consciousness is" has only to look into the eyes of a living creature to read the answer. It is that which takes on the mental forms of images, sounds, sensations, thoughts, and feelings. Consciousness is not just the awareness of having a body, emotions or thoughts, but of something beyond. It is our inner Being.

Essential to the understanding of the structures and dynamics of stakeholder participation (which is regarded here as a prerequisite for achieving socially just flood mitigation practices) is that one is concerned with *a community of caring*, and such a community is itself held together by a mutual concern on the one hand and, on the other hand, by the very special kind of relationship with one another that this communality of concern induces. This kind of relation-with-one-another, this Being-with-one-another, creates *states of solicitude*, as guided by *considerateness* and *forbearance*, as the "glue" that holds the body of stakeholders together and stabilises their relation to such an extent that, as exemplified shortly, they can do what would otherwise be impossible, so that when it is done it again appears as a 'miracle'.

In situations of this kind we can speak at the most basic level of a *transcendental æsthetic*. In order to see the generality of this term we may recall its use in the first of the Kantian Critiques, where moreover we can see more clearly how it can take on a further vital attribute again. Kant did not, of course, approach his subject in the present way, proceeding instead through the notion of experience, as a mode of cognition that must certainly make use of the understanding, but then immediately observing that the mode of understanding that is so employed is itself given to us a priori: it is again in itself pre-predicative, and in the more general biological case pre-ontological. This is most memorably expressed in Kant's most famous aphorism: "Sie begriffen, dass die Vernunft nur das einsieht, was sie selbst nach ihrem Entwurfe hervorbringt"//"Reason perceives only that which it produces after its own design" in the (more mellifluous) Meiklejohn translation (Kant 1787, 1924, 1979//1855/1934/see also 1929/1987).

The now-classic examples are those of space and time, both of which we know very well indeed, testing their nature from birth onwards and repeating our experiences of them through all our waking hours, but the experience of which is so much a consequence of our own nature that we cannot understand them, and indeed we see no need to understand them, such an "understanding" being experienced as superfluous (see again Abbott (2002), and, especially for the case of time (Fraser *et al.* (1978)). In this transcendental state, then, we have to do with things which, although we may know them very well, we once again can scarcely ever understand them. Put another way and even more simply for almost every such case: although we know that it is, we cannot understand why it is. It just 'is'.

6.3 A STATE OF SOCIAL JUSTICE AS A TELEOLOGICAL SUSPENSION OF THE ETHICAL

Although the transcendental æsthetic is important within the presentational aspects of hydroinformatics (e.g., 2D and 3D presentations of model results), much more important to us again is the transcendental ethic, and this priority is established by the entire Abrahamic tradition that is at the foundation of the Judaic, Christian and Islamic faiths. In this case we have no choice concerning our paradigm case! Correspondingly again, Kierkegaard had no choice either (Kierkegaard 1843/1976//1983/1985).

In his case the repetition, the repeated failure and overcoming of failure that prepared this event was that of Abraham's repeated efforts to produce a child on his lawful wife. He proceeded with his faith in this endeavour, in which repetition he persisted into an advanced age before she finally gave birth to Isaac, thus providing him with a foundation of faith from which to take the ultimate "leap of faith". We may then recall that Abraham was subjected by his God to the hardest test of faith that could be conceived in his case, namely that he should proceed with his so long-awaited and so dearly beloved son, Isaac, to an appointed place and should then kill him by way of a sacrifice. Abraham obeyed this command unquestioningly, but still kept his faith that his God would, in fact, not allow this sacrifice to happen. If Abraham had proceeded after his God's will without such a hope of redemption he would simply have resigned himself, reflecting a lack of faith, while if he had sacrificed his son only to reconcile an angry god in the Greek manner, he would have been no more than a Greek tragic hero, as exemplified by Agamemnon, who sacrificed his beloved daughter Iphigenia to this purpose. But Abraham did not resign himself and neither did he act to reconcile an angry god, but he acted in his faith in a merciful God. If he had acted for either of the first two motives it would be possible to understand him in an ethical manner, but in the face of the reality of his action it is no longer possible to understand him ethically.

Of this entire paradigm case, Kierkegaard spoke of *en teleologisk Suspension af det Ethiske* // a teleological suspension of the ethical. He explained his own feelings as follows (Kierkegaard 1843/1976, pp. 54, 59 //1980, pp. 33, 37, with italics added):

> Thinking about Abraham … I am shattered. I am constantly aware of the prodigious paradox that is the content of Abraham's life. I am constantly repelled, and despite all my passion, my thought cannot penetrate it, cannot get ahead by a hairsbreadth …
>
> Abraham I cannot understand; in a certain sense I can learn nothing from him except to be amazed. If someone deludes himself into thinking he may be moved to have faith by pondering the outcome of that story, he cheats himself …

We should observe that *this same property can be identified as injustice* within the context of mainstream theology, or at least in that branch that is commonly called "negative theology" where objects are defined through the absence of certain attributes. Within this context, injustice is clearly a product of that which is called in English translations "nothingness" (Kierkegaard's *Intethed*; Sartre's *le néant*, Barth's and Heidegger's *das Nichtige*). Thus, with Barth (1938–1950, III, 3, p. 354, again with italics added):

> Nothingness is absolutely without norm or standard. The explicable conforms to law, nothingness to none. It is simply aberration, transgression, evil. For this *reason it is inexplicable*, and can be affirmed only as that which is inherently inimical. For this reason it can be apprehended in its aspect of sin only as guilt, and in its aspect of evil and death only as retribution and misery, but never as a natural process or condition, never as a subject of systematic formulation … Being hostile before and against God, and also before and against His creature, *it is outside the sphere of systematisation*. Its defeat can be envisaged only as the purpose and the end of the history of God's dealings with His creature, and in no other way.

This notion of that which is transcendental, whether within the positive sense of social justice or the negative sense of social injustice, may perhaps be most familiar to hydroinformaticians from the inspiration that it gave to Cantor in his momentous investigation into the nature of transfinite numbers.

Jourdain introduced his translation of Cantor's memoirs (1895, 1897//1915) by mentioning an earlier one of Cantor's contributions, published as *Zur Lehre vom Transfiniten* in 1890 (see Cantor 1915, pp. 73, 74):

> A greater part of this little book is taken up with detailed discussions about philosophers' denials of the possibility of infinite numbers, extracts to and from philosophers and theologians, and so on.

From the theologians, however, Cantor must have come across the *credo quia absurdum* the "I believe it because it is absurd" of such as Irenaeus, Tertullian and Augustine and this inevitably led him to his presentation of infinite sets of numbers and the vital difference between those that were enumerable and those that were not, of which the most familiar of this latter kind were those which were "of the order of the continuum". Thus (Cantor 1915, p. 74):

> All so-called proofs of the impossibility of actually infinite numbers are, as may be shown in every particular case and also on general grounds, false in that they begin by attributing to the numbers in question all the properties of finite numbers, whereas the infinite numbers, if they are to be thinkable in any form, must constitute quite a new kind of number as opposed to the finite number, and the nature of this new kind of number is dependent on the nature of things and is an object of investigation, but not of arbitrariness or prejudice.

Thus, one finite set of elements is said to "have the same number" of elements as another finite set if a one-to-one correspondence can be established between each of the individual elements of the one set and each of the individual elements of the other set and in this case we say that "the sets are equivalent". In such a case, no set can be equivalent to a proper subset of itself. In the case of infinite sets, however, it is easy to construct, for example, a one-to-one correspondence between the natural numbers and the rational numbers through a process of diagonal counting, even though all our intuition tells us that the natural numbers are a proper subset of the rational numbers and indeed that there "should be infinitely more rational numbers than there are natural numbers". Such observations as these provide the foundations of the Functional Analysis that provides, in its turn, the basis for the computational hydraulics and the computational fluid dynamics that are situated at the very foundations of hydroinformatics (e.g., Kolmogorov & Fomin 1953, 1961; Abbott & Minns 1998, pp. 349–396, especially p. 352). For his part Kierkegaard used the expression "*i Kraft af det Absurde*", which, although it is usually translated as "*by virtue of the absurd*" can be rendered more literally as "by the power of its absurdity": we are back again with the dogmatic, and indeed more generally religious, paradox! Thus the entire theory of transfinite numbers rests upon this insight, while the power of this transcendence can be judged from Manin's observation (1977, p. 106) that: "In the hundred years since the introduction of transfinite induction, not a single new method of constructing sets has come into common use". Thus, to take one notable example that should be familiar to hydroinformaticians, it is easy to prove from the continuum equations of almost-two-dimensional fluid motion with bed resistance terms, but no diffusion terms, that no eddies can be produced in a two-dimensional flow, whereas it is observed that eddies are produced in the numerical computations that satisfy these same conditions and that these eddies are very similar to those observed and measured in nature and in physical models. In this case, the application of some innovations in Fourier transform theory to the analysis of the nonlinear discretised governing equations concerned led to an explanation of this phenomenon, but this explanation transcended the bounds of classical consistency theory, necessarily passing way beyond the conventional Banach-space assumptions of that theory (Abbott *et al.* 1985: see also Abbott & Basco 1989, pp. 357–367; Abbott & Minns 1998, pp. 438–457). In this case it could be shown that, so long as there was a sufficient level of resolution of the higher wave number components and the scheme used was non-diffusive, what was demonstrably impossible in the one language, such as the second-order language of real numbers, L_2Real, became perfectly possible in a first-order language, such as the first-order language of arithmetic, L_1Ar (see Manin (1977) for this distinction). A somewhat similar situation arises in the theory of weak solutions of conservation laws (see Abbott & Minns 1998, pp. 370–378) for which Peter Lax, its originator, received the million dollar 2005 Abel Prize of the Norwegian Academy of Science and Letters *on the fiftieth anniversary of its original publication*! (Fortunately for Lax, he was still alive to collect the prize!)

It is important to what follows to observe also that in this case it is the finite model that provides the transcendental state, which might at first sight appear to be in conflict with the notion of infinite numbers as transcendentals, but once again, and indeed as always seems to be the case, Kierkegaard was ahead of us. He observed that a complete transcendence consisted of two "movements", first an infinite (*uendelig*) movement that took the self into a relation with what he called "The Totally Other", as a first transcendence, and then a finite movement that brought this transcended self back into the nominally "everyday world" of the finite (*endelig*) movement, to function there through a second transcendence, and so in a "twice-transcended" form (Kierkegaard 1844/1851/1960//1980). It is precisely this double movement that has to be performed here, where we observe that what was impossible in one, once-transcended world, became possible in another, twice-transcended world. Thus, when we say, as we have above, that "each of the individual stakeholders attains not only to a broader, but also to a higher level of collective understanding, and this leads not only to a higher state of *collective consciousness* (a greater state of being aware) but also to a higher state of *collective conscientiousness* (a state of being motivated by caring) that can decide issues that are in the abstract unresolvable" we are speaking of just this kind of double movement that makes possible that which was otherwise impossible.

Associating the observations made above on the quiddity of stakeholder participation, which appears to proceed at just this transcendental level, it must appear that any authentic active stakeholder participation must be situated at this level if it is to meet its obligations towards a social justice that is situated at this same level. Concerning the force that drives the authentic and active stakeholders, just as it drives the hydroinformaticians involved in this process that brings the whole project together, we can then do no better than to return to an earlier work, probably written by H. H. John Paul II himself, where (1994, p. 104) he echoed a theme that also passed from Kierkegaard to Jaspers, and further again to the great Jewish thinker, Emmanuel Levinas (e.g., 1995/2006), namely that "Man, perhaps subconsciously, waits for such a challenge, indeed, man feels the inner need to transcend himself. Only by transcending himself does man become fully human (cf. Blaise Pascal, *Pensées*, Ed. Brunschvicg, (1670/1953) p. 434, "apprenez que l'homme passe infinement l'homme″ [// "understand that the man infinitely surpasses the man"])".

Necessarily putting to one side here the related insights of Karl Jaspers, as well as of Martin Heidegger, we may conclude this section with Wahl (1959//1969, pp. 74, 75):

> Jaspers' interpretation of the idea of repetition is different from Heidegger's and a little closer to Kierkegaardian repetition: our failures, he believes, are nothing but the affirmation of transcendence. The fact that we fail is an indication that there is some greater thing, a sphere higher than existence. We become aware of this sphere in what Jaspers calls 'boundary situations'.
>
> Thus failure makes it possible for us to mount from existence towards transcendence, and for this reason, in so far as it conducts us to repetition, it is the supreme 'cipher', the supreme symbol of transcendence.
>
> It is thus that the two ideas of failure and repetition are not so opposed as they seem at first, since in both Heidegger and Jaspers it is failure that brings about the possibility of repetition, it is failure that leads us to triumph and to authenticity.

The realisation of social justice is the reification of this process; it has been realised in some hydroinformatics applications (see, for example, the case study at the end of this chapter) and can be further developed in the work concerning flood risk mitigation, as one of the most significant applications of hydroinformatics. We have to give expression to it even though we can never really understand what it is "in itself", so that social justice is that which, even as it is experienced, surpasses our understanding.

An example in what may at first sight appear as being in an entirely other field again is that of the unprecedented youth riots across the UK in August 2011, against which the UK government called for draconic acts of revenge and suppression, even planning for a transformation of a police force traditionally based on care and respect into a US style of policing based primarily on fear. The New Orleans flooding provided a paradigm case of the dangers of this US approach This illustrated the effects of frustrating the will to transcendence of a large part of its younger people, and repeated flooding and the suffering and deprivation that this incurs must have had a similar effect in this paradigm case.

6.4 THE REPRESENTATION OF SOCIAL JUSTICE

It is of the essence of social justice that it should be seen to be done, and even if we cannot understand it, we still want to be able to recognise it when we see it. But how is this process of recognition to be represented, portrayed, instaurated, or "brought to life"? Clearly the language of ethics is much more one of actions than it is one of words, and it is then essential that these actions are displayed alongside the dialogue and the background of events: no amount of words alone could ever do justice to such actions and that which they symbolise. The deeper grounds for this, however, were explained as follows by Kathleen Raine (Keeble 2004, pp. 51, 52):

> Symbols speak to the soul. The psyche doesn't think in words; it is temporal man, the *ego*, that uses a verbal language. But a symbol which is finally grounded in nature reaches back to regions of experience that are far deeper than verbalisation can ever reach. These symbols touch us at a much deeper level; they come from a deeper level and they speak at a deeper level. They strike these resonances. A symbol must be grounded in physical nature because that is where the form of the symbol comes from, but it also resonates in a vertical level, in the level of the soul, and possibly beyond that, at a metaphysical level. Therefore, while words are, as it were, a horizontal level of discourse, symbols are a vertical level, calling into play all the different levels of our humanity: life, spirit, the whole range of our human being. Of course, dead symbols are not doing so; they are then like any other kind of verbalisation. But when a symbol is living, it speaks to the whole being. And again, to return to Jung, whom I must defend, the soul receives its oracles in symbolic form, as for example in dreams, very seldom in verbal form, and more often visual than verbal.

We should just mention in passing that in categorical theoretical terms, the sign and the symbol proceed 'in opposite directions'. In simplistic terms (see Abbott 2002) a sign (in the strict sense) is a token that, when given up to the other *is replaced* by that which it evokes (and in this case *signifies*) in the mind of the other. Thus, to use a familiar metaphor, the sign is that object which introduces the entirely other object which it evokes (and in this case signifies) onto the stage of consciousness, and then, having done its job, steps back into the wings of the stage, leaving the stage entirely to the signified object. Its category thus has as the simplest form of sign function:

Sign function: betokening object → betokened object

A symbol is a token that, when given up to the other, *replaces* that which it evokes (and in this case *symbolises*) in the mind of the other. In this case then, to continue with the previous metaphor, the symbol is that which, after introducing that which it evokes (and in this case symbolises) onto the stage of consciousness of the other, sends this, now symbolised object into the wings and takes over the role of this object itself. Thus, the symbol function proceeds in the opposite sense to the sign function. Its category takes the form:

Symbol function: betokened object → *betokening object*

We may associate this insight with the more mundane (as contrasted with the much more exalted, Platonic) formula of Locke (1690/1990, p. 208), that "words, in their primary or immediate signification, stand for nothing but the ideas in the mind of him that uses them", in which case we may continue with Locke to define an idea, albeit in a modern idiom, as "any object that has become a subject of the understanding" and here, we repeat, we are concerned with the outer limits and indeed the transcendence of the understanding. We may indeed recall that the goddess of justice is commonly depicted symbolically as a blindfolded woman holding a weighing scale, and wielding a sword: Figure 6.1.

Roman goddess of Justice Greek goddess of Justice

Figure 6.1 Lady Justice (Latin: *Justitia*, the Roman goddess of Justice, who is equivalent to the Greek goddess, Dike) is an allegorical personification of the moral and physical force inherent in judicial systems (Accessed on 12th October 2011 from: http://en.wikipedia.org/wiki/Lady_Justice)

Corresponding to this situation we need new didactic instruments that will transmit the nature of the qualitative changes that are occurring during the realisation of social justice, and these must accommodate all their manifestations as actions, words, facial expressions and even, passing above and beyond these again, their music. For the purpose of portraying the processes that culminate in social justice it is been found convenient to use videos of a BBC production entitled *Judge John Deed*, the nature and content of which is thoroughly covered in Wikipedia and which has been readily available from Amazon and other retail outlets until the two most relevant examples were well and truly suppressed. In other environments, a US film entitled *12 Angry Men* is frequently used to much the same effect. In the first case a particular interest accrues to one production in which the teleological suspension of the ethical occurs in such a way as to attain to social justice and correspondingly avert what would otherwise have been an extreme case of social injustice. This same production also excels in providing the means to inculcate in the mind of the viewer a particular form of knowledge, commonly called *tacit knowledge*, by following the words, facial expressions, gesticulations and other bodily movements and human interactions generally within a jury of a court of law, and thereby observing how an increasingly augmented state of collective consciousness comes to transmute into a transcendental state of collective conscientiousness within that particular environment – and so again how what starts out as impossible becomes possible. What is in play here is indeed much less knowledge itself, but that which is a necessary adjunct to it, which is imagination. Following Corbin's great work on Ibn 'Arabi (1958//1969/1996, p. 80) we may then say that "the active imagination guides, anticipates, moulds sense perception; that is why it transmutes sensory data into symbols". These are typical of the new ways of using the new kinds of references that must be employed in this kind of enterprise, and they have the added advantage that they are widely available in other languages than the original, either through sound-synchronisation or subtitling, while the invaluable role of background music, often serving as Leitmotiv, and in the John Deed case played by the London Symphony Orchestra and the Royal Philharmonic Orchestra no less, is then readily accommodated. As Corbin observed: "Something in the nature of harmonic perception is needed to perceive a world of many dimensions" – see Cheetham (2005, p. 32). It is through these devices, correspondingly, that the means now become available to express what would otherwise be inexpressible.

If we should ask what kinds of emotions are finally being aroused in such productions we are obliged to return to the sensation of "a miracle", such as may be exemplified in the instance just cited by the portrayal of the amazed relief of the accused when declared not guilty, and this sensation must also be induced within these environments as authentically as possible with the means available. For example, for such a person faced with disaster unless social justice is done, its actual appearance may well be experienced and projected as "a miracle": see the famous (or as one of his editors called it, "notorious") tenth section of Hume's Enquiry Concerning Human Understanding (Hume 1748, 1999). The last sentence of this section especially bears recording here:

> Mere reason is insufficient to convince us of its veracity. And whoever is moved by Faith to assent to it, is conscious of a continued miracle in his own person, which subverts all the principles of his understanding, and gives him a determination to believe what is most contrary to custom and experience.

Such a behaviour as this thus corresponds to an act of faith, wherewith we must then always add, thinking reflexively with Kierkegaard, "and faith is a miracle, otherwise it is not faith".

But still this should not excite too much euphoria, for as Pascal observed (Pascal/Lafuma 1670//1962, 2003, p. 161):

Les miracles ne servent pas à convertir mais à condamner
(miracles do not serve to convert, but to condemn)

which is to say that we must then be lacking in faith if we have to be looking for miracles (e.g., John 11:4). Correspondingly, the miracle is viewed in theology not so much as a product of the *faith* as of the *doctrine*.

We can otherwise speak of this as a revelation, that is, as an experience of a higher truth, or, to cross the line into theology, as a moment of epiphany. The dangers involved in this process should not need any emphasis here and the role of the hydroinformatician in avoiding at least some of these dangers is correspondingly taken up towards the end of this part of the book.

Chapter 7

Realising social justice in the context of flood risk mitigation

7.1 A PARADIGM SHIFT

As discussed in Part I of this book, substantial steps have been made in the flood risk mitigation practice over the past thirty years. However, in view of the limitations of modern science which have now become so evident, if we want to achieve a proper integration with the society and to truly incorporate those *primary values* (to repeat ourselves again, these are human values and those of nature) into the decision making processes we should then consider making a *paradigm shift*. A shift towards a *holistic paradigm* in which perceptive mode of consciousness is an important component of our worldview (see also Section 2.1, Chapter 2, Part I). A new paradigm would need to address the inner (or subjective) experience of people through *qualitative phenomenological enquiry* within a wider context whatever the nature or degree of that experience. This refers to how we take into account different values, perceptions, sensations, memories, thoughts and feelings, which we can refer to as qualities of perceptive consciousness. The need for such shift comes along with the major changes taking place in the world in which most of us live and which are reflected by the move from modern to post-modern society, from nationalism to globalisation, from cultural supremacy of one group over others to the concept of multi-culturalism and the acknowledgement of cultural diversity, from an understanding of one faith and its dominance in society to acceptance of multi-faiths. Since the transition from one paradigm to the next has never been so smooth we should spare a glance at the origin of the term paradigm shift and what is understood as its meaning.

The term *paradigm shift* itself was coined by Thomas Kuhn (1922–1996) who wrote about it in his book *The Structure of Scientific Revolutions* (Kuhn, 1970). The word paradigm is used to refer to the beliefs and assumptions underlying a particular way of thinking. The main paradigm of modern science is the belief in the primacy of the material universe. It states that if we fully understand the world of space-time-matter-energy we shall be able to account for everything in the cosmos. The paradigm of modern science is very much concerned with the *outer world* and as such it does not have much to say about the *inner world* (i.e., the world of qualities) which concerns human values and the inner world of subjectivity. Since these qualitative aspects cannot be weighed, measured, or otherwise pinned down they are either ignored by modern science or in some cases regarded as anomalies.

One theme in Kuhn's book is that scientific communities do not embrace radical new ideas without resistance. He showed that when anomalies first appear they are usually overlooked or rejected. This is true even when the new ideas are clearly more consistent with the data than the status quo. New ideas that challenge the status quo are scrutinized much more carefully and sometimes openly ridiculed. They are in some cases even likely to be ridiculed by some very important scientists. For example, in 1672, when Laurent Cassegrain proposed the design used in most modern research telescopes, he was publicly ridiculed by one the greatest scientist of his time, Isaac Newton.

Arthur Schopenhauer (1788–1860) characteristically observed:

> *All truth passes through three stages. First, it is ridiculed. Second, it is violently opposed. Third, it is accepted as being self-evident.*

The same reality was also described in a more radical way by Max Planck (1858–1947), the famous quantum physicist:

> *A new scientific truth does not triumph by convincing its opponents and making them see the light, but rather because its opponents eventually die, and a new generation grows up that is familiar with it.*

Returning to the subject of our book, as developments over the years have shown that the human dimension cannot be overlooked, many researchers and practitioners concerned with flood-related problems have made various attempts to account for it. However, only a limited success has been achieved and further efforts in finding the right approach to accessing and understanding the human dimension are needed. Whatever is put forward, we should be developing a new paradigm in which a human dimension is a fundamental component of reality rather than a component that can be placed within the purely materialistic world. That paradigm must be based upon active stakeholder participation and the creation of environments in which social justice principles such as fairness, maximising welfare, respecting individual rights, equality and virtue are continuously sought within new contexts. As mentioned earlier, the essential ingredients for this new paradigm are already in place as they already exist in philosophy and theology and they must be now carefully placed together in order to explore the new picture of reality that emerges.

Figure 7.1 Some of the key requirements for a socially-just flood risk mitigation practice

In what follows we discuss various aspects of a new flood risk mitigation paradigm, a paradigm which is based on social justice principles and values and in which stakeholders are truly in charge of the entire process, Figure 7.1. Therefore, we introduce stakeholder participation as a key means for realising such practice. We also discuss institutional, organisational and legal arrangements that are necessary to enable this process. We then review the key ethical principles and theories which have a direct relation to the subject of the present book and which have a strong foundation in both Western and non-Western worlds of thought. Certainly, as these principles and theories have their subjective elements they can be debated at length and even misused if they are guided by the negative side of human character, such as by greed, idolatry of money, the desire for power, self-centredness and egoistic ways of thinking generally. Therefore, one of the answers to this problem is to be sought by turning to a true Self as it is to be found in a sound, well-educated, well-informed, and above all *honest and thus authentic leadership*. This leadership will need to strive to minimise (if not eliminate) the negative influences and so provide the capacity to carry out changes at all levels and scales of the society.

7.2 STAKEHOLDER PARTICIPATION AS A MEANS FOR REALISING SOCIAL JUSTICE

Increasingly many projects in the water sector of the human economy have consequences for so many persons that some of them insist on participating directly in the making of the most vital decisions affecting the executions and the outcomes of these projects. Some of these persons then come to press the interests of specific human-social interests while others again press the claims of many creatures situated in the natural economy, so that they act as surrogates for these creatures. Such persons see themselves as active stakeholders in a project in which they feel that they have an interest, so that they press a claim to participate in the processes of decision-making within the project, and their interests may be so widely shared and their integrity and competence so respected that their claim to representation is upheld as legitimate within the society concerned. In this case we speak of an active stakeholder participation in the decision-making process that may or may not lead to social justice. At its most pragmatic level, the purpose of stakeholder participation is to induce a change in the built and managed environment that aligns with a positive change in the social environment. This change is, however, only recognised as positive when it corresponds to a sense of social justice on the part of the stakeholders, preferably as a whole or, in extremis, for the greater part. Correspondingly again, to use a term that is popular in the world of architecture, this gives this built environment the quality of an *attractor*, as a quality that will attract and filter the persons and businesses that set a store upon this environment, thus promoting such an environment even further.

Stakeholder participation is then justified on the basis of a claim that it is a means of realising social justice, whether this be a justice in human societies or a justice in the societies of nature within which human societies are in turn embedded. Since it is water which, with earth and air, is the most basic requirement for the sustainability of both the human societies and the societies of nature, these claims to social justice are among the most important of all claims. Thus, Abbott (2007) spoke of "an understanding that the greatest crime of all is to deprive people of the possibilities to exercise their own deep-rooted knowledge and understanding"; and correspondingly it is often observed that the greatest of all social injustices is to drive a people off its lands and to deny this people its waters, as those environments where their knowledge and understanding most essentially belong (see again, Roy 1999). As Machiavelli long ago explained (1516 +/-/ 1929 // 1970), this is commonly experienced as a social injustice that is something even more cruel and terrible than the murder of the members of one's own family. This has been evident in many places around the globe, and especially in India and China when the rich interests of urban expansion overcome the traditional but poor interests of farmers and their knowledge of traditional agricultural practices.

The present book is directed to an analysis of how hydroinformatics, as the study of how communication and information technologies can best be applied within human and other natural societies, can contribute to the realisation of social justice in the area of flood risk mitigation and, if only by surrogation, in the same area for other, natural, societies. It is thus directed towards identifying the social arrangements concerned and the mutations that these undergo during the participation process, the sociotechnical tools and environments that such arrangements can best use in order to catalyse these mutations, the ways that these tools and environments can best be employed to reinforce the corresponding processes and the new demands placed upon the producers, policies and users of such means. These requirements in turn have consequences for the more general social arrangements, as environments, best suited to the production and developments of the tools and vice versa. In the same vein, the ways in which these innovations are employed have to be brought into conjunction with the social mutations that are occurring within the societies concerned, and especially following the ways that these societies are themselves mutating from modern to post-modern conditions, that is, to repeat ourselves again, from societies composed primarily of knowers to societies composed primarily of consumers of knowledge. Correspondingly, the tools and environments are now for much the greater part web-based and make an ever more intensive use of the mobile internet using the increasingly ubiquitous resources of mobile telephony and working within a Software-as-a-Service (SaaS) paradigm. Beyond all this again, the nature of the person associated with flood risk mitigation work has to change too, and the nature of this change has also to be identified so that its potentialities can be augmented. This section again follows the insight of Latour (1991//1993; see also Abbott 1999) that a post-modern science then reverts in most cases to a pre-modern science, such as alchemy, thereby displacing modern science in such cases.

In the present book, the notion of transcendence has been used to describe a state of positive change in qualitative terms. In the stakeholder participation context, the same term would imply achievements of a higher state of awareness and active participation. Therefore, following the need to minimise, if not entirely eliminate, the level of risk for those exposed while maximising the use of available resources, and serving the most vulnerable first, we could then define *social justice as a state of transcendental praxis*. In this definition, the notion of praxis is used to describe the relationship between efforts to develop and implement effective measures on one side and the needs and concerns on the other. This is to say that instead of starting from a theory, then moving to practice to test that theory, praxis refers to the social needs and concerns as the basis for theorising, and then returning to the practical with a view to social transformation, and indeed to authentic transmutations in society. In doing so, the thinking and doing are very much tuned towards a *qualitative* change (which is also referred-to here as a transcendence). With this in mind, it is obvious that the traditional engineering thinking has to change into one where ideas emerge from social (and nature-related) concerns and which serve humanity (and the planet Earth generally).

Furthermore, the means of transforming technocratic and materialistic cultures into the sociotechnical culture that is based on *self-transcendent values of benevolence* (being honest, ethical, helpful and loyal) and *universalism* (caring about environment, a peaceful world and social justice) can only be brought to presence by engaging in an ongoing process which strives for repeated processes of transcendence. A representation of such a process as this proceeds through time is given in Figure 7.2.

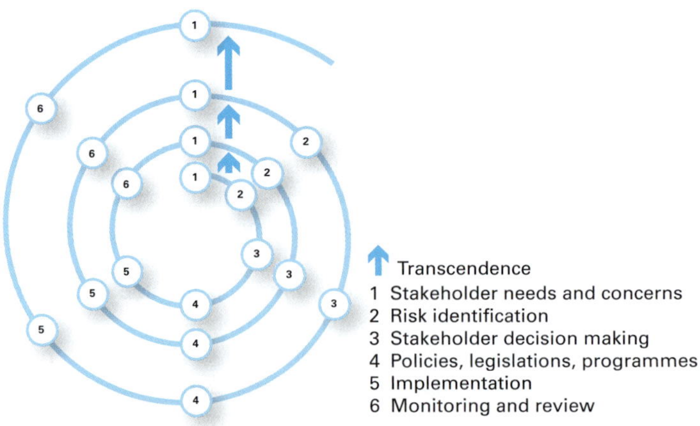

Figure 7.2 Stakeholder participation as a means for realising social justice within the flood risk mitigation practice

Active stakeholder participation in the flood risk mitigation practice is a complex process touching immediately and necessarily upon a variety of social, institutional, economic, cultural, political, environmental and technical issues. The effectiveness of this process depends largely on mobilising and enabling (i.e., empowering and providing the means for active participation) a potentially greater number of stakeholders from the very beginning and throughout the entire cycle. At the beginning of this process it is necessary to identify all parties that are affected or concerned and to develop the necessary means for facilitating their engagement and participation. Some examples of stakeholders are: various government representatives (from national, regional and local levels), water utilities, disaster prevention and mitigation officers, planners, all those at risk of flooding, technical experts, private sector interests, environmental agencies, NGOs, environmental law professionals, and, most important of all, those members of the general public who really care, and indeed in many cases care passionately about the causes at stake. Their role and the mechanism of their involvement (e.g., public hearings, interviews, workshops, consultations, delegation, etc., all supported by a *hydroinformatics environment* that we shall describe later here) need to be transparent and carefully designed so that the entire process is sustainable and just. It is to be expected that different parties will have different needs and requirements, based on their interests, their corresponding concerns and their cares, and as such they should be balanced and carefully introduced into the resulting flood mitigation policies, legislations and programmes. The flood mitigation policies, legislations and programmes should reflect the needs and concerns of those at risk, their perception of risk, considerations of ethical and other issues (e.g., technical, environmental, economic, legal, etc.) as well as the roles and responsibilities of those institutions and organisations which are in charge of their implementation and monitoring, all conditioned by the general public who care deeply and are correspondingly strongly engaged.

Overall, the objectives of stakeholder participation can be summarised as follows:

- Involvement of all parties;
- Empowerment of all parties;
- Ability to express, share and communicate needs, concerns, cares, views, information, knowledges and ideas;
- Achievement of a consensus and support for the outcomes;
- Direction for the development of policies, legislations and programmes as these form the decision on the issues at stake in the outer worlds of those so engaged;
- Sustainability and long-term commitment;

Figure 7.2 illustrates a mechanism whereby a group of stakeholders assigns benefits or burdens (or flood risks) across the society. Therefore, the aim of this process is to minimise, if not entirely eliminate, the level of risk for those exposed while maximising the use of available resources and serving the most vulnerable first. Certainly, the outcome of this process will greatly depend on the available resources that government can afford (human, financial, etc.) as the flood-related budgets will need to be allocated in conjunction with the needs from other sectors (e.g., education, health, etc.). In this respect, and building upon the previous discussions, active engagements in this process must strive towards *social justice* by following continuous cycles in a search and a realisation of justice within new contexts and higher values. As indicated earlier, this engagement has indefinite positive feed-back cycles and it starts from the creation of an environment that is open for a positive change. The overall process, as illustrated in Figure 7.2, has a form of a *spiral* rather than a circle as our states of reflective actions continuously transcend to higher levels following the ever changing standards, social needs, values and technological advances which come to presence.

We can also add that a process similar to the process depicted in Figure 7.2 also occurs within the group of stakeholders itself during their participation and interaction, again taking forms of successive failures and overcomings of failures, as repetitions, that drive the process of transcendence whereby the higher levels of consciousness (i.e., intuitions, feelings and other forms of subjective experience) mutate into these higher levels of conscientiousness (i.e., the states of inner wisdom, concern, respect, responsibility, equanimity and care) that then occur "spontaneously", so to speak, within the active stakeholder community. We observe that the first movement is modern and the second is pre-modern, and indeed it is in its expression alchemical[1]. What is occurring here is a transcendence within the character of the individuals and a transcendence within the character of the group, which is thereby a social transcendence, and it is this process that "leads us to triumph and authenticity", whereby, if conducted properly, where "conducted properly" has still to be defined in our present case, justice prevails and injustice is exposed. *Correspondingly, at the stakeholders' level, social justice can be defined as a transcendental state of social being.*

Within this context we continue to refer, as in common speech, to "the chemistry of the interaction" when the elements, the characters of the persons participating in the interaction, do not themselves change, so that only transformations can occur, whereas when we speak of "the alchemy of the interaction" we refer to situations in which the elements, the characters of the persons themselves, are changing, so that we then have to do with transmutations: see again Jung (1944//1952). We then commonly say that "if the chemistry is right" then people will "bond with one another" so as to "create a good team", but it is not then necessary that the individual persons concerned, the "elements" of the team, need themselves change very much, if at all. In the case of the alchemical transmutation the situation is totally different, in the sense that the "members of the team" come to change their attitudes, and even their "views on life", their "horizons" in the language of Husserl, as a consequence of their working together in the team: they are "not the same persons" after working together as they were before they worked together; they have changed qualitatively, attaining to another kind of Being-in-the-world (*In-der-Welt-sein*). At the risk of passing over into hyperbole, one might even speak with the ancient Greeks of a collective metamorphosis. One might correspondingly go so far as to say that without this higher-level-of-consciousness to higher-level-of-conscientiousness transmutation, social justice would be for the most part unrealisable. The task of the hydroinformatician within this virtual environment is then to mobilise the *activate imaginations* of the individual stakeholders in such a way that, at some points in their own virtual times and spaces, their *creative imaginations* come into action, and then with an increasing synchronicity, so as to transmute into *a collective movement of the creative imagination*, and thereby come to constitute a *collective force* (i.e., including a greater consensus). Clearly, the means to realise the first of these movements, of activating the primarily creative imagination, calls for more customised and individualised material contained in 2D and 3D coloured animations and transmitted over the web, while the second movement of the imagination, as it becomes increasingly collectively creative, is more concerned with supporting interactions between the stakeholders as these *communicate with one another in words and symbols as they seek out their transcendental states of collective action.*

We should now turn to the Denmark-Sweden Connection case study, a part of which is enclosed at the end of this part of the book, to see better how the introduction of certain kinds of hydroinformatics environments in processes of stakeholder participation have the capacity to change the ways in which individual stakeholders think and feel about the world in which they live and function, even as the stakeholder group as a whole then mutates correspondingly so as to provide one collective voice, as a voice that is experienced as one of social justice – just so long as this process is "conducted correctly".

[1]Alchemy is both a philosophy and an ancient practice that seeks to accomplish the transmutation and to attain ultimate wisdom.

7.3 ETHICS AS A CENTRAL PART OF RISK ANALYSIS

In Part I of the present book, we discussed the nature of urban flood risk as approached from different points of view without making any statement about the acceptability of the risk. In philosophy, such an approach is said to be *descriptive*. Descriptive statements tell us something about how the world (or in our case, the risk of flooding) is. However, if we want to address the acceptable level of flood risk, then we need to make a statement of how we *want* the world to be. These are called *normative* or *prescriptive* statements. Contrary to descriptive statements, normative statements are not value-free (i.e., they are not objective). They contain norms and values about how people should behave, about what is considered as right or wrong; or in short: how things *ought* to be.

For many engineers, making normative statements seems to go against their intuition and they would tend to respond by undertaking a form of calculation. In flood risk assessment, one such form is a Cost-Benefit Analysis (CBA) which is often used as a primary means to decide whether the safety level should be improved or not. As long as the cost of implementing safety measures to reduce the level of flood risk is lower than the cost of the damage that is prevented, such measures are regarded as worthwhile, or otherwise they are not considered as such. CBA resides on a principle which suggests that advantages should be weighed against disadvantages and costs weighed against benefits. However, notwithstanding its *prima facie* appeal, CBA is certainly not uncontroversial for at least two reasons (see also Anderson 1988; Sagoff 1988; Hansson 2007). First, CBA tends to place a monetary value on (the loss of) a human life and other "non-marketable goods" such as the environment and the cultural heritage. The "Ford Pinto Case" is often used as an illustration of a bad example of CBA in most Business Ethics texts. Second, CBA involves a great deal of value judgment and it is not such a value-free method as it is often perceived. This gives rise to the following questions: What are legitimate considerations in judging the acceptability of risk? Does cost-benefit (or risk-benefit) analysis provide an adequate method to come up with a decision or should we introduce an additional point of view which cannot be addressed within the spirit of modern science? To address these undeniably normative questions we have to turn to what is called "moral philosophy" and whatever happen to be the current ethical norms in the society concerned.

In the field of philosophy, then, the study of such norms and values is known as *moral philosophy* or *ethics* as these are accepted within the concerned society and problems that involve normative questions are referred to as *ethical or moral issues*. Ethical theories and principles are the foundations of ethical analysis and they emphasise different points of view that can be highly decision-relevant. For an ethical theory to be useful (i.e., socially acceptable) it is important that such a theory is directed towards a set of principles. Examples of such principles are a restricted liberty principle and the principle of reciprocity. Van Velsen (2000) has even argued that these two principles are necessary and sufficient for peaceful coexistence within the society that is at issue (see also Zandvoort 2008).

The field of ethics becomes an important aspect in a socially just risk assessment process as it prevents us from being little more than blind users of statistical methods and numerical models. Instead of uncritically calculating risk in terms of probabilities multiplied by unwanted outcomes and aggregating preferences we should also use ethical and social considerations to answer what are good reasons for or against options for action. Therefore, we argue that ethics should form a central part of the risk analysis process.

7.3.1 The restricted liberty principle

This principle is also called a "no harm principle" stating that everyone is free to do what he/she pleases as long as he/she does not harm others. This principle dates back at least to the philosopher John Stuart Mill and his essay *On Liberty* (Mill 1859). An equivalent formulation of the restricted liberty principle is the right to be safeguarded (Van Velsen 2000, 2003): *"Everyone has the right to be safeguarded from the consequences of another person's actions"*. An implication is that actions are right if and only if: either there are no (possible) consequences for others; or those who will experience the (possible) consequences have consented after having been fully informed.

We can than relate this principle to the flood risk situation: *"Every person has the right to be safeguarded from the negative impacts of floods that are caused by another person's actions"*. This principle also satisfies the principle of equal rights. Certainly, such a principle has its subjective element and the term "negative impacts" (or harm), which is often followed by the requirement of *informed consent* needs careful consideration and input from various kinds of experts.

7.3.2 Reciprocity principle

The ethical principle of reciprocity is a necessary addition to restricted liberty: *"He/she who violates a right of another may face a reciprocal reaction"*. This means that somebody who infringes a right of another will him/herself lose that same right insofar as is necessary (and no more than that) in order to correct the original violation or to compensate for it – and in order to, if necessary, prevent further infringement.

We can also relate this principle to the flood risk situation: *"All flood mitigation activities must be given permission by those who are at risk through their informed consent. Those who engage in flood-related activities without consent from those at risk can be held to full and unconditional liability for the negative impacts that these activities may cause to those at risk who did not give their consent"*.

Obviously, these and some other similar principles should be incorporated into the decision making framework of socially just flood risk mitigation practice. Given the qualitative nature of such principles, they cannot be addressed by the probabilistic cost (or risk) benefit analysis alone as they require much deeper philosophical and social considerations. Hence, we shall now discuss the four main *ethical* theories and we shall show their capacity to shed a different light on different moral questions, and therefore raise some apparently highly relevant points.

7.4 AN OVERVIEW OF ETHICAL THEORIES

In making a normative statement we can always have different points of departure. We may focus on the person who is acting, on the action itself, on the outcome, or on the effect of the act on other persons. This can be illustrated by the following example.

Example 1: Breaking a promise for the sake of helping the neighbour?
Carly promised Peter, her youngest son, to teach him mathematics. Although Peter is good at mathematics, he needs the help of his mother in order to achieve an excellent mark in the exam tomorrow. Ten minutes before the time that they agreed to start with Peter's preparation for exam, the next door neighbour knocks on the door and asks Carly to bring his daughter to the hospital as she swallowed a little piece of something or the other and urgently needs first aid. What should Carly do?

In this example, there are several ways of approaching the problem posed. We can bring the focus on Carly and try to analyse what should be the right thing to do for a virtuous mother? Should Carly, as a good mother, let the interest of her son prevail over the interest of the neighbour? Or should she set an example to her son by showing readiness to help those who are in the need of help? We can also look at the action itself at a more abstract level. In this case, bringing a child to the hospital would involve breaking the promise that Carly has made to her son. Is she allowed to break a promise? Or we could look at the consequences and judge which action would lead to the best consequences. In that case, the outcome "good grade for the math exam" should be compared with the outcome "preventing someone from suffocation." These different points of departure (person, action, or consequence) can be coupled to three different moral theories: virtue ethics, deontological ethics and consequentialist ethics, respectively. As a fourth category of ethical theories, we could also distinguish rights-based theories. Rights-based theories take the right of people who are affected by an action as the point of departure. Rights are to be protected and given highest priority. In this example, the right to be taken to hospital will probably have to receive the highest priority.

In what follows we shall discuss these four ethical theories in detail and show how they can be used to address some of the important question that typically faced in the flood risk mitigation practice. We start our discussion with consequentialism, the theory that is most akin to the engineers' practice of calculation.

7.4.1 Consequentialism

As the name of this theory already suggests, the theory of consequentialism takes the consequences as the primary point of evaluation. In a very rough definition, consequentialism refers to those moral theories which hold that the consequences of one's acting are the basis for any evaluation of the rightness or goodness of those actions. From a consequentialist standpoint, the action that leads to the best consequences is the morally preferred action. What counts as "the best consequences" is evaluated in terms of one value. Accordingly, all consequences are expressed in terms of the selected value. This value should then be maximised. Hence, for consequentialists, determining the morally preferred action is to calculate which action leads to the highest outcome of the selected value, which could be seen as an optimisation problem.

Consequentialist theories can be divided according to the value to be maximised and for whom it is to be maximised. Here, we limit ourselves to the most well-known variant of consequentialism, viz. utilitarianism. The term utilitarianism is derived from the Latin word for useful: *utilis*.

As mentioned earlier, classic proponents of utilitarianism were Jeremy Bentham (1748–1832) and John Stuart Mill (1806–1873), Figure 7.3. Jeremy Bentham can be considered as the founder of utilitarianism. According to Bentham, the value to be optimised should be something that has value in itself and not as something that has value to achieve something else. In philosophical terms, something that has value in itself is called *intrinsic value*. Examples of intrinsic values are beauty, love, health and friendship. The opposite is instrumental value, meaning that something has value in order to achieve something else which has some higher value. Money, for example, is for most people an instrumental value with which to buy things.

Figure 7.3 Jeremy Bentham (1748–1832) – left; John Stuart Mill (1806–1873) – right; (Accessed on 12th October 2011 from: http://en.wikipedia.org/wiki/Jeremy_Bentham)

According to Bentham (1948 [1789]), the intrinsic value to be maximised is human pleasure. In the end, all human beings strive for pleasure and try to avoid pain (seen as the opposite of pleasure). Hence, that which provides pleasure or avoids pain is good, and that which provides pain or reduces pleasure is bad. By focusing on pleasure and pain, experience gets a prominent place in Bentham's theory since he sees it as an elementary fact of experience that people strive by nature for pleasure and avoid pain (Van de Poel and Royakkers 2011). Moreover, people know what provides pleasure and what results in pain, and also how pleasure can be realised. Accordingly, people can form a moral judgment on the basis of their experience without the need for intervention by an external authority (such as "the legislator of God").

Once we have established that we need to maximise pleasure, the next question is: whose pleasure is to be maximised. According to Bentham, we should sum all the "values of pleasure" of all people on one side and subtract all the "values of pain" of all people and see whether a particular act increases or decreases this sum. Acts can then be ranked on the basis of the amount of pleasure they produce. Bentham was aware that it may be difficult to quantify pleasure and pain, but to overcome this problem he proposed the so-called *hedonistic calculus*, a list of features of pleasure to which attention must be paid in order to assess how great it is: intensity, duration, certainty, propinquity, fecundity, purity, and extent. Artificial as this calculus may sound, it is important to remember that Bentham's view does not allow an individual to place special weight on his or her own pleasure. Instead, the calculus is about maximising the pleasure of all people (captured by Bentham's famous phrase: "the greatest good for the greatest number"). Notwithstanding Bentham's hedonistic calculus, it seems difficult to measure pleasure and pain objectively. First of all, pleasure is a subjective term which may vary between different people. What counts as pleasure for one, can be pain or indifferent to others (e.g., watching television, going shopping, or exhausting oneself at the gym). Second, it is also difficult to compare different kinds of pleasure: does going to the gym give more pleasure than playing music? In a professional setting, we can think of improving food production by using chemicals, as illustrated in Example 2. In such an example, it seems nearly impossible to draw a quantitative balance sheet in terms of "pleasure" for different options.

Example 2: Hedonistic calculus and the use of chemicals in food production

Let's suppose that bananas can be brought to the Western world at a relatively low price if certain chemicals with low or no bad effects on consumers are applied. Those who love eating bananas in the Western world would certainly gain pleasure by having bananas available at any time in their local stores even though these chemicals are used. However, if such chemicals would have bad effects on store employees then it would be questionable if the people working in those stores would be happy about their working conditions. If you were the intermediary between the stores in the Western world and the local producers, what would you do? Would you continue importing bananas with chemicals, or would you change to a fair trade, including environmentally-friendly production (with the risk that the higher price will lower the demand and consequently the local producers' turnover)? The hedonistic calculus does not provide a clear-cut answer in this situation.

John Stuart Mill pursued the idea of Jeremy Bentham, but his idea is nevertheless also quite distinct from Bentham's original idea of utilitarianism. Whereas Bentham did not make a distinction between different kinds of pleasure, Mill (1979 [1863]) argued that qualities need to be taken into account in determining a valuable pleasure as some pleasures are more valuable than others. Although Mill gave some broad indication of what kind of pleasure can be valued higher (e.g., intellectual ones) and what kind of pleasure can be valued lower (e.g., physical or animal desires) he did not give a clear and universal answer. Also, an important distinction compared to Bentham's theory is the introduction of the freedom principle. Bentham's utilitarianism was criticised for taking no account of how pleasure is distributed or what is called *distributive justice*. Taken to the extreme, in the utilitarian view the ends justify the means. This implies that the pain of one person may be acceptable – or stronger: morally required – if that would give pleasure to many other people. In other words, the position of individuals cannot always be protected if the hedonistic calculus indicates that the pleasure of the majority outweighs the pain of the minority. This could easily lead to exploitation of individuals: bullying one person would become acceptable if it would make 5 other people happy. As the example of the production of bananas shows, the hedonistic calculus would probably indicate that the happiness of banana consumers, which we assume here to be the majority, would outweigh the health concern of local store employees. Mill was aware of this weakness in Bentham's theory, which he labelled the "tyranny of the majority," and he therefore introduced his freedom principle. This freedom principle states that everyone is free to strive for his or her own pleasure, as long as he or she does not deny or hinder the pleasure of others. Hence, people must seek to maximise pleasure, but under the condition that they do not inflict harm on other people's human nature and dignity. Mill also discussed his freedom principle in relation to the state power. In his famous book *On Liberty* he argued that "the only purpose for which power can be rightfully exercised over any member of a civilised community, against his will, is to prevent harm to others" (Mill 1859). Nowadays, this principle has become known as the *no harm principle*. In the context of public policy making, this principle is very relevant. One could think of the forced migration of population groups in order to implement a large infrastructure project, for example a large dam. According to Mill, this would be allowed only if the dam is necessary to prevent harm (i.e., flooding or extreme drought). It would be then questionable whether the construction of a dam for the generation of hydroelectric power justifies forced migration.

Although Mill improved Bentham's theory in terms of distributive justice, his theory is still vulnerable from this same point of criticism. Even if people are not directly harmed, should we allow activities that only benefit the pleasure of the happy few? Furthermore, we could easily observe that welfare is not equally distributed over the world's population. The question that we could ask is: should we try to increase the aggregated welfare of all people (with the risk that some measures would only benefit the people who are already well-off) or should we try to improve the welfare of those people who are worst off and try to make the distribution of welfare as equal as possible? In the discussion of rights-based theories, we shall see that some philosophers aim at the latter.

Notwithstanding these points of criticism, consequentialism is nowadays the most dominant theory in public policy making. After all, most decisions are made on the basis of CBA. CBA seeks to implement policies that can achieve greater social benefits while minimising social harms, which is a direct translation of consequentialism. Nowadays, the value that is usually optimised is not pleasure, as it was the case in Bentham's and Mill's utilitarianism, it is the cost. Hence, the same line of criticism that is applied to consequentialism could also be applied at CBA. Furthermore, most forms of CBA do not take distributive issues into account. Example 3 illustrates how CBA can be used to choose between different flood mitigation strategies.

Example 3: Building a flood wall or strengthening of dykes?

Let's suppose that you were asked to give an expert advice to a middle sized town which frequently suffers from floods. The town has areas A and B that are at risk of flooding. In area A, the current risk of flooding is 5% AEP (or a 20-year ARI event) and the number of people living in this area is 2000. Since this is a rich area the value of houses and belongings is relatively high. In case of a major flood, the likely number of affected individuals is 2 and the expected material damage is 14 million Euros. Area B is in the poor part of the town with a squatter settlement. The current risk of flooding in this area is 20% AEP (or a 5-year ARI event). The number of people living in this area is 5000 and the expected number of affected individuals for a major flood is 25. The expected damage is very little as the houses and belongings are cheap. In order to apply a CBA analysis to this case, an insurance company placed a monetary value of 500,000 Euro (or 0.5 MEuro) on each life that may be lost. Hence, the expected yearly damage D has been calculated as follows:

Area A: $D_{A;\,initial} = 0.05 * (2 * 0.5\,\text{MEuro} + 14\,\text{MEuro}) = 0.75\,\text{MEuro}$

Area B: $D_{B;\,initial} = 0.2 * (25 * 0.5\,\text{MEuro}) = 2.5\,\text{MEuro}$

Expected yearly damage (total; initial): $D_{T;\,initial} = 3.25\,\text{MEuro}$

Two engineering options are identified for mitigating flood risk:

(1) The first option is to strengthen the dykes around area A. This will improve to a greater extent safety in area A and to a smaller extent area B. The yearly probability of flooding in area A will decrease from 5% to 1%. In area B, the yearly probability of flooding will decrease from 20% to 10%. Investment costs (including operation and maintenance costs over a period of 30 years) are 30 MEuro. Hence, the average annual costs of this option are 1 MEuro.

(2) The second option is to build a flood wall. This option would only improve the safety in area B; the yearly probability of flooding will decrease from 20% to 6%. Investment costs (including operation and maintenance costs over a period of 30 years) are also found to be 30 MEuro and the average annual costs are also 1 MEuro.

The available annual budget for implementing flood mitigation measures is 1 MEuro. Since the combination of two options is not affordable the choice for a preferred option needs to be made. Which measure would you advise the town to implement?

If you choose strengthening of dykes, the net benefit can be calculated by comparing the total expected yearly damage before and after dyke strengthening for both areas A and B, minus the yearly costs:

Area A (after strengthening): $D_{A;\,strengthening} = 0.01 * (2 * 0.5\,\text{MEuro} + 14\,\text{MEuro}) = 0.15\,\text{MEuro}$

Area B (after strengthening): $D_{B;\,strengthening} = 0.1 * (25 * 0.5\,\text{MEuro}) = 1.25\,\text{MEuro}$

Expected total yearly damage (after strengthening): $D_{T;\,strengthening} = 0.15\,\text{MEuro} + 1.25\,\text{MEuro} = 1.4\,\text{MEuro}$

Benefit (yearly) $= 3.25\,\text{MEuro} - 1.4\,\text{MEuro} = 1.85\,\text{MEuro}$

Costs (yearly) $= 1.0\,\text{MEuro}$

Net result $=$ benefit $-$ costs $= 1.85\,\text{MEuro} - 1.0\,\text{MEuro} = 0.85\,\text{MEuro}$

Similarly, the net benefit of the flood wall can be calculated by comparing the damage before and after its implementation (note that the expected damage for area A does not change):

Area B (after building flood wall): $D_{B;\,floodwall} = 0.06 * (25 * 0.5\,\text{MEuro}) = 0.75\,\text{MEuro}$

Benefit (yearly) $= D_{B;initial} - D_{B;floodwall} = 2.5\,\text{MEuro} - 0.75\,\text{MEuro} = 1.75\,\text{MEuro}$

Costs (yearly) $= 1.0\,\text{MEuro}$

Net result $=$ benefit $-$ costs $= 1.75\,\text{MEuro} - 1.0\,\text{MEuro} = 0.75\,\text{MEuro}$

Which option would you choose? If we would follow a blind application of CBA (as per the calculation above) we would probably choose the dyke strengthening option over flood wall option as the numbers that came out from the analysis appear to be more favourable. This is a classic example of an unethical and immoral practice where the greed for a positive bottom line is more important than the value of human life.

This hypothetical case brings another yet problematic issue which is how to estimate the value of a human life? Could a monetary figure be assigned to a human being?

With the discussion of this case we hope to have shown how consequentialism (in the form of CBA) is the dominant framework for decision making in public policy. However, as the case shows, a strictly consequentialist way of reasoning cannot capture all relevant considerations, such as the fair distributions of risks and scarce resources. Additionally, converting everything (including the value of human life) into one monetary unit is nothing less than insane. In the remainder of this section, we will discuss three other ethical theories that approach the morally right thing to do from a different angle.

7.4.2 Deontological ethics

Of the four theories discussed in this book, deontological ethics is probably the most difficult one to grasp. This is partly related to the very formal language of the classical texts that stand in this tradition, but also to the complex theory that lies behind. In order to do justice to the theory, we follow the reasoning behind and justification of the main ideas. However, we try to explicate the reasoning as clearly as possible and make some simplifications where necessary.

The word deontological (adjective) or deontology (noun) is derived from the Greek work *deon* which means duty or obligation. Deontological ethics is therefore also known as *duty ethics*. The most well-known defender of deontological ethics remains the German philosopher Immanuel Kant (1724–1804), Figure 7.4. Although there are other variants of deontological ethics, we limit our discussion to Kant's theory, because this is the most developed and well-known variant, even as we do return briefly to this subject later here.

Figure 7.4 Immanuel Kant (1724–1804) (Accessed on 12th October 2011 from: http://en.wikipedia.org/wiki/Immanuel_Kant)

In his famous *Critique of Pure Reason* (In German: *Kritik der reinen Vernunft* using the present-day spelling), Kant argued that philosophy can be summarised as addressing three fundamental questions: "what can I know?", "what ought I to do?" and "what may I hope?" (Kant 1787 [1781]). The first question belongs in the domain of knowledge theory (or, to some extent, in the philosophy of science), the second in moral philosophy and the third in the domain of religion. By formulating these three questions,

Kant indicates a particular view on human persons. A human person is a being with cognitive abilities and a capacity for moral judgment. These cognitive or rational abilities and the capacity for moral judgment are closely related. Even more, it is for these capacities that, for Kant, humans are to be valued.

Before Kant, the basis of ethics was mostly sought in strivings for human happiness or welfare. Kant changed this view radically. For him, the basis of ethics cannot be sought in happiness because happiness is an individual matter which varies between persons. Moreover, striving for happiness can lead to immoral behaviour if people only strive for their own happiness. But if it is not on the basis of happiness or human wellbeing, how then can we judge whether something is good or bad, right or wrong? According to Kant, nothing can be conceived as good without qualification except the will or the intention of the person acting. Hence, if we want to judge whether an action is good or bad, we have to look at the intention with which the action is done. Not the consequence of an action determines its goodness or rightness but simply the will or intention of the person acting.

But what turns a will into a *goodwill*? For Kant it was important that ethical judgments should be *universal*; that is, applicable to all situations and not varying from situation to situation or from person to person. Kant states that a rule, in order to have moral force, should be absolutely necessary and not conditional on other factors. An action is morally right if it is done in agreement with (or with the intention directed at) a *universal* moral rule (law, norm, or principle), independent of the consequences (Kant 1786).

So far, Kant's theory sounds very abstract. An action is right if it is done with the right intention and if it is in conformity with a universal moral rule. But how should we understand this moral rule? The answer to this question can be derived from Kant's portrayal of mankind. Kant's refers to man as a human being with rational and moral capacities. According to Kant, people should themselves be able to determine what is morally correct through reasoning. This should be possible independently of external norms, such as religious norms. People should impose on themselves the moral rules. This is called autonomy, from the Greek words *auto* (self) and *nomos* (law). People should be their own lawmakers. Hence, as an intermediary conclusion, we could say that moral rules should be universal and they should be set by an act of a goodwill. A will can become a *goodwill* if it acts according to a moral rule for no reason other than out of a respect for that same rule.

The next question is, then, what are these moral rules? According to Kant, the rules that are set by the human will can be formulated in so-called maxims. These are commands that express the intention of an agent. Kant makes a distinction between two types of maxims: hypothetical maxims and categorical maxims. According to Kant, if people use their capacity for reasoning, they will all come to the same moral laws or rules. These rules are valid independently of contextual circumstances and they are universally applicable (in Kant's terms, these norms are *categorically applicable*). These rules have the following form: "thou shall …" or "thou shall not …" (e.g., "thou shall not kill" or "thou shall respect thy parents"). The categorical rules can be contrasted with hypothetical or conditional rules, which are only valid if one wants to achieve something. For example, "if you want to pass your examination, you have to study." This rule is valid under the condition that you want to pass the examination. Moral rules are categorical maxims, that is, maxims that are universally valid.

How can we derive these moral rules or maxims? As stated above, Kant sees a central role here for the human capacity of reasoning. He argues that there is one universal principle from which all moral rules can be derived. This principle for choosing maxims is referred to as the *categorical imperative*. An imperative is an obligatory rule. Hence, the categorical imperative is an obligatory rule that is unconditionally valid independently of circumstances. Since all people share the capacity for reasoning, any rational person, by arguing reasonably, should be capable of judging whether an optional action is morally right. And the outcome will be the same for all people. Hence, the categorical imperative is a principle that selects those maxims that are moral rules. And because all people have the capacity for reasoning, the outcome of this principle is the same for all people.

The categorical imperative was formulated by Kant in different ways. The first formulation of the categorical imperative, the universality principle, is as follows: "Act only according to that maxim whereby you can, at the same time, will that it should become a universal law."

Above we described the maxim of an action as the rule that is issued by the will. According to Kant, we have to act according to those maxims that should be able to serve as a general law, applicable to each and every one, without giving rise to contradiction. Subsequently, we must oblige ourselves to follow that maxim. Let us return to the example of a certain person called Carly, who had to choose between breaking a promise she made to her son (and helping a neighbour), or keeping the promise (and not helping the neighbour). Most people would probably say that in this particular situation, Carly was allowed to break the promise she made to her son. However, for Kant the answer would be different. Kant would then argue as follows. Say that we allow Carly to break her promise. The maxim of the action to be undertaken would then be "I am allowed to break my promise." On the basis of the categorical imperative we can now judge whether this maxim is a reasonable one. This is the case if we could reasonably wish everyone to break their promise without contradiction. For Kant, it is evident that this cannot be the case. If breaking a promise would become a general law (i.e., if it would become accepted), no one would trust another person to keep their promise. Promises would lose their meaning and the whole point of making a promise would disappear. It would be of no use any more to make a promise in the first place. Hence, we would contradict ourselves if we wish to break a promise and want the breaking of promises to become a general law. Accordingly, Kant would argue that breaking a promise is not allowed and Carly should help her son with his preparation for his examination rather than helping her neighbour.

It is obvious that Kant has a very formal ethical theory. So far, he has not given any content to his ethical theory, other than a procedure for deriving the rightness or wrongness of an action. Kant realised this and he gave a second formulation of the categorical imperative, in which he provided more "content." This formulation reads as follows: "Act as to treat humanity, whether in your own person or in that of any other, in every case as an end, and never as means only." The principle underlying this formulation of the categorical imperative is called the *reciprocity principle*. At first glance, this formulation of the categorical imperative sounds very complex as well, but it has a more mundane equivalent in the golden rule. The golden rule states "Do unto others as you would have them do unto you." Although this rule is often attributed to the Bible, it can be found in many more and also older philosophies, such as Hinduism, Buddhism, and Confucianism.

In this second formulation of the categorical imperative, "reason" or "rationality" is seen as the equivalent of "humanity." It is for their reasoning capacity that human beings are different from inanimate objects and (supposedly other) animals and it is exactly for this reason that humans are to be respected. In more concrete terms, we are not allowed to mislead the rationality of other human beings, either by treating them as means to achieve our own ends or by constraining their rational nature as free human beings. If we return to the example of Carly, we should again conclude that breaking the promise is not allowed because Carly would then mislead (the rationality of) her son. Hence, breaking the promise is not allowed.

As stated at the beginning of our discussion of deontology, this theory is often considered the most difficult of all ethical theories. We therefore summarise the most important characteristics as follows:

(1) Deontology applies to all human beings under all conditions (*universally* applicable);
(2) Human beings are to be respected for their reasoning capacity. Any infringement on this capacity is therefore wrong (respect for *autonomy*);

(3) Not the consequence of an action but the intention of the acting person determines whether an action is right or wrong (*good will*).

Kant's theory has been criticised on several points. According to Kant, all moral laws can be derived from the categorical imperative. However, Kant does not give any indication on what to do when the outcomes conflict. How to act, for example, if one has to choose between telling a lie and being faithful to someone? If we use an example of a professional situation where an employee has a confidentiality contract with a client but also discovers that the client is in fact building a dangerously unsafe breakwater. Both breaching the confidentiality contract and being silent on a potential risk seem to be forbidden by the categorical imperative. Kant, possibly mistakenly, assumes that the outcome of the categorical imperative forms an unambiguous and consistent system of norms, but it is questionable whether this will always be the case.

A second point of criticism is that Kant does not allow for any exceptions. There is no such thing as bending a rule (Van de Poel and Royakkers 2011). Kant is aware that his theory may sometimes lead to unexpected outcomes. He gives the example of murderer who knocks on the door in search for a particular person. Even in such a situation, one is not allowed to lie, according to Kant. For many of us, this is not an acceptable outcome. If we can save an innocent person by telling a lie, most of us would say that it is preferable from a moral point of view to tell this lie. The evaluation of our behaviour cannot be reduced to intention or good will, most people argue. Sometimes the consequences of one's actions *do* matter, contrary to Kant's view.

A third point of criticism relates to the formal character of Kant's theory. As we showed earlier, Kant mainly provides a formal "procedure" for deriving what ought to be done. The contemporary philosopher Alasdair Macintyre (1984) argued that the categorical imperative, exactly because of its formal nature, can be used to derive anything and is therefore not as universally applicable as Kant says it is. By formulating exceptions, it is possible to manipulate the categorical imperative towards one's own self-interest, Macintyre argues:

> "I can without any inconsistency whatsoever flout it; 'Let everyone except me be treated as a means' may be immoral, but is not inconsistent and there is not even any inconsistency in willing a universe of egotists all of whom live by this maxim" (MacIntyre 1984, p. 46)

Although Macintyre has a point in that Kant's theory is sometimes too formal, his example clearly contradicts the second formulation of the categorical imperative ("Act as to treat humanity, whether in your own person or in that of any other, in every case as an end, never as means only"). Hence, it may be wise to keep this second formulation in mind if one wants to judge an action in terms of Kant's ethical theory.

In addition to criticism phrased above, Macintyre also questions Kant's rational justification of his ethical theory. According to Macintyre, the ambition to make ethics "universally applicable" does not do justice to the diversity of human beings and their own particular lives (in short: to context). Reason alone is not sufficient to act morally, Macintyre argues. The next ethical theory, of which Macintyre is a proponent, places a greater emphasis on the context.

7.4.3 Virtue ethics

Virtue ethics stems from a long tradition and was the dominant form of ethical theory in ancient and medieval times. The ancient Greek philosopher Aristotle (Figure 7.5) was the first to define virtue ethics as a field of inquiry in itself (Van de Poel and Royakkers 2011). After some centuries of only little popularity, virtue ethics has experienced a revival in the last few decades and it has become the third big ethical theory besides consequentialism and deontological ethics (Kuusela 2011).

Figure 7.5 Aristotle (384 BC–322 BC) (Accessed on 12th October 2011 from: http://en.wikipedia.org/wiki/Aristotle)

Similar to utilitarianism, proponents of virtue ethics argue that human beings should strive for something that has value in itself. For utilitarians, this highest value is pleasure. For Aristotle, the highest good is *eudaimonia*, the Greek word that is usually translated as happiness. However, in the context of Aristotle's ethics, *eudomania* is better understood as "the good life." It does not refer to some happy circumstance that brings momentary pleasure (as with the utilitarians) but a state of being a good person. For Aristotle, leading "the good life" means that one should live one's life in a way that is characteristic for human beings and that one tries to excel in doing so. The old Greek word for "excellence" is the same as the word for "virtue." Hence, virtue ethics refers to a kind of ethics which promotes the notion that people excel in their actions. But what actions are virtuous actions? Aristotle argued that virtuous actions are those actions that a virtuous person would take. This sounds of course circular or at least arbitrary: what counts as virtuous actions are whatever actions the virtuous person would do. However, if we take virtue to refer to excellence, we could say that a virtuous person tries to excel in being a human. It means leading a life as is meant to be the case for a human; one should excel in the things that are part of being human. Hence, we have to look at what is characteristic for humans. Since humans were regarded as the only living beings that could reason at the level of humans, people should use this superior reasoning capacity to the fullest extent. However, whereas Kant had a very cognitive interpretation of man's rationality, for Aristotle this reasoning capacity means being responsive for certain features or aspects of reality. What is distinctive about a virtuous person is that he or she knows how to act in a particular situation. As such, virtuous persons provide a kind of ideal or role model for how we ought to live. Virtue ethics is often explained on the basis of sports. If someone wants to become a good sports man (say a good soccer player), we mostly respond with something like "try to play like Pele, Johan Cruijff, or Lionel Messi" rather than "try to follow this or that rule." Being an excellent soccer player means knowing how and when to play a ball, how to pass an opponent, and when to aim at the goal. This cannot be taught by listing some rules, but primarily by practicing and looking at good examples. Just like a prospective soccer player should practice in order to develop his or her talents, so should human beings develop their character traits in agreement with the virtues necessary to realise the good life. By doing so, acting morally becomes one's nature. Moral virtue is therefore also referred to as character virtue by Aristotle. Hence, where deontological ethics and consequentialism take actions and the consequences thereof as the primary basis for moral evaluation, adherents of virtue ethics look at the person's moral character. This capacity is associated in this book with *transcendence* as introduced earlier here and as will be refined later, rather than moral virtue within the field of ethics, this last term having nowadays fallen into disrepute, at least when the word "ethical" is taken in the malleable secular sense that is so widely accepted and employed in the West. Thus, within this last context, ethics is merely concerned with pragmatic choices based on reasoning within whatever social arrangements happen to be acceptable at the time. Thus what was considered as most unethical and was correspondingly severely punishable by law at one time, such as abortion and homosexuality, can become ethical at another time: such an ethics is temporal through and through. Such changes are of course justified in the name of *progress*. Correspondingly, social justice can be characterised in very different ways within such societies, and even more between different societies, even though it may be experienced in the same kind of way at the level of the emotions. However, it suffices to observe our fellow creatures here on earth to see that social justice must in their case at least be pre-temporal, supra-temporal and post-temporal; correspondingly, this is the position concerning humanity too that is adopted in almost all religions and spiritual practices and so in most places outside the West (See Barth, II. 1, pp. 638, 639). This is also so when we have to work within what we can perhaps best describe here as a *perennial ethics*, such as is expressed in the analects of Confucius.

Although virtue ethics does not give specific rules for acting morally, we can still derive some guidelines for moral behaviour. According to Aristotle, a moral virtue is always a middle course between two extremes (which are always wrong, according to Aristotle): courage is centred between cowardice and recklessness, generosity between stinginess and prodigality, pride between vanity and submissiveness, etc. Finding the right balance between two extremes in a given situation requires a certain ratio that is essential to humans. The character traits that help us find this balance are not given to us at birth but they can be developed by our deeds. Hence, by acting morally our moral virtues become more "instilled" in our character. In other words, moral virtues can be practiced just like all arts: "For the things we have to learn before we can do them, we learn by doing them, e.g. men become builders by building and lyre players by playing the lyre; so too we become just by doing just acts, temperate by doing temperate acts, brave by doing brave acts" (Aristotle 1980 [350 BC]: 1130a).

Finding the right middle course between two extremes is not always easy. It partly depends on the particular context and the problem at hand. In other words, what counts as good in one situation may not be the best thing to do in some other situation. Aristotle used the term "practical wisdom" (in Greek: *phronèsis*) to refer to the capacity concerned with how to act in particular situations. Practical wisdom allows us to make the right choices for actions concerning what is good and useful for a successful life. According to Aristotle, a wise man can see what he has to do in the specific and often complex circumstances of life. This practical wisdom is therefore a precondition for finding the middle course. As such we can say that moral virtues and the intellectual virtue of practical wisdom together allow us to do the right thing (i.e., to act morally).

In the thirteenth century, the Christian philosopher Thomas Aquinas (1225–1274) developed a Christian version of virtue ethics. Aquinas distinguished seven virtues: four cardinal virtues that are natural and revealed in nature and three theological virtues, which have God as their object. The cardinal virtues are prudence, temperance, justice, and fortitude. The theological virtues are faith, hope, and charity. The cardinal virtues are binding on everyone and they tell us how to behave towards other people. The theological virtues give instructions on how to relate to God. From 1600 onwards, consequentialism and deontological ethics replaced virtue ethics as the dominant doctrine, but in the last decades virtue ethics has gained in popularity again. This is partially due to the influence of the British philosopher Alasdair Macintyre (born in 1923), mentioned in the previous paragraph. MacIntyre emphasises the importance of people's practice. By practice he means a particular community or context in which people act, live, work, etc. By participating in a practice, people form their opinions of what constitutes "the good life". Moral deliberation should therefore not be separated from the practice itself (MacIntyre 1984). Hence, what professional behaviour amounts to can only be judged in the context of a particular profession. For engineers, professional behaviour would probably best be described by virtues such as expertise, accuracy, creativity, cooperation, striving for social justice, etc.

It should be clear that virtue ethics does not give specific rules to follow, other than finding a virtue that is the middle way between two extreme character traits. People who want to have a "recipe-like" instruction for moral behaviour may be left disappointed. However, especially in the context of professional ethics, virtue ethics has become a popular doctrine. In commercial businesses or governmental agencies, people are often confronted with conflicting demands (stemming from time or money constraints, institutional constraints, etc). Virtue ethics can be helpful for finding a right balance and making an acceptable compromise.

Professionalism

In their work, professionals are often confronted with dilemmas that place high demands on their standards of integrity. How to act, for example, in a situation where bribery and corruption are commonplace? Corruption is wrong, no matter what, and the same holds for full-fledged bribery. But when does negotiation become bribery? One may be inclined to simplify the case to a simple black-and-white situation: either to reject categorically any presents given by a potential client or to go along with it uncritically. However, not accepting certain offers may in some situations be considered disrespectful. Is rejecting a cultural visit organised by a potential client a matter of moral courage or is it a lack of sensitivity for context. In other situations, it may simply be an unavoidable obstacle for achieving otherwise morally desirable ends. Especially in totalitarian countries, working on projects that benefit the poor is sometimes only allowed after one has finished a project for some higher government official. Would you reject the whole project at the price that the poor will also be deprived of their benefits? Or is it perhaps better to negotiate an even better outcome for the poor? These questions are very hard to answer on the basis of abstract moral theories. Sometimes a very principled position is the only right thing to do. In other situations, it may be necessary to make compromises. Because of its sensitivity to context, virtue ethics may be helpful for responding to delicate situations by analysing what character traits the different reactions amount to and see whether it is possible to find an acceptable middle ground between two extremes. Similarly, virtue ethics may be helpful in shaping one's own professional practice in a *morally acceptable* way. Let's return to the problem phrased in Example 2 (the use of chemicals in food production). If you are the CEO of a consultancy firm, would you require that the people working in your company's canteen use biological fair-trade products only? Or how would you deal with people who use the company's printer for private purposes? Would your answer be different if these people made working days of 11 hours? As one runs into those kinds of situations, it is important to recognise the dilemmas in the first place and, consequently, to have an "ethical toolbox" available to reflect on these issues and decide what to do. Keeping in mind that what counts as a desirable middle ground may have changed between Aristotle's and our present time, virtue ethics may be helpful in this regard.

Virtue ethics has been criticised for not giving any concrete clues about how to act in a particular situation. Adherents of virtue ethics would probably respond to this criticism that it is their strength rather than their weakness that they do not give concrete rules. Moral behaviour is about knowing how to act *in a particular context* and trying to capture this in a pre-set rule does not necessarily do justice to the particularities of the situation. Having the right virtues, in contrast, does facilitate moral behaviour because people will strive for a middle-ground between extremes. Kant rejects this latter argument by giving the example of a cold psychopath who uses the virtues "self-control" and "deliberation" to prepare wilful murder. In this case, his so-called moral virtues facilitated his immoral behaviour. The final word has not been said yet in this discussion. For now it is important to recognise that virtue ethics – more than any of the other theories discussed – looks at the context to see what is the right thing to do.

7.4.4 (Egalitarian) rights-based theories

Although they are not considered as a distinct category of ethical theories, rights-based theories deserve a special place and discussion. The reason for this is that rights-based theories are, more than any of the other ethical theories, attentive to distributive issues.

In contemporary understanding of morality, rights play a crucial role. People claim to have certain rights, for example the right to freedom of speech and assembly, property rights, the right not to be discriminated, the right to education and basic medical care, the right to privacy, or – relevant for our book – the right to have a safe place to live, etc. We judge institutions (i.e., our governments, businesses, the content of laws, etc.) and other people's actions to the extent that they respect these rights.

But what exactly are rights? And do we all have the same rights? A distinction is usually made between so-called *liberty rights* and *welfare rights*. Liberty rights are rights to exercise one's own freedom. These rights place a duty on other people not to interfere with these people's rights. For example, if one has a property right, other people have a duty to refrain from interfering with this property right. These rights are therefore also called *barrier rights*. Barrier rights indicate that other people are barred from acting in certain ways to that person. Welfare rights, in turn, refer to a particular state of affairs to which a person is entitled. If we say that one has a right to education, this means that other people are required to make this education accessible to that person. Hence, both liberty rights and welfare rights are accompanied with duties for other people (either the duty to refrain from interfering or the duty to bring about a certain state of affairs). Some people therefore say that rights-based theories are in a way a type of deontological or duty ethics.

The extent to which people are entitled to welfare rights is controversial, especially when they enter into the law or when public money is spent to promote these rights (Martin and Schinzinger 2005). Adherents of libertarian rights ethics deny the existence of welfare rights and claim that only barrier rights exist (cf. Locke 1960 [1690]; Nozick 1974). However, so-called egalitarian libertarians claim that welfare rights *do* exist, but they differ in their interpretation of these rights. Based on the idea that all humans are equal in fundamental worth or social status, they promote some sort of equality of welfare (Dworkin 2000). Egalitarians try to remove economic inequalities among people. However, equality can be promoted in different ways and a famous question is therefore "equality of what?": Should people have equal access to health care, irrespective of their wealth and their life style? And should people all have the same chance to get a job, irrespective of their prior education? These are very delicate questions and they cannot be easily answered, especially if we realise that redistribution of welfare may infringe on other people's liberty rights. Some people claim, for example, that to levy tax on citizens' income is already an infringement on their property right. Hence, if a government uses tax money to implement measures to mitigate flood risk, people's welfare rights (the right to a safe living environment) may conflict with other people's liberty rights (the right to property).

We shall now discuss the two principles that are used to distribute scarce resources, one that aims at efficiency and the other that aims at equality.

- *Pareto optimality* (also known as Pareto or allocative efficiency): This principle is named after the economist Vilfredo Pareto (1848–1923), Figure 7.6. Although this principle does not say anything about equality or social justice, the principle is worth discussing because it is often applied in engineering and social sciences. Pareto (1966 [1909]) developed this principle in his studies of economic efficiency and income distribution. The principles states that, given an initial allocation of goods among a set of individuals, a change to a different allocation that makes at least one individual better off without making any other individual worse off is called a Pareto improvement. An allocation is defined as "Pareto efficient" or "Pareto optimal" when no further Pareto improvements can be made. Distributions that are not Pareto efficient are to be avoided, for the simple reason that they do not get the most out of the scarce resources. In economic systems and public policy, Pareto efficiency is an important criterion for evaluation. Since Pareto optimality does not look at equality, the outcome does not necessarily lead to a distribution of resources that is desirable from the perspective of social justice. In the context of flood risk mitigation, measures that increase the safety of people without harming the interests of other people should be implemented from a Pareto efficiency point of view. It is clear, though, that most safety measures involve costs, mostly public money. The public money often involves tax money paid by people who do not live in the area that is exposed to flood risk. It is therefore difficult to think of an example where a flood risk mitigation measure does not harm the interests of other people if these interests are understood in a broad sense.

Figure 7.6 Vilfredo Pareto (1848–1923) (Accessed on 12th October 2011 from: http://en.wikipedia.org/wiki/Vilfredo_Pareto)

- *Maximin rule* (also known as a difference principle): This rule is renowned from the twentieth century political philosopher John Rawls (1921–2002). Rawls developed a theory of justice in which "fairness" played a central role (a theory, which he aptly labelled "Justice as Fairness" (Rawls 1999 [1971]; Rawls 2001)). Rawls aim was to develop a criterion of justice that would be agreed upon by all people under conditions that are fair to all people. Rawls therefore introduced the hypothetical "original position." Imagine that people do not know in which layer or place of society they live, how would they want society to be organised? Rawls thought that people would try to arrange society in such a way that the person who is worst off will get the *relatively* best position (or the "least bad" position). A simple example can be illustrated by the following. If you have to divide a cake in four pieces and you know that you are the last to choose a piece, you will probably try to make the pieces as equal as possible, such as to *max*imise the size of the smallest (*min*imum) piece (hence, the name *maximin*). The same holds for arranging our society, Rawls argued. If you do not know in which country you live, whether you belong to the upper or the lower classes, whether you are man or woman, black or white, you will probably make the society as equal as possible; that is, equal opportunities and chances, equal distribution of resources, equal distribution of burdens (such as risks), etc. Hence, according to Rawls, justice has all to do with fairness (fair distributions). The principle in which he elaborated this idea of "justice as fairness" is the so-called difference principle. This principle or rule states that social and economic inequalities should be arranged in such way so that "they are to be of the greatest benefit to the least-advantaged members of society." To illustrate how this works in practice, let us go back to Example 3 (Building a flood wall or strengthening the dykes). Whereas a CBA indicated that dyke strengthening is the preferred option, the difference principle would indicate that we have to focus on the population in area B first, because these people are the "least-advantaged" (or the most vulnerable) members of society. Hence, even if the considerations of efficiency indicates differently, it may be fair to spent tax money unevenly if it maximises the benefits of those who have little welfare resources.

With the discussion of Rawls' *maximin* rule, we hope to have shown that there are alternative distributing procedures than those solely based on efficiency. If we want to realise social justice in the context of flood risk mitigation, it might be necessary to look at these other distributing rationales and try to reconcile the demands from efficiency with those stemming from considerations of social inequalities.

Further to the above discussion of different ethical theories and their relevance for dealing with flood risk mitigation issues (see also Table 7.1) it should be noted that this discussion is not exhaustive, as we have not covered all possible theories. Moreover, the discussion was limited to ethical theories that stand in the tradition of Western philosophy. It is increasingly recognised that Eastern philosophies may well be better suited to answer particular questions. However, it is beyond the scope of this book to go into discussion of non-Western ethical theories. The interested reader is referred to the following books and papers for a discussion of ethical issues in the Islamic (Hamed 1993; Wersal 1995; Foltz 2002), Confucian (Chan 2008; Wong forthcoming), or Eastern philosophy tradition in general (Koehn 1999).

Table 7.1 Ethical theories.

Ethical theory	Characteristic focus	Relevant considerations
Rights-based theories	Distributive issues	Intragenerational justice Intergenerational justice Availability of alternatives
Consequentialism	Consequences	Efficiency/efficacy
Deontology	Autonomy	Informed consent Participatory approaches

In any event, it must be faced that *these are only theories* and in most cases they are very difficult to be applied in 'the real world' in which the great majority of us live. They provide various levels of desiderata, but are all subject to constraints, and even severe constraints, in our present-day, Heideggerian, "mundane and average world". Examples appear every day and today's FT (02.08.11, p. 7) has an article by Philip Stephens that includes (with italics added):

> They knew when he was elected that Mr Cameron hailed from wealth and privilege. If the prime minister is to be criticised it is for pretending otherwise. Pace George Osborne, the chancellor, *we have never been all in this together*. Sure, ministers have taken a pay cut. They have also capped the pay of senior officials. This is *gesture politics …*
>
> As my FT colleague Brian Groom has detailed, the chief executives of FTSE 100 companies saw their pay rise by 32 percent. That compares with 2 percent for most workers. This was not a blip. In 1998, the average pay of these business leaders was 47 times higher than that of employees. By 2010 the multiple had soared to 120 times. Yet the real value of their companies hardly changed at all. So much for claims that pay matches performance.
>
> The upshot has been a new plutocracy whose interests and incentives are wholly detached from the rest of society.

In this grotesque situation there can be no structure of common ethics, but only a variety of ethics, with each one special to a specific social group. This situation had, at the same time as Philip Stephens' commentary, reached such a state in the USA where opposing interest groups produced such differences concerning the revaluation of permissible US debt that there could be no community of interests at all, but only deep differences that reflected largely conflicting ethical positions. This is a situation that one commentator has described as 'insane'. We may return to one of Blaise Pascal's typically pithy commentaries that generalises ethics within the broader framework of Justice (Pascal/Le Guern 1977/2004, p. 89):

Justice.
Comme la mode fait l'agrément, aussi fait-elle la justice.

Justice.
Just as the current fashion is agreed, so also it makes the current justice

7.5 INSTITUTIONAL AND LEGAL REQUIREMENTS

The present book emphasises that a socially just flood risk mitigation practice is an imperative to deal with flood risk in a sustainable manner. Being a multi-disciplinary activity, its successful implementation greatly depends on the seamless interaction between various sectors of society and institutions. As discussed earlier, social justice is very much a property of active and continuous stakeholder participation but it is also a property of procedures and mechanisms whereby institutions can fairly assign benefits or burdens (or in this case risks) across the society. Correspondingly, there is a need for a better integration and creative synergies between the actions of various stakeholders and their effective coordination within the existing legal and institutional arrangements. The related decisions must take into account not only their effect on the mitigation of flood risk, but also of all other resulting impacts (i.e., social, economic, environmental, land development, etc.). Similarly, the decision–making processes of other authorities and organisations whose activities may have an influence on hydrological and other related responses must be carefully integrated and coordinated with those decisions that relate to flood risk mitigation. Such integration and coordination is needed at different levels and scales (e.g., horizontal, vertical, national, regional, local). In doing so, it is important that the appropriate mechanisms that can enable effective participation of stakeholders are in place. Figure 7.7 presents a framework whereby the stakeholders are involved in the development of various policies, legislations and other concrete mechanisms and instruments that can put the necessary change in place for mitigating flood risk issues.

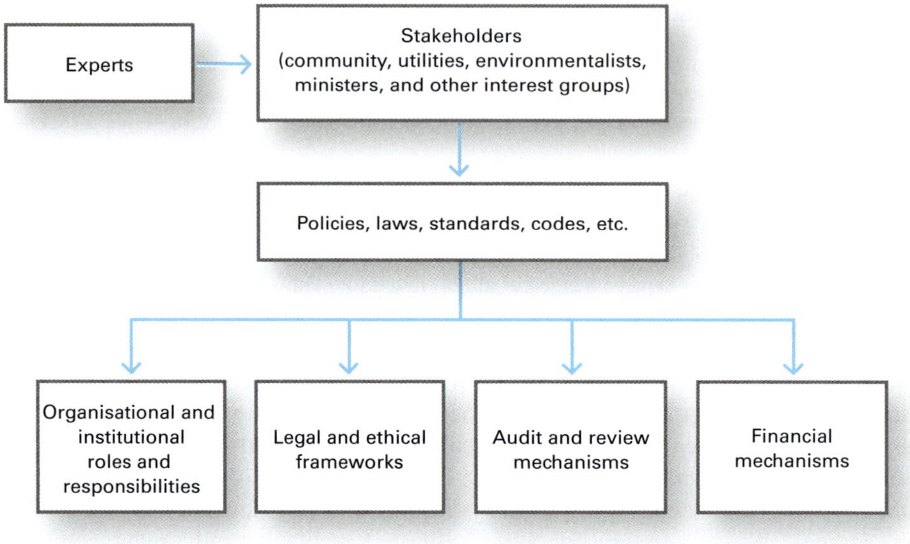

Figure 7.7 Stakeholder involvement in flood risk mitigation

The main objectives of the framework given in Figure 7.7 can be summarised to the following:

(1) To enable active stakeholder participation through:
   (a) coordination and cooperation between various organisations, institutions, sectors and community;
   (b) availability and accessibility of the basic data and information for informed decision making;
   (c) creation of hydroinformatics-based environments that can enable active and continuous stakeholder participation through raising the awareness and mobilising the imaginations of the stakeholders, initially individually in terms of activating their personal imaginations, and then proceeding interactively across the participating stakeholder population as a whole, so as to raise their collective creative imaginations, with of course equal rights being given to every such participant;
(2) To integrate the planning of flood mitigation measures (concerning all potential kinds of floods: pluvial, fluvial, coastal, groundwater, etc.) in conjunction with other planning activities such as water/wastewater, land use planning and other sector planning activities:
   (a) to adopt a best mix of strategies taking into account the relevant issues from all other areas and sectors;
   (b) to search for synergies between different water-related areas and different sectors;
(3) To exercise virtue and ethical principles in all decision making processes;

The experience to date has shown that some flood related decisions (e.g., what is the acceptable level of risk or which mitigation measures should be adopted) are often political and/or serve interests of a particular group of people or organisational body. It is then essential that all stakeholders are involved in this process in order to ensure that political interference is minimal and that a wide consensus is achieved. Some governments have addressed such issues by adopting the principles of precautionary limits (see e.g., *Communication from the European Commission on the Precautionary Principle* 2000, 1, 19, which is available at the Eurlex website at http://europa.eu.int/eur-lex/en/). In such a case their obligations are derived either from international agreements or more directly through domestic legislation.

Precautionary Principle

The precautionary principle can broadly be defined as the imposition of "controls in advance of complete scientific understanding" (see also Salter, J. and Howsam, 2002, and Sands, 2000). In the context of environmental protection, Principle 15 of the Rio Declaration provides that:

> *"where there are threats of serious or irreversible damage, lack of full scientific certainty shall not be used as a reason for postponing cost-effective measures to prevent environmental degradation."*

A precautionary approach could equally be taken with respect to the protection of human life or property. Article 3u3) of the United Nations Framework Convention on Climate Change14 states that:

"The Parties should take precautionary measures to anticipate, prevent or minimize the causes of climate change and mitigate its adverse effects. Where there are threats of serious or irreversible damage, lack of full scientific certainty should not be used as a reason for postponing such measures, taking into account that policies and measures to deal with climate change should be cost-effective so as to ensure global benefits at the lowest possible cost."

(*UNECE* Guidelines on Sustainable Flood Prevention, supra note 2, para.13(c)).

The *UNECE*, in its *Guidelines on Sustainable Flood Prevention*, suggests that flood prevention measures should be governed by the above mentioned precautionary principle. However, the same document does not provide further guidance as to how this should be applied in practice. Therefore, a clear and unambiguous legal and institutional framework is required to ensure seamless interfaces between different layers of government and different sectors of society in dealing with flood related decisions.

7.6 ENHANCING STAKEHOLDER PARTICIPATION WITHIN LEGAL AND INSTITUTIONAL ARRANGEMENTS

As stated earlier, stakeholder participation plays a critical role in socially just flood risk mitigation process. The practicality of enabling effective participation requires careful consideration as it has to deal with the following issues:

- Identification of all stakeholders (e.g., stormwater/flood defence authorities, water resources authorities, community representatives, disaster managers and civil defence/security authorities, environmental agencies, nature conservation, forestry, agriculture, spatial and land use planning, public health, transport, etc.);
- Recognition of all parties including those exposed to risk, minorities and other disadvantaged groups;
- Equal rights and equal dignity between the group representatives;
- Empowering marginalised groups to advocate and defend their interests;
- A more equitable distribution of power and opportunity;
- Realisation of civic, social and economic rights;
- Taking-up a vision of a more inclusive and understanding society in which the basic needs of all people may be met;

- Taking a comprehensive view of risk (initial, residual, technological, etc.);
- Taking into account all sources of floods (e.g., pluvial, fluvial, coastal, groundwater, etc.);
- Changing power relationships, and even power structures, between citizens, government, business and civil society;
- Identifying and changing the causes of social inequalities;
- Testing innovative approaches to social change;
- Bringing stakeholders together to understand their role in an unjust system;
- Encouraging people who are part of the problem to become part of the solution;
- Access to assets and capabilities to reverse injustice;
- Maximising welfare – maximising the benefits of flood mitigation measures by giving the priority to those that are most vulnerable;
- Fairness – public resources should be used fairly and efficiently by taking into account the needs and concerns of present and future generations;
- Respecting individual rights – everyone has the right to be safeguarded from floods and/or another person's actions in relation to floods;
- Promoting virtue – public policy should encourage sharing risks and working together to protect the most vulnerable members of the community first;

The outputs from stakeholder participation in the pre-flood preparedness and planning process should be followed with the introduction of necessary legislations, policies, laws, standards, building codes and so forth. These should be then taken at the governmental level and integrated so that the government's strategy, implemented through different departments, is coherent and harmonised. An appropriate mechanism for sharing the costs of risk mitigation based on providing the basic protection against a minimum level of flood that addresses the vulnerability and maximises the welfare should be established: see for example WMO (2009b).

Appropriate measures should be applied at all levels of public planning, whether national, regional or local, and involve all relevant public agencies. The corresponding actions will vary between those responsible for planning, for operation and maintenance of flood defences and for forecasting and warning services, and those in charge of disaster preparedness, response and recovery. Therefore, apart from involvement in decisions regarding the definition of acceptable levels of risk and planning of mitigation measures that will minimise flood risk, the participation of stakeholders in defining the activities during and after the flood emergency is also necessary. An appropriate institutional framework is required to manage the interfaces between different layers of government in flood emergency situations and to enable a more effective response at the appropriate levels.

Particular attention should be paid on how to maintain the longevity of the participatory process term. Another critical issue is how to build trust. This is typically done through information sharing and repeated interactions. However, the task of developing and installing the true *sociotechnical arrangements* that can enable these facilities remains for the most part as a challenge that has still to be faced.

Hence, enhancing stakeholder participation in the process of dealing with flood risk mitigation may be achieved through:

- Mobilising the necessary sociotechnical arrangements that can enable longevity and transparency of the process;
- Raising the awareness and understanding of flood risk (i.e., stakeholder capacity building);
- Provision of mechanism to manage various kinds of conflicts (e.g., relationship conflict, data conflict, values conflict, interest conflict, organisational and institutional conflict, etc.);
- Better sharing of information and resources between different sectors;

- Better procedural and legislative mechanisms and instruments that can put the necessary change in place;
- Use of incentive schemes for risk sharing;
- Commitment to *authentic leadership*;
- Obtaining political willingness;
- Building trust;

7.7 INTEGRATING PUBLIC PLANNING PROCESSES

In whatever way the society is organised, decisions of various other sectoral planning activities will always have a considerable effect on flood risk characteristics and it is important that such decisions are properly coordinated and synchronised with activities considering the risks from all possible floods (e.g., pluvial, fluvial, coastal, groundwater, etc.). For example, the influence of spatial and land use planning on flood risk is enormous. The location of key components of the human and the natural economies on flood plains may provide economic benefits but at the same time these components may also create another potential for flood risk. There are two major aspects that connect land- use planning and flooding (see also WMO/WGP 2007):

- The location of values and key components of the economy on flood plains may provide economic benefits but at the same time it may also create and enlarge the flood risk.
- The development of land change from impervious to pervious surfaces has a potential not only to accelerate runoff but also to reduce the infiltration capacity of soils or to obstruct the natural drainage system, as well as to increase sediment and pollutant transport.

Taking these aspects into consideration, legislation or policy documents may identify the bodies that must be consulted with respect to certain types and magnitudes of proposed development, along with those that have a right to have their views considered.

There are also direct links between deforestation, agriculture and transport activities and floods (see e.g., FAO 2003; FAO and CIFOR 2005). These activities have proved to have their negative effects on the flow of water and creation of flood risk and as such their influence should be appropriately reflected in legislation or policy documents.

The broader role of drainage in the context of wastewater, industrial development, agricultural and irrigation activities, water resources management, transport and communication development, as well as disaster preparedness, response and recovery, should be addressed in an integrated manner by involving all related authorities. All existing sources of pollution on active flood plains, such as industrial premises, wastewater treatment plants, car parks and domestic central heating oil storage tanks, must be identified and appropriate confinement measures should be implemented. A particular consideration must be given to those situations where floodwaters might potentially come into contact with these and other contaminants. Toxic substances from diffused pollutants (pesticides or sludge in arable farming) should be also taken into account. Therefore, planning of all activities that may have their direct impact on the creation of new or amplification of existing hazards (and therefore risks) should be subject to flood risk assessment and as such they should be reflected within the related management plans.

For example, in the UK, the Planning Policy Statement 25 (PPS25) and CIRIA 624 (2004) set out the national policy on development and flood risk. The aims of these documents are to ensure that flood risk is taken into account at all stages in the planning process to avoid inappropriate development in areas at risk of flooding, and to direct development away from areas of highest risk. Where new development is particularly necessary, the policy aims to make it safe, without increasing flood risk elsewhere, and, where possible, reducing the overall flood risk.

Besides the institutional, legislative, legal and other arrangements that are necessary for implementation of a socially just practice a form of stewardship or *leadership* that can carry out the change is also very much needed. For this purpose it is necessary to provide a raft of theory and a rudder of philosophy whereby what we shall now introduce as the leader can navigate the rapids of social justice activities more effectively. We shall now discuss different forms of leadership and describe their features and characteristics within in the first place the world of active stakeholder participation.

Chapter 8

Leadership and social justice

8.1 WHY LEADERSHIP?

Leadership (together with some measure of what passes as serendipity) is at the heart of every successful change. Engaging in social justice work of any kind requires constant change and transcendence towards higher states of consciousness. In academia, leadership is charged with directing the research and entrepreneurial activities in the direction that is holistic, socially just and ecologically sustainable. Similarly, in practice (or in industry generally), leadership is charged with directing and implementing such changes in various industrial, organisational and team (or group) activities. One of the true answers to the problems associated with successful realisation of socially just flood risk mitigation is then to be found in providing sound, well-educated, well-informed, and above all *honest*, and thus *authentic, leadership*. This is a leadership that will continually strive for transcendence of industrial, organisational, team and personal consciousness. At the personal level, this may require substantial self-leadership efforts which will at the first place let go various kinds of fear-based beliefs that we all carry throughout our lives. As soon as such fears start to diminish, the mind of the leader becomes free to take on the higher values of the *true self* and those of the world. Therefore, being in touch with the *inner Self* is the fundamental quality of the authentic leadership.

This section aims to provide the readers with the necessary material on this subject and to shed some light through the verbiage of confused thinking, disinformation and deception that is such a feature of what has become an *ideology of management*. We observe that as a result of the virulent spreading of this ideology, our present-day societies are choking under a surfeit of essentially parasitic persons who are really only masquerading as managers. To use an old English expression, one would say that most of these persons "could not even run a whelk stall". We all recognise them as they circulate long, meandering and entirely pointless 'memoranda', call endless 'meetings' to discuss every kind of nonsense. We recognise them also by their being the only ones in the organisation who try to take at all seriously the pretence that they are doing anything other than wasting everyone else's time and energy – for they must always make themselves 'felt' in the organisation and so they constantly interfere in the work of others, and in particular in the work of those who are truly productive.

One might think that the answer to this problem, which will surely be familiar to almost everyone, is simply to introduce 'better management', and indeed many business schools are claiming, more or less discreetly, to be doing just that, and we like to believe that some really are doing it. Our own position however is that in most cases these managers are entirely redundant, and that the persons who are actually involved in the directly productive work of the organisation can much better manage their own affairs in their own time, and that this arrangement now becomes ever more effective due to advances in the communication technologies and the understanding of the right employment of these technologies. We argue that the introduction of self-managing environments, in which power structures came into proper coincidence with knowledge structures as *endostructures*, whereby these structures emerge from the work processes and market conditions directly, rather than, as in far too many cases, from the whims and impositions of pretending 'managers', can provide far higher efficiencies and flexibilities than can most existing, *exostructured* arrangements. A few but really talented and well educated managers will always be necessary, but the incompetent and inauthentic managers in the numbers that we see today are totally unnecessary for the purposes of true, or *authentic* management. This is not to say that this mass of incompetent and inauthentic managers does not serve a social, or rather antisocial purpose, namely that of allowing persons appointed as leaders who in fact have no leadership qualities to stay in control of organisations. The purpose of incompetent and inauthentic management is to enable persons who have gained leadership positions without themselves being leaders, essentially by 'politicking', to maintain distance control over an organisation through a self-serving power structure, as will be described later in this section. At the present time, reductions in the numbers of persons employed as 'managers' of some eighty percent seem to be in order and indeed have been the norm in those situations where properly communication-technologically-supported and well-organised self-management teams have been introduced and maintained by a sound leadership.

Correspondingly, developments of this positive kind, and of the many other kinds, all necessitate something that is essentially different even to good management, which is *sound leadership*. Management is concerned with *order*, while leadership is concerned with *direction*. Developments in the information and communication technologies and the proper means to apply these technologies in society have now come so far that the persons working directly in industry can for a large part best order their working processes between themselves, so that far less 'hands-on' or, even worse, 'proactive' management is required, and indeed this kind of management for the sake of managing something becomes increasingly counterproductive.

The objective of this section is then to develop a capacity to recognise and analyse situations where leadership is required and to evolve the modes of leadership best suited to each situation. Clearly, it is then necessary to provide an education in leadership suited to a wide variety of kinds of leadership, and then within each kind to promote the innate capacity of each individual for independent thought and action in the exercise of that particular kind of leadership.

8.2 ON LEADERSHIP IN GENERAL

The quality of leadership is widely recognised even though it appears in different individuals, within different contexts and in different forms. This capacity to recognise one and the same essential quality within many different manifestations of the quality marks leadership as a *Form* (and so with an upper-case 'F') in the sense initiated by Plato in his *Theory of Forms*. Such a capacity is then something inherent in many of our fellow creatures as well as ourselves, being a capacity to map many different things that appear to us separately in our consciousness into just one thing which is common to all of these manifestations even as it is not itself any particular one of them. This general capacity to intuit one immutable and eternal Form behind many mutable and transitory forms might be supposed to be something that is genetically programmed into us. Indeed, as just one example, observing leadership so widely among our fellow creatures, we may suppose that the propensity to develop Forms is more archaic than *Homo sapiens* himself and thus more archaic than language and logical ratiocination. It is accordingly still experienced as much, or even more, at the level of the limbic system than it is experienced at the level of the cerebral system itself: it is experienced as much, or even more, at an emotional level than it is experienced at an intellectual level. There is then always something ineffable, and something of a mystery, about our every individual experience of leadership and this necessarily arises as we relate any one of these experiences back to the underlying Form.

Accordingly, from a more theological standpoint, leadership corresponds to *a natural form of bonding*, where we use the word 'natural' in the Pascalian sense of 'as given at the Creation', or as 'that which we cannot lose'. As such it corresponds to 'a law of nature' and so to a law that cannot be violated so long as it truly is what it claims to be.

Despite the fact that our experience of the Form originates at a level that is more basic than that of language and of logic, it is still possible to explain it and to make it intelligible, so that it is possible to define it, but then only so long as we accommodate the ineffable, or mysterious, element as an indefinable component. This was achieved already by the Scholastics of the Middle Ages, who applied the notion of *intention*, and the more precise concept of *intentionality*, for this purpose: in the present context, it is the intentions of the leader that drive a process of leadership by inducing a just intention within a social group. This, however, is always realised through a process that is associated with the presence of an ineffable, 'totally other', element that provides, so to say, the binding force of leadership that makes the induction possible. Thus, at a level of generality that itself necessarily verges on the incomprehensible:

> *The leader is one who imbues a social group with a common intention by linking this social group with an ineffable, totally other element, called leadership, as that which mobilises the Form which makes this process possible.*

This element called *leadership* thus transforms the intentions of the one, the leader, through establishing a relation between the leader and the group, into a something which structures the actions of the others, and so introduces, or even imposes, a *structure of intentionality*, or an *intentional structure*, upon the group. It is because of the ineffability of this 'totally other' that leadership cannot be taught, although it can be identified, encouraged, developed, refined and polished. Thus one cannot 'teach leadership', but one can provide a course in leadership studies: leadership, like mathematics in this respect, "must be caught rather than taught", and that is of the essence of this section in the present book.

To every Form there corresponds a specific system of ratiocination, which we call *a logic*, and in the case of leadership this is a *logic of obligation*, that is, it is a *deontic logic*. By this we are not of course saying anything like 'leadership per-se *is* an obligation', which would be incorrect even if it were true in some particular instances, but that:

> *The logic of* leadership is *the logic of* an obligation to strive for social justice.

By way of simple exemplification suited to the engineer, we may consider the difference between a structure or a mechanism, which in each case is a physical object, and the logic of a structure or the logic of a mechanism, which is a product of the mind that may have a representation, such as in a set of marks on a piece of paper. In the case of the logic of a structure this is a structural analysis that has a central formalism representable through successive differentiations with respect to the distances along the directions of the members of the structure as represented by an algebra of ordinary differential equations, while in the case of the logic of a mechanism it is a logic of geometrical conformity between the links of the mechanism, such as may be represented through a geometry of instantaneous centres.

The serious study of deontic logics goes back at least to the scholastics, but the systematic search for a satisfactory axiom system for deontic logic was initiated only in our own times by the Finnish logician and philosopher Georg Henrik von Wright in 1951, and since that time several systems have been elaborated, and through this effort the understanding of such logics has developed extensively. Thus the possibility of mapping patterns of leadership behaviour into graphical and algebraic representations and thence into computer codes has also been considerably advanced. We shall not enter into such formalisms at all here, only observing that, being deontic, the logic of leadership *is not in itself a logic of preference*, even though it may in its processes of leading appear to proceed through logics of preference – and even though von Wright also wrote a classic on this class of logics too. For example, leadership is almost always transitive, in that if *A* leads *B* and *B* leads *C*, then *A* effectively leads *C*, while preference is commonly not transitive: if *A* prefers *B* for a certain position or a certain task and *B* prefers *C* for a certain subordinate position or task, it does not follow that *A* has such a preference for *C* and indeed he may not accept the preference of *B*. (Since neither leadership nor preference are symmetric on the other hand, neither satisfies the axioms of an equivalence relation.)

In short: to the extent that the leader is obliged to lead at all, it is in the last analysis through an obligation issuing from his or her own inner self, and those who are led are obliged by their own inner selves to follow *through the power of leadership in itself.* It is this leadership and nothing else that provides the relation between the Self of the leader and the Selves of the led: there can be no experience of preference in the case of true leadership. When seen from without, we experience this power as *commitment towards social justice*.

Thus, to proceed at once to the sublime in exemplification, the historical Jesus of Nazareth, as the archetypal leader, did not take upon himself the obligation of leadership as the Messiah as any kind of preference, and his twelve true disciples did not take on their obligations as a preference for one way over others: there was throughout one and only one way, experienced, essentially gladly, as *an obligation to one's own self* which came to presence in the outer world of name and form as commitment. The thirteenth disciple was the only one subject to a logic of preference, and he chose the pieces of silver, with consequences that are well recorded. It is this property of obligation to one's own self that subsumes the essential mystery of leadership.

We can take this search for the origins of and essential nature of leadership to a greater depth again if we return to the theological, and even more strictly 'spiritual' level of understanding, which necessitates that we try to understand those who have used their faculties in promoting their understanding, which in turn necessitates that we learn how they earlier used certain faculties that are nowadays very little employed, if at all. It appears as "doing what comes naturally". We have just introduced the natural as 'That which was given at the Creation', which we may follow, albeit repeating ourselves, with Blaise Pascal's definition:

Il n'y a rien qu'on ne rende naturel.
Il n'y a naturel qu'on ne fasse perdre.

(There is nothing that we can render natural.
That which is natural is that which we cannot lose.)

However, we have now also to introduce what theologians and other students of the spiritual existence of humans call 'the supernatural'. This has of course nothing whatsoever to do with the common or vulgar use of this term in everyday speech, but with "that which proceeds beyond the Creation, as the intervention of the Divinity in the present world". This is obviously open to all manner of misinterpretations and misrepresentations and it requires a very special knowledge, and kind of knowledge, to 'read' such interventions and thus to interpret them correctly. So rare and so uncertain has this 'spiritual knowledge' now become that it is generally discounted in our so-called 'Western' societies – perhaps only a Teilhard de Chardin could still pass muster in our own times – but it is certain that it did exist even in Europe among some of those whom we still read, so that they become 'the immortals', and that it still persists quite widely even today in societies outside of those of the European peoples. Accordingly, we can best pass outside our present predominantly Christian context if we are to follow this thread of understanding. Thus, from the *Īśā Upaniṣad*, in what still remains its most popular, Purohit and Yeats, English translation:

> Then hope for a hundred years doing your duty. No other way can prevent deeds from clinging, proud as you are of your human life … .
>
> The Self is everywhere, without a body, without a shape, whole, pure, wise, all knowing, far shining, self-depending, transcending; *in the eternal procession assigning to every period its proper duty*

At this level then, leadership is *the recognition* in the first place, and thence *the natural acceptance*, in the second place, of the obligation of leadership, as a being-true-to-one's-own-Self in one's own situation in one's own time, and the mystery of leadership resides in the leader recognising him-self or her-self in the places and times – and the places and times recognising themselves, so to say, in the leader.

At this level of understanding then, the mystery of leadership is that which assigns to the leader his or her proper duty at the appropriate place and time as the natural expression of his most inner self, as that which we have written as 'Self'. It is, again so to say, 'a mystery of integration', and thus belongs to the Form for which the paradigm case is the entire biosphere in all its natural state. Evidently, only someone possessed of supernatural knowledge could ever attain to any kind of understanding of such a mystery.

8.3 THE VARIETY IN THE FORMS OF LEADERSHIP

One Form commonly provides the one archetype of many different (lower case) forms, but these forms can in turn be divided into classes in terms of certain similarities that are exhibited within each such class. It is usual to choose one, or at most a few, exemplars from among the class as a way of characterising and even defining the class as a whole. We call these exemplars *paradigm cases* and the class that corresponds to any paradigm case in this way is called a *paradigm*. Since leadership has, or is, a Form, it provides several different paradigms of leadership which in turn characterise the different classes of leadership. We may list some of these in a roughly historical order, as follows:

- The paradigm of the leader who intends (and in this case is intended) to change the behaviour of a social group which is currently governed almost exclusively by ethical norms without true religious faith by himself exemplifying the power of faith over conventional ethical behaviour. The paradigm case here has to be the Judaic, Biblical and Koranic archetype of Abraham and his 'teleological suspension of the ethical' by his willingness to sacrifice his only and dearly-loved son Isaac at the Word of the Totally Other which was his God.

  This Word was not then accepted as one preference over others, but as an obligation to the own Self of Abraham. Concerning the mystery of the connection of this exemplar to its archetypal Form, and indeed to their integration, as pointed out earlier, Kierkegaard exclaimed: "Abraham I cannot understand; in a way all that I can learn from him is to be amazed".
- The leader of a people adapted to subjugation in a foreign land who inculcates in this people the intention to revolt against their subjection and leave for a 'Promised Land', and who then leads them to this land. The exemplar is of course the Biblical Moses. The position of Moses as a link between the forms and the Form in this case was expressed by Moses himself when he said that he had attained to *emunach*, which was to say that, following the great Jewish religious existential

philosopher and vehement defender of the human rights of Arabs in the State of Israel, Martin Buber, "He was the reliable and consistent messenger for the expression of the commandments and the confirmation of their power. *Emunach* means in this context: "reliable signal". To the extent that the people of the Biblical Israel experienced this as an obligation, they were assured of leadership, and to the extent that this sense of obligation wavered and preferences appeared within this people this intimation of leadership within them wavered in its turn.

- The messianic leader who imbues his disciples and through them whole peoples with the faith to proceed beyond the accepted interpretation of a prevailing religion and to embrace its total reinterpretation: the figure of the historical Jesus of Nazareth is again the obvious Christian exemplar, but every religion has its own exemplar again of this paradigm. The mystery surrounding the activation of the Form is here most evident.
- The paradigm of a leader who changes the intentions of a wide class of persons in pursuing a particular occupation, as exemplified by the change in attitude towards reading that occurred in the Europe of the XI and XII centuries, from a *monastic reading*, in which the text was sounded either externally by the voice or within the mind internally, to a *scholastic reading*, in which the text was intuited directly without passing through its bodily incorporations. An exemplar in this case would be Hugh of St Victor, who clearly delineated the mystery that intervened in this process and related the act of reading as synonymous with the act of eating. Just to quote from Cheetham (2003. p. 130):

> Reading is experienced by Hugh as a bodily motor activity. In a tradition of one and a half millennia, the sounding pages are echoed by the by the resonance of the moving lips and tongue … the sequence of letters translates directly into body movements and nervous impulses. By reading the page is literally embodied, incorporated.

For example, the definitions of 'reality' and 'truth', as introduced earlier here, have not only to be defined again in a single sentence, but have also again to be defined vocally in a single breath. Otherwise, although we can still more or less understand how we come to use a language, but we can no longer understand how a language uses us.

- Following closely upon this, in the XIII century, the paradigm of the leader who changed the intentions of persons seeking to understand the world of nature, as exemplified by the individual scholastics who introduced a monotheistic cosmology in place of the polytheistic cosmology that had been bequeathed to European thought by the ancient Greeks, as already introduced earlier. Henceforth the intention was to find laws that were true of all processes in all places at all times, as revealed by a one and only God, rather than describing the effects of a variety of laws imposed by a variety of gods and goddesses, as again already introduced earlier here. Only such universal laws could be 'natural'. In this case the intrusion of the 'Totally Other' was assured by the predominant influence of the theologians in this process that led to the formation of modern science through the condemnation of the Greek polytheistic cosmology and, way beyond all this, its Islamic and Judaic ramifications, in March 1277, as introduced and discussed earlier here.
- The form of the leader who imbues a group to share his intention to describe a particular world of thought in another way: Galileo's development at the beginning of the XVII century of algebraic manipulations in place of geometric constructions in the scientific description and prediction of physical systems provides a paradigm case. Descartes' almost simultaneous publications of his *Discourses*, his *Meditations* and his *Principles* indicated his awareness of the mystery prevailing when reaching back to the Forms behind the forms of modern science, so that he also could be

described as a leader in researching the ways of expression of scientific truth even though he is seen more as an adept of Modern Science.

- The paradigm of the leader in promoting qualities of leadership in a student body. Franz (von) Brentano shines as an exemplar as he resurrected and refurbished scholastic theories of intentionality within the (Aristotelian) Theory of Forms and introduced other innovations, to develop such students as:
  - *Tomas Masaryk*, who designed, implemented and presided over the world's first multinational democratic republic, in the form of Czechoslovakia;
  - *Sigmund Freud*, who established and developed psychology as an empirical science;
  - *Edmund Husserl*, who researched, established and led the philosophical movement of Phenomenology, as one of the two principle branches of philosophy world-wide in the XX century;
  - *Alexius (von) Meinong*, who researched, developed and taught the world's first, and still only, unified object-value theory.
  - And others – great leaders, even if less famous – in several other fields.
- The paradigms of leadership in sociotechnology, with such exemplars as Edison and Bell in the USA. In such cases we have to do with situations in which one and the same Form, of leadership, is in play in two or more environments at one and the same time, providing enabling archetypes for two or more knowledges that have to operate simultaneously and interactively. Thus, from one of the classics on sociotechnology (Law 1991, p. 9):

  > Edison was a 'heterogeneous engineer'. He worked not [only] on physical materials, but [also] on and through people, texts, devices, city councils, architects, economics and the rest. Each of these materials had to be moulded to his design if the system as a whole was to work. And, as a consequence, he travelled between these different domains, weaving an emergent web which constituted and reconstituted the bits and pieces that it brought together.

- A leadership in artistic production that predicts, and even prophesises, the ethos of a future epoch, leading us to the deepest of insights into the future consequences of our continuing human, and in this case all too dehumanised, modes of existence, for which the painter Francis Bacon provided the paradigm case. His cold and pitiless depictions of the extremities of human agony, of the mutilation of the soul expressed through the mortification of the body, of the depths of an utter and complete despair, always of isolated individuals tortured without the presence of any torturer, always enclosed within walls and in some connotations also within cages: all this together burns a searing shaft of awareness through our collective consciousness and into our individual souls. And way beyond and above even all this, albeit proceeding through it, there is the sensation of the terrible generality of its prophetic import. The writer and painter John Berger (*Le Monde Diplomatique*, June 2004) saw the world without pity that Bacon evoked and tried to exhercise it as a prophetic revelation, whereby the personal drama of the artist could reflect within half a century the crisis of a complete civilisation. "And how? Mysteriously".

> This is then *a leadership in understanding*, which in this case is *a comprehending of a future condition of humanity*. The less that compassion is expressed in Bacon's work, the more the consequences of its absence are impressed upon us and the more we come to comprehend these consequences. Sir Isaac Newton, one of the most outstanding alchemists of his age, predicted the end of human existence in 2060, and when we view Bacon's works we comprehend how this now becomes possible, and we may even regard this time estimate as optimistic!

It will then be clear already that there are many forms (with a lower-case 'f') of leadership.

8.4 LEADERSHIP, MANAGEMENT AND ORGANISATIONS

We have observed that leadership is about providing direction and management is about providing order. 'Direction' and 'order' are clearly two very different things and should never be confused. Nature provides plentiful examples of both, in some of which leadership predominates strongly over management, as occurs in some packs of predators when hunting, and others where management predominates over leadership, as is the case most of the time in ant nests. However, leadership and management differ at another level again, in that one cannot have 'self leadership' – which would be a *contradictio in adjecto* – but one can have self management, and most cases of management in nature involve self-management processes which are processes of cooperative action. As has been observed in ant nests, for example, most cooperative activity arises from individuals carrying out simple, repetitive tasks that are genetically programmed, and what we call 'natural selection' has ensured that these combine spontaneously to provide ordered relations and structures of activity. Such relations and structures are then said to be *emergent*. Nature has singularly few managers in the way that human society has, and certainly not in the way that our so-called 'first world' societies appear to be developing, into 'societies composed mainly of pseudo-managers'.

Relations and structures are among the most basic Forms of order. In everyday speech, we regard relations in natural and human societies as for the most part pregiven, inherent and even spontaneous arrangements that are as a consequence usually only hierarchised by custom and usage. The paradigm case is the relation of hereditary, of family, as we normally experience this in our so-called 'Western' societies today. In this particular case it is clear that, in more technical terms, the relations are reflexive, symmetric and transitive: *A* is related to *A*; if *A* is related to *B* then *B* is related to *A*; and if *A* is related to *B* and *B* is related to *C* then *A* is related to *C*. This then constitutes what we have mentioned earlier here as an equivalence relation. On the other hand, other forms of relation do not constitute equivalence relations: for example in the case of ancestry, *A* cannot be an ancestor of *A*, if *A* is an ancestor of *B* then *B* cannot be an ancestor of *A*, but still if *A* is an ancestor of *B* and *B* an ancestor of *C* then *A* is an ancestor of *C*. Social structures, on the other hand, are commonly regarded as more codified, more socially organised and more directed arrangements which are often strongly hierarchised. Thus, for example, in a company hierarchy, it is common that *A* reports to *B* and *B* reports to *C*, but *A* does not report to *C*, *C* does not report to *A* or *B* and *B* does not report to *A*: to the extent that we can apply these more technical terms at all, structures are normally irreflexive, asymmetric and intransitive.

Among the most important of the forms of these two Forms from the present point of view are those of power relations and power structures on the one hand, and knowledge relations and knowledge structures on the other hand. In many situations power and knowledge are so intertwined that it is usual to speak, following Foucault, only of knowledge/power relations and knowledge/power structures. The forms of power are however often still further divided into a power derived from an established authority (where we might perhaps better speak of 'power/knowledge') and a power derived from influence based upon superior knowledge (where 'knowledge/power' appears to be more appropriate). Most, and indeed nearly all, social organisations possess all of these forms to a greater or lesser degree, functioning simultaneously. However, the predominance of a power derived from authority over a power derived from influence based upon superior knowledge is then associated with an organisational structure that can be described as an *exostructure*, while a power derived more from an influence based upon superior knowledge provides what can then be described as an *endostructure*. Exostructures are then defined by an authority, such as a board of directors and its executive officers, while endostructures are scarcely 'defined' at all, but emerge spontaneously from the work processes proceeding within the organisation. Then not only do almost all organisations have both structures functioning simultaneously, but also functioning interactively. If, however, the exostructure comes to dominate over the endostructure to such a degree as to inhibit the operations of this endostructure, *the organisation will demands the constant attention of a larger number of managers that will in turn feed the exostructure further*. If, on the other hand, the endostructure dominates over the exostructure, *the organisation will become more autonomously managed*, or *'self-managed', requiring an even smaller exostructure and ever fewer managers.*

Organisations that are predominantly exostructured are ordered by the *fiat* of their management, usually with relatively rigid organisational structures. The reactions to a changing environment of such organisations must be planned, transmitted, codified and organised by their management and, to the extent that these often involve changes in the management structure through reorganisations and restructurings, these changes must again be implemented by their managers. Organisations that are predominantly endostructured, on the other hand, are constantly reordering and reforming themselves so as to adapt to the changing environment of the organisation. They are constantly *evolving* new working relations that can better react to the exigencies of the organisation's environment. In predominantly exostructured organisations, working arrangements are *imposed* by management, while in more endostructured organisations working arrangements *emerge* from the interactions of the persons who are actually carrying out the work. Endostructures are essentially emergent structures, and it is in this sense that they are, in the traditional-social sense already introduced earlier here, *anarchistic*, and even, but in a very different and specifically modern-scientific sense, *chaotic*. We have only introduced predominately exostructured and predominately endostructured organisations here and we shall return to discuss the manner in which most organisations contain both later on.

Endostructured organisations necessitate the constant attention of good leadership, while exostructures, once established, can continue to exist for some time even in the absence of any real leadership at all. As just observed above, it is indeed very common that organisations fall into the hands of persons who, although occupying the positions and titles of leaders, have little or no real leadership capabilities, and organisations most commonly fail through this process. We have also introduced the view that the process itself usually occurs through what is colloquially called 'politicking', which may proceed in many ways, such as through ingratiation, flattery, obsequiousness and other such means on the one hand and through the withholding of vital information, the encouragement of incorrect decision-making on the part of 'competitors' and subsequent behind-the-back denigration on the other hand, all directed towards those deciding upon the succession. In order to avoid an exposure of the resulting ongoing process of failure, the resulting pseudo-leaders are obliged to establish exostructures even where there were previously none, and for this purpose they need to introduce persons who are entirely under their control within all decision-making parts of the organisation. For this purpose, competent managers would be a nuisance, since they would rail against the incompetence, indifference, dishonesty and generally destructive behaviour of such pseudo-leaders, while true leaders would certainly subvert whatever the pseudo-leaders decided in order to save what could still be saved of the productions and good name of the organisation – just so long as they had any power to do so. Thus the only possibility is to 'parachute-in' incompetent managers *above* the potential leaders and to displace or otherwise remove competent managers, most of whom will anyway usually leave of their own accord in such situations. Being so patently incompetent, the pseudo-managers are totally dependent on the pseudo-leaders for their jobs and will therefore do whatever is required of them unquestioningly.

We can now return to the point that what has changed so decisively over the last decade is the possibility to strengthen self-organising relations that constantly adapt themselves to changing environments and other conditions by using greatly improved computational and communication facilities. Without these new facilities, the possibilities inherent in endostructured organisations would not be realisable, since their 'anarchistic' and even in a certain restricted sense 'chaotic' behaviour would make their behaviour unpredictable and inherently unstable. The task of developing and installing the *sociotechnical arrangements* necessary to instantiate these facilities, however, remains for the most part as a challenge that has still to be faced. This task will then call in the very first place for a much improved leadership, being one that leads the organisation in ways to provide higher levels of *self-management* of productive, distributive and other functions, that is, a management *by* the persons who are actually carrying out these functions. In the limit – a limit scarcely to be attained in practice today however – such an organisation would come to employ no full-time 'professional' managers at all. Several writers on management practice have presaged this development, such as Warren Bennis (as with Bert Nanus 1985, and separately, 1989) while there have been some outstanding works decrying the excesses of pseudo-management, but all of these appear to look at leadership, and management also, 'from the outside in', while the purpose here is more directed to looking 'from the inside out'. Thus Bennis defines leadership as "the capacity to create a compelling vision and to translate this into action and sustain it", which is often the way that leadership presents itself *to us when we observe it*, while here we are concerned with the sources of leadership; its 'why' and its 'how', *as it comes to presence within the person of the leader himself* and, inseparably from this, *within the persons he leads*. As Husserl long since demonstrated, such an investigation cannot be treated psychologically, but necessitates the application of a phenomenology and, as became increasingly clear over the last century, this must proceed through a study of the nature of human existence itself, as an existentialism, whether of the philosophical or the theological kind.

The exploitation of advances in information and communication technologies, and of the developments in sociotechnology necessary to realise endostructured organisations, constitutes the greatest opportunity and the greatest leadership challenge of our present age.

8.5 THE CONCEPTS OF DIRECT AND INDIRECT LEADERSHIP AND THEIR REGULAR AND IRREGULAR FORMS

Direct leadership proceeds primarily through *direct person-to-person(s) communication of intentions* (through orders, persuasion, actions, etc.) while indirect leadership proceeds primarily through the leader *creating situations* whereby other persons are led in a direction desired by the leader. Direct leadership and indirect leadership can well be, and even in many cases should be, provided by one and the same person. Further to that we may now introduce the following principles of direct and indirect leadership and their further subdivisions:

(1) Direct leadership is most immediately associated with organisational structures that are defined and regulated from within the organisation itself. They are often hierarchical and have already been introduced as *exostructures.* We now further observe however that, functioning rather like exoskeletons in natural organisms, it is these that give form to the body of the organisation. Like exoskeletons in this analogy, once established they provide genetically-defined and thus totally-ordered biological forms. Exostructures then, like exoskeletons, can only be changed, or 'reengineered', by more or less painful destructive/creative processes of 'restructuring'. We sometimes describe them as 'brittle', being hard but easily fractured by shocks from their environment.

(2) The kind of leadership that functions most obviously in exostructured organisations, being one that is regulated by fiat from within these organisations and proceeds through direct communication, may now be introduced as a *regular direct leadership.* As introduced already, it commonly proceeds on the basis of an authority conferred upon the leader by the organisation's hierarchy and it proceeds through the giving of orders whereby anyone disobeying these orders will

be disciplined through the actions of the hierarchy. As constantly recalled here, however, alongside this kind of leadership there is quite commonly another kind of leadership again which is at work within even the outwardly most exostructured of organisations, this being one that is established more by influence than by authority, which we may now introduce as an *irregular direct leadership* which, if suppressed, may become an *irregular indirect leadership.* This is, again, associated with a structure of influence within the organisation that is based more upon knowledge than upon administrative power, so that it is usually less hierarchical, less well defined and less ordered. This is then, of course, also an endostructure.

(3) The greatest exostructures of all time were those of the former Soviet Union and its fellow 'command economies' in Central and Eastern Europe. These were so tightly organised in most places and among one another that their endostructures came to work primarily against the economical interests espoused by these same exostructures. Some parts of the regular direct 'leadership' of these exostructured organisations often at one and the same time came together in endostructures to form 'cliques' and worked to subvert the declared objectives of the exostructure, thereby becoming simultaneously something more in the nature of irregular indirect leaders. Under the stress of exploiting rapidly developing technologies and the resulting need for rapid sociotechnical transformation, these exostructures could not restructure with sufficient rapidity and consequently fractured through whole economies and across these economies, leading to what is called in complex-system theory 'a common-mode failure'. This was a failure of regular direct leadership on the largest scale ever, even as some of the individual leaders had also become irregular indirect leaders at the time of failure, and it was a failure that was inherent in the system itself. From a point of view that we shall develop later here, this failure can be associated with an active suppression of entrepreneurial initiatives, whereby when the common-mode failure did finally occur, the entrepreneurial forces that were latent within the societies, and had self-organised themselves into endostructures that had simultaneously penetrated into the exostructures, were capable of taking over almost all the industries of the societies concerned as their own private property. Their leaders then transformed back from irregular indirect leaders to regular direct leaders again.

(4) Then, to return to the biological analogy, very much as organisms with endoskeletons are more flexible and nimble, both in their everyday existence and in their adaptation to changing environmental conditions, endostructures are generally more adaptable, and thus more resilient, to their changing sociotechnical environments. Endostructures are forever restructuring themselves. In the same vein, much as organisms with endoskeletons depend upon more complex and highly interactive nervous systems for their survival, so endostructures depend for their sustainability upon highly developed internal and external communications systems. Their basic mode of communication is then that of *broadband*, that is, that which provides *multiply connected* communication. No one necessarily 'reports to' anyone in this kind of structure: there is a natural interdependency between all the parts. Leadership here is regular and indirect, and this naturally places a greatly increased emphasis on human understanding, on empathy and compassion, on solicitude and on other values that are of a spiritual nature. We repeat: endostructures are self-organising structures that emerge spontaneously within almost every organisation but which demand very different qualities of leadership than do exostructured organisations.

As a first example of a specific irregular indirect leader in the field of military sociotechnology, we may take the case of (the later Sir) Basil Liddell Hart, who became the principle British military historian and contemporary critic of military affairs over several decades. Although not actually ‘in’ the organisation – as one British general acerbically remarked, “Captain Hart does not occupy a room in the War Office” – he led most thinking in military matters, even though his books and articles had the greatest influence in Germany despite his every effort to influence military opinion in Britain. He saw the effect that a combination of technical advances in armoured vehicles, aviation, radio and other fields would have on the prosecution of war and proposed new organisational forms, tactics and strategies correspondingly. As adopted in Germany, his ideas led to the rapid disintegration of the enemies of Germany in 1939 and 1940 in continental Europe, leaving Britain alone and isolated after the collapse of her last ally, Greece, in May, 1941 – until the ‘Great Coalition’ led by Germany attacked the then Soviet Union on 22 June 1941 and the outcome of the war was effectively decided elsewhere. In fact, a major part of British preparations for war in the 1930s proceeded through irregular indirect leadership, such as in the development of much more advanced aircraft, new kinds of aircraft engines, new kinds of propellers, entirely new radar equipment and other advances, and without these Britain would have almost certainly succumbed.

Laurence of Arabia was another and more colourful example, being distinctly irregular in the eyes of most of his British military contemporaries and having scarcely any leadership role at all. For the Arabs, on the other hand, his leadership was regular and direct. Both Hart and Laurence were capable of great compassion and human understanding – and spoke their minds correspondingly – so that they could never become regular in the British army.

We can now make a clearer distinction between *regular and irregular indirect leadership.* The first is that which originates within the exostructure of the organisation, even though it generates an endostructure outside the organisation. The form of regular indirect leadership also has the property that *it can lead the competition into ruinous courses of action*, and then not at all by deceit, but by exploiting weaknesses in the competitor’s own leadership.

We are left with irregular indirect leadership: the stuff of spontaneous mass movements, of mass protests, of revolts and revolutions – and the very apotheosis of endostructure. Leadership is always an emergent phenomenon – it is in some sense genetic but only comes to presence in certain life situations – but in the case of irregular indirect leadership this coming to presence is often spontaneous, and even quite unexpected by those who become leaders. Although unexpected, it is always accepted. Abraham provides the archetype of the irregular indirect leader in his own time and place, just as Moses provided the archetype of the regular direct leader. The true disciples of Jesus who subsequently became apostles and thus the prototypes of the regular direct leaders of the Christian Church, provide a range of twelve very different examples of irregular indirect leaders when seen within the context of the time of their ‘apprenticeship’ as disciples. As introduced above, irregular indirect leadership in such an exalted sphere is never sought, but it is always accepted, essentially gladly, as an obligation and as an article of faith indivisibly. This is something that most of us cannot hope to comprehend, and neither to understand, but only to accept. The actions of individuals fighting voluntarily for democracy in Libya has provided further examples of this kind of what the mundane and average world can only comprehend as ‘sacrifice’.

Irregular indirect leadership is observed at the Kierkegaardian ‘level of the aesthetic’ also: many poets, painters, musicians, novelists and others may serve as examples. The example of the painter Francis Bacon has been presented as a paradigm case earlier here. More generally speaking, the ‘leading of taste’ commonly proceeds from persons who are in some sense ‘outsiders’ to the artistic and general cultural establishment of their times and ‘only’ influence their environment through their works, and even then quite commonly posthumously. Schopenhauer, Kierkegaard and Nietzsche provide specific paradigms of irregular indirect leadership in philosophy and theology that developed its endostructure, of phenomenologist and existentialist philosophical and theological movements, long after their deaths.

Industrial development also has had its share of 'unsung heroes' who were irregular within their own time and only had their say much later. Some subsequently became regular, such as Whittle in the case of the jet engine and Watson Watt in the case of radar, but most remain forgotten outside of a few small circles of industrial historians.

8.6 ANARCHISTIC SYSTEMS AND CHAOTIC SYSTEMS

We have seen earlier that the very word 'anarchy' has had all manner of pejorative connotations from the seventeenth century onwards, even as it has long had deeper meanings and influences, but any treatment of leadership today is obliged to treat anarchistic systems seriously and dispassionately in their own right. This is because these systems have become particularly relevant over the last two decades as a result of the current and still rapidly ongoing developments in information and communication technologies and, even more significantly, their sociotechnologies. It is these developments that are again making anarchistic arrangements, as sociotechnical constructs, viable and increasingly attractive, and not least in industry. The so-called 'open source' movement in software production, such as has produced the Linux Kernel (that in turn still drives the most powerful challenge to the Microsoft quasi-monopoly of operating systems) provides a paradigm case. Clearly, such anarchistic systems pose their own opportunities as well as their own problems for future social justice leaders.

As we have seen earlier, 'anarchy', as derived from a Greek word signifying 'without a rule', was first applied within and to Stoic philosophy at the time of the ancient Greeks. From its origins it then also had the connotation of 'without a ruler', so that the essence of anarchism has always been that it is *antiauthoritarian.* Now although authority is usually regarded as something inseparable from leadership in the context of military affairs, it is remarkable that in his highly influential *On the Psychology of Military Incompetence*, Dixon (1976) has shown, following well established precedents in psychology generally, that it is precisely the character trait of *the authoritarian personality* that has been mostly associated with *incompetent* leadership. Possibly more surprisingly still to those uninitiated in such psychological studies, Dixon showed that authoritarianism was at variance not only with humanitarian, but also with disciplinarian traits. Thus (*ibid*. p. 275):

> Certainly such great leaders as Wolfe, Wellington, Shaka, Lawrence, Monash and Montgomery not only displayed a general absence of authoritarian traits but also showed a lively regard for the prime responsibility of a commander: conservation of his forces and the psychological and physical welfare of his troops. As Trevelyan wrote of Wellington, 'It is fortunate for Britain that Wellington was at once a great humanitarian and a great disciplinarian'.

In order to view anarchism more clearly within this context, we may follow the excellent *Encyclopaedia Britannica* entry on this subject in identifying its basic tenets as being those of *mutualism* between participants (i.e., the recognition of mutual interests and the provision of mutual support), of *federalism* (i.e., the formation of a federal structure between the participants, such as may be provided by a syndicate) and of *direct action* (i.e., the direct involvement of this federation/syndicate in processes occurring in society and the world of nature). As introduced earlier, civil rights movements, human rights movements, environmental movements, peace movements and most non-governmental organisations then fall under this category. The traditional university, which endured in Europe from the twelfth into the second half of the twentieth century and still survives in a few places even today, provides an example of an anarchistic system in this broader sense.

One of the most determining features of any anarchistic organisation, as a self-organising and self-managing system, is that *it is held together by a shared ethic*. Now, among most so-called 'higher' animals – those with endoskeletons and highly developed nervous systems – codes of social behaviour which approximate ethics are inculcated for a large part by the playing of games, and of course this applies to human societies as well. Current research into means to inculcate ethical norms into sociotechnical systems accordingly attaches great importance to inducing habits of secondary reflection and other promoters of intersubjectivity by introducing Internet-based games (e.g., Jonoski 2003). Now the essential point about such present-day anarchistic movements as that of open source software development, as exemplified by the GNU/Linux system development, is that *such a development is itself a game, with all of its participants functioning as players*. Thus, *the ethics necessary to support such a mode of development are self-generating and self-sustaining* (see Jonoski 2002).

Another, but already well established feature of anarchistic systems, and particularly of anarchosyndicalist systems, is that they are abhorrent of bureaucracy. The most cited example is that of the Spanish *Confederacion Nacional del Trabajo*, which had some 2,000,000 members in 1936 while employing only one paid secretary!

If they are to be at all sustainable, anarchistic systems necessitate what the poet and critic Herbert Read (1968) described as "a process of individuation, accomplished by general education and personal discipline". However, whatever their general education and personal discipline, such systems have been regularly destroyed by authoritarian regimes and their ideologies: established bolshevism was at one with fascism and Nazism in its uncompromising and thoroughly brutal suppression of anarchistic movements. An analysis of the success of these suppressions shows however that they only succeeded because the authoritarian regimes had definite socio*technical* advantages over the anarchistic organisations. What is changing in our own time is that the technical means to maintain anarchistic organisations is strengthening by the day through the development of communication technologies, so that, to the extent that the sociotechnology can keep pace with this development, this suppression must meet with an ever stronger and more tenacious resistance – and this resistance to repression must increasingly prevail. Again we return to the revolutions in North Africa as current paradigm cases at this time of writing.

Writers on the subject of anarchism and anarchistic movements have strongly emphasised that anarchism is the antithesis of chaos, and in this they draw upon the wellspring of anarchism, which is the understanding that creativity is always driven by that covenant of the spirit that we call *love*. This theme runs through all the great writers on anarchism: Winstanley, Godwin, Proudhon, Bakunin, Kropotkin, Tolstoy, Ghandi, and so many others (see Section 1.5). All religions are at one in declaring that it is only with love that anything of any lasting value can be created: "Unless the Lord buildeth the house, its builders work in vain". Thus, what we call 'anarchism' is an essential component in any creative endeavour, particularly within the context of social justice and it is surely for that reason that works on anarchism have appealed so strongly to artists of all kinds. To return to the *Encyclopaedia Britannica* article:

> [The principles of anarchism] lingered on in the traditions of English Protestant sects and reached their ultimate flowering in the masterpiece of … William Godwin, who in 1793 published his *Social Justice* …
>
> Goodwin enjoyed great celebrity in the 1790s and influenced such writers as Percy Bysshe Shelley (whose Queen Mab and Prometheus Unbound are virtually anarchistic poems). William Wordsworth, William Hazlitt, and Robert Owen … .
>
> During the 1890s, especially in France, Anarchism was adopted as a philosophy by avant-garde painters and writers. Gustave Courbet had already been a disciple of Proudhon; among those who in the 1890s accepted an Anarchistic philosophy were Camille Pissarro, Georges Seurat, Paul Signac, Paul Adam, Octave Mirabeau, Laurent Tailhade, and, at least as a strong sympathiser, Stéphane Mallarmé. At the same time in England, Oscar Wilde declared himself an Anarchist and, under Kropotkin's inspiration, wrote a libertarian essay: 'The Soul of Man under Socialism' in 1891.

Thus that most misused and abused of all words, *love*, was recognised in anarchism as the source of all creation and thereby the antithesis of all chaos, but this was essentially also a rediscovery of a profound truth of the kind that is always of the essence of the ancient Greek myths, which in this case tells us that *Eros*, the god of love and fertility, was the son of the god *Chaos*, as his compensating progeny. As explained earlier, such forms of language as the myths are archetypal and recur incessantly.

We cannot enter further into such depths here, although we can recall for example, that in a biology founded upon information theory and thermodynamics, natural creation itself is regarded as the formation of 'entropy-bending structures' providing ever longer 'detours to death', where entropy is itself employed as a universal measure of disorder, including that associated with chaotic phenomena. Creation, as the struggle between *Eros* and *Thanatos* in the Greek myths, then becomes resurrected in and through the language of quantum chemistry (e.g., Monod 1970).

It must then appear that although anarchistic organisations appear as antithetical to chaos, they must also have a close, and indeed the closest of relations to chaos, as the overcoming of chaos, and certain more recent developments within chaos theory also support this view, suggesting that this theory is able to provide the means to study how organisational structures may be expected to emerge and dissolve in self-managing, anarchistic systems within certain sociotechnical environments and at what rates these processes of coalescence and dissolution may occur, or may already be occurring. Chaos itself is then understood in the modern-scientific sense as covering situations in which, although we cannot ourselves discern any order in these situations directly, we may process measurements to arrive at meaningful descriptions of underlying orders that have emerged through interactions occurring within the processes. Thus the possibility arises to identify organisational structures as they are emerging and transforming and passing away again within anarchistic systems. This possibility opens other perspectives again for analysing, designing, constructing and leading self-managed working environments which can lead to the states of social justice. Of course, the techniques of chaos theory so far provide no more than an instrumentalism – they contain no knowledge-discovery environments – but even with this restricted functionality, chaos theory still has its own utility (e.g., Ababarnel 1996).

8.7 AN EXISTENTIAL-PHILOSOPHICAL VIEW OF THE PREHISTORIC ORIGINS OF LEADERSHIP

From an existentialist-philosophical point of view, leadership is a calling, and this is a calling from within and a calling from without, so that it is only when there is a conjunction between these callings, including their synchronisation, that true leadership becomes possible.

We have earlier here identified the origins of leadership in the observed existing behaviour of animals, and especially in the behaviour of our fellow primates, but it then follows from this that leadership must predate the very existence of *Homo sapiens*; it is older than human Being so that, in respect to humanity, it is in its essence *pre-ontological*. These 'primitive' forms of leadership already serve to link categories of leadership with types of social structures, including hierarchies. Salient features are those of divisions between tactical and strategic leadership of both territorial and migrating populations, between leadership by virtue of strength and agility and leadership by virtue of experience, knowledge and wisdom, and between leadership in tactics of phenotypal survival and strategies of genotypal survival, including sacrificial strategies. Such characteristics of leadership as its circumspective concern (the Heideggerian *umsichtig Besorgen*), its empathy (*Einfühlung*) and its solicitude (*Führsorge*) also appear clearly here. In this sense we might even speak of a *palaeontology of leadership*.

There is then further to this an *archaeology of leadership* which, among other findings, emphasises how authoritarian leadership has always been associated with, and indeed legitimised by 'superhuman' accomplishments, often associated with magical practices that were increasingly subsumed under religious practices which imitated the magical in order to transcend it (e.g., Moses' staff-cum-snake that ate up all the staff-cum-snakes of Pharaoh's magicians).

The *myths* of ancient times were not intended to project the *realities* of leadership, but their *idealities*: we repeat, "They never were, but always are". These myths probed the depths of the collective unconscious of the times of the ancient Greeks in their search for essences, including the essences of leadership in a world whose religious forces were faltering, and which was therefore seeking new, but still necessarily divine, inspiration. Prometheus, as he stole fire from the gods for the benefit of man, provided the tragic-heroic ideal of such a leader, providing the ancient Greek prototype of what has already been introduced as irregular indirect leadership. This is also, as already adumbrated through Shelley's great poem, the prototype of the anarchic leader. Since the collective unconscious of the times of the ancient Greeks was not so very different to that of our own times in Europe, these myths maintain their relevance to this day: "They always are".

From our current standpoint of leadership, Homer stands supreme in his delineations of leadership in the form of *legends* and *sagas*, and especially in *The Trojan Wars*, and then, within that work again, in *The Odyssey*. The leaders in Homer, and especially Ulysses, become steadily more real and less ideal: in the legends and sagas the story takes increasingly the form of the *epic*, as the celebration of the adventures of some heroic personage of contemporary history or tradition. The increasing relevance and refinement of astrology later in this period provided new insights into the forces guiding the 'destiny' of the heroic leader, whereby this leader became increasingly a *tragic-heroic leader*. This model of leadership has never been entirely supplanted by the modern scientific point of view, as demonstrated by the histories of such *inauthentic leaders* as Napoleon and Hitler, where the nature of this inauthenticity will be introduced later here.

Histories of leaders are legion, covering all possible fields of human endeavour and all manner of forms of leadership. Although, unlike the myths, legends and sagas, histories ostensibly proceed exclusively in chronological time, histories of leadership still necessarily contain the residues of myth, legend and saga written, so to say, 'between the lines' of the historical text, and so out of chronological time: they come, so to say, to mimic the legends and the sagas in this respect. Some histories, regarded as 'Grand Narratives', will be used in the course of this book in order to illustrate the limited historicity of leadership alongside these historiographical aspects.

8.8 THE TAXONOMIES OF LEADERSHIP AND THEIR ASSOCIATION WITH THE TAXONOMIES OF POWER STRUCTURES AND KNOWLEDGE RELATIONS

Let us start here with some recapitulation. We have earlier distinguished between *direct* and *indirect* leadership. We saw that direct leadership proceeds primarily through *direct person-to-person(s) communication* (orders, persuasions, actions, etc.) while indirect leadership proceeds primarily through the leader *creating situations* whereby other persons, usually outside the organisation itself, are led in a leader-desired direction.

Just to recapitulate further, we saw that direct leadership is most immediately associated with organisational structures that are defined and regulated from within the company itself and we called such an *organisationally empowered* leadership *regular direct leadership*. We saw also how indirect leadership can originate from within the organisational structure, as just indicated, and in this case we speak of *regular indirect leadership*. Even in the most strictly organised and hierarchical of organisations, however, we observed how knowledge relations develop between groups of individuals within the organisation such that some individuals come to exercise influence within these groups: they thus come to exercise *regular indirect leadership* if they are tolerated by the hierarchy and *irregular indirect leadership* otherwise within the organisation as a whole. In other cases again, however, we exemplified how a leader who does not originate from within the organisation itself, but from outside it, may have a major and even decisive *influence* on an organisation – such as by writing about it and so 'creating a situation' – while being himself no part of that organisation, thereby providing *irregular indirect leadership.* As we saw, however, this form of leadership may well transform into a *regular direct leadership* as this leader's theories and prophetic utterances transform into realities.

As also already introduced, specific categories of leadership are associated with specific kinds of organisational structures. We have seen how regular direct leadership is almost invariably associated with what we have called *exostructures.* By using an analogy with exoskeletons in natural organisms, we saw how these define the form of the organisation and its internal workings through establishing what are essentially *power structures* that commonly take the forms of hierarchies. Thus, exostructures, once established, provide institutionally defined and totally-ordered organisational arrangements. Exostructured organisations then, like creatures with exoskeletons, can only be changed to meet new environmental conditions by more or less painful destructive/creative processes of 'restructuring'.

By way of contrast we have indicated how irregular leadership, and especially the irregular indirect form of leadership, are associated much more with what are called *endostructures*, these being less hierarchical, less well defined and less ordered. Very much as organisms with endoskeletons are more flexible and nimble, both in their everyday existence and in their adaptation to changing environmental conditions as we saw earlier, endostructures are more adaptable, and thus more resilient, to their changing sociotechnical environments. Endostructures are forever restructuring themselves in response to the changing demands placed upon them. Rather as creatures with endoskeletons usually depend upon highly developed nervous systems for their survival, so endostructures depend upon highly developed internal and external *communication systems* for their sustainability. Having emergent structures defined by their own interactions subject to their external environment, they are for the most part *self-management systems.* As we also intimated, even the most highly exostructured, hierarchical, organisations also have endostructures, so that an irregular, and usually indirect, leadership based upon knowledge/influence coexists with the 'official' regular direct leadership of the organisation.

On this basis, we can proceed to introduce orderings that can be spatialised so as to introduce a *leadership space* as a place where we can trace the changing nature of forms of leadership as trajectories within the space, and in this way also we can construct a *taxonomy.* Proceeding through its legendary and Biblical-historical development, the space of leadership mapped within this taxonomy is *spanned* between Abraham, situated in the one corner in both legendary and Biblical-historical terms, as the archetypal irregular indirect leader, and Moses, situated in the other corner, and again in both legendary and Biblical-historical terms, as the archetypal regular direct leader.

The space defined by this simplest of taxonomies of leadership can be conveniently represented in the form of a table with two rows, corresponding to the regular and irregular forms of leadership, and two columns, corresponding to the direct and indirect forms. All the above features, and more, can be assigned places in a table of this kind and the locus of a particular leader can be traced in time by points associated with particular times and by arrows joining these points (Figure 8.1). The corresponding power structures of exostructures and knowledge relations of endostructures can be introduced similarly.

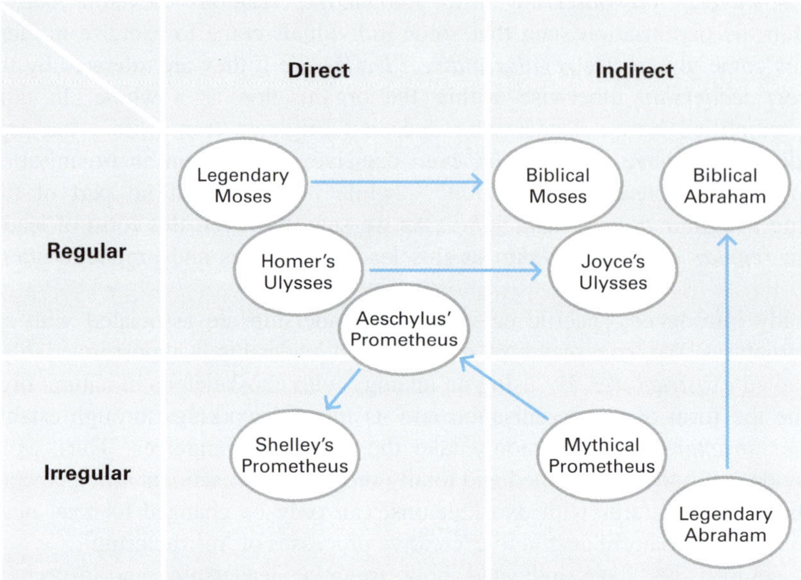

Figure 8.1 Different forms of leadership

Clearly other and much more extended taxonomies can be generated by introducing further divisions in categories of leadership. Basic differences between tactical and strategic leadership and oppositions between authentic and inauthentic leadership can be introduced into another or an extended leadership space. In particular, the role of *ideology* in the formation of *inauthentic leadership*, whereby *it is the ideology that really does the leading and the nominal 'leader' is merely a cipher possessed by this ideology*, may be introduced. The 'great dictators' of the first part of the twentieth century provide paradigm cases, but much more cogent today are the still ongoing subversions of *modern science* and of *management* into ideologies, and a particular attention needs to be given to these inauthentic forms. Once again, however, it is the exceptional business opportunities presented by such subversions for well-led organisations that need to be emphasised, discouraging any morbid interest in the pathologies of the processes of subversion themselves.

There remains one further subdivision however which we have not so far so much as mentioned, which is that between *male leadership* and *female leadership*. This division cuts, so to say, orthogonally across every other coordinate in the space of leadership. It is for this reason that we could so far use only the masculine form, ubiquitously, so that while referring to 'he' and 'his' explicitly, we were always simultaneously referring to 'she' and 'her' implicitly: there is little difference between men and women in the forms so far considered. However, although it may be impolitic to say so, we shall argue that there is a major and deeply-founded difference between the tasks of leadership best suited to men and those best suited to women, and that this difference is of an essentially genetic nature. This is the subject of the a paper (Abbott 2000) that demonstrated the different ways in which men and women regard two kinds of psychic operations, the one kind of operation being that which carries the mind from an object to its intrinsic or personal value and the other kind of operation being that which carries the mind from this same object to its extrinsic or social value. In our rather deviant form of category theory the mind-carrying objects were introduced as *functors*. Thus the first kind of operation/functor may carry the mind of a parent from a photograph of his or her new-born child to the strength of the emotion of love that this experience engenders as an intrinsic or personal value, while accepting that the second kind of operation/functor, that takes the mind from the same photograph to its extrinsic or social value – such as might be measured by what anyone else would pay for such a photograph – might well be accepted as being little or nothing. In a very different situation, the first kind of operation might take the experience of holding a finely machined piece of highly refined steel into a negligible feeling of intrinsic or personal value while accepting that the second kind of operation may transform the experience of this object into an intimation of a quite substantial extrinsic or social value. We all of us have such value measures of all objects. The value of using category theory was that it demonstrated that the functor that mapped the object into its intrinsic value and the functor that mapped this same object into its extrinsic value could be mapped, the one into the other, and the name of this mapping operator was a *natural transformation*.

An analysis of this mathematics of the qualities was provided and further investigated in a remarkable essay *Sur l'idée de concept mathématique 'naturel'//On the idea of the mathematical concept of 'natural'*, in which the French mathematician Gilles-Gaston Granger (1994, pp. 157, 158) explained this significance of the word "natural" that is used here, as follows:

> Mathematicians, who do not usually show much repugnance about using metaphors, have not made much use of the word 'natural'. In more recent times, in French, one could at most make a case for the expression 'fundamental transformation' in Category Theory. The word seems then to refer to a univocal determination of a correspondence that is transposed from one universe of mathematical objects to another; but in this sense it is generally the adjective 'canonical' that is utilised … It is however a more ancient, traditional usage of this adjective 'natural' that serves as the starting point for our commentary; [such as is used when] one speaks of 'natural numbers' …
>
> The first characteristic that appears to justify at least provisionally the introduction of the concepts of 'natural' is that these are immediately present before any mathematical elaboration, at one and the same time as tools and objects of thought at the centre of our spontaneous activity …
>
> And this that we call their 'presence' signifies the fact that they have to be shown and employed in an experience, which we shall try to show does not demand anything of the senses but must be recognised as an intellectual intuition [as an intellectual experience].

The remarkable thing for us is that these mathematical theories, or ways of thinking mathematically, need not necessarily be associated with conventional mathematical objects, but can be applied to objects, such as the special aptitudes of women and the conventions of words, that are not "mathematical objects" in any normative sense. Thus, the word "natural" refers, again following Pascal, to "that which was given at the creation" and so "that which we cannot lose" in the religious view of the world, and this is a precedence that it shares only with the natural numbers, which again cannot be defined, being intellectual intuitions or experiences "given at the creation", and so "natural" (see Manin 1977, p. 17). We are of course using these terms in the sense of foundation-mythical components of a religious discourse, and there can be no suggestion here of any kind of popular "creationism" or other such current debasements of religious experience: we may indeed here once more recall Jaspers' celebrated "*Nicht Vernichtung, sondern Wiederherstellung der mythischen Sprache ist der Sinn*"//"*Not the destruction, but the restoration of the mythical language is the objective*" (see Jaspers & Bultmann 1954, p. 22).

It can accordingly be observed that women experience the *balance* between the intrinsic values of objects and their extrinsic values in quite another way than do men, and this suits women much better than men as leaders in certain occupations and even in certain industries, even as men are more privileged, *qua* men, in other occupations and other industries again. Thus, as more publicised examples, women are often better suited to industrial activities involving highly refined social-aesthetic values, such as in several branches of architecture, clothing, furnishing and publishing, as well as to social-ethical modes of leadership, such as may be concerned with the provision of many public services. Simply by reading this summary it will be clear that the conception and the employment of this construction is greatly aided by some degree of formalisation, and can be realised by drawing upon a few elements of a unified object-value theory (as originated by the Alexius (von) Meinong that we have earlier introduced here as a student of Franz (von) Brentano) together with a few fragments of Category Theory, as one of the more popular of the current 'mathematics of non-mathematical objects'.

The most cogent aspect in our own case is that of leadership in active stakeholder participation, where in various fields of application, such as in legal aspects, women appear to be far more capable than men.

Chapter 9

On sociotechnology

9.1 LEADERSHIP AS A SOCIOTECHNICAL ACTIVITY

The theory and philosophy of leadership, as introduced earlier here, is essential to the understanding of the nature of the individual leader, but beyond this again the leader needs a theoretical apparatus that will facilitate and enhance his understanding of critical events and situations and promote the transformation of this understanding into well-formed judgements and decisions that can enable states of social justice. In many cases the leader will have to construct the relevant theories personally, or at best with the help of others, and for this purpose some 'theory of theorising', which is a pragmatic description of philosophy, will be required. The following discussion concerns the development of theory through philosophical understanding and its application within the context of *leadership for social justice*. It will be demonstrated that *this is a sociotechnical endeavour through and through*: that very few leadership problems nowadays are any more purely social or purely technical, but they are for much the greater part sociotechnical. Sociotechnology constitutes a particular species of what we have repeatedly described as *conjunctive* knowledges, being knowledges that join together two or more different and otherwise separate knowledges, thereby providing a new field of knowledge with qualitatively different properties than those of its constituent parts. The study and practice of sociotechnology is essential if one wishes to understand how technical innovations have social consequences and how the resulting social changes react back again on the technical developments, and so on and on, again and again, in what become sequences of 'positive feedback' cycles. It will be shown that many subjects which are rarely if ever regarded sociotechnically, take on another aspect when they are so considered, even as in the currently predominant modernistic ideology 'progress' continues to be mostly defined in a specific technocratic way that is anathema within the present context. The ethos of active stakeholder participation is to escape from the false embraces of this paradigm by first accepting that knowledge is not achieved by the senses alone. Thus, with Bortoft (2010, p. 68):

> There is always a nonsensory element in knowledge, and this must be so whether the element is verbal-intellectual or intuitive. The difference is that, whereas the verbal-intellectual mind withdraws from the sensory aspect of the phenomenon into abstraction and generality, the intuitive mind goes into and through the sensory surface of the of the phenomenon to perceive it in its own depth. It is by first going into the full richness and diversity of sensory detail that the intellectual mind is rendered ineffective, so that we escape from its prison into the freedom of intuition … .

It was said that there is no intellectual equivalent to this experience, and the reason for this is now clearly because it is an intuitive experience which depends on a change of consciousness.

It is this freedom that we seek through engaging the attentions, interests and passions of the active stakeholders as they engage individually within the correctly, truthfully and beautifully illustrated hydroinformatics environments that we provide, primarily over the Web and further through the means of mobile-communications technologies.

9.2 CHARACTERISATIONS OF SOCIOTECHNOLOGY

Sociotechnology is concerned with the study of processes in which social aspects cannot be separated in thought from technical aspects: the social and the technical are so related to one another in the process that the presence of the one makes no sense without the presence of the other. A common metaphor is that of a weaving together of two different kinds of thread, each of which is an essentially a one-dimensional object, so as to form a piece of cloth, which is an essentially two-dimensional object. This object formed by the weaving process is then a quite different kind of object to the threads from which it is materially constituted and it has quite other applications from these. In particular, one cannot very well cloth oneself with pieces of thread, but one can do so with pieces if cloth. A new value is then added through the weaving process. As introduced in the sociotechnical terms of a Meinongian unified object-value theory, the two composing objects are then said in such a case to be *implected* or *implexively contained*, the one in the other. Another metaphor is that of the beam of the sociotechnical supported on the columns of the social and the technical: the beam is a quite different kind of structure to the columns upon which it is supported and has to be analysed and designed quite differently accordingly. Another analogy again is with the bridge of sociotechnology joining together the one river bank of the social to the other river bank of the technical.

The sociotechnical knowledge that is involved is in all such cases of the kind that we have introduced earlier reappears as a conjunctive knowledge. Thus in the first metaphor the conjunctive knowledge that is used in the process of weaving together the two kinds of threads is a knowledge of weaving processes, in the second metaphor the conjunctive knowledge is a more extended knowledge of structural analysis, and in the third metaphor it is simply the bridge. Sociotechnologies are always composed of conjunctive knowledges and *sociotechnology is the study of processes in which the social and the technical are indivisibly combined.*

Since it is observed that in our present-day societies there are fewer and fewer purely social problems and even fewer purely technical problems; today most problems are of a sociotechnical nature that necessitate the deployment of conjunctive knowledges. With this observation goes the increasing realisation that many historical developments that have long been regarded either as essentially social or as essentially technical were in fact interacting and interdependent sociotechnical developments and have now to be studied as such if they are to be properly understood. We shall proceed through the institutions that are of the most vital importance to the decision making of the leader in our own times by tracing the principle developments that have occurred in these institutions historically from a sociotechnical point of view.

9.3 EARLY POLITICAL AND ORGANISED-RELIGIOUS DEVELOPMENTS SEEN AS SOCIOTECHNOLOGIES

The entire social development of Europe after the end of feudalism cannot be understood without taking account of the technological development – together with the increasingly modern-scientific basis for this development – that occurred simultaneously in Europe. At the same time, and as was also adumbrated earlier, this technical and increasingly supportive scientific development would, in its turn, have been nullified if the social developments that supported and nurtured it had been frustrated. The fact that these social developments were in fact supportive of the technical developments can be seen as a consequence of a particular deployment of political power, just as this particular deployment of political power was strongly supported by the technical developments. There then occurred in certain places a synergy between the exercise of political power and the development of industry through the application of an increasingly science-based technology, and it was this synergy that provided the motor that drove Europe to its 'pinnacle' of world domination – and then just as surely sent it down to self-destruction and penury by 1945.

As the influence of technological advances upon politics increased, so the influence of organised religion declined. Feudalism, as sanctioned by the Church, slowly became ever more incompatible with the advance of technology and this had in turn its effects on organised religion, whereby technological advance became increasingly associated, largely through the Reformation, with what Max Weber called the *Protestant Work Ethic*. In politics, the cliques formed around princes, dukes, counts, or marquises – who were kept to some extent in check by the princes of the church and their underlings who were responsible for actually administering the state – were joined by other cliques, composed of merchants, industrialists, businessmen and bankers. As the latter grew to challenge the dominance of the former through the strongly interactive development of commerce and industry, so these two kinds of cliques or 'power brokers' came to form what are commonly called *cadre* or *brokerage parties*. Thereby two kinds of cadre or brokerage parties arose, the one composed for the most part of landowners who drew their wealth from the labour of an illiterate peasantry who were in turn held in check by a traditionalist clergy, and the other composed of industrialists, merchants, tradesmen, bankers, financiers and professional people who increasingly needed more literate and enterprising workers to staff their ever more extensive and sophisticated offices, factories, ships, warehouses and all manner of other organisational and technical constructs. The rise of this second political caste was in fact associated in many Catholic European societies with the rise to prominence of Freemasonry, and especially in Belgium and in France, as an almost anticlerical movement – but this aspect cannot possibly be taken up here.

The leading of people is an exercise of power and within the consolidating nation states of much of Europe a contention arose in civil society between the cadre or brokerage parties over who was to articulate and exercise leadership and thus who was to articulate and exercise power. The process of consolidation into nation states was itself a sociotechnical process of course, since it depended not only upon military and financial means, but just as well upon the development of means to improve communications within the territories of the consolidating state faster than communications were improving in neighbouring states that had pretensions to this same territory. Since the industry-based cadre parties had an advantage not only in providing, but also in utilising these improvements in military might, finance and communications, they tended to gain the upper hand during these processes of national consolidation. The development of inland communication associated in a large part with the development of canals was important in many European nation states and led, in particular, to the rise of the Netherlands as a major player in European geopolitics, but then through this technical strand of development being interwoven with a Protestant mercantile and processing-industrial ethic to form a sociotechnical system.

The sociotechnical nature of politics becomes much more apparent when we trace the transition from political cliques and factions into political parties, such as occurred in Britain in the eighteenth century and in France after the post-Napoleonic restoration. This necessitated a more active intervention in the affairs of state on the part of the citizenry, which then took on the form of an *electorate*. Of course the suffrage was restricted to men of sufficient means to be worth counting as taxpayers, and also landowners even if they did not pay tax, it being argued that these were the only persons with any real stake in the political process: others in society would have no interest in 'the common good' and would simply sell their votes to 'real' stakeholders even if given them.

It was against this background that Burke, in Britain, which was the first nation state to enter the industrial revolution, argued for a new justification for *party*, which he defined as "a body of men united on public principle which could act as a constitutional link between king and Parliament, providing consistency and strength in administration, or principled criticism in opposition". The means to link the electorates of the contending parties to their representatives over the territory of the nation state were however technical, comprising on the one hand the improvements in road-building and road transport technologies generally that proceeded through the eighteenth century and accelerated rapidly at the turn of the eighteenth and nineteenth centuries and, on the other hand, the almost simultaneous industrialisation of printing processes. Both of these developments were aspects of a more general industrial revolution that was occurring in *communication technologies* at that time. Without these technical advances in communication, which culminated in the relatively cheap production and widespread distribution of news through newspapers that could influence the opinions of the electorate so as to suit the interests of 'their' representatives, the constitutional-representative political structure could not have been maintained. At the same time, without this representative-democratic political structure, the industrial revolution itself would have faltered and been endangered. For example, the legislation of copyright and patent laws that protected intellectual property was a necessary condition for a sustainable technological development at that time. As with all sociotechnical systems, the social and technical 'aspects' were interdependent: the one without the other would have been unsustainable. In such situations, the social and the technical are then commonly said to be 'in system'.

As is well known, the level of representation of the citizenry in Britain, which pioneered this development, was initially highly uneven, with some boroughs being open constituencies and others being mere 'pocket boroughs', which could be bought by money or otherwise acquired through patronage. The extension of a more representative democracy in Britain, unlike for example in France and Germany, had to await the widespread introduction of the railway and the electric telegraph. We then say that the technical development 'catalysed' the social development, but then always in such a way that the social reacted in its turn to promote the technical again so as to constitute a *sociotechnical system*.

Among the most common of sociotechnical *constructs* more generally – which may or may not constitute *systems* in the strict (Kantian) sense – are *networks*. These then collectively provide a specific type of sociotechnical development that has been particularly extensively studied. Thus we may trace fresh water supply and waste water disposal networks from the earliest historical times as the *conditio sine qua non* of every sustainable urban development, while it has been and continues to this day to be just these urban developments that in turn made and continue to make such constructions technically and financially feasible. Irrigation networks have similarly provided the technical foundations of many, and even most, of the earliest of those great civilisations that were in their turn the only kinds of societies that could mobilise the labour, organisation and financial resources to construct and maintain such systems. We may similarly identify sea-trading networks as providing some of the first sociotechnical foundations of trading and banking organisations and other social activities that in turn provided the distribution and financial systems that serviced the trading infrastructure. Road networks, postal networks, canal networks, gas supply networks, rail networks, telegraph networks, steamship networks with their networks of coal-bunkering stations, electricity supply networks, telephone networks, radio and television transmission networks and many others can be identified, all of which were of course sociotechnical constructs in themselves, but all of which had this further property in common: that all had their individual influences on political processes, leading generally to an ever wider extension of the suffrage as they made possible an ever more intensive influence and control over the opinions of the electorate that could be employed to promote these same networks.

To the 'material' or 'outer-world' technological developments were thus added the so-called '*technologies of persuasion*' to which we must now turn, and which provided 'inner world' changes, or the formation of 'opinions' within the minds of the electorate, without which representative democracy could not have extended its sway so extensively, even as this extension was in its turn essential to the promotion of the technological advances that, once again in their turn, made political process itself sustainable. In order to grasp the nature of this new class of technologies we need first however to introduce a few more notions and definitions.

So far we have only considered technology as a process of bringing something new to presence in our outer world, but clearly we can also consider another process again, which is that of bringing something new to presence in our inner world. This something new may be a new thought or way of thinking, a new opinion, a new knowledge, a new understanding, a new attitude, a new belief system, a new ideology, a new doctrine and even a new faith – and any number of other such mental and more general psychic constructs.

Reviewing the previous examples we see again that sociotechnological constructs arise when two processes, the one proceeding in the inner world and the other proceeding in the outer world, occur not only interactively but also in synergy, and correspondingly synchronously.

Clearly the technologies of persuasion are the principle tools of the trade of the politician. Since they are always directed by the persuader to *influence the behaviour* of another person, so that this other person becomes the persuaded and behaves accordingly, *the technologies of persuasion are devices for mobilising power through communication*. Since we are again dealing with power we are again led, following Foucault, to start any investigation of these technologies by investigating how they function at their 'lowest' level, and so in this case at the level of those who are to be persuaded.

There is now a very large and extensive literature on this subject which does indeed start at this level, dealing for example with the inculcation of 'consumerist attitudes' in advertising, 'feel-good factors' and 'promoters of upward social mobility' in party-political election campaigns and within many other contexts, and not least in most 'Business Schools'. Common to nearly all of this is its play upon apparently irrational impulses, so that it is said to function primarily at the level of the unreasonable, the subversion of logical rules, or, to be precise, *the paralogical*, while at the same time playing upon the doubts and uncertainties of those to whom it is directed, so as to induce states of total perplexity, such as by simultaneously supporting incommensurable beliefs, as states of *aporia*, to which it in turn 'opposes' its own states, of *euphoria*. Much of this is obviously manipulative, such as by way of advertisement for this or that product or service, but much is not so obviously manipulative and much too much is not so obviously so at all. Thus, for example, almost all the contents of newspapers and television programmes that purport to report on events, developments, personalities and so on are in fact not only or even so much directed to informing their readers or viewers, but also and even principally concerned to influence the opinions and attitudes of these readers and viewers. Their reports are not primarily *constative*, directed to informing their audience about facts, but much more *performative*, and so directed to persuading their audience about something and so gaining power over them. Thus a performative statement refers to a reality that it intends to create itself, because it is stated under circumstances which are intended to make it into an act (See Manin 1971, p. 81). It is one consequence of this that nowadays one often only asks in the second place *what* the newspaper or journal article or radio or television programme is saying, while asking *why* they are saying this in the first place. In the same vein, before reading an article nowadays one almost automatically looks for the name and affiliation of its author – and then, as often as not, one no longer needs to read the article itself, for the thrust of its presentation and its arguments are already completely predictable. All of this is of course eminently political and in particular it points to the close symbiosis between the party-political machinations of 'government' and 'opposition' and the productions of the 'mass media' with their own oligarchies of 'press barons', as now expanded to "television kings".

Writing in 1847 on this same phenomenon but within the context of his own time, Søren Kierkegaard observed that, in so far as we are led to reflect upon real objects, so we have to do with a *communication of knowledge*, while when, as for the most part nowadays, we are led to reflect upon unreal, or more strictly non***ex***istent objects, we always have to do with the *communication of power*. He saw how, in the limit, there would finally be no communication of knowledge at all, but only a communication of power, and that the absurd situation would then arise that everything that should be communicated directly as power would instead be communicated indirectly, masquerading as knowledge. We should then exist in a *totally manipulated society*, such as George Orwell came to delineate in his '1984'. The result could only be a total confusion and an inability to make any sound decisions at all. Kierkegaard described the behaviour of the communicators in this case as one of *impropriety*, and so as something improper, unseemly and morally reprehensible. The Kierkegaardian "purveyors of opinions" and "merchants of drivel" remain among the greatest menaces of our age, even as this age has become, and now on the largest possible scale, the age of gold for such miscreants.

What the technologies of persuasion are aiming to influence are our *judgements*, and what we have to protect ourselves best from their enticements is our *experience*. Through the 1930s, as Europe advanced towards its last grand military denouement, its final throw of the dice with its destiny, Edmund Husserl prepared the last of his great works, which he entitled *Erfahrung und Urteil* (1948)//*Experience and Judgement* (1973). Husserl emphasised that 'experience' is itself something active – and indeed *poietic*. Thus (p. 52):

> Our life-world in its originality … is also the world of experience in the wholly concrete sense which is commonly tied in with the word 'experience' … Thus this commonplace, familiar and concrete sense of the word 'experience' points much more to a mode of behaviour which is practically active and evaluative than to one that is cognitive and judicative.

This is to say that the deconstructions of the communications of the politicians and the mass media that are forever forced upon the critical mind must be based upon an experience of the reality of the world, that is to say upon the oneness of the written and screen-projected world with the inner world of accumulated personal experience.

Some further explanation is still in order, even if only by way of repetition. We may first regard the simplest form of a predicative judgement as being a component in a string of inferences, as follows:

(belief system, facts(data))→attitude→position→judgement→decision→action (3.1)

where in the usual way of things it is only the belief system and the set of data that constitute a priori observables and so are underlined accordingly. Experience is that which alters the belief system, as the basis of the inner world of an individual, whereby the same facts as drawn from the same data may lead to the individual taking a different attitude and thence a different position than others and correspondingly making a different judgement. On this basis a different decision will be made as a result of the experience and this will lead to a different action. If now we reverse the direction of the arrows in this inference string, so that the string of inferences proceeds in the opposite direction, and may even play the role of a dual category to 3.1, and we take the action also as an observable, we may work our way back to an elicitation of the belief system of the judging individual:

action→decision→judgement→position→attitude→ (facts (data, belief system)) (3.2)

In the event that both (3.1) and (3.2) are known, we say that the judgemental process is *transparent.* What we are then seeking when we are evaluating the pronouncements of politicians and others who are in such compromised and compromising positions is *transparency* (see also Jonoski 2002, pp. 91–94).

With this minimum of conceptual machinery in place and up and running, we can move on to consider sociotechnology and sociotechnical arrangements outside of the immediate sphere of the political and so enter into the space in which the Hydroinformatician has to work so far as possible with the politician.

9.4 THE SOCIOTECHNOLOGY OF NETWORKS AS THE HALLMARKS OF MODERNITY AND THE RISE OF PARTICIPATIVE DEMOCRACY

The development of representative democracy was associated with the development of networks that provided new and much enhanced channels of communication. This development is closely associated with the rise to dominance of what has already been introduced as *modernity*. However, modernity has led in its turn to new sociotechnical constructs again, and these may be conveniently introduced quite generally here as *superpositions of networks*. The paradigm case here is the superposition of the Internet on a telephone system which is in its turn superimposed upon both a postal network and an electricity supply network which is in its turn again superimposed upon a coal, oil and gas transport network. Among the feedback loops that are superimposed on this simple superposition, we immediately identify railway, road and pipeline transport systems that depend upon electricity that may in its turn be dispatched through the uses of electronic communication networks, including the Internet. This process of superimposing ever more layers of new networks with ever more complex feedback loops between the earlier superposed layers and their networks is often regarded nowadays as *the culmination of modernism* as has been intimated earlier, where it was seen to be closely related to, or to be 'in system with', the drive towards producing ever more persons as 'knowers'. Now, however, under the influence of the new kinds of networks associated with the Internet (reinforced by intranets and extranets, iPhones, iPods and iPads and so much else besides) and mobile telephony and all its productions, we have to do more with the *post-modern* project of turning ever more persons into 'consumers of knowledge', and this development is more 'in system' with forms of participative democracy – which in turn support the claims to autonomy of these new communication media. Similarly, the move from the *broadcast modes* of communication at fixed times, which have reached their apogee in real-time-transmission of colour television, to *broadband modes* of communication operating at all times indifferently of the kind that advanced communication networks and mobile telephony now provide, are already changing social relations markedly, increasingly challenging the decisions and even the legitimacy of representative government, and increasingly successfully. As adumbrated earlier, these last sociotechnical developments open up the possibility of stabilising otherwise unstable self-managing, 'anarchistic' or 'endostructured' organisations by the provision of appropriate technical equipment – which is once again equipment that would have no purpose outside of the evolution of these new self-organising or 'emergent' social structures that could (possibly) lead towards states of social justice. It is then often said that in such systems the social and the technical components are 'mutually reinforcing'. Of course these same devices can be used to fulminate discontent and rioting, hooliganism and vandalism, but this is outside of our immediate concerns.

The kinds of networks that are of the greatest interest to the leader that strives for social justice are those composed of closely-interacting technical equipment and people working both as individuals and in teams, such as are exemplified by Internet-connected working environments. Within the increasingly unified European context, these networks often cannot function using natural languages because there are several teams working closely together, interactively and so concurrently, with each team having its own natural language as well as its own technically-specialised expressions. For example, a Finnish team may be installing the communication and signalling system together with a German team installing the electrical system together with a Polish team installing the surface finishing panels inside one and the same tunnel element deep under water at one and the same time. This concurrent mode of working is necessitated by the contract conditions that are directed to completing the tunnel and other works six months before schedule in order to claim the generous bonuses that have been offered for early completion of the project and thus the most rapid return on investment for the owner. Clearly such an operation necessitates the use of an extensively analysed, well designed and competently installed communication system that necessitates in its turn an expertise in graphics and semiotics generally in order to communicate with means, such as portable computers and advanced mobile telephony, that are simple, relatively cheap and reliable. This system in turn however necessitates not only a knowledge of its immediately enabling technologies, such as may include several Java and Java-based technologies and their successors, but also an intimate knowledge of all the work processes of all the teams who are employing the system – and then of course the system will need to accommodate many more than just the three teams that happen to be working in this particular tunnel element if it is itself to be cost-effective. In this case there are several different human networks associated with several different languages describing several different technologies that have to be conjoined and synchronised through networks of technical equipment. This is then a typical sociotechnical construct of our present-day World.

This development is however connected to, and in fact inseparable from another, which is that of a legislated right to public access to all relevant information, backed by appropriate knowledge-promoting activities allowing this public to interpret the information provided, and a corresponding legislated right of this now well-informed public to intervene in actions that could have deleterious effects upon their living and working environments. In the case of the Denmark-Sweden traffic link that will be used to conclude this chapter, this again required both extranets and intranets supplying information and knowledge services to the general population in such a way that presentations were customised to the specific areas of concern and the specific intentions of each of the interested parties. This flow of knowledge and information was usually mediated by specific non-governmental organisations representing the main interest groups. This so-called 'mass-customisation of knowledge' emanating from a 'knowledge provider' provides a typical example of a system providing the resources to support a movement towards *participative democracy*, as both a complement and a supplement to, and even a replacement of, representative democracy.

Chapter 10

Data – Information – Knowledge – Understanding – Wisdom

10.1 WHY DO IT AT ALL?

From out earliest years we have been taught that before we embark upon any undertaking we should always ask three basic questions, in this order:

(1) Why do it at all?
(2) Why do it now?
(3) Why do it this way?

In the case of a theory and a philosophy of leadership we should then in the first place pose the questions: Why have a theory of knowledge at all? Why have a philosophy of knowledge at all? And in this case, where we are dealing with knowledge, which is itself the very stuff of both theory and practice, we can proceed even further again and ask: Why have a theory at all? Why have a philosophy at all?

In the case of a theory we might suppose that our need follows directly from the definition: "A theory is an *explanation* of observed facts and actions", whereby we might be simply tempted to take it as given that a competent leader must be able to *explain* what he observes in the world around him, as facts, and what he observes of the actions of himself and others and the way in which others and himself act in relation to the facts. But, we may insist, *why is this*? We then observe, however, that an explanation is itself an answer to a question of 'why?' so that we are then brought back, recursively, to posing the question of 'why should we pose the question "why?" in the first place?' Or, which is the same thing, of 'explaining the need for an explanation'.

Theory is often set up in opposition to practice, but of course these are two different things, the first being itself concerned with the 'why?' of acting or being oneself the recipient of actions and the second being concerned with the 'how' of the acting or the action to which one is subject. However, it is the 'why?' that determines whether the question 'how?' will ever need to be posed at all and, if so, in what way, so that this 'why?' has precedence over the 'how?' At the same time, of course, without an answer to the question of 'how?' this precedence has no practical significance. As Karl Marx most memorably expressed this matter (and as we have mentioned earlier): "Practice without theory is blind; theory without practice is sterile".

Thus theory and practice are indivisible in what we refer to as *praxis*, but theory has a certain precedence, being concerned with whether a certain thing should be done at all, whether it should be done at this time or another time and whether it should be done in the one or another way - whereas practice is concerned with actually doing it. This section is concerned with both the theory and the practice of knowledge as such and the application of knowledge itself to the application of theory and practice.

Obviously both theory and practice require this something that we call ‘knowledge’, but then theory requires something more again, which is what we call ‘understanding’ and even ‘wisdom’. Through knowledge we learn how to act in our own better interests. Will this decision lead to greater well-being, or greater suffering? What is the right way to respond in a particular situation? Is it coming from the state of our consciousness, or from the egoistic mind?

Wisdom reflects the values and criteria that we apply to our knowledge and understanding. Its essence is ability to distinguish justice form injustice, right from wrong, helpful from harmful, truth from delusion.

10.2 THE REALITIES AND THE IDEALITIES OF KNOWLEDGE – AND THE NOTION OF A ‘KNOWLEDGE SOCIETY’

The study of strings such as ‘data’, ‘information’, ‘cognisance’, ‘knowledge’, ‘meaning’ and ‘understanding’ dates back through time immemorial, and certainly by the times of Socrates and Plato and Aristotle. It was already clearly understood that the very study of these strings must itself depend upon our own exercise of ‘data’, ‘information’, ‘cognisance’, ‘knowledge’, ‘meaning’ and ‘understanding’. But we must be pragmatic here and enter this circle as best we can, and for this purpose it is most convenient to enter it at the point of ‘knowledge’ and then to proceed, so to say, backwards to define ‘cognisance’ and ‘information’, and thence ‘data’, and then to proceed forward to define ‘meaning’, ‘understanding’ and perhaps, possibly ‘wisdom’, Table 10.1.

Table 10.1 Knowledge terminology.

Data	That which has the capacity to provide information.
Information	That which has the capacity to provide knowledge.
Knowledge	The intangible which provides a change in one’s state of awareness, leading to a change in intentions and thus behaviour (as actionable knowledge’).
Understanding	The faculty of comprehending or apprehending something or somebody clearly.
Wisdom	The arbiter and executor of knowledge and understanding; that which governs the way in which knowledge and understanding are applied.

We look backwards correspondingly only to the scholastics of the eleventh to fourteenth centuries, who introduced the broad and powerful concept of *intentionality* in order to proceed further in such investigations. We believe that *intentions* are things existing only within ourselves and, we must suppose, within other living creatures, but which we believe to be absent in all inanimate objects, and it is these intentions which drive our own searches for data, information, cognisance, knowledge, meaning and understanding. In this book, intentionality is only traced in the merest outline from Brentano in the nineteenth century, and then principally through Husserl to more recent, twentieth-century authors such as Heidegger, Merleau-Ponty, Foucault and Derrida, while other directions, such as those proceeding through the works of Carnap and Quine, are usually no more than mentioned, while the influence of Bergson, passing also through Foucault, is scarcely mentioned at all. Even on this basis, however, we can still define knowledge pragmatically *as that which, by its acquisition, changes our behaviour and thus has the capacity to change our intentions.* The following, from Karl Barth, from the *Kirchliche Dogmatik/Church Dogmatics* (Part 1 of Volume 1, p. 188), extends this to define knowledge at its highest, theological, level:

> By the knowledge of an object by men we understand confirmation of their acquaintance with its reality in respect to its existence and its nature. But the confirmation of their acquaintance means that the reality of the object concerned, its existence and nature, being true in themselves, now become in some way, and in some degree of clarity and distinctness, true for men too. Their acquaintance with it, instead of being a contingent and outward determination of their own existence, now becomes a necessary and inward determination. They no longer exist without it, but with it. In so far as they think of it, with the same confidence with which they dare to think in general, they must think of it as a true reality, as true in its existence and nature. Whatever else and however else they think of it, they must begin by thinking of the truth of its reality. Face to face with the truth they can no longer withdraw into themselves in order to affirm, question or deny it thence. Its truth has come home to them, has become their own. And in the process they themselves have become the truth's. This event, this confirmation, in contrast to mere cognisance, we call knowledge. Cognisance becomes knowledge when man becomes a responsible witness to its content.

We are at this point using an *encyclopædic model of a definition* in order to break into the hermeneutic circle whereby one definition always presupposes and is presupposed by another in the dictionary model. With this one piece of the circle in place, we can then proceed to construct the other pieces by extending upon it. Thus, *cognisance is that which becomes knowledge when man becomes a responsible witness to its content*, so that *information is that which has the potential, or capacity, to induce cognisance and thus to inculcate knowledge.* Thence again, *data is that which has the potential, or capacity, to provide information.* Or, proceeding in the other direction, *the meaning of an item of knowledge is the significance to the knower of that item of knowledge* and *an understanding of someone or something is a comprehension or apprehension of the meaning of that someone or something to the knower.* For Dilthey, for example, "understanding is the process of recognising a mental state from a sense-given sign by which it is expressed" (Dilthey-Rickman 1976, p. 248). Clearly to understand someone is something more, and usually much more, than merely knowing someone and this distinction carries over further to our understanding, as opposed to our mere knowledge, of things more generally. The greater effort (which in turn excites the will to action through acts of judgement) that is involved in understanding as compared with acquiring knowledge is expressed in the construction of the corresponding words for 'understanding' in language. Thus, just keeping to a few European languages, already in English we may decompose the word into 'under-standing' or, as other examples, in French 'comprendre' = 'comme-prendre', similar to the Dutch 'begrijpen' = 'be-grijpen', or the German 'verstehen' = 'ver-stehen', similar to the Danish 'forstår' = 'for-står'.

It is usually accepted as a part of the modern condition of our societies that, to the extent that man has knowledge of something at all, he may in principle attain to some understanding of it. Following Husserl, however, this is really only a tenuous assumption that has been nurtured through the modern condition and one that has become increasingly questioned as our societies move into a postmodern condition, where societies of 'knowers' are increasingly displaced and even largely replaced by 'consumers of knowledge'. It is this sea change in the mode of application, and thus the very nature of knowledge that has driven the communication revolution of recent times, as exemplified by the internet and the mobile telephone. On the one hand most knowledge becomes increasingly 'commoditised', being ever more available at market-established rates, and on the other hand it becomes increasingly fragmented between 'markets', to provide a multitude of 'knowledge*s*', so that we nearly always nowadays speak and write of it in the plural.

It is often rather acerbically observed by those who obstinately remain as knowers that just as it becomes so much easier and convenient to access and dispose of knowledge 'at the click of a mouse', so the understanding of the knowledge thereby accessed decreases correspondingly. It must then appear, at least to these 'dinosaurs', that as humanity as a whole will come collectively to know more and more, so its individual members will come to understand less and less, so that in some limit, *humanity as a whole will know absolutely everything, just as each one of its individual members will understand nothing whatsoever!* (But then it is assumed in this somewhat jaundiced view that humanity will have destroyed itself some time before this limit is reached.)

What has been lost in our post-modern world in this view, and what was already on its way to extinction during the descent of the world through the modern era, was the quality of *wisdom*. It is usual to place wisdom, somewhat capriciously, as the arbiter and adjudicator of knowledge, as the source of all true understanding and as the supreme exercise of judgement. It is often considered that wisdom is not, like knowledge and even understanding, something that is attainable entirely through the efforts of the individual self, but that it is always experienced, at least to some extent, as something that is given to, or conferred upon the *wise person* from outside his or her own Self. Since wisdom is clearly so rare in our present-day world, and it has anyway been said that only the wise can discourse upon wisdom (which they have usually done, most wisely, by remaining silent!) there seems to be little purpose in entering further into its nature in the present work. When we read Augustine, or Anselm of Canterbury or William of Ockham, let alone the Bible, the Koran and other Holy Scriptures, we experience voices of true wisdom, but it seems that all that we can learn from these about wisdom itself, as Kierkegaard said when he read the story of Abraham, is to be amazed.

The theory of knowledge, or *epistemology*, is pursued in philosophy through investigations into the nature and scope of knowledge, its presuppositions and their foundations, and the reliability that can be attached to various kinds of claims to knowledge. The art or science of interpreting representations of knowledge, such as may in principle take the form of written texts, musical compositions, sculptures, buildings and many other artefacts, as this was intended to be interpreted, is called *hermeneutics*. Wilhelm Dilthey restricted this however to "the methodology of the interpretation of written records" (*ibid*, p. 249). The systematic understanding of recorded expressions, as a way of teasing out their essential meanings, is called *exegesis*.

As our world has descended ever deeper into the (mis)use of language and other signs, primarily as a means of influencing its listeners and so gaining power over them, and even as it often does this by making a pretence of informing them, much stronger means for uncovering the intentions of those using the language and thence uncovering their deeper motives – their so-called 'hidden agendas' – have become necessary, and the devices used for these purposes include the instruments of *deconstruction*. In principle, deconstruction is by no means limited to such pathological applications, but in practice today it is most widely employed, and needs to be employed, for this purpose. Following Kierkegaard, it is not efficient to contradict the falsifiers and destroyers of meaning by confronting them directly with contrary evidence, for these miscreants bear with them all the powers of nothingness, and are armed with all the ruses and cunning needed to confuse and mislead their audiences. Only an indirect approach can be effective and this must proceed from *within the same structure* that produces the lies and the manipulation. Thus, with the 'father of deconstruction', with Jacques Derrida (1991, p. 41):

> The movements of deconstruction do not destroy structures from the outside. They are not possible and cannot take accurate aim, except by inhabiting those structures. Inhabiting them in a certain way, because one always inhabits, and all the more when one does not suspect it. Operating necessarily from the inside, borrowing all the economic resources from the old structure, borrowing them structurally, that is to say without being able

to isolate their elements and atoms, the enterprise of deconstruction always in a certain way falls prey to its own work.

Following the unified object-value theory established by Alexius von Meinong, knowledge is composed entirely of non-*ex*istent objects, in that we can nowhere perceive knowledge in our outer or *ex*ternal world as the world that we experience with our senses. Thus, just as in the case of numbers and money, as mentioned earlier, although we may have any number of signs in our outer worlds of sense experience that direct our attention to and inculcate knowledge, such as the worlds of nature, of books and of films, we can never actually encounter knowledge itself outside of our own individual minds. For the sake of simplicity we have introduced, an '*ex*ists' in order to emphasise this connotation of '*ex*ists in our *ex*ternal world' even though standard Meinongian theory, being so much more rigorous, has no need of this device. Of course we can imbibe knowledge from entities that ***ex***ist, such as speech acts, newspapers, television programmes, etc., but these sources only provide knowledge to the extent that their content enter our minds and remains there, *sub*sisting in our conscious and ***sub***conscious minds.

As a non-***ex***istent object, so one that is not encountered in our *ex*ternal world, knowledge belongs to the Platonic world of Ideas, even as it is itself concerned with the ordering and structuring of *id*eas in the most general sense. It is thus a prime candidate for transformation into an *id*eology, and even further again into an *id*olatry. This transformation is now however increasingly catalysed by our ongoing drift into a post-modern condition, whereby the value placed upon knowledge comes to overtake the values placed upon numbers and money so as to become the dominant value in our increasingly post-modern societies. It is then widely argued that in these societies a superior level of knowledge can be translated into a higher level of money income, as expressed in greater numbers of accumulated monetary units. A greater rate of consumption of knowledge can be translated similarly into a greater rate of accumulation of monetary wealth. Thus an ideology of money, descending to an idolatry of money, is displaced – although of course not replaced – by an *ideology of knowledge*, descending to *an idolatry of knowledge*. The earlier aim of making a society more moneyed is transformed into the current aim of making it more knowledgeable: through both, society itself becomes increasingly hedonistic. Thus the notion arises that Europe, for example, should become *a knowledge society*, a notion that has now been given some credence by being adopted by European heads of state as the so-called *Lisbon Agenda*, whereas what we are calling for here is a Europe becoming *an understanding society*. But what then are the realities behind this? And what are its idealities?

10.3 THE ENCAPSULATION OF KNOWLEDGE IN PRODUCTS AND SERVICES: THE CONCEPT OF 'KNOWLEDGE CONTENT'

Knowledge is in the mind of the knower, even if perhaps only temporarily in the case of the consumer of knowledge, as a non***ex***istent object, as something intangible, but it can be used, such as through applications of technology, to provide ***e****x*istent objects, such as ***e****x*istent products and ***ex***istent services, understood as products and services that are tangible objects, ***e****x*isting in our ***ex***ternal world of name and form. The knowledge that was then in the mind of the knower or the consumer of knowledge has then entered into these products and services and in this case this knowledge is said to be *encapsulated* in these products and services.

According to the extent that knowledge is encapsulated in this way, so we may speak of lower or higher levels of *knowledge content*. The knowledge content is then high when the value of the knowledge that has been used to manufacture a product or provide a service is large as compared with the sum of the values of this knowledge and the work that is done to transform this knowledge into the corresponding product or service, and low in the contrary case. The value of the knowledge that is used may still be expressed in terms of the money value of accessing this knowledge, which may then correspond to the expenditure on maintaining this knowledge in the manufacturing or service-providing organisation or the cost of accessing it through other means, such as through the employment of an outside agent, possibly over the internet and possibly already in a partly encapsulated form. Although the work that is done to transform this knowledge into a product or service obviously itself demands some further knowledge, this is not usually included in the 'knowledge content' in the view of the current ideology.

In a society where the costs of transforming knowledge into products, for example, is high, the knowledge content must be high correspondingly if the producing and/or servicing organisation is to remain competitive. This implies the introduction of high levels of automation of work processes, flexible computerised tooling processes, intensive robotisation and the use of other knowledge-intensive equipment, all of which in turn necessitates the employment of a well-educated, highly-skilled and properly-motivated operatives. In another situation, where the employment costs are much lower, the knowledge content may be less in standard, quantifiable terms, but it may be higher in the less standard, qualitative terms involved in analysing, designing and implementing manufacturing processes with much lower investments in knowledge-intensive equipment. Just as the early industrial revolution was marked by an increased application of mechanical, and subsequently also electrical power to work processes, such as might be measured in terms of the number of horse-power or kilowatts of power at the disposal of each person employed in the production process, so the current information/knowledge revolution is driven by the increasing amounts and levels of encapsulated knowledge placed at the disposal of each such person. Similarly, just as the producing person at the time of the industrial revolution needed new and generally higher skills, a better education and an increased motivation, such as through the provision of increasing levels of security, so the producing person today needs higher levels of skill again, considerably higher levels of education and a much enhanced system of social services to maintain not only his/her immediate motivation, but also his/her direct, active and creative participation in the larger productive process. The knowledge that is necessarily present in the producing person or persons and their equipment (which is already regarded as encapsulated knowledge) in order to encapsulate further knowledge is often called the *knowledge frame* provided by the person, persons and their equipment.

10.4 THEORIES OF KNOWLEDGE FLOWS, KNOWLEDGE GRADIENTS AND RATES OF CIRCULATION OF KNOWLEDGE

Much the greater part of all knowledge flows in the world today are in the form of encapsulated knowledge. Thus knowledge nowadays flows increasingly in encapsulated forms in clothing, in computers, in farm animals, in television sets, in mobile telephones, in motor vehicles, in aircraft, in weapons and indeed, quite literally in the ideology, in *everything* else. In some cases the knowledge content may be small, as in the case of the provision of readily available minerals, while in other cases it may be very high, as is the case in many scientific instruments, in some types of aircraft and, perhaps most of all, in weapons systems.

Correspondingly, trade in goods nowadays is increasingly in the form of encapsulated knowledge and ever less in the form of the raw materials, labour and transport that make up most of the rest of the values of the goods concerned – which we conventionally distinguish from 'encapsulated knowledge'. There are however certain very special features of knowledge that make its behaviour markedly different to these other components that contribute to the value of goods, and of services also.

The first of these features is that the gradients in knowledges that drive the flow of goods and services from a place with a generally higher level of knowledges to a place with a generally lower level of knowledges are of an essentially different nature to the gradients in raw-material costs and manufacturing and servicing costs that also contribute to driving these flows. This is because, whereas labour and material costs are still quite strongly constrained geographically, *knowledges are less, and indeed much less constrained geographically, and indeed in many situations they are not so constrained at all.* As one result, it has become common to locate centres of production of large-volume goods in areas with lower raw material costs, with lower labour rates relative to the associated levels of skills and as close to the market as possible in order to economise on transport and its related costs.

We have used the notion of 'knowledge gradient' very loosely above, but in some cases it is necessary to be more precise about its definition. The expression itself is metaphorical, with 'gradient' as its subject and 'knowledge' as its modifier. However, a gradient is in its origins concerned with something that changes over *distance*, so that we are obliged to define what we mean by 'distance' in the present metaphor. Since in this metaphor we might reasonably say that 'knowledge flows along the direction of its gradient from the giver of the knowledge to the receiver of the knowledge and with an intensity that is proportional to the magnitude of this gradient', we must then define what we mean by the 'distance between the giver and the receiver of knowledge'. Clearly the very fact that knowledge flows between the giver and the receiver establishes both of these as subjects as opposed to mere objects, so this distance is something between subjects and thus assumes a certain intersubjectivity; we can accordingly describe it as *an inter-subjective distance*. Thus when there is a high level of intersubjectivity between the giver and the receiver, such as commonly occurs between parents and their offspring, the inter-subjective distance may be very small and the gradient correspondingly large, with a correspondingly rapid and strong flow of knowledge – even if not understanding – from the parent to the offspring. Thus, to take a popular example, if an enamoured young man introduces the lady of his affections to his mother and his mother responds with an "Over my dead body!", then we are clearly in the presence of a very strong knowledge gradient. We commonly refer to such knowledge as is transmitted along strong knowledge gradients as *advice*, so that in our popular example the mother is conveying a maximum of advice with a minimum of semantic resources. The young man, the son, may not *understand* at all the thinking and, more especially, the feeling that support his mother's words, but, to the extent that he has *faith in his mother's judgement*, so that he *trusts her judgement*, so he does receive the intended knowledge. By the same token of course, the mother may herself not *understand* why she feels the way she does about the lady concerned, but her whole life experience is still behind her words.

The second such feature that is vital here is that knowledge, once encapsulated, is *almost infinitely renewable at a negligible cost*, while raw materials, labour and transport are not. This corresponds to the theory of the rentability of knowledge more generally, whereby the unit cost of using a once-established entity of knowledge falls almost linearly with the frequency with which it is employed. This was the very principle of mass-production from its earliest times and paradigm cases are provided today by the publishing and pharmaceutics industries. Corresponding to this, the *rentability of encapsulated knowledge within an enterprise and between this enterprise and its environment is proportional to the rate of circulation of this knowledge*. The purpose of such developments as those involving the electronic encapsulation of knowledge in software has been precisely that of accelerating the circulation of existing knowledge by promoting the development of new *knowledge frames* and their environments.[1]

[1]We are here using the word 'frame' in the Heideggerian sense (Heidegger 1977/tenth edition, 2002, p. 19) where *Enframing* is used as a translation of the German *Ge-stell* where it is used to denote "a 'calling-forth', a 'challenging claim', a 'demanding summons', that 'gathers' so as to reveal" in the words of the English-language translator.

The third feature of encapsulated knowledge is less obvious than these, but still most important. It has to do with *the more limited predictability of knowledge requirements* as compared with the predictability of supply of materials, labour and transportation. This commonly occurs when knowledge is, so to say, 'de-commoditised', such as when more specialised and esoteric knowledges that were not initially considered at all come necessarily to be involved. Some examples are those of the knowledge that enters specialised modelling software. In these examples the profit margins on the original investment have been very large indeed, but these margins were usually necessary, being devoured at least for some considerable initial period in order to finance the rapid and intensive development of the new product that was necessary as it came to 'track its market' and 'learn its way around' correspondingly, discovering new and unforeseen markets, and thereby throwing up all manner of faults and lacunæ in the original concepts underlying the system. We then commonly say that during the introduction of such a new product "the investment levers are locked on".

If only because all the objects of knowledge are non***ex***istent objects, subsisting only in the minds of persons, even as some of these objects of knowledge come to drive the technology of existing objects that is at the source of all industry, *the study of the functioning of knowledges is always a sociotechnical study*. Every theory of the functioning of knowledge in industry is correspondingly a sociotechnical theory. This is then to say that the study of the ways in which any combination of knowledges comes to permeate and transform industrial processes cannot be separated from the ways in which this or other related combinations of knowledges come to permeate and transform the society within which the technology comes to presence, commonly with further consequences again for the technology, and often so on again for some further cycles. The social 'context' may be simply the factory or construction site or design office or whatever other place where the knowledges first come to presence and are applied, whereby it is this context that is transformed, or it may extend to broader, and possibly much broader, sociotechnical environments. *Knowledge engineering*, even as it may be initiated by studies of applications of knowledge within technologies, must always extend its scope similarly to become an essentially sociotechnical endeavour.

Sometimes the social components of the sociotechnical transformation can be separated out into a single 'object', such as occurred in the transformation of numerical modelling from an entirely in-house activity into a world-wide activity as realised through the 'packaging' of the encapsulated knowledge into modelling-system products and the marketing of these products through large numbers of local agents. In this case the 'object' that caught the attention of the social components was that of transforming at least a part of the organisation "from a performer of projects into a provider of products".

In most situations of this kind there is bound to be resistance to such innovations and often this resistance exhibits all the symptoms of aporia. In this last case, for example, it was on the one side argued that by these means all the knowledge accumulated in the original organisation would be sold off once and for all for everyone in the world to use, while on the other side it was argued that few if any persons outside of the organisation would ever really come to understand this knowledge and so would misuse it even if they ever used it at all. The second argument was marginalised by accepting that many persons outside the producing organisation would indeed encounter problems in some cases and in fact might find some projects too onerous, but that in just these cases they would return to the original product-producing organisation to help them solve their problems and even perform their projects for them. Thence arose the *'mantra-like object'* that *"the projects follow the products"*, which worked itself out in practice, albeit with some time lags that were of no real consequences to the organisations concerned.

A final point that should be mentioned, although it may be rather obvious, is that although human knowledge is accumulative because it is accompanied by a process of selection that is associated with forgetting, encapsulated knowledge is accumulated without any such compensating device. Associated with this is a divorce that commonly, and nowadays increasingly, occurs between human knowledge and encapsulated knowledge, in that over time there are fewer and fewer persons who truly understand the encapsulated knowledge, even as at the same time there are more and more persons using it that have never really understood how it functions in the first place. It is then inevitable that mistakes are increasingly made when using knowledge that is encapsulated, and especially when this is encapsulated in software. There is an extraordinary illusion abroad that just because we live in societies where human knowledge as a totality is increasing, human understanding is increasing correspondingly, but outside of a few specialisations this is surely incorrect. But then industrial leadership is already trying to tackle these problems, and must increasingly tackle them sociotechnically.

Chapter 11

The role of hydroinformatics in active stakeholder participation

11.1 THE RELATION OF HYDROINFORMATICS TO COMMUNICATION TECHNOLOGY

The predominance of "modern science" as it spread across the world emanated however much more from its marriage with a technology that developed also in Europe. Indeed, as already remarked, technology itself came to be identified in the West with "applied modern science", thereby raising up this science, as a pseudo-myth, to the level of an ideology – and even for some persons to an idolatry – that was geographically unconstrained, and indeed was regarded as "universal". As Heidegger (1963/2002//1977) observed, such a technology becomes itself mathematical in the quantitative sense even before it is cast into mathematical form. It was this restricted view and its exclusively quantitative objectivity that was castigated and rejected already in the *Hydroinformatics* of 1991, where the ideological foundations of this view were already exposed following earlier such denunciations by other authors. These denunciations have now become increasingly strident (e.g., Newcomen Society 2009). In order to further motivate these reactions, let us return for a moment to the post-modern/pre-modern view on technology, retracing a further development of the much publicised Heideggerian vision.

We then speak again about industry, conventionally, as systematic work or labour, but then in the first place as intelligent and skilful employment necessitating diligence and assiduity in the performance of its tasks. For the purposes of performing this kind of work satisfactorily, those who engage in it are said to *draw upon technology*. The processes through which they do this however are by no means simple and straightforward, and indeed the whole notion of technology itself needs to be more carefully examined and thoroughly understood again if we are to proceed anywhere in comprehending the essential nature of industrial activity as it "draws upon technology". Such a comprehension is of course fundamental for every leader of industry.

As is usually the case in such matters, the first obstacle that has to be overcome in arriving at any deeper understanding is the belief that *we already understand enough*, that is, enough for us to perform the work that we have in hand, so that no further investigation is required. Thus, in the present case, it may at first appear that everyone knows what technology is, such as "technology is the sum of the means directed to an end through human activity". It is then understood that technology is not the human activity itself, but the means directed to an end that are drawn upon during the human activity. For example, glass-blowing for the manufacture of vases and other household articles is a craft that draws upon certain elements of glass technology for informing the processes that it employs; just as cabinet-making is another craft that draws upon other and generally very different elements of technology for informing the operations of its craftsmen. For the craftsman such as the glass-blower, such a definition is entirely adequate, for he does not need to *explicate what it is* that is being drawn upon by his mind in conjunction with his face and throat muscles, his breath, his hands and arms and indeed his whole mind and body as he goes about his productive activities. He simply "knows it intuitively" as he "does the job". Accordingly, since technology is not itself the actions of the mind and body of the craftsman that makes the production *possible*, but the means for guiding all the activities of the mind and body in this *productive process*, technology is that which prepares and informs the craftsman for and during his work: technology is then not the act itself, *but that which makes the act possible*.

On the one hand, technology then appears as a kind of knowledge, but on the other hand, to the extent that it prepares and guides the coordinated mental and bodily processes of the craftsman so as to produce a craft, it is something that also transcends knowledge in the ordinary sense. Thus, to the extent that it is admitted as a kind of knowledge at all, it is an *extra*ordinary knowledge. It is certainly not intellectual in most cases and it is only existential to the extent that it is inseparable from the existence of the person or persons possessing just this specific knowledge.

The above simple way of defining technology, still widely current, implies that those who engage in any such particular activity can do so by drawing upon some technology in ways chosen by themselves and directed to an end of their own choosing. Accordingly, in this view, technology can serve as a mere instrument, in that those who are using it can *select it, master it and use it within well-established limits*, implying in this view that *it is entirely in the hands and under the control of those who are using it*. In this view, then, technology is entirely subject to their will and their control. This view remains so prevalent and is so widely accepted that it is taken for granted as something eminently realistic and always attainable. It is then for just this reason that when we observe that a particular application of technology threatens to slip out of our control so as no longer to be subject to our will, as is currently exemplified by the case of so-called "weapons of mass destruction" and "terrorist attacks", *our will to be the master over the technology becomes ever more pronounced and urgent, and ever more compulsive*. Thus, to the extent that we can use this characterisation of technology as a definition at all, we must circumscribe its realm of application, and we accordingly refer to such a definition as *an instrumental and anthropological definition of technology*.

In his short but seminal work, *The Question Concerning Technology*, which first introduced these and several consequent notions, Martin Heidegger pointed out that, although such a mode of definition might be adequate for the popular and everyday view of technology, it was not at all adequate for those who have to work and live with technology, who have to understand *technology in its essence*, and so are obliged to define *technology as such*. He thereby posed the question of what technology was *in its essence*. Thus: what is technology, essentially?

For the purposes of his investigation, he began by employing the classical philosophical distinction between *what is correct* and *what is true*. To start with an example, in everyday speech we may well say "I think that is correct", or even just "That is correct", even when we understand that there is the possibility that we might be wrong about it, so that it is not necessarily true. Accordingly philosophy makes a very sharp distinction between what is correct and what is true. This becomes most evident nowadays when we say ironically that some statement is "politically correct", where we imply that this is a statement that is just as likely as not to be untrue!

Heidegger accepted that the instrumental and anthropological definition could be regarded as *correct* within certain circumscribed contexts, but he then further demonstrated that *it must be untrue for technology as such*. He showed that technology, in its essence, was certainly no mere means, but had, so to say, "a way of being of its own", albeit one in which human activity was intimately connected. It is vitally necessary that we understand this, and we shall follow Heidegger's demonstration accordingly, introducing some of his vocabulary that we shall have frequent occasion to use later in this book.

In order to arrive at this deeper understanding of this-that-makes-all-industry-possible that we call technology, we can then best follow *how technology functions*, with a special concern with how it enters into industry. We can then say at once that every industrial activity *changes* as technology enters into it, so that technology is in some way *at the source*, or even is *the cause*, of what changes in industry. Technology is thus in some way *responsible* for what comes to presence in industry. The introduction of technology into industry *starts something moving towards a new way of functioning*, so that something new *comes into being*, is *occasioned*, or *comes to presence*, in industry, and it is in this way that technology is responsible for something in industry to *move forward, to bring something forth*. Observing this, Heidegger was led to refer back to a particularly relevant observation of Plato, from the *Symposium* upon which we have already drawn earlier here:

> Every occasion for whatever passes over and goes forward into presencing from that which is not presencing is *poiēsis*, is bringing forth [*Her-vor-bringen* in Heidegger's specially hyphenated German].

It is really only at this point in Heidegger's analysis that technology begins to appear in a new light, as something that is called in English *poietic*, that is, *as something creative, formative, productive and active*, and then indeed *as such in its own right*, even as we have introduced this word earlier here without properly providing its origin at this depth.

For his part, Heidegger emphasised further that, for the Greeks, *poiēsis* applied not only to handicraft manufacture, and even not only, continuing further, to visual-artistic and poetical expressions, but it applied just as well to the acts of nature where something new came to presence, such as the bursting into blossom of a flower.

This in turn, however, inevitably recalls Goethe's celebrated aphorism (see Dilthey 1976, p. 60) that "Just as in the sphere of morality we are meant to approach the highest good so, in the intellectual sphere, contemplation of an ever creative nature is intended to make us worthy of participating in its productions". In this exalted view, then, we return to the definition given earlier, but hopefully with a better understanding behind it:

Our technology is nothing less than the continuation of the Creation by human means.

11.2 WHAT DOES HYDROINFORMATICS CURRENTLY PROVIDE AND WHAT DOES SOCIETY WANT?

The answer to the first part of this question was provided in Hydroinformatics: Information Technology and the Aquatic Environment (Abbott 1991), even as, at the same time, the limitations of this response were also presented. Hydroinformatics at that time was taken up almost entirely with numbers and their calculations, making use of computational hydraulics and drawing upon computational fluid dynamics the first an established technology and the second a putative science to provide numerical models; so that it was using numbers in ways that were essentially new, as expressed in the aphorism that "the numbers are beginning to function in another way". At the same time, of course, these models were themselves being applied within a new environment, being one in which they were becoming the central elements in a much more far-ranging activity employing measuring equipment, its radio links and SCADA systems, weather radars, meteorological forecasts and early warning systems generally, remote sensing equipment and its interpretation facilities, and many other such accessories as are introduced later here. It was then this whole complex of equipment and activities that was brought together under the one rubric of hydroinformatics (Abbott 1991).

Already within this founding document, however, the limitations of a world that was being reduced only to collections of numbers was criticised as something undesirable in itself and dangerous in its potentialities. It accordingly drew heavily, as we have seen again here, upon the insights and analyses of Martin Heidegger. However, one could not stop with Heidegger, but was obliged to deepen his analysis by placing it within the more general environment of a rapidly spreading and specifically European philosophical nihilism and theological nothingness, following such authors as Kierkegaard, Nietzsche and Barth. This, however, might be said to have been the book's undoing, for it was launched for the most part into a Western world which was rapidly losing its foundations in spiritual values, and especially those that were mediated by the established churches as expressed in their church dogmatics. Correspondingly, such a world could not, for the most part, understand the onto-theological language in which the second part of the book was written, so that this fell for the greater part on deaf ears.

However, as Heidegger (1963/2002, p. 32//1977, p. 32) had predicted [2002//1977]:

So birgt denn, was wir am wenigsten vermuten, das Wesende der Technik
den möglichen Aufgang des Rettenden in sich
(Thus the coming to presence of technology harbours in itself what we least suspect, the possible arising of the saving power)

We are constantly reminded that our present-day world is crying out for something quite other than mere numbers: this world is crying out for *social justice*!

11.3 THE HYDROINFORMATICIAN IN THE SERVICE OF SOCIAL JUSTICE

Even in hydroinformatics as it exists today, the work of the hydroinformatician remains centred upon the model, whether a physically-based numerical model or a more empirical or instrumental data-driven model. Currently only the numerical model is used in the kind of operations considered here and all the other resources of hydroinformatics all manner of instrumentation; SCADA systems; radio transmission links; geographical information systems; geographical positioning systems; fixed and mobile internet, increasingly with World Wide Web access; satellite-based positioning and radio and television-transmission facilities, and so on are, so to say, distributed around and connected to a central modelling activity. However, as we have emphasised repeatedly in this chapter, what is now needed is something quite other than mere numbers: we must have a hydroinformatics that can deal with *the qualities*. We call then the organisation that brings all the knowledge of this kind together and combines it in such a way as to be accessible to the stakeholder group, a knowledge provider. The stakeholders then become our "knowledge consumers". The hydroinformatician is the one who brings the productions of the knowledge provider together, currently through the agency of the model, and processes this knowledge into forms that are recognisable to the stakeholders, these being forms such as animations and graphical expressions that can remove polarisation that tends to arise between experts and the public. Typically, experts regard public responses as being emotional and biased, to which they oppose their own views which are predominantly based on "scientific evidence". People are said to perceive risk using their feelings and imagination, whereas experts assess and evaluate risk by applying some form of scientific knowledge or method. We should certainly not ignore the perception of risk of the public, and we should do our utmost to include psychological and emotional factors as part of the risk assessment process. The hydroinformatician's role is then to help in achieving a greater consensus between stakeholders by stimulating their processes of imagination. This is of a particular importance as ethical judgements already crucially depend on the use of the imagination, while *teleological judgements depend essentially upon the various levels of the imagination, making of it a creative imagination in itself.*

The person that we call "the modeller" is then in the first place the person who possesses the ocular system and brain functions that are capable of combining the productions of all of the technical facilities through the agencies of the numerical model, and then of computing using such images as may be relevant in order to provide the metonymic structures that will appear on the screens of the user interfaces of the individual, specifically caring and concerned, stakeholders. Each of the productions of this initiating modeller then has the capacity to convey information that is specifically relevant to the inner and the outer worlds of each one of these stakeholders. These stakeholders then become *terminal modellers* in the sense that they possess the qualities of eye and mind that enable them to transform the information contained in the metonymic structures that the *initiating modeller* specifically directs to them into metaphors that are meaningful within the ambit of their own specific concerns, so that they come to experience what we might now describe as *symbolic knowledge*. In the language of a Husserlian phenomenology, the initiating modeller is one who has the capacity to use all the indicative signs that are available to create the metonymic structures, and these have in turn the capacity to provide the different expressive signs that are specifically appropriate to each and every one of the individual terminal modellers (compare Husserl 1900, 1901/1913//1970, with Husserl 1945//1973). These terminal modellers then proceed, individually and in interaction with other such modellers, to transform these expressive signs into symbols, recalling again from Kathleen Raine that "These symbols touch us at a much deeper level; they come from a deeper level and they speak at a deeper level. They strike these resonances". It is then a conditio sine qua non of all such stakeholder participation that all the content produced in this way must be available to all the stakeholders and that the means can be mobilised to connect them with other stakeholders who can also interpret the symbolic meaning of the content, albeit often in different, and even at some stages in conflicting, ways. In the same vein, the terminal modellers may well be provided with the possibility to introduce changes in the parameters of existing models over the Web, and possibly to introduce other changes again, so as to arrive at other analyses and insights, frequently transmitting the consequences to other terminal modellers as well as to the initial modeller. It should be emphasised that we are here using the word "model" in the sense of Abbott (2002) as that which captures and encapsulates the full range of data, numerical computations, expertise, physical knowledge and all else that is accumulated in hydroinformatics and projects the results so accumulated in the forms of the metonymic structures that appear on the screens of the persons who must (and thereby can) transmute this into meaningful metaphors within their own more specific and circumscribed fields of concern. (We of course use the longstanding definition of a model [Abbott 2002] as the source of a collection of indicative signs that create an expressive sign in the mind of the beholder, using the terms "indicative sign" and "expressive sign" again in the manner of Husserl (1900, 1901).)

Thus, all the modellers, initiating and terminal, are persons who receive structured collections of indicative signs and, in the case of the terminal modellers, whose eyes and minds transform these into expressive signs and thence, either individually or, more commonly, through interaction with others, into symbols by employing their resources of active and creative imaginations which the hydroinformatician promotes. In the kinds of environments that are being advanced within the present context, increasingly functioning with possibly serious time constraints, most of these operations proceed over the mobile internet and should be increasingly catalysed by improved satellite-based communication coverage and, in distributed applications, more advanced mobile telephone technologies.

(As one of the leading nations in these respects, and as one where extensive public participation has become an accepted means of preparation for decision-making, China has a particular relevance to this development. Thus, in an editorial entitled "Transparency: it's the law" (Yao Bin 2008, p. 3), after castigating the many examples of social injustice that so much concern the peoples and the governments at all levels of that nation, it is observed that: "Clearly, a complete change in the above-mentioned situation lies in the establishment of a strict and transparent legislative system that features the broadest public participation". This has been strengthened by new legislation on property and water rights and responsibilities that had become increasingly urgent and which involved the active participation of many tens of thousands of persons, even though this participation itself necessarily led to considerable legislative delays in many cases. It depends of course on the omnipresence of the Web and the mobile internet. With the rapidly ongoing sinification of Africa and so many other places, the spread of public participation, as may be specialised to active stakeholder participation using the mobile internet, can be expected to expand into those parts of the world where it is needed even more desperately: see, for example, Michel & Beuret (2008).)

Whatever the geographical context, however, and further to this kind of process again, the views of each individual stakeholder may well be influenced by what first appear as "outsiders", even though this should not in principle be allowed in a court of law, so that these are, de facto, other operators again. In this case we arrive in a world of knowledge supply chains that may be quite complex and convoluted and which may span the entire space between the initiating and the terminal modellers. In the knowledge supply chain analyses of hydroinformatics, although still at an early stage, there are already two distinct models, namely one which is called colloquially (because the names are so familiar also in China!) a "Walmart–Carrefour" model, which is a centralised, exostructured "collect, process and disperse" model, and the other, called a "farmers market" model, in which all relevant contributors come together of their own volition in largely unstructured, decentralised and endostructured knowledge markets, proceeding as a "knowledge Agora". The possible attempts to subvert social justice that are inherent in both of these structures must, of course, be taken into account by both the initiating modeller and the terminal modellers interactively. The first of these models is better suited to mass-customised advice-serving systems (Abbott & Jonoski 2001) and environmental learning games (Jonoski 2002). The sociotechnical arrangements necessary in the present case, which are by far the most preferable in the situations described here, and such as can be realised by providing the means to promote the spontaneous foliation of an outer halo through the introduction of open-source software components, can only be adumbrated using as a paradigm case the introduction of what is commonly called middleware, following the use of this term by IBM.

The manner in which this middleware can be implanted from its technical side is through introducing what we have called *sites* into parts of the modelling code that are open-source and which will attract and bind a potential halo of persons and organisations. Abbott & Vojinović (2009) explain this as follows:

The analogy and associated metaphors that are introduced have been taken from elementary quantum chemistry and molecular biology, where patterns of free hydrogen ions on a living organism are steriospecific with complementary patterns on molecules that need to be bonded for this organism to grow and prosper. Consequently, in the random, chaotic movements of molecules, the free hydrogen ions on the living mega-molecule that is the living cell come to attract and bind with those of the so-far free-ranging molecules, bringing these into the life systems and temporal cycles of the living cell. In "IBM-speak", such interfaces are called *service-orientated platforms*. These can then be configured so as to provide bonding sites for outside developers, entrepreneurs and others in the water sector of industry generally. [We then speak of processes of *steriospecific recognition*.] Obviously, just as in the molecular-biology analogy that is further described later here, the persons and organisations that are so attracted and come to bond may, and in many cases will, provide further such sites so as to enrich their own operations again. They may of course use these facilities to bond with their own legacy software and the IBM paradigm case and our own experience suggest that such applications may be of great significance in our own field too, opening up even wider perspectives.

It is in this way that the environments that we are concerned to construct in this context acquire again the properties of attractors, attracting and maintaining extended halos around them. It is this process that has come to be identified with knowledge supply chain formation as a process that must be not only accommodated but also actively promoted in any post-modern organisation, understood as one that combines and integrates knowledge-provision and knowledge-brokering functions through its intentional creation of an extended halo around it. Such an organisation has then to understand that it must accordingly and simultaneously create intentionally an environment that may, and at some stage must, compete with at least some of its own productions. This in turn corresponds to the essential nature of such an innovation, whereby it also becomes a disruptive (socio-) technology (Abbott *et al.* 2005).

There are clearly circumstances in which the initiating operator, functioning as a leader in the indirect but regular sense, must identify and participate in at least some of these intervening influences and may then come to serve them also, even as they in turn serve the entire stakeholder complex. Among the most influential of these participating influences are, of course, the media, but in this place we can better eschew the contentious issues that are raised when treating with the media. In the same vein, we abjure any consideration here of the role of what are now widely called "media studies".

11.4 FOR WHAT, THEN, IS THE HYDROINFORMATICIAN SEARCHING?

What the initiating hydroinformatician is constantly looking for is cause and causality: what it is that is causing the issues that are of concern to the active stakeholders to lead to a failure to agree on a mutually acceptable arrangement, or verdict. From the first part of this paper it is clear that he or she is looking for the cause of failure, so that for the initial modeller, just as for the psychoanalyst described by Lacan (1964/1973//1977/2004, p. 22): "In short, there is cause only in something that does not work". When it can be brought "to work", through the processes of repetition, then we arrive at "The Great Work", the magnum opus of the alchemist, as both an individual transcendence of the Self and as a group transcendence of the community of Selves of the active stakeholders. Clearly, no such process is realisable in the case of passive stakeholders. In the same vein, only in the case that the initiating hydroinformatician can overcome the temptations of exercising only "deficient modes of solicitude" in the now-classical Heideggerian sense (whereby the stakeholders are maintained in a passive state, such as happens when these stakeholders are only being "consulted", "informed" and "directed", rather than being challenged-out to exercise and develop their own inherent knowledges, imaginations and judgements, and to exercise these both independently and interactively) can "The Great Work" succeed. We shall return to this concept again, but in a broader sense, in a moment.

The second immediate consequence is that the function of the hydroinformatician changes, and indeed drastically, in such situations, as corresponds to this reversion from the chemical back to the alchemical. Whereas in the pre-internet era of hydroinformatics the emphasis was on the model, in its new web-connected and mobile era the emphasis of hydroinformatics is much more on the modeller: the emphasis changes from the operand to the operator. In the words of Lacan (1964/1973//1977/2004, p. 9):

> What is it that makes us say that, despite the dazzling character of the stories … from ages past, alchemy, when all is said and done, is not a science? Something, in my view, is decisive, namely, that the purity of the soul of the operator was, as such, and in a specific way, an essential element in the matter.

In hydroinformatics similarly, the purity of the soul of the operator, which we can better here describe as the quality of the character of the modeller, becomes inseparable from the quality of the model in the quality of the total production. Thus, in modern science, the outcome of an experiment should be independent of the nature of the experimenter, whereas in a pre-modern (and thereby, as in the present case, post-modern) science this is not the case. Correspondingly, the central question now comes to be posed of "what is the modeller's desire?" It is this in turn that establishes the economy of the modeller's libidinal resources.

Thus the proper employment of the web-based SaaS paradigm in hydroinformatics depends vitally upon an authentic answer to this question, and thus, so far as is possible, an authentic understanding. Our question then becomes: with what kind of human activity, and correspondingly with what kind of participating Selves, are we concerned here? To put this question in the terms appropriated by Lacan for his field of psycho-analysis: "What grounds it as praxis?"

But then we must pose the question of what we now mean by praxis? We follow Lacan, again adding italic, when he says that "It is the broadest term to designate a concerted human action, whatever it may be, which places man in a position to treat the real by the symbolic". What we then do is to take our praxis with us, as interconnected applications in our outer world, and we let our observations of the applications "direct us at once towards some fairly well-located, specifiable points of praxis". It is then necessary, continuing to follow Lacan, to avoid a profound misunderstanding, which in our case is the misunderstanding that hydroinformatics is some form of research. We can do no better than to quote Lacan (1964/1973//1977/2004, p. 7) when answering from a similar position in his own field:

> Well, allow me to say quite clearly c in particular to the public authorities for whom this search has seemed, for some time now, to serve as a shibboleth for any number of things – that I am a bit suspicious of this term research. Personally, I have never regarded myself as a researcher. As Picasso once said, to the shocked surprise of those around him – I do not seek, I find.
> Indeed, there are in the field of so-called scientific research two domains that can quite easily be recognised, that in which one seeks, and that in which one finds.
>
> Curiously enough, this corresponds to a fairly well defined frontier between what may and may not qualify as a science. Furthermore, there is no doubt some affinity between the research that seeks and the religious register. In the religious register, the phrase is often used – You would not seek me if you had not already found me. The already found is already behind, but stricken by something like oblivion. Is it not then a complaisant, endless search that is then opening up?

Hydroinformatics provides a paradigm case of something that was found while following a closely woven web of technological innovations that were observed, during the praxis, to be brought together around one single theme even when viewed initially only technologically. The theme itself that was thereby found was not itself modern-scientific, however, or at least it could not be described adequately in the language of modern science, so that it had to be described in terms of something else again, and in this case in terms of the onto-theological categories and their corresponding languages.

The study of an apparently alternative approach, based upon the Derridian concept of *différance* (which does not function as a word, but only as a sign), showed that this onto-theological mode of expression might possibly be avoided. However, the conceptual difficulty of Derrida's (1967//1975/1989) linguistic innovation, that appeared to exemplify another approach to language than that of the onto-theological, in turn appeared as excessively abstract and thus too difficult conceptually for most of our present purposes. However, in one application at least it might be admitted here, namely when expressing the complexity of the processes that are occurring in such an enterprise and the corresponding didactic simplicity that must characterise their realisation. Even at the level of the computational systems that were in use in the first paradigm case, as described in conclusion of this chapter, these were mostly the *MIKE 21* and the *MIKE 3*, and in the second paradigm case as its successor that will be described later, this is currently the *MIKE Urban* coupled seamlessly to the *MIKE 21* with standard *DHI* dynamic viewing facilities there are some trillions of different digital operations in all possible applications of the code that might simulate the phenomena. Correspondingly, the digital user interface has at least 750 000 pixels and each of these can have a range of possible intensities and possible colours, while all of these again may change, through combinations and permutations, in an immense number of ways over time, so that we can have, at least in principle, some trillions of possible different graphical representations. Only one of the possible sequences of numerical operations is, in fact, performed and only one sequence of graphical representation also. This leads to the problem of choosing the computation and the dynamic graphical representation that are best suited to the aims of the enterprise, and in this case of promoting and indeed realising states of social justice.

11.5 THE HYDROINFORMATICS OF THE QUALITIES

The realisation of social justice in the flood risk mitigation practice of the human economy, and thus by surrogation also in the natural economy, necessitates a new kind of hydroinformatics that has already been introduced here as the *hydroinformatics of the qualities*. In this environment it is essential to understand and accept that:

The things that really count cannot be counted.

Its corresponding mathematics is then a mathematics of the qualities rather than of the quantities, for which Category Theory, with its extension so far unexplored within the present perspective, to Allegory Theory, provides a paradigm case (e.g., Freyd and Scedrov 1990). The question of how to judge whether a risk is acceptable or not requires such a mathematics of the qualities and much deeper social and philosophical considerations rather than simple calculations of probabilities multiplied by unwanted outcomes and their balancing between costs and benefits.

It is within this perspective that we are obliged to mention the totally other view of social justice that then appears when the existence of this Third Economy is accepted, and so in most societies other than those of the West. It is then usual to speak of *The City*, instead of "The Urban Conurbation", as being at the centre of any such Economy, so that we must then relativise, and even compromise, everything that we have written above by giving a moment of expression to this other view, which we hope may find a resonance in the West. We shall follow here Rowan Williams, the current Archbishop of Canterbury, when he writes (Williams 1978/1990/2002):

> The city of man, the *civitas terrena*, is built in self love, out of the human desire to make order and intelligibility and rest upon their achievements. And so two kinds of love have constructed the two cities: the earthly city, built by love of self leading to contempt of God, the heavenly city by love of God leading to contempt of self … The earthly city is built by those who are confident of their control, not only over their own biographies but over the whole human world. And this attitude Augustine rejects as fiercely in the social as in the individual sphere. Social good *is* good, peace and justice *are* to be sought with passion; but with … the recognition that they are not to be considered as signs of human glory, human authority, human capacity to tame the world … If sought and 'enjoyed', loved for their own sake rather than … directed to a greater end … , their precariousness is at once obvious, and destructive misery follows for those who put their trust in them … Those who look for a home of their own making secure only their own deepest alienation and homelessness.

It is such considerations as these that indicate that, insofar as a hydroinformatics of the qualities develops, there must occur a division, even if initially only at the *socio*technical level, between a hydroinformatics that "begins as a European possibility" (Abbott 1991, p. 76) and, in the first place, an Asian hydroinformatics. This is because in Asia, in particular, the social–spiritual environment is so different from that of the present-day European peoples, and this difference, contrary to so much Western wishful thinking, is becoming ever greater by the day (see Abbott 2008b, 2009).

The notion of *social justice*, which has been discussed at length in this part of the book, is to enforce the sociotechnical point of view within a flood risk mitigation practice in which technical parts cannot be separated from social parts and in which ethics and social justice play an important role. It is within this context that the term social justice has been defined as a state of *transcendental praxis* and within the context of stakeholder participation it has been defined as a *transcendental state of social being*. Such definitions call for a theory of social justice that takes the human values and those of nature as primary values. The theory that is based upon active stakeholder participation and environment in which fairness, maximising welfare, respecting individual rights, equality and virtue are sought by following continuous cycles in a search and realisation of justice within new contexts. Obviously, this can never be realised by the simple probabilistic cost-benefit analysis as the engagement in socially just practice requires much deeper philosophical and social considerations. Such realisation in the context of flood risk mitigation practice poses a formidable challenge as it requires the hydroinformatics of the qualities to enable such process.

11.6 CASE STUDY: THE BUILT ENVIRONMENT AS A MANIFESTATION OF SOCIAL JUSTICE: THE CASE OF THE DENMARK-SWEDEN CONNECTION

11.6.1 A first paradigm case of a stakeholder participation process

The pith of every great discovery as well as of every world plan of historical importance is the product of a propitious moment when, through favourable outer and inner circumstances, *complicated causal series or concealed causes of phenomena* – whether seen already a thousand times or proceeding along obscure paths never previously followed – *suddenly reveal themselves to the understanding.*

Arthur Schopenhauer, *The Fourfold Root of the Principle of Sufficient Reason*, 1813//1974, p. 115

A first paradigm case of a stakeholder participation process which can be regarded as a socially just process is the construction of the road and rail link across the Sound between Denmark and Sweden (Figure 11.1), a project that was concluded in 2000 without any deleterious consequences for the environment, within budget and six months ahead of schedule (Thorkilsen and Dynesen 2001).

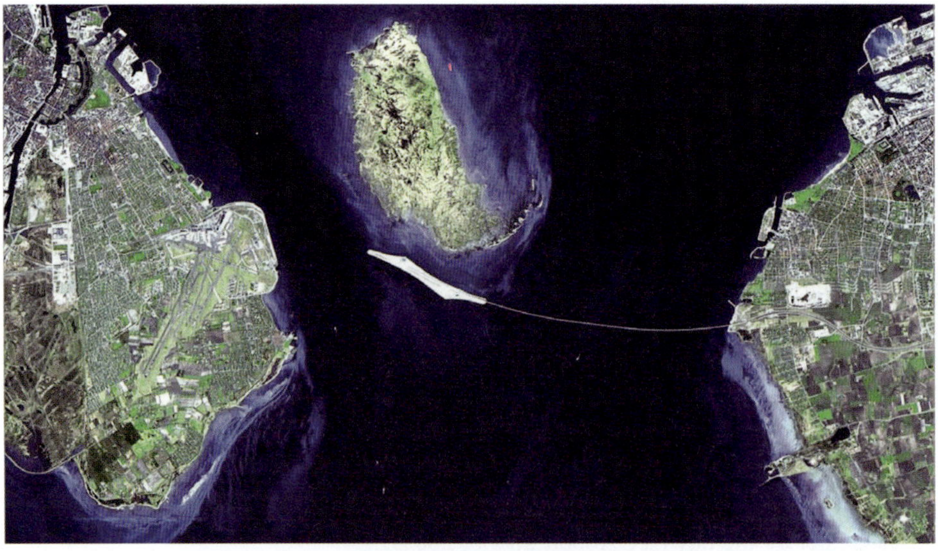

Figure 11.1 Satellite image of the completed Øresund fixed bridge and tunnel link between Malmo in Sweden (right) and Copenhagen in Denmark (left) – a world-class example of stakeholder participation (NASA/GSFC/METI/ERSDAC/JAROS and U.S./Japan ASTER Science Team)

The fixed traffic links between the Scandinavian peninsula and western Europe pass over the Størebaelt, which is entirely within Danish territorial waters even though it is also an international waterway, and across the Øresund, which is divided between Denmark and Sweden while also guaranteeing uninterrupted passage to navigation (Figure 11.2).

Figure 11.2 Locations of the Øresund and Størebaelt fixed links which now connect the Danish and Swedish peninsulas

Completed in 2000, the total value of the two projects was around £6 billion. The Størebaelt project, being within Danish waters exclusively, was the easier to legislate and the principal concerns were primarily of a technical nature. However, both projects needed the acceptance of all nations riparian to the Baltic and this required the Danish Government to ensure that the exchange of salt and oxygen between the North Sea and the Baltic should not be changed by the works. The requirements of unrestricted navigation also necessitated in both cases that the crossings should consist for one part of a tunnel, allowing unrestricted headroom for oil and gas exploration and exploitation rigs, and a bridge for vessels drawing great depths of water – such as very large oil tankers and semi-submerged nuclear submarines in the case of the Størebaelt, and cruise ships in the case of the Øresund. The tunnels and bridges needed to be connected by islands that would allow road and rail traffic to pass without major speed reductions between the base levels of the tunnels and the approach levels of the ramps of the bridges (Figure 11.3). Because of these constructions, however, major blockages of water exchange were unavoidable unless extensive compensatory dredging works were undertaken in the still-open waters. In the case of the Størebaelt crossing, the compensatory dredging did not cause serious environmental problems and so aroused only limited opposition. However, in the case of the Øresund – where much of the dredged material was chalk and clay – the environmental problems were of another order again and the public opposition was intense (Figure 11.4). It was for this reason that, in the case of the motorway and fast-train connection between Denmark and Sweden, it was essential to engage all stakeholders in a positive and meaningful way if the link was to be so much as legislated, let alone built. This necessitated the genuine empowerment of all the interested stakeholders in the project.

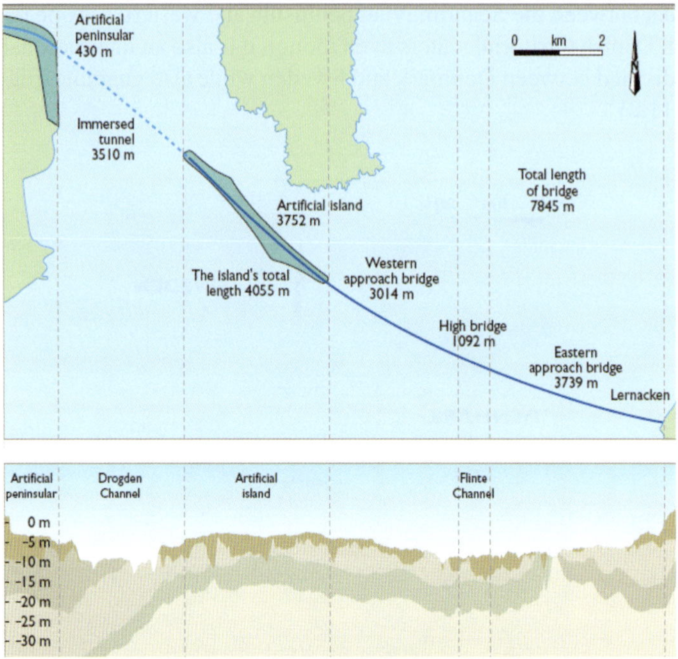

Figure 11.3 Plan and geotechnical section of the Øresund fixed link – the tunnel and bridge sections needed to be connected by a 4 km long artificial island

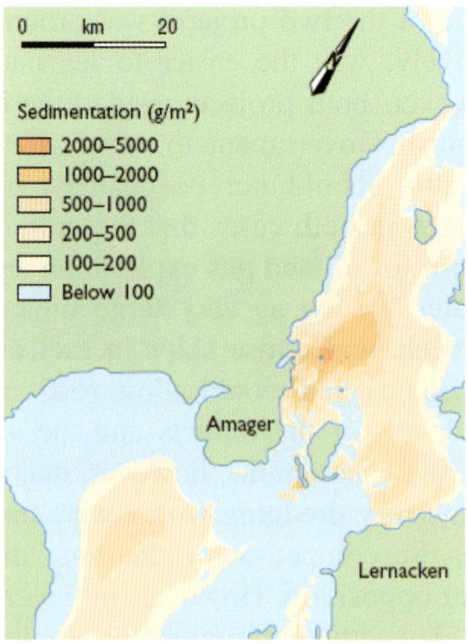

Figure 11.4 Chalk and clay sediment accumulated on the Øresund seabed at the time of the project opening in 2000. The discharge of sediments only under the most favourable (usually stratified) flow conditions, as predicted by numerical models and supported by measurements, alleviated the concerns that had led initially to such intense public opposition

Thorkilsen and Dynesen (2001) observed that:

> In effect, this amalgamation of different technologies provided such synergies that a new confidence was generated in the capacity of this new kind of integrating technology, called hydroinformatics, to provide solutions not only to unprecedented technical problems, but also to hitherto intractable social problems concerning the integrity of the natural environment. As and by themselves, of course, these earlier developments did not suffice to justify the construction of the road and rail link between Denmark and Sweden. However, without these developments it would have been much more difficult, and perhaps impossible, to persuade the various parties involved to proceed with this project. It was at this point of persuasion of the various stakeholders to grant their consents, and to incorporate these into the enabling legislation and contractual agreements, that the social side of the sociotechnical equation of hydroinformatics first became satisfied …
>
> A common feature of all the different phases of the project was the widespread uses of ICT solutions, not only functioning in the role of technical problem solvers, but also as facilitators for communicating complex decision problems to all interested persons and parties. This was done, for example, by simulating work procedures, providing visualisations of design features and especially of information about environmental aspects related to the flow of water. The task of informing the politicians, the responsible authorities and the public as a whole in a truthful and realistic way about the consequences of the construction and, inseparably from this, the empowering of these to act as genuine stakeholders in the management of the natural environment, was a task of hydroinformatics over and above its tasks within the physical project itself …

When concluding, these authors observed that:

> This overall tendency away from an exclusively representative democracy and towards direct participatory democracy (itself now being more and more catalysed by the technologies of the Internet) made it clear that public and political support could only be regained through allowing a much wider dissemination of environmental information and, in particular, information on the potential consequences in a readily understandable form.

This process accordingly introduces new forms of what, following the teachings and writings of Michel Foucault (e.g., 1972–1977//1980) are widely called knowledge/power relations, such as lead to the formation of new knowledge/power structures. These new structures are widely seen as attractive alternatives to the longer established knowledge/power relations and structures of what is commonly called "representative democracy", such as is associated with the practice of politics. Indeed, they are often regarded as reactions to what is nowadays perceived as an ever-increasing politicisation of nearly all social activities, including those of the water sector. It is in fact a mark of the strength of this reaction that the word "politicisation" (replacing the earlier "politicalisation") has entered so widely into the currency of everyday discourse, and that it now appears regularly in translations into English from other languages, not the least of which is Chinese. In nearly all cases this term is used in a deprecatory, and often even in a derogatory, sense and it is commonly contrasted unfavourably with what is called "participatory democracy", often also called "participative democracy". It is claimed here that the principal reason for this distinction and preference arises from the increasingly widespread view among many stakeholders that "representative democracy" is failing to promote social justice, in the present sector as in so many others, and that certain forms of participatory justice can do much better in this respect. Of course the politicians are still represented in participatory democratic arrangements, but they now become only one part, and usually even a small part, of a much more extended population of stakeholders.

For the present purposes, however, the concluding observations of this paper were among the most significant (see also Abbott 2007, p. 29):

> There was complete transparency in all data and the means to interpret it by its often very disparate users. This procedure was followed with particular care throughout the project in relation to the media, with the consequence that an originally dismissive and often antagonistic tone was transformed, step by step, into a more accommodating and finally quite supportive and even congratulatory attitude.

This may then be seen as a first indication that this project was perceived by a major section of the public as a manifestation of social justice in the water sector. Given the scale of the investment in this case together with the very similar Great-Belt link that complemented it (of some 8 billion Euros in present-day values) this public appreciation was highly valued within the political circles of the European Union.

Already, in Abbott (2007), it had been explained that:

> There could be no way to "manage" the stakeholders in the Øresund project and that instead it was essential to engage them all actively in the project as equal partners with a stake in its success. Although this activity thus had little or no need of formal management, it did call for a leadership by persons with influence at the highest executive level. Accordingly, a major part of top executive time on the Øresund project was taken up with this function. The empowering process started when the project was first conceived and continued until some time after it was completed. It necessitated complete honesty, transparency of decision making and the exercise of excellent communication skills, while using properly integrated technologies.

In social-organisational terms the stakeholders together functioned as what is described precisely as an anarchistic arrangement, whereby (Abbott 2004, drawing again on the *Encyclopædia Britannica* article by George Woodcock 1967):

> In order to view this category more clearly, we may recall the three principle components of all anarchistic movements, namely those of mutualism between participants (the recognition of mutual interests and the provision of mutual support), of federalism (the formation of a federal structure between the groups of participants, such as may be provided by the structure of a syndicate) and of direct action (the direct involvement of this federation/syndicate in processes occurring in society and in the world of nature).

The critical importance of a shared ethic in any such movement was similarly related already at that stage to the observation that "such a development is itself a game with all of its participants functioning as players". Thus, referring again to the work on gaming in hydroinformatics of Jonoski (2002), "the ethics necessary to support such a mode of development are self-generating and self-sustaining", a matter to which we shall return shortly.

Since the present book is concerned with identifying the new social arrangements that are required in order to realise such expectations and to relate them to the technologies that can promote and stabilise these arrangements, we are scarcely concerned here at all with the devices and procedures employed for realising such developments as, for example, Integrated River Basin Management Systems or in Integrated Urban Water Management, while the ways in which expert knowledge and understanding are employed within the knowledge/power structures of participatory democracy are very different from their employment when applied within a representative democracy (e.g., Evers 2008). Many of the current tools and facilities can be taken over from the representative approach to the participative approach, but these are then used in markedly different ways and within entirely different social environments to provide results that, as will be explained, would be unrealisable within the representative approach. Thus, although it may be claimed that many, and possibly most, of the required scientific knowledges and modern technologies have already been developed within hydroinformatics, the social arrangements necessary to employ these tools must be completely rethought and new conjunctive knowledges and tools developed accordingly. This is clearly a very demanding and difficult task, and correspondingly it is primarily these aspects that need further developments.

In the case of the Denmark–Sweden connection (see Abbott 2007), which can be regarded as a paradigm case of social justice, it was clear from the very beginning that it would be essential to engage all the stakeholders in a positive and meaningful way if the link were to be so much as legislated, let alone built. This necessitated the genuine empowerment of all interested stakeholders in the project. The key stakeholders in this case were composed of representatives of the owner–where the very form of ownership was strongly conditioned by the stakeholder participations–the contractors, Danish and Swedish parliaments and governments, their respective ministries (environment, fishing, transport, and so on) and their departments, various organs of the European Commission, representatives of governments riparian to the Baltic and certain of their ministries, the general public, represented so far as possible by their non-governmental organisations, Danish and Swedish mussel-bed farmers, Danish and Swedish eel fisheries associations and herring fishery associations, the society for the protection of the eider duck, swan and seal protection interests, and of course the media. Then (from Abbott 2007, with some italics added):

> This example illustrated that such an *infrastructural project* has two sides: it has on the one side the visible and tangible *physical infrastructure* of steel, concrete, stone, and whatever other materials, as shaped by various machines and formed by surveying equipment, GPS systems, fixed and towed measuring devices and their radio links, SCADA systems, numerical modelling systems with their seamless interfaces to a GIS, aerial photographic records and their interpreting equipment, satellite imaging devices, cartographic coordinate transformation systems and so much more again. But on the other side it has an invisible and intangible *social infrastructure* sustained by the intentions, beliefs and codes of behaviour of people, most of whom are interested in, and some of whom are actively engaged in, *this same project*. This 'inner world' of the minds of socialised people requires means to transform the data, information and knowledge provided by the physical equipment into forms that these minds can easily assimilate, each after an own individual interest and ability. The consortium charged with constructing and subsequently operating the connection accordingly set up facilities for transforming the content provided by the equipment into forms that could be assimilated by the various stakeholders, each in its own special ways. Thus the data needed by those concerned with respiring mussels was not the same as that required by those concerned with feeding swans and such differences had to be addressed correspondingly. The Internet-based system, called the *Eagle*, that carried this material out to the different stakeholders and accommodated their reactions, was a central bearing instrument in the success of the project as a whole.
>
> Thus, such an infrastructural project has to comprehend two very different aspects that often appear as two different worlds. On the one side it has to do with creating in the outer world of physical forms and activities, while on the other side it has to do with the creating in the inner world of human minds as these come together through communication subject to specific social arrangements. Being creative processes, both are technologies, the one being a technology of the outer world and the other of the inner world. The first is constantly advancing of course, but it is the second that is advancing so much more rapidly today due to the catalytic effects of advances in the Internet and the World-Wide Web and, much more again, by the rapidly ongoing advances in applications software and supporting services that are enabled by these facilities. Previously, what happened in the outer world was what was obvious because it was immediately visible and tangible, while what happened in the inner world was mostly hidden and usually unheard and unfelt in a wider circle. Now this situation changes as the inner world broadcasts its observations, inspirations and aspirations through the developments made possible by advances in the computation and communication technologies and the proper understanding of their appropriate applications. The most immediate consequence is that the media loses its quasi-monopoly in the technologies of persuasion, as these are increasingly appropriated by the web-sites, 'blogs' and other instruments of the environmental and other, less regular,

> bodies. The nature of leadership increasingly shifts correspondingly from direct and regular forms to indirect and irregular forms.
>
> Stakeholder participation changes the way in which the individual stakeholders think and feel about the world in which they live and function. This begins already with the behaviour of persons in government agencies who are otherwise so constrained in the forms of their thinking and their very being by their subjection to political imperatives. Thus, to extrapolate from an expression of Findley (1977), one could say that during stakeholder participation and interaction there emerges "a bustling Agora [a boisterous market or bazaar] where such forms are involved in endless transactions and conversations; and even though it is by the intermediation of such forms that there is a reaching-out to their individual instances, they none the less enjoy a relative independence there, a detachment in the thought-ether, that they never enjoy elsewhere". Thus even those who must otherwise be most beholden to their political masters find here an opportunity to do what they know to be right rather than what is most currently politically expedient, or 'correct'.
>
> Similarly, by their intercourse with those who identify with specific creatures, such as certain species of mammals, birds, plants and others in the world of nature, participants in this stakeholder endeavour come to attain to an awareness that the knowledge of modern science is in fact exceedingly restricted and severely circumscribed as compared with the depth of knowledge and understanding exhibited by the world of nature, and this is one of the most salutary of all experiences …
>
> Thus, through stakeholder participation and involvement, each of the individual stakeholders attains not only to a broader, but also to a higher level of collective understanding, and this leads not only to a higher state of *collective consciousness* (i.e., to experience feelings and wakefulness) but also to a higher state of *collective conscientiousness* (i.e., to act with care) that can decide issues that are in the abstract unresolvable. There thus arises a much clearer sense of responsibility and duty generally that, although it may appear to proceed ethically, in fact transcends the ethical. This is the essential nature and ultimate form – the quiddity – of stakeholder participation.

This characterisation then indicates that, to the extent that stakeholder participation can be employed to realise social justice at all, it must have a transcendental character and that this must necessarily have the capacity to transcend the level of the ethical, at least when the word "ethical" is taken in the malleable secular sense that is so widely accepted and employed in the West, as outlined earlier here.

We repeat that ethics today is merely concerned with pragmatic choices based on reasoning within whatever social arrangements happen to be acceptable at the time.

Clearly those involved in flood risk mitigation work have to take account of these aspects even while striving to create an environment that will cause the present-day stakeholders to rise above their temporalities: he or she must ensure that the evidence is presented in as clear and unadulterated a way as possible within his or her fields of competence, and especially when this evidence is conflictual. He or she can never even so much as try to impose a personal view upon whatever is the current ethic: he or she can only try to create situations in which the living stakeholders themselves are confronted with the consequences of their own opinions and their own actions in their own time.

The greatest danger in this case is that the whole discourse will descend to a state that is commonly described by its German name, of *Gerede*, mainly because it can only be translated much too innocuously as "gossip, chatter and idle talk", while it is in point of fact much more dangerous in its consequences than these translations might suggest (see Heidegger 1927, pp 167–170//1962, pp 211–214). Accordingly, those participating must be provided not only with the data, the information and the knowledge relevant to the project, but also the understanding of the reasons why the operations in the outer world are deemed necessary to the project as a whole. In some cases, at least, this necessitates following the very thought processes, including the trains of reasoning, that underlie the engineering decisions that are being taken during the project. The process of creating an environment of transparency is a sociotechnical process in which technical developments in such areas as the World Wide Web and mobile telephony are woven together with the emergence of group and more general social structures that may in some cases lead to the creation of new facilities, new institutions and new legal frameworks as already exemplified by the case of the Denmark–Sweden fixed link (see again Thorkilsen & Dynesen 2001). This process is observed to provide also the means to arrive at deeper and broader conflations of understandings within the stakeholder community as a whole. This exemplifies only one of the many "technologies of the inner world" that this discipline requires.

Another "technology of the inner world" expresses itself in terms of the coherence and the intelligibility of the participative discourse, and the leader of such a participative endeavour must be well versed in the exigencies of attaining to these desiderata. It should, however, also be mentioned that it is these developments which, when combined with the above conflations of understanding, lead to the exposure of any attempts by certain special-interest groups to promote their more selfish and otherwise hidden agendas, while corruption can also be more quickly identified and exposed in this way. Thus, to the extent that such states of coherence and intelligibility are induced throughout the stakeholder community, so there occurs a positive opening-up in the inner world of the project as it progresses in time, and then in such a way that this is congruent with the positive opening-up that is occurring simultaneously, and indeed in this case synchronously, in the outer world of the project.

11.6.2 The possibilities for active stakeholder participation in the 'Third World'

(First published as MANAGING THE INNER WORLD OF INFRASTRUCTURE, 32, PROCEEDINGS OF THE INSTITUTION OF CIVIL ENGINEERS – CIVIL ENGINEERING, 2007, 160, No. CE1 ISSN 0965 089 X)

Despite the consumption of natural resources by a more prosperous part of human society at a rate that is clearly unsustainable, in another and larger part of human society the number of persons living in abject poverty and who are correspondingly subject to the most extreme miseries of hunger, illness and early death, continues to increase. This is the greatest scandal of our age and one for which the civil engineer must experience a direct and even a personal responsibility. Of course many civil engineering works have been undertaken and are still in progress to redress the situation, but their success is clearly marginal and often unsustainable. There is correspondingly a great campaign to increase expenditures on civil engineering works in this so-called third world, especially in the areas of infrastructure and particularly in the water sector. But these efforts have little overall effect and, indeed, are seen by many with a long and direct experience of working in the world as fundamentally misguided and consequently futile.

At the same time, there is a consensus among some persons with a long and close experience of working on infrastructural projects in the third world that the situation arises because of the absence of any real, or genuine, stakeholder participation. This is in a large part because most of the true or genuine stakeholders – the great majority who have the most direct stake in the outcome of the project concerned – are not represented at all, having been disenfranchised by a richer and more influential minority. There have, of course, been many attempts to change this situation by direct interventions on the part of international and non-governmental organisations, primarily financed by first-world societies but directed towards the enablement of specific communities in the third world.

Whatever the good intentions of community enablement, however, the autonomy of the stakeholders concerned is necessarily compromised. The introduction of a genuine stakeholder participation – in which the poor and oppressed could take a major and indeed leading part in a project – necessitates the taking of a socio-technical position that is way beyond the capacities of most of the local community. This situation can in turn be traced to the long-ongoing struggle for the control of the development agenda between the various organisations involved, which has actively discouraged the development of any serious capacity in the socio-technical leadership of civil engineering projects within the this world. Instead there reigns a technocratic attitude that is largely dismissive of such initiatives. There are singularly few persons who are prepared to grasp the nettle of genuine stakeholder participation and, inseparable from this, stakeholder empowerment, in the third world.

More specifically, the relations between those promoting such projects and those who are realising and operating them are afflicted by technocratic attitudes that Heidegger characterised in terms of "deficient modes of solicitude". The manner in which even an apparently positive mode may in fact be deficient is often experienced where an aid recipient is only "receiving instructions", basically only being trained, rather than being challenged to develop his or her inherent capabilities and to exercise these independently.

From a socio-technical point of view, the poor and oppressed have been for the most part disenfranchised within projects – even when they have been to some extent enabled–because they lack power, even as the success of the projects depends fundamentally upon the knowledge of just these persons. This disenfranchisement occurs in turn because the knowledge that this great majority possess is for the most part not accepted as such by those who are nominally responsible for the projects concerned, who see themselves and the more privileged minorities with which they cooperate as the repositories of "the only true knowledge", which they like to associate with "modern scientific knowledge".

Foucault exposed "the inhibiting effect of global, totalitarian theories" of this kind. The presently ongoing, even if only partial and ephemeral, insurrections of subjugated knowledges that we observe in many places today have the capacity, if properly led, to provide the only realistic means to overcome the present impasse in third-world development. Correspondingly, this is currently the most relevant aspect of stakeholder participation, even if many, and probably most, of these stakeholders are currently dispossessed, terrorised, driven from their lands, deprived of their water and abused in every other way, and all of this on a massive scale. To the very extent that they succeed, movements of this kind can come to constitute a rejection of the mandate claimed by the respective governments to exercise violence and the threat of violence within this sphere – and this, in turn, often leads into other forms of insurrection again. This process can only be mitigated when democratic means exist or can be imposed, either to prescribe limits to the mandate, as in some parts of India, or to promote genuine stakeholder participation against entrenched privileged minorities, as in several parts of Latin America.

A socio-technical perspective that accepted this and worked out from it, understanding that the greatest crime of all is to deprive people of the possibilities to exercise their own deep-rooted knowledge and understanding, could quite easily overcome this present untenable situation, and then at a small fraction of the costs currently being incurred in the largely euphemistic name of "development aid".

Part III

Scientific and Technical Aspects of Flooding

Chapter 12

Floods and drainage systems

12.1 OVERVIEW OF PART III

As noted earlier, flood behaviour and therefore risk is influenced by a range of factors that can vary significantly from one location to another and need to be understood and managed locally as well as in a wider context. In this part of the book we shall review some of the key scientific and technical aspects that are necessary to attain to a sustainable flood risk mitigation. We describe how floodplain shape, slope, storage, development, vegetation and flood controls, both natural (receiving waters) and man-made (roads, buildings and other structures), all have a significant influence on the routing of flood flows (i.e., hydraulic processes) and thereby the derivation of flood behaviour from both hydraulic and hydrological point of view. We note that physically-based computational modelling is invaluable for this purpose. With such models, it is possible to explore the generation of floods and to simulate the consequent effects in response to any control actions. Therefore, particular attention will be given to the modelling theory and issues that are important for practical applications.

We start with the review of different types of collection systems and describe their key elements. We then proceed with discussing the key concepts of urban hydrology and hydraulics. We then move to modelling concepts, data collection and modelling practice. Certainly, a thorough discussion of these topics could easily fill up this book on its own. Therefore, we can only mention some of the key aspects and more interested readers are encouraged to seek further information from our references and other sources.

12.2 TYPES OF FLOODS

We have already noted elsewhere that the amount or volume of runoff (and the risk of flooding) generated by a rainfall event in an urban area, and the corresponding rate of rise of water levels, depths and velocities will depend on a number of factors. Some of these factors are: rainfall characteristics (intensity, duration, etc.), land characteristics (percentage of developed area, topography, slope, permeability, degree of moisture already present in the soil, etc.), surface vegetation characteristics (extent, type, etc.) and drainage system characteristics (conveyance, efficiency, capacity, etc.). Another factor that can also increase the potential of urban floods is the lack of appropriate planning and development regulations and policies. Impacts from such factors will vary in relation to different types of floods and their mechanisms. In this respect we can distinguish five major types of urban floods:

- *Fluvial floods* which result from rivers breaching or overtopping flood defences and inundating urban areas;
- *Coastal floods* which result from tidal or storm surges in cities close to the coast or in deltas;
- *Flash floods* which are caused by the rapid response of ephemeral streams to heavy rainfall, related, *inter alia* to steep slopes. Recent work on flash floods (Ruin *et al.* 2008) and their rapid response during storms requires a reexamination of the central assumptions concerning the role of flood protection during flash floods and the importance of flash floods compared with other types of flooding.
- *Groundwater floods* which are caused by the groundwater table rising. This tends to occur in low-lying areas, and especially those that are underlined by permeable rocks (to provide aquifers) after much longer periods of sustained high rainfall. Compared to other types of floods (e.g., flash floods) flooding due to groundwater has much longer response times. It also takes longer time to dissipate and flow away underground because of its slow movements;
- *Pluvial floods* which are due to heavy rainfall directly on the urban area such that the runoff exceeds the capacity of the drainage systems;

Certainly, flooding can be also experienced form lakes through inflow or wind induced water level rise. Furthermore, there could be a case where two or more of the above mentioned flood types are combined. Examples of different types of urban floods are illustrated in Figure 12.1.

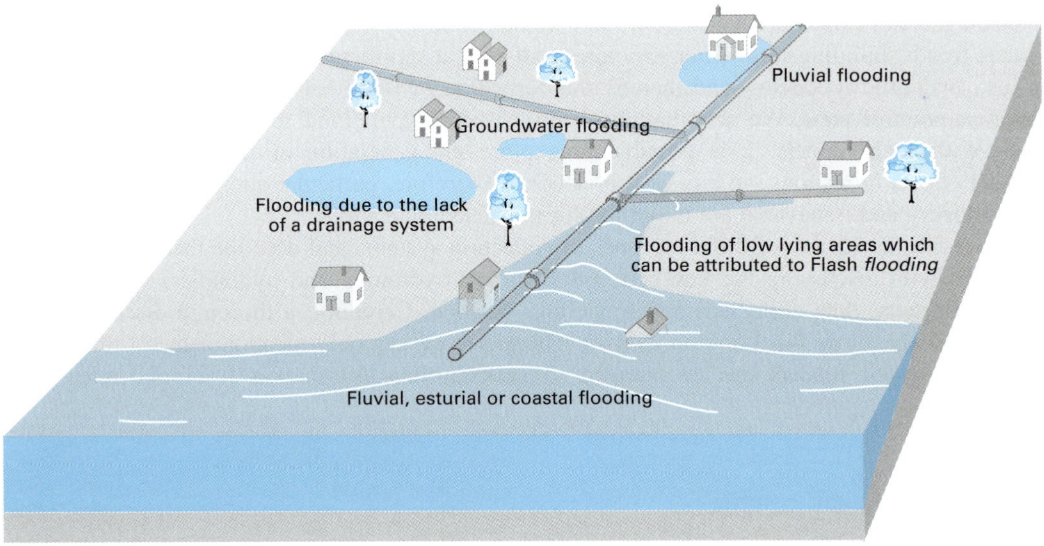

Figure 12.1 Illustration of different types of floods

In addition to the distinction between different types of floods, the magnitude and severity of floods can be also classified into localised (minor) flooding, larger-scale (major) flooding and catastrophic (extreme) flooding. Such a classification is usually done on the basis of the population and number of structures affected.

12.3 IMPACTS OF FLOODS

Experience to date has shown that flooding can produce widespread impacts in urban areas. Almost any type of industrial, commercial, or residential development located on a floodplain is vulnerable to flooding. In flood incidents, buildings can sometimes experience small damages that can be easily repaired, but sometimes they will incur significant damage that is beyond repair. Household contents, furnishings and business inventories, among other entities, can easily be ruined if they are not removed to a safe location before the flood strikes. Very often urban floods cause extensive damage to public utilities, disruptions to the delivery of services, to community life and to the economy. Loss of electric power and failure of communications due to floods are also very common. In some cases, drinking water and wastewater treatment facilities may be temporarily put out of operation. Other impacts, such as erosion, sedimentation, land slides and relocation of people from their homes, are also experienced during certain flood events. In addition to these, inhabitants can face the threat of such risks as electric shocks, gas poisoning or even explosions, the outbreak of waterborne diseases, danger to traffic on roads, etc. Furthermore, severe floods can also cause psychological traumas or even loss of human lives due to incidents of physical instability and drowning.

Public perception of floods is subjective, whereby it may differ between individuals. This is due to the fact that despite the negative effects of floods, there may also be some positive consequences. For example, recharge of groundwater, provision of water for crops, supply of fresh water for irrigation and domestic use, increased fishery resources, improved navigation transport, economic opportunities for provision of services to those affected by the flood water, etc. Overall however, the economic losses due to floods are much more significant than the benefits since they usually affect a much larger area than the area directly inundated and they may endure for a much longer time than the flood incident. This is not only due to the physical damage to buildings and infrastructure but also due to the loss of working hours as the working population may not be able to get to their jobs. Also, depending on the service and products provided, some commercial businesses and industries may be more severely affected than others.

There is also a wide range of flood impacts on public health. According to the WHO (WHO 2002) in terms of occurrence, the flood impacts can be categorised as:

- those happening during or immediately after the flooding;
- those developing in the days or early weeks following the flooding; and
- longer-term effects, which may appear after and/or last for months or years.

All these categories can be divided into direct and indirect health effects. Direct health effects occur during the flood itself and are caused by the direct contact with the flood water. They include the following:

- Mortality from drowning, heart attacks and injuries. The number of deaths associated with flooding is closely related to the life-threatening characteristic of floods (rapidly rising waters, deep and cold flood waters, objects carried by the rapidly flowing water) and the behaviour of victims.
- Injuries (sprains/strains, lacerations, contusions, etc.). These are also likely to occur in the aftermath of a flood disaster as residents return to their homes to clean up damage and debris.

Indirect health effects can be the consequence of damage to major infrastructure and property. They include the following:

- Infectious diseases (gastrointestinal diseases, dermatitis, conjunctivitis) and some rare cases of vector borne diseases. Infectious diseases are not common and are normally confined to illnesses endemic to the flooded region. The risk of introducing new diseases, such as vector borne diseases, is negligible.

Nevertheless, the rates of diseases that were present before a flood may increase because of a reduction in sanitation or overcrowding among displaced persons.
- Poisoning caused by the rupture of underground pipelines, dislocation of storage tanks, overflow of toxic waste sites, or release of chemicals stored at ground level.
- Post-traumatic stress disorder, including anxiety and depression, psychosocial disturbances and suicides. Aside from the trauma caused by the flooding itself, many mental health problems stem from geographical displacement, damage to the home or loss of familiar possessions, and often lack of insurance. They may continue for months or even years after the event itself.

The effects of flooding can be particularly devastating to already vulnerable groups such as children, the elderly, the disabled, ethnic minorities and those with a low income who, because of social, political and economic constraints, experience special health care needs. Consequently, these groups may suffer more from flooding and may need special consideration during the response and recovery periods.

In terms of the environmental impacts, the quality of runoff has a direct influence on pollution problems. Since the stormwater runoff has the effect of cleaning urban surfaces and drainage channels, its quality depends on the land use, waste disposal and sanitation practices. The so-called 'first flush' phenomena is related to the initial surface runoff of a rainfall event and the fact that the pollutant concentrations tend to be much higher at the beginning of an event then in the middle or at the end of the event. This term is often also used to address the first flood after a dry period, which is supposed to contain higher such concentrations than a subsequent one. Consequently, these high concentrations of urban runoff result in high levels of pollutants discharged from the stormwater system to the receiving waters. High rainfall intensities may cause severe erosion to the land and there is a wide range of pollutants that can enter receiving waters via the stormwater system. They typically range from gross pollutants, particulates, oxygen demanding material, heavy metals, pesticides, hydrocarbons, suspended solids and other soluble toxins.

12.4 COLLECTION SYSTEMS

The urban drainage practice based on scientific methods was established in the second half of the nineteenth century. Development of collection systems was initiated due to the need to mitigate public health issues. Initially, drainage of stormwater and groundwater was uniformly viewed as part of the more general issue of sewerage. In 1880s, the concept of separate systems was developed and professional engineers have started to address urban drainage and sewerage separately. Rome was one of the first cities to develop an extensive collection system. However, the beginning of the modern era of drainage and sewerage collection systems began in Paris and spread to other cities in Europe and then to the US. Since then, there has been an ongoing issue of whether it is better to collect the stormwater and wastewater together in a single network of conduits, or to do so separately in two networks. Some of their key advantages and disadvantages are discussed later here.

Before we explore different aspects of collection systems (a.k.a. urban drainage systems) we need to differentiate between different types of water that require collection and transport from urban areas and to conceptualise their different phases. The supply and use of potable water in urban areas usually leads to the contamination of that water with significant pollutants (domestic or nondomestic) that are dissolved or in suspension: see Figure 12.2. This contaminated water is referred to as *wastewater*. The wastewater needs to be returned to the environment and should be disposed of safely. Similarly, the extensive paving of the ground surface in urban areas means that the infiltration of rainfall is limited compared to the former natural catchment, and therefore excess rainfall-runoff (i.e., *stormwater*), which has the tendency to pond on the ground surface, has to be collected and transported in order to limit the frequency and magnitude of surface flooding. In both cases there is a need to collect the unwanted water and to convey it for processing or discharging to an appropriate receptor. Whereas in some cities around the world open drains along streets have been and continue to be a primary means of evacuating wastewater and stormwater from the urban area, the health risks posed to the community are considerable. In dry weather conditions, the dry weather flow component is collected from domestic, public and industrial sources via underground pipes, and conveyed under gravity either to a suitable receiving water, such as a stream, river or coastal waters, or to a treatment works where a proportion of the pollutants are removed before the effluent is discharged back to the environment. In wet weather conditions, as the surface runoff inflow and infiltration increase, the excess flows are allowed to spill into the receiving waters via overflows: see Figure 12.2.

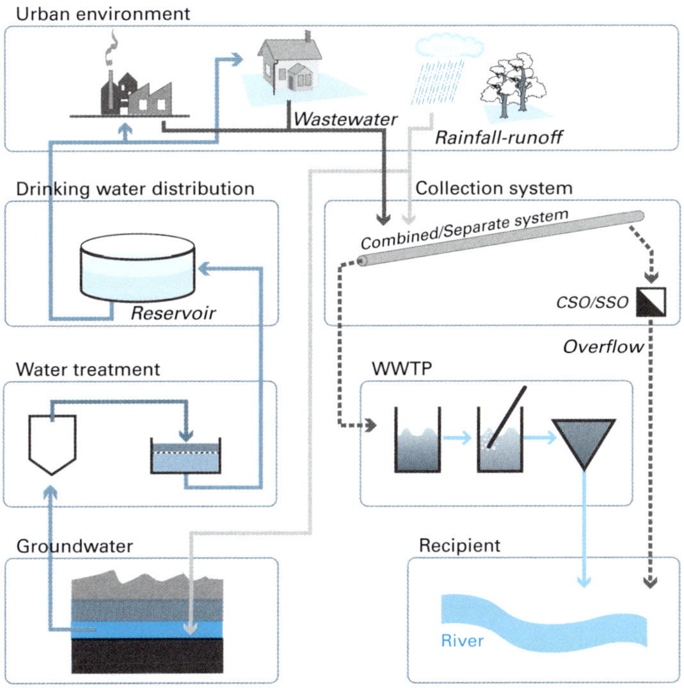

Figure 12.2 Urban water phases

The excess rainfall-runoff forming the stormwater component may similarly be collected from roofs, roads and other paved areas and conveyed through surface channels which may discharge to underground pipe networks. The objective is then to dispose of the stormwater from the more frequent rainfall events to the nearest appropriate stream, river or coastal water under gravity. Of course, this will not necessarily be possible for rarer events. Recognising this, some countries make provision for these events by designing routes (or overland flowpaths) through an urban area for the excess flows and effectively creating a dual drainage system. In such cases the pipe network is regarded as the 'minor' system and the above-ground channel network and associated storage areas are treated as the 'major' system.

Typically, collection systems can be in a form of open channels or closed conduits. The term 'sewers' usually refers to the network of closed conduits (i.e., piped systems) which are used to convey the flow from a certain area, whereas the term 'sewerage' refers to the entire wastewater/stormwater system infrastructure in the particular city.

There are two main types of collection systems: combined and separate. The following Figures 12.3 and 12.4 illustrate these two types of systems.

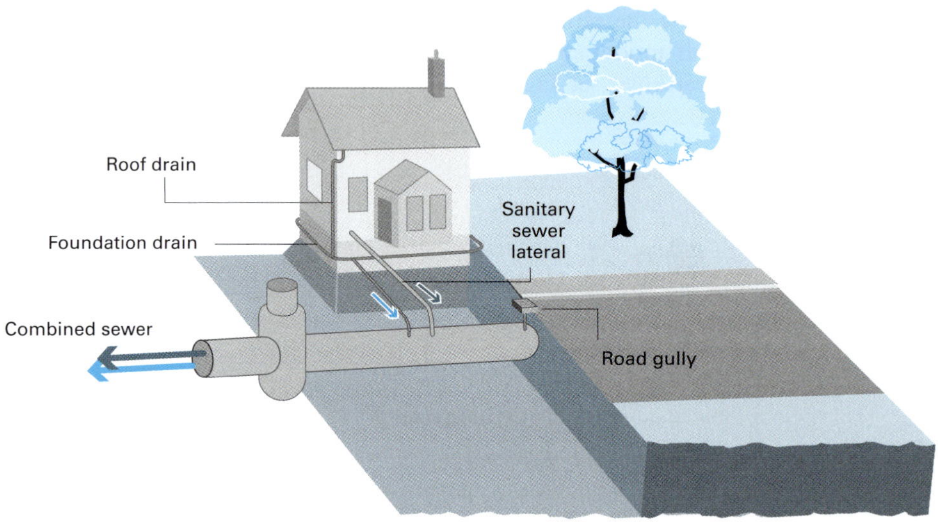

Figure 12.3 Combined system

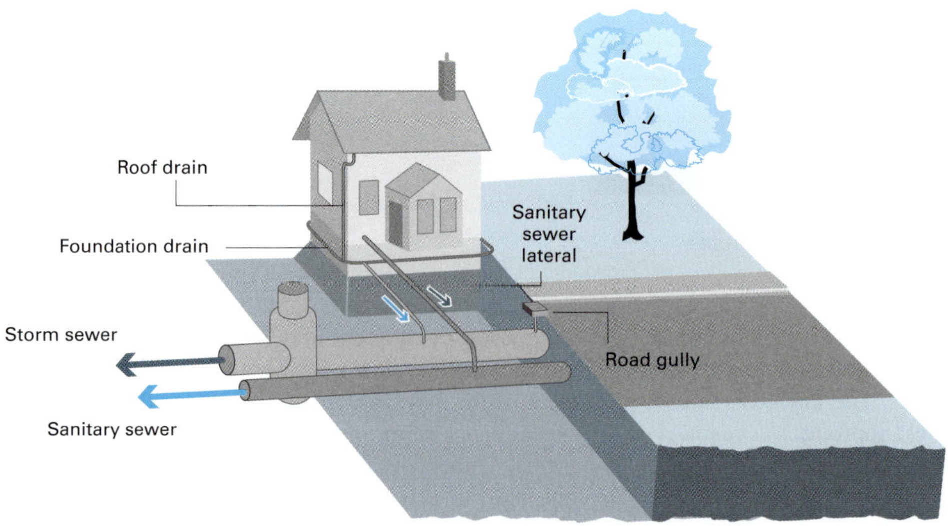

Figure 12.4 Separate system

In terms of effectiveness, the two types of systems (i.e., separate and combined systems) have their own respective advantages and disadvantages and there is no clear cut view upon which system is the better. It has always been at first sight an attractive proposition to collect both stormwater and wastewater in the same system. Generally, the collection systems introduced during the industrial revolution in Europe followed the natural drainage paths so that the full effect of gravity could be taken into account. Drainage engineers have tended to avoid the need for moving parts in their systems wherever possible, and to rely on pumping primarily to lift sewage up into a treatment works at the lowest point in the urban catchment, usually close to a river, estuary or coastal water. It is only as systems have aged, and engineers have striven to get the most out of the systems, that they have resorted increasingly to moving parts such as gates and additional pumps with or without real time control.

A system which contains elements of both of these two systems is known as a hybrid system. In addition, there are also small collection systems (which use pressure or vacuum and they are known as vacuum systems) and simplified (i.e., condominial or communal) systems. Vacuum sewerage systems consist of a vacuum station (where the vacuum is generated), the vacuum pipeline system, collection chambers with collection tanks and interface valve units. Simplified sewers are constructed using smaller diameter pipes laid at a shallower depth and at a flatter gradient than conventional sewers. Apart from small diameter pipes, they also contain simple inspection chambers or flushing points (instead of standard manholes) and interceptor tanks to prevent settleable solids and trash from entering the sewer.

Overall, key elements of collection systems (Figures 12.5, 12.6 and 12.7) can be summarised as follows:

- reticulation or collection networks (pipes, rising mains, channels, culverts, manholes, etc.),
- gully inlets (a.k.a., catchpits),
- ventilation shafts,
- chambers and tanks,
- combined sewer overflows and sanitary sewer overflows,
- outfalls,
- controls (orifices, gates, penstocks, vortex regulators, flap valves, etc.),
- pumping stations, bifurcations, weirs, inverted siphons, treatment and other ancillary structures,
- storages (offline, online, open, closed, above-ground, below-ground),
- watercourses,
- overland flow paths and designated floodplains,
- dams.

Figure 12.5 Elements of collection systems

Combined sewer overflow (Pattaya, Thailand) | Pumping station (Pattaya, Thailand) | Stormwater retention pond (Belo Horizonte, Brazil)

Figure 12.6 Elements of collection systems (Photographs: Zoran Vojinović)

Natural stomwater channel | Stepped chutes | Stormwater culvert and lined channel

Figure 12.7 Elements of collection systems (Photographs: Zoran Vojinović)

Most of the time collection systems are designed to collect and drain (or transport) wastewater and stormwater by gravity. However, in some cases it is not feasible to achieve efficient operation of the system only by gravity. In such a situation, the use of pumping facilities is inevitable. Such facilities operate on a stop-start basis with respect to the water level in the wet well. The hydraulic performance of a pump is determined by its characteristic curve which is expressed as a head-flow relationship.

An important element of collection systems are overflow connections, which mainly serve as relief points for the separate wastewater systems (a.k.a., sanitary or separate sewer overflows – SSOs), or as a means for conveying excessive flows from combined pipes (a.k.a., combined sewer overflows – CSOs) into the receiving waters or the separate stormwater system. Overflow structures are built into systems enabling discharge into detention basins or designated locations in the environment, primarily where the effect on human health is minimised. The objective is to ensure that sewage does not flow back up through the system into properties in the event of blockages, mechanical or electrical failures, or when excessive stormwater enters the system. Hydraulic design of overflow structures requires a special care in order to avoid frequent (or premature) spills (in case of too low weir crest level) or excessive surcharge upstream (in case of too high weir crest level). To minimise impacts form sewer overflows, structures (or facilities) such as stilling ponds, vortex separators storages and screens are often designed in conjunction.

In terms of the system layout, collection systems usually consist of a number of pipes connected in a dendritic network (Figure 12.8).

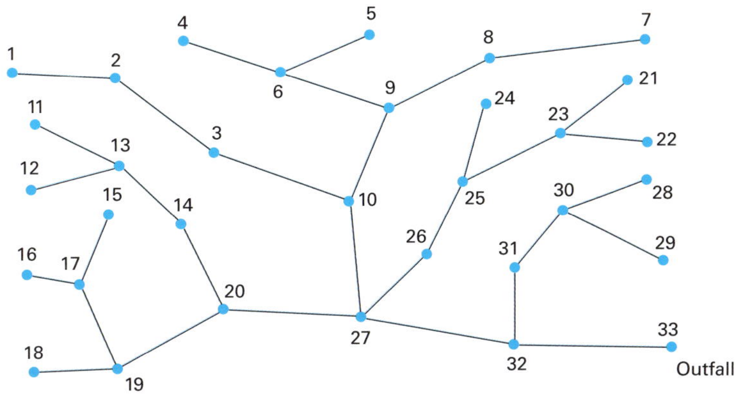

Figure 12.8 Dendritic network

A collection network is more suited to a dendritic structure because the aim of the drainage engineer is to design a system that drains under gravity, so far as that is possible. Connections between different branches may be introduced, especially in the larger networks, in order to balance the loading on the network generated by spatially varying rainfall-runoff, or to bypass pipes of limited capacity. Diversion structures are also common in sewerage networks, so that excess loads on particular parts of a network may be limited. For example, overflows are used to divert excess flows from entering a wastewater treatment works.

12.5 WASTEWATER SYSTEMS

As described in the previous section, wastewater systems are needed because of the requirement to dispose of water that has been used for washing, cleaning, food preparation, and flushing toilets, whether in domestic or in industrial processes. The main sources of wastewater can be grouped into: domestic, non-domestic (commercial and industrial) and infiltration and inflow. The domestic component of wastewater, which refers to the wastewater from residential areas, has a strong link with the water usage in terms of the patterns as well as the quantities involved. Usually, the percentage of wastewater discharged with respect to the water consumption is between 70 and 95% depending on local climate conditions, availability of water, life style, culture, and so on. The pattern of daily (i.e., *diurnal*) production of domestic wastewater usually varies between working and non-working days and between different seasons. Therefore, the nature of wastewater flow is continuous with a repeatable pattern.

Typically, the wastewater contains dissolved substances and particulate matter that have various pollutants attached. Much of the particulate matter will normally be carried in suspension, but the coarser, denser material may be carried as a bed-load along the bottom of the channel or pipe. The range of chemicals in wastewater includes chemicals such as nitrites, phosphates, dissolved oxygen, carbon, heavy metals, as well as materials that have chemical and biological oxygen demand, bacteria, and a sophisticated range of trace elements originating in the various modern substances that are incorporated as ingredients in modern washing and other products. Water is a very good solvent and dissolves most substances, but there are certain chemicals that it is unable to dissolve, such as various oils. These are carried along in wastewater as colloidal globules in suspension or on the surface.

In most urban areas the aim is to provide wastewater collection from every significant building where people live or work. Pipes with small diameter run from kitchen sinks, baths and toilets, and possibly surface drains in back-yards to sewers along the adjacent street. Such sewers are connected in a network that is aligned mainly along the urban roads or the paths of the natural stream network identified by the topography, and convey the wastewater essentially under gravity to a treatment works, Figure 12.9.

Figure 12.9 Wastewater treatment plant in Tallin, Estonia (Photograph: D. Brdjanovic)

The pipes are sized on the bases of estimates of the total wastewater flow draining to each pipe. This is done on the basis of product of the number of people living in an area and the volume of generated wastewater per person per day, plus allowances for infiltration and industrial effluent discharges. Furthermore, a peaking factor needs to be included in order to account for peak flows. Another factor may be included to take into account the standard of living in the area, as this is known to affect the amount of water used, and therefore the amount of wastewater produced. Where the time variation of the wastewater flow is important, such as planning the daily operation of a treatment works, then a standard daily unit hydrograph can be used that has peaks in the morning and evening when people are getting up or having their evening meal. Typically, the hydrograph will have small flows during the night: see Figure 12.10 and also Chapter 16, Section 16.3.

Traditionally, the sewer systems are designed to carry sediments by specifying the minimum 'self-cleansing' velocity (e.g., 0.6 m/s). This approach has proved to be only partially efficient due to the fact that it does not relate to the characteristics and concentrations of sediments as well as the hydraulic capacity of sewer flows to transport sediments. Alternatively, the new practice has taken into consideration methods that determine both local bed shear stress values and mean bed shear stress values (the first based on the logarithmic velocity profile and the other on the Reynolds shear stress distributions using a method based on the energy slope).

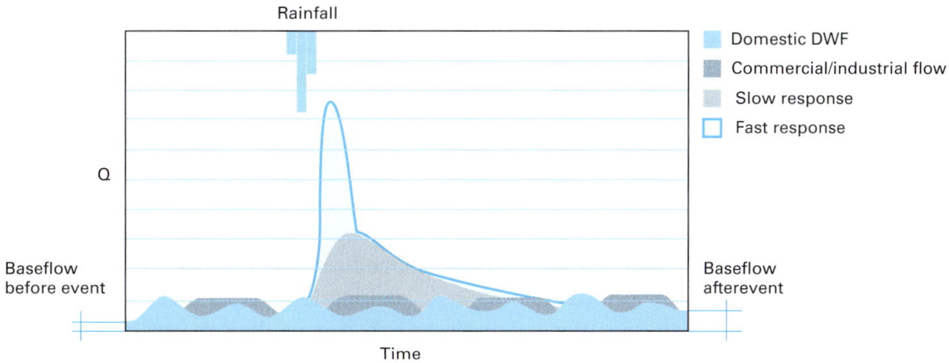

Figure 12.10 Different components of wet weather flow

Generally a treatment works is designed to take, say, 6 to 8 times the average dry weather flow. This is such that the treatment works can accommodate the daily flow variations comfortably. Because the wastewater draining from housing and industry is reasonably consistent over a long period of time some storage upstream of a treatment works is normally sufficient to balance any extreme wastewater flows due to unusual events. However there are other factors at work which can make even the storage in the sewerage system insufficient to prevent flooding from the network before reaching the treatment works. Often the most important factor is the illegal connection of stormwater drains to the sanitary sewerage system. Developers sometimes make mistakes about which is the separate sanitary sewer, or which is the separate stormwater drain, and connect the collectors from buildings to the wrong service. This portion of water that enters the pipes through lateral connections (which are often referred to as 'wrong' or 'illegal' connections) is referred to as *inflow*. Typically, in such situations, smoke surveys are carried out to detect the source of such connections.

Another factor is the *infiltration* of groundwater from around the pipes (or the water from the water supply/distribution pipe leaks), such that the pipe network is acting like a land drain. The extent of infiltration usually depends on the quality of the design and construction works, the material and the age of the system, the surrounding soil characteristics and frequency of surcharge. This is particularly difficult to manage because, although grouting and pipe lining can improve the integrity of the main pipe network, it is not so easy to repair the smaller pipes taking the wastewater from buildings to the main sewers.

Due to both inflow and infiltration, the wastewater flow can be augmented to become several times the normal dry weather discharge, so that the pipe network becomes more of a storm drain than a wastewater collector, Figure 12.10. If the treatment works cannot cope with the peak flows that occur, and the storage facility upstream of the works is inadequate, then provision has to made for excess flow to be discharged directly from the sewerage network to the environment, whether to urban streams or to coastal waters. This can generate difficult water quality conditions in the receiving water.

In addition to inflow and infiltration processes (which are often abbreviated as 'I and I', or 'I/I'), the reverse process, where the wastewater leaks out from pipes, also known as *exfiltration*, gives rise to yet another concern as it creates the potential for groundwater contamination. The extent of exfiltration usually depends on similar factors to those that have impacts on infiltration process.

The key to planning and implementing effective strategies for designing or rehabilitating wastewater networks are dynamic models, and not just the steady flow models that are traditionally used for sizing the pipes in such networks. The reason for using a dynamic model is to take into account the short term responses of the flow in a network due to the stormwater component. A number of authorities around the world have developed procedures based on the use of a dynamic model for wastewater systems analysis to identify and manage illegal (stormwater) inflows and unwanted groundwater infiltration and to assess the impacts of combined and sanitary sewer overflows. Such procedures depend on the strategic use of short term flow surveys that take advantage of technological developments in velocity and depth monitors in pipes.

12.6 STORMWATER SYSTEMS

An urban area is characterised by an extensive sealing of the land surface using concrete, asphalt, stone and other construction materials in order to provide buildings, parking lots, roads, pavements/side-walks and cycle paths. Even where some of these surfaces are reasonably porous their existence means that rainfall (or snow melt) which would normally infiltrate to the subsurface is limited or prevented from doing so. As a consequence the runoff over the paved surface tends to collect in surface depressions giving rise to the need for special provision to be made in order to remove the collected water. This is done through adjusting or adapting to the surface topography by providing a network of carefully positioned and sized surface channels or through the construction of a network of underground pipes that accept the stormwater through specially designed inlets (e.g., road gullies). In both cases, stormwater is conveyed to a convenient urban stream, pond or lake often following the natural drainage paths, or for urban areas adjacent to the sea to a coastal outfall. Most of the time, separate stormwater systems will remain empty during dry weather periods and they will only operate during and immediately after the wet weather periods. Care has to be taken to ensure that the conduits, including the receiving waters are large enough to convey a certain flow. There is a limit to the flow that can be conveyed due to the cost of constructing the channel or pipe network. This cost has to be balanced against the damage that would ensue if the generating storm created a runoff with a potential discharge in excess of the capacity of the conduits, leading to unacceptable surface flooding. In other words, the drainage network has a limiting capacity which defines a return period of flow beyond which damage due to flooding may occur. A consequence is that any storm drainage system can only limit the frequency of surface flooding and cannot prevent the phenomenon under all circumstances, Figure 12.11.

Figure 12.11 Construction of a rectangular stormwater section in the Central Business District in Auckland, New Zealand (Photograph: John Tettoroo)

Flooding in the networks of channels or pipes can occur in one of two ways: water can emerge from the network to flow overland, or the runoff locally may be too large to enter the network and so ponds-up on the surface, possibly flowing away from any inlet or channel. Where the management of excessive surface flooding has been recognised as important, separate provision has been made for overland flood flows as well as the traditional above ground channel or below-ground pipe networks. In such cases, urban drainage systems are divided into *minor* (or primary) and *major* (or secondary) systems; see Figure 12.12. Minor systems therefore refer to the underground pipes and open drains, culverted watercourses and highway drains, whereas major systems refer to the above ground drainage and consist of overland flow paths and the floodplains associated with urban streams. In some cases, the major system utilises the roads as part of the drainage system. In such cases, the multiple use of roads calls for special design and maintenance considerations.

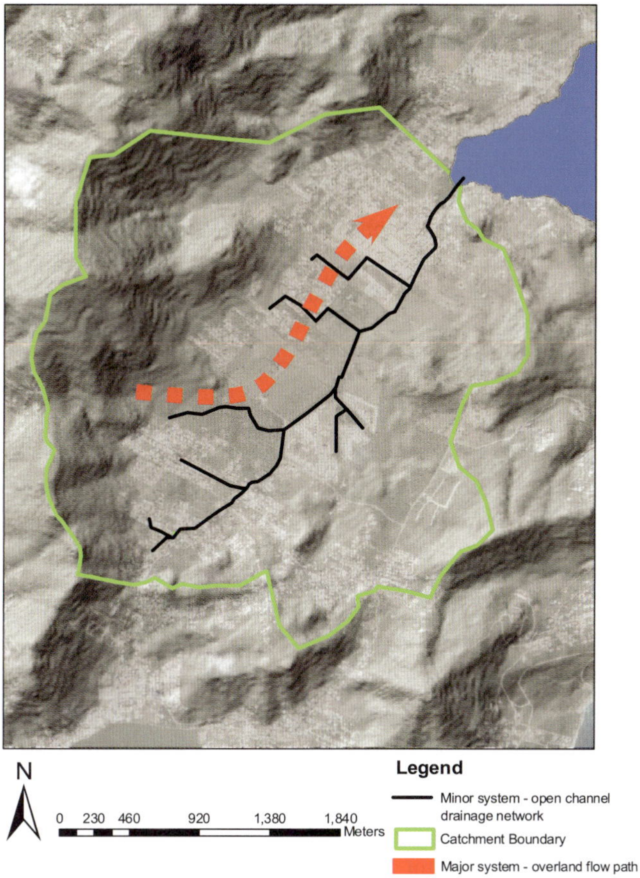

Figure 12.12 Example of minor and major systems

The stormwater runoff over the urban surface tends to collect detritus and sediment deposited by human activities, such as vehicle transport and construction sites, and by the wind. Street cleaning attempts to limit the accumulation of sediment and detritus, but each significant storm event carries with it a certain quantity of pollutants, including rubbers and heavy metals, and decaying vegetation. This wash-off of pollutants from the urban surface can lead to a degradation of water quality in the receiving waters for the storm runoff.

Therefore the service provided by an urban storm drainage system is to collect excess rainfall-runoff from an urban area and to discharge it to an adjacent water course, or to convey stormwater originating in rural catchments outside the urban area safely in (natural) channels passing through the urban area. Such a service is provided in order to establish and maintain a safe living environment by limiting the frequency of surface flooding and of inundation from urban runoff and streams flowing through the urban area. This may involve improving and maintaining the drainage of low-lying swampy areas.

The drainage service fails when there is a too frequent surface flooding in the urban area, which leads to a deterioration in the acceptability of the environment. This could be caused by an inadequate capacity of the drainage network, but could also occur temporarily because of blockages in the conduits. There is also a failure when the pollution of the stormwater impacts unacceptably on the receiving waters. Once again, this could arise because of the lack of street cleaning and garbage collection or even extensive construction in the urban catchment. Additionally, wastewater pipes, which should have been connected to a separate wastewater sewerage network, could have been introduced illegally to the storm drainage network. This again would exacerbate any pollution problem in the receiving waters, damaging such waters as a resource for, say, downstream supply.

As with wastewater systems, strategies for improving or extending stormwater drainage can benefit considerably from simulation modelling. However, in this case the modelling is more complex than for water distribution networks, in that different phases in the drainage process have to be identified and modelled separately, including rainfall, runoff from urban catchments and possibly rural catchments, loading and wash-off from urban surfaces, groundwater, flow and quality in receiving waters, and above ground-overland flow. All of these phases are also included in the case of combined sewerage, and therefore discussion of their modelling is given in the subsequent sections.

12.7 COMBINED SYSTEMS

The combined system has all the features of the separate wastewater and stormwater collection systems, except that illegal connections are no longer such an important issue (Figure 12.13). There are certain design criteria that need to be selected from those for the separate systems. For example, the sizing of the pipes is based primarily on the peak stormwater discharge plus the dry weather flow, whereas the gradient of the pipes has to be sufficient to move sediment that may be deposited under dry weather flow conditions.

The service provided by a combined system is for the collection of both wastewater and stormwater in the same conduits and the delivery of anything up to a predetermined part of the cumulative wastewater flow to treatment, with the discharge of any excess flow proceeding through combined sewer overflows to an adjacent water course. There are several reasons for providing such a service. One objective is to convey waste material contained in water (sewage) from housing and industry away from the urban area to minimise the exposure of human beings to dangerous bacteria in wastewater-causing infectious diseases. A second complementary reason is to limit the frequency of flooding and of inundation in the urban area. In both cases the aim is to establish and maintain a safe living environment. Where the combined system conveys sewage to treatment there is the additional service of removing waste material from the sewage before discharging the effluent to the receiving waters. This has the benefit of helping to establish and maintain acceptable water quality in the receiving waters. There can also be the additional benefit of being able to reutilise sludge produced by the treatment works and even to reclaim the treated water where water is scarce.

Figure 12.13 Combined sewer system in Manchester at the inlet to the wastewater treatment plant Davyhulme (Photograph: D. Brdjanovic)

The service provided by a combined system can fail for a number of reasons, such as when wastewater does not drain away from the generating source because of blockages in the system downstream caused by sediment, congealed fats or a sewer collapse, or because the hydraulic capacity of the system downstream is inadequate due to increased loading of the system from developments upstream. Such blockages or lack of capacity may increase the frequency of surface flooding upstream. Alternatively, despite a treatment works at the end of the combined system there may be too frequent or too polluted discharges to receiving waters from combined sewer overflows. This leads to a general deterioration of the water quality in the receiving waters and the possibility of (hopefully short term) fish kills.

Obviously, the management of a combined system with its integration of a variety of different processes and potential problems presents the greatest challenge for urban hydroinformatics at the purely technical level, which is principally one of numerical modelling. For this reason we consider the implications for simulation modelling in some detail. In particular, we identify the following phases for such modelling: rainfall (precipitation), rainfall-runoff for urban catchments, surface pollutant loading and wash-off, dry weather flow, groundwater, pipe network flow, above ground-overland flow, pipe network flow, treatment works, rainfall-runoff for rural catchments, receiving waters (urban streams, lakes, coastal waters); see Figure 12.14.

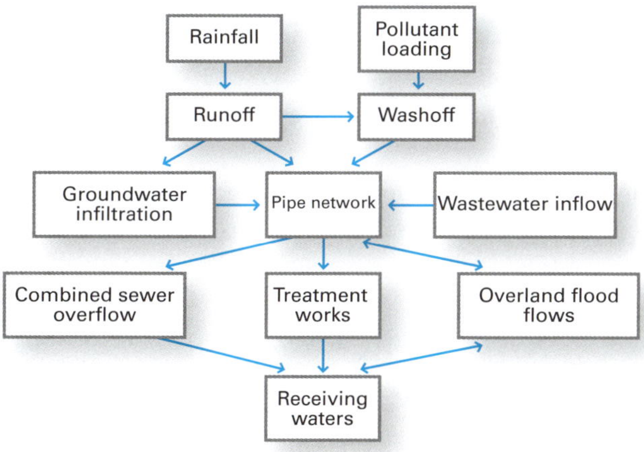

Figure 12.14 Phases for urban drainage. (After Price and Vojinović 2011)

Table 12.1 Advantages and disadvantages of separate and combined systems. (After Price & Vojinović 2011).

Separate system		Combined system	
Advantages	**Disadvantages**	**Advantages**	**Disadvantages**
Lower frequency and smaller volume of overflow spills.	Additional costs due to the construction of two pipe systems.	Lower construction costs.	Greater frequency and volume of overflow spills.
Less inflow to the WWTP.	Possibility for illegal- or cross-connections.	Occupation of a smaller public space.	Larger inflow into the WWTP.
Smaller variation in flows.	The two pipe systems together with their other elements occupy a larger public space.	Smaller maintenance costs.	Greater variation in flows.
Apart from the first flush, the stormwater is relatively less polluted.	Greater maintenance costs (including possible additional pumping).	The pipe slope may be lower.	In the case of flooding, flood water will also contain raw sewerage.
Flooding is by stormwater only and as such it can be used for irrigation purposes.	Smaller flows may require that the wastewater pipes are constructed with higher slopes in order to maintain self-cleansing characteristics.	Some portion of the stormwater will be treated at the WWTP.	Design and construction difficulties associated by including the two constituent flows in one pipe system.
The two systems can be laid separately following their own optimal route and depth.	Difficulties due to the need to construct pipes with higher slopes.	Less planning and asset management activity prior to the production of a single proposed asset management plan for active stakeholder participation.	Stormwater cannot be used for irrigation purposes and this must be made clear to active stakeholders.

A summary of the advantages and disadvantages of separate and combined sewer systems is given in Table 12.1.

Chapter 13

Quantifying urban processes

13.1 URBAN HYDROLOGY

The consequences of floods in both developed and developing countries have become increasingly devastating in recent decades, and the rising trend could reach dramatic proportions with an accelerated urbanisation and more rapid population growth. In many cities in the world, new development activities often take place in the floodplains of urban streams and rivers and along natural drainage pathways, and the problems of achieving sustainable protection measures are often limited. Along with the increase in the population living in urban areas, the increase in the number of cars and other forms of road transportation will inevitably cause an increase in the paved surfaces of the roads and parking areas. A combination of these factors has a direct influence on the hydrologic cycle, Figure 13.1.

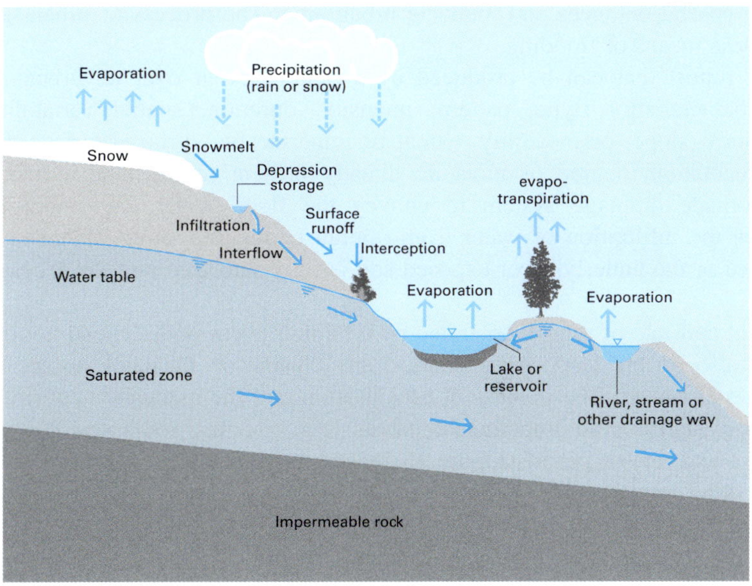

Figure 13.1 Hydrological cycle

The water moves from one place to another within the hydrologic cycle. It moves from river to ocean, and from the ocean to the atmosphere, and then once again to the ground, by the physical processes of evaporation, condensation, precipitation, infiltration, runoff, and subsurface flow, Figure 13.1. The urban part of the hydrologic cycle refers to natural and artificial inflows to the urban area from urban streams and water supply, together with precipitation, and subsequent runoff in the urban streams and rivers and groundwater away from the area. As far as precipitation is concerned the marginally higher temperatures within urban areas tend to intensify convective rainfall, leading to more intense thunderstorms with high intensities and short durations. Urban areas adjacent to mountains and the coast can be subject to devastating flash floods generated by orthographic lifting of warm moist air. A similar effect can occur due to frontal lifting as the warm moist air is forced upwards over a colder drier air mass. In both cases uplifting causes condensation to occur and the possibility of precipitation along the frontal boundary. Cyclones, tropical depressions or hurricanes are intense areas of low pressure often associated with high winds which gain energy while moving over warm ocean water. Such depressions can cause torrential rains and heavy storms with high intensity and long durations. Extreme depressions that occur in the Caribbean are commonly referred as hurricanes, in the Bay of Bengal as monsoon depressions, in the Far East as typhoons and in Southern Africa and Pacific as cyclones. The associated winds, together with the low pressure centres of the storms can produce significant surges in coastal areas that devastate low lying cities.

Flooding in urban areas can be therefore be caused by a variety of means. Intense rainfall over an urban area can lead to an excess surface runoff which is too much to manage for the natural or man-made drainage network. The water that collects on the surface finds its own routes by gravity above ground, generating flooding, especially in low-lying areas. Alternatively, water may be discharged from overloaded drainage networks so as to flood a low-lying area. Similarly, intense rainfall on natural catchments above the urban area, such as on adjacent hills or mountains, can overload streams that flow through the urban area leading to a spillage of flood water from the channels into the developed area. Again, urban areas adjacent to coastal waters can be subject to storm surges created by intense meteorological depressions which can overtop coastal defences and inundate urban area. The process of urbanisation can certainly accentuate all of these means of flooding.

The amount of runoff that can be produced by a rainfall event over an urban area thus depends on the rainfall characteristics (type, pattern, intensity, duration, spatial variability, etc.) and the surface characteristics (slope, permeability, extent of impermeable surface, surface depression storage, type of surface vegetation, degree of moisture already present in the land surface, efficiency of a natural or a man-made drainage system to convey the flood water, and more besides.). Surface runoff occurs when the infiltration of water from rainfall or melting snow is limited due to the paved nature of the surface or the underlying or exposed soil is fully saturated and the excess water flows over the surface.

Since the amount and rate of runoff generated by a rainfall event both depend not only on the rainfall characteristics but also on the land characteristics, any change on the land surface will influence the amount and the rate of runoff. The process of urbanisation and the associated construction of buildings may cause dramatic increases in impermeable areas (e.g., paving, roofs, etc.) which will in return increase the amount and rate of runoff (Figure 13.2), and as such it may have disastrous consequences. Paving, roofs and other impermeable surfaces do not allow percolation of the water down through the soil to the aquifer. It is instead forced directly into the surface runoff process. In this case, flooding is the obvious problem, but then not the only one. Increased runoff also reduces groundwater recharge and therefore lowers the water table and makes droughts worse, especially for farmers and others who depend on water wells.

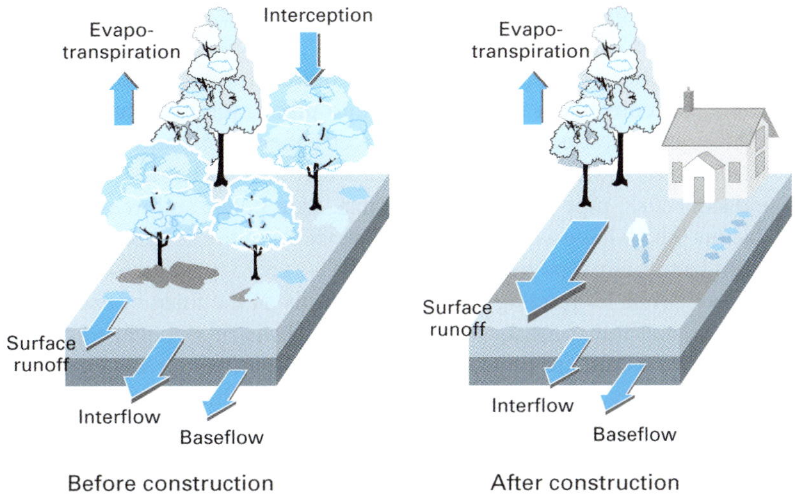

Figure 13.2 The hydrologic cycle (before and after urbanisation)

As shown by the schematisation of runoff hydrographs in Figure 13.2, the process of urbanisation can be reflected by the two major effects on surface runoff. Firstly, the replacement of natural vegetation with impervious areas such as pavements, roads and roofs will lead to an increase in the volume of surface runoff reaching the drainage channel. Secondly, the urban drainage system of gutters and pipes will have much less resistance to flow than the rough natural drainage system, which will shorten the catchment response time. The combination of these factors will produce a runoff hydrograph from the developed area which has a higher peak discharge and occurs earlier than the peak of the natural channel hydrograph. Finally, it will cause hydraulic overload of the drainage system and consequently flooding.

In addition to the increase in frequency and magnitude of urban flooding due to urbanisation, another associated problem that often arises is pollution of urban streams and other receiving water bodies. The cause of pollution is often attributed to the level of discharges coming from the wastewater systems and the concentration of pollutants that are carried within the wastewater flows. Also, a significant amount of pollutants comes from the roads and other paved areas themselves (e.g., suspended solids, zinc, copper, lead, etc.) due to the first flush phenomena. This is often described as the initial surface runoff after the dry period. During this process, water pollution entering drainage systems in areas with high proportions of impervious surfaces is typically more concentrated compared to the remainder of the storm. Consequently these high concentrations of urban runoff result in high levels of pollutants discharged from drainage systems to receiving water bodies.

Besides the damage that can be incurred by the quantity of the water generated during a flood event, another problem is the pollution in the flood water directly affecting public health of the local population and the pollutant loading of receiving waters.

13.2 URBAN MORPHOLOGY

Since urban hydrology is concerned with the study of water and its relationships with urban activities, understanding of the structure of the city and its physiological process of urbanisation represents yet another important aspect. Through such understanding the growth that can alleviate urban runoff processes can be planned and encouraged.

Urbanisation is usually defined as a process in which the land use occupancy changes as a result of concentration of people and conversion of rural lands into urban, suburban and industrial communities. Certainly, the way that a city is structured with its street pattern, lot pattern and building pattern will have a significant influence on various street components that have to be designed and constructed (e.g., pavements, curbs and sidewalks, pipe networks and drains, light poles, and trees, etc.).

There are many factors which have a direct influence on the structure of urban areas or a city as a whole. Some of these important factors are: topographic characteristics, geology, street form, political and economic forces, etc. In this respect, urban morphology refers to the study of the form of human settlements and the process of their formation and transformation. It involves analysis of physical form of street pattern, lot pattern and building pattern, which are collectively referred to as urban grain. Generally, there are three approaches to urban morphology: Italian, British, and French. The Italian approach revolves around the work of Saverio Muratori and dates back to 1940s. The British approach revolves around the work of M.R.G. Conzen, who developed a technique called 'town-plan analysis.' The key aspects of this technique are: the town plan, pattern of building forms and pattern of land use. The French approach is primarily based on the Versailles School of Architecture.

Sociologists, economists, and geographers have developed several models of urban structure, explaining where different types of people and businesses tend to exist within the urban setting. Key examples of such models are: grid model, concentric zonal model (i.e., Burgess model), sectoral model (i.e., Hoyt's model), multiple nuclei model (Harris & Ullman's model) and irregular pattern model, etc. These models, which in essence represent generalisations of urban structures, have been cause for much debate, Figure 13.3. Some cities display a structure which has elements of two or more of these models.

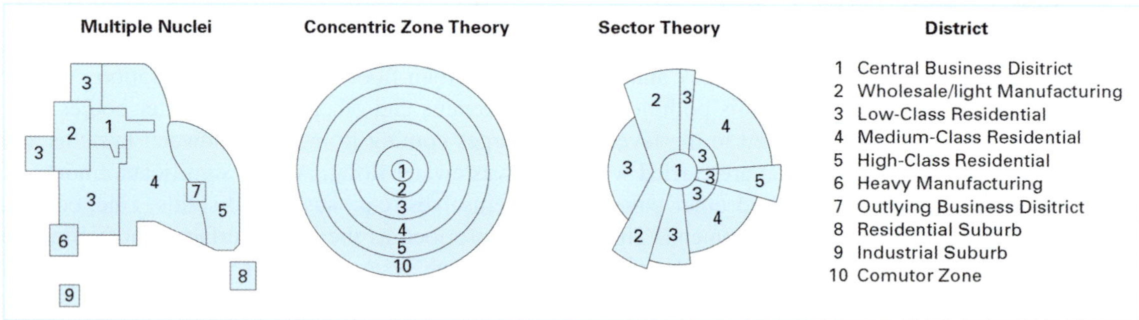

Figure 13.3 Illustration of the internal structure of cities

Understanding urban structure is an important aspect that needs to be taken into consideration while planning flood mitigation measures as its evolving patterns will have an influence on what can be achieved. This refers not only to the implementation of structural measures but also to the implementation of non-structural measures such as development planning and disaster prevention, response and recovery. In this respect, the capabilities of transportation and the degree of trade relationship with the surrounding region will have an influence on the effectiveness of such measures.

Following along similar lines, effects of climate change will also have their influences on the future spatial patterns, growth and development of cities.

13.3 CLIMATE CHANGE

It is becoming more and more evident that the effects of global warming and climate change represent yet another factor which has an impact on urban areas. Even though these changes are likely to be due to a combination of natural and human activities it is believed that the main cause for global warming is the increase of concentration of greenhouse gases (GHG) (IPCC 2001). Gaining an understanding of these phenomena and evolution of their potential future scenarios requires significant modelling efforts.

Climate models are complex mathematical descriptions of the climate system and its interactions with different aspects [see e.g., May *et al.* (2005)]. The key input into climate models is the level of emissions of important gases. The UN Intergovernmental Panel on Climate Change (IPCC) has defined 40 Special Report on Emissions Scenarios (SRES) scenarios which together encompass the current range of uncertainties of future GHG emissions arising from different drivers (IPCC 2000); see Figure 13.4.

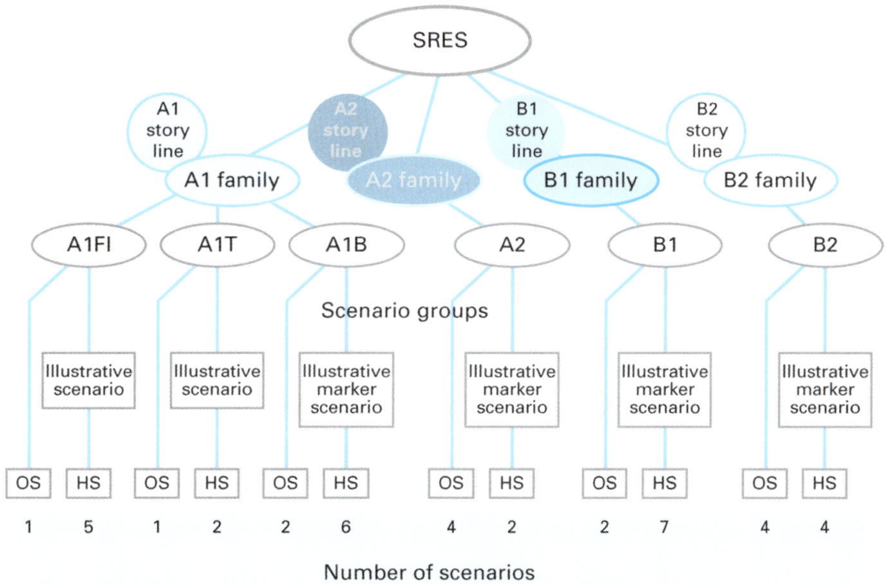

Figure 13.4 Schematic illustration of SRES scenarios, After IPCC (2000)

To assess impacts on urban areas climate forecasts need to be developed by using Global Circulation Models (GCMs). Current GCMs typically link dynamic models of the atmosphere and ocean so that the interactions between these two aspects of the climate system can be captured. Like any other model, these models are also adjusted (or calibrated) by tuning them until they can predict past climate conditions with a publically-acceptable accuracy. Downscaling climate data is a strategy for generating locally relevant data from Global Circulation Models (GCMs); see Figure 13.5. A typical strategy is to connect global scale predictions and regional dynamics to generate regionally-specific forecasts.

Downscaling can be done by applying various methods such as nesting, statistical regression, etc. Typically, local climate projections are derived from regional atmospheric climate model data which are derived from the global circulation model data (see e.g., Berggren *et al.* 2007).

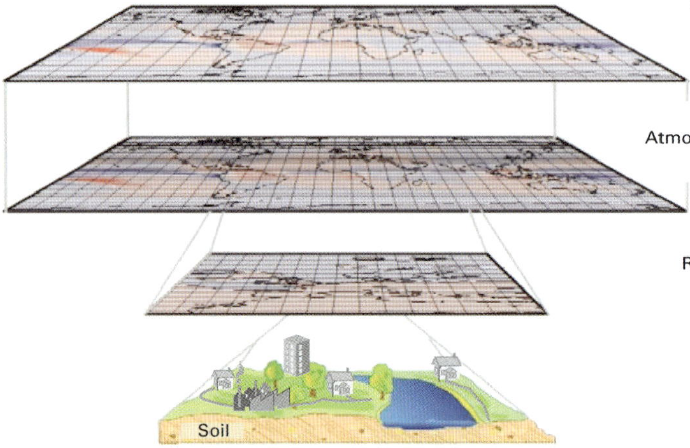

Figure 13.5 Downscaling of climate data

The study of Grum *et al.* (2006) discussed the impacts of climate change on local weather conditions in Copenhagen, Denmark. They derived the local rainfall data from the regional climate model HIRHAM, developed by the Danish Meteorological Institute, and discovered potential for significant change in return periods of events; see Figure 13.6.

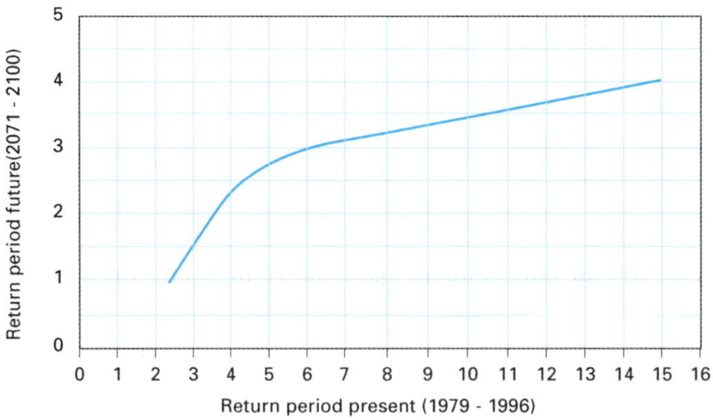

Figure 13.6 Effects of climate change evaluated in terms of changes in return periods of events for Copenhagen, Denmark (Source: Grum *et al.* 2006)

The evidence of climate change on urban areas is reflected not only in the expansion of cities' spatial patterns, growth and development but also in the increase of flood incidents due to more precipitation and storm surges, landslides, drought, saltwater intrusion, typhoons (or hurricanes) and other hazards. The International Red Cross and Red Crescent have reported an increase in the intensity and frequency of disasters that climate change will only make worse. It has already been reported that the climate-related vulnerability of urban areas will only increase with the time if the necessary measures are not taken in advance. The potential implications can be summarised to the following:

- More frequent floods due to increases in flows which will exceed the capacity of existing drainage and sewerage systems (as they were designed to meet standards which may no longer be applicable),
- Faster deterioration of systems as they will be exposed to more frequent surcharges, increase in the pollution of receiving waters, and
- Less efficient operation of wastewater treatment plants due to the increase of influent loads.

With the view on the events over the past decade it can be observed that climate change is making weather less predictable, rains more uncertain and heavy storm rainfalls more likely. The unpredictability of rainfall is shown both by observations, such as the large fluctuations in the water levels of receiving waters, and by the experiences of urban communities, who report much more frequent storm-producing floods since 1990. Heavy thunderstorm-associated rains appear to have increased in frequency. In many places, urban areas may help to increase thunderstorm activity because their built-up surfaces attain higher temperatures than surrounding areas and create a local air circulation that produces an 'urban heat island'. Dust particles caught up in such a circulation act as nuclei on which moisture in clouds condenses, forming rain droplets that eventually may coalesce into the large rain drops of a major thunderstorm. Climate change also works in an indirect way to aggravate urban flooding. Droughts and floods in rural areas, such as the African Sahel droughts and major floods in Mozambique, have forced many rural people to migrate to towns and cities, adding large new populations to existing slum communities. These rural refugees add further to the urban activities that increase the flow of rainwater to rivers and thus the intensity of local flooding.

With the threats of climate change, increasing urban population and economic development, great uncertainty exists about the necessary adaptation of the urban environment to flood risk. This requires the development and employment of new measures (or perhaps extension of traditional measures, which mainly have an impact on the flooding pathways) that can enable the communal management of the flood risk in a more flexible and holistic way. Such measures will need to have a positive impact on the source, the pathway, the receptor and the consequences of any flood event. There must be greater synergy and synchronisation between structural and non-structural flood mitigation and adaptation measures.

In terms of the communal management of urban drainage systems, the following potential impacts due to climate change can be identified:

- Increased number of frontal rain systems, particularly in winter months – This could give rise to increased river flooding, which would impact drainage assets on or near the floodplain. Also, drainage systems may become increasingly overwhelmed, resulting in increased risk of surface-water flooding.
- Occurrence of increased intensity rainfall events all year round: this can result in surface flooding (or pluvial flooding) in many urban locations that may not be on the fluvial floodplain.
- Combined impacts of rainfall-derived and increased sea level rise – Increased sea level due to climate change (e.g., mean level rise and/or increased storm surge) combined with heavy rainfall events could place at risk of flooding low-lying locations in coastal areas. For example, high sea levels may cause backing up of stormwater sewer flow or overflow of storage within the sewer system close to the sea.

13.4 URBAN DRAINAGE HYDRAULICS

The subject of hydraulics deals with different properties and laws governing the pressure and flow of fluids and the application of these laws to engineering practice. All fluids are compressible to a certain extent, in a way that the changes in pressure or temperature will result in changes in density. However, in many situations the changes in pressure and temperature are sufficiently small that the changes in density are negligible. In such cases the flow can be considered as an incompressible flow. This greatly simplifies the formulation of the equations governing the flow of water which we assume to behave as if it is perfectly continuous in structure (*continuum hypothesis*).

The flow of water may be classified in different ways (e.g., turbulent, laminar; real, ideal; steady, unsteady; uniform, nonuniform; open channel, pipe; one-dimensional, two-dimensional and three-dimensional). *Turbulent* flows are the most prevalent in engineering practice. In turbulent flow, the particles of water move in very irregular paths which results in greater shear stresses and more irreversibilites (or energy losses). In *laminar* flow, the particles of water move along smooth paths in laminas, or layers, with one layer gliding smoothly over an adjacent layer. *Steady* flow occurs when conditions at any point in the fluid do not change with time. In the opposite case we speak of *unsteady* flow. *Uniform* flow occurs when along the flow the velocity vector remains identical in magnitude and direction over distance. Flow such that the velocity vector varies from place to place at any instant is described as *nonuniform* flow. *Pipe flow* is usually regarded as a flow in which water flows under pressure. In an *open channel flow* water flows under the influence of gravity, with a free surface at atmospheric pressure. *One-dimensional* flow (1D) considers the variation in velocity in only one direction (e.g., pipe flow). *Two-dimensional* flow (2D) considers the variations in velocities in two directions. *Three-dimensional* flow (3D) is the most general flow in which all three velocity components are taken into account.

The total energy of a particle of water at any instant is the sum of its *potential energy*, its *kinetic energy* and its *pressure energy*. Assuming an ideal flow, for any mass of water in which there is a continuous connection between all the particles along a continuous trajectory, the total energy of each particle is the same. This is known as Bernoulli's theorem. When water flows in a pipe or a channel, some energy (or head) is 'lost' from water. The energy grade line (EGL) is a line that represents the total head available to the fluid. The hydraulic grade line (HGL) is a line representing the total head available to the fluid minus the velocity head. The hydraulic grade and the energy line are graphical presentations of the Bernoulli equation. The energy (or head) losses in flow in a pipe are subjected to friction losses and local losses. Friction losses are caused by forces between the liquid and the solid boundary. Local losses are caused by disruptions to the flow at local features (e.g., bends, junctions). Total head-loss is the sum of these two components. The basic representation of friction losses is the Darcy-Weisbach equation. The Darcy–Weisbach equation contains a dimensionless friction factor, known as the Darcy–Weisbach friction factor or Moody friction factor. This factor can be determined numerically from the Colebrook-White equation and graphically from the Moody diagram, Figure 13.7.

The *density* ρ of a fluid is defined as its mass per unit volume. All real fluids resist any force tending to cause one layer to move over another while movement is taking place. Water can be treated for most purposes as a real Newtonian fluid; that is, it resists deformation such that the shear stress is proportional to the velocity gradient normal to the stress. The constant of proportionality is the *viscosity* (i.e., resistance of the movement of one layer over an adjoining one). The resulting time dependent three dimensional flow equations are called the Navier-Stokes equations. Given that the water particles immediately next to a rigid boundary do not move (the no-slip condition) the shear stress in the flow is locally large near to the boundary. This generates a strong rotation, or vorticity, in the fluid. Physically, vorticity is generated along the boundary, and it can be shown that the vorticity diffuses away from the boundary into the flow where it is convected further away. The region near the boundary where the vorticity is large is termed the boundary layer. The thickness of this conceptual layer is proportional to the viscosity and inversely proportional to the square root of the Reynolds number, UL/ν, where U is the scale of the velocity, L is the length scale and ν is the *kinematic viscosity*.

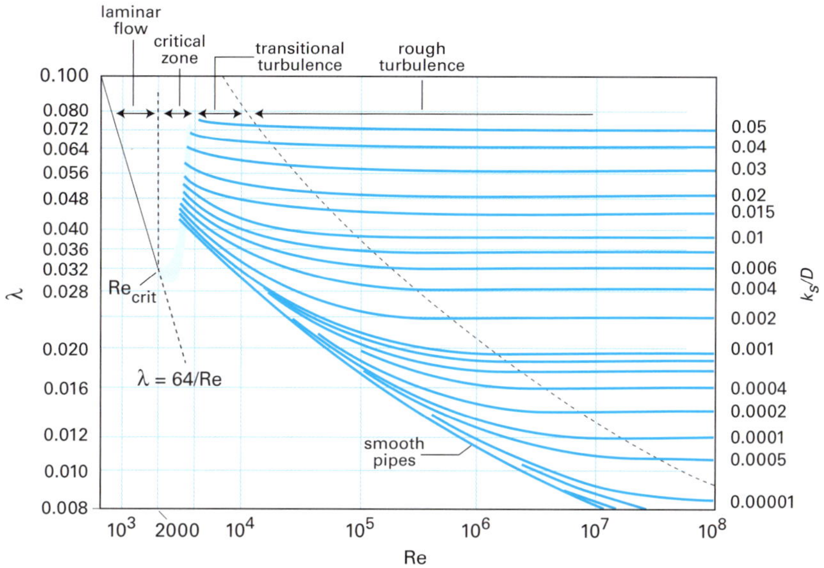

Figure 13.7 Moody diagram

The boundary layer is important for sediment transport and the introduction of random fluctuations, or turbulence, into the flow. Whereas it is possible to formulate the Navier-Stokes equations for laminar flow rigorously from first principles, turbulence introduces further dependent variables, namely the Reynolds stresses, with the need for additional equations to generate a unique solution for the mean flow variables. The closure problem is circumvented in open channel hydraulics by assuming that the energy loss in the flow is proportional to the square of the velocity averaged over depth (in two spatial dimensions, 2D) or over the cross section (in one dimensional flow, 1D).

The equations derived in this way for 2D or 1D flow are termed the Saint Venant equations. The additional assumption made in the derivation is that the flow has a gradually varying free surface such that the vertical accelerations of the water particles are negligible, and the pressure in the flow is then essentially hydrostatic, that is, the pressure is the same as if there were no flow. Although corresponding equations can be derived when the pressure deviates to first order away from being hydrostatic, these equations do not feature in the modelling of urban water systems considered in this book. Besides free surface flow, equations are needed for surcharged flow under pressure in pipes. The equations are essentially similar to the free surface flow equations with the exception that the discharge is uniform along a given pipe at any instant due to the assumed incompressibility of the fluid.

The fact that water can have a well defined surface with the atmosphere is due to the bonds between the hydrogen atoms. The surface tension that is the result of these bonds enables insects to walk on water, a meniscus to form in the contact of the free surface of water with a hard surface (which benefits osmosis in plants), and raindrops to form on dust particles.

Finally, water is an excellent solvent, which means that it dissolves and carries with it a range of substances, as well as eroding and transporting sediment. The concentration of a particular substance is advected with the mean velocity of the flow, diffused due to the turbulence (and the Brownian movement) and dispersed according to the three dimensional structure of the flow. The interaction between various organic and inorganic substances and the processes of sediment transport introduce the complex analysis of water quality in urban water systems.

In urban drainage hydraulics we are mainly concerned with two types of flow: pipe flow and open-channel flow. The flow in sewers is most of the time hybrid of these two, that is, part-full pipe flow, which is in theory treated as a special case of open-channel flow, Figure 13.8. Figure 13.8 gives dimensionless relationships for part-full pipe flow where maximum flow velocity and discharge do not occur when the pipe is running full: in practice they occur when it is slightly less full.

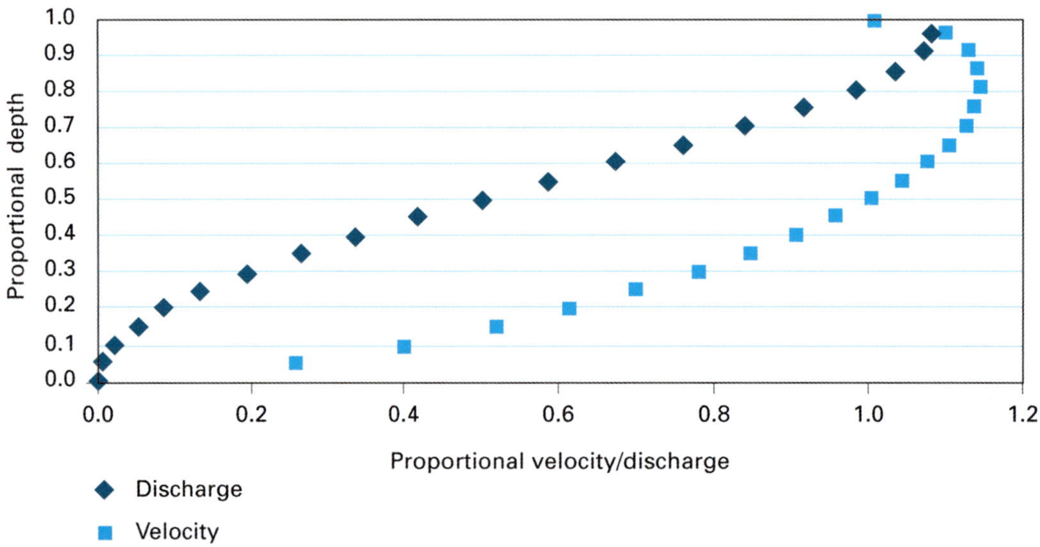

Figure 13.8 Pipe flow

In what follows we deconstruct the physics of the flow of water in conduits (1D), over a 2D surface and in 3D groundwater aquifers or treatment processes, as this is appropriate for hydraulic modelling. Our concern is to formulate the analytical equations based on the physical processes involved. Once these have been defined we can then address the issue about how to solve the equations numerically in different contexts.

13.5 BACKGROUND TO MODELLING

13.5.1 Introduction

As described in Price and Vojinović (2011), the purpose of modelling is about quantifying the physical and social processes in the physical world through the use of an underlying physics and mathematics. We assume that all such processes can be represented individually or in an integrated manner by some form of symbolic equations or relationships, whether based on good science and deduction or working with measured data and induction. This means that with models we can capture the behaviour of a physical world in order to explore the possibilities for improving the performance of the physical systems for urban flood risk assessment, and for identifying and working with the social systems that initiate the demand for safer environment.

We could raise a number of questions concerning the purpose of the modelling work. Do we want to investigate an intervention in the physical world, such as extending a network, redeveloping an urban area, or rehabilitating an existing system? Or is our purpose non-structural: do we want to produce a future development policy, to assess the feasibility of an early warning system, to forecast the operation of a stormwater outlet or combined sewer overflow? In the first case, we need to work with a model that represents the physics of the situation accurately, whereas in the second case we may be more dependent on a model that learns directly from relevant data. So we appear to have at least two model paradigms: one that is deductive from physical (and possibly, social interaction) principles, and the other inductive from data. However, not all models turn out to be purely deductive or inductive: deductive models have inductive components, and vice versa. So, just as in Price and Vojinović (2011) we adopt here an alternative way of categorising models as one that broadly supports the distinction between the example applications above.

13.5.2 Model categories

We recognize that much of our modelling is normally concerned with water interacting with physical objects in physical space (though we also have reason to address below the modelling of social processes in social 'space'). In particular, we can talk about the 'model space'. This is the physical geometric domain potentially occupied by the water: the pipe or channel network, the groundwater strata containing an aquifer, the above-ground surface topography for overland flow, the sub-catchment draining to a pipe, the regions of a city with a demand for water, or the bathymetry of the coastal waters. Are we concerned with the motion of the water in the particular space, or are we more interested in the space containing the water? Such questions lead us to consider two categories of model (see Table 13.1): in one, we define mathematical equations formulating the physical laws for the flow of water *at a point* (in 1D, 2D or 3D space), we then adopt a one-to-one mapping of the physical space onto the model space and discretise the model space consistent with finding a numerical solution to the analytical equations; in the other, we compartmentalise the model space according to logically-deduced physical boundaries and introduce appropriate mathematical constructs to relate the output to the input for each compartment (see Tables 13.2 and 13.3). In the former case the focus is on how the actual physical geometry affects flows at all points in the fluid, and how changes in the geometry affect these flows. In the latter case, there may be little if any interest in the detailed flows with respect to the actual geometry, but only in the relationship of the output to the input. The mathematical constructs may be derived from quasi-physical principles, resulting in analytical equations that apply to the compartment as a whole. Alternatively, the constructs may have little if anything to do with the physics, but are designed to provide a consistent connection between the output and the input; that is, the construct learns about the nature of the connection from data; see Table 13.1.

What is common to both categories of models is that we have to identify the model space in terms of the physical objects that make it up. The model space may be defined by a network of sewer pipes. For the first category of model we have to include in the model space each significant object that contains the flow. All of the objects play some part in contributing to the whole of the sewer system.

Table 13.1 Categories of model (Adopted from Price and Vojinović, 2011).

Categories of model	Description
Physically (or process) based (or white box)	Numerical solution of equations describing the physical (and chemical or biological) processes for the fluid as a continuum in the model space, which is a one-to-one mapping of the actual physical space and is deconstructed into cells or elements appropriate for the numerical solution
'Conceptual' (lumped or grey box)	Numerical solution of equations describing the physical (and chemical or biological) processes for the fluid as a continuum in the model space which is a simplification of the actual physical space or a lumped quantity in one or more conceptual objects linking output to input in the model space or in compartments for the model space which are identified on physical principles
Data driven (or black box)	Numerical solution of equations derived for a particular mathematical construct that includes a number of parameters, whose values are trained to replicate the connection between the output and input observed for the model space or for compartments for the model space which are identified on physical principles
Rule based	An inference rule is a statement that has two parts, an IF clause and a THEN clause.
Deterministic	The model predicts the same output each time it receives the same input
Stochastic	The model predicts a different output each time it receives the same input

Table 13.2 Physically-based modelling.

Physically-based modelling
Deconstructs the model space into compartments (cells/elements) consistent with the numerical solution of the algorithmic equations

Table 13.3 Compartmentalised modelling.

Compartmentalised modelling
'Conceptual'
Lumped
Black box
Grey box
Deconstructs the model space into compartments consistent with geophysical concepts and connects the output from each compartment with the input

Meanwhile, for the second category of model, we are not so interested in the details of the model space, instead the model space is a simplification of the physical space, such as an equivalent uniform reach corresponding to an irregular natural reach, one or more reaches along an open channel in the

Muskingum routing method (see below) or a sequence of storage tanks in a tank model for rainfall-runoff from a catchment, or a *zero dimensional* model consisting simply of a time series for the inflow to the network.

As an alternative, consider an urban sub-catchment contributing to a given outlet or conduit in a drainage network. The sub-catchment may include its own sub-network of pipes. What we are interested in is the relationship between the rainfall falling on the sub-catchment and the runoff. Whereas we could use the first category of model in simulating the local rainfall losses and the runoff over the 2D surface, the high resolution required makes the model cumbersome and accuracy is prejudiced. In practice, only the second category of model is applicable to this situation, especially as we are focused on the outfall of the sub-catchment and not on the details of the flows in the sub-catchment. There are different types of model that may be used, depending on whether some physical information about the sub-catchment is needed or not. Those models dependent on physical information require the sub-catchment to be defined by the geographical boundary that forms the (approximate) watershed of the sub-catchment. The sub-catchment then has its own unique values for certain parameters that may be important for the model, such as the total area, the percentage paved area, and the characteristic slope. There are however, models that are purely driven by the data and simply need the time series for the rainfall and the runoff (if available). There is no need for any other physical information. Obviously this last type of model is heavily dependent on adequate data for the rainfall and runoff.

As another example of the first category of model consider a pipe network defining the model space. Analytical equations can be derived for the time dependent flow at a point in 3D. But it can be argued that the flow in the pipes can be approximated as being one dimensional in that the length scale along the pipe is considerably greater than the scale of the cross section, and therefore the three dimensional equations are integrated over the uniform cross section of a pipe to deduce a 1D version of the equations. We then solve these equations numerically within the geometric domain for the flow in each pipe. In particular, the model space in a pipe is discretised into equal lengths according to the size of the pipe, with the dependent variables for the flow defined at the sections between each length; see Figure 13.9.

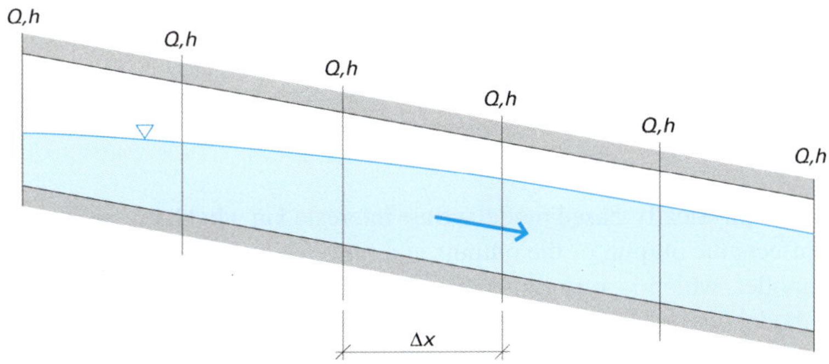

Figure 13.9 Discretisation of a 1D model space for a conduit

In line with normal practice we call the first category, 'physically-based' modelling. However, the second category does not have a universally accepted nomenclature. For convenience we call it 'compartmentalised' modelling. Other terms used under the category of compartmentalised modelling are: conceptual, lumped, black box, grey box, and data-driven. 'Conceptual' is not used here as a separate category because of the generic use of the word to identify processes and objects. 'Lumped' reflects the word 'compartmentalised', so despite its popularity it is not used here as an independent category in this book. The primary sub-categories of compartmentalised modelling recognised here are: 'data-driven' of 'black box' and 'grey box' modelling. Data driven modelling uses mathematical constructs that have a minimal dependence on knowledge about the processes and objects in the model space. Included in this sub-category are machine learning methods such as artificial neural networks (ANNs) (Hassoun 1995), model trees (MTs) (Quinlan 1992), nearest neighbour methods (Toth *et al.* 2000), fuzzy logic generators (Lobbrecht and Solomatine 1999), evolutionary programming (Savic *et al.* 1999) etc. Also included under data driven modelling are regression (Yevjevich 1972), unit hydrograph (Sherman 1932) and ARMA-type (Box *et al.* 1976) methods. In contrast, grey box modelling includes all those approaches that include some knowledge about the processes and objects in the model space. For example, hydrological methods for catchment runoff that are based on the notion of a linear (or non-linear) reservoir would be included in this sub-category, as would the rational and time-area methods for determining the peak discharge and time dependent runoff respectively from a catchment; see Figure 13.10. Also, included in the compartmentalised category is the Muskingum method for flow routing referred to above; see McCarthy (1938). This method deconstructs a length of conduit or a reach of river into one or more routing compartments for which it is assumed that the storage in each compartment is dependent on a linear combination of the input and output discharge. Interestingly, provided the linear combination is correctly formulated, the Muskingum method can be shown to be consistent with an approximation of the 1D gradually varying free surface flow equations as the length of the compartment goes to zero (Price 2009).

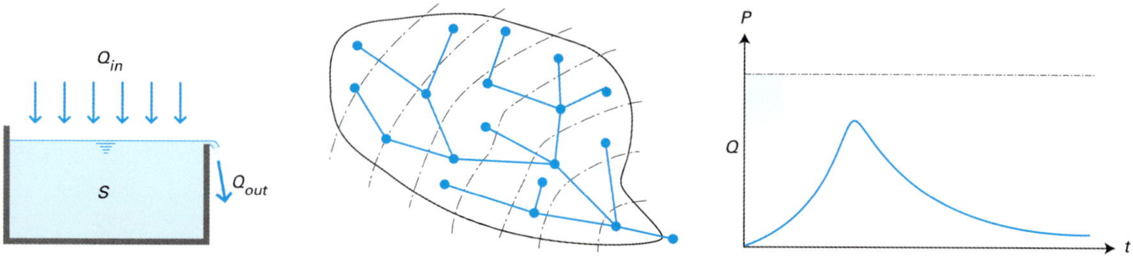

Figure 13.10 Alternative rainfall-runoff model concepts

To summarise, with a physically-based model we are interested in what happens in the physical geometric domain and how it affects the output, or the outputs at a range of locations within the model space. With a compartmentalised model, which is zero dimensional, we are only interested in the output and not in the changes within the model space. Our focus in this book is on the physically-based models.

13.5.3 What is a model?

We have defined a model in the most general sense earlier here, as that which provides a collection of indicative signs that provide an expressive sign in the mind of the observer, where we use the words 'indicative sign' and 'expressive sign' in the now long-established sense of Husserl (1900, 1901): see Abbott and Minns (1998).

Before we proceed further, however, we need to be more precise about what we mean by a model within the various contexts within which it is most currently employed in our own field; and we shall be focused on 'physically-based models'. As described at the beginning of this chapter, modelling is about quantifying the physical and social processes in the real world through the use of an underlying mathematics. For us, a model is essentially a mathematical object that generates output data from input data. And yet it is more than that. The object represents an aspect of reality, and must necessarily be a simplification of that reality. Further, the model is designed such that when operating properly it provides insights into particular attributes of a system, whether it be physical, biological, economic or social (USEPA 2008).

All of the models described in this book are deterministic in that the same input always produces the same output; in other words, there is no random component in the models. Necessarily, assumptions are made in the derivation of the mathematical algorithms for all physically-based models. In making these assumptions we introduce the notion of parameters for the model, which are usually closely tied to particular physical processes that have to do with energy loss, such as the frictional effect of a boundary surface, or the highly turbulent and contorted flow through structures, including gates, orifices, weirs and pumps. The parameters may acquire fixed numerical values through an analysis of experimental laboratory or field data, or their values may be left open in order to be defined separately for each model application. The values of these open parameters may be defined in terms of qualitative descriptions of physical structures, such as the roughness of a boundary surface. But there may be situations where guidelines are given for a range of appropriate values even as the precise value is unknown. In these circumstances the modeller has to go further in 'calibrating' the parameter, possibly by adjusting the output of the model for the prescribed input such that the output closely resembles observations of the output variables.

13.5.4 Calibration of a model

Calibration is a process that can easily be abused. The desire of the modeller is to have a physically-based model that reproduces the values of an observed variable as closely as possible, and sometimes the parameters will be adjusted unreasonably in order to achieve this. Such an adjustment fails to recognise that a model is always going to be an idealised version of the real world, and therefore incorporates a number of approximations that generate errors in the model output and then even with the correct values of the parameters. Therefore, setting aside the fact that there are likely to be errors in the data for the observed input and output variables, it is important that the modeller learns to live with the limitations of the model, and to avoid falling into the trap of force-fitting the model by adjusting the parameters outside their normal ranges of values. The penalty of not observing this limitation is the probable prediction of incorrect results in applying the force-fitted model to design and prediction. Of course there are situations where there is little experience available in identifying values for certain parameters in particular contexts. This is often the case for diffusion coefficients in advection-diffusion processes for pollutants or rate coefficients in chemical and biological processes; even rainfall-runoff coefficients for a catchment may be difficult to quantify. Even then the modeler should endeavour to seek published values of the parameters for similar models. In such cases calibration becomes more meaningful as the modeller adjusts the parameter values to replicate the observed variable data.

Calibration should therefore be handled with care, and in general optimisation should not be used to refine parameter values with physically-based models. Where calibration is involved, many modellers like to verify the parameter values that they have selected by running the model with another set of input data and demonstrating that the results continue to be reasonable. Unfortunately, modellers tend to use words like 'verification' and 'validation' somewhat loosely. As pointed out by Oreskes *et al.* (1994), such words have to do with truth, and it is very difficult to justify claims to truth in modelling. The authors suggest that a better word than either 'verification' or 'validation' is 'confirmation'. This book uses the word 'confirm' as the act of gaining confidence that a model is fit-for-purpose, often through peer review. Nevertheless, verification and validation will continue to be referred to because of their common use by the modelling community.

13.5.5 Confirming a model

As stated, the words 'validate' and 'verify' have been generally used as a sequel to calibration. 'Verification' has also been used in the UK to refer to the process of confirming that the integrity of the objects in the model space for urban drainage networks in particular, which replicate the structural objects that make up the physical space, is complete; that is to say, there are no additional or missing objects, they are correctly connected and the geometric data for each object is acceptably accurate. In most hydraulic modelling situations this form of verification is not an issue. However, it happens that in urban drainage especially, there are numerous occasions when there is missing information about the below-ground network, and even with expensive manhole and CCTV surveys it is difficult to be sure that all the details of the pipe network have been recorded accurately. It is necessary therefore for the integrity of the network structure replicated in the model space to be 'verified'. This is done in the normal way by running the model for observed events and comparing the results of the model with the observed flows. Through both a quantitative and/or a qualitative analysis it is possible to identify deficiencies in the structural data and then to confirm or disprove the existence of such deficiencies through field-survey data. A difficulty with this approach is that there may also be the need to calibrate the model parameters at the same time, and therefore confusion can arise between what is justifiably a problem with the structural data and what is the consequence of an adjustment to one or more model parameters.

Another note of caution about physically-based models should be raised at this point, and this concerns the fact that almost all such models depend on numerical techniques for the solution of the mathematical algorithms. This implies the development of modelling software to implement the solvers of the algorithms in a solution space that is based on a particular generic deconstruction of the model space. In other words, the modelling software is generic and can be applied to any situation that satisfies the class of problem for which the software is designed, whether for distribution or collection networks of pipes or open channels, groundwater, river networks, and so on. A physically-based model is the instantiation of the appropriate modelling software with the geometry and input data for the given situation. The modelling software on its own is not 'the model', as is abundently clear from the definition with which we started this section

So far as data-driven models are concerned there are strict rules for training and verifying them. A data-driven model is entirely dependent on training a selected mathematical construct that intercedes between observed output and input data. The particular underlying mathematical algorithm forming the connection has nothing to do with the physics of the situation, and therefore there is no dependence of the model on the physical structure of the model space. Where the physics becomes important is in the choice of the input variables and the use of the data values. Such a choice may be affected by factors such as the shape and characteristic geometry of the 'compartment' (the catchment, the wastewater treatment process, etc), the nominal 'lag time' between the input and output response, or the decision to deconstruct the physical process into components and to model them separately, such as separating the base flow in the runoff for a catchment from the response to rainfall. The instantiation of different types of data driven model is discussed at the end of this chapter.

Obviously, the development of a model can be highly subjective and is dependent on the individual choices made by the modeller on what to include and what not to include, the selection of time dependent data for calibrating and verifying the model, and the way in which the model should be applied. Therefore it is important that there should be a peer review of the model in order to confirm the approach adopted by the modeler and the outcome.

Finally, all users should be aware of the limitations of such models. They are designed to replicate the physical world; and whether we are concerned with physically-based or compartmentalised models, they are only as good as the underlying mathematics, the data for the model space, the temporal input (boundary) data, the calibration or training of the model parameters, and the user's expertise. They are therefore limited in their capability to explain and predict, and the wise user will always bear that in mind when interpreting the output from such models.

Chapter 14

Data collection for modelling

14.1 PREPARING FOR A DATA COLLECTION CAMPAIGN

A number of steps can be identified in the process of collecting the data for modelling purposes, Figure 14.1.

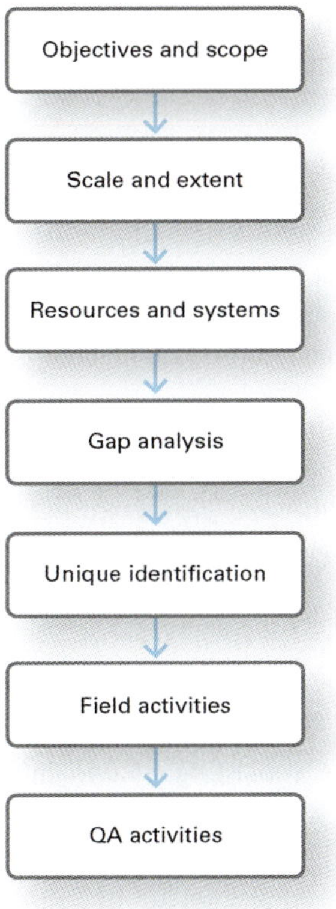

Figure 14.1 Data collection

Defining a set of clear objectives is the most important aspect of the data collection campaign. Collecting the data without clear purpose and justification can result in collecting potentially unnecessary data and/or missing essential data. Once the scale of a data collection campaign is defined, details such as existing data gap analysis, asset identification, site and equipment selection, selection of variables, budgeting and other financial possibilities and decisions, together with more general social and institutional requirements should be considered. The objective of gap analysis is to identify which of the data needs of the project can be met by the existing data and which data will have to be collected through new data collection activities. Aspects to consider are availability, quality and compatibility of existing data.

Having defined the needs for new data, the next step in designing a data collection programme is to specify a unique identification for each asset of element for which data is to be collected. In doing this, it is important to develop a system of numbers which is logical and has meaning on the ground. Furthermore, it has to allow for extension in order to accommodate newly created assets or the development of more sophisticated databases.

For temporal data collection it is important to have an understanding of the drainage system infrastructure within the area under consideration and the characteristics of different site locations. Sometimes it is necessary to undertake a site investigation prior to the data collection programme in order to obtain necessary information and minimise the potential waste of resources and time. For example, a careful selection of a raingauge site is important when planning a flow survey. Data from an inappropriately selected raingauge sites will affect the success of any model calibration. Furthermore, the geometry of assets must be measured accurately and measurements from design drawings should not be assumed. In the case of a buried infrastructure, a specific survey may be required. In the case of active stakeholder partcipation as introduced earlier in this book, there must be a high level of flexibility in this programme as the stakeholder group succesively transcends itself and may consequently require extra data as this process proceeds.

14.2 SPATIAL DATA COLLECTION

An overview of some key spatial data that is required for management of different urban water systems is given in Table 14.1.

In terms of the pipe network details, the following data is needed:

- Cover level (or ground level) of the manhole,
- Invert level of the manhole,
- Manhole geometry,
- Incoming pipe length,
- Outgoing pipe length,
- Incoming pipe invert level,
- Outgoing pipe invert level,
- Incoming pipe geometry,
- Outgoing pipe geometry.

A typical drainage manhole structure consisting of two incoming pipes and one outgoing pipe is presented in Figure 14.2.

For open channel networks (either artificial or natural), the following details are needed:

- Cross section geometry (distances and elevations); see Figure 14.3,
- Chainages, and
- Long sections.

Table 14.1 An overview of spatial data requirements.

Storm water and wastewater networks and wastewater treatment plants	Receiving waters (streams, rivers, watercourses, lakes, wetlands, coastal waters)
Terrain and landuse data;	Network layout of streams, rivers, watercourse, lakes, dams, groundwater aquifer;
Catchment and sub-catchment characteristics boundaries and their aerial coverage;	Contributing catchment areas, their characteristics and aerial coverage;
Pipe/open channel network layout;	Coastal zone layout;
Manhole data (IDs, X and Y coordinates, elevation);	Coastal bathymetry
Channel cross sections (IDs, X and Y coordinates, geometry, chainage);	Channel cross sections (IDs, X and Y coordinates, geometry, chainage);
Storages (IDs, X and Y coordinates, bottom elevation, storage area);	Location of flow, level, pressure and water quality measurement devices (ID, X and Y coordinates);
Gravity pipes, rising mains and inverted siphons (IDs, start node, end node, length, diameter, inlet and outlet elevations);	Stormwater outlets (sources and sinks);
Pumping stations (IDs, X and Y coordinates, wet well geometry and elevation);	Connecting CSOs and SSOs;
Catch-pits and gully pot data;	Structures;
CSOs and SSOs (IDs, X and Y coordinates, geometry);	
Weirs, Orifices and other ancillary structures (IDs, X and Y coordinates, geometry);	
Culverts/bridges (IDs, X and Y coordinates, geometry);	
Outlets (IDs, start node, end node, length, diameter, inlet and outlet elevations);	
Wastewater treatment plants and its elements (IDs, X and Y coordinates);	
Location of flow, level, pressure and water quality measurement devices (ID, X and Y coordinates);	
Cadastre and property data;	
Population data (per census area unit);	
Ponds, storages, tanks, soakaways, infiltration wells and basins, swales, (IDs, X and Y coordinates, geometry);	
Asset condition data;	

In addition to the pipe/channel network, details of the ancillary structure geometry are also necessary. Ancillary structures typically include constructed weirs/overflows, bifurcations, pumping stations, storage tanks flow control devices and inlets/outlets; see Figure 14.4.

In addition to the above mentioned spatial data, there are also other data which are needed irrespective to the particular system under consideration. Such data includes digital terrain model data, land use

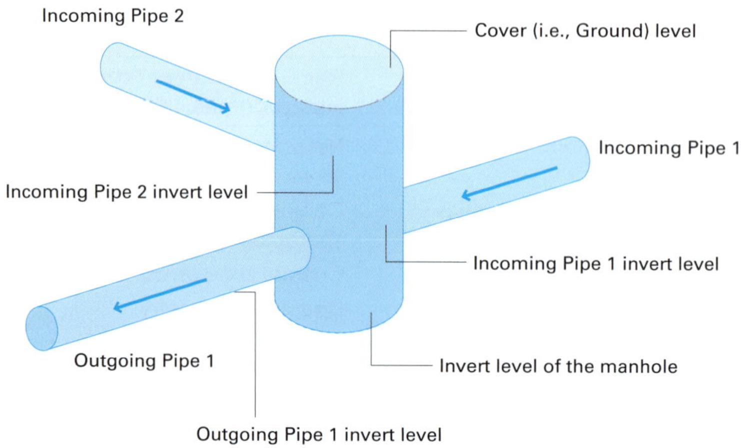

Figure 14.2 Manhole/Pipe details

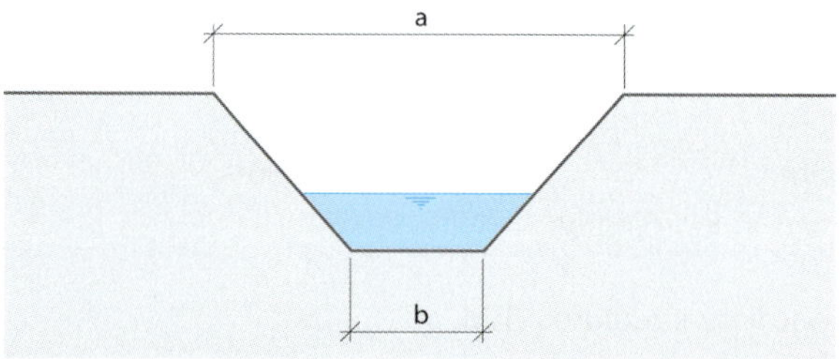

Figure 14.3 Example of a channel cross section

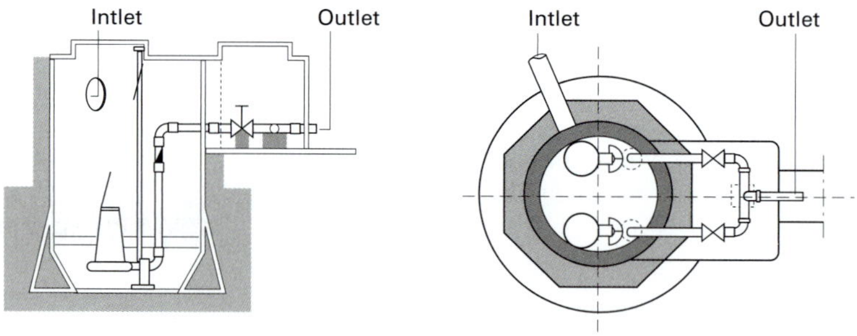

Figure 14.4 Pumping station geometry (cross-sectional view left and top view right)

characteristics, layout of property boundaries, road network layout, geology and soil distribution, administrative boundaries, layout and boundaries of census areas, etc.

14.3 TERRAIN DATA COLLECTION

A digital terrain model (DTM) refers to a topographic map which contains terrain elevations, and as such it is used to represent the terrain (or land) surface and its properties. Correspondingly, the term DEM (or digital elevation model), although usually associated with the land surface, refers to the elevations of any surface. For example, a DEM may be used to describe the water surface of a lake or the water elevations in a groundwater aquifer. There are various structures of DEMs and each of them have their own merits and short comings. Typical structures are: line model, triangulated irregular network (TIN) and grid network. The line model describes the terrain elevations by contours; see Figure 14.5.

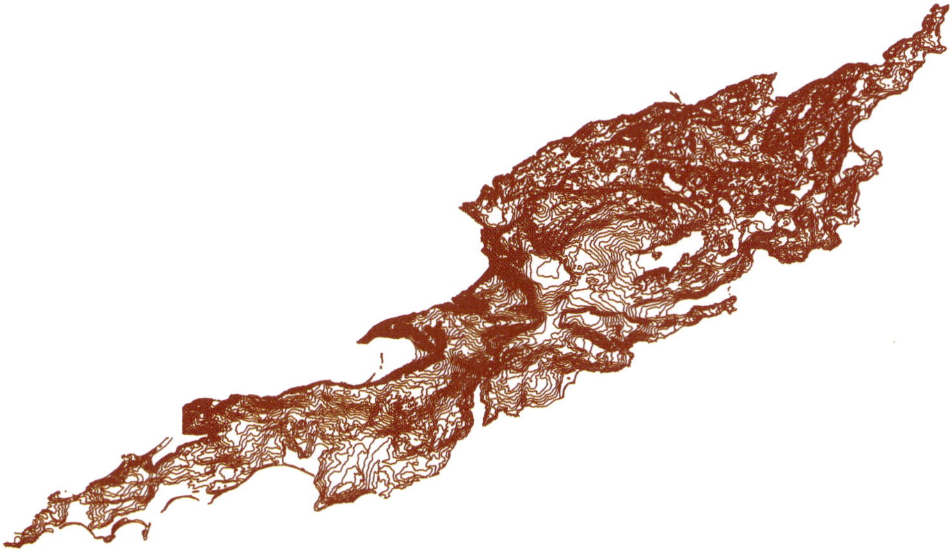

Figure 14.5 Example of a line model – contour map

Figure 14.6 Example of a grid structure DTM processed in a raster-based GIS

The grid structure usually contains regularly spaced square grids. Each grid (or cell) is bounded by three or four adjacent grids (or cells).

Typically, DTM data (Figure 14.6) can be obtained from ground surveys (e.g., total stations together with Global Positioning System – GPS), aerial stereo photography, satellite stereo imagery, airborne laser scanning or by digitising points/contours from analogue format, such as transferring a paper map to a digital format, so that they can be stored and displayed within a GIS package and then interpolated.

Aerial Stereo Photography is one of the oldest and most versatile forms of remote sensing[1] of the Earth's surface. It is taken from a flying object (airplane, kite, balloon etc.). Most professional Aerial Stereo Photographs are taken from special planes which are equipped with cameras at the bottom pointing straight down to the earth; see Figure 14.7. These (large format) photographs are then used to calculate the elevations of the terrain (and buildings) through photogrammetry.

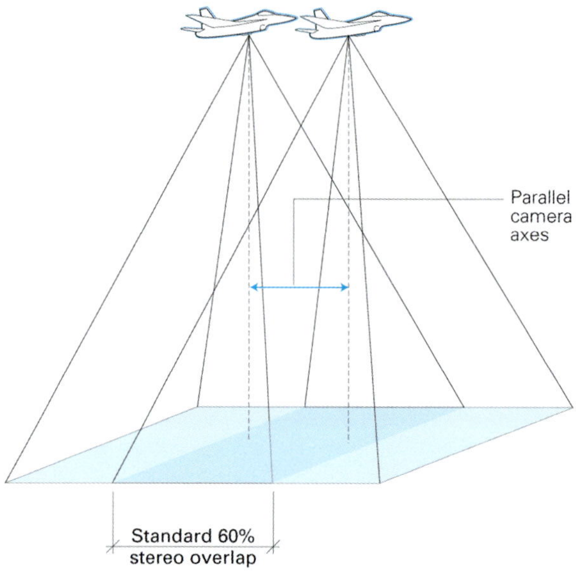

Figure 14.7 Aerial stereo photography

Satellite stereo imagery is another remote sensing technique which consists of two satellite images of the same location on Earth, taken from two different perspectives during one orbital pass; see Figure 14.8a. A Multispectral image is one that captures image data at specific frequencies across the electromagnetic spectrum, Figure 14.8b. The pair of images is collected in-track, or on the same ground path just moments apart, to maintain the tonal consistency between each image, enabling better interpretability. Typically, each stereo pair contains an image collected at a low elevation angle (above 60 degrees) as well as an image collected at a higher elevation angle (above 72 degrees) with 30°–45° convergence (0.54 to 0.83 base-to-height ratio).

[1]Remote Sensing can be regarded as the process of acquiring, processing and interpreting data or information about objects without being in contact with them.

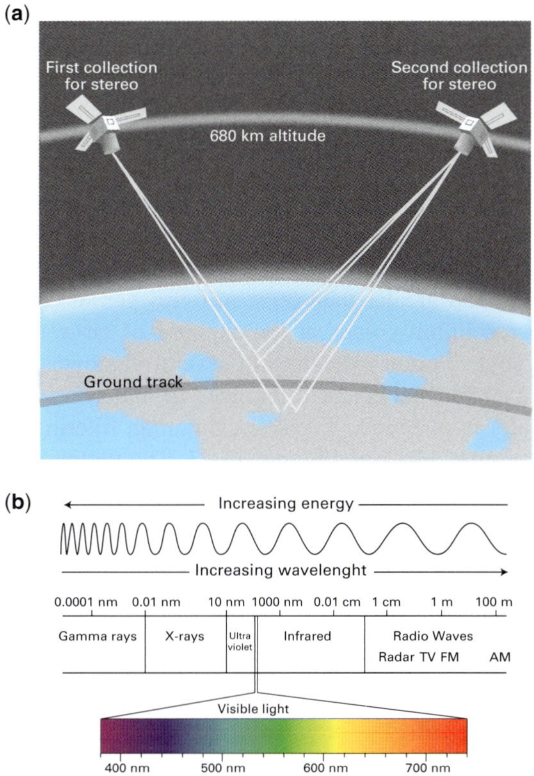

Figure 14.8 Satellite stereo imagery (2), Electromagnetic spectrum (b)

Airborne Laser Scanning (ALS) or Light Detection and Ranging (LiDAR) is yet another remote sensing technique that is used to measure the elevation of an area in an accurate and economical manner. LiDAR is an optical remote sensing technology that measures the properties of scattered light in order to find the range and/or other information of a distant target. The prevalent method to determine the distance to an object or surface is the use of laser pulses. Similar to radar technology, which uses radio waves, the range to an object is determined by measuring the time delay between the transmission of a pulse and the detection of the reflected signal. LiDAR is a surveying method, which can give us accurate x, y, and z (height) positions. LiDAR collects highly accurate and dense points (which typically range from tens of millions to even a hundred million points) thoroughly from the surface or terrain to a desired resolution. This data is collected using aircraft-mounted lasers, capable of recording elevation measurements at a rate of 2,000 to 5,000 pulses per second and has a vertical precision of 15 centimetres. Collected sample points represent the height of a terrain. The points collected from LiDAR, are highly accurate and dense, which helps when constructing a Digital Elevation Model (DEM). In order to construct the DEM from raw data, pre and post-processes need to be performed, including the modelling of systematic errors, filtering, and thinning. Within the overall data process, filtering poses the greatest challenge.

LiDAR can deliver information on terrain levels to a desired resolution. The end result from such a survey is a large number of spot elevations spread across the area. Typically, spot heights are provided at a spacing of 0.5 to 5 m with a horizontal accuracy of 0.3 m and vertical accuracy of 0.15 m. A LiDAR scanner is mounted in an aircraft together with a GPS positioning device, and an inertial navigation system is used to capture the raw data; see Figure 14.9. The laser emits a pulse of light along a vector which is scanned from side to side in the direction of the flight path resulting in a set of zigzag scan lines; see Figure 14.10. Each pulse returns some reflected light to the device from soft objects such as vegetation and eventually a hard object such as the land or a building. The distance is calculated by measuring the time of travel of the pulse, and the spot point determined from the position of the aircraft (by an onboard GPS unit), the distance and the angle of the light vector. Pulse scanners have a range of up to a kilometre, and the rate of data capture which is limited by the physics of light travel and reflectance is between hundreds and thousands of points per second. Depending on flying height and speed the scan width can vary from 200 m to 1 or 2 km with 500 or more points per scan line and greater than 66 scan liners per second. The requirement to obtain ground level data at a higher resolution comes at the expense of large data files which are typically hundreds of thousands of megabytes, often extending to several gigabytes. The complexities associated with manipulating such large data files has created the need to undertake further processing of the raw data and to deliver a manageable set of points. Typically, thinning, filtering and interpolation are part of this process; see Figure 14.11, Figure 14.12 and Figure 14.13. The thinning (or reduction of data points) is usually attained by removing neighbouring points which are within a defined elevation tolerance. Filtering is a process of automatic detection and interpretation of bare earth and objects. Recently, a number of algorithms have been developed, in order to enhance the extraction process that separates surface points from the point clouds, obtained by LIDAR. Most of these filtering algorithms classify data into at least two levels, namely object and surface; by using local neighbourhood operations on points (Sithole and Vosselman 2004). Typically, there are four filtering approaches:

(1) The first group is the so-called, mathematical morphological filtering;
(2) The second group can be described as progressive filters;
(3) The third group of algorithms are those algorithms that progressively increase the density of the DTM, in order to approximate the bare earth;
(4) The fourth group are filters based on segments.

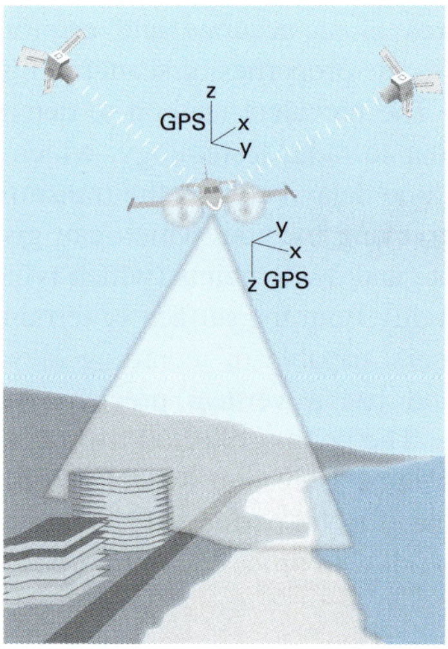

Figure 14.9 Use of LiDAR for collection of digital terrain data

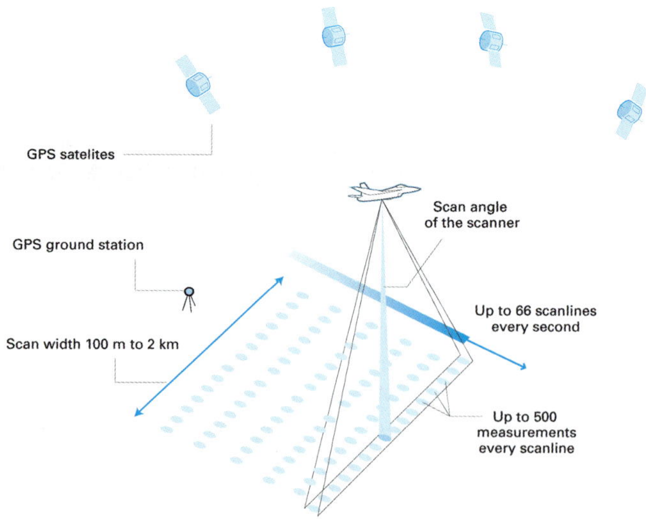

Figure 14.10 LiDAR scanning

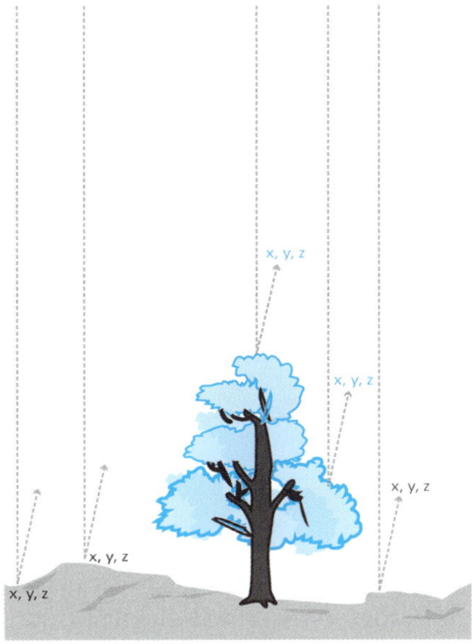

Figure 14.11 Obstruction by morphological objects

The main difference between the above mentioned approaches is the strategy that they employ for estimating the planimetric and height differences between objects and surface points. Additionally, until now, most of these algorithms have been built for a rural environment and very few have focused on separating urban objects from its complex surface. The filtering process for an urban environment is essential in order to produce an accurate DTM; especially when the DTM is used as an input in the urban flood model (see e.g., Abdullah 2011a and 2011b).

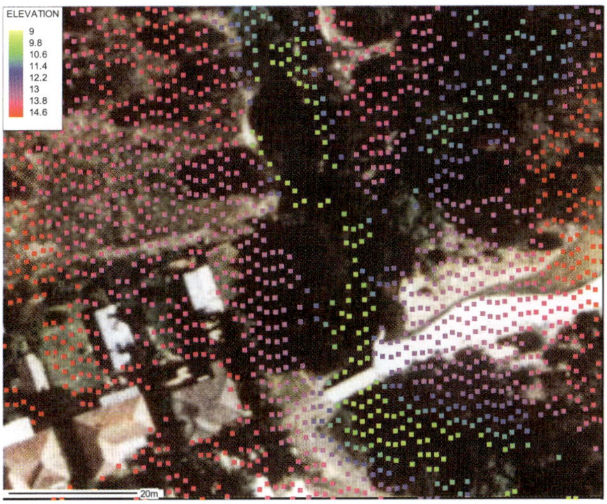

Figure 14.12 Filtering of ALS points

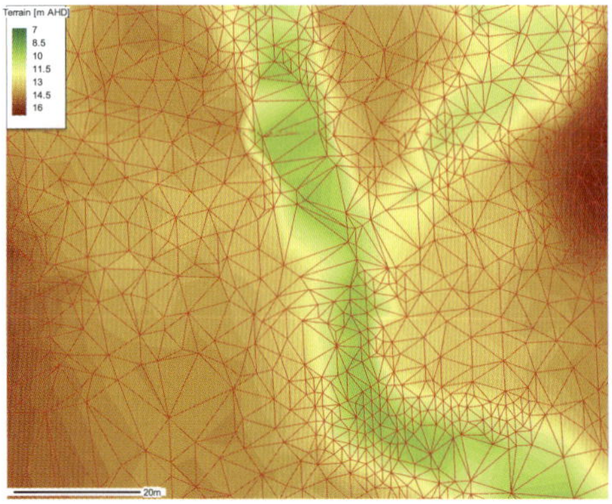

Figure 14.13 Interpolation of filtered data

There are several filtering algorithms which have been developed based on the above mentioned four approaches. Some of them are: progressive morphological filter, Morph and Morph2D, elevation threshold with expand window (ETEW), maximum local slope, iterative polynomial fitting, polynomial 2-surface filter, adaptive TIN (ATIN), etc. For further elaboration of different filtering algorithms and their application to drainage system modelling see Abdullah *et al.* (2011a). Morphological filtering is done by classifying whether a point is on the ground, on a piece of vegetation (e.g., bush or a tree) or on a building. This ground spot height data is then saved for the creation of a digital elevation model (DEM).

Interpolation refers to the process of constructing new data points within the range of a discrete set of known data points. There are many interpolation methods available. The typical ones used in most GIS packages are: inverse distance weighted (IDW), spline, kriging and natural neighbour; see Figure 14.14.

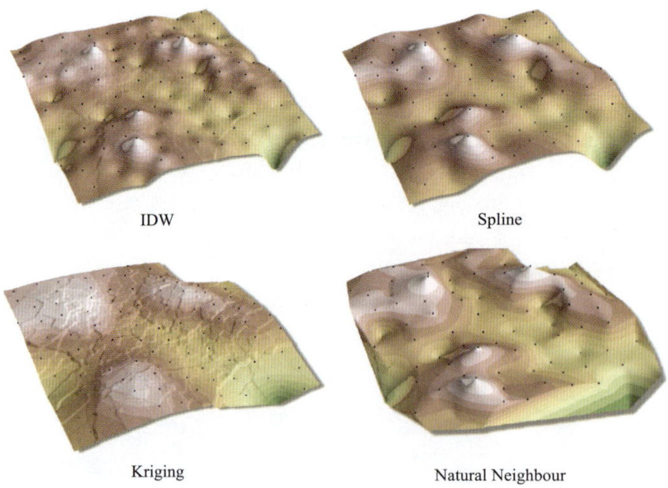

Figure 14.14 Illustration of common interpolation algorithms

IDW determines cell values using a linear-weighted combination set of sample points. The weight assigned, is a function of the distance of an input point, from the output cell location. The greater the distance, the less influence the cell has on the output value. For this reason, the IDW function should be used when the set of points is dense enough to capture the extent of local surface variation needed for analysis.

Spline estimates value using a mathematical function that minimizes the overall surface curvature. This results in a smooth surface that passes exactly through the input points. There are two variations of spline, which are regularized and tension. A regularized spline incorporates the slope (first derivative), rate of change in slope (second derivative), and rate of change in the second derivative (third derivative), into its minimization calculations. A tension spine uses only the first and second derivatives, but includes more points in the calculations.

Kriging assumes that the distance or direction between points reflects a spatial correlation that can be used to explain the variation in the surface. It fits a function to a specified number of points or all points within a specified radius, in order to determine the output value for each location. Kriging is most appropriate when a spatially correlated distance or directional bias in the data is known. The predicted values are derived from the measure of relationship in samples using a weighted average technique. It uses a search radius that can be fixed or variable.

Natural neighbour can be used for interpolation and extrapolation. It is another weighted method that generally works well with clustered scatter points. The basic equation used is identical to IDW. This method can efficiently handle large data sets of input points.

In addition to LiDAR data collection technology, the airborne interferometric synthetic aperture radar (IFSAR) for digital mapping is another technology that can be used to obtain DTM data. IFSAR is an aircraft-mounted sensor designed to measure surface elevations. The data collected is used to produce topographic imagery. Radar pulses are aimed at targets on the earth's surface, and the return signals from the ground are received by two antennae that record the elevations at specific ground coordinates. Similar to the LiDAR technology, the ground coordinates are determined by the GPS and inertial measurement unit (IMU) technology. Post-processing of these data produces topographic information in the form of orthorectified radar imagery (ORRI).

14.4 COMBINING MULTIDIMENSIONAL VIEWS OF TOPOGRAPHICAL INFORMATION TO IMPROVE DETAILS OF URBAN FEATURES

The advantage from using airborne or space borne remote sensing images and airborne laser scanning (ALS) information is that it provides details of the terrain and land cover from a top viewpoint, which is suitable to represent a larger surface area. However, from the top viewpoint in a dense urban area, some details of building ground-footprints as well as narrow passages (e.g., alleys, etc.) may be lost or hidden by roofs or other structures. Therefore, to ensure that all pathways are properly delineated and incorporated in a DTM (or in a 2D model) an approach that combines multidimensional views obtained from remote sensing images (e.g., satellite images) and photographs taken from the ground has shown promising results, Figure 14.15. This approach uses Structure from Motion technique (SfM) as a way to generate the accuracy and detail in space and time of topographical information from a number of photographs.

Figure 14.16 shows an example of 319 photographs which have been taken in an urban area using a compact digital camera. With the use of SfM technique, the point cloud results can be created by detecting particular features, matching them and estimating their geometry from a number of photographs. By using ground control points, the re-projecting point cloud into the real world geometry can be made, Figure 14.17a and Figure 14.17b.

Figure 14.15 Ponding and runoff of flood water in an urban area: direction of runoff based on ALS data without alleyway ('original exit' – red arrow), actual direction of runoff based on eyewitness reports ('original exit' – yellow arrow) (Boonya-aroonnet, 2008)

Figure 14.16 Terrestrial-viewpoint photographs taken in one of the streets in Delft, the Netherlands. (Photograph: Vorawit Meesuk)

Figure 14.17 Terrestrial viewpoint of point cloud generated by using SfM technique (a). Top-viewpoint showing break-line footprints of buildings (b)

14.5 BATHYMETRY DATA COLLECTION

Bathymetry estimation is usually referred to as the study of the "beds" or "floors" of water bodies, and hence it represents the record or data of seafloor topography. Bathymetric data is used for many purposes, which include studies of ocean biology, management of coastal zones and setting up of numerical models that can be used for modelling the effects of tsunamis and storm surges.

There are several remote sensing techniques that can be utilised to derive bathymetry data in coastal areas. The key techniques are: sonar (sound navigating and ranging), LiDAR (light detection and ranging) and high-resolution satellite images.

Sonar is the underwater measuring technique which uses the pulse of sound to detect the seabed and objects in the water column, see Figure 14.18 (a). This technique is based on the measurements of the speed of sound in water. By determining the time lapse between the emission of the sound pulse and its reflected reception, the detector can determine the range and orientation of the object.

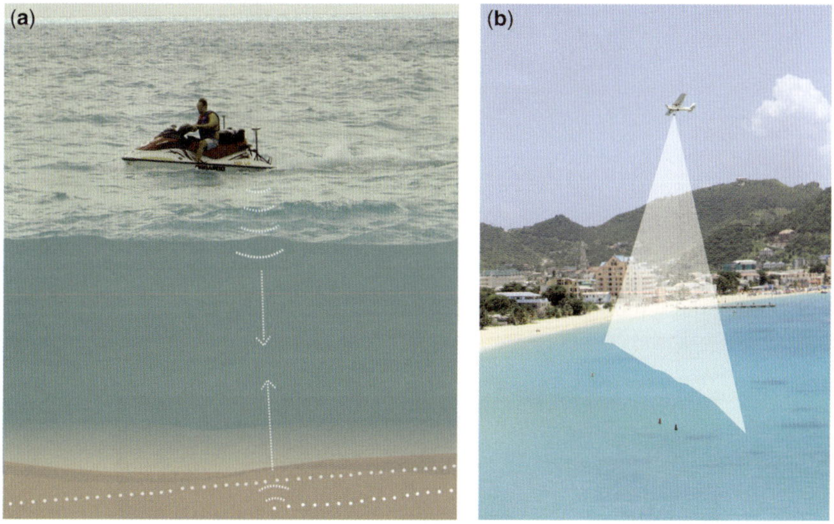

Figure 14.18 (a) Measurements with echo-sounder, and (b) Measurements with LiDAR

Sonar data is usually collected with a Hydro-box Single Beam Echo-sounder (SBES) which can be mounted either on the boat or on the JetSki (Figure 14.19). The role of SBES is to transmit the acoustic signal through the water. In addition, the boat or the JetSki is equipped with a GPS receiver to identify the position in which the measurements are taken.

Sonar measurements, if done properly, can obtain a good quality dataset. However, in certain situations, performing a complete and detailed survey of a coastal zone using sonar techniques can be a very expensive and time consuming task. As a result of this, many researchers and practitioners nowadays combine less intensive sonar measuring with satellite data in order to obtain bathymetry data sets, Figure 14.20.

In this respect, some researchers have used either LiDAR data (see e.g., Stumpf *et al.* 2003) or sonar data (see e.g., Corucci and Masini 2011) in order to evaluate the performance of the bathymetry that can be obtained from satellite imagery. The modelling of water depth from satellite images can be established using theoretical, empirical or statistical models based on the relationship between pixel values and measured depths. These methods include a linear transform algorithm (Lyzenga 1978), ratio transform method (Stumpf *et al.* 2003), a back propagation artificial neural network (Wang *et al.* 2007) and a Support Vector Machine (UNESCO-IHE 2011).

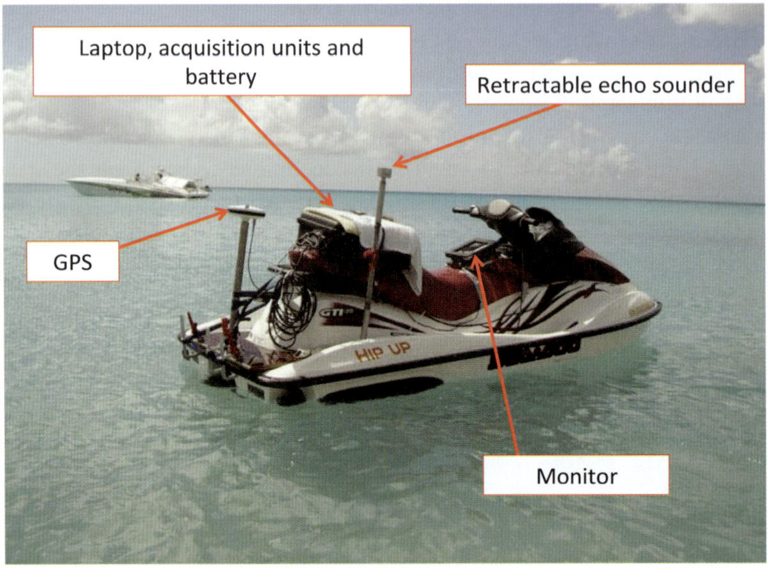

Figure 14.19 Sonar equipment mounted on a Jet Ski

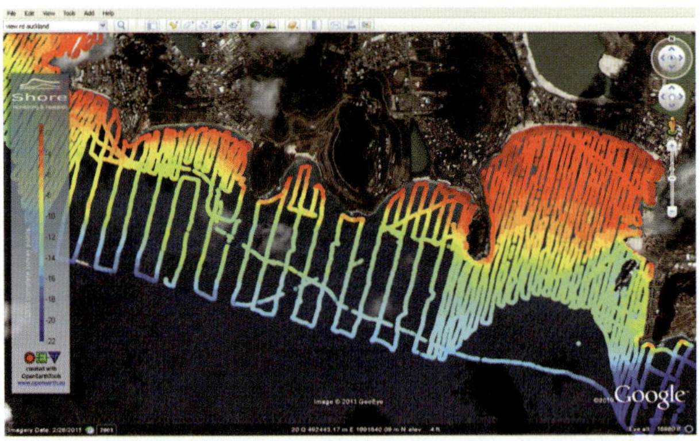

Figure 14.20 Example of sonar measurements taken in 50 and 100 meter offsets

The following flow chart (Figure 14.21) illustrates the methodology that can be used to obtain the bathymetry from sonar data and satellite imagery.

A data driven technique such as an SVM technique focuses on the reconstruction of unknown functions from known data, which in this case could be in this case the ratio of blue band to green band and ground true information obtained from sonar technique. This model has shown to be capable of handling non-linearity inherent in the data sets.

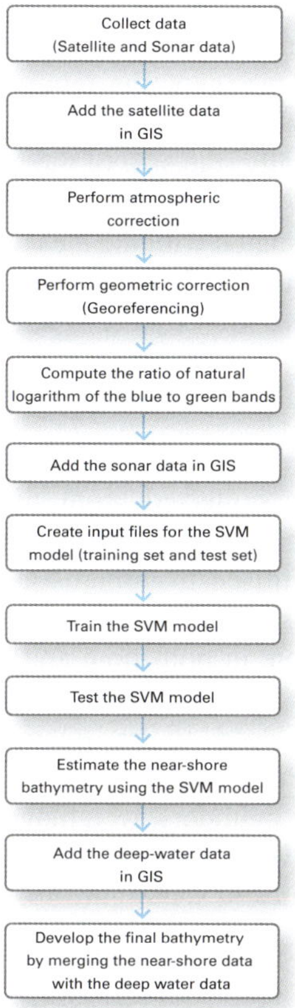

Figure 14.21 A flow chart for bathymetry estimation from sonar data and satellite imagery

14.6 TEMPORAL DATA COLLECTION

Temporal data collection refers to the measurement of a physical variable over a period of time and can be defined as the process by which a numerical value is assigned by the sensor at specific time instances (or time steps). A sensor is a device that can detect a signal or a physical condition.

Typically, different quantities such as depth, velocity, rainfall, pH, suspended solids, and total coliforms will require different kinds of sensors. It is important to note that there is a wide range of sensors that can be found on the market and their accuracy (the closeness of the agreement between the result of a measurement and the true value at the site of the measurement) and time of response, range of measurement, repeatability (ability to give the same result when used to measure the same quantity several times in succession) and robustness vary according to the application and the environment. There is always a trade-off between the accuracy, repeatability and robustness that can be achieved and a given cost.

Selection of the appropriate measurement equipment for a site may not be simple or easy. Difficulties can arise due to presence of vibrations, direct sunlight, high humidity, aggressive environment, high or low temperature, difficult structural conditions, etc. Furthermore, special consideration must to be given to sites which have a high potential for vandalism, in which case appropriate housing arrangements may need to be constructed.

In designing a temporal data collection campaign it is important to be aware of potential uncertainties associated with the measuring equipment as well as with the nature of the phenomena under consideration. In this respect, careful definition of the temporal and spatial scales are necessary. The temporal scale refers to two dimensions such as frequency and duration of measurements. The spatial scale also refers to two dimensions, nominally, the size of the area under consideration and the density of the gauges.

To minimise potential uncertainties associated with the measurement process it is necessary to undertake the calibration of all instruments before their installation and use. In this way, we can minimise the error of the measurements, remove the potential bias, identify dominant sources of error, quantify their effects and estimate uncertainties.

14.6.1 Meteorological data

Typical meteorological data that can be collected from urban environments involves temperature, precipitation, (rain, snow and hail), evapotranspiration, humidity and wind speed. Out of these five quantities, precipitation is the one most utilised by urban water engineers and it is often measured not only daily but also at much shorter time scales. Depending on the climate, topography and other local conditions, such data can be highly variable in both time and space. Typically, this data is collected by raingauges which in effect represent point measurements of rainfall and, in the case of multiple stations, the readings between different stations need to be interpolated in order to obtain some spatial representation of the precipitation (e.g., using Thiessen polygons, IDW, kriging, etc.).

Types of rain gauges include: graduated cylinders, weighing gauges, tipping bucket gauges, and simple buried pit collectors. Each type has its advantages and disadvantages for collecting rain data. The most commonly used raingauge is the tipping bucket; see Figure 14.22.

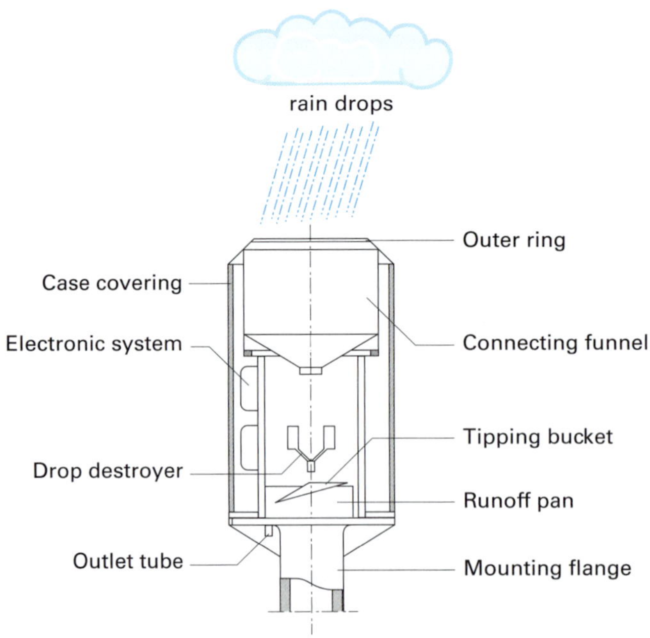

Figure 14.22 Tipping bucket rain gauge

The tipping bucket raingauge consists of a large copper cylinder which is typically set into the ground. At the top of the cylinder is a funnel that collects and channels the precipitation. The precipitation falls onto one of two small buckets or levers which are balanced in the same manner as a scale. After a certain amount of precipitation is accumulated the lever tips and an electrical signal is sent to the recorder. The recorder consists of a pen mounted on an arm attached to a geared wheel that moves once with each signal sent from the collector.

If the rainfall data is to be used for model calibration purposes it is very important that the sites are selected in such a way that they will provide a good representation of the rainfall across the area of the catchment.

In principle, the raingauges should be placed twice as far from an obstacle as the obstacle is high off the ground. For example, if an object is 3 meters high, the rain gauge should be at least 6 meters from that object. The top of the rain gauge should be at least 0.5 m off the ground in order to minimise the effects of the wind near the ground and to increase the capture of the rainfall.

Apart from precipitation measurements that can be undertaken using raingauges, recent developments in weather radar techniques have made radar a very valuable tool for collecting precipitation data in urban areas: see Figure 14.23. Such techniques have the advantage of measuring rainfall at all points over a catchment rather than at a few locations. This creates a special advantage over the use of raingauges, especially where spatial effects are important.

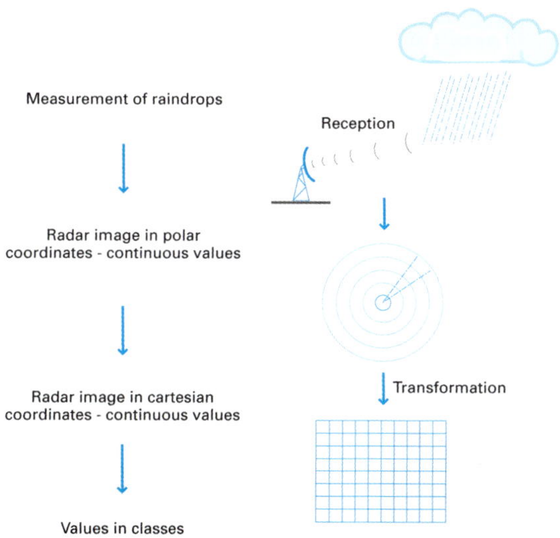

Figure 14.23 Radar measurement processing (After Einfalt *et al.* 1990)

The method of measurement involves a radar beam being reflected from falling raindrops; see Figure 14.23. The intensity of the reflected beam is related to the intensity of the rainfall. Before use it is necessary to make several corrections to the data:

(1) Reflections from ground based objects such as hills and tall buildings (called ground clutter) must be removed.

(2) High intensity signals caused by ice crystals at the bottom of the cloud layers (bright band) must be corrected for.

(3) Other corrections are required for the size of the raindrops, which can be made using the Doppler frequency shift to measure their velocity.
(4) It is also usual to calibrate the signal against rainfall intensities measured at rain gauges within the catchment. This may involve constructing a rainfall intensity map from the rain gauge results and comparing this to the rainfall intensity map from the radar.

All of these effects vary during the passage of a storm, and continuous recalculation of the corrections is required.

For urban water management applications, the results are typically produced as a series of rainfall intensities on a rectangular grid with a mesh size of between 0.5 and 1 km. The temporal frequency is usually provided at 15- or even 5-minute intervals. The appropriate mesh point is chosen for each contributing area in the catchment.

14.6.2 Wastewater and stormwater systems and treatment plants

As discussed previously, wastewater and storm water systems typically include pipe and open channel networks, ancillary structures, above-ground or under-ground storages and wastewater treatment plants. The quantities of interest for measurement are flow rates, velocities, levels (or depths), overflow rates, inflow and exfiltration rates and various biological and chemical parameters. The flow monitoring units normally include the depth, pressure and velocity sensors, the data loggers, the mounting bands, cabling, data modems, power supply, meters (or rulers), structures in the pipes (weirs, flumes, etc.), portable devices and other ancillary equipment; see Figure 14.24 and Figure 14.25. Depth measurements can be obtained from meters (or rulers, "depth of flow stick") readings, pressure sensors or ultrasonic sensors. Meters or rulers are effective for steady and correspondingly slow flows and since they require a person to read them they are only good for spot or single readings. They are not efficient for closed conduits because it is difficult to find a suitable location (without debris) and because by standing in the flow and placing the ruler in the flow an effect of a dam will be created and as such this may misrepresent the measured flow depth.

Figure 14.24 Pressure sensors (Photograph: Zoran Vojinović; AIT Hydraulic Laboratory, Bangkok)

Another way of measuring flow depths is to use a pressure sensor which measures the pressure of the flow on the top of the sensor: see Figure 14.24. As the pressure changes, it records a change in the depth of the flow. The device works with barometric, that is, atmospheric pressure. The sensor must be fully immersed in the flow in order to give accurate readings. Barometric sensors may have vent tubes that have to be kept out of the flow at all times in order for them to function properly. Some of the problems that can be encountered with pressure sensors are: too low a flow for measurement, a blocked vent tube or a loss of power to the sensors (low battery of electrical failure). The depth sensor is normally capable of measuring the depth of the flow over a range from 30 mm up to and including 6 m above the invert of the conduit and with a minimum instrument reading resolution of 2 mm increments. Typical accuracy is in the order of +/−1 – 10% and the accuracy tends to be worst at low flow conditions. To prevent loss of data from damage or fouling of the instrument, multiple ultrasonic level sensors are usually installed to provide redundancy.

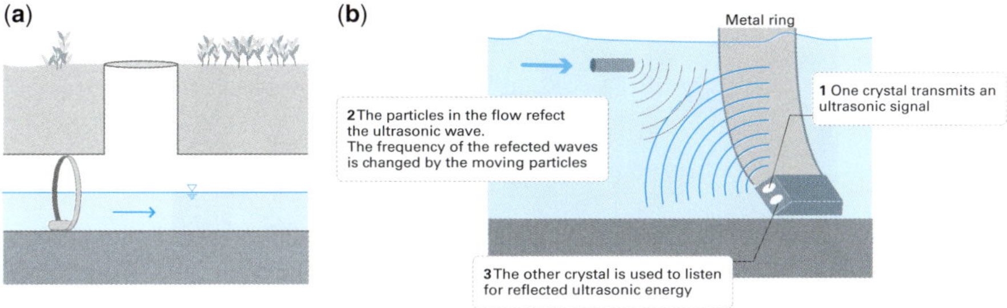

Figure 14.25 Operation of the velocity sensor

Ultrasonic sensors are also used to measure the depth of the flow. They measure sound reflections through the air space above the flow. The sensor must both transmit and receive an ultrasonic pulse. The ultrasonic signal frequency is usually 40 KHz which does not cause any harm to people. Some of the typical problems associated with ultrasonic sensors are: low charge in the batteries that result in weak signals and therefore shallow depth readings, installation difficulties (due to the fact that the sensors need to be installed level to the surface of the flow), a dead band (which is the minimum distance from the face of the transducer at which the sensor can measure) and loss of power.

The velocity sensor is normally capable of measuring the velocity of the flow in a drainage system over a range from –1.0 m/s to +4.5 m/s inclusive, for depths of flow of 30 mm or greater above the invert of the drainage pipe. The typical installation of such sensor is in the upstream pipe of a manhole such that it faces into the flow; see Figure 14.25 (a). It operates using doppler sound waves; see Figure 14.25 (b). The sensor is mounted on a metal ring which is inserted at the downstream end of the pipe; see Figure 14.26 and Figure 14.27.

The sensor devices incorporate a means of measuring, recording and storing the velocity and depth of flow at the required location and at a time interval specified in increments from 1 minute and larger. They are also capable of and suitable for discrete measurements at regular sampling intervals: 5 minutes for temporary monitoring and 15 minutes for strategic gauging under both open channel and surcharge conditions.

For discrete velocity profiles in sewers a suitable hand held velocity measurement instrument (i.e., pygmy or propeller meter) is normally used for in-situ calibration. The depth of flow measured in the sewer at its deepest point is needed to determine what velocity profile to use.

Field calibration of the measuring instruments is an integral part of the measurement process because it confirms the accuracy of the field measurements. An initial calibration is required during the installation of the instrument, and thereafter it is important to ensure a programme of field calibrations which validate sensor readings across a representative range of flows, that is, the minimum, average and peak day flows as well as other elevated flows due to wet weather, whereever practicable.

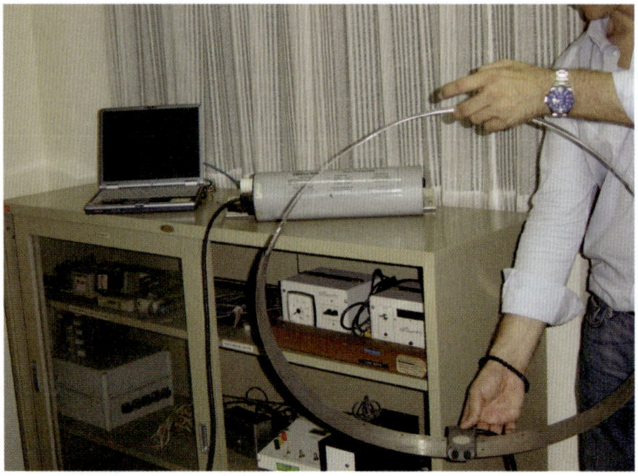

Figure 14.26 A typical flow monitoring installation (metal ring, sensors and flow transmitter connection to the computer – Courtesy of AIT Hydraulic Laboratory, Bangkok)

Figure 14.27 Installation of a velocity sensor in a pipe (Photograph: Sutat Weesakul; AIT Hydraulic Laboratory, Bangkok)

The dimensions of the sewer in which the flow monitor sensors are installed are always measured and recorded at the time the flow monitor is installed and again upon removal or relocation of the sensors.

The major vertical and horizontal dimensions of regular conduits are measured and recorded. Where conduits are of an irregular shape, sufficient measurements are taken and recorded to allow the cross sectional area to be adequately calculated by accepted mathematical means. The industry practice is that the accuracy of measured sewer dimensions is ± 5 mm up to a dimension of 1 m and ± 10 mm for greater dimensions.

In the case of low water depths at a gauging site, a weir structure is usually installed. This structure is normally located perpendicular to the direction of the flow in plan view, with a vertical wall in section view, such that the top of the V-notch weir plate is horizontal. In plan view, the weir is placed in the manhole such that the flow through the calibrated weir plate falls freely onto the manhole invert. The weir pool should incorporate at least two thirds of the manhole area. The notch of the "V" is positioned such that it is vertically above the centre line of the pipe, and so that it is at least 50 mm higher in level than the soffit of the inlet pipe(s) entering the manhole. The weir structure is normally fixed to the manhole such that the manhole's structural integrity is not affected. The weir is installed and maintained such that it remains in position during dry and wet weather flow conditions without affecting its structural integrity as a measuring device.

The level or pressure sensor is placed in or above the pool created by the weir such that it measures the level of the pool unaffected by draw-down near the V-notch. The level sensing equipment is used to measure accurate levels in the range from the notch of the V-notch weir up to the overtopping level of the weir structure.

Independent level measurement calibrations are usually performed on at least three occasions during the gauging period. Additional calibrations may be required to confirm the accuracy of the recorded data.

In principle, regular audits are usually required, and they include the analysis of the following:

- Scattergraphs of Depth vs. Velocity,
- Verification of the flow using a dual equation calculation of the flow at each site,
- Flow balancing,
- Statistical analysis of the field verification results.

The following deliverables are normally obtained as the final output from the flow gauging process:

- Site geometry and schematics of the incoming and outgoing lines;
- Edited depth, velocity and flow data recorded at a specified time interval in an electronic form,
- Graphical plots of edited depth, flow and velocity data for each monitor to a scale such that individual storms can be easily examined,
- A quality flag of any missing data and an assessment of the accuracy of the data presented, so that basic quality coding of the data may be completed,
- Details of any equipment failure or location problems (e.g., silting, excessive ragging, high, low or layered/variable velocity), and an indication of the "drift" in the depth measurement for each monitor,
- A summary of the month's results including comments on the performance of the monitor sites,
- Copies of the check sheets of velocity and depth,
- An installation sheet and monitoring summary for each site,
- Project progress reports delivered at the agreed frequency,
- Processed data, quality coded, in ASCII, Access or any other database format,
- Inflow/infiltration related monitoring (if required) would include:
  - Average dry weather hydrograph for each site,
  - Comparison of the theoretical average dry weather flow and the actual dry weather flow,
  - Typical wet weather hydrographs for the rain events,
  - Quantification of the four components of flow – wastewater; high groundwater; rain-related inflow and rain-induced infiltration,

   - Ranking of the subcatchments in terms of severity for each of the components of inflow and infiltration (I/I),
   - Recommended source detection programme.
- A summary of quality control measures including flow balancing; review of scattergraphs and a statistical assessment of the set of calibrations for each site,
- Commentary of any hydraulic or system features apparent from the flow data,
- Final report discussing all the foregoing points.

Overflow weirs are monitored with simple rectangular or V-notch weirs or pressure/ultrasonic level sensors.

If the intention is to use the data for model calibration purposes, the placement of flow gauges should:

- avoid downstream backwater effects (e.g., be located at the furthest point upstream of the major confluence),
- choose straight sections, both in plan and grade,
- avoid drop connections;
- choose leafy catchments where possible in such a way as to avoid subtraction errors,
- avoid sections with low velocities because the velocity sensors are less accurate below 0.6 m/s and such sites are prone to siltation.
- avoid steeper section and situations where the likely flow depth is below 40–50 mm.

The monitoring of pumping stations should be accurate in circumstances where a particular station has sufficient wet weather capacity, i.e., peak wet weather inflows are lower than the duty flow rate of the pump (s). If a pumping station floods regularly during wet weather (in which case overflow is suspected) and if there are only one or two major incoming sewers, it may be better to gauge these sewers, being aware of ay drawdown effects from the wet well level fluctuations. An overflow may be deduced from the level data, but it may require a gauge or level sensor at the inlet to the overflow. It should be possible to estimate the overflow potential from the pump start/stop records during selected rainfall events prior to the gauge selection.

In terms of the water quality parameters (such as BOD, COD, DO, TSS, TN, TP, heavy metals, etc.) within the pipe/open channel network, these are measured by sampling, which can be either grab, composite or continuous. Such samples are typically taken at any instance in time (in a dry weather or wet weather) and at any point within the system depending on the objectives of the monitoring programme. Typically, grab samples are used for dry weather flow sampling, whereas continuous auto-samplers are used for wet weather flow sampling.

The sampling interval can be different for different parameters. Recent developments in gauging equipment have enabled higher sampling frequencies using sensors even though most of the parameters require laboratory analyses. Measurements with higher sampling frequencies are used in situations where the purpose is to construct 'pollutographs'. However, if the purpose is to obtain event-mean concentrations (EMC), the same samples can be combined to produce the mean values. Whatever sampling equipment is chosen (auto-samplers, gravity-fed samplers, probes, etc.) careful consideration needs to be given to the site selection and time interval in order to ensure that representative samples are collected.

Sediment transport, deposition and re-entrainment through pipe and open channel networks can be characterised over time by regular sampling a few times a year. Such samples are then taken to a laboratory for further analyses (particle size, density, etc.).

In terms of the wastewater treatment plant monitoring, apart from the regular monitoring and sampling activities throughout different treatment processes, monitoring at the inlet (i.e., influent) and the outlet (i.e., effluent) to the treatment plant are essential in order to assess the removal performance.

Since the flow gauging and water quality sampling is a very expensive and time consuming task, before embarking on such programme it is important to consider whether particular gauge/sampling information provides the necessary information and is therefore worth installation. Furthermore, the proposed site should be analysed in order to ascertain whether it is appropriate in terms of the slope, drop junctions, depth and potential major access problems.

14.6.3 Receiving waters

Receiving waters in urban areas typically refer to various types of water bodies which often provide a range of goods and services that are highly valued by the community (e.g., rivers, lakes, coastal waters, wetlands, etc.). With an ongoing process of urbanisation these water bodies tend to receive increased impacts from storm water runoff, overflow discharges and effluent. Typically, legislative and regulatory requirements are used as standards for protection, enhancement and restoration of receiving waters. For example, in Europe it is the European Water Framework Directive (CEC 2000) which prescribes such requirements and in the United States there are different ecosystem monitoring and protection programmes managed by state and regional EPA offices. With respect to the measurements of coastal systems, tide gauges and various Acoustic Doppler Current Profiler (ADCP) instruments are normally utilised. Tide gauges are used to measure tides and quantify the size of tsunamis: see Figure 14.28. They are usually placed on piers to measure the sea level relative to a nearby geodetic benchmark. Tide gauges also monitor meteorological factors that affect sea levels, such as barometric pressure and wind-speed distributions, so that these variable factors can be eliminated from long-term assessments of sea level change. Typically, with ADCP one can map current structures along the measurement track with high vertical (10–20 m) and horizontal (2 km) resolution. A more complete and realistic picture of the currents can be obtained by accessing both components (horizontal and vertical) of the velocity vectors.

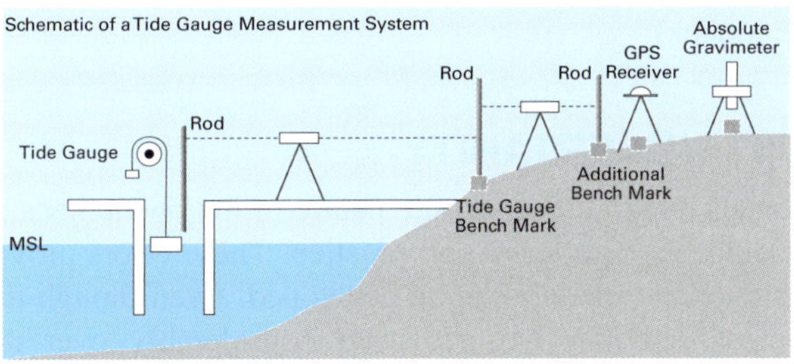

Figure 14.28 Illustration of a tide gauge

In terms of the water quality measurements, tracer studies are often carried out in order to detect contaminants within or coming from wastewater systems. Tracers are compounds, usually dyes or salts used for measuring, mapping, and monitoring wastewater systems. Fluorescent tracers are often chosen because they are cost-effective and can be accurately measured on-site with a portable, field ready fluorometer. The ideal tracer is nontoxic, usable in small quantities, cost effective, easy to measure at very low concentrations, and stable during the course of the study. In such studies the travel time and distribution of the tracer is of interest.

In order to determine the status of receiving water bodies at different spatial and temporal scales, various quantities need to be monitored. For such water bodies, indicators such as temperature, pH, turbidity, carbon and oxygen levels are of particular importance. These are usually measured using different sampling programmes and collected samples are processed in laboratories. Furthermore, bacterial concentrations, unpleasant odours and visual disturbances are also monitored. Sampling from receiving waters can be of grab, composite or continuous natures just as for pipe and open channel networks. Typical environmental monitoring systems are multi-parameter water quality measurement systems. They contain various sondes, which are placed in the water to gather water quality data. Sondes may have multiple probes, with one or more sensors that read water quality data. Typical parameters that can be measured are: dissolved oxygen, conductivity, specific conductance, salinity, total dissolved solids, resistivity, temperature, pH, depth, level, flow, turbidity, nitrate-N, ammonia-N, ammonium-N, chloride, chlorophyll, etc. An example of a sonde is shown in Figure 14.29.

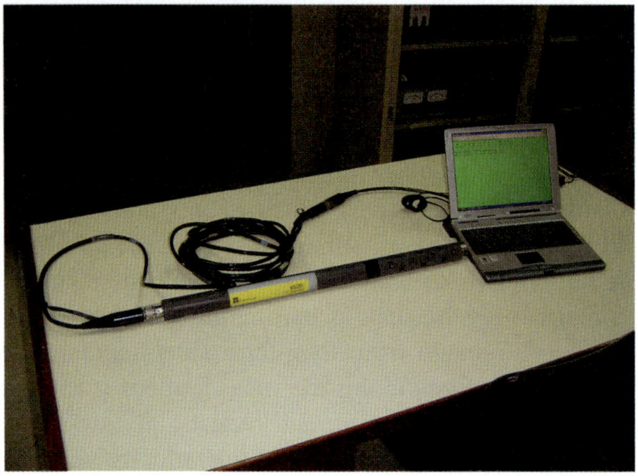

Figure 14.29 Example of an environmental sonde connected to the computer (Photograph: Zoran Vojinović; AIT Hydraulic Laboratory, Bangkok)

14.7 MEASUREMENT UNCERTAINTY

Measurement uncertainty arises due to the stochastic nature of the phenomena that has been measured as well as errors due to the measuring equipment and practice. These errors provide a fundamental limit to the accuracy with which subsequent models can be developed. Even though the science and technology concerning data measurement activities has advanced considerably over the past two decades the associated uncertainty has not been fully eliminated. Therefore, understanding the sources of uncertainty which are inherent within the data measurement process is an important aspect of urban water management. The process of quantifying errors and identifying their dominant sources is often referred to as *uncertainty analysis*. Measurement uncertainty can be defined as the sum of the following uncertainties:

- sampling errors: for example, measurements taken at the point in the sewer pipe which does not give the maximum flow;
- instrument errors: for example, sensor readings are biased consistently below the true value;
- practice error: for example, the rainfall not being measured accurately due to a poor placement of a rain gauge.

To date, certain engineering fields have made attempts to categorise different error types based on their effects on measured values (e.g., AIAA (1995); ANSI/ASME (1985); ASME (1998)). They divide the total measurement error into two components: a bias (fixed) component and a precision (random) component. Some other standards (ISO 1993) have grouped uncertainties into two categories according to the method used to estimate their numerical values: Type A (which refer to the use of statistical methods and which are founded on frequency distributions) and Type B (which refer to other means and which are founded on a priori distributions). Most of the international standards which are currently available are based on the following assumptions:

- uncertainty sources are normally distributed;
- elemental uncertainty estimates for individual variables are combined using the root sum of squares (RSS) method;
- propagation of the elemental errors to the final measurement results are made using first order Taylor series approximations;
- systematic and random uncertainties are RSS combined and multiplied by a coverage factor to obtain confidence in the results;

Since there will be always a degree of uncertainty associated with the collection of data from urban water systems it is necessary to develop a set of QA procedures (concerning choice of the measuring equipment, site selection, instrument calibration, data processing, etc.) which will ensure that the errors arising from the measurement process are minimal.

14.8 OTHER DATA

In addition to the above data, we may also need to collect other kinds of data such as:

- infrastructure condition data,
- geological and soil data,
- customer-related data,
- flooding and surcharge incidents,
- system design and environmental standards,
- economic indicators,
- construction and maintenance cost data,
- etc.

In the presence of active stakeholder participation, much of this data needs to be accessed for initiating and supporting the insights of the actively participating stakeholders: there is a process of transcendence that proceeds in this direction also.

14.9 DATA VALIDATION, PROCESSING, HANDLING AND STORAGE

Upon the completion of data collection activities, all collected data needs to be validated and processed before it is used and stored; otherwise, it should be treated as raw and inaccurate. This will ensure that all outliers in the data which may occur due to a poor sensor reading or disturbances at the gauging site are detected and removed. This also refers to the removal of biased values and other data anomalies. Various statistical methods such as variance and principal component analysis can be applied in this context, where by short-term missing or invalid data can be replaced by carefully calculated estimates. Certainly, every detail of such validation and processing work must be documented and discussed.

Once the data is collected and processed it is then necessary that it be stored adequately in a database so that it can be easily shared among different users. Building up appropriate databases using the latest relational or object-oriented database technology can produce considerable cost savings in terms of planning, design, maintenance and operations. Having the data 'to hand' is also extremely important for job organisation, customer service, and the satisfaction of regulatory requirements – not to speak of the new requirements of other, and potentially much more influential, active stakeholders. Furthermore, an appropriate database can facilitate the coordination of all activities required for providing the desired level of service in a cost-effective manner.

A comprehensive urban water system database uses data from a variety of sources, which can be both spatial and non-spatial. The classes of data needed include the following:

- asset description
- customer related data
- ageing and condition data
- performance data (CSOs, SSOs, flood incidence, blockages, collapses, inflow/infiltration, exfiltration, leakage, intermittent water supply, etc.)
- domestic and industrial discharges,
- condition related data
- historic data
- policy related data
- environment related data
- financial and economic data
- measurements (flow, level, pressure, rainfall, climate, water consumption, effluent, etc.)
- land use
- population
- policy changes on levels of service
- changes in design policy: source control, CSO performance standards
- etc.

Once the data is validated, processed and stored in appropriate databases, its manipulation and maintenance is usually undertaken within a system that can handle spatial and temporal data, and this is essential in the case of public, and especially active stakeholder, participation.

14.10 GEOGRAPHIC INFORMATION SYSTEMS

An important aspect of urban water systems management is to determine the process of collecting and inserting the data into the system that ensures cost-effective data storage and retrieval. This is usually done within a data management framework which links various types of databases with a Geographic Information System (GIS). Within such a framework, users can better understand the way in which the asset condition and performance change over time and what is their relation with the provision of services to customers.

Geographic Information Systems (GIS) are a constituted of versatile and intuitive technologies which have a broad application in urban water management in general, and especially when the management proceeds through public participation. The use of a GIS system within the water/wastewater (river basin or coastal) authority can provide benefits at different organisational levels and different users (e.g., all citizens who care so much as to participate actively, such managers and planners as remain, as well as scientists, engineers, technicians, operators, etc.), Tables 14.2 and 14.3. Overall, some of the benefits are:

(1) More efficient information sharing between different stakeholders, increasingly changing the balance of power between the organised active stakeholders on the one hand and the managers, planners, etc. on the other;

(2) Ability to offer an improved map service that will promote this change;
(3) Improved cooperation and integration between different departments/social organisations;
(4) Opportunity to reduce the sets of manual maps held and the associated storage costs;
(5) Developing and implementing practices for public supervision of land development, including reassesed land value assessments, and ensuring that such information is readily available;
(6) Greater efficiency resulting in increased staff capacity and savings on the one hand and promoting social justice on the other;
(7) Faster and more extensive access to geographic information throughout the stakeholder group as well as between departments and other authorities;
(8) The ability to track and monitor socially desirable growth and development over time;
(9) Improved analyses e.g., the extent of flood prone areas, emergency response planning, quick provisions of underground and above-ground infrastructure information, housing issues, use of public facilities, etc;
(10) Improved productivity in providing public information that promotes the interests of this public;
(11) Better communication of information throuout the public as the ultimate decision-makers;
(12) Consequent muc improved decision-making;
(13) Improved ability to aggregate data for specific sub-areas of interest;
(14) Improved quality of services for and by each stakeholder;
(15) Correspondingly better targeting and coordination of services;
(16) Ability to perform and display different types of professional analyses; and
(17) Improved policy formulation.

Table 14.2 Different users of GIS.

Kind of user	Kind of data	Kind of analysis/tools
Data providers	Raw data, qualitative, quantitative	Flexible, large range of tools
Researchers	Raw data, pre-processed, quantitative	Exploratory, explanatory models
Policy preparers	Pre-processed, quantitative, qualitative	Decision support
Policy makers	Strategic information, quantitative, qualitative	Decision support and communication
Public	Information merging with on-site access to public knowledge and experience, both quantitative and qualitative	Communication and decision making promotion

Table 14.3 Some of the common commercial GIS software packages.

System Name	Developer	Web Extension	Geometric Storage
ARCMAP	ESRI	ArcIMS/ArcGIS Server	Vector and Raster
MAPINFO	Mapinfo Corporation	Mapinfo MapX	Vector and Raster
GEOMEDIA	Integraph	GeoMedia WebMap	Vector and Raster
IDRISI Taiga	IDRISI	IDRISI Taiga	Vector and Raster
MANIFOLD	Manifold Net Ltd	Manifold IMS (Internet Map Server)	Vector and Raster

Geographic stands for spatial data (i.e., the data that can be referenced in space, or *georeferenced*), and displayed in the form of a map. *Information* may be viewed within this special context as *a communication of value-added information. System* may refer, again within this special context, to an integrated interaction of hardware, software and users. Particularly, such spatial technology is invaluable for analysing spatial data, in producing thematic maps as well as in presenting numerical model results.

GIS tools have been developed to geo-reference various data sets into a spatial framework (or to a grid which includes a model of the terrain). With the use of GIS technology, asset data can be linked to the cadastre and property databases and as such used very efficiently in the overall urban planning process.

Geographic data can be characterised by three major aspects: *position, attributes* and *time.* Position usually refers to the location within the coordinate system, attributes refer to the characteristic of an entity contained in the database (pipe diameter, condition, material, etc.) and time refers to the point (or period) of time when the geographic data were collected. There are four geographic data types: *points, lines, polygons* and *continuous surfaces.* Their spatial distribution is represented in two basic types of spatial model: *vector* and *raster* (or *tessellation*) models. The approach used in the vector model is to specify precisely the position of points, lines and polygons with either one or a series of X-Y location pointers. For raster models, the approach used is based on a grid (or cellular) organisation of the spatial data.

Most of the water/wastewater assets are geographically referenced (i.e., *georeferenced*). Therefore, it makes sense to store all these data along with their geographical locations within a spatial information system. This is an obvious step for the networks and the ancillary structures, the above ground catchments, rainfall gauge locations, and so on. The value of the GIS comes when the data can be interrogated from a relationship perspective. So, for example, a query can quickly determine the houses served by a particular water main, or identify the nearest sewers to a factory. The initial value of the GIS comes however, in relating such services to each other, along a main road for project and job management, or quickly determining what part of a water distribution network needs to be shut down for servicing.

Overall applications of GIS systems can be summarised to the following:

- utility networks and cadrastal mapping
- topographic mapping and thematic mapping
- surveying and photogrammetry
- remote sensing and image processing
- civil engineering
- mining engineering
- geography
- soil sciences
- military
- public services
- tourism (etc.)
- land use planning and management
- mineral exploration
- environmental impact studies
- management of water resources
- natural hazard mapping
- forestry and wildlife management
- soil degradation studies
- monitoring desertification

There are four basic functions of GIS:

- data input and pre-processing,
- data management (storage and retrieval),
- manipulation and analysis,
- data output (reports and publications).

Some of the common applications of GIS in urban water management are:

- Combination of different maps with different scales,
- Classification of features,
- Creation of digital elevation models (DEMs),
- DEM interpolation,
- Combination of different factors so that new spatial contexts for different urban water management measures can be visualised,
- Manual and automatic creation of thematic maps,
- Discrimination of various types of pervious and impervious areas,
- Analysis of the DEM to identify possible flow patterns and routes and determine their elevations and cross-sections,
- Cross-referencing the underground infrastructure against the surface terrain features,
- Delineation of the catchment into subcatchments and assignment of land features,
- Flood risk and flood hazard mapping,
- Processing of radar rainfall data,
- Creation of model input files that take account of all of the above in so far as these concern the active stakeholders in particular and the general public more generally,
- Presentation of model results in ways that are required by the needs of the active stakeholder community in particular, and thence to the public interest more generally.

It is still questionable however, how much can be done by interacting with a GIS and how much is better left to the judgment of an engineer. Certainly, the identification or delineation of catchment boundaries and streamlines is sometimes done best by the engineer on the ground. However, for most cases, a combination of a GIS-based delineation followed by a manual adjustment is likely to be the most sensible approach; see Figure 14.30 and Figure 14.31.

A Geographic database refers to the collection of spatially referenced data that acts as a model of whatever is conceived by the stakeholders as 'reality': for the definition of 'reality' as used in this book see Chapter 4, Part I. Within this environment, two main components of the geographic data can be defined:

- attributes or properties
- geographic position

Each entity usually has a number of attributes (properties). Attributes may be:

- nominal (identity e.g., manhole, pipe, maize, granite, lake)
- ordinal (ranking e.g., class1, class2 etc.)
- scalar (value e.g., ground level, invert level,water depth, elevation, BOD)

There are four common categories of analysis procedures:

- retrieval, (re)classification and measurement operations
- overlay operations

- neighbourhood operations
- connectivity operations

Neighbourhood operations include:

- retrieval functions
- interpolation functions
- topographic functions

Connectivity operations include:

- contiguity functions: connected areas that share common characteristic(s)
- proximity functions: connected areas having the same distance to a given point, line or area
- spread functions: connected areas that result from spreading, dilution or accumulation of phenomena from a given point, line or area.

Figure 14.30 Example of catchments/sub-catchments delineation initially pre-processed in GIS and then manually post-processed

Nowadays, access to spatial data as well as to advanced mapping and spatial analysis over the Internet is becoming more common. In the past, an individual would have to buy an expensive software package to use and manipulate the data needed for GIS. With the advent of Java based programming, software applications for web-based GIS work are now available. Some of these programs require the user to buy some additional software, and others require plug-ins to be added to web browsers, but some programs require no special additional software at all. With fast Internet connections the amount of information that can be transferred over the Internet is immense. A typical web-based system is a computerised system that delivers information to the user using a "thin-client" Web browser like Internet Explorer or Netscape Navigator. The computer server that hosts the system application is linked to the user's computer by a network with the Transmissions Control Protocol/Internet Protocol (TCP/IP) protocol. Example of a web-based decision support system is given in Choi *et al.* (2005). The same system can be applied to the operation of water systems where a web-GIS system can be utilised for catchment delineation, map interfaces, data preparation routines, model instantiation and web communications for Internet-based system operation.

Figure 14.31 Example of flow paths definition initially pre-processed in GIS and then manually post-processed

Chapter 15

Rainfall data analysis and catchment delineation

15.1 USE OF STATISTICS AND PROBABILITY

Hydrologic processes are often regarded as stochastic as they evolve in space and time in partly deterministic and partly random manners. In this sense, the use of probability and statistics is beneficial in the analysis of hydrological data. Statistics deal with methods for analysing data sets and for drawing inferences about the properties of a population based on the properties of a sample from that population. They are also used in addressing the uncertainty in knowledge about the processes involved. Together with statistics, probability theory provides a range of standard methods used for 'handling' uncertainty mathematically. In this respect, there are two broad views on modernistic probability theory for representing uncertainty: a *frequentist view* and a *Bayesian view*. The frequentist approach is to evaluate observed samples and to infer the characteristics of the source population. The concepts of populations and samples are fundamental to statistical and probability. A population that may be either finite or infinite is defined as the entire collection of all possible occurrences of a given quantity. An example of a finite population is the number of possible outcomes of the throw of the dice, a fixed number. An example of an infinite population is the number of different peak monthly discharges possible for a given stream. Therefore, the measurements that hydrologists often work with are only a small sample from a large and potentially indefinite population.

Probability refers to as a relative frequency of occurrence such that a confidence interval is translated into a percentage of a confidence interval, as estimated from different samples from the same population. It is also interpreted as a degree of belief such that the confidence interval translates into a level of confidence that the true value falls within the calculated interval. It is a standard practice that flood risk assessment and design of flood protection measures take into consideration likelihood or probability of flood events. The Bayesian approach is that of starting by explicitly stating the up-front assumptions about the population (i.e., prior probability distributions), to confront these assumptions with sample information (i.e., likelihood) and to arrive at modified assumptions (i.e., posterior probability distributions). In case where flooding is caused by the cumulative effect of two or more sources of flood (e.g., pluvial, fluvial and coastal) then joint probability methods needs to be used for the analysis. Joint probability analysis gives the probability of the relevant source variables taking high values simultaneously and thus creating a situation where flooding may occur (see e.g., Defra 2005a and Defra 2005b).

15.2 SPATIAL DISTRIBUTION OF RAINFALL

The spatial distribution of rainfall and its intensity are known to have substantial influences upon the modelling of flooding events (Andréassian *et al.* 2004; Saulnier & Le Lay 2009). A large part of the rainfall-runoff modelling errors can be explained by the uncertainties on rainfall estimates (Moulin *et al.* 2009; Sangati & Borga 2009). The shape, timing and peak of a rainfall event are significantly influenced by spatial and temporal variability across the catchment. An example of a spatial variability of rainfall for an urban catchment is given in Figure 15.1.

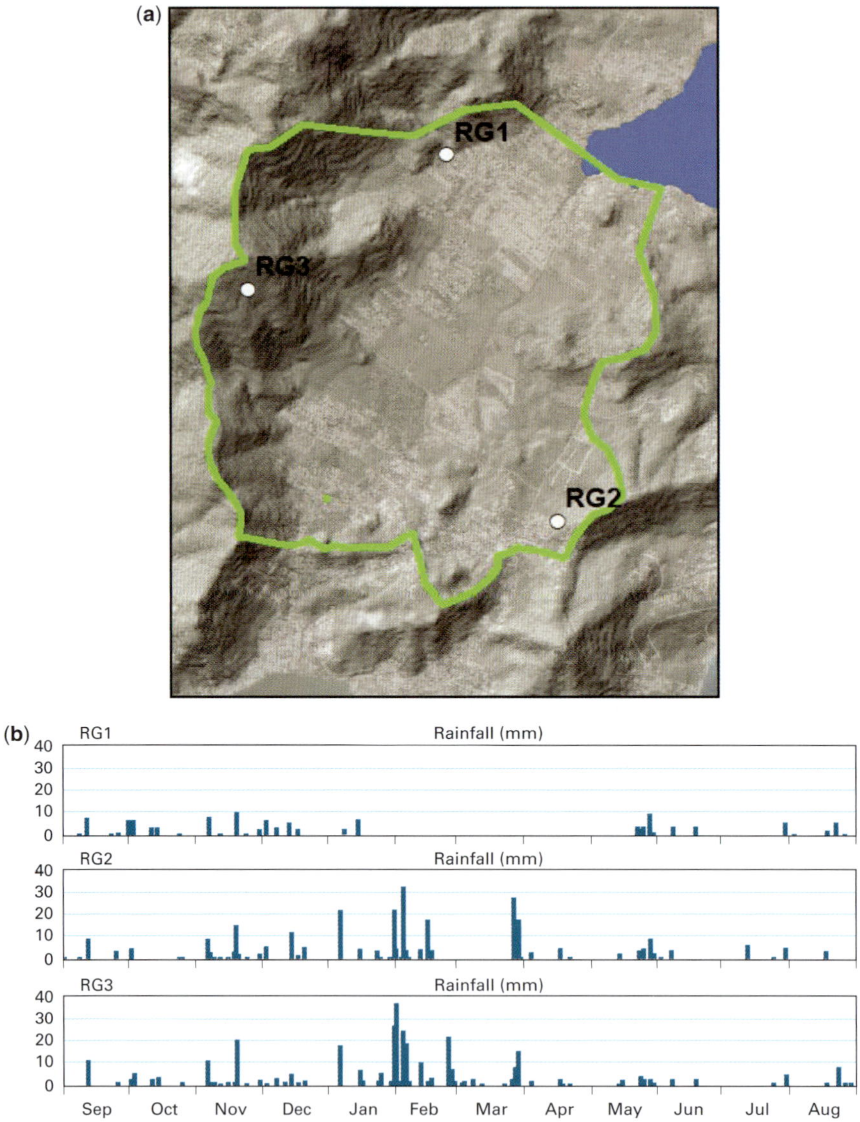

Figure 15.1 Location of raingauges in a catchment area (a) and rainfall measurements (b)

There are various methods that can be used to estimate the average rainfall over an area (i.e., areal rainfall) from point measurements. Some of the commonly used methods are: average depth method, Thiessen method, Isohyetal method, Kriging, reciprocal distance method and Inverse Distance Weight (IDW).

The average depth method is a method based on the arithmetic mean of the rainfall amounts measured across an area. The arithmetic method uses only those gauging stations within the catchment area:

$$\bar{P} = \sum_{i=1}^{n} \frac{P_i}{n} \tag{15.1}$$

where

P_i is the average precipitation over cell (mm/hr)
n is the number of gauges within the catchment.

For example, if the measured precipitation by gauge 1 is 10 mm, gauge 2 is 12 mm and gauge 3 is 16 mm then the average precipitation is simply: $(10 + 12 + 16)/3 = 12.67$ mm.

The Thiessen polygon method adjusts for the non-uniform location of gauging stations by attempting to determine the areas of influence for each gauge. The procedure consists of connecting each precipitation station with straight lines and constructing perpendicular bisectors of the connecting lines to form polygons with these bisectors. In effect, the precipitation surface is assumed to be constant and equal to the gauge value throughout its polygon area of influence (see also Figure 15.2). Therefore, the average precipitation over a polygon area can be calculated as:

$$\bar{P} = \sum_{i=1}^{n} W_i P_i \tag{15.2}$$

where

$W_i = A_p/A$,
A_p is area of the polygon (ha)
A is the total catchment area (ha)
n is the number of polygons

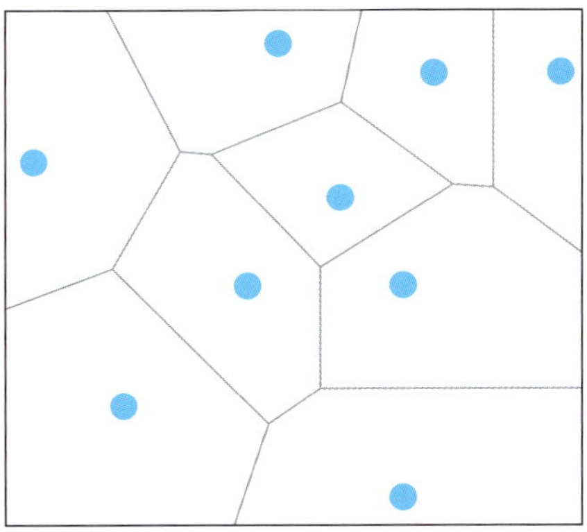

Figure 15.2 Thiessen polygons

The Isohyetal method is also regarded as an accurate method for estimating average precipitation within a catchment area; however, this average is difficult to reproduce because of the complexity of drawing the isohyets; see Figure 15.3.

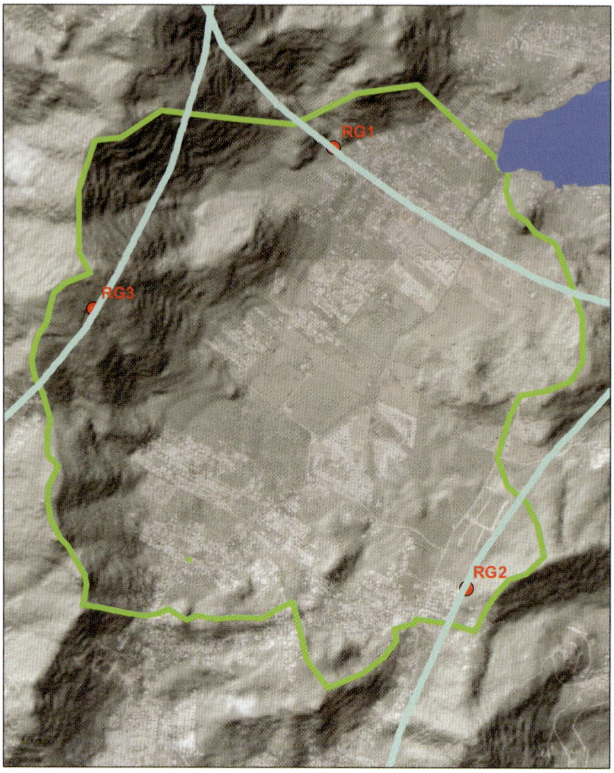

Figure 15.3 The Isohyetal method

Typical parameters that define the statistics for the rainfall are:

- Intensity – the rate of rainfall (usually expressed in mm/hr),
- Duration – a period of rainfall event (usually expressed in minutes or hours),
- Depth – thickness of water layer on the surface (usually expressed in mm),
- Area – geographic extent of the rainfall (usually expressed in km^2), and
- Frequency – the number of repetitions of a particular rainfall event per unit time (usually expressed in terms of 'return period' values).

15.3 FREQUENCY OF RAINFALL EVENTS

The Average Recurrence Interval (ARI) and the Annual Exceedance Probability (AEP) are usually used as a measure of the frequency of a rainfall event. ARI can be defined as the average, or expected, value of the periods between the exceedances of a given rainfall total accumulated over a given duration (e.g., 10-year ARI). AEP can be defined as the probability that a given total rainfall accumulated over a given duration will be exceeded in any one year (e.g., 10% AEP). The relationship between the ARI and AEP can be expressed as:

$$AEP = 1 - \exp\left(\frac{-1}{ARI}\right) \tag{15.3}$$

This relationship yields the values given in Table 15.1:

Table 15.1 Annual Exceedance Probability

ARI (years)	AEP	AEP (%)
1	0.632	63.2
2	0.393	39.3
5	0.181	18.1
10	0.095	9.5
20	0.049	4.9
50	0.02	2
100	0.01	1

When dealing with collection systems (i.e., the design of a new system or the evaluation of the performance of an existing system), different events are often used for the analysis. It is important to remember that this is based on statistics and probability and there is no guarantee that once an event of a certain duration and intensity has occurred, for example a 10 year event, it will not happen again for another 10 years. In 100 years of record there should be ten such rainfall events (an average of one every 10 years). These events could be in consecutive years or even multiple events in a single year! There is also no guarantee that a nearby catchment will not experience such an event in, say, consecutive years.

This concept may be misleading as people may consider that, after a 10-year event, another event of the same magnitude will occur again in 10 years time. Indeed, such conclusions are incorrect as even relatively low probability floods have a significant likelihood of occurring. Several researchers and practitioners have suggested avoiding such terms (see e.g., Fleming 2001; CIRIA C624 2004). Therefore, an alternative terminology could be that of a percentage probability of a flood of a given magnitude being equalled or exceeded in any year (e.g., a 10 year flood event is an event which has 10% of annual probability).

15.4 DESIGN RAINFALL

Much of the concern of urban drainage engineers is with limiting the frequency of flooding. This leads to the analysis of observed rainfall data in order to derive synthetic (or design) rainfall data in terms of the frequency of events. There are different forms of design rainfall that need to be considered:

- Intensity-duration-frequency curves, derived specifically for return periods greater than a year, and developed separately for return periods of less than a year
- The rainfall-event profile (hyetograph)
- Time series of critical events
- Stochastic time series

Design storm events are therefore particularly useful for analysing the frequency of a range of phenomena in a drainage system, such as surface flooding, surcharging in pipes, filling of storm tanks, operation of overflows and other diversion structures, and impact on bathing beaches. In particular, design storms are used for the prescriptive sizing of pipes in a new drainage system. A design storm is normally derived with a rainfall depth for the event, together with a corresponding duration and a profile. Each of these parameters may have their own statistical distributions.

15.4.1 Intensity-duration-frequency curves

Three parameters are needed to describe the statistics of rainfall:

- The *duration* of the period in which the rainfall occurs;
- The rainfall *intensity* during that period;
- The *frequency* with which the rainfall intensity over the particular period is likely to occur, or the probability of it occurring in any particular year. (Note that the inverse of the probability is the return period).

These three parameters are linked in an intensity-duration-frequency (IDF) relationship. Generally, as the rainfall duration increases the average intensity reduces, and the rarer the event the greater the intensity; see Figure 15.4. Note that an actual storm rainfall event can be analysed for a range of durations while each such duration will have its own (average) intensity and frequency.

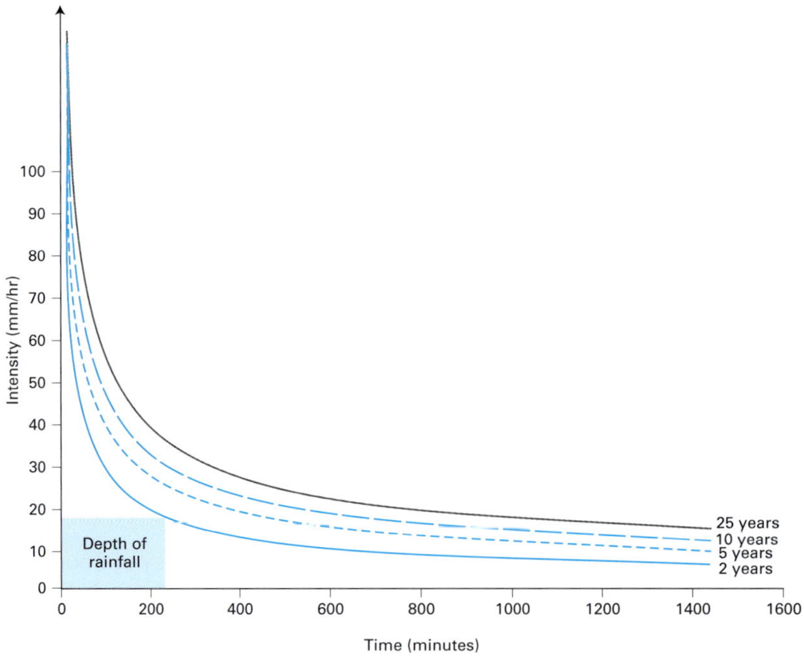

Figure 15.4 Intensity – Duration – Frequency curves

IDF relationships are formally derived by analysing a long set of rainfall records for a large range of durations to determine the greatest depth of rain for each duration in each year of the record. These depths form an annual maximum depth series for each such duration, from which the depth of rain for various return periods can be derived.

Bilham (1936) carried out such an analysis for 10 years of data from 12 stations situated in lowland England and Wales with an average annual rainfall of 889 mm. He came up with the formula:

$$n = \frac{1.25D}{(I/25.4 + 0.1)^{-3.55}} \quad (15.4)$$

where

n is the number of times the rainfall occurs
I is the rainfall depth (mm)
D is the duration (h)

The analysis is valid for durations between 5 min to 2 h.

15.4.2 Storm profile

The rainfall intensity during an event is not uniform in time, and it is therefore necessary to define a storm profile to describe the variation. The profile can be characterised by two parameters, although the details of the shapes used require further parameters. The two basic parameters are:

- The *peakedness* of the storm, which defines the relationship between the peak intensity and the average intensity of the design storm relative to recorded events.
- The *skew* that defines the position of the peak within the duration of the storm.

These are illustrated in Figures 15.5 and 15.6.

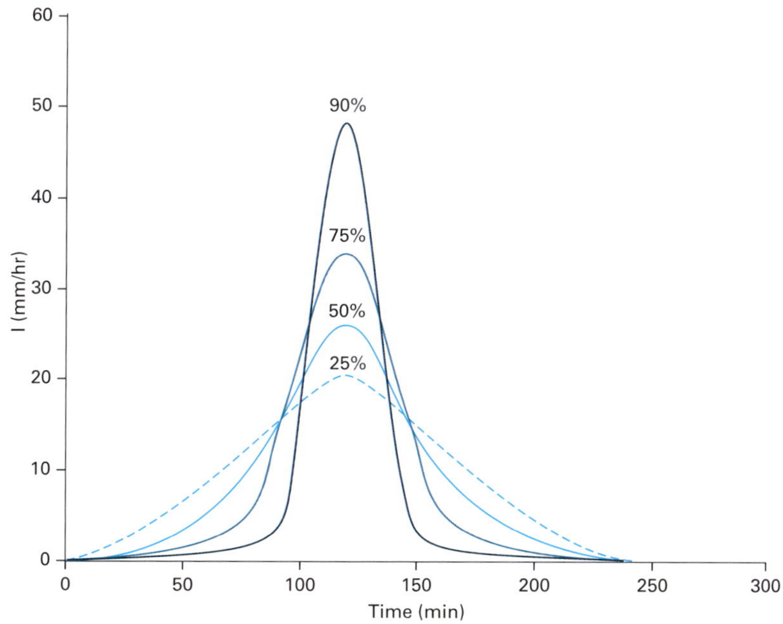

Figure 15.5 Storm peakedness profiles

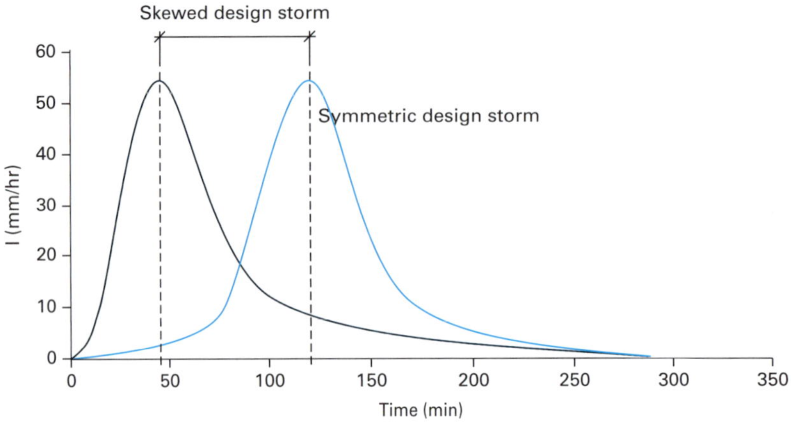

Figure 15.6 Storm skewness profiles

Design rainfall events (or hyetographs) for use in simulation modelling can be derived from IDF data using the Chicago method or by analysing an extended data set of recorded autographic rainfalls. The Chicago method derives a time varying storm that has the same frequency for each rainfall duration occurring during the storm: see Figure 15.7. Obviously, this does not happen in practice, but it simplifies the analysis of the return period of flooding by running a single event defined by the Chicago method. Whole events are therefore derived for specific return periods. They can be assumed to be symmetric or skewed. Different countries or states have adopted alternative ways of defining such profiles, but their aim is to provide a synthetic storm that has an observed peak intensity and also the correct rainfall depth. Where possible the locally recommended design hyetograph should be used to preserve consistency in the calculations.

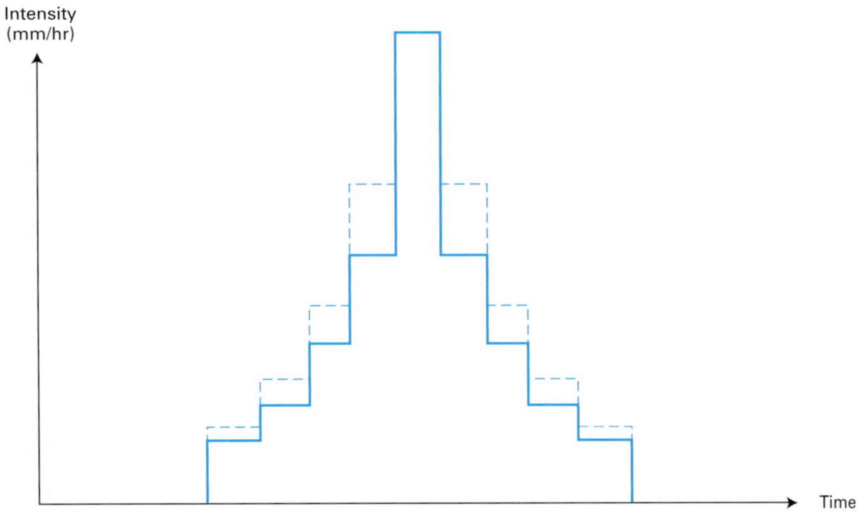

Figure 15.7 Chicago profile

15.5 SELECTION OF DESIGN STORM

The return period of the storm is not necessarily the same as the return period of the flow. This is because there are several independent, statistically defined variables contributing to the flow. These include: rainfall depth, storm profile, duration and the antecedent wetness of the catchment. Given the rainfall depth of given return period, the duration equal to the time of concentration to the point in question, and the 50% summer profile, then the antecedent wetness can be determined as that which leads to the generation of a peak flow of the same return period as the rainfall depth. The summer profile is usually chosen because in urban areas the critical storms usually occur in the summer rather than the winter, and the rarer events are typically thunderstorms. This is discussed further in the section below on rainfall-runoff.

For design storm analysis it is necessary to identify the critical duration of the rain (i.e., critical rainfall) that generates the largest discharge at a point in the system. Sometimes, depending on the rainfall profile and the system characteristics, rainfalls of a shorter duration (and higher intensity) may cause flooding upstream, whereas longer storms will not produce a flow in excess of the maximum peak flow of the critical duration storm.

15.6 ANNUAL TIME SERIES

The IDF analysis is carried out using an annual maxima series for which only the single largest event is taken from each year of the rainfall data. This means that the analysis cannot be used to define storms with a return period of less than one year. However, such storms can be very important when analysing the frequency with which overflows occur. The IDF relationship needs therefore to be complemented by an alternative approach to deal with frequencies of an event occurring more than once a year. Such an approach can be provided by an annual rainfall time series, which includes events in a typical year with frequencies greater than once a year.

15.7 SYNTHETIC TIME SERIES

A criticism of the annual time series is that it is biased towards short storms, and tends to be catchment-specific. Also, its use in model simulation is time consuming and is not suitable for the assessment of drainage solutions to meet the bathing beaches consent criteria of 3 spills a season, which requires a simulation of a 10 or 20 years time series duration. To overcome these problems a stochastic rainfall model has been produced by Cowpertwait (1991) generating a continuous rainfall series for a period of up to 30 years, which is representative of the rainfall climate for the catchment. The rainfall generator is based upon the Neyman-Scott Rectangular Pulses rainfall model; see also Entekhabi *et al.* (1989). The model was developed using five main stations, each located in one of the five regions.

The accuracy of the rainfall generator depends to some degree on the location in which it is being applied, but the generally held view is that the accuracy is adequate for most project activities and it reflects the rainfall character of the area in a reasonable way.

15.8 DELINEATION OF CATCHMENTS AND SUB-CATCHMENTS

An urban catchment area can be defined as an area where water is collected by the natural landscape. In a catchment area, the surface run-off eventually flows to the receiving environment or into the groundwater via the gullies (or catchpits), soakholes and pipe network system. For modelling purposes, after defining the overall catchment boundary it is important to determine the relevant contributing areas of each individual subcatchment covering the key features that are present in the catchment. Figures 15.8 and 15.9 illustrate the schematics of a delineation process for the stormwater catchment and individual subcatchments boundaries.

(a)

(b)

Figure 15.8 Example of a delineated catchment boundary (green line) in a 3D view (a) and 2D view (b)

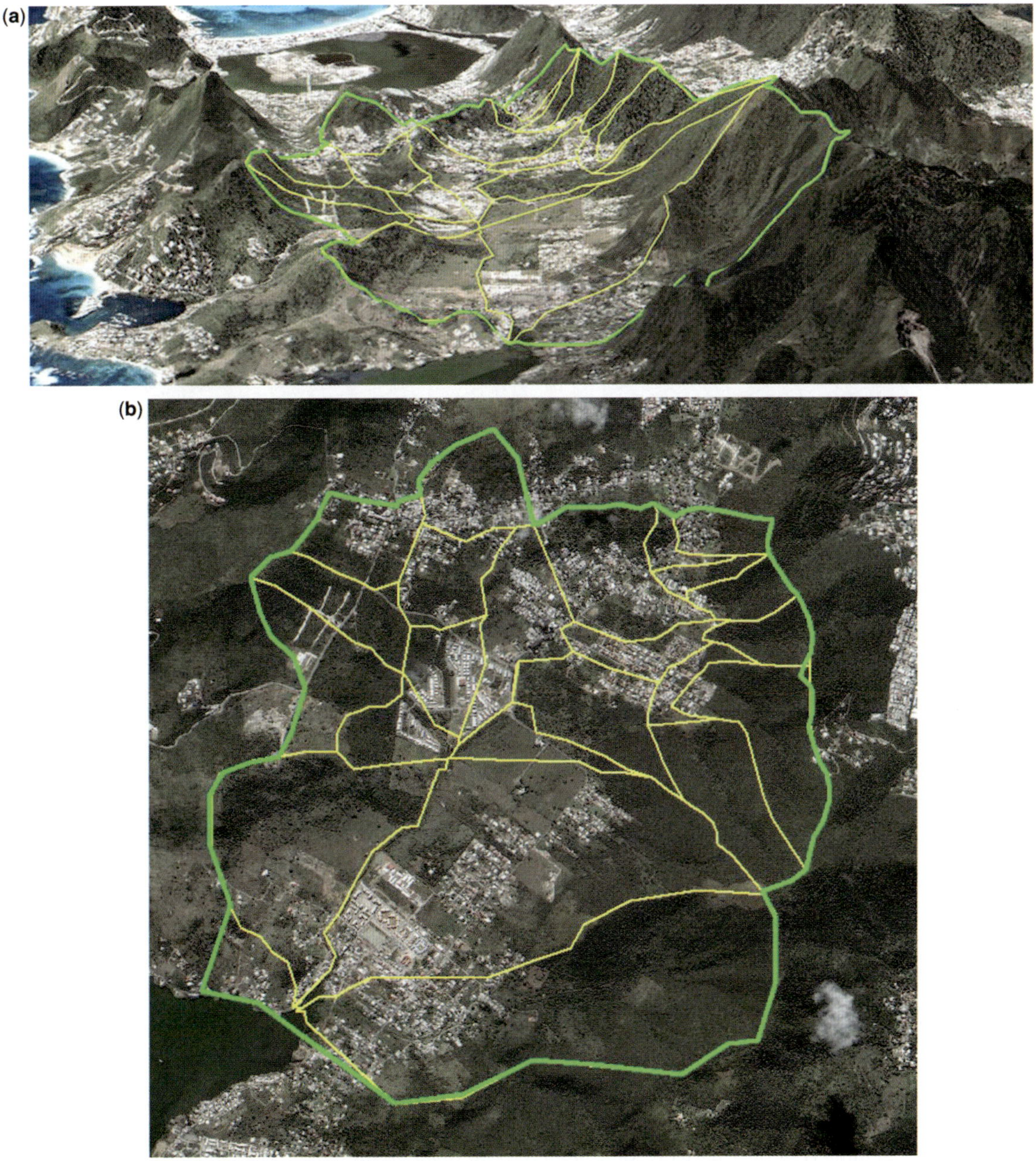

Figure 15.9 Example of delineated subcatchment boundaries (yellow line) in a 3D view (a) and 2D view (b)

In the case of combined sewer system modelling, the wastewater catchment boundary (and also related subcatchments) may not necessarily follow the ridgeline of a natural landscape and as such may not coincide with the stormwater catchment boundary (and its subcatchments): Figure 15.10. Usually, the wastewater subcatchments follow the property lines (as per the household connections) and not the natural landscape. From the modelling point of view, this does not pose a problem since the catchments can be separated with rainfall applied to the stormwater subcatchments and the equivalent population (for residential areas) or area (for commercial and industrial areas) applied to the wastewater subcatchments.

Figure 15.10 Separate wastewater subcatchment schematisation/delineation. The closed rectangles show the subcatchment boundaries. The lines and dots represent pipes and manholes

The process of catchment and subcatchment delineation is normally initiated using built-in GIS functions (also referred to as automatic delineation) and is further refined manually. After this process, inflow loading points (i.e., a manhole of a pipe network or a branch of an open channel network) must be defined. These loading points then effectively define the upstream extents of the model. In principle, the subcatchment loading point should be defined at least one manhole upstream from the known problem areas or ancillaries (i.e., overflows, pumping stations, bifurcations, control devices and inlet structures). Loading points should be defined at an appropriate manhole (or a branch) within the subcatchment such that the resultant flows entering the model result in representative flows and surcharge levels in the system. This is particularly important in the vicinity of flow monitors, bifurcations and overflows. Special attention should be paid to the constructed overflow locations to make sure that subcatchment boundaries do not include flows downstream of the overflow.

Subcatchment definition should also take into account the location of flow monitors to ensure that subcatchment boundaries coincide with monitored (sub)catchments. Large areas that exist within an urban catchment such as playing fields, parkland and major roadways which provide no dry weather sanitary flow contribution to the drainage system should be differentiated from land use maps during the model subcatchment delineation. The land use maps should clearly identify these areas such that wet weather flow parameters can be assigned appropriately. Certainly, the entire area within the catchment boundary needs to be divided into subcatchments without omitted areas. This also includes areas that do not drain directly to the drainage network. Furthermore, consideration should be given to the location of major problem sites identified from the incident/complaint records, and indeed has to be given in the cases of active stakeholder participation.

There are several methods that can be used for GIS-based (or automatic) catchment delineation. Some typical examples are: the watershed method, the basin method and the proxy method. The watershed method uses flow direction grid and loading point grid as inputs into the watershed function to delineate the contributing area above a set of cells in a grid. The basin method is similar to the watershed method with the difference being that the basin function delineates contributing area by identifying ridge lines between basins. The proxy (or proximity analysis) method attempts to mimics the manual techniques which is used to delineate contributing area in separate sewer areas. This method dissolves parcel boundaries based on each parcel's proximity to a generalized sewer network. Therefore, this method uses parcels, modeled sewers and loading points to delineate contributing areas.

Chapter 16

Modelling wet weather and dry weather flows

16.1 MODELLING RAINFALL-RUNOFF FROM URBAN AREAS

The purpose of rainfall-runoff (hydrologic or wet weather) modelling is in the transformation of rainfall over an urban area into a hydrograph which can be used as an input into a pipe or (open channel) flow model. Once rainfall reaches the urban surface the water proceeding from the rain undergoes a far-reaching transformation. Some of the rain falls on buildings and vegetation and is therefore intercepted before it can reach the ground surface or level. Other rain reaches the ground surface directly. In both cases a small amount of water is used to wet the different surfaces. Some water begins to infiltrate through the ground surface, whether it is through cracks and voids in a concrete or asphalt layer or directly into the soil. If the rainfall intensity exceeds the rate at which water infiltrates the local surface, or the absorption capacity is filled, water begins flowing over the surface and fills any remaining small voids or depressions in the surface. Once a void or depression is full, water moves under gravity to an adjacent depression, and therefore moves further and further, creating streams and sheet flow over the three dimensional surface. Necessarily the urban surface is extremely complex, consisting of roads with their camber and drainage gulleys or channels each side, pavements or side walks, paved parking areas, buildings with roof drainage, trees, and pervious areas with shrubs and grass. Consequently the shallow flows over the urban surface are controlled by very small differences in topography, and especially by steps formed by curbs and low walls.

Modelling the detailed flow over this complex urban surface is too time consuming for practical purposes at present, and it would be difficult to guarantee a level of accuracy in the results for, say, the depths of flow over the surface. Therefore the modelling of rainfall-runoff over an urban catchment follows the compartmental approach and treats the catchment contributing to a pipe or surface channel as a whole, forming deterministic equations to relate the volume of runoff to the volume of rainfall taking into account the characteristics of the catchment, and similarly the rate at which the excess rainfall-runoff flows from the catchment.

Rainfall-runoff from an urban catchment is very different from the runoff from a natural rural catchment. The rate of runoff from the urban catchment is usually much faster than a corresponding natural catchment. In addition, there is (or should be) very little infiltration into the conduits from groundwater in the urban catchment, so that almost all runoff occurs over the surface, whereas the groundwater component in the natural catchment feeding the stream network can be considerable.

There are two major concepts associated with rainfall-runoff from an urban catchment draining to a node of a drainage network. The first is the notion of a 'loss' of rainfall that fails to reach the outfall of the urban catchment. A basic assumption is that the rainfall event has a certain volume, which is reduced by an initial amount before the water starts flowing over the urban surface, and thereafter it has a constant or variable continuing loss depending on the nature of the losses computed in the model that is used. The losses model can however, take one of several different forms. Given that the initial loss is removed from the total rainfall volume to leave an effective rainfall volume, the simplest approach is one of using a runoff coefficient that determines the factor of the total effective rainfall that runs off. However, runoff can be said to occur from three or more types of surface, including roofs, paved areas and pervious areas. A different continuing losses model can therefore be assumed for each surface type. Whereas runoff from a paved area is subject to a small constant continuing loss, the continuing losses on the pervious area might be subject to the variation in the infiltration rate. Yet another losses model lumps the continuing losses from all surface types and then reallocates them to each surface type according to some prescribed rule.

The second major concept associated with the rainfall-runoff concerns the rate at which the remaining surface flow runs off the catchment and into the outfall or node. This is a matter of routing the runoff to the outfall.

The following sections describe some of the most common rainfall-runoff models used in modelling urban catchments.

16.1.1 Runoff Coefficient Model

The continuing runoff losses after initial losses can be defined as being at a fixed rate and therefore independent of antecedent conditions and changes taking place during an event. This type of representation is only advised for use with impervious areas: the runoff coefficient from pervious areas is known to vary significantly with antecedent conditions.

Typical values for runoff coefficients are given in Table 16.1. These are conservative values. It is advised that the use of these coefficients be supported either by field observations or by expert judgement.

Table 16.1 Typical values of the runoff coefficient.

Surface type	Description	Coefficient
Paved	High quality paved roads with gullies <100 m apart	1.00
Paved	High quality paved roads with gullies >100 m apart	0.90
Paved	Medium quality paved roads	0.85
Paved	Poor quality paved roads	0.80
Mixed Paved/Permeable	High density housing	0.55
Mixed Paved/Permeable	Medium density housing	0.45
Mixed Paved/Permeable	Low density housing or industrial areas	0.35
Permeable	Open areas	0.0–0.25

16.1.2 The Horton infiltration model

If the runoff coefficient model deals with runoff from paved areas, we need a model for runoff from permeable surfaces. A suitable model for this is the Horton model, which assumes that runoff from permeable surfaces increases as an event takes place; see Horton (1933). The paved runoff factor is assumed to be constant while the permeable runoff is a function of the conceptual wetting and infiltration processes. The Horton equation is derived from infiltrometer/small-catchment studies and is usually expressed as a function of time:

$$f = f_c + (f_0 - f_c)e^{-kt} \tag{16.1}$$

where

f_0 is the initial infiltration rate (mm/hr)
f_c is the final (limiting) infiltration rate (mm/hr)
k is the exponential term coefficient (1/hr)
t is the time (hr)

The cumulative infiltration F is given by the integration of Eq. (7) as a function of time

$$F = \int_0^t f\,dt = f_0 t + \frac{f_0 - f_c}{k}\left(1 - e^{-kt}\right) \tag{16.2}$$

The Horton equation, Eq. (16.2), represents the potential infiltration as a function of time when the supply rate (rainfall rate) is unlimited, that is, when the rainfall is higher than the potential infiltration rate. Expressed as a function of time, it is not suitable for use in a continuous simulation model. As a result the equation needs to be transformed to be a function of a soil moisture storage, which can be accounted for continuously:

$$f = f_0 - k\theta \tag{16.3}$$

Here θ is the antecedent moisture parameter given by

$$\theta = \int_0^t (f - f_c)dt \tag{16.4}$$

Eq. (16.4) provides a direct mechanism for updating the Horton curve during a rainfall event if the rainfall intensity falls below the potential infiltration rate. Furthermore, the outflow q can be initialised as a function of soil moisture.

Figure 16.1 provides an overview of the calculations. First, the infiltration rate f at time t is calculated. This value is compared with the supply rate of rainfall, i.

If $i > f$ then ponding occurs and the volume of rainfall losses due to infiltration is calculated over the time interval Δt.

If $f > i$ then two cases can exist:

If f is greater than i at the end of the interval Δt then all the rainfall infiltrates into the soil.
If f is less than i at the end of interval Δt then ponding occurs during the interval Δt.

The condition is tested by calculating a new infiltration rate f' at the end of interval Δt assuming all the rainfall infiltrates. If f' is less than i then the time to ponding t_p is calculated from

$$t_p = \left(\frac{i - f}{f' - f}\right)\Delta t \tag{16.5}$$

Following this time the rainfall is allowed to infiltrate. A new f at time t_p is than calculated. The infiltration volume is calculated from Eq. (16.5), while substituting $(\Delta t - t_p)$ for Δt.

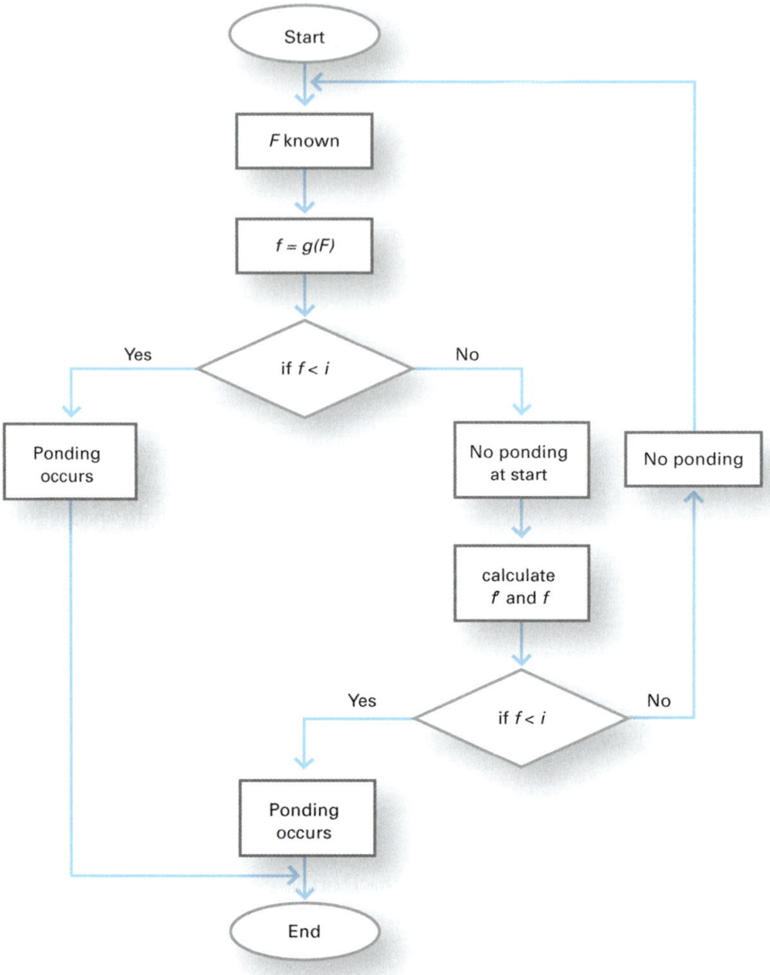

Figure 16.1 Flow chart of Horton model infiltration algorithm

At the end of these calculations a new storage term q is calculated. Various values for Horton's infiltration model are available from the literature: see Table 16.2.

Table 16.2 Various values for Horton's infiltration model.

SCS Soil Group	f_0 (mm/hour)	f_c (mm/hour)	k (1/hour)
A	250	25.4	2
B	200	12.7	2
C	125	6.3	2
D	76	2.5	2

16.1.3 Conceptual framework for rainfall-runoff models (UK)

The rainfall-runoff models in the UK Wallingford Procedure consist of a set of models which enable engineers to represent the transformation of rainfall into runoff for both urban and rural catchment areas contributing to piped or channelled drainage systems. Irrespective of the model option selected, there is a basic conceptual framework underpinning the use of the model. First, the overall catchment model is divided into a series of sub-catchments. Each sub-catchment can then be subdivided into three surface types. It is normally assumed that the first two surface types represent impervious areas, whereas the remainder of the sub-catchment represents pervious surfaces. The surface runoff model is divided into two major process models, representing rainfall losses and runoff routing, each of which can act independently on the surfaces in each sub-catchment; see Figure 16.2.

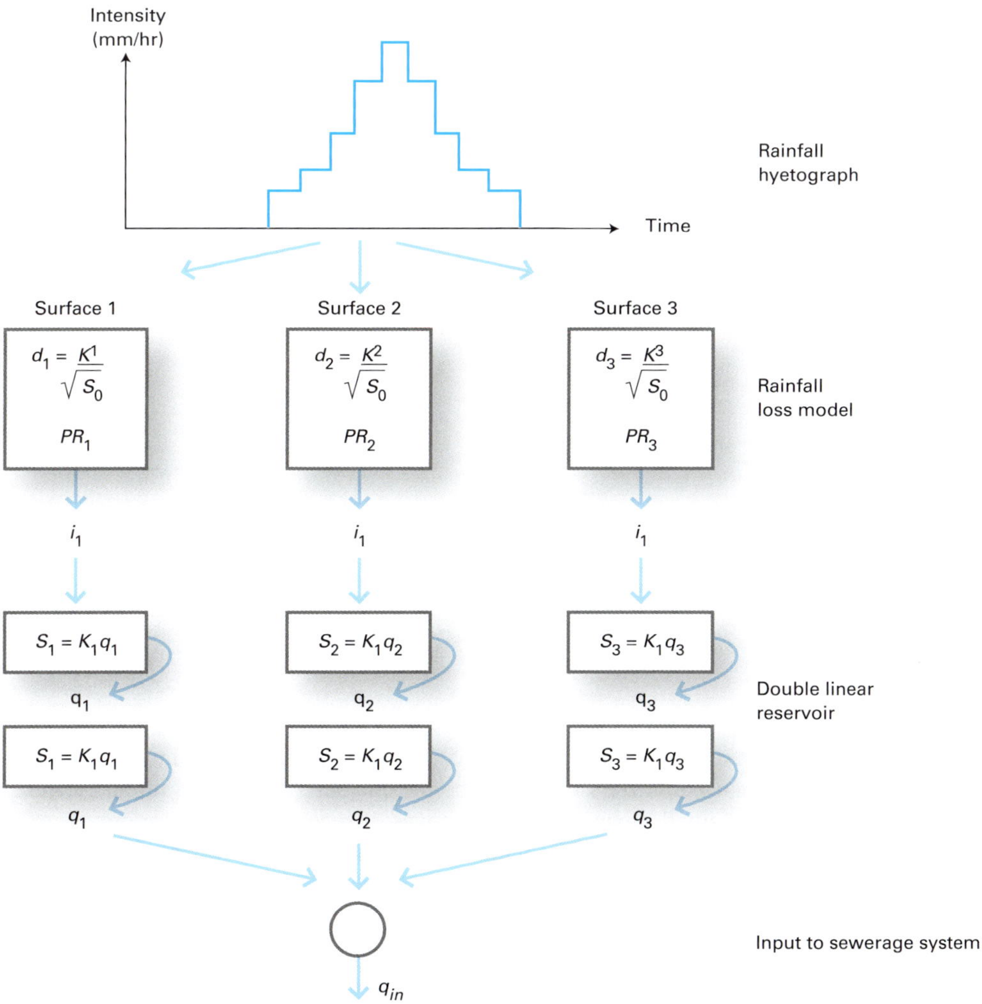

Figure 16.2 Wallingford Procedure rainfall-runoff model

The runoff model for the UK Wallingford Procedure (HR Wallingford 1981a) is dependent on experimental studies carried out largely in Europe but including some studies from North America. First, 34 studies were carried out in the UK, Denmark, France, Italy, Norway, Sweden, Yugoslavia, USA, Canada and Australia where measurements were taken in the sewer, monitoring the response of a collection of individual areas and surfaces. The resulting experimental data enabled a percentage runoff to be determined for each catchment, and subsequently a generic regression equation to be derived for general application. Secondly, 20 experimental studies in the UK, The Netherlands and Sweden based on runoff from individual urban surfaces (such as roads and roofs) measured at the inlets into the sewerage system were carried out in order to calibrate a generic 'rate of runoff' model.

16.1.4 Rainfall-losses models (UK)

There are a number of processes that have been explicitly represented in the Wallingford Procedure runoff models for the losses:

(1) Depression Storage Model for the initial losses
(2) Percentage Runoff Model including an Infiltration Model for pervious areas

Depression storage model (UK)

Although on paved surfaces this is dominated by depression storage, other losses such as surface wetting, infiltration and evapotranspiration all have an impact on the value of the initial loss. On pervious surfaces interception is a further complicating process. Initial losses are defined as the quantity of rainfall required just before overland runoff begins. The experimental data provides

$$D = \frac{k}{\sqrt{s}} \tag{16.6}$$

where

D is the average depth of initial losses (m)
s is the ground slope (m/m)
k is a coefficient dependent on the surface type (0.000071 for impervious surfaces and 0.00028 for pervious surfaces)

Percentage runoff model (UK)

A better model for the continuing losses is provided by an equation that relates the percentage runoff PR equation to the more important parameters that describe the characteristics of an urban catchment. From data collected for a number of small urban catchments it was deduced that the more important catchment characteristics are the soil type and the antecedent wetness. At first a regression equation for PR was derived from the data for discrete storms using the percentage impervious area (PIMP) for the catchment, the soil index defined by the Flood Studies Report (NERC 1975) and an urban catchment antecedent wetness index (UCWI). UCWI was based on a 5-day antecedent precipitation index (API5) and the soil moisture deficit (SMD). The equation was found to under-predict the runoff volume for catchments with relatively low PIMP and with light soils, as also for long duration storms. The equation was not easy to use with time series rainfall. Consequently, an alternative equation was developed, which calculates the runoff from the paved and pervious surfaces separately; see Eq. (16.7).

$$PR = IF^* PIMP + (100 - IF^* PIMP)^* \frac{NAPI}{PF} \tag{16.7}$$

where

PIMP is the percentage impervious area,
IF is the effective impervious area factor; see Table 16.3,
PF is the moisture depth parameter (mm) calibrated to be 200 mm,
NAPI is the 30 day antecedent precipitation index (*API*30) derived from the net rainfall after the subtraction of running depression storage.

Table 16.3 Recommended values of IF.

Surface condition	Effective impervious area factor, *IF*
POOR	0.45
FAIR	0.60
GOOD	0.75

The losses on pervious surfaces and non-effective impervious areas are represented by the second term in the equation. *NAPI* is defined as a 30-day *API* with evapotranspiration and initial losses subtracted from rainfall. API_{30} is given by

$$API_{30} = \sum_{n=1,30} P_{-n} C_p^{n-0.5} \tag{16.8}$$

The value *C* of the *API* has been made dependent on the soil type to reflect the faster reduction of soil moisture on lighter soils; see Table 16.4.

Table 16.4 Relationship between C and soil type.

Soil type	C
1	0.1
2	0.5
3	0.7
4	0.9
5	0.99

16.1.5 The US Soil Conservation Method SCS model

The SCS method provides a widely accepted model for predicting storm flows from rural catchments. It can therefore also be used as the permeable component in a semi-urban environment. This is a simple runoff model which allows for variations in runoff depending on catchment wetness. The model is described by two parameters: soil storage depth and catchment wetness index. The basis of the method is the continuity equation

$$P = I_a + F + Q \tag{16.9}$$

where

I_a is the initial abstractions (initial losses) (mm)
F is the cumulative actual retention (mm)
Q is the total runoff (mm)
P is the total rainfall (mm)

together with a relationship between rainfall, runoff and actual retention of the form

$$\frac{F}{S} = \frac{Q}{P - I_a} \tag{16.10}$$

where S is the maximum potential retention (storage) (mm)

These equations combine to give the SCS model:

$$Q = \frac{(P - I_a)^2}{P - I_a + S} \tag{16.11}$$

for $P > I_a$. It is normally assumed that S does not include I_a (Hjelmfelt 1990).

The form of this model can be adapted to be suitable for continuous simulation. The differential form of Eq. (16.11) is given as

$$\frac{dQ}{dt} = \frac{(P - I_a)(P - I_a + 2S)}{(P - I_a + S)^2}\frac{dP}{dt} \tag{16.12}$$

Numerical representation of the derivative terms leads to

$$\frac{q}{p} = \frac{(P - I_a)(P - I_a + 2S)}{(P - I_a + S)^2} \tag{16.13}$$

where

q is the runoff over time interval (mm)
p is the rainfall in time interval (mm)

This equation is used incrementally enabling the runoff coefficient (q/p) to change during the event. The two parameters S and I_a are related by

$$I_a = kS \tag{16.14}$$

where $0 < k < 0.2$. The original SCS approach recommended is $k = 0.2$. However other studies suggest k values between 0.05 and 0.1 may be more appropriate.

The storage parameter itself is related to an index known as the runoff curve number (CN) representing the combined influence of soil type, land management practices, vegetation cover, urban development and antecedent moisture conditions on hydrological response. The coefficient i varies between 0 and 100, 0 giving no runoff and 100 giving 100 per cent runoff. The storage parameter S is related to CN by

$$S = \frac{25400}{CN} - 254 \tag{16.15}$$

The CN value can either be defined globally for the catchment model or can be associated with specific surface types. CN values for different conditions are available from various sources (see e.g., USDA NRCS 1986).

16.2 RAINFALL-RUNOFF ROUTING MODELS

The rainfall losses model is a compartmental model that converts the rainfall into an effective rainfall for the runoff. Similarly, the runoff routing model is another compartmental model that transforms the effective rainfall into the runoff. Necessary features of the latter model are that it should provide a time delay and attenuation between the peak effective rainfall (times the catchment area) and the surface runoff. The approaches that can be readily adopted to model the surface runoff are: (design) unit hydrograph, time-area diagram, linear reservoir or kinematic wave.

16.2.1 Design unit hydrograph

The design unit hydrograph of Harmes and Verworn (1984) for the HYSTEM model takes the form shown in Figure 16.3 such that

$$\begin{aligned} Q &= \frac{t}{t_p} Q_p \quad 0 < t < t_p \\ Q &= Q_p e^{-\frac{t-t_p}{k}} \quad t \geq t_p \end{aligned} \tag{16.16}$$

where

Q is the flow rate (m^3/s)
Q_p is the peak flow rate (m^3/s)
t is the time (s)
t_p is the time to peak (s)
k is an exponential decay constant (s^{-1})

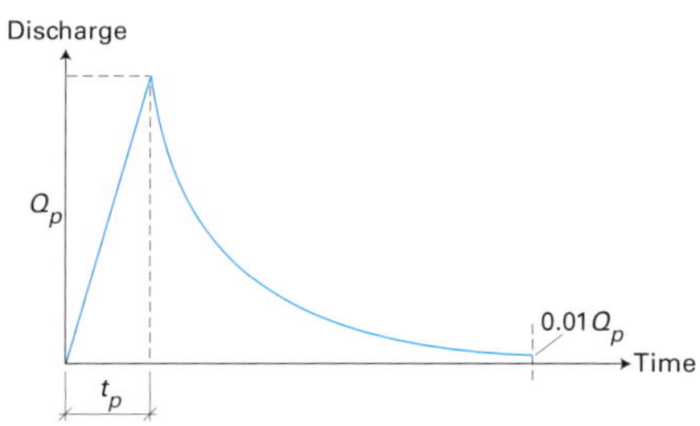

Figure 16.3 HYSTEM design unit hydrograph

The three parameters are related to catchment characteristics by

$$Q_p = \frac{0.96A}{t_L}$$

$$t_p = 0.49t_L \tag{16.17}$$

$$k = \frac{\frac{A}{Q_p} - \frac{t_p}{2}}{0.99}$$

where

t_L is a lag time (hr) calculated from catchment characteristics
A is the catchment area (ha)

The USA-SCS method uses a design unit hydrograph defined by

$$Q_p = \frac{484A}{t_p}$$

$$B = 1.67t_p$$

$$t_p = t_r/2 + t_L \tag{16.18}$$

$$t_L = \frac{L^{0.8}(S+1)^{0.7}}{1900y^{0.5}}$$

$$S = 1000/CN - 10$$

Here

Q_p is the peak discharge (cfs) for a one inch rainfall excess
t_p is the time to peak (hrs)
A is the catchment area (sq.mi)
B is time of fall (hrs)
t_L is the lag time (hrs)
t_r is the rainfall duration (hrs)
L is the main stream length (ft)
y is the average basin slope (%)
CN is the curve number

See Figure 16.4.

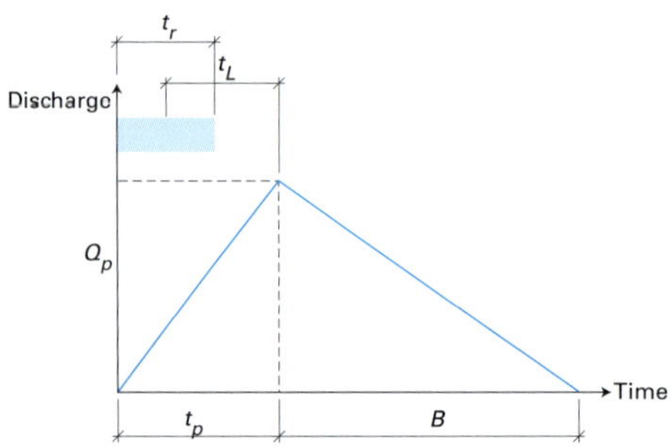

Figure 16.4 Design unit hydrograph for the SCS method

16.2.2 Time-area method

Typically, in time-area method the runoff is controlled by the initial loss, the size of the contributing area and the continuous hydrological loss. The shape of the runoff hydrograph is controlled by the concentration time and by the time area (T–A) curve: see Figure 16.5.

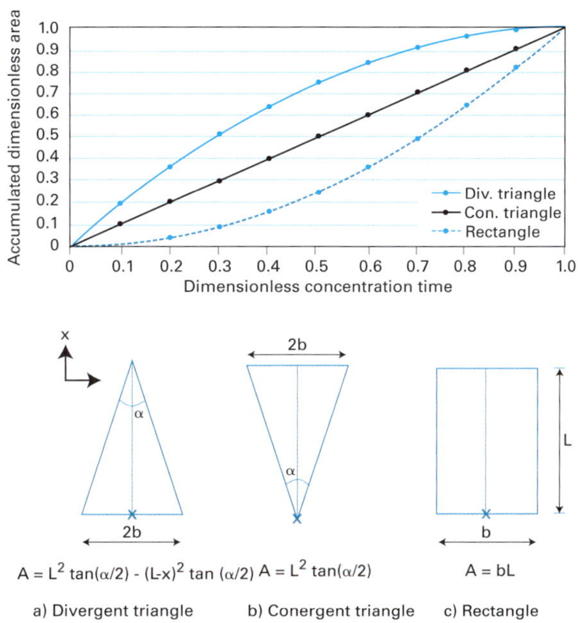

Figure 16.5 Time-area method for different shaped catchments

If the rainfall is given by the sequence of rainfall intensities $\{P_n\}$ then the flow at the outfall is given by

$$Q(N\Delta t) = \sum_{j=1}^{n} \Delta A_j P_{N-j} \tag{16.19}$$

where

Δt is the time increment (s)
ΔA_j is the area contributing to the flow at the outfall after j time increments (m^2)

16.2.3 Linear reservoir

The translation and attenuation can be simulated through the storage concept, which is introduced as a reservoir, Figure 16.6. In fact, rainfall-runoff models based on the tank or reservoir concept are very popular with hydrologists. In its simplest form, the reservoir can be linear, that is, the storage is proportional to the outflow:

$$\frac{dS}{dt} = Q_{in} - Q_{out}$$
$$S = kQ_{out} \tag{16.20}$$

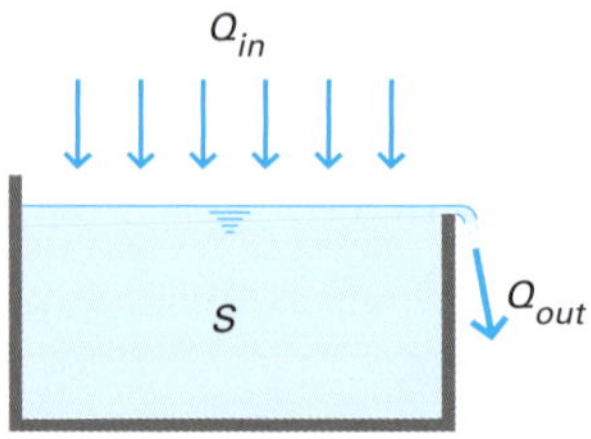

Figure 16.6 Linear reservoir

One important feature of the linear reservoir solution is that the peak discharge for the outflow coincides in time with a point on the recession of the inflow hydrograph. This does not in general occur in practice. Therefore an extension of the simple linear reservoir model is called for. Nash (1957) suggested that a sequence of identical linear reservoirs removes the need for the peak of the outflow to be on the recession of the inflow hydrograph. Such a sequence is termed the Nash cascade:

$$u(t, k, n) = \frac{1}{k\Gamma(n)}\left(\frac{t}{k}\right)^{n-1} e^{-\frac{t}{k}} \tag{16.21}$$

where

$\Gamma(n)$ is the gamma function.
k is the storage constant (1/s)

So the Nash cascade is described by two parameters: namely k and n. The number of reservoirs k and n can be chosen to match the peak output and the lag between the peak input and output.

16.2.4 Kinematic Wave (Nonlinear Reservoir)

In this method the surface runoff is computed as flow in an open channel, taking into account only the gravitational and friction forces and ignoring the contribution from the water surface gradient. The runoff amount is controlled by the size of the contributing area as well as the hydrological losses. The shape of the runoff hydrograph is controlled by the catchment parameters such as length, slope and roughness of the catchment surface. The surface runoff at time t is calculated as:

$$Q(t) = MBI^{\frac{1}{2}} y_r(t)^{\frac{5}{3}} \tag{16.22}$$

The depth $y_r(t)$ is calculated from the continuity equation:

$$I_{eff}A - Q(t) = \frac{dy_r}{dt}A \tag{16.23}$$

where

M is Manning's number,
B is the flow channel width (m), calculated as: $B = A/L$
I is the surface slope,
$y_r(t)$ is the runoff depth at time t (m)
I_{eff} is the effective rainfall (mm/hr)
A is the contributing catchment surface area (m^2)

dt is the time step (s)
dy_r is the change in runoff depth (m)

16.2.5 Runoff routing models (UK)

The overland flow routing model in the UK Wallingford Procedure (Wallingford 1981b) is based on the use of a double quasi-linear reservoir applied in series for each surface type with the same storage-output relationship for each reservoir:

$$S = kq \tag{16.24}$$

where

$$k = C\bar{i}^{-0.39} \tag{16.25}$$

in which

$$\bar{i} = 0.5(1 + i_{10}) \tag{16.26}$$

where i_{10} is the running ten minute average of rainfall intensity.

Combination of the two storage relationships with the continuity equation leads to a second-order ordinary differential equation of the form:

$$k^2\frac{d^2q}{dt^2} + 2k\frac{dq}{dt} + q = \bar{i} \tag{16.27}$$

C was derived by optimisation for each of the experimental catchments. Then a generic functional form for C was derived from the characteristics of the catchments. An overall relationship relating this value to catchment characteristics was then developed and the final relationship is:

$$C = 0.117S^{-0.13}A^{0.24} \tag{16.28}$$

where

S is the slope (m/m)
A is the area (m^2)

The parameters are subject to the following limits:

if ($S < 0.002$) $S = 0.002$
if ($A < 1000$) $A = 1000$
if ($A > 10000$) $A = 10000$

16.2.6 Extension for large sub-catchments

The routing model is developed for small catchments contributing to a pipe. For larger, possibly sewered catchments an alternative is needed. When systems are in surcharge, the value and use of the big catchment model is much reduced.

$$K' = 0.03A^{-0.022}S^{-0.228}L^{0.46} \tag{16.29}$$

$$T' = 4.334A^{-0.009}S^{-0.173}L^{0.46} \tag{16.30}$$

These terms are applied individually to each subcatchment. K' is the multiplier of k and must be greater than 1.0, whereas T' is a time shift applied to the nearest whole time step used in the rainfall file and effectively applies a shift in the rainfall file so that it reads $T–T'$ at each time step. Where figures for length and slope arc not provided, these are assumed to be:

$L = 2A^{0.5}$
$S = 1.5$ times the slope of the downstream sewer link.

16.3 DRY WEATHER FLOWS

In cases of modelling separate or combined sewer system, in addition to the surface runoff component (which is generated from the hydrological model computation) the dry weather flow component needs to be applied to each modelled subcatchment as well. A typical wastewater production per person per day should be defined together with the diurnal dimensionless profile to represent the 24-hour cycle: Figure 16.7. The wastewater production values can vary significantly due to the various reasons already cited (e.g., climate, culture, etc.).

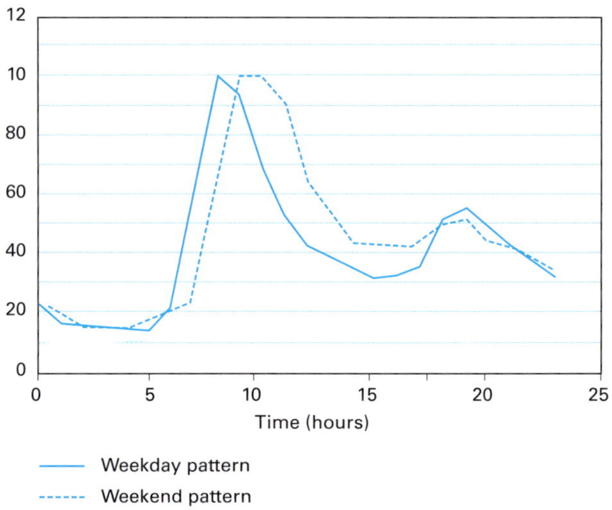

Figure 16.7 Example of a typical weekday and weekend diurnal patterns

The average daily water consumption for each residential property should be summed for each modelled subcatchment and as such it should be used as a comparison and a reality check of the final per capita discharge rates used in the calibrated model, while it may assist in identifying areas with significant exfiltration.

The average daily water consumption for each commercial and industrially designated property should be also estimated. However, if available, detailed information regarding the discharge pattern from any major commercial or industrial facilities can be used to assign the most appropriate profile from the standard profile list, or in exceptional circumstances create a new dimensionless curve following discussion with the system operator.

16.4 POLLUTANT LOADING AND WASHOFF

In addition to the modelling of flows in collection systems, the pollution-related aspects can be also incorporated. Typically, the following aspects are modelled:

- wash-off and build-up of sediments and attached pollutants (inputs into the model are: size, density and porosity of fine and coarse sediments);
- sediment transport (transport of cohesive and non-cohesive sediments and deposition/erosion processes);
- advection and dispersion (transport of dissolved and suspended material; inputs into the model are: dispersion coefficients, initial conditions, boundary conditions and decay rates);
- water quality (processes in the pipe network system considering chemical compounds and biological constituents; examples are: BOD, COD and H_2S).

The modelling of water quality in a sewerage system assumes various sources of pollution: domestic wastewater, surface washoff, industrial inflows, and erosion of sewer sediments. Both dissolved and particulate pollutants are transported through a sewerage system. The dissolved pollutants normally include ammoniacal nitrogen NH_4^+ (total ammonia) and oxygen demanding substances represented by Biological Oxygen Demand (BOD). These are normally treated as being conservative, that is, they do not degrade with time. Sediments with different characteristics (size and density) are also transported through the system. These are classified into sediment fractions. Typically, a minimum of two sediment fractions are used, one coarse high-density material (grit) and one fine (organics). In addition to being dissolved, pollutants such as BOD and COD are also assumed to be permanently attached to sediments. (Ammonia is normally only present in the dissolved phase).

Each source of sediment is categorised by the same sediment fractions (i.e. size and density). However the pollutants associated with a given sediment fraction may be different for each source. As the sediments are routed through the system those from different sources are mixed together, and therefore the concentrations of associated pollutants change during a simulation. The sediments from each source are assumed to have a characteristic amount of pollutants attached to them. This amount is assumed to be fixed and the ratio is referred to as the *potency factor* of the pollutant, expressed as kg of pollutant per kg of sediment.

All surface sediments (sometimes described as suspended solids) are defined by one value of size (d_{50}) and one value of density. These values apply to sediments derived from any source, including household and industrial inflows, and sediments eroded from and around pipes.

The build up of sediment on the catchment surface is a function of time and an accumulation rate coefficient. Sediment "build up" on the catchment surface during periods of dry weather starts from an initial quantity present at the end of the previous event. The basic hypothesis is one of a linear accumulation of pollution on the ground, which is related to the land use of that catchment. There is a limiting value for surface mass (B_s/K_s) as the number of days tends to infinity.

$$\frac{dM_r}{dt} = B_r - K_r M_r \tag{16.31}$$

where

M_r is the mass of deposit per surface unit (kg/ha)
B_r is the build up factor (kg/ha · day)
K_r is the decay factor (/day)

Therefore the mass of sediment built up at the start of an event is:

$$M_0 = M_i e^{K_r N \Delta t} + \frac{B_r}{K_r}\left(1 - e^{K_r N \Delta t}\right) \tag{16.32}$$

where

M_0 is the mass prior to the event (kg)
M_i is the initial mass of sediment (kg)
N is the duration of dry weather (days)

The rate of sediment washoff is dependent on an erosion coefficient and the quantity of available sediment. As the event proceeds and sediment is removed from the catchment, the quantity of available sediment, and hence the rate of sediment washoff, is reduced.

The washoff model can be linked, for example, to a single linear reservoir coefficient. The mass of sediment or dissolved pollutant mobilised is described by:

$$M_s(t) = K \cdot f_s(t) \tag{16.33}$$

where

$M_s(t)$ is the mass of pollutant mobilised (kg/ha)
$f_s(t)$ is the pollutant flow (kg/s)
K is the linear reservoir coefficient.

The mass of sediment eroded from the bed is a function of the rainfall intensity based on a 5 minute moving average:

$$\frac{dM_r}{dt} = -K_a \cdot M_r(t) \tag{16.34}$$

where

$$K_a(t) = c_1 i_5(t)^{c_2} - c_3 i_5(t) \tag{16.35}$$

Here

K_a is rainfall coefficient
c_1, c_2, c_3 are coefficients
$i_5(t)$ is the effective 5 minute rainfall

The mass of sediment that is carried from the catchment in suspension is governed by

$$\frac{dM_s}{dt} = K_a \cdot M_r(t) - f_s(t) \tag{16.36}$$

The outflow of sediment mass for each time increment is:

$$f_s^{n+1} = f_s^n e^{-\Delta t/K} + (e^{-\Delta t/K_a} - e^{-\Delta t/K})\frac{K_a^2}{K_a - K} M_r^n \tag{16.37}$$

where $f_s(t)$ is the TSS flow per unit of active surface (kg/s/ha)

This leads to an effective flow of sediment from the catchment of

$$F_s(t) = C \cdot A \cdot f_s(t) \tag{16.38}$$

where

C is the proportion of the catchment that is impermeable
A is the catchment area (ha)
F_s is the sediment flow (kg/s)

16.4.1 Attached pollutants

The quantity of pollutants attached to sediments is determined by means of potency factors. A potency factor describes the strength of a pollutant attached to a particular kind of sediment, and is defined as the ratio of the volume of pollutant to the volume of sediment. The potency factor does not change during the rainfall event. However, when sediments with different potency factors mix, the overall potency alters.

The relationship between total suspended solids (TSS) and attached pollutants is defined as:

$$P_{k,n} = a_1(i_5 - a_2)^{-b_1} + a_3 \tag{16.39}$$

where

$P_{k,n}$ is the potency factor
i_5 is the 5-minute rainfall intensity (mm/hr)
a_1, a_2, a_3 and b_1 are coefficients

Thus the calculation of washoff for the particular attached pollutant is:

$$F_{s,n}(t) = P_{k,n}(t) \cdot CA \cdot f_s(t) \tag{16.40}$$

where $F_{s,n}(t)$ is the pollutant flow (kg/s)

16.4.2 Dissolved pollutants

Dissolved pollutants build up and are removed from each subcatchment via the "gully pot" model. With regard to build up, there are assumed to be no dissolved pollutants in the gully pot at the end of the previous event (although this does not necessarily mean that all dissolved pollutants are removed during a rainfall event). Starting from zero, dissolved pollutants build-up in a linear manner according the equation

$$P = \frac{(C + MN)V_g}{1000} \tag{16.41}$$

where

C, M are parameters
N is the number of days since the previous event
V_g is the volume of the gully pot calculated from the depth of the gully pot and the catchment area (m^3)

The rate of washoff of dissolved pollutants is a function of the time for which it has been raining and the quantity of available pollutant.

In the calibration of the washoff model, the only term found to be significant is the one for erosion caused by rainfall impact.

A major input of the UK-UPM Research Programme has been the assimilation of data describing the nature and behaviour of sediments found in combined sewers and storage tanks. Crabtree (2007) presented a 5 category classification of the types of combined sewer sediments; see Table 16.5.

Table 16.5 Combined sewer sediments.

TYPE A	Coarse, predominantly granular mineral material found in the invert of pipes.
TYPE B	Similar in composition to Type A but concreted by the addition of mineral cements, fats and tars.
TYPE C	Mobile, fine grained, largely organic sludges found in quiescent zones.
TYPE D	Organic pipe wall slimes and biofilms.
TYPE E	Fine grained organic and mineral deposits found in storage tanks.

As explained by Butler and Davies (2000), the pollution processes take place in and between the wastewater, the atmosphere, the biofilm on the sewer wall and the sediment both in suspension and on the bed. The biodegradation of the organic material is caused by the micro-organisms in the biofilm and the wastewater, and is dependent on the presence of dissolved oxygen (DO) to facilitate these aerobic processes. Often water quality models for sewer networks assume that the pollutants are conservative, but degradation modelled by simple decay laws are appropriate in systems with long retention times.

Chapter 17

Hydraulic modelling

17.1 THE FUNDAMENTAL LAWS

In Section 1.4 (Urban drainage hydraulics) we have reviewed some of the important properties of water and in this section we shall introduce the fundamental physical principles, or laws, that describe the fluid flow. Regardless of the nature of the flow, the important laws that govern fluid flows are: *Newton's laws*, *the first and second laws of thermodynamics* and *conservation laws*. The Newton's laws of motion are:

(1) A body either remains at rest or continues in a state of uniform rectilinear motion unless acted upon by another body or force;
(2) The instantaneous force applied to a body is equal to the product of the instantaneous mass and the instantaneous acceleration of the body; and
(3) Action and reaction are equal and opposite.

These laws are usually called *Newton's laws of motion* even though they had several distinguished forbearers as observed in Part II of the present work. In addition to these laws the other important laws are the *first and second laws of thermodynamics*. The first law postulates the existence of walls that prevent all interactions (and so are, in particular, "impermeable to the flow of heat") and states that:

(1) In a system isolated by such walls, the work done in taking one state A of a system into another state B is entirely determined by the terminal states A and B. The internal energy difference between state A and state B is defined as the mechanical work done in taking either state A into state B or state B into state A, whichever is possible;

The second law of thermodynamics states that:

(2) There is a tendency on the part of nature to proceed towards a state of greater disorder.

The second law of thermodynamics says that when energy of parallel or near parallel motion (i.e., horizontal and nearly horizontal flows) is converted into energy of less parallel motion, essentially short wave motion and turbulent motion, the energy so converted can never be entirely reconverted into energy of parallel motion. According to this law, in all physically observable processes on Earth only one of the transformations $A \rightarrow B$ and $B \rightarrow A$ mentioned in the connection with the first law of thermodynamics is in fact possible. This process is then said to be *irreversible*. In the event that a process allows both transformations A → B and B → A, at least to some close approximation or assumption, the process is said to be *reversible*.

In order to establish relations between hydraulic variables such as depths, velocities, pressures, etc. we need to consider physical laws which express the conservation of certain quantities or combination of variables and the simplest and the most common of these quantities (or variable combinations) are: *mass*, *momentum* and *energy*. Derivation of these laws for our present purposes is best followed using the concept of a *control volume*, which is a fixed frame in space through which the fluid passes, and this is called the *Eulerian* view of motion (see Abbott 1980).

The law of conservation of mass states that the mass within a closed system remains constant with time. In other words, conservation means Inflow – Outflow = Change in storage. For an infinitesimal small control volume of fluid of density ρ and velocities u, v, w along the axes x, y, z the *continuity equation* can be expressed as:

$$\frac{\partial \rho}{\partial t} + \frac{\partial}{\partial x}(\rho u) + \frac{\partial}{\partial y}(\rho v) + \frac{\partial}{\partial z}(\rho w) = 0 \tag{17.1}$$

For an incompressible flow (i.e., ρ = constant) the continuity equation becomes:

$$\frac{\partial u}{\partial x} + \frac{\partial v}{\partial y} + \frac{\partial w}{\partial z} = 0 \tag{17.2}$$

The law of conservation of momentum states that in a closed system, such as one without friction, the amount of momentum remains constant; momentum is neither created nor destroyed, but only changed through the action of forces as described by Newton's laws of motion. Therefore, *Newton's second law of motion* is used as a basis for developing the *momentum equation*:

$$\mathbf{F} = \frac{d}{dt}(m\mathbf{V}) = m\mathbf{a} \tag{17.3}$$

Since the velocity at a given point in a moving fluid must generally be considered to change with time, and since the velocity at a given instant generally changes from point to point, an elementary particle of the fluid may be expected to undergo acceleration owing to passage of time and to displacement along the flow path. Thus, in the direction initially tangent to the path line passing through the point in question, the component of total acceleration will be the sum of *local* (*or temporal*) and *convective* (*or spatial*) *accelerations*:

$$\begin{aligned} a_x &= \frac{du}{dt} = \frac{\partial u}{\partial t} + u\frac{\partial u}{\partial x} + v\frac{\partial u}{\partial y} + w\frac{\partial u}{\partial z} \\ a_y &= \frac{dv}{dt} = \frac{\partial v}{\partial t} + u\frac{\partial v}{\partial x} + v\frac{\partial v}{\partial y} + w\frac{\partial v}{\partial z} \\ a_z &= \frac{dw}{dt} = \frac{\partial w}{\partial t} + u\frac{\partial w}{\partial x} + v\frac{\partial w}{\partial y} + w\frac{\partial w}{\partial z} \end{aligned} \tag{17.4}$$

where the terms: $\frac{\partial u}{\partial t}, \frac{\partial v}{\partial t}, \frac{\partial w}{\partial t}$ represent the local (or temporal) acceleration and terms $u\frac{\partial u}{\partial x}, v\frac{\partial u}{\partial y}, w\frac{\partial u}{\partial z}$, etc. are convective acceleration terms which describe velocity changes by movements in space. The local terms represent the degree of unsteadiness, while the convective terms represent the degree of non-uniformity. If the flow is steady, the local terms will obviously disappear, whereas if the flow is uniform the convective terms will be equal to zero. Only if the flow is both steady and uniform there will be no acceleration of any sort. In accordance with the Newton's laws of motion, such acceleration of an elementary fluid volume will be equal to and in the direction of the net force per unit mass which acts upon it. There are different types of force that may affect the motion of a fluid (volume force or weight, pressure force, friction force). If we include only the force which is due to a change in pressure intensity and which is equal to the pressure gradient in the corresponding direction we can obtain the following:

$$\begin{aligned}\frac{\partial u}{\partial t}+u\frac{\partial u}{\partial x}+v\frac{\partial u}{\partial y}+w\frac{\partial u}{\partial z}&=-\frac{1}{\rho}\frac{\partial p}{\partial x}\\ \frac{\partial v}{\partial t}+u\frac{\partial v}{\partial x}+v\frac{\partial v}{\partial y}+w\frac{\partial v}{\partial z}&=-\frac{1}{\rho}\frac{\partial p}{\partial y}\\ \frac{\partial w}{\partial t}+u\frac{\partial w}{\partial x}+v\frac{\partial w}{\partial y}+w\frac{\partial w}{\partial z}&=-\frac{1}{\rho}\frac{\partial p}{\partial z}\end{aligned} \tag{17.5}$$

These equations are the differential form of Newton's second law of motion for a fluid flow in a pressure field without friction. They describe relationship between forces and accelerations and they are also referred to as *Euler* equations.

Now, by adding the effect of viscosity (which can be defined as the ratio of shear intensity τ in the x–y, y–z and z–x plane) to Euler equations to account for frictional forces and shear stresses we obtain the *Navier-Stokes* equations:

$$\begin{aligned}\tau_{xy}&=\mu\left(\frac{\partial u}{\partial y}+\frac{\partial v}{\partial x}\right)\\ \tau_{yz}&=\mu\left(\frac{\partial v}{\partial z}+\frac{\partial w}{\partial y}\right)\\ \tau_{zx}&=\mu\left(\frac{\partial w}{\partial x}+\frac{\partial u}{\partial z}\right)\end{aligned} \tag{17.6}$$

$$\begin{aligned}\frac{\partial u}{\partial t}+u\frac{\partial u}{\partial x}+v\frac{\partial u}{\partial y}+w\frac{\partial u}{\partial z}&=-\frac{1}{\rho}\frac{\partial p}{\partial x}+\frac{\mu}{\rho}\left[\frac{\partial^2 u}{\partial x^2}+\frac{\partial^2 u}{\partial y^2}+\frac{\partial^2 u}{\partial z^2}\right]\\ \frac{\partial v}{\partial t}+u\frac{\partial v}{\partial x}+v\frac{\partial v}{\partial y}+w\frac{\partial v}{\partial z}&=-\frac{1}{\rho}\frac{\partial p}{\partial y}+\frac{\mu}{\rho}\left[\frac{\partial^2 v}{\partial x^2}+\frac{\partial^2 v}{\partial y^2}+\frac{\partial^2 v}{\partial z^2}\right]\\ \frac{\partial w}{\partial t}+u\frac{\partial w}{\partial x}+v\frac{\partial w}{\partial y}+w\frac{\partial w}{\partial z}&=-\frac{1}{\rho}\frac{\partial p}{\partial z}+\frac{\mu}{\rho}\left[\frac{\partial^2 w}{\partial x^2}+\frac{\partial^2 w}{\partial y^2}+\frac{\partial^2 w}{\partial z^2}\right]\end{aligned} \tag{17.7}$$

where

τ is the shear stress
u, v, w are the velocity components (m/s) in the x, y, z directions (m)
t is the time (s)
p is the pressure ($kg/s^2/m$)
ρ is the density (kg/m^3)
μ is the dynamic viscosity (kg/m s)
g is the acceleration due to gravity (m/s^2)

These equations cannot be solved in their general form because they are non-linear and second order and they have to be simplified in order to be solved. For laminar, incompressible flow of a Newtonian fluid treated as a continuum they are the Navier-Stokes equations; see Batchelor (2000) and Temam (2001).

These hyperbolic equations are symbolic of Newtonian laws of motion, namely that the acceleration of the water is proportional to the applied force made up of gravity and the internal friction losses, and that the mass of water is conserved. In deriving the Navier-Stokes equations it is assumed that the fluid is Newtonian, that is, the shear stress in the flow is proportional to the velocity gradient normal to the direction of flow. If we add the continuity equation to Eq. (17.7), then there are four equations in four dependent variables u, v, w and p, so it is possible to determine a unique solution in the solution space (x, y, z, t). Note that we distinguish between the model space which is formed by the physical boundary to the flow, namely the interior of the pipe, say, and the corresponding solution space.

A major difficulty with these equations is in extending them for turbulent flow. The classical way of doing this is to introduce turbulent fluctuations in the dependent variables and averaging over the random fluctuations. This generates a similar set of equations but with six additional unknown "Reynolds stresses" arising from the non-linear inertial terms in the momentum equations. A number of different assumptions have been made in order to define these Reynolds stresses in terms of the mean dependent variables; see Speziale (1991). The assumptions become important in large scale 2D and 3D flows, such as in coastal waters or reservoirs, and in computational fluid dynamics applied to flow in overflow chambers or chlorine contact and activated sludge tanks. Turbulence is characterized by eddies of different sizes and randomness in the flow. Energy is dissipated in the flow as the larger eddies break down into smaller eddies, and so on, eventually converting into heat. There are three primary length scales associated with turbulence.

The length scale of the larger eddies is the integral length scale, l_0, given by

$$l_0 \propto \frac{k^{3/2}}{\varepsilon} \tag{17.8}$$

where

k is the turbulent kinetic energy (m^2/s^2)
ε is the energy dissipation rate (m^2/s^3)

The turbulence Reynolds number associated with these large eddies is defined as

$$\mathrm{Re}_L = \frac{k^{1/2} l_0}{\nu} = \frac{k^2}{\varepsilon \nu} \tag{17.9}$$

The Kolmogorov length scale, η, is for the size of the smallest eddies that are responsible for dissipating energy when the Reynolds number is 1, that is,

$$\eta = \left(\frac{\nu^3}{\varepsilon}\right)^{1/4} = l_0 \mathrm{Re}_L^{-3/4} \tag{17.10}$$

The Taylor length scale, λ, is for eddies in the inertial sub-range, which are intermediate between the large and small scale eddies:

$$\lambda = \left(\frac{10\nu k}{\varepsilon}\right)^{1/2} = \sqrt{10} l_0 \mathrm{Re}_L^{-1/2} \tag{17.11}$$

The different length scales are used to determine the resolution of the numerical grid for the solution of the equations in a given boundary domain.

Boussinesq followed the assumption of the shear stress in laminar flow being proportional to the velocity gradient normal to the shear stress and proposed that the turbulent Reynolds stresses can similarly be assumed proportional to the mean velocity gradient normal to the stresses, with the coefficient of proportionality being the eddy viscosity. Kolmogorov and Prandtl suggested that the turbulent viscosity is given by

$$\nu_{\text{turb}} = C_{\mu}\frac{k^2}{\varepsilon} \tag{17.12}$$

where

C_{μ} is an empirical constant and k and ε are given by

$$\frac{\partial}{\partial t}(\rho k) + \frac{\partial}{\partial x_j}(\rho k u_j) = \frac{\partial}{\partial x_j}\left[\left(\mu + \frac{\mu_t}{\sigma_k}\right)\frac{\partial k}{\partial x_j}\right] + P_k - P_b - \rho\varepsilon - Y_m + S_k$$

$$\frac{\partial}{\partial t}(\rho \varepsilon) + \frac{\partial}{\partial x_j}(\rho \varepsilon u_j) = \frac{\partial}{\partial x_j}\left[\left(\mu + \frac{\mu_t}{\sigma_\varepsilon}\right)\frac{\partial \varepsilon}{\partial x_j}\right] + \rho C_{1\varepsilon}\frac{\varepsilon}{k}(P_k + C_{3\varepsilon}P_b) - C_{2\varepsilon}\rho\frac{\varepsilon^2}{k} + S_c$$

$$\mu_t = \rho C_{\mu}\frac{k^2}{\varepsilon} \tag{17.13}$$

$$P_k = -\rho\overline{u_i u_j'}\frac{\partial u_j}{\partial x_i} = \mu_t S^2 = 2\mu_t S_{ij}S_{ij}$$

$$P_b = \beta g_i \frac{\mu_t}{\text{Pr}_t}\frac{\partial T}{x_i}$$

where

β is the volume expansion coefficient
Pr_t is the turbulent Prandtl number (~0.85)
g_i is the gravitational acceleration in the ith direction (m/s^2)
T is the temperature
σ_k, σ_ε, $C_{1\varepsilon}$, $C_{2\varepsilon}$ and $C_{3\varepsilon}$ are empirical constants

In these equations, μ_t is the turbulent viscosity, P_k represents the generation of turbulence kinetic energy due to the mean velocity gradients, calculated in same manner as standard k–ε model. P_b is the generation of turbulence kinetic energy due to buoyancy, calculated in same way as standard k–ε model. Some of the coefficients take the following values:

$C_{1e} = 1.44$, $C_{2e} = 1.92$, $C_{\mu} = 0.09$, $\sigma_k = 1.0$, $\sigma_\varepsilon = 1.3$

The coefficient of thermal expansion, β, is defined as

$$\beta = -\frac{1}{\rho}\left(\frac{\partial \rho}{\partial T}\right)_p \tag{17.14}$$

The interested reader should consult Mathieu and Scott (2000).

If we allow for effects of vertical accelerations into the momentum equation we can then obtain the so-called *Boussinesq* equation which is the same as Euler equation except for the last third order Boussinesq term:

$$\frac{\partial u}{\partial t} + u\frac{\partial u}{\partial x} + g\frac{\partial h}{\partial x} - \frac{1}{3}h^2\frac{\partial^3 u}{\partial x^2 \partial t} = 0 \tag{17.15}$$

The linearised Euler equations in 2D can be given as:

$$\frac{\partial^2 h}{\partial t^2} - c^2\left(\frac{\partial^2 h}{\partial x^2} + \frac{\partial^2 h}{\partial y^2}\right) = 0 \tag{17.16}$$

where, $c^2 = gh$

For steady-state flows we obtain an eliptic form of partial differential equation which is used for groundwater flow problems:

$$\frac{\partial^2 h}{\partial x^2} + \frac{\partial^2 h}{\partial y^2} = 0 \tag{17.17}$$

Besides mass and momentum we can also consider conservation of energy. By recalling the first law of thermodynamics we can follow that the net energy (e.g., heat and potential energy) supplied to the system equals the increase in energy of the system plus the energy that leaves the system (i.e., accumulation = inflow – outflow). The energy conservation can be expressed as:

$$\frac{\partial}{\partial t}\left[\rho h\left(\frac{u^2}{2} + \frac{gh}{2}\right)\right] + \frac{\partial}{\partial x}\left[\rho uh\left(\frac{u^2}{2} + \frac{gh}{2}\right)\right] = 0 \tag{17.18}$$

Or with constant ρ:

$$\frac{\partial}{\partial t}\left[h\left(\frac{u^2}{2} + \frac{gh}{2}\right)\right] + \frac{\partial}{\partial x}\left[uh\left(\frac{u^2}{2} + \frac{gh}{2}\right)\right] = 0$$

$$h\frac{\partial}{\partial t}\left(\frac{u^2}{2} + \frac{gh}{2}\right) + \left(\frac{u^2}{2} + \frac{gh}{2}\right)\frac{\partial h}{\partial t} - \frac{gh}{2}\frac{\partial h}{\partial t} + uh\frac{\partial}{\partial x}\left(\frac{u^2}{2} + \frac{gh}{2}\right) + \left(\frac{u^2}{2} + \frac{gh}{2}\right)\frac{\partial}{\partial x}(uh) = 0 \tag{17.19}$$

$$uh\frac{\partial u}{\partial t} + uh\frac{\partial}{\partial x}\left(\frac{u^2}{2} + \frac{gh}{2}\right) = 0$$

So:

$$\frac{\partial u}{\partial t} + \frac{\partial}{\partial x}\left(\frac{u^2}{2} + \frac{gh}{2}\right) = 0 \tag{17.20}$$

This is sometimes known as the *Bernoulli* equation that for steady state flow leads to:

$$\frac{u^2}{2} + gh = const \tag{17.21}$$

17.2 SAINT VENANT EQUATIONS

Given a gradually varying free surface to the flow such that the vertical accelerations of the water particles are negligible compared to the acceleration due to gravity, and assuming that the viscous terms in the third equation above can be ignored, the momentum equation in the vertical direction becomes

$$0 = -\frac{1}{\rho}\frac{\partial p}{\partial z} + g \tag{17.22}$$

Integrating with respect to z we obtain:

$$p = p_0 + \rho(h - z) \tag{17.23}$$

where p_0 is the atmospheric pressure ($kg \cdot m^2/s$) at the free surface. The pressure is therefore a linear function of depth, and is identical to the pressure when the water is at rest (the hydrostatic pressure). The viscous terms involving the second order derivatives of the horizontal velocities in x and y can be regarded as negligible compared with the remaining viscous term, as can the derivatives of the Reynolds stresses in x and y. Integrating the two remaining momentum equations over depth and assuming that the Boussinesq coefficients formed by the integration of the non-linear convection of momentum terms are approximately unity:

$$\begin{aligned} h\left(\int_0^h u^2\,dz\right)\Big/\left(\int_0^h u\,dz\right)^2 &\approx 1 \\ h\left(\int_0^h v^2\,dz\right)\Big/\left(\int_0^h v\,dz\right)^2 &\approx 1 \\ h\left(\int_0^h uv\,dz\right)\Big/\left(\int_0^h u\,dz\right)\left(\int_0^h u\,dz\right) &\approx 1 \end{aligned} \tag{17.24}$$

gives

$$\begin{aligned} \frac{\partial u}{\partial t} + \frac{\partial(u^2)}{\partial x} + \frac{\partial(uv)}{\partial y} + g\frac{\partial h}{\partial x} - \frac{1}{\rho}\left(\tau_{x,\mathrm{surf}} - \tau_{x,\mathrm{bed}}\right) &= 0 \\ \frac{\partial v}{\partial t} + \frac{\partial(uv)}{\partial x} + \frac{\partial(v^2)}{\partial y} + g\frac{\partial h}{\partial y} - \frac{1}{\rho}\left(\tau_{y,\mathrm{surf}} - \tau_{y,\mathrm{bed}}\right) &= 0 \\ \frac{\partial h}{\partial t} + \frac{\partial(uh)}{\partial x} + \frac{\partial(vh)}{\partial y} &= q \end{aligned} \tag{17.25}$$

where

$\tau_{x,\mathrm{surf}}$ is the viscous and Reynolds stress in the x-direction at the free surface ($kg/s^2/m$)
$\tau_{x,\mathrm{bed}}$ is the viscous and Reynolds stress in the x-direction at the bed ($kg/s^2/m$)
$\tau_{y,\mathrm{surf}}$ is the viscous and Reynolds stress in the y-direction at the free surface ($kg/s^2/m$)
$\tau_{y,\mathrm{bed}}$ is the viscous and Reynolds stress in the y-direction at the bed ($kg/s^2/m$)

Assuming that the stress at the surface is zero (e.g., there is no wind) and that the bed stresses can be replaced by

$$\begin{aligned} \tau_{x,bed} &= f\rho\frac{u\sqrt{u^2+v^2}}{4h} \\ \tau_{y,bed} &= f\rho\frac{v\sqrt{u^2+v^2}}{4h} \end{aligned} \tag{17.26}$$

where f is the Darcy friction factor (see below) then

$$\frac{\partial u}{\partial t}+\frac{\partial(u^2)}{\partial x}+\frac{\partial(uv)}{\partial y}+g\left(\frac{\partial h}{\partial x}+f\,\frac{u\sqrt{u^2+v^2}}{4gh}\right)=0$$

$$\frac{\partial v}{\partial t}+\frac{\partial(uv)}{\partial x}+\frac{\partial(v^2)}{\partial y}+g\left(\frac{\partial h}{\partial y}+f\,\frac{v\sqrt{u^2+v^2}}{4gh}\right)=0 \qquad (17.27)$$

$$\frac{\partial h}{\partial t}+\frac{\partial(uh)}{\partial x}+\frac{\partial(vh)}{\partial y}=q$$

Here q is the inflow per unit area (m/s).

Note that the relationships in Eq. (17.27) are only approximate, but they are good enough for most practical purposes within the present context.

17.3 1D SAINT VENANT EQUATIONS

In previous sections we have described in a very general way the equations governing unsteady flow. We can then note that for a description of one-dimensional unsteady flow we need two dependent variables and two equations, each of which must represent a physical law. Since we can formulate three physical laws in such flow (conservation of mass, momentum and energy) the choice of two laws is important from a theoretical as well as the practical points of view. As shown by Abbott (1980) wherever the flow is varying smoothly at every point in a continuum, and so remains nearly horizontal, the mass-momentum couple and mass-energy couple are equivalent concepts. However, as a discontinuity such as hydraulic jump and bore appear, these two concepts are no longer equivalent and their equations will produce different answers. Therefore, since the mass-momentum couple of conservation laws is applicable to both discontinues and continuous situations and the mass-energy is not, as described by Abbott (1980), we shall base our derivations on mass-momentum conservation.

The 1D Saint Venant equations for gradually varying free surface flow in uniform open channels can be deduced from Eq. (17.13) with the adjustment that h in the friction term is replaced by R, the hydraulic radius:

Conservation of momentum:

$$\frac{\partial u}{\partial t}+u\frac{\partial u}{\partial x}+g\left(\frac{\partial h}{\partial x}+f\,\frac{u|u|}{4gR}\right)=0 \qquad (17.28)$$

Conservation of mass:

$$\frac{\partial h}{\partial t}+u\frac{\partial h}{\partial x}+\frac{A}{B}\frac{\partial u}{\partial x}=0 \qquad (17.29)$$

where

A is the flow area (m^2)
B is the surface flow width (m)
R is defined as A/P (m), where P is the wetted perimeter (m)

See Figure 17.1.

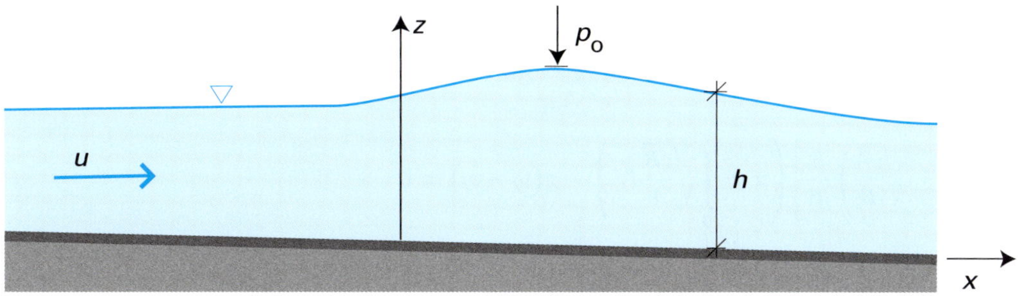

Figure 17.1 Flow in a channel

The assumptions underlying these equations include:

(1) The fluid, namely water, is a continuum.
(2) The water surface gradients are gradually varying in space and time such that the pressure is hydrostatic; that is, the vertical accelerations of water particles in the flow are negligible compared to the acceleration due to gravity.
(3) The Boussinesque coefficients modifying the convection of momentum terms in the momentum equation, and arising from the integration of the 3D Navier-Stokes equations over a cross-section to obtain the 1D averaged equations, are approximately constant and equal to unity.
(4) The fluid is incompressible.
(5) The resistance to the flow due to the shear stress at the boundaries and the dissipation of energy due to turbulence is proportional to the square of the mean velocity at a section, and the coefficient of proportionality can be derived in terms of the geometry of the cross-section, a measure of the boundary roughness and the Reynolds number.
(6) The forms for the resistance (or friction) force, which are derived from laboratory experiments under steady flow conditions, are applicable to unsteady flow.

The Saint Venant (or shallow water) equations can be written in different ways. They have different forms and different approximations. Consider first alternative forms.

17.3.1 Algorithmic form

This is written as:

$$\begin{aligned} u\frac{\partial u}{\partial t} - g\frac{\partial h}{\partial t} + \left(u^2 - g\frac{A}{B}\right)\frac{\partial u}{\partial x} + gus_f &= 0 \\ -\frac{A}{B}\frac{\partial u}{\partial t} + u\frac{\partial h}{\partial t} + \left(u^2 - g\frac{A}{B}\right)\frac{\partial h}{\partial x} - g\frac{A}{B}s_f &= 0 \end{aligned} \tag{17.30}$$

provided the Froude number, $Fr \equiv u/\sqrt{gA/B} \neq 1$. This formulation is important when forming the 6-point implicit finite difference scheme below.

17.3.2 Characteristic form

Another important formulation depends on the hyperbolic nature of the equations, giving the characteristic form for a uniform channel:

$$\left[\frac{\partial}{\partial t}+\left(u\pm\sqrt{\frac{gA}{B}}\right)\frac{\partial}{\partial x}\right]\left(u\pm\int_0^y\sqrt{\frac{gB}{A}}dy'\right)-g\left(s_0-s_f\right)=0 \tag{17.31}$$

where

y is the depth (m) defined by $h=y-s_0x$
s_0 is the bed slope (m/m)

There are two characteristic curves passing through any point in the solution space (x, t) given by the lines:

$$\frac{dx}{dt}=\left(u\pm\sqrt{\frac{gA}{B}}\right) \tag{17.32}$$

The characteristic curves carry information into the solution space from its boundaries; see Figure 17.2. This stresses the importance of the boundaries in determining uniquely what happens in the space. Note that here a "boundary" is denoted by what happens in time to the flow at the upstream and downstream ends of the pipe, and the initial state of the flow along the pipe: we are not referring to the physical boundary of the flow formed by the interior of the pipe.

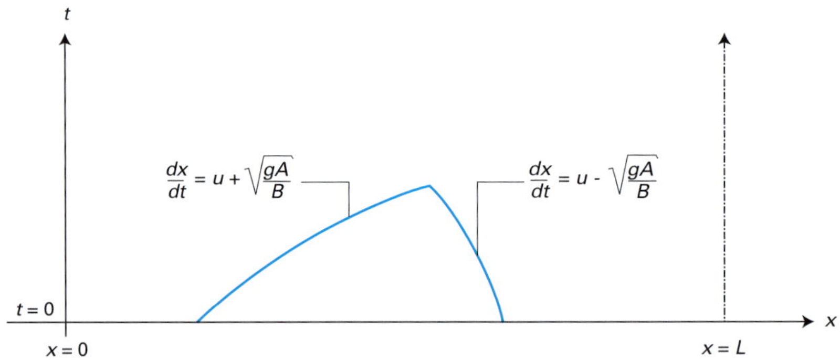

Figure 17.2 Characteristic curves in solution space

The two equations in Eq. (17.31) refer to the propagation of two sets of dynamic waves. Because $u+\sqrt{gA/B}$ is always positive for $u>0$, every point in the domain receives information from the upstream boundary (or from the initial condition at time $t=0$). However, $u-\sqrt{gA/B}$ can be either negative or positive for $u>0$. When it is negative, that is $Fr<1$, the corresponding characteristic curve conveys information from the downstream boundary (or from the initial condition at time $t=$ 0). If $Fr=1$, information does not come from upstream or downstream, in which case we have a "critical" flow condition. For $Fr<$ 1the flow is said to be sub-critical, and for $Fr>1$ the flow is super-critical. In the latter case, no information comes from downstream: all the information comes from upstream. At points where $Fr = 1$ there is a discontinuity in water levels. This is referred to as a hydraulic jump; see Figure 17.3 and Figure 17.4. Momentum and mass are conserved across the jump, and according to the second law of thermodynamics energy is converted into other forms such as through the formation of short-period waves, heat and sound. Conditions on the upstream and downstream velocity and depth can be found such that Eq. (17.31) apply upstream and downstream of the trajectory of the jump in the solution space. In other words, the trajectory of the jump marks the boundary where Eq. (17.31) has a singularity but where conditions can be imposed at the trajectory to ensure continuity of the solution across it.

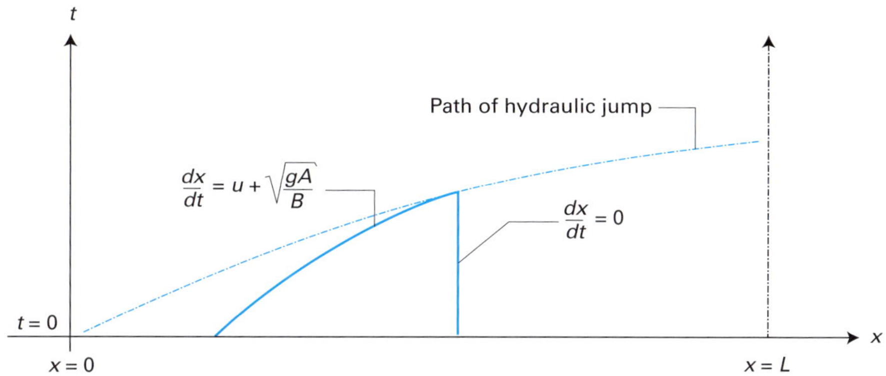

Figure 17.3 Path of hydraulic jump in solution space

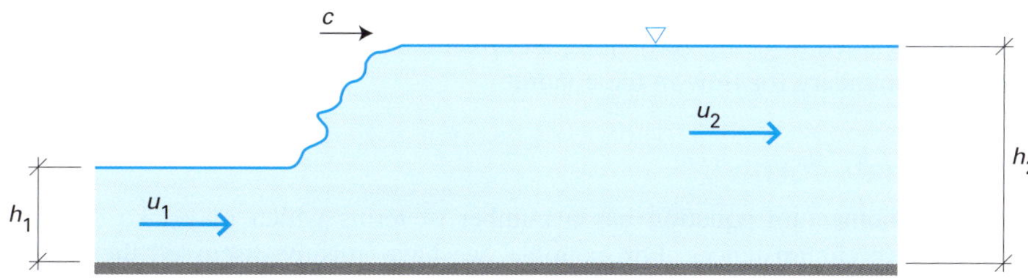

Figure 17.4 Hydraulic jump

17.3.3 Discharge form

The momentum equation, Eq. (17.28), with u replaced by Q/A, where Q is the discharge, and the continuity equation Eq. (17.29), become

$$\begin{aligned} \frac{\partial Q}{\partial t} + \frac{\partial}{\partial x}\left(\frac{Q^2}{A}\right) + gA\left[\frac{\partial h}{\partial x} + f\frac{Q|Q|}{4gRA^2}\right] &= 0 \\ \frac{\partial A}{\partial t} + \frac{\partial Q}{\partial x} &= q \end{aligned} \qquad (17.33)$$

Typically, in pipe applications for drainage, say, f is given by the Darcy friction factor, where

$$\frac{1}{\sqrt{f}} = -2\log_{10}\left(\frac{k_s}{14.8R} + \frac{2.51}{\mathrm{Re}\sqrt{f}}\right) \tag{17.34}$$

where

R is the hydraulic radius ($= A/P$) (m)
Re is the Reynolds number
k_s is the Nikuradse equivalent sand roughness height (mm)

k_s can be defined for the internal surface of a pipe, though it may be difficult to prescribe accurately when the pipe is in poor condition with cracks and badly aligned joints, or when the pipe is penetrated by tree roots. In such cases, k_s may need to become a calibration parameter.

When the Saint Venant equations are applied to rivers or coastal waters, modellers prefer to use simpler forms for f including Manning:

$$f = \frac{4n^2}{R^{2/3}} \tag{17.35}$$

or Chezy:

$$f = \frac{4}{C^2} \tag{17.36}$$

Here

n is the Manning coefficient (taking a typical value of 0.03 for an urban river)
C is the Chezy coefficient (taking a corresponding value of about 35).

The conservation of volume equation normally has a source term q. Unlike lateral inflow along a river reach q is generally regarded as being zero along an urban conduit, and any lateral inflow is assumed to enter at the upstream boundary for flow in the conduit.

17.3.4 Approximate forms

The conservation of momentum equation has a number of terms which may take different orders of magnitude in alternative circumstances. For example, we have already discussed the case of gradually varying flow in open channels where the vertical acceleration of water particles is small compared to the acceleration due to gravity, resulting in the assumption of hydrostatic pressure in the flow. A careful scale analysis of the terms in Eq. (17.33) shows that the momentum equation becomes

$$\varepsilon\gamma\frac{\partial Q}{\partial t} + \varepsilon\gamma\frac{\partial}{\partial x}\left(\frac{Q^2}{A}\right) + gA\left[\varepsilon\frac{\partial y}{\partial x} - s_0 + f\frac{Q|Q|}{4gRA^2}\right] = 0 \tag{17.37}$$

where ε and γ are inserted into the equation to denote the magnitudes of the associated terms and are such that ε is the characteristic surface slope number given by $\varepsilon \approx \left|\partial y/\partial x\right|/|s_0|$, and γ is the square of the Froude number given by $\gamma \approx \left|u^2B/gA\right| \equiv \left|Fr^2\right|$.

If $\varepsilon \approx 0$, the momentum equation becomes:

$$Q = \frac{1}{\sqrt{f}}A\sqrt{4gRs_0} \tag{17.38}$$

When combined with the conservation of mass equation this gives:

$$\frac{\partial Q}{\partial t}+\frac{1}{B}\frac{dQ}{dy}\frac{\partial Q}{\partial x}=\frac{1}{B}\frac{dQ}{dy}q \tag{17.39}$$

Here B is the surface width (m).

Eq. (17.39) describes a kinematic wave, which translates with speed $c(Q)=(1/B)\,dQ/dy$ downstream but without any change to the peak (assuming there is no lateral inflow). If we assume that $0<\varepsilon\ll 1$ then in place of the kinematic wave equation we obtain:

$$\frac{\partial Q}{\partial t}+c(Q)\frac{\partial Q}{\partial x}+\varepsilon c(Q)\frac{\partial}{\partial t}\left(\frac{a(Q)}{c^2(Q)}\frac{\partial Q}{\partial x}\right)=c(Q)q-\varepsilon c(Q)\frac{\partial}{\partial t}\left(\frac{Qq}{2gAs_0}\right)+O\left(\varepsilon^2\right) \tag{17.40}$$

where $a(Q)$ is termed the attenuation parameter.

$$a(Q)=\frac{Q}{2s_0B}\left\{1-\frac{B}{gA}\left(c-\frac{Q}{A}\right)^2\right\} \tag{17.41}$$

Eq. (17.40) is the analytical equation for the Muskingum-Cunge method of flood routing. The second derivative term in this equation modifies the kinematic wave, especially in giving a decrease in the peak discharge. This attenuation is induced by gravity, which tends to dampen the wave disturbance. If $\gamma\approx 0$ then we obtain the non-inertial equations for a uniform channel:

$$\begin{aligned}
&\frac{\partial Q}{\partial t}+c(Q,h)\frac{\partial Q}{\partial x}-a(Q,h)\frac{\partial^2 Q}{\partial x^2}=0\\
&B\frac{\partial h}{\partial t}+\frac{\partial Q}{\partial x}=q\\
&\frac{\partial h}{\partial x}+\frac{Q^2}{K^2}=0\\
&c(Q,h)=\frac{Q}{BK}\frac{dK}{dh}-\frac{a(Q,h)}{B}\frac{dB}{dh}\\
&a(Q,h)=\frac{K^2}{2BQ}
\end{aligned} \tag{17.42}$$

Note that the equation for the discharge is an advection–diffusion equation, namely a mixed hyperbolic-parabolic equation, which occurs in a variety of other circumstances, such as the movement of pollutants in water and the diffusion and advection of vorticity away from a fixed boundary.

17.4 BOUNDARY CONDITIONS FOR PIPE FLOW

Where the Saint Venant equations are applied to a pipe it is important to establish the appropriate boundary conditions for the solution space. Given that the pipe has a bed gradient $s_0\neq 0$ then we can define upstream and downstream nodes for the pipe. Because of the hyperbolic nature of the two equations, and because they have first-order derivatives in x and t, they introduce four boundary conditions: one upstream and one downstream (for sub-critical flow) or two upstream boundary conditions (for super-critical flow), and two initial conditions along the pipe. Note that Eq. (17.40) only needs one upstream condition and one initial condition, whereas Eq. (17.41) needs an upstream condition and a downstream condition as well as an initial condition. At the upstream boundary the flow can be sub-critical (a) or critical (b); see Figure 17.5. The water level in the upstream manhole can be above the outgoing soffit, but, where the pipe is not flowing full air may be entrained in the form of bubbles and brought in from the manhole (c). Additional energy is therefore lost as the flow is forced to contract into the pipe downstream.

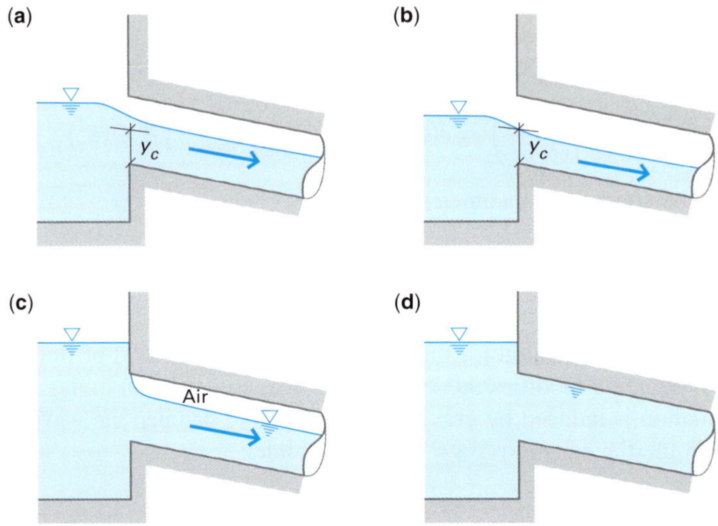

Figure 17.5 Upstream boundary conditions for flow in a pipe (y_c is the critical depth) (after Kellagher (1993))

Similar conditions apply at the downstream boundary: see Figure 17.6. In this case the flow in the pipe can be sub-critical or super-critical. If it is sub-critical there can be a critical flow point before the outflow reaches the water level in the downstream manhole (a), or alternatively, the water level is high enough to submerge the critical flow condition (b). If the flow is super-critical in the pipe then the flow at the downstream boundary will be at normal depth and there may or may not be a hydraulic jump in the manhole depending on the water level there (c). There is one other condition to note, and that is the case when the discharge is greater than the pipe full discharge. In this case the downstream boundary condition is such that the water level is at or above the outgoing soffit; see (d) and (e).

Note that head-losses can occur both at the upstream and downstream boundaries. There is generally a contraction of flow into the pipe at the upstream boundary and an expansion of flow at the downstream boundary. These head-losses need to be included in the analysis, and are defined by:

$$\Delta H = K\frac{u^2}{2g} \quad (17.43)$$

where K is an energy loss coefficient. See Figure 17.7. The actual head-losses that occur across a manhole vary according to the shape of the manhole, any benching designed to reduce head-losses, the alignment of the incoming and outgoing pipes, shaping of the entrance to the pipe, the height of the water level above the outgoing soffit, possible air entrainment, and so on. In other words, there is a large combination of possible factors affecting the calculation of the head-loss at the upstream and downstream ends of a pipe. A considerable number of laboratory experiments have been done to identify the head-loss coefficients for ranges of different manholes, but the experiments are by no means complete; see Yen (2004).

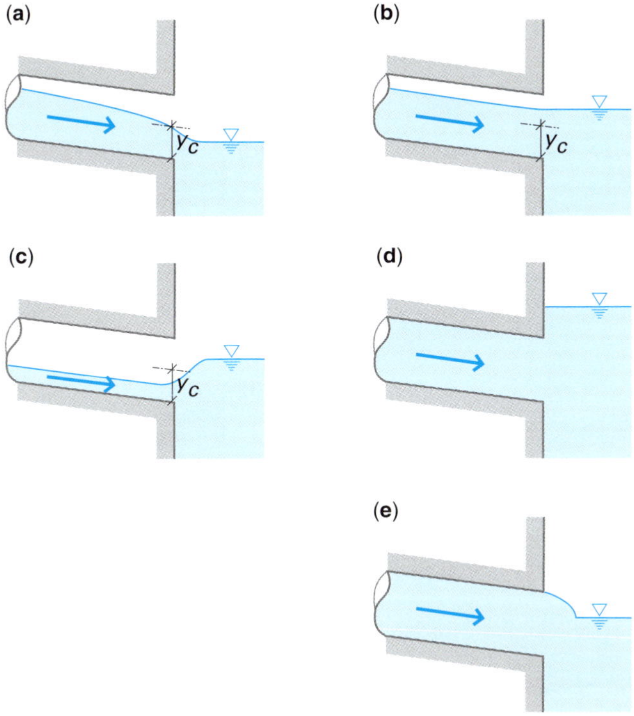

Figure 17.6 Downstream boundary conditions for flow in a pipe (y_c is the critical depth) (after Kellagher (1993))

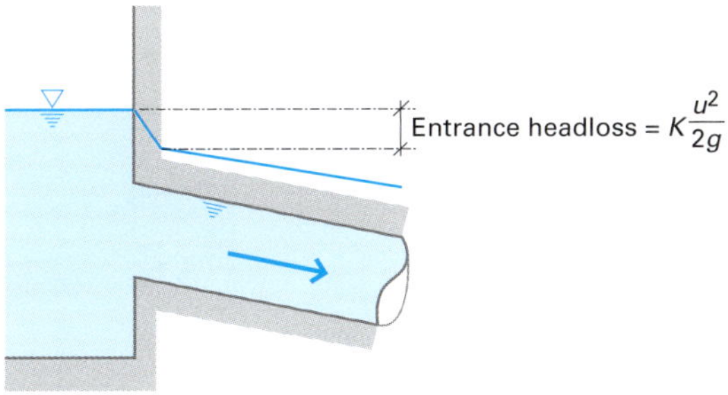

Figure 17.7 Head-loss at a manhole

17.5 PRESSURISED FLOW

The pipe can flow partially and completely full. When the transition between free surface and pressurized flow takes place at some point along the pipe, there are a number of possibilities; see Figure 17.8. Of the 10 options, three of them: (e), (f) and (h), cannot exist under steady flow conditions.

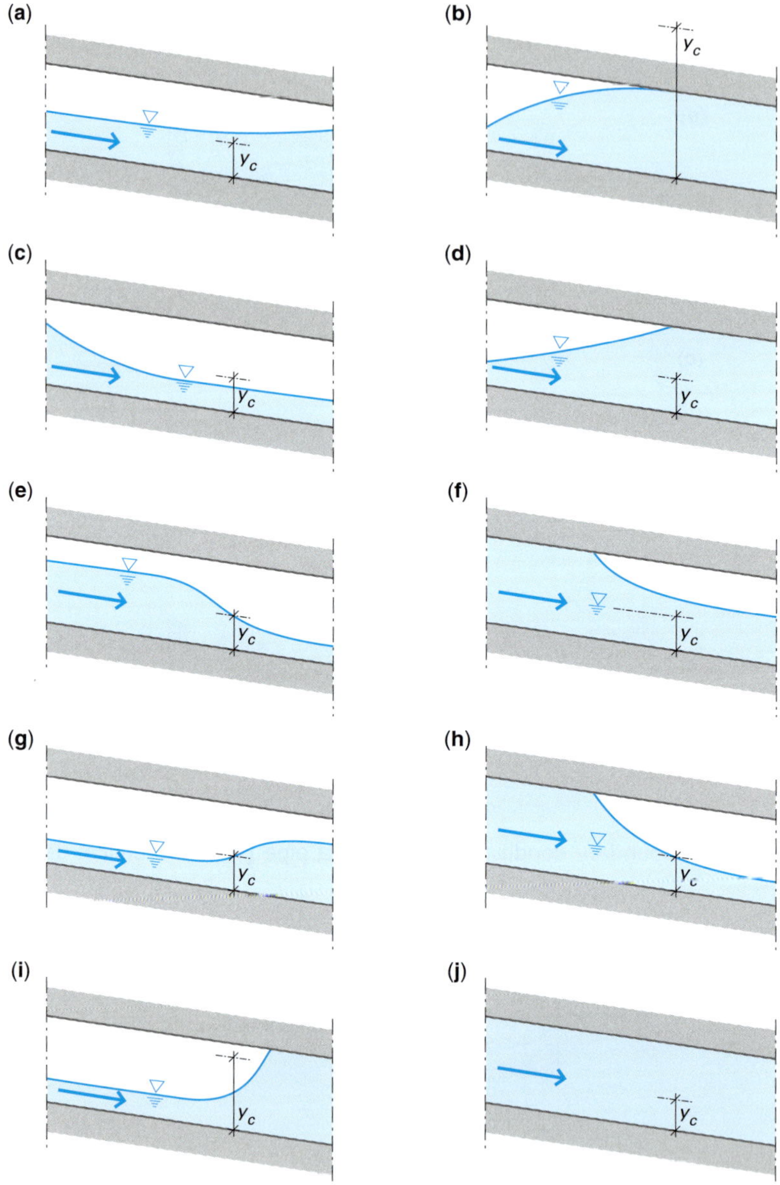

Figure 17.8 Free-surface and pressurised flow conditions in a pipe (y_c is the critical depth) (After Kellagher 1993)

The transition between free surface and pressurised flow needs to be considered separately. Assume that the water surface is parallel to the bed of the pipe, that is, normal depth flow applies. The relationship between the discharge and the water level for free surface flow in a circular pipe is as shown in Figure 17.9. Here, the proportional discharge and velocity are defined in terms of the pipe-full discharge and velocity. An interesting fact is that the velocity takes the same value at the middle depth in a circular pipe as at the top (soffit/crown) of the pipe. As the discharge is increased (which is difficult to conceive in practice because we are talking about steady flow) then the water depth must "jump" and fill the pipe suddenly. Sufficient head has then to be built up at the upstream end of the pipe in order to drive the flow through the pipe. Similarly, as the discharge is decreased from a condition that the pipe is fully surcharged the water level suddenly "drops". This is descriptive of a hysteresis effect: see Figure 17.10. The pipe completely surcharged is shown in Figure 17.11. In practice it is assumed that the depth-discharge relationship is unique and follows the rising limb of the hysteresis loop. Note that when the flow into a pipe is increasing such that the discharge exceeds the pipe full capacity there will be a shock travelling along the pipe downstream and the water level in the upstream manhole increases rapidly. Provided the discharge keeps on increasing (or at least not decreasing) the shock travels quickly down the pipe and the whole pipe becomes surcharged. Examination of the time series for the water level in the upstream manhole can reveal a sharp oscillation, which is made complicated by air entrainment and the trapping of pockets of air in the pipe. Surcharging can also occur when the water level in the downstream manhole increases above the outgoing soffit. There is backup into the pipe. Due to instabilities of water surface at nearly full pipe, flow may become surcharged in the middle section of the pipe. This is more likely to happen if air circulates in the direction opposite from water flow.

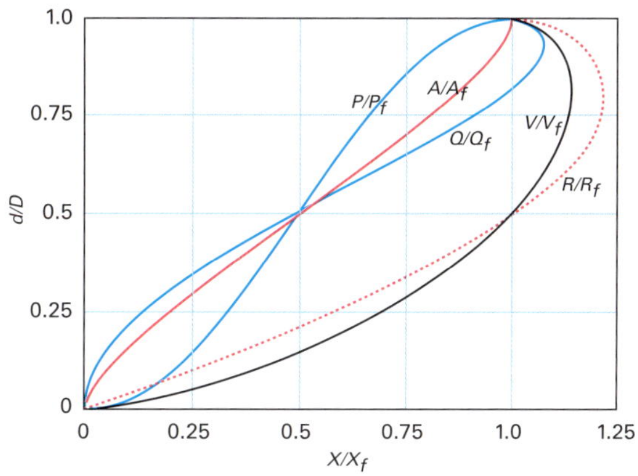

Figure 17.9 Proportional discharge and velocity curves for a circular pipe (after Butler and Davis 1993)

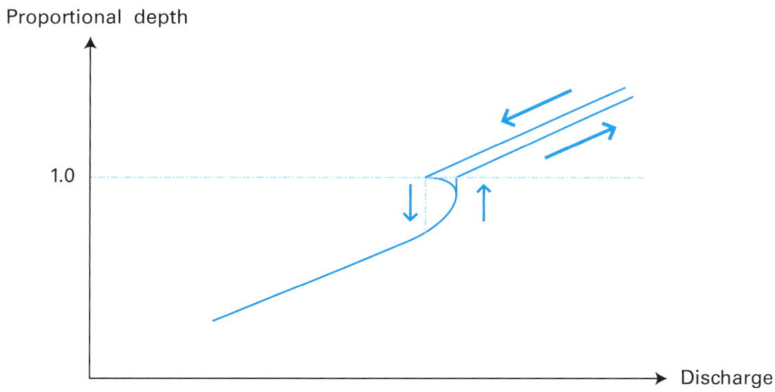

Figure 17.10 Hysteresis effect between free-surface and pressurised flow in a pipe

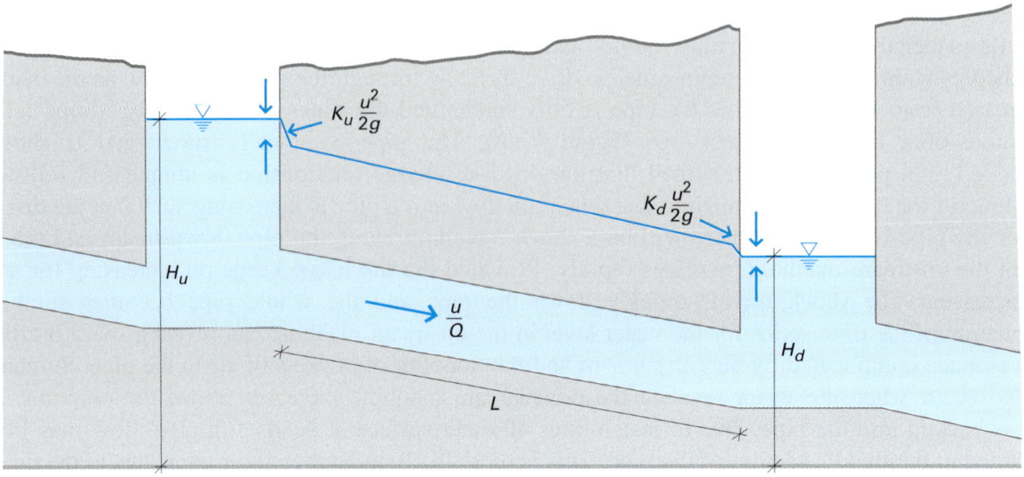

Figure 17.11 Flow in a surcharged pipe (after Kellagher 1993)

The issue now arises about how to switch between the two modes: free surface and pressurised flow, in a pipe. Free surface flow can be represented by the 1D Saint Venant equations. Assuming similar equations apply to the pressurised flow along a pipe, they become:

$$\begin{aligned} &\frac{\partial u}{\partial x} = 0 \\ &\frac{\partial u}{\partial t} + \frac{g}{L}(H_d - H_u) + \left(\frac{f}{D} + \frac{K_u + K_d}{L}\right)\frac{u|u|}{2} = 0 \end{aligned} \tag{17.44}$$

where

L is the length of the pipe (m)
H_d is the water level in the downstream node (m)
H_u is the water level in the upstream node (m)
K_u is the head-loss coefficient upstream
K_d is the head-loss coefficient downstream
D is the pipe diameter (m)

The dependent variable u is independent of x. This would not matter if the flow in the pipe always switches from having a free surface along the whole length of the pipe to being pipe full at the next instant. But this does not happen: at any one time part of the pipe can be surcharged and the flow in the remainder may have a free surface. Although surcharging due to the discharge exceeding the capacity of a closed conduit is very significant in identifying throttles in a network, conduits also surcharge because of backup from downstream. In this case a conduit can have the steady situation of a free surface flow upstream and pressurised flow downstream. (A steady situation with pressurised flow upstream and free surface flow in the pipe downstream cannot occur).

Preissmann slot

When it comes to solving the two categories of equation numerically, the solution algorithm has to be switched between free-surface and pressurised flow. A considerable amount of book keeping has to be done to keep track of the segments of the pipe that are surcharged or not. To get round this problem, Preissmann (Cunge & Wegner 1964) came up with the ingenious idea of assuming that the flow always has a free surface by assuming a conceptual slot of small width in the top of the pipe along its entire length; See Figure 17.12.

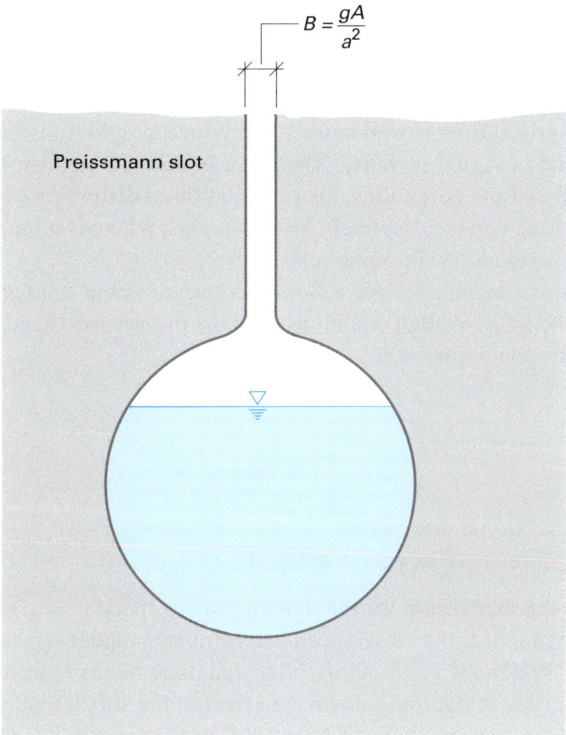

Figure 17.12 Preissmann slot

A major issue is the width of the slot. Obviously it has to be small to reproduce what happens in a surcharged pipe, but it cannot be too small, otherwise the Courant number would be very large and as a consequence the time step would need to be extremely small to preserve accuracy. The equations for pressurised, compressible flow in an elastic pipe are:

$$\begin{aligned}\frac{\partial H}{\partial t} + u\frac{\partial H}{\partial x} + u\sin\theta + \frac{a^2}{g}\frac{\partial u}{\partial x} &= 0\\ \frac{\partial u}{\partial t} + u\frac{\partial u}{\partial x} + g\frac{\partial H}{\partial x} + f\frac{u|u|}{D} &= 0\end{aligned} \tag{17.45}$$

where

H is the piezometric head (m)

and a is defined by

$$a = \sqrt{\frac{K/\rho}{1 + \frac{D}{e}\frac{K}{E}}} \tag{17.46}$$

where

K is the fluid bulk modulus (kg/m/s^2)
e is the thickness of the pipe wall (m)
E is the elastic modulus (kg/m/s^2)

See Chaudhry (1987). Pressurised flow is well known to produce a rapid transmission of disturbances along a conduit related to the speed of sound in water. Therefore in marked contrast to the propagation speed of disturbances under free surface flow conditions, the propagation of disturbances under pressurised flow in a 1 m circular conduit 100 m long can be effectively instantaneous, whereas it may take 60 s for a disturbance under free surface flow to travel along the same pipe.

The unsteady flow equations relating transient flow of a compressible fluid within an elastic pipeline can be seen to be analogous to the Saint Venant equations with the piezometric head replacing the depth and the surface width, B, replaced by the expression:

$$B = \frac{gA_{pf}}{a^2} \tag{17.47}$$

where

A_{pf} is the pipe full cross-sectional area (m^2)
a is the speed of the pressure wave in water (m/s)

Eq. (17.47) is derived from the expression for the dynamic wave speed $c = \sqrt{gA/B}$. Typically, values of a are greater than 100 m/s. In practice, the wave speed can be much smaller due to pressure relaxation effects introduced by lateral pipes (Watanabe *et al.* 1990). Note that there needs to be a smooth transition from the pipe to the slot. This is important in damping down the effect of the shock that is known to propagate along the pipe when the discharge is increasing beyond pipe full. Care has also to be taken so that the storage in the slot is compensated for by adjusting the storage in the manholes. This concept of Preissmann was, so to say, "turned on its head" by Abbott (see Abbott 1980) who introduced a slot descending through a smooth transition from the bottom of the pipe so as to accommodate the local drying-out of a pipe while keeping the same numerical scheme.

17.6 MANHOLE STORAGE

Normally, in a distribution network, there is no storage at the nodes representing the junctions between the pipes. In the collection systems, however, manholes are provided for inspection at all junctions. These provide a certain amount of storage in the system. The water levels in the manholes change in time according to the net inflow or outflow:

$$\frac{dV_j}{dt} - \sum Q_{ij} + Q_{jk} - Q_{j,in} + Q_{j,out} = 0 \tag{17.48}$$

where

V_j is the volume in manhole j (m^2)
Q_{ij} is the discharge into manhole j from conduit ij, where manhole i is upstream of manhole j (m^3/s)

Q_{jk} is the discharge from manhole j to manhole k downstream (m^3/s)
$Q_{j,in}$ is the inlet discharge from the surface runoff to manhole j (m^3/s)
$Q_{j,out}$ is the discharge out of manhole j such as to surface flooding, if any (m^3/s)

Note that Eq. (17.48) is the finite difference form of the continuity equation Eq. (17.29).

The manhole storage equation Eq. (17.48) is straightforward to use, except that the volume is normally a function of depth. Generally, there will be a storage chamber below ground with a narrow access shaft leading up to the ground surface; See Figure 17.13. Another issue to be considered at this point is the destiny of floodwater discharging from the manhole if the surcharging in the conduits is extreme. If pipe network model is not coupled with the surface flow model, then it is customary therefore, to include a model of the above-ground storage to accommodate excess water issuing from the manhole. This can, for example, be defined as a variable plan area with depth, though in practice the scarcity of data on the ground profile and the lack of knowledge on what happens to emerging flood water means that fairly simple above-ground storage models are adequate in the first instance: see Figure 17.14. We discuss the integration of 1D pipe network models and 2D over ground models below.

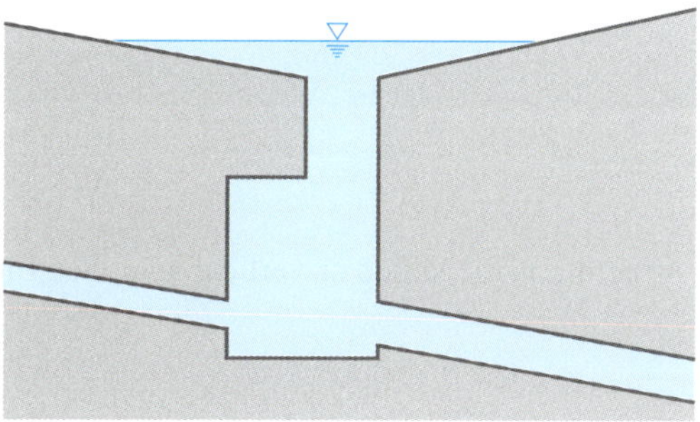

Figure 17.13 Manhole storage

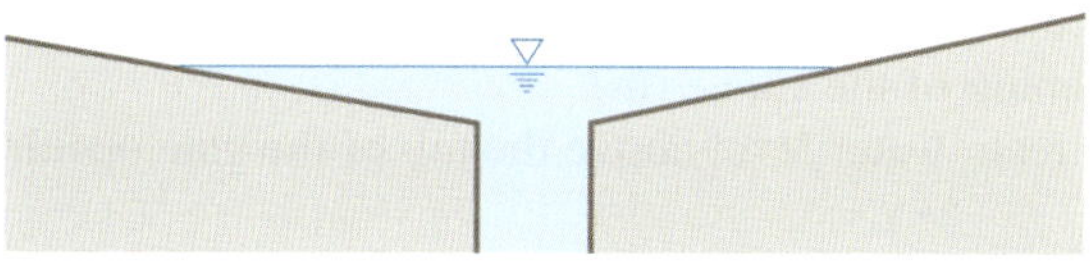

Figure 17.14 Above ground storage

17.7 ANCILLARY STRUCTURES

Besides pipes and manholes there is a range of other structures, termed ancillary structures, that are associated with urban drainage systems. One of the most common structures is the weir, which comes in different forms such as a side weir, a leaping weir, and the traditional weir across the direction of flow: see Figure 17.15. The basic equation for modular flow over a weir normal to the flow is given by:

$$Q = C_d L \sqrt{g} h^{3/2} \tag{17.49}$$

where

C_d is the discharge coefficient

L is the length of the weir (m)
h is the height of the water level upstream above the weir crest (m)

Eq. (17.49) assumes that the flow is modular, that is, the flow over the weir has a critical flow at some point. This implies that the water level downstream has no influence on the water level upstream and therefore on the flow over the weir. If the water level downstream is sufficiently high to submerge the flow in the cross-section where otherwise critical depth would occur, then the flow may no longer be supercritical and the discharge over the weir is a function of the downstream water level as well. In this case the discharge given by Eq. (17.49) is modified by a factor that is a function of the height of the downstream water level above the weir crest, as well as the height of the upstream water level above the weir crest. The equation for a side weir is more complicated than Eq. (17.49). See Butler and Davies (2000) for further details.

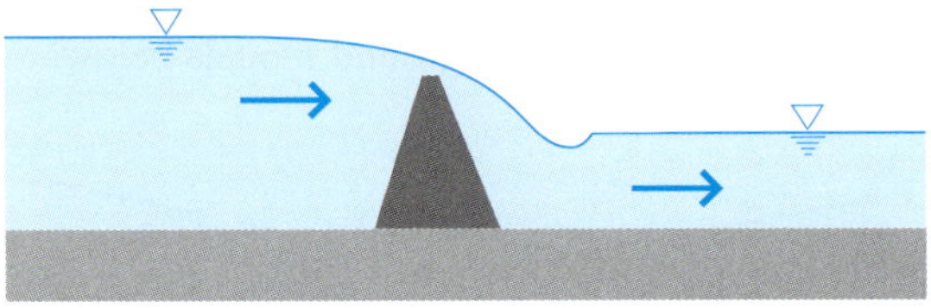

Figure 17.15 A simple weir

Another important ancillary structure is the sluice or vertical gate; See Figure 17.16. Here the discharge through the sluice is given by

$$Q = C_d L h_d \sqrt{g} h^{1/2} \qquad (17.50)$$

where

C_d is the discharge coefficient
L is the width of the sluice (m)
h_d is the height of the opening of the sluice (m)
h is the height of the upstream water level above the top of the gate opening (m)

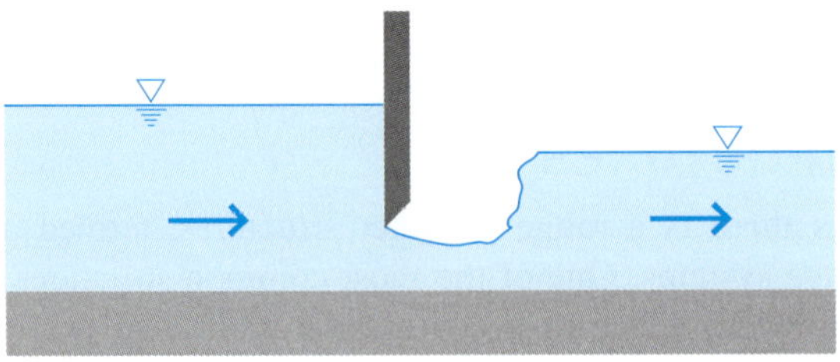

Figure 17.16 A sluice

Note that the water level downstream is such that the flow immediately downstream of the sluice is supercritical. If the downstream water level is sufficiently high then the downstream water level drowns out the sluice reducing the discharge through the structure. Again, the discharge given by Eq. (17.50) is reduced by a factor which reflects the magnitude of the downstream water level above the top of the gate-opening as well as the height of the upstream water level above the top of the gate-opening.

Orifices (or valves, Figure 17.17) constitute yet another common structure. The basic equation for an orifice is

$$H_u - H_d = k_v Q^2 \tag{17.51}$$

where

H_u is the head upstream (m)
H_d is the head downstream (m)
k_v is the head-loss coefficient for the flow through the orifice (s^2/m^5)

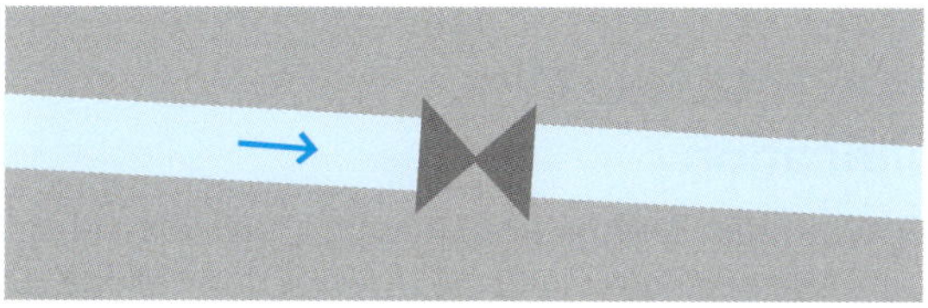

Figure 17.17 A valve

As introduced earlier, pumping stations are needed for cases where it is not feasible to achieve efficient operation of the system only by gravity. Pumping facilities operate on a stop-start basis with respect to the water level in the wet well, Figure 17.18.

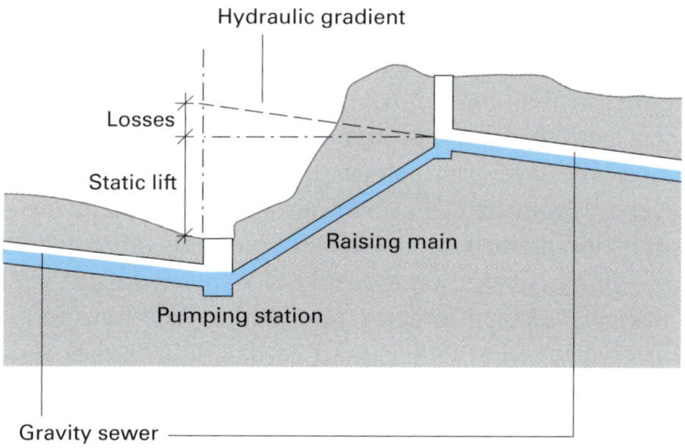

Figure 17.18 Pump (after Butler and Davies (2000))

The hydraulic performance of a pump is determined by its characteristic curve which is expressed as a head-flow relationship, Figure 17.19. The pump characteristic curve gives the head that a pump is capable of producing while producing a certain flow rate. The system characteristic curve gives the head that would be required for the system to deliver a certain flow rate. The point where these two curves cross is referred to as the operating point or duty point.

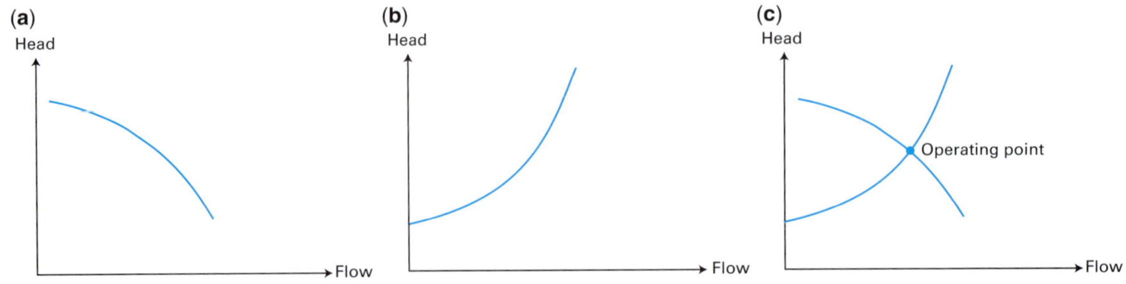

Figure 17.19 Pump and system characteristic curves: a) pump characteristic; b) system characteristic; c) operating point

17.8 MODELLING WATER QUALITY

17.8.1 Process simplification

The modelling of water quality in pipes and natural channels provide substantial problems whereby it is particularly difficult to quantify and regulate water quality because of the diverse nature of the pollutant inflows, the development of pollutants such as bio-films in water distribution pipes, and the intermittent and varied nature of overflow discharges. Because the chemical and biological processes are so very complex, simplifications have to be made in order to model the pollutant behaviour, albeit with some level of confidence.

17.8.2 Sediment transport

It is assumed that we can distinguish between sediment and chemical and biological pollutants. Fine sediment occurs in two layers: an active layer in which the sediment is un-consolidated and available for pick-up when the bed shear stress exceeds a certain threshold, and a passive layer with more permanent deposits on the bed of the sewer. Given a sufficient velocity, fine sediment deposited in the active layer during quiescent periods can be brought into suspension. This is related to what is termed the "first foul flush". The passive layer can contribute to a decrease in the pipe's capacity, though in most cases the passive layer is assumed to have a uniform depth along the pipe. The capacity of the flow in a conduit to transport fine-grained or low-density particles in suspension can be calculated from the Ackers (1993) equation, which is based on the Ackers-White sediment transport equation (Ackers & White 1973) for alluvial channels adapted to apply to conduits. The transport capacity can be expressed as the volumetric sediment concentration as the ratio of the flow rate of sediment to the flow rate of the water:

$$C_v^{max} = J\left(\frac{W_e R}{A}\right)^{\alpha}\left(\frac{d_{50}}{R}\right)^{\beta}\lambda_c^{\gamma}\left[\frac{|u|}{\sqrt{g(s-1)R}} - K\lambda_c^{\delta}\left(\frac{d_{50}}{R}\right)^{c}\right]^{m} \tag{17.52}$$

where

W_e is the effective width of the deposited sediment bed (m)
λ_c is the composite Dacry-Weisbach friction factor of the flow
A is the cross-sectional area (m^2)
R is the hydraulic radius (m)

u is the velocity of the water (m/s)
s is the specific gravity
d_{50} is the mean particle size (mm)
α, β, γ, δ, c and m are constants

The coefficients depend on the sediment size:

$$D_{gr} = d_{50}\left[\frac{g(s-1)}{\nu^2}\right] \quad (17.53)$$

where ν is the kinematic viscosity of the water. For fully suspended sediment transport, D_{gr} should not be greater than unity.

The effective width W_e is related to the actual width, W_b of the sediment bed by

$$W_e = \left[0.2 + \frac{10}{3}\left(\frac{y_b}{D} - 0.01\right)\right]W_b \quad (17.54)$$

where

D is the diameter of the pipe (m)
y_b is the depth of the deposited bed (m)
W_e has a maximum value of $0.5W_b$ and a minimum value of $0.02D$.

It is assumed that the suspended sediment is well mixed with the water and that the concentration C is advected by the flow with negligible dispersion according to

$$\frac{\partial C}{\partial t} + u\frac{\partial C}{\partial x} = 0 \quad (17.55)$$

The model described here does not include the transport of coarser granular sediment as bed-load, but the hydraulic effects of the deposited materials can be taken into account. The roughness of the bed can be determined as a function of the shear stress for which $d_{50} = 1.0$ mm and $s = 2.6$. The composite friction factor λ_c takes into account the different roughnesses of the pipe walls and the deposited sediment bed according to their relative wetted perimeters.

The sediment will be suspended and transported according to Eq. (5.52). If the suspended load exceeds the transporting capacity of the flow, it is assumed that the particles settle at their fall velocity onto the invert of the pipe or any permanent layer of deposited sediment.

17.8.3 Chemical pollutants

Pollutants such as are characterised as Total Suspended Solids (TSS), Biochemical Oxygen Demand (BOD), and Total Nitrogen (TN) are either dissolved in the water or attached to the sediment according to a potency factor defined separately for each pollutant. (Note that Ammonia only appears in the dissolved state). Consequently, a pollutant is advected with the water and diffused according to the turbulent structure of the flow. This diffusion is made up of molecular diffusion (Brownian movement), turbulent diffusion and dimensional effects. The net effect is a degree of mixing of the pollutant. In addition, there are processes of dilution, flocculation as electro-chemical forces bring sediment particles together, deposition of sediment when the shear stress is sufficiently low, erosion when the shear stress exceeds a threshold, thermal effects and re-aeration. The dynamic concentration can be described by the advection-diffusion equation:

$$\frac{\partial C}{\partial t} + u\frac{\partial C}{\partial x} - D\frac{\partial^2 C}{\partial x^2} = S \tag{17.56}$$

where

C is the concentration of pollutant (kg/m^3)
u is the speed of the water (m/s)
D is the diffusion coefficient (s^2/m)
S is the source term which includes the range of processes above (kg/m^3/s)

The source term may take a variety of forms such as

$$\begin{aligned} &\text{(i)} \quad S = KC \\ &\text{(ii)} \quad S = KC^n \\ &\text{(iii)} \quad S = \frac{KC}{a + C} \end{aligned} \tag{17.57}$$

(iii) is the Michaelis-Menton form.

The biochemical processes include aerobic and anaerobic oxidation, nitrification, adsorption and desorption of metals and other toxic compounds. In addition, we may have microbiological processes, including the growth and die-off of species and toxicant accumulation. The oxygen balance in sewers is given by:

$$\frac{\partial C_O}{\partial t} + u\frac{\partial C_O}{\partial x} - D\frac{\partial^2 C_O}{\partial x^2} = K_{LA}(C_{O,S} - C_O) - (r_w + r_b + r_s) \tag{17.58}$$

where

$C_{O,S}$ is the saturation value of the dissolved oxygen concentration (kg/m^3)
C_O is the actual dissolved oxygen concentration (kg/m^3)
K_{LA} is the volumetric re-aeration coefficient (1/s)
r_w is the oxygen consumption rate in the bulk water (kg/m^3/s)
r_b is the oxygen consumption rate in the biofilm (kg/m^3/s)
r_s is the oxygen consumption rate in the sediment (kg/m^3/s)

There are various important chemical processes involving oxygen, nitrogen, phosphorus and carbon. Oxygen is dissolved through the free surface from the atmosphere into the water and is diffused through the flow. There is a saturation level for oxygen in the water. In sewers Pomeroy and Parkhurst (1972) showed that

$$K_{LA} = 0.96\left(1 + 0.17\left(\frac{u^2}{gd_m}\right)\right)\gamma(s_f u)^{3/8}\frac{1}{d_m} \tag{17.59}$$

where

u is the mean velocity (m/s)
d_m is the hydraulic mean depth (m)
γ is the temperature correction factor (1.00 at 20 °C)
s_f is the hydraulic gradient

There are a variety of different chemical and ecological processes going on in a river or pipe. They have different spatial and temporal scales; See Figure 17.20.

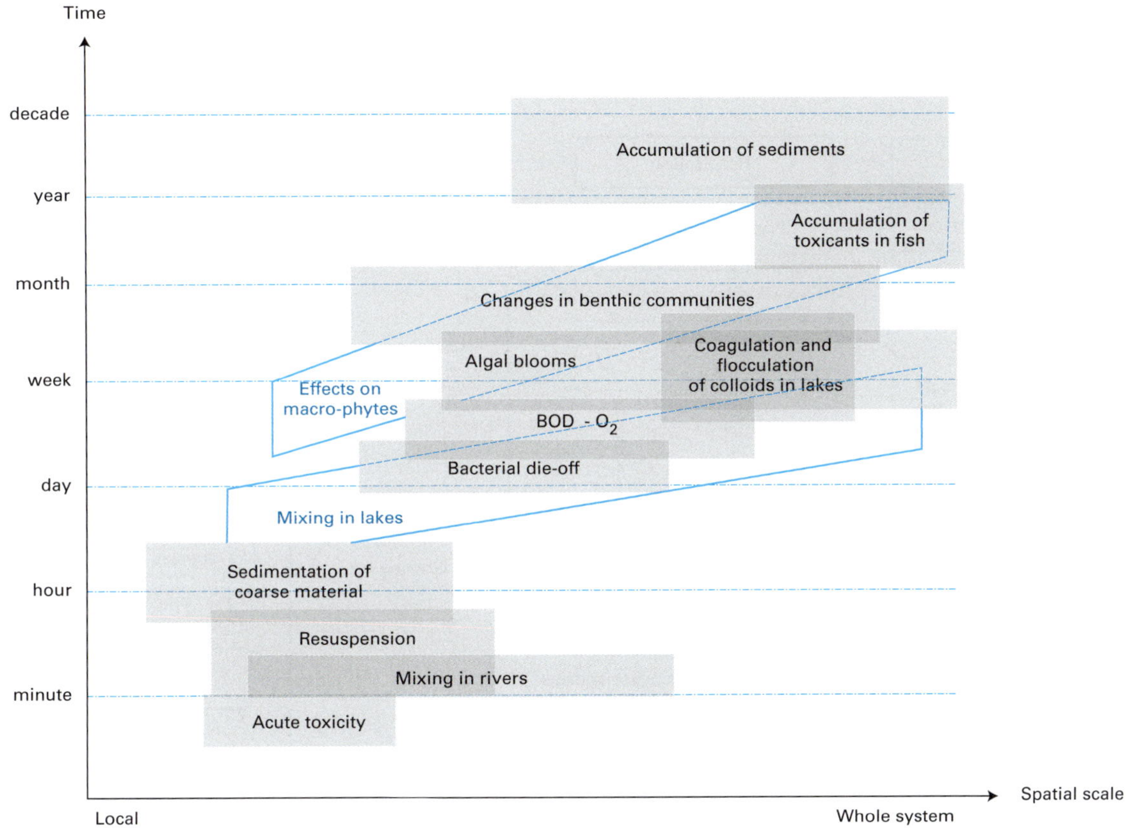

Figure 17.20 Temporal and spatial scales for water quality processes (adapted from Butler and Davies (2000))

Similarly there is a wide range of biological processes: see Figure 17.21. Typical ordinary differential equations for these processes were developed by Lotka (1925) and Volterra (1926), essentially to describe food chains in nature, such as:

$$\frac{dX}{dt} = \alpha X - \beta XY$$
$$\frac{dY}{dt} = \gamma\beta XY - \delta Y \qquad (17.60)$$

where

X is the concentration of the prey species (1/l)
Y is the concentration of the predator species (1/l)

α is the prey species' growth rate (1/s)
β is the predation rate of the predator species on the prey species (1/s)
γ is the assimilation efficiency of the predator species
δ is the mortality rate of the predator species (1/s)

Obviously, the models for both the chemical and biological processes can be extremely complex. In practice the processes have to be limited to those that are perceived as being important – and sufficiently calculable.

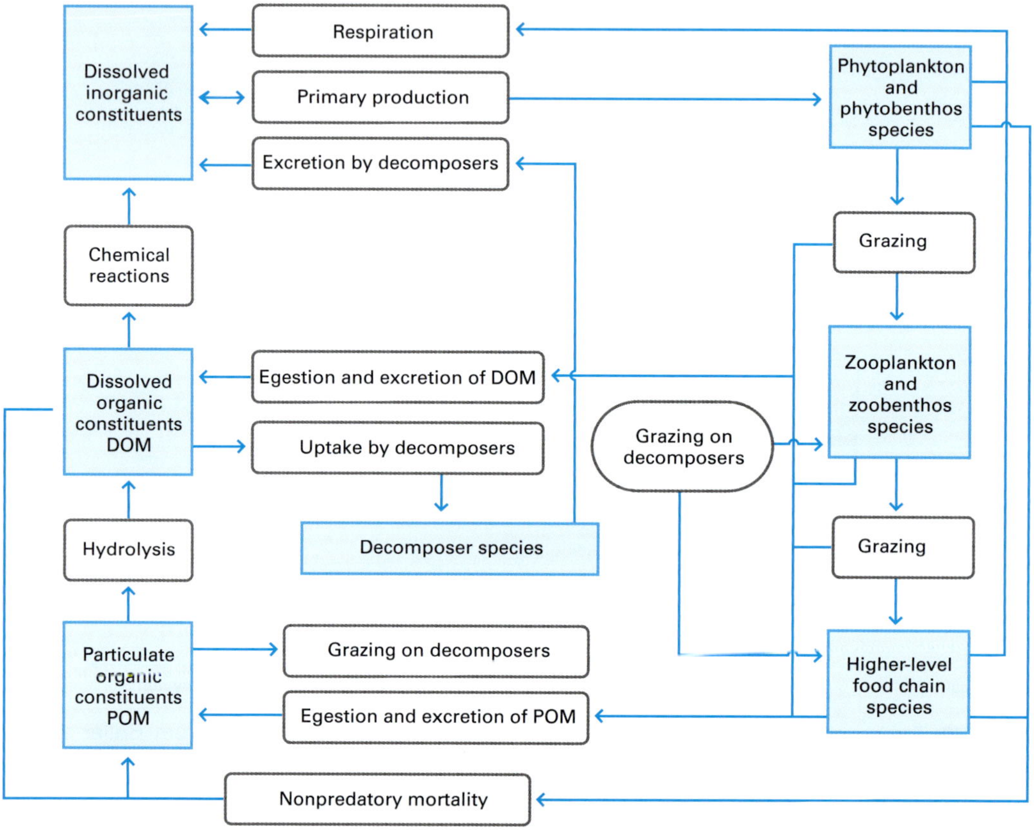

Figure 17.21 Ecological processes

17.9 GROUNDWATER

The Richards equation for flow in an aquifer is given by:

$$S_s \frac{\partial h}{\partial t} - \frac{\partial}{\partial x}\left(k_x \frac{\partial h}{\partial x}\right) - \frac{\partial}{\partial y}\left(k_y \frac{\partial h}{\partial y}\right) - \frac{\partial}{\partial z}\left(k_z \frac{\partial h}{\partial z}\right) = -G \tag{17.61}$$

where

S_s is specific storage (volume that can be released from a unit volume of a saturated aquifer by a unit reduction in hydraulic head)

h is the groundwater potential (m)
k_x, k_y, k_z are the hydraulic conductivities (m^2/s)
G is the source term (m/s)

The assumptions made in the derivation of Eq. (17.61) include:

- the aquifer material is incompressible,
- the water is of constant density (incompressible),
- any external loads on the aquifer (e.g., overburden, atmospheric pressure) are constant,
- the groundwater is flowing slowly (Reynolds number less than unity), and
- the hydraulic conductivity (k) is an isotropic scalar.

The two-dimensional form of the equation for an unconfined aquifer is

$$S_y \frac{\partial h}{\partial t} - \frac{\partial}{\partial x}\left(kh\frac{\partial h}{\partial x}\right) - \frac{\partial}{\partial y}\left(kh\frac{\partial h}{\partial y}\right) = -N \tag{17.62}$$

where

N is the recharge (m/s)
S_y is the specific yield of the aquifer
H is now the head (m)

The interaction between an aquifer and a river can be derived as:

$$q = 5\alpha kP(h - h_r) \tag{17.63}$$

where

q is the lateral inflow (m^2/s)
α is a constant which is approximately 1.0
P is the wetted perimeter formed by the aquifer in contact with the river channel (m)
h_r is the stage in the river (m)

See Miles (1985).

Chapter 18

Numerical solutions of equations

18.1 NUMERICAL SOLUTIONS OF THE SAINT VENANT EQUATIONS

18.1.1 Discretization

The solution space of the 1D Saint Venant equations for a pipe occupies the domain in (x, t) space bounded by the lines $t = 0$, $x = 0$ and $x = L$. A numerical solution of the equations is based on a discretization of this domain. The discretization can be done using finite differences, elements or volumes in different coordinate systems (Cartesian, curvilinear, cylindrical, radial). A finite difference representation is the most popular method, though the finite volume technique is being used more widely in some contexts.

There are a number of different ways of forming a finite difference representation of the basic equations, but only a limited set provide a satisfactory solution. In particular, the solution should be stable (i.e., it should not exhibit rapidly growing oscillations that have a period of twice times the time increment), be consistent with the original equations, and converge to the right solution as the space and time increments tend to zero (as can be investigated through what happens to the truncation error between the numerical and analytical equations). Also, the solution should be economical to obtain. A special attention is given to schemes that satisfy mass, momentum and energy conservation conditions. In general, provided consistency and convergence are satisfied, attention is focused on the stability and economy of the scheme used. Note that almost all numerical solutions of simulation equations for water flow focus on finding values for the dependent variables (such as water level or discharge) in physical space for a time step rather than calculating the dependent variables for every time step at a succession of space steps. In other words, the solution is found throughout the physical space at successive time levels with increasing time.

There are two major categories of finite difference schemes: explicit and implicit. There are however several other types of schemes, but these can be reduced to the one or the other of these two types. An explicit scheme calculates the dependent variables at the new time level solely in terms of the dependent variables at the previous time level (or even at a previous time level or recursively at a 'half time level'). The Lax-Wendroff scheme (see Morton and Mayers 2005) is a well known explicit scheme; see Figure 18.1. To explore this scheme, consider the advection equation:

$$\frac{\partial u}{\partial t} + c\frac{\partial u}{\partial x} = 0 \qquad (18.1)$$

where c is the advection speed (m/s)

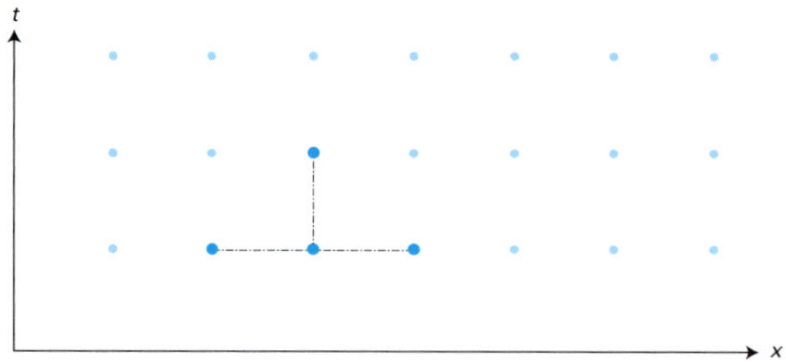

Figure 18.1 Lax-Wendroff explicit scheme

The Lax-Wendroff scheme uses the following finite difference representations for the first order differentials:

$$\begin{aligned}\frac{\partial u}{\partial t} &\approx \frac{u_{j+1}^{n+1} - u_{j+1}^{n}}{\Delta t}\\ \frac{\partial u}{\partial x} &\approx \frac{u_{j+1}^{n} - u_{j-1}^{n}}{2\Delta x}\end{aligned} \tag{18.2}$$

See Figure 18.1. Suppose we take the Taylor expansion of $u(x, t + \Delta t)$ to second order:

$$\begin{aligned}u(x, t + \Delta t) &= u(x, t) + \Delta t\frac{\partial u}{\partial t} + \Delta t^2\frac{\partial^2 u}{\partial t^2} + O(\Delta t^3)\\ &= u(x, t) - \Delta tc\frac{\partial u}{\partial x} + \frac{1}{2}\Delta t^2\left[-\frac{\partial c}{\partial t}\frac{\partial u}{\partial x} + c\frac{\partial}{\partial x}\left(c\frac{\partial u}{\partial x}\right)\right] + O(\Delta t^3)\end{aligned} \tag{18.3}$$

We can write the finite difference form as:

$$\begin{aligned}u_j^{n+1} &= u_j^n - \Delta tc_j^n\left(\frac{u_{j+1}^n - u_{j-1}^n}{2\Delta x}\right) - \frac{1}{2}\Delta t^2(c_t)_j^n\left(\frac{u_{j+1}^n - u_{j-1}^n}{2\Delta x}\right)\\ &\quad + \frac{1}{2}\Delta t^2 c_j^n\left[(c_x)_j^n\left(\frac{u_{j+1}^n - u_{j-1}^n}{2\Delta x}\right) + c_j^n\left(\frac{u_{j+1}^n - 2u_j^n + u_{j-1}^n}{\Delta x}\right)\right] + O(\Delta t^3)\end{aligned} \tag{18.4}$$

If we now write $u_j^n = \alpha^n e^{i\beta j\Delta x}$ and assume that c is constant then

$$\begin{aligned}\alpha &= 1 - \frac{c\Delta t}{2\Delta x}\left(e^{i\beta\Delta x} - e^{-i\beta\Delta x}\right) + \frac{c^2\Delta t^2}{2\Delta x^2}\left(e^{i\beta\Delta x} - 2 + e^{-i\beta\Delta x}\right)\\ &= 1 + \frac{c\Delta t}{\Delta x}i\sin(\beta\Delta x) + \frac{c^2\Delta t^2}{\Delta x^2}\left[\cos(\beta\Delta x) - 1\right]\end{aligned} \tag{18.5}$$

Therefore

$$|\alpha| = \left\{1 - \frac{c^2\Delta t^2}{\Delta x^2}\left[1 - \cos(\beta\Delta x)\right]\right\}^2 + \frac{c^2\Delta t^2}{\Delta x^2}\sin^2(\beta\Delta x) < 1 \tag{18.6}$$

The Lax-Wendroff scheme is unconditionally stable if $|\alpha| < 1$, that is, if $c\Delta t/\Delta x < 1$; otherwise, any perturbations in the solution will grow uncontrollably. The dimensionless expression $c\Delta t/\Delta x$ is termed the *Courant number*.

An implicit scheme, on the other hand, calculates the dependent variables at the new time level in terms of other adjacent, dependent variables at the new time level as well as dependent variables at the old time level. There are two well known implicit schemes: 6-point and 4-point schemes.

18.1.2 Method of characteristics

The following description of the method is based on Abbott (1966), Cunge *et al.* (1980), Garcia-Navarro and Burguete (2006), and Price and Vojinović (2011).

The characteristic method has two important fields of application: as a standard for other methods and as a means of representing the boundary conditions in the methods which cannot compute all flow variables at exterior or interior boundary points of the model (Cunge *et al.* 1980).

The characteristic formulation in 1D leads naturally to the method of characteristics. It can be seen as a way to move from a problem governed by partial differential equations to another problem based on ordinary differential equations (Garcia-Navarro and Burguete 2006).

Consider the characteristic form of the Saint Venant equations for prismatic channel with a resistance shear stress $S_f \neq 0$ and constant longitudinal bottom slope S_o (Garcia-Navarro and Burguete 2006):

$$\begin{aligned} \frac{\partial}{\partial t}(u + 2c) + (u + c)\frac{\partial}{\partial t}(u + 2c) &= -g(s_f - s_o) \\ \frac{\partial}{\partial t}(u - 2c) + (u - c)\frac{\partial}{\partial t}(u - 2c) &= -g(s_f - s_o) \end{aligned} \tag{18.7}$$

For frictionless and horizontal channel,

$$u + 2c = \text{constant} = J^+ \tag{18.8}$$

along the C^+ characteristics which is defined as

$$\frac{dx}{dt} = u + c \tag{18.9}$$

whereas

$$u - 2c = \text{constant} = J^- \tag{18.10}$$

along the C^- characteristics which is defined as

$$\frac{dx}{dt} = u - c \tag{18.11}$$

According to (Cunge *et al.* 1980), the characteristic curves carry information into the solution space (x, t) from its boundaries. They represent the speed of propagation of the perturbations and hence are the convective wave velocities. Any disturbance which occurs at any point Q in Figure 18.2 (a) at time $t = 0$ in open channel flow propagates along the channel in time and both downstream and upstream directions. Hence, the perturbations at point Q have a range of influence on the region limited by C^+ and C^- characteristic curves originating at point Q. However, taking a point P in Figure 18.2 (b), the domain of dependence shows the region in which disturbances can influence the flow at P. This region is formed by C^+ and C^- characteristics that meet at point P. Anything happening outside the domain of dependence will not influence the state at P.

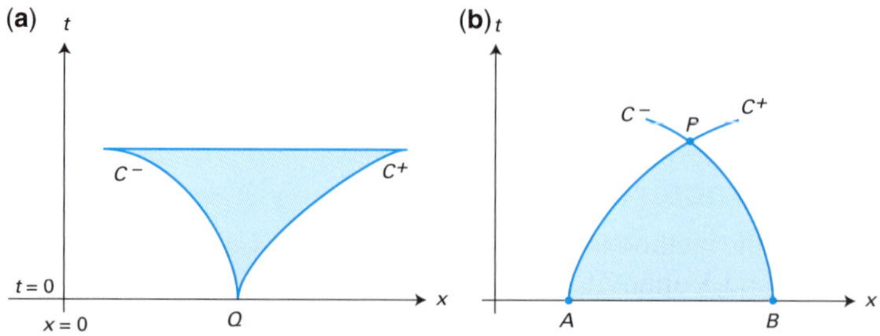

Figure 18.2 Propagation of a disturbance in channel flow. (a) Range of influence of disturbance at point Q (b) Domain of dependence of point P (Cunge *et al.* 1980)

As shown in Figure 18.3, assuming that initial data are known at points L and R in the solution space (x, t), the characteristics issuing from these points and defined by Eq. (18.9) and Eq. (18.11) intersect at point M.

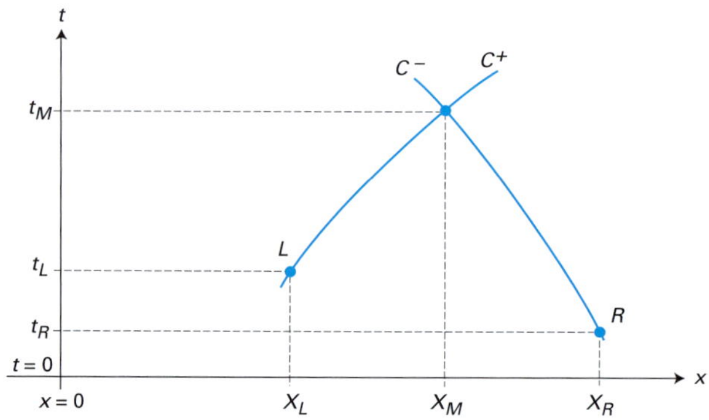

Figure 18.3 Basis of numerical solution using the characteristic method (Cunge *et al.* 1980)

Since Eq. (18.9) and Eq. (18.11) express derivatives along the characteristics and since all variables are known at points L and R, it is possible to integrate along the characteristics as follows:

$$
\begin{aligned}
x_M &= x_L + \int_{t_L}^{t_M} (u + c)\,dt \\
u_M + 2c_M &= u_L + 2c_L + \int_{t_L}^{t_M} g(s_o - s_f)\,dt \\
x_M &= x_R + \int_{t_R}^{t_M} (u - c)\,dt \\
u_M - 2c_M &= u_R - 2c_R + \int_{t_R}^{t_M} g(s_o - s_f)\,dt
\end{aligned}
\tag{18.12}
$$

The four unknowns – coordinates x_M and t_M, and flow variables u_M and h_M – at point M can be solved from the four relations above. If we implement the trapezoidal rule of integration between points L and M, and R and M leads to:

$$\begin{aligned}
x_M - x_L &= (t_M - t_L)\left(\frac{u_M + c_M}{2} + \frac{u_L + c_L}{2}\right) \\
u_M + 2c_M &= u_L + 2c_L + g(t_M - t_L)\left(\frac{s_{O_M} - s_{f_M}}{2} + \frac{s_{O_L} - s_{f_L}}{2}\right) \\
x_M - x_R &= (t_M - t_R)\left(\frac{u_M - c_M}{2} + \frac{u_R - c_R}{2}\right) \\
u_M - 2c_M &= u_R - 2c_R + g(t_M - t_R)\left(\frac{s_{O_M} - s_{f_M}}{2} + \frac{s_{O_R} - s_{f_R}}{2}\right)
\end{aligned} \tag{18.13}$$

The above four non-linear algebraic equations may be solved by iteration, and the accuracy of the values of x_M, t_M, u_M and h_M is dependent on the accuracy of iterations and the truncation error in the trapezoidal rule of integration. Increasing the number of iterations and decreasing the interval $x_R - x_L$ helps in reducing both sources of error.

18.1.3 6-point implicit scheme

One is the 6-point staggered grid scheme with h and u identified at alternate computational points along a pipe: see Figure 18.4. In this case the algorithmic equations are written as:

$$\begin{aligned}
&u\frac{\partial u}{\partial t} - g\frac{\partial h}{\partial t} + \left(u^2 - g\frac{A}{B}\right)\frac{\partial u}{\partial x} + gus_f = 0 \\
&-\frac{A}{B}\frac{\partial u}{\partial t} + u\frac{\partial h}{\partial t} + \left(u^2 - g\frac{A}{B}\right)\frac{\partial h}{\partial x} - g\frac{A}{B}s_f = 0
\end{aligned} \tag{18.14}$$

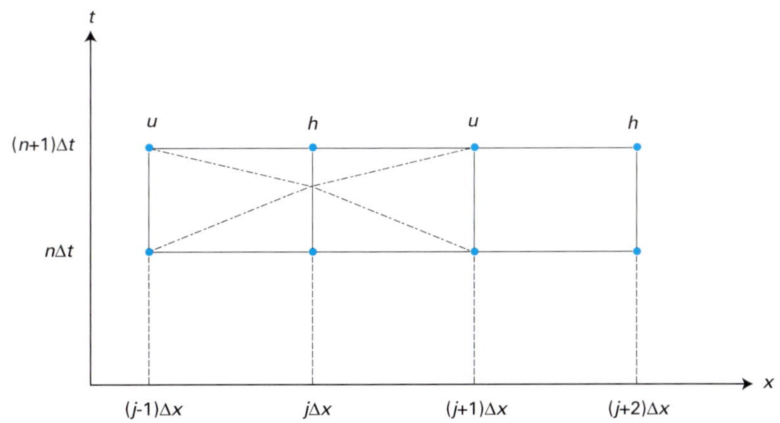

Figure 18.4 6-point Abbott-Ionescu scheme

The corresponding finite difference expressions for the derivatives are given by

$$\begin{aligned}
\frac{\partial h}{\partial t} &\approx \frac{h_j^{n+1} - h_j^n}{\Delta t} \\
\frac{\partial h}{\partial x} &\approx \theta \frac{h_{j+2}^{n+1} - h_j^{n+1}}{\Delta x} + (1-\theta)\frac{h_{j+2}^n - h_j^n}{\Delta x} \\
\frac{\partial u}{\partial t} &\approx \frac{u_{j+1}^{n+1} - u_{j+1}^n}{\Delta t} \\
\frac{\partial u}{\partial x} &\approx \theta \frac{u_{j+1}^{n+1} - u_{j-1}^{n+1}}{\Delta x} + (1-\theta)\frac{u_{j+1}^n - u_{j-1}^n}{\Delta x}
\end{aligned} \tag{18.15}$$

where

h_j^n refers to the depth (m) at the point $(j\Delta x, n\Delta t)$
Δx is the space step (m)
Δt is the time step (s)
θ is the weighting factor for stability ($0.5 < \theta < 1.0$)

This scheme is consistent with a 1D subset of the standard 2D finite difference scheme on a rectangular grid; see Figure 18.5 and Abbott and Ionescu (1967). The focus of the scheme is on evaluating the derivatives consistently near the centre point in terms of space and time. The space derivative is in some cases deliberately defined off-centre and biased to the new time level according to the value of $\theta > 0.5$. This device introduces a small numerical diffusion that dampens down any oscillations that might occur, and therefore preserves the stability of the calculations even when otherwise disturbing influences arise.

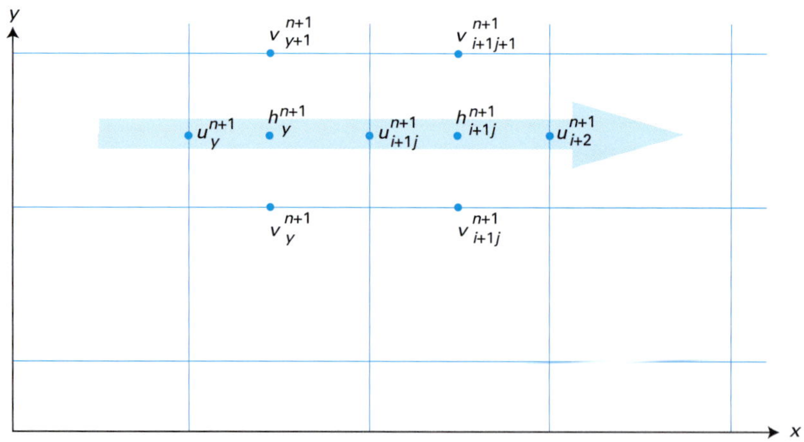

Figure 18.5 2D finite difference grid

As shown in the original publication of Abbott and Ionescu (1967), the scheme may be solved with $\theta = 0.5$ while conserving all three consistency conditions, that is, it conserves mass, momentum and energy simultaneously. Correspondingly; it is unconditionally stable. This does not of course exclude the creation of dispersion effects that arise with higher Courant numbers, although additional third-order terms can be introduces to reduce these to an acceptable level. This scheme is easily extended to two- and three-dimensional flows that are also unconditionally stable. Other third-order terms can also be introduced to simulate higher-order terms in the governing partial differential equations, such as those that account for accelerations in the vertical and so with short-period wave simulations and the flows that these generate as well as Tsunamis waves. These are not used within the context of this book outside of providing early warnings of Tsunamis propagations and their land-fall effects through real-time simulations. These approaches have been discussed in Abbott and Minns (1998).

There are other devices that are introduced to deal with small depth flows and the tendency of the standard finite difference formulation for the friction term to introduce more than one solution. These details are discussed below.

18.1.4 4-point implicit scheme

The other major implicit scheme is the 4-point scheme with h and u defined at each of the computational points along a pipe; see Figure 18.6. In this case the finite difference expressions for the derivatives are given by:

$$
\begin{aligned}
\frac{\partial h}{\partial x} &\approx \theta\frac{h_{j+1}^{n+1}-h_j^{n+1}}{\Delta x}+(1-\theta)\frac{h_{j+1}^{n}-h_j^{n}}{\Delta x}\\
\frac{\partial u}{\partial x} &\approx \theta\frac{u_{j+1}^{n+1}-u_j^{n+1}}{\Delta x}+(1-\theta)\frac{u_{j+1}^{n}-u_j^{n}}{\Delta x}\\
\frac{\partial h}{\partial t} &\approx \frac{h_j^{n+1}-h_j^{n}}{2\Delta t}+\frac{h_{j+1}^{n+1}-h_{j+1}^{n}}{2\Delta t}\\
\frac{\partial u}{\partial t} &\approx \frac{u_j^{n+1}-u_j^{n}}{2\Delta t}+\frac{u_{j+1}^{n+1}-u_{j+1}^{n}}{2\Delta t}
\end{aligned}
\tag{18.16}
$$

Again, $\theta > 0.5$ for the stability of the scheme when subject to destabilising influences. The same precautions in dealing with small depth flows and non-unique solutions as for the 6-point scheme are also needed here.

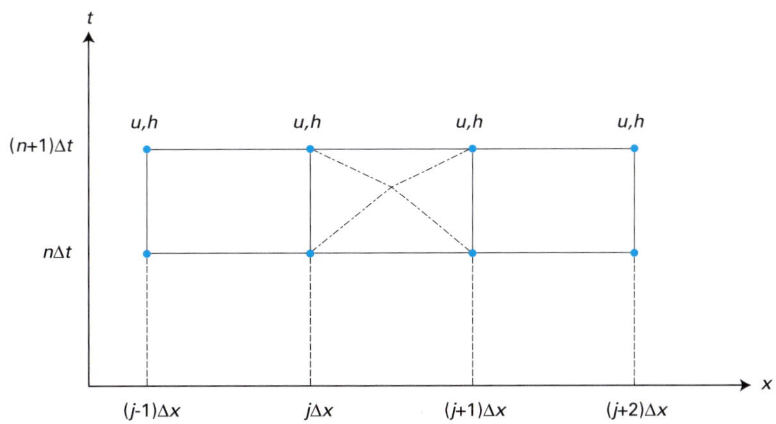

Figure 18.6 4-point scheme

18.1.5 Double sweep algorithm

Obviously, in both cases the finite difference equations are non-linear. Therefore, to solve them they need to be linearised in terms of the increments to the estimates for the dependent variables. Such an approach opens up the possibility of using an iterative Newton-Raphson approach (Tjalling 1995) and Abbott and Minns (1998) whereby a solution is obtained for the increments to the estimates for the dependent variables; the estimates of the dependent variables are then updated, and new increments are calculated. The process is repeated until there is acceptable convergence for the solution of the dependent variables at the new time level $(n + 1)\Delta t$. Given a single pipe the solution for the increments is obtained by inverting a tri-diagonal or four-diagonal matrix. This involves a double sweep algorithm (Abbott and Minns 1998); the first sweep down the diagonal involves the elimination of certain elements in the lower triangle of the matrix and then solving for the most downstream increment, before returning back up the diagonal and solving for the remaining increments; see Figure 18.7. This process is called a *double-sweep algorithm.*

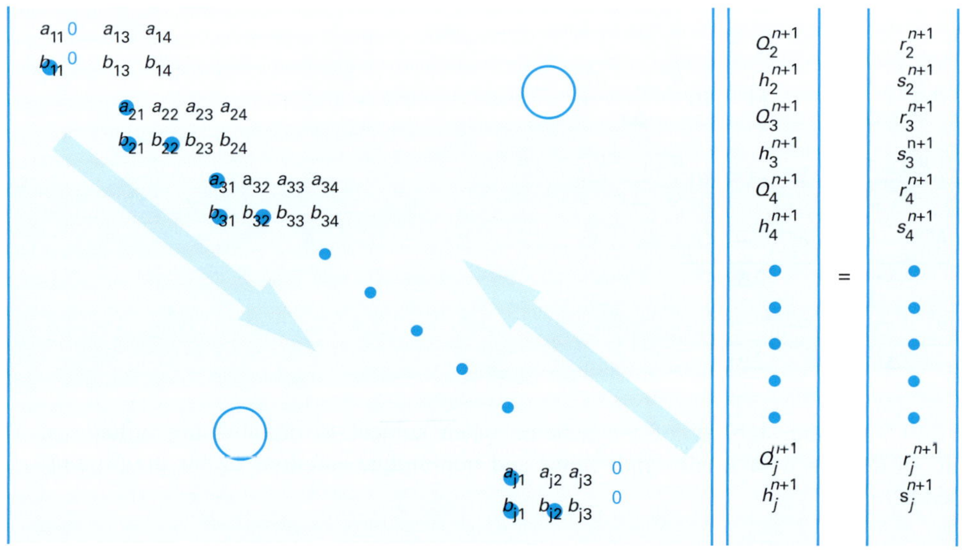

Figure 18.7 Double sweep algorithm (r and s are the residues of the two equations)

If there is failure to converge due to external perturbations then the time increment can be reduced and the process repeated. Alternatively, if convergence has been achieved for an adequate number of successive time increments then the time increment can be increased.

18.1.6 Network of pipes or channels

Where there is a network of pipes then more work has to be done in order to get a solution. This is because the solution for one pipe can be dependent on the solution for adjacent pipes. Before we introduce the solution method however, we need to indicate how head-losses at manholes and at ancillary structures (such as pumps, weirs, sluices, valves, etc.) are included.

The head-loss at, say, the exit of a manhole into a pipe is defined explicitly as an energy loss between two sections zero distance apart. The same concept is used at the exit of the pipe into the (downstream) manhole.

For the head-loss for the flow into pipe *ik* (as with the 4-point scheme):

$$h_{ik,0}^{n+1} = h_{ik,1}^{n+1} + K_{ik}\frac{\left(u_{ik,0}^{n+1}\right)^2}{2g} \qquad (18.17)$$

$$u_{ik,0}^{n+1} = u_{ik,1}^{n+1}$$

where the index 0 and 1 with h refers to two sections at the entrance to the pipe.

An ancillary structure is treated in a similar manner. It is assumed that there are computational nodes either side of the structure and the head-loss generated by the structure occurs between the two nodes. In effect, this defines a dummy pipe (or link). The distance between the two nodes can be non zero. This can affect the mass conservation equation. So far as the relationship between the water levels is concerned, this is given by

$$\frac{1}{2}\left(Q_{ik,1}^{n+1} + Q_{ik,2}^{n+1}\right) = F\left(h_{ik,1}^{n+1}, h_{ik,2}^{n+1}\right) \qquad (18.18)$$

(again for the 4-point scheme) with the mass conservation equation providing a second equation linking the values for the discharge and water level.

Given the difference equations for each pipe then it can be shown that partial elimination using the double sweep process along the pipes results in a system of equations for the increments in the water levels at the nodes of the network; see Abbott and Minns (1998). This sparse matrix can be inverted and the solution found for the water levels at the nodes. Once the water levels (or the increments in the water levels) are determined at the nodes, it is a simple matter to calculate the water level and discharge increments along each pipe, and therefore to update the water levels and discharges.

18.1.7 SWMM

The most significant of the explicit methods used in urban drainage modelling is the modified Euler method adopted by SWMM-EXTRAN; see Roesner *et al.* (1983). The momentum equation is rewritten as

$$\frac{\partial Q}{\partial t} = -\frac{gn^2}{1.49^2 R^{4/3}}Q|u| + 2u\frac{\partial A}{\partial t} + u^2\frac{\partial A}{\partial x} - gA\frac{\partial h}{\partial x} \qquad (18.19)$$

where the Manning form of the friction slope is used. The corresponding finite difference form of this equation for the whole pipe is written as:

$$Q_{jk}^{n+1} = Q_{jk}^{n} - \frac{k\Delta t}{R^{4/3}}Q_{jk}^{n}\left|u_{jk}^{n}\right| + 2u_{jk}^{n}\frac{\Delta A}{\Delta t}\Delta t + \left(u_{jk}^{n}\right)^2\frac{A_2 - A_1}{L_{jk}}\Delta t - gA\left(\frac{h_2 - h_1}{L_{jk}}\right)\Delta t \qquad (18.20)$$

Similarly, the finite difference form for the continuity equation is

$$h_j^{n+1} = h_j^n + \frac{1}{P_j}\sum Q_{jk}^n \Delta t \qquad (18.21)$$

where P_j is the plan area of the jth manhole or node.

Eqs (18.20) and (18.21) are solved sequentially to determine the discharge in each conduit and the water level in the manholes over a time increment. Roesner actually calculates values for the water level and pipe discharges at the half time step:

(1) Compute the discharge at $t + \Delta t/2$ in all conduits based on the water levels in the nodes at time t.
(2) Compute the water levels at all nodes at $t + \Delta t/2$ based on the average of the discharges in the conduits at time t and time $t + \Delta t/2$.

(3) Compute the discharge at $t + \Delta t$ in all conduits based on the water levels in the nodes at time t.
(4) $t + \Delta t/2$.
(5) Compute the water levels at all nodes at $t + \Delta t$ based on the average of the discharges in the conduits at time t and time $t + \Delta t$.

For stability this scheme requires

$$\Delta t \leq \frac{L_{jk}}{\sqrt{g}D_{jk}}$$
$$\Delta t \leq C^* \frac{P_j h_{\max,j}}{\sum_k Q_{jk}} \tag{18.22}$$

where C^* is a dimensionless constant = 0.10
$H_{\max,j}$ is the maximum water surface rise (m) in time step Δt
Note that the finite difference scheme is applied to a complete pipe, and therefore the pipe length must be neither too long or too short. Long pipes can be divided into a number of shorter pipes.

Roesner *et al*. (1983) states that time increments the order of 10 secs are typical for applications of the method. Some adjustments are made to the way in which the flooded surface area is assigned to the nodes in special cases, such as when there is back-up from the manhole downstream and there is no flow in the pipe from the upstream manhole.

Other than in SWMM, explicit schemes have not been widely used, though there is growing interest in schemes such as those proposed by McCormack (Garcia-Navarro and Saviron 1992) and Roe (Priestley 1989) because of their ability to model discontinuous flows accurately.

18.1.8 Roe scheme

The first order Roe scheme is a numerical method used to solve Riemann problems. It has been proved that the solver can be applied to the non-linear hyperbolic SWEs (shallow water equations). The Roe scheme is an upwind scheme which is based on the idea of discretizing the spatial derivatives so that information is taken from the side from which it comes (Garcia-Navarro and Burguete 2006).

Ignoring the source term, 1D unsteady flow equation can be written in vector form as (Garcia-Navarro and Burguete 2006):

$$\frac{\partial \mathrm{U}}{\partial t} + \frac{\partial \mathrm{F}}{\partial x} = 0 \tag{18.23}$$

where

$$\mathrm{U} = \begin{bmatrix} A \\ Q \end{bmatrix}$$
$$\mathrm{F} = \begin{bmatrix} Q \\ \dfrac{Q^2}{A} + g\dfrac{A^2}{2B} \end{bmatrix} \tag{18.24}$$

If $\mathbf{F} = \mathbf{F}(\mathbf{U})$, it is possible to rewrite the conservative system in the form

$$\frac{\partial \mathrm{U}}{\partial t} + \mathrm{J}\frac{\partial \mathrm{U}}{\partial x} = 0 \tag{18.25}$$

where **J** is the exact Jacobian matrix of the 1D Saint Venant equations, and defined as

$$J = \frac{\partial F}{\partial U} = \begin{bmatrix} 0 & 1 \\ c^2 - u^2 & 2u \end{bmatrix} \tag{18.26}$$

with corresponding eigenvalues Eq (18.27) and eigenvectors Eq. (18.28) given by

$$\lambda_1 = u + c \quad \text{and} \quad \lambda_2 = u - c \tag{18.27}$$

$$e_1 = (1, u + c)^T \quad \text{and} \quad e_2 = (1, u - c)^T \tag{18.28}$$

where $u = Q/A$ is the cross section averaged water velocity and $c = \sqrt{(gA/b)}$ is the celerity of the small amplitude surface waves in a channel of width B.

The Jacobian matrix is diagonalisable, whereby

$$J = P\Lambda P^{-1} \tag{18.29}$$

where **Λ** is the diagonal matrix with the eigenvalues in the main diagonal

$$\Lambda = \begin{bmatrix} u + c & 0 \\ 0 & u - c \end{bmatrix} \tag{18.30}$$

and **P** is the matrix made of the right eigenvectors.

$$P = \begin{bmatrix} 1 & 1 \\ u + c & u - c \end{bmatrix} \tag{18.31}$$

For non-linear cases, Roe's approximate Riemann solver performs a local linearization. At every cell limited by nodes L (left) and R (right), an approximate Jacobian ($\tilde{J}_{RL}$) is defined satisfying the following conditions (Roe 1981):

(1) $\tilde{J}_{RL}$ depends only on the $\mathbf{U_R}$ and $\mathbf{U_L}$ states,
(2) $(\mathbf{F_R} - \mathbf{F_L}) = \tilde{J}_{RL}(\mathbf{U_R} - \mathbf{U_L})$,
(3) $\tilde{J}_{RL}$ has real and distinct eigenvalues and a complete set of eigenvectors,
(4) $\tilde{J}_{RL} = \mathbf{J}(\mathbf{U_R}) = \mathbf{J}(\mathbf{U_L})$ if $\mathbf{U_R} = \mathbf{U_L}$

The expressions for the eigenvalues and eigenvectors of the approximate Jacobian are similar to those of the original Jacobian but evaluated at the average values in (L, R) (Garcia-Navarro and Burguete 2006).

$$\tilde{\lambda}_{1,2} = \tilde{u} \pm \tilde{c} \tag{18.32}$$

$$\tilde{e}_{1,2} = \begin{bmatrix} 1 \\ \tilde{u} \pm \tilde{c} \end{bmatrix} \tag{18.33}$$

where the average velocity and celerity at the interface (L, R) are

$$\tilde{u} = \frac{Q_R\sqrt{A_L} + Q_L\sqrt{A_R}}{\sqrt{A_L A_R}(\sqrt{A_L} + \sqrt{A_R})} \tag{18.34}$$

$$\tilde{c} = \frac{1}{2}(c_L + c_R) \tag{18.35}$$

At the same time, equations Eq. (18.30) and Eq. (18.31) will be updated as:

$$\tilde{\Lambda} = \begin{bmatrix} \tilde{u} + \tilde{c} & 0 \\ 0 & \tilde{u} - \tilde{c} \end{bmatrix} \tag{18.36}$$

$$\tilde{P} = \begin{bmatrix} 1 & 1 \\ \tilde{u} + \tilde{c} & \tilde{u} - \tilde{c} \end{bmatrix} \tag{18.37}$$

Then, the vector **U** at the next time step is defined as:

$$U_i^{n+1} = U_i^n - \frac{\Delta t}{\Delta x}(F_{i+1/2}^* - F_{i-1/2}^*) \tag{18.38}$$

where the flux is given by

$$F_{i+1/2}^* = \frac{1}{2}(F_{i+1} + F_i) - \frac{1}{2}\left[(\tilde{P}\left|\tilde{\Lambda}\right|\tilde{P}^{-1})\right]_{i+1/2}(U_{i+1} - U_i) \tag{18.39}$$

where using Eq. (18.30) and Eq. (18.31), and going through some steps,

$$\tilde{P}\left|\tilde{\Lambda}\right|\tilde{P}^{-1} = \begin{bmatrix} \dfrac{\tilde{c}^2 - \tilde{u}^2 + (\tilde{u} + \tilde{c})|\tilde{u} - \tilde{c}|}{2\tilde{c}} & \dfrac{\tilde{u} + \tilde{c} - |\tilde{u} - \tilde{c}|}{2\tilde{c}} \\ \dfrac{\tilde{c}^2 - \tilde{u}^2 + (\tilde{u} + \tilde{c} - |\tilde{u} - \tilde{c}|)}{2\tilde{c}} & \dfrac{(\tilde{u} + \tilde{c})^2 - (\tilde{u} - \tilde{c})|\tilde{u} - \tilde{c}|}{2\tilde{c}} \end{bmatrix} \tag{18.40}$$

$F_{i-1/2}^*$ refers to the previous value of the flux.

18.1.9 McCormack scheme

The McCormack scheme is an explicit finite difference scheme which is second order accurate in space and time. It is one of the shock-capturing techniques. Its main advantage is, therefore, its ability to simultaneously handle calculations of slowly varying flows as well as rapidly varying ones, containing shocks or discontinuities. The scheme is a two-step scheme whose main advantage is the non-centered character of the spatial differentiation (Garcia-Navarro and Saviron 1992). The scheme uses unilateral (forward and backward) differences to approximate the spatial derivatives instead of the central approximations. The numerical algorithm is based on a second order Runge-Kutta method which enables the time stepping of the solution of a hyperbolic equation in conservative form by means of a predictor-corrector iteration (Garcia-Navarro and Burguete 2006).

From the system of equations given in Eq. (18.23), the solution at a temporal level $n + 1$ is obtained in two steps (Garcia-Navarro and Burguete 2006). The predictor step provides a first approximation to the solution, and these values are used in the corrector step.

$$\begin{aligned} U_i^P &= U_i^n - \frac{\Delta t}{\Delta x}(F_{i+1}^n - F_i^n) \\ U_i^C &= U_i^n - \frac{\Delta t}{\Delta x}(F_i^P - F_{i-1}^P) \end{aligned} \tag{18.41}$$

Finally, the average of the two steps leads to the updated solution

$$U_i^{n+1} = \frac{1}{2}(U_i^P + U_i^C) \tag{18.42}$$

The numerical flux of this scheme is given by

$$F^{*}_{i+1/2} = \frac{1}{2}(F^{n}_{i+1} + F^{P}_{i}) \tag{18.43}$$

18.1.10 Small depth problem

When we come to form the finite difference expressions for the momentum equation, there are two difficulties. One is that without care in formulating the finite difference expression for the friction slope two solutions for the water depth can be obtained for a given discharge. The numerical solution can therefore tend to oscillate between two flow states; see Cunge *et. al.* (1980). This can be avoided by modifying the finite difference representation of the conveyance term in the friction slope. The average of the upstream and downstream values is replaced by a simple upstream weighting. In addition, a small base flow can be introduced throughout the pipe network using a synthetic slot in the bottom of the pipes as introduced earlier here (e.g., see MOUSE) or alternatively a small artificial flow can be injected at the upstream end of a pipe and removed at the downstream end (e.g., see InfoWorks CS). Such a concept avoids the issue about what to do with the wetting of a previously dry bed.

18.1.11 Treatment of sub- and super-critical flows

The remaining numerical problems with the solution of the Saint Venant equations for gradually varying flow in conduits concern the transition from sub- to super-critical flow and the instantiation of the initial condition through a pipe system. As shown above in Section 17.3, the end boundary conditions for sub-critical flow include an upstream boundary condition (preferably the discharge) and a downstream boundary condition (preferably the water level or a functional relationship between the downstream water level and the corresponding discharge). The functional relationship at the downstream boundary could be the critical depth condition in case of free outflow. For super-critical flow the situation is very different: there need to be two upstream boundary conditions. Inevitably, this means that if the numerical solution is faithful to the basic analytical equations then the numerical solution algorithm has to be changed and rigorous book keeping is needed to track the propagation of the critical point(s). This is complicated in the extreme, so most modelling software systems make the assumption that the algorithm for sub-critical flow does not change its structure. To achieve this, the non-linear acceleration of momentum terms in the conservation of momentum equation are reduced to zero as the Froude number approaches unity. The resulting non-inertial equations can be applied to super-critical flow conditions in a conduit.

18.1.12 Reduction of the convective momentum term

This approach is based on the method first suggested by Havnø *et al.* (1985) and elaborated by Kutija (1993) who showed that the Saint Venant equations can be written by excluding the gravity term due to the bottom slop and the friction term for the time being, and expressed in the following form:

$$\begin{aligned} &\frac{\partial h}{\partial t} + \frac{\partial (uh)}{\partial x} = 0 \\ &\frac{\partial (uh)}{\partial t} + \frac{\partial (u^2 h)}{\partial x} + gh\frac{\partial h}{\partial x} = 0 \end{aligned} \tag{18.44}$$

where uh is discharge per unit width and u is average flow velocity.

The slopes of the characteristic lines of Eq. (18.44) are given by

$$\frac{dx}{dt} = u \pm \sqrt{gh} \tag{18.45}$$

For supercritical flow conditions, where the magnitude of u exceeds $\sqrt{gh}$, the slopes of both characteristic lines have the same sign and therefore enter the domain from one side only. This allows the use of a single sweep algorithm which is different from a double sweep algorithm provided for subcritical flows.

To avoid this problem the author suggested that it is possible to split up the convective momentum term in the momentum equation and reduce only a part of it. Hence, expanding the convective momentum term further gives:

$$\frac{\partial(uh)}{\partial t} + hu\frac{\partial u}{\partial x} + u\frac{\partial(uh)}{\partial x} + gh\frac{\partial h}{\partial x} = 0 \tag{18.46}$$

If the second term is removed, previous equation takes the form:

$$\frac{\partial(uh)}{\partial t} + u\frac{\partial(uh)}{\partial x} + gh\frac{\partial h}{\partial x} = 0 \tag{18.47}$$

The slopes of the characteristic curves of the de Saint Venant equations are then given by:

$$\frac{dx}{dt} = \frac{u}{2} \pm \sqrt{\frac{u^2}{4} + gh} \tag{18.48}$$

From expression Eq. (18.48), for the slopes of the characteristic lines, it is apparent that these are not independent of the flow velocity. The slope of the C^- characteristic remains negative no matter what the relation between of u and $\sqrt{gh}$; but its magnitude decreases with the increase of the Froude number. When flow is supercritical the downstream boundary condition is, strictly speaking, physically unrealistic, and therefore completely artificial, but this property has the consequence that the influence of this departure from reality should not be spread very widely through the domain. By the very steepness of the C^- characteristic (with flow taken in the positive direction), its influence is limited to a very narrow region.

18.1.13 Generation of the initial condition

It is desirable to have a consistent initial condition for flows in a pipe network before simulating the response to driving forces such as rainfall or extraneous time dependent inflows. In theory, arbitrary water levels and discharges can be defined in each conduit at every computational point and the simulation can be allowed to run until a steady state is achieved. In practice this can take some time, and if the system contains pumps controlled by water levels in their wet-wells then there will probably be no steady state solution. Consequently, two particular actions may be taken. One is to increase the numerical diffusion by temporarily weighting the space derivative to the forward time level. This accelerates convergence to the (possible) steady state. The second action is to modify the pump discharges artificially, again temporarily, to preserve a steady state condition at the pumps.

18.1.14 Groundwater

The solution of the Richards equation, Eq. (17.61), is normally done using a fully forward-difference scheme with the second spatial derivatives evaluated at the forward time level and the time derivative expressed in terms of the difference between the groundwater potential, h, at the new and the old time levels. This is the approach adopted in MODFLOW, which is one of the most widely used physically-based modelling software package for groundwater problems.

18.1.15 Solving the pollutant transport equations

The solution of the advection-diffusion equation is complicated by the transport term. Generally, it is convenient to solve for the transport (advection) term using the Galerkin approach along the characteristic (see, e.g., Hughes *et. al.* (1989)) and to solve for the parabolic diffusion terms separately using the Crank-Nicolson method; see Crank and Nicolson (1947) and Crank and Nicolson (1996).

18.2 2D ABOVE-GROUND FLOW MODELLING

Besides the 1D modelling of flow in networks of conduits, significant modelling effort is also needed for overland flood flows originating from excess rainfall-runoff but prevented from entering the 1D network, discharges emerging from the surcharged 1D network, or over-bank spillage from urban streams or rivers. Such overland flows are typically shallow and can therefore be treated as being two dimensional. Necessarily, the flows are controlled by a complex topography formed by the streets and buildings of the urban area. Simulating flood flows on inundated flood plains is difficult enough, let alone the artificial features of curbs and cambers on roads, the flows round or through buildings, and the natural flow paths generated by the general ground terrain as if there were no buildings or structures at all. In addition, there may be a large number of discrete points or lines at which there are local head-losses. It would take an enormous modelling effort to include such head-losses explicitly in any 2D model, and the normal practice is not to include them directly, but to assume that they are incorporated implicitly in the surface roughnesses (such as represented in the Manning or Chezy forms) and variation over the domain.

The equations describing the flow are:

$$
\begin{aligned}
&\frac{\partial u}{\partial t} + u\frac{\partial u}{\partial x} + v\frac{\partial u}{\partial y} + g\left(\frac{\partial h}{\partial x} + f\,\frac{u\sqrt{u^2+v^2}}{4gh}\right) = 0 \\
&\frac{\partial v}{\partial t} + u\frac{\partial v}{\partial x} + v\frac{\partial v}{\partial y} + g\left(\frac{\partial h}{\partial y} + f\,\frac{v\sqrt{u^2+v^2}}{4gh}\right) = 0 \\
&\frac{\partial h}{\partial t} + \frac{\partial (uh)}{\partial x} + \frac{\partial (vh)}{\partial y} = q
\end{aligned}
\tag{18.49}
$$

These equations are typically applied to flows in coastal waters or wide lakes. Because the urban overland flows are typically shallow and can be super-critical, the standard finite difference solution can become extremely complex. Therefore the inertia terms (or the non-linear inertia terms) in the momentum equations are often neglected in such special circumstances.

18.2.1 Numerical solution of the 2D equations

As with the 1D equations, finite differences, elements and volumes can be used to solve the 2D equations numerically. Such a solution is needed increasingly to model flood flows in urban areas as well as coastal waters receiving overflow and outfall discharges from drainage networks. A straight forward finite difference approach is to use a rectangular grid as shown in Figure 18.8. The ground levels can be defined at the centres of the resulting cells. An alternative is to define the ground levels at the mid-points of the boundaries for each cell. This has an advantage that a natural channel crossing the cell from one side to the opposite side can be accommodated by the geometry. Further, if the ground topography is defined at a finer resolution than the computational grid then the *A*-*h* and *B*-*h* relationships have to be defined at the boundaries of each cell and similarly the *V*-*h* relationship for each cell.

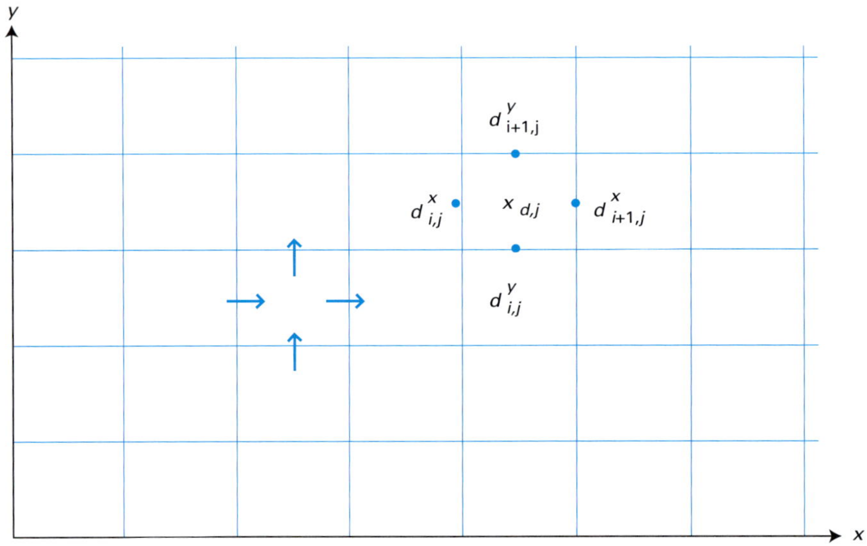

Figure 18.8 2D finite difference grid

It is straightforward to set up an explicit finite difference form of Eq (18.49) for a rectangular grid. So far as urban situations are concerned there is little value in retaining the non-linear inertial terms in the momentum equations, so Eq. (18.50) shows an implicit finite difference form of the reduced equations. The resulting finite difference equations are however, still non-linear, so the Newton-Raphson iterative method can be used to solve them at each time step. This results in a set of linear equations in the increments to the water levels in each cell. The resulting matrix can be inverted and the solution for the increments to the water levels can be found. Iteration can be carried out if the magnitude of the maximum increment is too large. The iterations are concluded if the magnitude of the maximum increment is acceptable. The calculations then proceed to the next time step.

A problem with this solution is that the number of arithmetical operations involved in inverting the matrix, and therefore the computational time, is at least proportional to N^3, where N is the smaller number of cells in the *x*- or *y*-directions. An alternative numerical solution can be derived using the alternating direction (ADI) method whereby the velocities in the *x*-direction are calculated at the half time step, and the velocities in the *y*-direction at the full time step; see Figure 18.9.

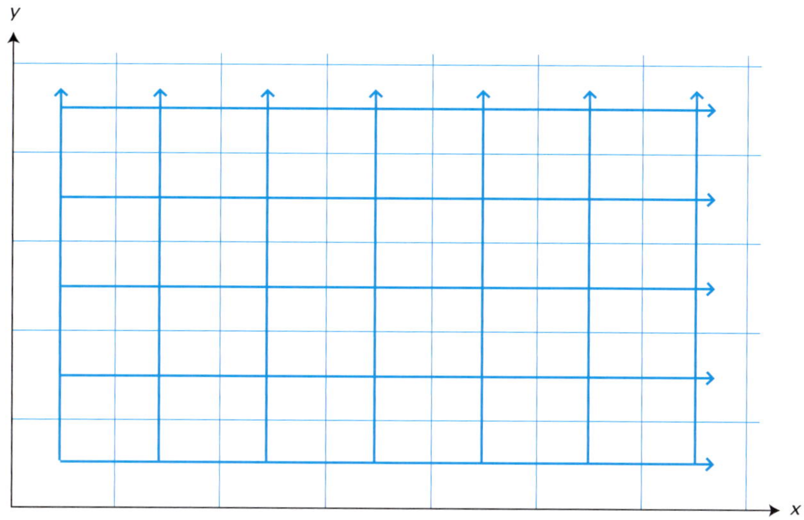

Figure 18.9 ADI approach

$$\frac{1}{\Delta t}\left(\frac{Q_{i+1,j}^{n+\frac{1}{2}}}{A_{i+1,j}^{n+\frac{1}{2}}} - \frac{Q_{i+1,j}^{n-\frac{1}{2}}}{A_{i+1,j}^{n-\frac{1}{2}}}\right) + g\theta\frac{h_{i+1,j}^{n+\frac{1}{2}} - h_{i,j}^{n+\frac{1}{2}}}{\Delta x} + g(1-\theta)\frac{h_{i+1,j}^{n-\frac{1}{2}} - h_{i,j}^{n-\frac{1}{2}}}{\Delta x} +$$

$$\theta\frac{1}{K_{i+1,j}^{n+\frac{1}{2}\,2}}\frac{Q_{i+1,j}^{n+\frac{1}{2}}}{A_{i+1,j}^{n+\frac{1}{2}}}\sqrt{\left(\frac{Q_{i+1,j}^{n+\frac{1}{2}}}{A_{i+1,j}^{n+\frac{1}{2}}}\right)^2 + \bar{v}_{i+1,j}^{n\;2}} + (1-\theta)\frac{1}{K_{i+1,j}^{n-\frac{1}{2}\,2}}\frac{Q_{i+1,j}^{n-\frac{1}{2}}}{A_{i+1,j}^{n-\frac{1}{2}}}\sqrt{\left(\frac{Q_{i+1,j}^{n-\frac{1}{2}}}{A_{i+1,j}^{n-\frac{1}{2}}}\right)^2 + \bar{v}_{i+1,j}^{n\;2}} = 0$$

$$\frac{1}{\Delta t}\left(\frac{R_{i,j+1}^{n+1}}{A_{i,j+1}^{n+1}} - \frac{R_{i,j+1}^{n}}{A_{i,j+1}^{n}}\right) + g\theta\frac{h_{i,j+1}^{n+1} - h_{i,j}^{n+1}}{\Delta x} + g(1-\theta)\frac{h_{i,j+1}^{n} - h_{i,j}^{n}}{\Delta x} +$$

$$\theta\frac{1}{K_{i,j+1}^{n+1\;2}}\frac{R_{i,j+1}^{n+1}}{A_{i,j+1}^{n+1}}\sqrt{\left(\frac{R_{i,j+1}^{n+1}}{A_{i,j+1}^{n+1}}\right)^2 + \bar{u}_{i+1,j}^{n+\frac{1}{2}\;2}} + (1-\theta)\frac{1}{K_{i,j+1}^{n\;2}}\frac{R_{i,j+1}^{n}}{A_{i,j+1}^{n}}\sqrt{\left(\frac{R_{i,j+1}^{n}}{A_{i,j+1}^{n}}\right)^2 + \bar{u}_{i+1,j}^{n+\frac{1}{2}\;2}} = 0 \qquad (18.50)$$

$$\frac{1}{\Delta t}\left(V_{i,j}^{n+\frac{1}{2}} - V_{i,j}^{n-\frac{1}{2}}\right) + \theta\left(Q_{i+1,j}^{n+\frac{1}{2}} - Q_{i,j}^{n+\frac{1}{2}}\right) + (1-\theta)\left(Q_{i+1,j}^{n-\frac{1}{2}} - Q_{i,j}^{n-\frac{1}{2}}\right) + \left(R_{i,j+1}^{n} - R_{i,j}^{n}\right) = q$$

$$\frac{1}{\Delta t}\left(V_{i,j}^{n+1} - V_{i,j}^{n}\right) + \left(Q_{i+1,j}^{n+\frac{1}{2}} - Q_{i,j}^{n+\frac{1}{2}}\right) + \theta\left(R_{i,j+1}^{n+1} - R_{i,j}^{n+1}\right) + (1-\theta)\left(R_{i,j+1}^{n} - R_{i,j}^{n}\right) = q$$

Here

Q represents discharge across the cell boundary in the x-direction
R represents discharge across the cell boundary in the y-direction
V represents the volume.

The numerical solution then resolves into solving for the water levels in the cells for a series of 1D channels, first in the x-direction and then in the y-direction. The solution procedure for a single channel is outlined above. The non-linear difference equations can be solved using a Newton-Raphson methodology, but solving for the whole domain before testing the maximum increment in water level in order to decide whether or not to repeat the iteration in order to refine the estimates of the solution for the water levels. Just as with the 1D model, the 2D model can adjust its time steps, halving them, for example, if the increments in water level fail to converge within a required number of iterations, or doubling the time step if the convergence has been satisfied for a further number of time steps; see Seyoum *et al.* (2010).

The ADI approach has the number of arithmetic operations of the order of N^2, and is therefore considerably faster than the first approach. For this reason, almost all finite difference methods for solving 2D flow use either a straightforward explicit or implicit ADI approach. This is of course at odds with simulations of flows in extensive and deeper water bodies such as seas, estuaries, harbours and similar extensive regions, where implicit schemes that are unconditionally stable for two-dimensional and, if necessary, three-dimensional models are de rigueur. These depend upon what is called a *side-feeding* of convected terms that are orthogonal to the directions of the sweeps in the algorithm using data from the side that has just been established. This gives rise to four-level schemes, while in three dimensions it leads to eight level schemes in order to guarantee stability and satisfaction of all conservation laws.

18.2.2 Integrating 1D and 2D models

The integration of the 1D pipe network model with the 2D overland flow model needs to be done with care. Ideally, they should be integrated implicitly so that there is a consistency in the water level at the nodes of the network that are flooded, that is, where and when water emerges from the network nodes above ground level and can generate overland flow. A similar link can be constructed between flow spilling over the banks of an urban stream and flowing through the urban area.

One method for linking the two models is to assume that flow emerging from a manhole spills out radially into an above-ground 2D cell via a circular orifice, with the flow determined by the difference in water level between the manhole and the cell. The discharge emerging from the manhole during the time step is introduced as a source to the 2D model: see Figure 18.10. The 2D model then computes the resulting overland flow for a number of time steps for a duration equal to the 1D time step (normally the 2D model time step will be smaller than the 1D time step). The calculation then returns to the 1D model using the latest water levels at the manhole and in the cell containing the manhole. See Figure 18.11 for an illustration of the effectiveness of this integrated approach.

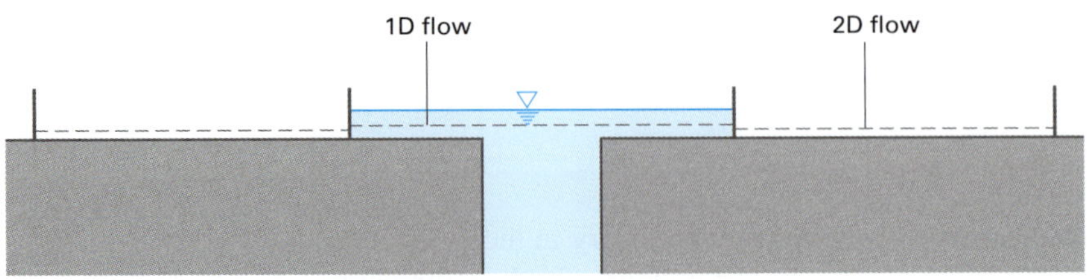

Figure 18.10 Interaction of 1D and 2D flow at a manhole

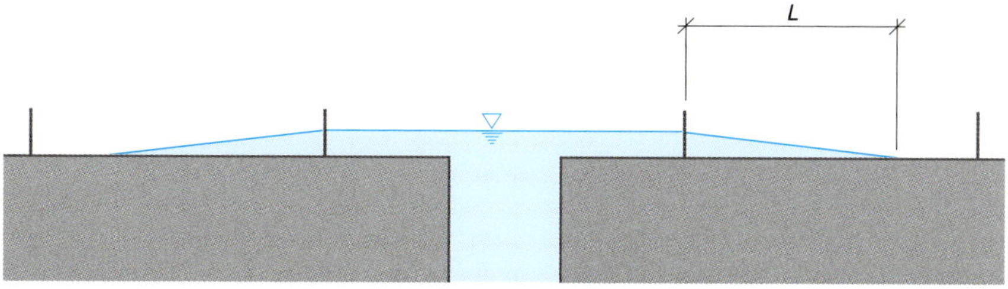

Figure 18.11 2D flow of water from a manhole on a flat surface

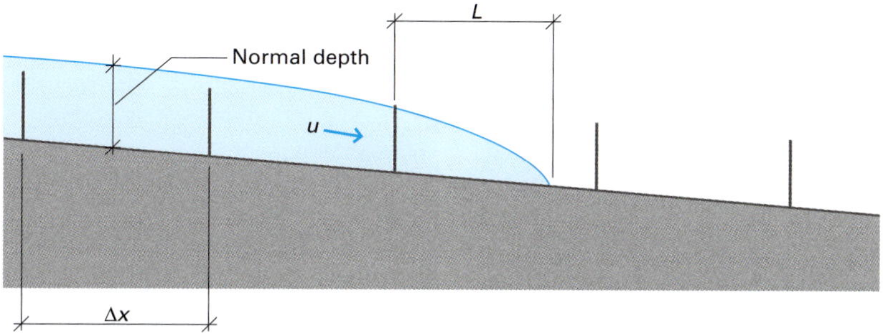

Figure 18.12 Propagation of a front over a dry bed on a sloping surface

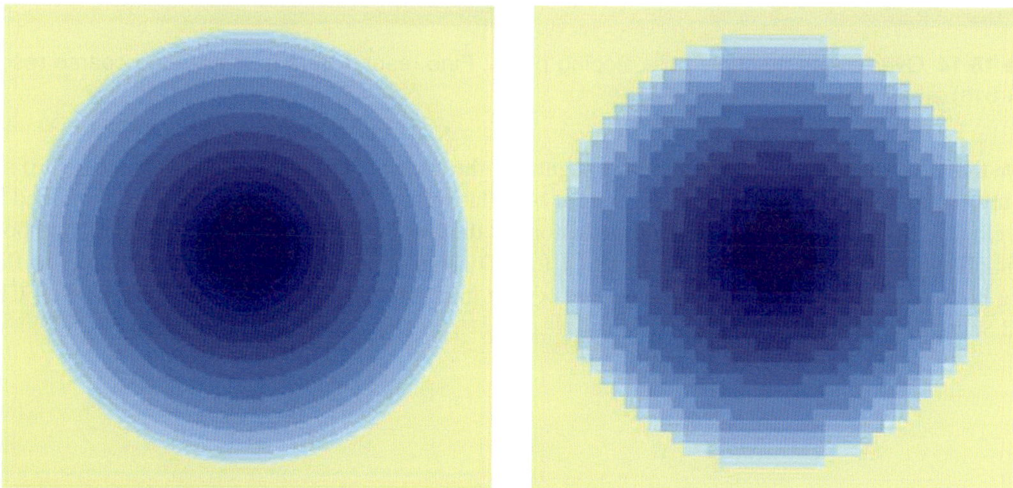

Figure 18.13 Overflowing manhole on a flat plain: Fine resolution (1 m x 1 m) and coarse resolution (5 m x 5 m)

18.2.3 Wetting and drying

The 2D model has to be able to cater for the flow emerging from a manhole and flowing over a dry surface. The condition at the front is important. Here it is assumed that the front propagates across a cell with a velocity at any instant equal to the velocity u of the inflow to the cell; See Figure 18.12. The cell is assumed to be wet (and the front can move to the next dry cell) when

$$L = \sum_{t} u\Delta t < \Delta x \tag{18.51}$$

where

L is the cumulative distance (m) that the front travels in the cell since first arriving

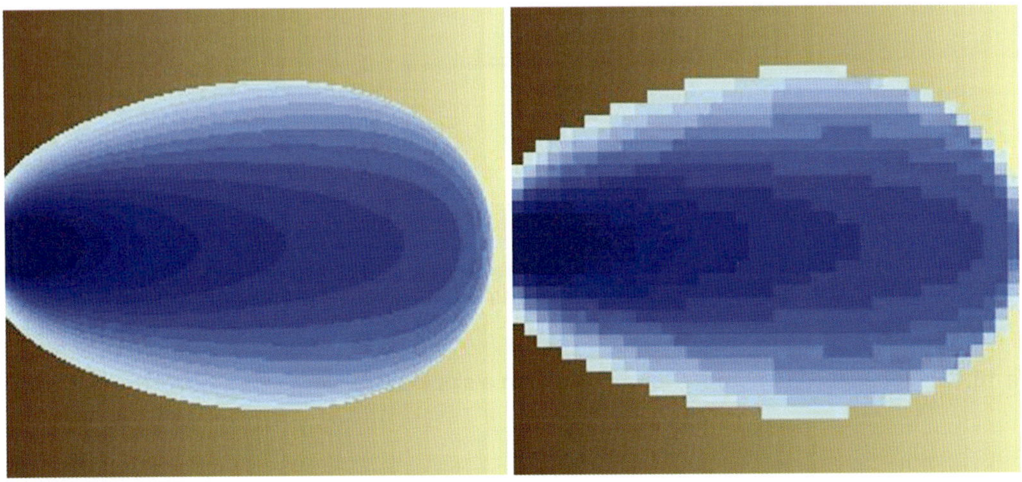

Figure 18.14 Overflowing manhole on a sloping plane: Fine resolution (1 m x 1 m) and coarse resolution (5 m × 5 m)

Note that on a sloping surface the front propagates at the speed associated with the (approximate) normal depth upstream. Such a normal depth cannot exist on a flat surface, but the propagation is determined by the local velocity, which is a function of the time history of the forcing flow from the manhole(s), and possibly the rainfall runoff. See Figures 18.13 and 18.14 for 2D calculations for flow from a manhole on flat and sloping beds with a 1 m *1 m and a coarse 5 m × 5 m grid. For further details see Section 19.9, Chapter 19 and Seyoum *et al.* (2009).

Chapter 19

Modelling practice

19.1 MODEL INSTANTIATION

Typically, the process of instantiating a physically-based model involves the following tasks:

(1) Decide on the purpose of the model;
(2) Review existing data and collect other necessary data;
(3) Build the model infrastructure and select initial values of the parameters;
(4) Carry out consistency tests of the model;
(5) Calibrate parameters and confirm the model;
(6) Produce the base performance against which to test improvements to the physical system;
(7) Document modelling conditions;

A fuller description of the model instantiation process is given in Price and Vojinović (2011).

The models are constructed by first identifying the physical geometric space in the physical world that contains the flowing water, and which the model is going to replicate. This physical space becomes the model space. So the next step is to deconstruct the physical space into (idealised) geometric objects that comprise the corresponding objects in the model space. Similarly, the flow of water is governed by particular generic processes. A *hydroinformatics model* in Abbott's sense is one that consists of the integration of a collection of indicative signs that point to the geometric objects and physical processes as the expressive signs that are the graphical output objects. Note that all of these are designed by the developer of the software (where the developer can of course be a team of people) who works from a unified point of view: see Figure 19.1. The developer deconstructs the physical world from an own perspective, which certainly is not necessarily that of the potential user. Consequently there are risks associated with the use of the software by person other than the developer. Nevertheless these risks can be mitigated by the user's understanding of the limitations and constraints of the software and so not using it outside the limits for which it was designed. It should be borne in mind that much is left up to the user concerning how to structure his/her model, what data to select, how to calibrate the model, how to interpret the results, and so on, such that two modellers working independently are more than likely to arrive at different (yet similar) models for the same situation. Generally, the user has behind him yet a third person or group of persons who is/are even more remote from the modelling process, but who are ultimately concerned with what the model produces in terms of data that will assist in the decision making process. All of these different types of people (and there are many kinds of such people) obviously play essential roles in the sociotechnical production of hydroinformatics.

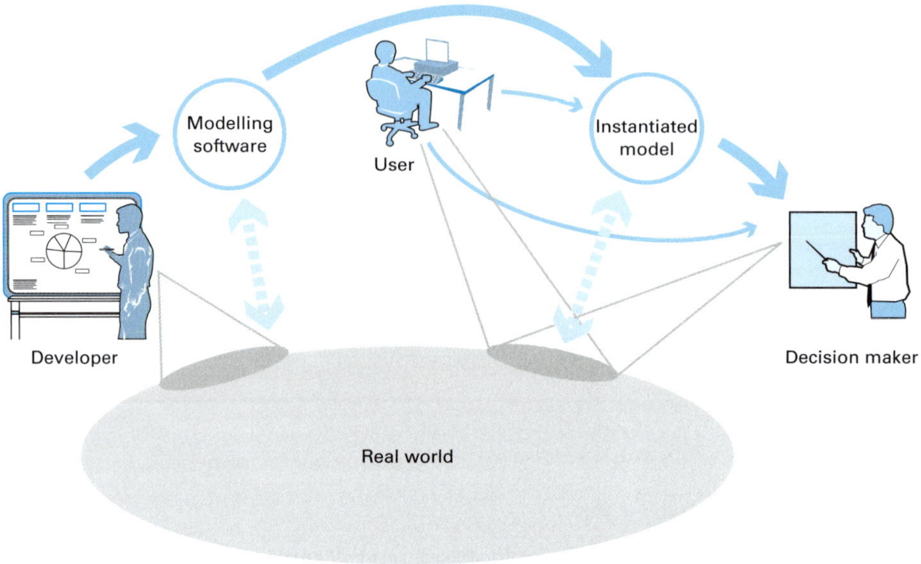

Figure 19.1 Modelling relationships (After: Price and Vojinović 2011)

19.2 MODELLING FLOW IN NETWORKS OF CHANNELS AND/OR PIPES

The modelling of flow in networks of pipes and channels has been described earlier here. It remains important to recall that generally speaking it is the full 1D Saint Venant equations that are solved, rather than using the non-inertial or kinematic wave equation approximations. Note that the latter approximation assumes no downstream boundary condition and therefore cannot take backwater effects into account. Such effects are very important in most sewerage network simulations. Care has to be taken with the dry bed condition, the hysteresis in the flow at the top of a pipe where there is a transition between the free surface and pressurised flow, the corresponding inclusions of the 'imaginary' Preissmann and Abbott slots to preserve the same numerical algorithm for pressurised flow and cessation of flow as for free surface flow, the specification of discrete head-losses at the upstream and downstream ends of each pipe, and the preservation of the sub-critical flow algorithm for super-critical flow. In addition, a range of ancillary structures should be anticipated as possibly being needed for incorporation in the model, such as various weirs (e.g., combined sewer overflows), gates, sluices, different pumps, outfall flaps and orifices. Allowance should be made for the emergence of sewage from a manhole. (All surface and wastewater inflows and outflows are assumed to be at the upstream end of a pipe rather than being distributed irregularly along its length). The emerging effluent from an underground pipe network or an above ground channel network can interact with an overland 2D flow model.

Traditionally, the collection systems have been modelled using a 1D approach which was limited to the modelling of flows within channels and/or pipes. A step forward has been the dual drainage concept, where urban surface is treated as a network of open channels and ponds (major system) connected to the channel/pipe system (minor system), commonly referred to as 1D/1D approach. More recently, coupled 1D/2D models have emerged, in which channel/pipe network flow models are tightly coupled with the flood flow model that treats the urban surface as a two-dimensional flow domain. In both approaches, complex interactions take place through surface/sub-surface links, which is also explicitly taken into account using adequate equations. The great majority of processes occurring within the collection systems can be adequately represented by the use of 1D, 1D/1D or 1D/2D modelling approaches. However, care must be taken particularly in the case of overflow structures where 3 dimensional effects of flow can significantly reduce or increase flows over weirs compared to values predicted by standard equations. In addition, the use of low side weirs, where supercritical flows and hydraulic jumps can take place, is rarely capable of being explicitly represented in standard 1D models. The ever-increasing power of computers is now allowing a 3D fluid dynamic analysis of such structures, and in time these may be integrated into 1D models where such locations require this kind of analysis. These complete models of the entire three-dimensional domain were long ago integrated into the *Système Hydrologique Européen//European Hydrologic System*, known as the *SHE*, and then at both the Danish Hydraulic Institute (DHI) and the University of Newcastle, and subsequently brought up to a fourth-generation standard as the *MIKE SHE* at DHI (Abbott *et al.* 1986).

Water quality is simulated using the advection-diffusion equation for the different concentrations of the primary pollutants with particular source terms identified for the pollutant and associated processes. The advection speed is the mean velocity calculated from the hydraulic model. Note that the full water quality model is dependent on having a safe and reliable hydraulic model.

The modelling of the interaction of infiltration to a pipe network from groundwater and infiltration from the surface rainfall and flood flows to groundwater can be done using a two layer groundwater flow model, with the upper layer representing the unsaturated zone and the lower layer the saturated zone.

19.3 1D MODELLING APPROACH

As discussed earlier, the hydraulic models used for the modelling of collection systems have a sound physical basis of shallow-water equations and their reliability increases with the amount of measurement data that can be used for their instantiation and calibration (or validation). For these and some other reasons, one or another form of hydraulic modelling, such as a one-dimensional (1D) approach has been used as a standard industry practice for quite some time: see Figure 19.2. Consequently there are still many researchers and practitioners who favour the use of a 1D modelling approach over other approaches. Some of the commonly perceived reasons are that these models are relatively easy to set up, calibrate and explain.

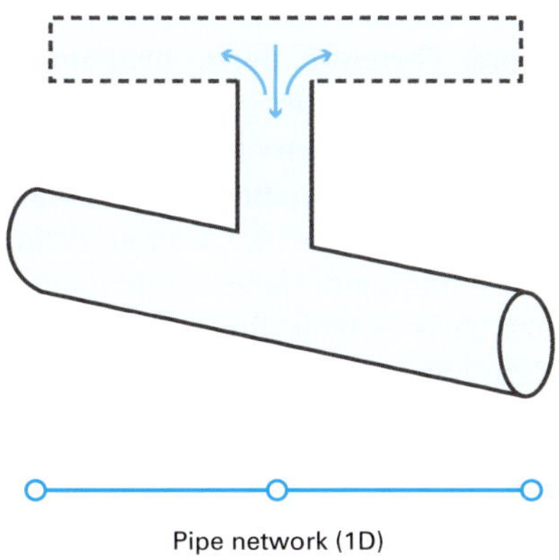

Figure 19.2 Longitudinal profile of the pipe network system and above-ground storage of the 1D modelling framework. When the water level exceeds the ground level an artificial inundation basin is created above the manhole. There is no spill above the ground level (i.e., street) and therefore no above-ground flow

A horizontal plan of the 1D model structure may be represented as exemplified in Figure 19.3.

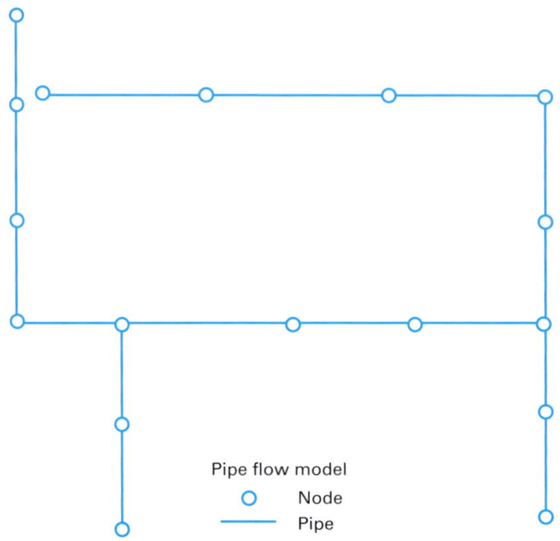

Figure 19.3 A horizontal plan, or model layout, of a 1D model

19.4 SIMPLIFICATION OF 1D MODELS

The 1D pipe flow (or open channel) models are either developed on a skeletal (or simplified) level or on a more detailed level depending on the number of pipes and nodes (or branches and cross-sections). The choice for the appropriate extent of the model (and the level of its complexity) is normally decided by the modelling objectives, the availability of data and the financial capacity of the modelling project.

The purpose for the development of a skeletal (or simplified, strategic planning) model is to gain the basic understanding of functioning of the system and to undertake a simple planning/management analysis. This can be also regarded as a cost-effective way of approaching the modelling task. The detailed model is usually developed in order to understand the detailed relationship between problems in the system and the hydraulics of the drainage infrastructure. Therefore, due to the complexity and capacity required in order to describe various characteristics of an urban drainage system and its functioning, detailed models are also used in operational and system maintenance activities.

Very often, the collection system modelling studies would commission with the development of a skeletal (or simplified) model and then continued with the more detailed model refinement in the critical (or problem) areas only. Typically, as a minimum, pipes (or branches of the open channel network) of a simplified network are selected on the basis of their dimension (i.e., trunk lines), proximity to the known problem area and location of ancillary structures (e.g., overflows, pumping stations, etc.). This process should take into consideration also the location of flow/level gauges.

The simplification of a model requires that the head-losses and storages from manholes/pipes/channels which are outside of the simplified model network are also incorporated. This is usually done in a way that the additional storage (or the storage due to the omitted network) is added to the modelled manhole downstream, and additional head-losses are added to the modelled manhole upstream. In doing this it is important to take into consideration also other local circumstances such as low terrain elevations and significant slope changes.

The process of computing the additional storage can be done in a GIS package by cross-referencing the ground level of the highest upstream node in the simplified network and invert levels of all of the other nodes that are above from that node (i.e., outside from the simplified network). Their values need to be subtracted in order to compute the additional storage volume; see Figure 19.4. Thereafter, for every pipe and node upstream the additional storage volume needs to be calculated.

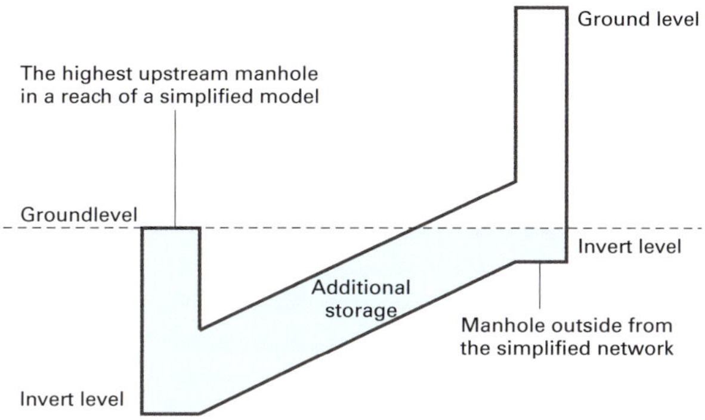

Figure 19.4 Schematisation of the additional storage area/volume

Furthermore, there could also be intermediate nodes which could also be omitted from the simplified model network. Their storage volumes must also be summed and added to the next model node downstream. For converging junctions, the storage should be summed along each incoming branch and added at the junction node.

In terms of the additional head-loss calculations, the angle of each link with respect to the direction of the flow needs to be calculated. Indeed, if there is a straight-line continuation between the links, then their angles can be ignored and only the angle of the conduit relative to North at the downstream node should be calculated. Also, head-losses due to directional and elevation changes (e.g., backdrop manhole), contraction/expansion and converging/diverging junctions need to be taken into account.

The head-loss through a node is a function of the angle at which the pipe section bends through and can be defined by the following equation:

$$K = 1 - \left(\frac{90^2 - \theta^2}{90^2}\right) \tag{19.1}$$

where θ represents an angle of the pipe bend.

Typically, the head-loss value for a manhole where the incoming flow enters and exits in a straight line can be given as: k = 0.1,

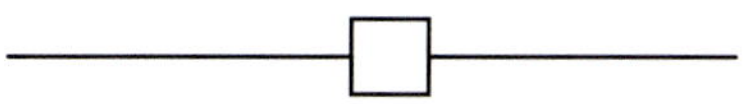

and for or the case of 90 bend: k = 1,

Head-loss at backdrop manholes

A backdrop manhole can be defined as one in which the incoming sewer invert is above the soffit of the outgoing pipe. For these manholes, a K value of 1.2 is usually applied.

Head-loss at converging junctions

At converging junctions the head-loss can be calculated as a weighted average of the angular head-losses of all the incoming pipes. This weighting can be based on the square of the pipe diameter. This is done because it is difficult to know the actual flow through the pipes at the junction: thus it is assumed that, the bigger the diameter, the bigger the flow will be.

$$K = \frac{\sum_{i=1}^{n} K_i \times H_i^2}{\sum_{i=1}^{n} H_i^2} \tag{19.2}$$

where $K_i =$ Head-loss for a particular incoming pipe.

Head-loss at diverging junctions

For diverging junctions or bifurcations the head-loss is taken to be the average of the head-losses at that point.

Head-loss at outfalls and head manholes

For head manholes and outfalls a K value of 0.1 is usually applied.

Example of a simplified network layout against the background of the total pipe network is given in Figure 19.5.

Figure 19.5 An example of a simplified model layout against the total drainage network

19.5 1D/1D MODELLING APPROACH

The *dual drainage* concept, where the urban surface is treated as a network of open channels and ponds (major system) connected to the pipe (or channel) network system (minor system), is commonly referred to as the 1D/1D approach; see Figure 19.6. This is because both systems, below-ground (minor) and above-ground (major), are modelled with sets of 1D equations for each of the two systems.

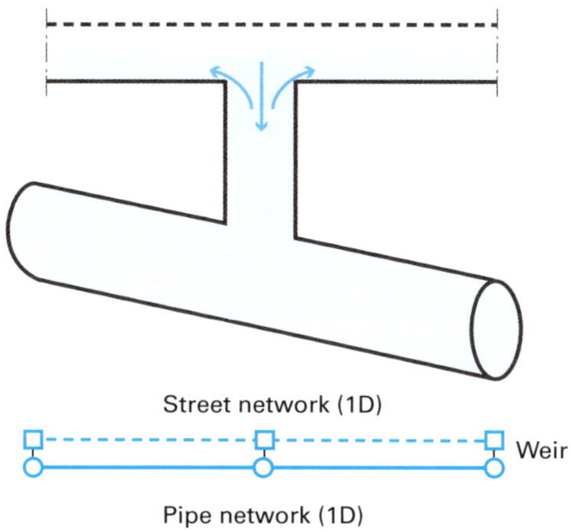

Figure 19.6 Longitudinal profile of the dual drainage model (i.e., 1D/1D modelling framework). It is important to note that there is a weir at the ground level which allows water to spill to the ground level (i.e., street) and create the above ground (i.e., street) flow

A cross-sectional view of the 1D/1D model configuration may be represented as in the example given in Figures 19.7 and 19.8.

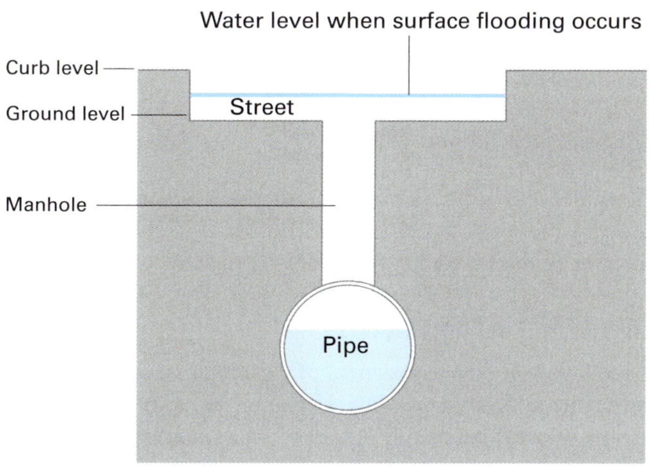

Figure 19.7 A cross-sectional view of the 1D/1D model set up

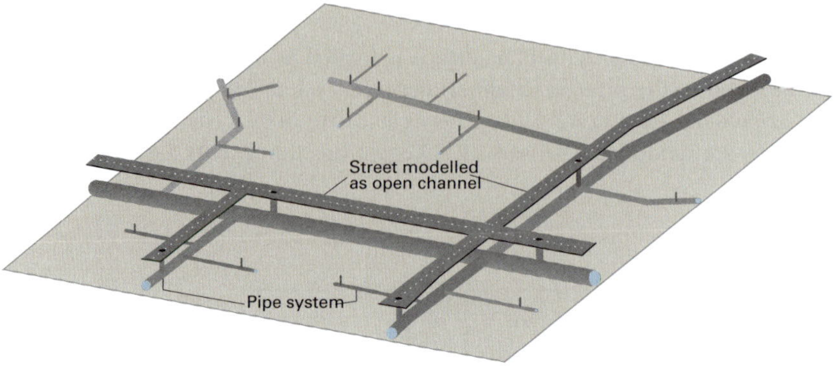

Figure 19.8 1D/1D model set up. Note that pipes and streets do not have to necessarily have the same layout. Furthermore, there could be also other forms of surface flow paths and not only streets (e.g., passages between buildings, etc.).

A horizontal plan of the 1D/1D model set up may be represented as an example given in Figure 19.9. Surface channels and pipe system may sometimes overlap (as given in Figure 19.9) but this is not always the case. The layout and cross-sections of the surface flow paths can be generated using GIS-based analytical tools (Maksimovic *et al.* 2009).

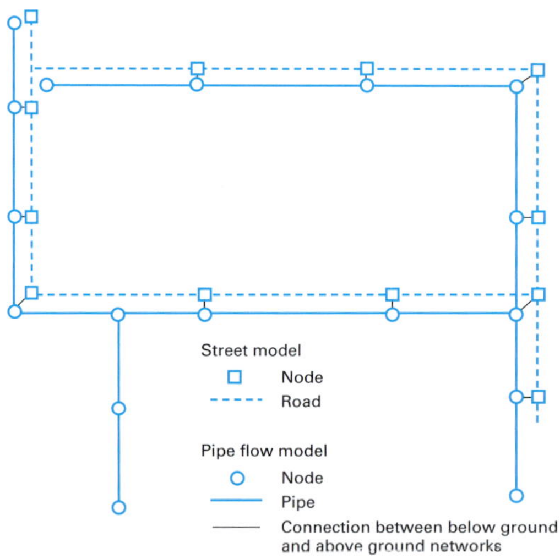

Figure 19.9 A horizontal plan, or model layout, of the 1D/1D model

19.6 1D/2D MODELLING APPROACH

The appropriate level of modelling for the assessment of particular issues depends crucially on the nature of the physical situation and on the availability of data. Urban topography, drainage network layouts and even detailed geometry of the networks can be readily surveyed using current technology. Where flood flows are confined to well-defined conduits, a robust 1D model can usually be instantiated and, once adequately calibrated and verified, its results can be considered sufficiently safe for decision-making. However, the flows generated along urban floodplains are normally highly complex because the morphology of the urban surface is eminently artificial with correspondingly highly irregular geometries, and the flows may run contrary to natural flow paths. Modelling of flows in such complex geometrical situations is difficult. Small geometric 'discontinuities' such as road or pavement curbs can play a significant role in diverting the shallow flows that are generated along roads, through fences and around buildings. Head-losses due to flows over or around such structures are particularly difficult to represent and accommodate. Such issues necessitate the coupling of simulations using 1D and 2D modelling systems; see, for example, Hsu *et al.* (2000), Mark *et al.* (2004), Djordjevic *et al.* (2005), Chen *et al.* (2005), Vojinović *et al.* (2006) and Vojinović and Tutulic (2009), Neal *et al.* (2009); Verwey *et al.* (2008); Mignot *et al.* (2006).

The simulation process in the case of coupled 1D/2D modelling is rather complex and it can become computationally very expensive (i.e., time consuming). It is based on complex numerical solution schemes for the computation of water levels, discharges and velocities. The surface model (i.e., 2D model) simulates vertically-integrated two-dimensional unsteady flow given the relevant boundary and other ancillary conditions (e.g., resistance coefficients, etc.) and bathymetry/terrain (as provided by a digital terrain model of the catchment area): see Figure 19.10. In this respect, the use of parallel computing seems to be one of the promising ways of reducing the computational times. Furthermore, several other approaches have proved to be beneficial in this respect. For example, an approach based on the adjusted conveyance and storage characteristics to account for fine-scale flow obstructions (i.e., change in storage and lateral fluxes) is an approach which is showing promising results (this approach and some other approaches will be discussed in Section 19.8).

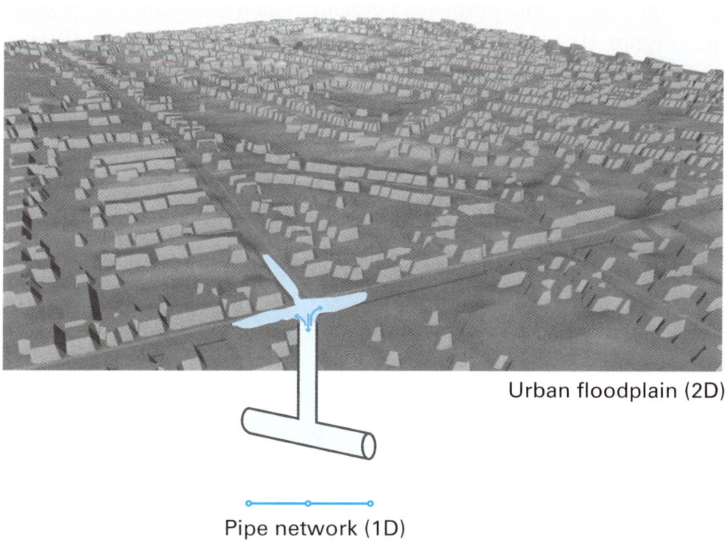

Figure 19.10 Illustration of the 1D/2D modelling framework

The interactions between the below-ground (or minor) and above-ground (or major) system are determined according to the type of link. For example, discharges generated by pumping stations, weirs or orifices are regarded as the lateral inflows to the 2D model. Also, if the pipe flow exceeds the ground level or bank levels (for open channels) then the discharge is computed by the weir (or orifice) discharge equation and it is considered as a lateral inflow to the 2D model (i.e., the source term F_s).

For setting up 2D models, regular grid representations are often the preferred choice of data format due to their ease in manipulation and representation within GIS packages. There are a large variety of methods available for generalising regular grids and one of the relatively easy ways that can be used for generalisation (besides re-sampling) is a low-pass filter (Burrough & McDonnel 1998). This is a global filtering technique which does not distinguish between geomorphologic characteristic and uncharacteristic features.

19.7 ISSUES CONCERNED WITH SETTING UP THE 2D OVERLAND FLOW MODEL WITH REGULAR GRIDS

In general terms, 2D flow simulation can be done on a structured or unstructured mesh – regular or irregular grid. The former approach may appear more convenient as it can directly use LiDAR DTM data, thus has a simpler (matrix) data structure and is consequently easier to code. The latter, however, captures surface features much more efficiently, using TIN which adapts computational mesh to match boundaries of buildings, curbs, etc. The following discussion concerns setting up of 2D models with regular grids.

Floodwaters flowing through urban areas follow different kinds of paths as the water moves through and around buildings, fences and other obstructions. Such obstructions tend to dissipate energy by forcing the water to change its direction and velocity, and by forming eddies behind them. Therefore, the rising and lowering of water levels together with the estimation of flow velocities, flow directions, flood durations and inundation extents are important aspects and need accurate representations. For such representations and consequent interpretations, sufficiently detailed and accurate representation of the topography plays a key role. Digital Terrain Models (DTMs) or Digital Surface Models (DSMs) created from ground surveys or LiDAR data are often utilised for this purpose. Typically, such data can be processed to represent either the land surface alone (Figure 19.11 (a)), land surface with the road network (Figure 19.11 (b)) or land surface together with the road network and buildings (Figure 19.11 (c)). Furthermore, in case of buildings with underground car parks, such objects can be modelled as storages, Figure 19.12.

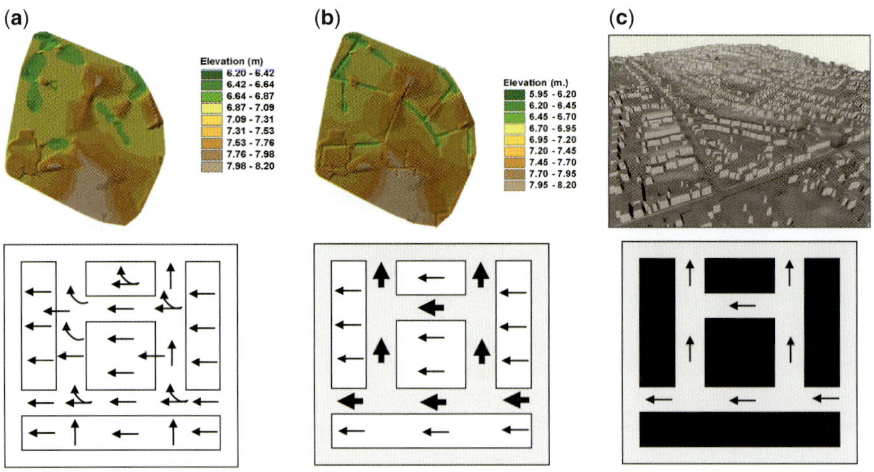

Figure 19.11 Schematisation of flow vectors in urban areas: (a) DTM only – buildings and roads are represented as hollow objects; (b) DTM with road network (this can be achieved by lowering the road network elevation relative to the land surface); (c) DTM with buildings and roads represented as solid objects (this can be achieved by lowering the road network elevation relative to the land surface and raising the building elevations above the land surface)

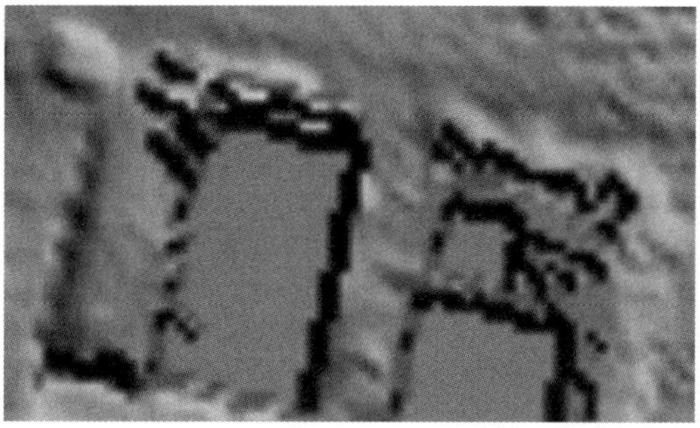

Figure 19.12 Example of an underground car park buildings represented as a storage within a DTM

It is fair to say that a more detailed 2D model (or a 2D model with higher resolution) will enable better description of physical processes over the urban floodplain. However, computing resources still restrict the applications when very detailed information is demanded over a very large scale area. Increasing the grid size can effectively improve efficiency, but also can necessitate reductions in details. For instance, elevations of buildings represented in fine-grid models would be smeared or completely removed when the grid is coarsened. For this reason, the straightforward 2D modelling technique may not reflect properly the local flow phenomena in coarse-grid applications (in which case one of the approaches described in Section 19.8 may be found suitable, see the following section). Illustration of impacts of a gradual increase of 2D model grid size on buildings and roads is given in Figure 19.13.

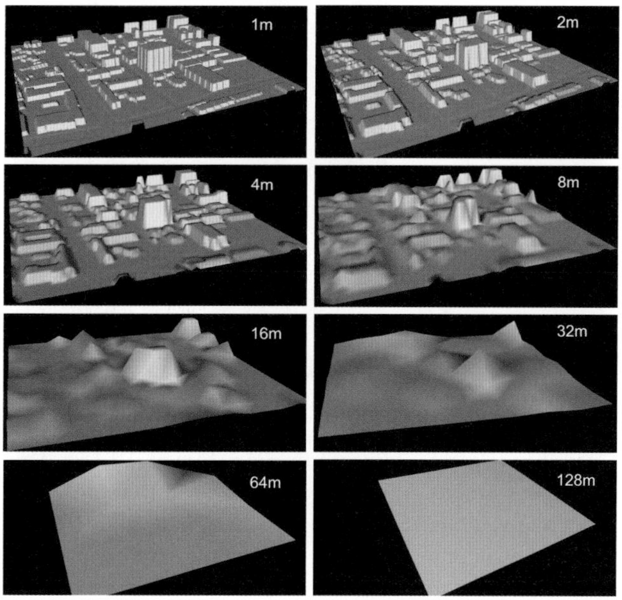

Figure 19.13 Effects of a grid resolution on capturing buildings and road features (catchment area size is 7 ha). The grid size is displayed in a top-right corner. The road features have completely disappeared at 8 m grid size and most of the buildings have disappeared at a 16 m grid size. Approximately 95% of the all floodplain features have disappeared at a 32 m grid size. All the floodplain features have disappeared at a 64 m grid size

In general terms, the work to date has shown that 2D models which are derived from a coarser DTM resolution tend to give rise to a wider spread of flood water with shallower depths when compared to those models which are derived from finer DTM resolutions, where the flood water tends to get trapped by local terrain depressions and thus gives rise to higher depths (Vojinović *et al.* 2011): see Figure 19.14.

Figure 19.14 Impacts of different DTM resolutions on model results (5 m, 10 m and 15 DTM resolutions) (See Vojinović *et al.* 2011)

Further to the issues raised above, several approaches have been devised to represent the influence of buildings on flood flow behaviour, including:

(1) *Increased model roughness for building footprints* (Figure 19.15); This method might be more suitable for coarse meshes where groups of cells are used to represent the bulk effect on flow of building groups (as opposed to individual buildings). It allows for storage effects and also allows for variation of the roughness parameter to apply calibration to field data (Syme 2008). In terms of practical applications, this approach can be found troublesome as it is difficult to come up with appropriate roughness values, and since the friction factor holds a strong relationship to velocity, calibration to a single event can be unreliable (Alcudro 2004). Furthermore, the bed friction used may be non-stationary with mesh resolution which results in a fixed calibration (Fewtrell *et al.* 2008; Yu & Lane 2006).

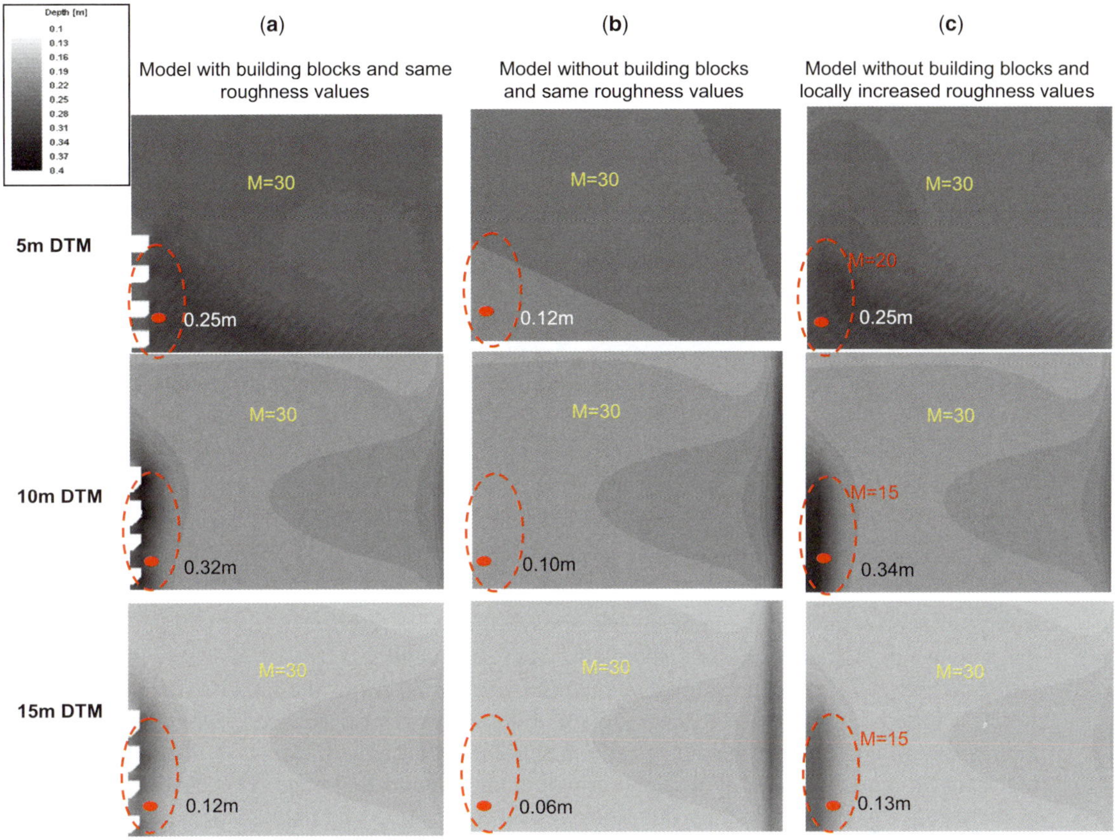

Figure 19.15 2D model results (depths) for different DTM resolutions representing buildings as solid blocks (a), not representing buildings at all (b) and buildings represented with locally increased roughness values. The roughness values are the same throughout the 2D model domain for cases (a) and (b), that is, Manning's value of 30 was used, whereas in case (c) the roughness values of cells within the building footprints have been increased (i.e., Manning's value of 20 was used); The flow was introduced in the direction from East to West.

(2) *Blocking out model elements for building footprints by raising the cell elevations to a height that is above the flood level or lowering the cell elevations for those buildings that may act as storage areas* (Figure 19.16); This approach can produce a flow pattern that is visually consistent with the conditions on the ground. It is important to elevate the bed of the building footprint cells to a certain height to ensure that the building is correctly represented (i.e., that the water level will not overtop the building) (see also, Brown *et al.* 2007). It should be noted that using abrupt bottom elevations changes violates the shallow water equation assumption that the bed slope is sufficiently small to ensure that the sine and tangent of the slope are close in value to an angle itself. However, as the flow is propagating around the buildings, this assumption may not necessarily be violated (see Alcrudo 2004). It should be also mentioned that the representation of buildings through increased bed elevations can destabilise the numerical solutions unless limiting conditions are introduced. With respect to the limitation necessary to account for

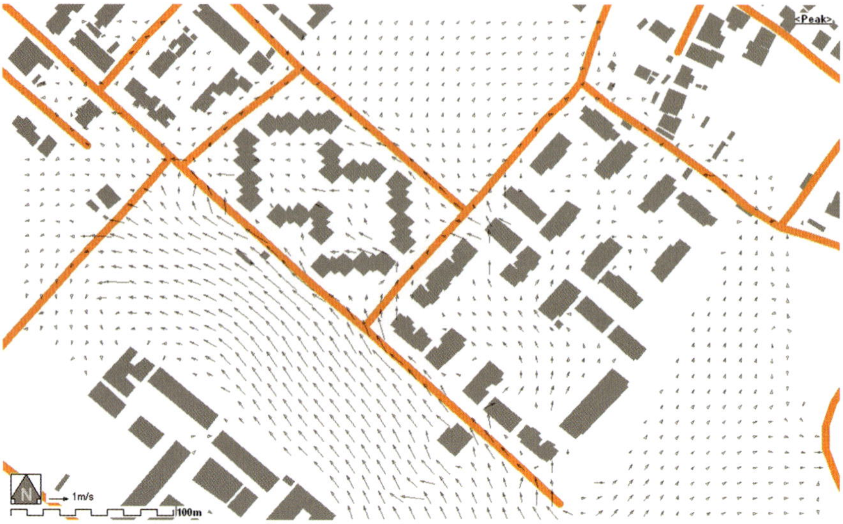

Figure 19.16 Example of 2D modelling of flows around buildings

storage effects on the flood volume in case of modelling underground car parks and other buildings that may act as storage areas, see Abdullah *et al.* (2011a and 2011b).

(3) *Using an energy-loss coefficient over the building footprints*; As an alternative to the first approach, which is based on an increased Manning's n value (i.e., i*ncreased model roughness for building footprints*), it is also possible to specify form- (and consequent energy-) loss coefficients to represent the fine-scale energy dissipation within and around buildings. This can be more correct as the energy-loss occurs mainly due to the water contracting and expanding as it flows through and around buildings. Whilst a 2D-scheme takes into account some of these form losses (e.g., the expansion of water downstream of the building), the fine scale losses that are not well represented, need to be included: hence the need for additional energy dissipation terms. Similarly to the first approach described here, that is, i*ncreased model roughness for building footprints*, this approach can also be troublesome as it is difficult to know what would be the appropriate form-loss values in advance.

(4) *Modelling buildings as 'porous' elements*; This method requires modification of the shallow water equations and it may not be found feasible for applications where the commercial software is used in the modelling work. For examples and further discussions, see Sanders *et al.* (2008) and Alcrudo (2004), Guinot and Soares-Frazão (2006), Soares-Frazão *et al.* (2008).

(5) *Representing buildings with adjusted conveyance characteristics*; This method requires adjustment of conveyances in x–y directions to account for effects of building blocks and similar to the previous method it may not be found feasible for applications where the commercial software is used in the modelling work. For examples and further discussions see Chen *et al.* (2008), Evans *et al.* (2009), Salum (2010) and Evans (2010).

In addition to the above mentioned methods it could be a method where the cells located on building footprints are simply removed from the 2D model domain. Furthermore, it should be noted that other urban features such as curbs, roads, stairs, fences, cars, morphological objects, etc. should be carefully considered as they also may have an important role in diverting the flow of flood water through urban areas. TIN which adapts computational mesh to match boundaries of buildings, curbs, etc. The following discussion concerns setting up of 2D models with regular coarse grids and representation of buildings as key features.

19.8 REPRESENTATION OF KEY FEATURES WITHIN COARSE GRID RESOLUTIONS

19.8.1 A method based on sub-grid scale porosity treatment and adjustment of storage characteristics

An example of this approach is an approach adopted by Yu and Lane (2006) which represents fine scale characteristics within a coarse grid using sub-grid parameterization. They introduced a "porosity" function within the coarse model; this was used as a means of maintaining the slope integrity of a surface within the coarse representation. McMillan and Brasington (2007) adopted a similar approach in their work, whereby fine scale topology is also represented within coarse grids in terms of its "porosity". Figure 19.17 shows an example of a 1 m grid cell topology within a 5 m grid cell. The fine scale topology provides information ascertaining to where water can enter the coarse grid and the direction that it can leave. This relationship is dependent on the depth of water within the cell and it is defined as a "porosity" and referenced from a lookup table. A volume–depth relationship is created based on the percentage of the cell volume above the ground surface and hence available for water storage. Such approach can enable modelling of urban features at coarse resolutions with little increase in computational costs (see also Soares-Frazão *et al.* 2008).

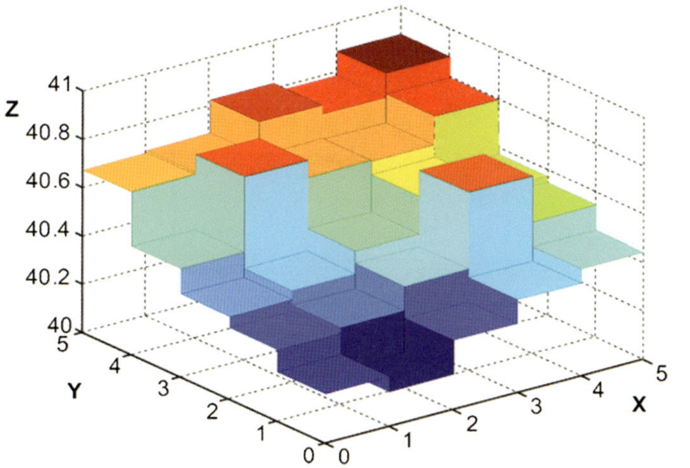

Figure 19.17 Schematic representation of micro-topography within coarse model grids. A 1 m resolution DEM embedded within a 5 m resolution model cell (McMillan & Brasington 2007)

19.8.2 A method based on adjusted conveyance and storage characteristics

A method based on adjusted conveyance and storage characteristics represents yet another methodology that can be used to reproduce similar flow characteristics along the floodplain between fine and coarse 2D model resolutions (Salum 2010; Seyoum *et al.* 2012; Vojinović *et al.* 2012). In the following demonstration of this method, different hypothetical case studies were set by changing the terrain topography and building positions. An automatic computation of depth-volume relationships for transition between fine and coarse resolutions was implemented within the code of a 2D non-inertia model,: see Figure 19.18. This is to say that model solves continuity and momentum equations over the cells of the coarse model while taking the minimum elevation as a surface level. For each depth increment, the volume stored in every cell is calculated as a volume-depth relationship. The same was also applied for the area-depth relationship, and for every increase in depth the area filled with water was calculated.

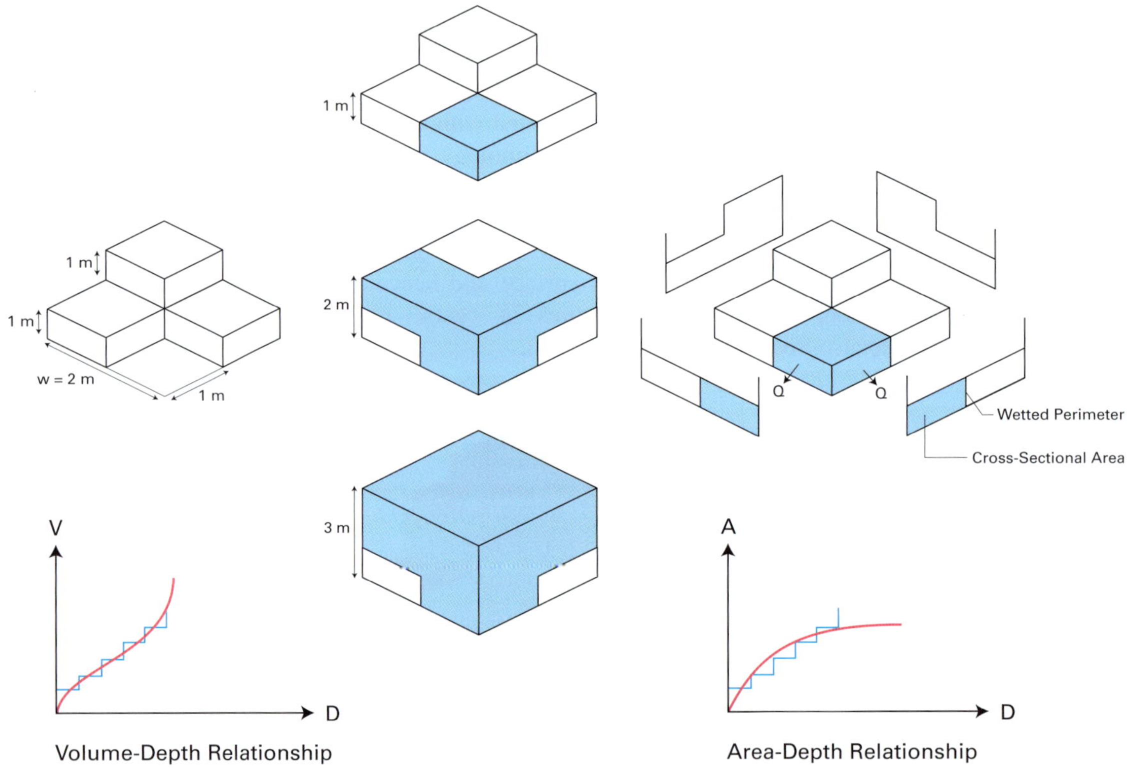

Figure 19.18 A volume-depth and area-depth relationship for a 1 m grid resolution with flood progression

In order to replicate restriction in conveyances in x–y directions of fine resolution models due to building blocks, the Chezy values in the coarse-resolution model were adjusted. Some of the results obtained from this demonstrational work are given in Figure 19.19. After such adjustment, it was observed that the coarse-model results were found to be very close to the fine-resolution model results (when compared in accordance to flood depths and flood extents). In the case given in Figure 19.19, the terrain slopes from left to the right and the source was placed close to the edge of the left side of the model domain. In this initial case, no building blocks were modelled.

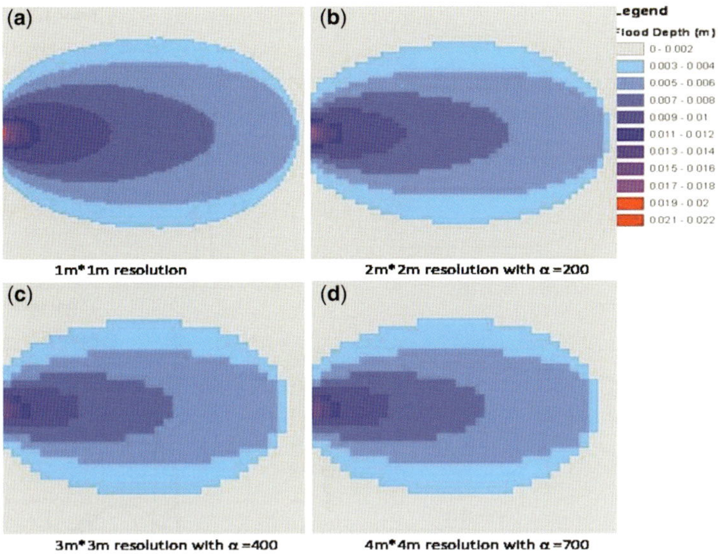

Figure 19.19 Example of model simulation results with 1 m, 2 m, 3 m and 4 m DTM resolutions using a method based on adjusted conveyance and storage characteristics

The following Figures 19.20 and 19.21 show model simulation results from a real-life case study where different grid resolutions were used (Seyoum, 2012).

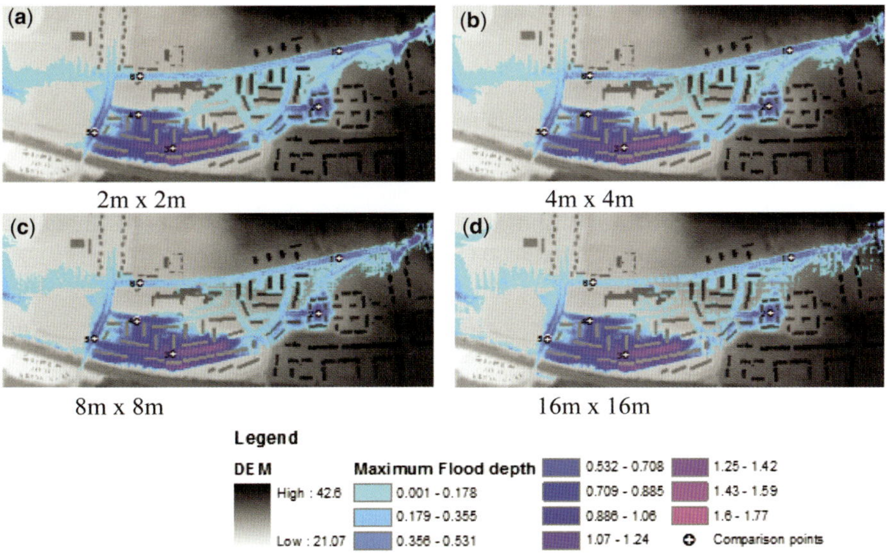

Figure 19.20 Model simulation results for different topographic resolutions (2m, 4m, 8m, and 16m grid resolutions) using a method based on adjusted conveyance and storage characteristics; Cockenzie Street and surrounding streets in Glasgow, UK (Environment Agency/Heriot-Watt University, 2010): see Seyoum *et al.* 2012

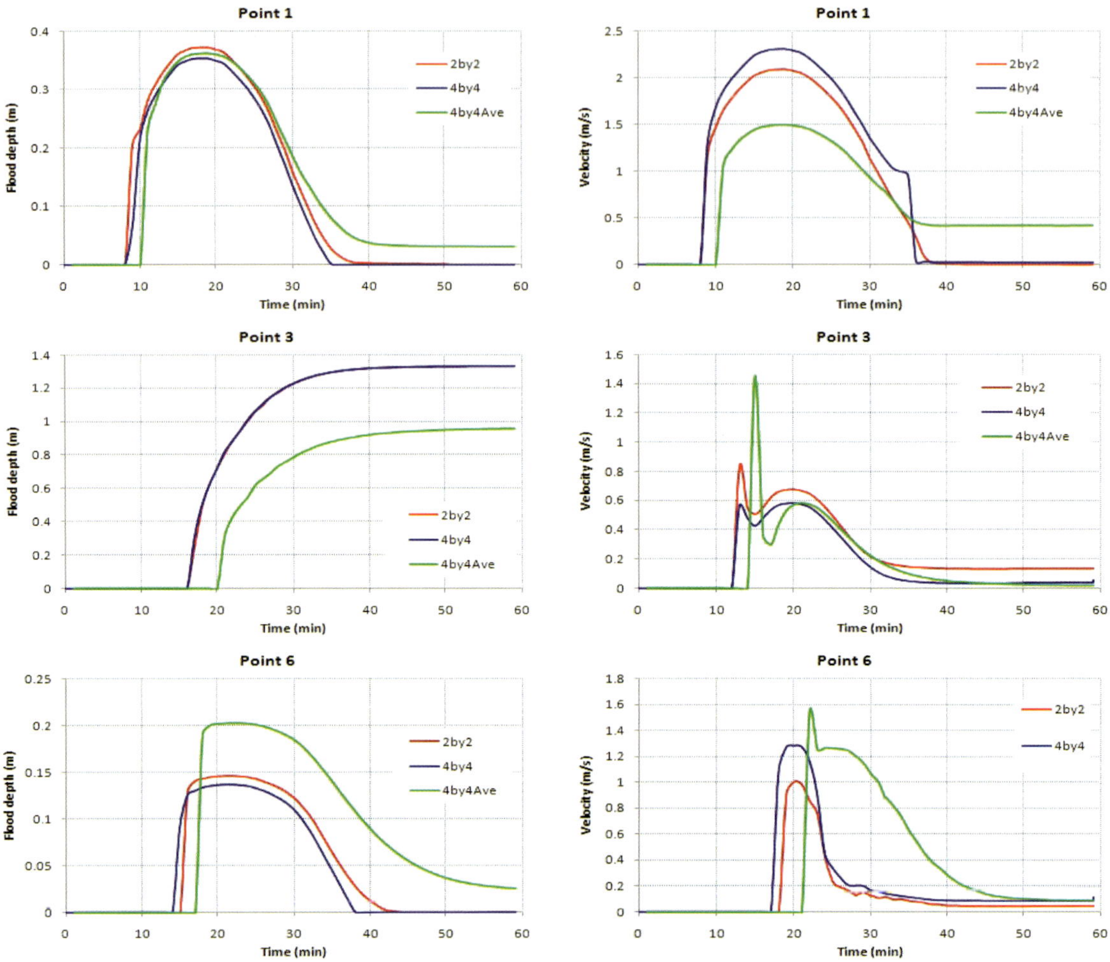

Figure 19.21 Time series of flood depths and velocities for points 1, 3, and 6 of the model results shown in Figure 19.20 for different topographic resolution (Seyoum *et al.* 2012)

Further to the above results, it can be observed that the coarse model with volume-depth and area-depth relationship has the potential to predict the depths and velocities that are similar to those of the fine grid model. Certainly, further studies are required to evaluate in detail strengths and limitations of this method.

19.8.3 A multilayered approach

In Chen *et al.* (2008), the area occupied by buildings within coarse grids is represented numerically with the use of a Building Coverage Ratio (BCR) and the restriction of surface flow propagation from one coarse cell to another was governed by the Conveyance Reduction Factor (CRF). This approach however allows for different BCR and CRF values to be assigned to each coarse grid cell. A multilayered approach is then achieved by representing effects of buildings numerically within a multiple layer grid format, Evans (2010) and Evans *et al.* (2009). In Evans (2010), the transformation from a fine-grid DTM into a coarse-grid multi-layered representation was initially done by generating a Boolean building grid representation of the terrain for reference, and then, using this information, coupled with a discretizing window, the DTM is then broken up into layers using a Cellular Automata methodology, Figure 19.22.

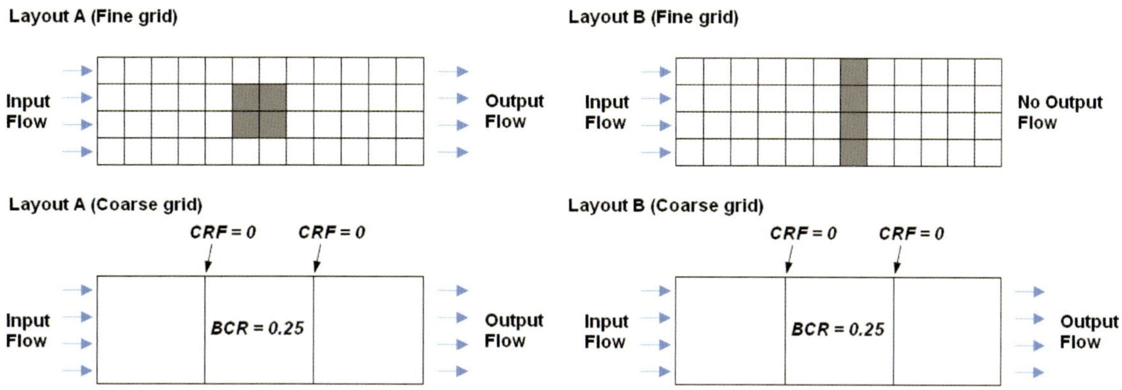

Figure 19.22 Comparison of flows across coarse grids in a single layer (Evans 2010)

19.8.4 An approach based on multi-cell finite difference solver as implemented in MIKEFLOOD and MIKE21

The multi-cell finite difference solver for overland flow has been implemented in MIKE FLOOD and MIKE 21 and it offers a number of advantages in comparison to the regular finite difference solver. This technique offers a reduction in model runtime by a factor of 5 to 10 when compared with a standard 2D finite difference solution to the Saint Venant equations. The Multi-cell Difference Solver uses a coarse scale topographical data to carry out time-varying water surface and flow velocity calculations while the fine scale topographical data is utilised when evaluating the governing equations for the 2D solver (Hartnack *et al.* 2009).

19.9 DECIDING ON A MODELLING APPROACH

Clearly there are several aspects that need to be considered before deciding on a particular modelling approach. Some of these aspects include objectives of the work, type of application (strategic analysis, engineering design, off-line, on-line, real-time, etc.), terrain topography, availability and resolution of data, computational time, commercial software costs, and so on. Table 19.1 outlines positive and negative aspects with respect to 1D and 2D modelling approaches and their application to urban floodplain modelling. Interested readers are also encouraged to see Vojinović and Tutulic (2009), Mark *et al.* (2004) and Allitt *et al.* (2009).

One crucial point has to be emphasised, namely that the above-mentioned modelling techniques are capable of predicting flood phenomena and it is not suggested that the use of the 1D modelling approach should be disregarded and the 1D/2D approach should be adopted exclusively. It is however suggested that since the 1D above-ground approach does not account for 2D flow processes, it places more responsibility on its users, and clearly any unthinking use of either of the two approaches will fail to produce satisfactory results. Still, a wealth of domain knowledge and spatial mapping techniques is required to apply these approaches in a way that can generate fruitful results. This applies to data

Table 19.1 Positive and negative sides of using 1D, 1D/1D and 1D-2D approaches.

Approach	1D and 1D/1D	1D-2D
Positive sides	• Easy to set up and manipulate/edit; • Suitable for design of minor systems and to some limited extent some major systems; • Only one type of model needs to be understood; • Less computational time required; It makes this approach more applicable to real-time control (or flood warning) applications; • There are many cases where such a model is capable of producing reliable results; • Calibration/validation is relatively easy; • Financially it is cheaper considering that the use of a single modelling system is needed;	• More detailed and realistic representation of surface geometry and overland flow physics; • Ability to set up different bed friction coefficients for each cell across the floodplain; • Extrapolation of results is not needed. The results can be presented directly on a 2D map. • Momentum transfer is explicitly accounted for; • It can simulate overland flow when flow is two-dimensional. Flow is directly routed over the surface and the actual flow path depending on terrain features can be determined by the model itself; No need to predefine overland flow paths;
Negative Sides	• Surface geometry and flow physics are simplified; Therefore, their use is limited to those cases where the nature of surface flow is close to one-dimensional, that is, where there is little uncertainly about drainage routes and where the flow outside larger ponds is mostly limited to the road width. • It is very difficult to produce sufficiently wide and accurate representation of cross-sections which can include the floodplains (therefore, the need for extrapolation and consequently there will be a possibility to obtain misleading results). • In the case of a two-layer approach (1D model for channel and separate 1D model for terrain), it requires extra effort to define the terrain geometry and overland flow paths; • Velocity is mean cross-sectional and unidirectional. The magnitude and directional changes are not accounted for. • Momentum transfer between the channel and floodplain is not accounted for; • Extrapolation of results across the floodplain for irregular terrains is limited and needs special attention.	• Two models need to be understood; • Setting up of 1D and 2D models requires more time for preprocessing and model instantiation; • This is computationally more demanding: it requires more computer power and memory; • When the coupling is done explicitly, the model can become unstable; • The model is very sensitive to stability. Usually, a time step of 1s is required; • Its suitability for real-time control (or flood warning) applications is accordingly limited; • Commercial software is more expensive considering that the use of two modelling systems is needed; • Calibration/validation of overland flows is more complex.

preprocessing, model instantiation, calibration/validation, the interpretation of results and the presentation of results within a GIS framework.

19.10 DETERMINING PEAK FLOWS IN A DENDRITIC NETWORK

The prescriptive design of stormwater pipes or channels in a dendritic network is normally carried out by determining a peak runoff for rainfall of a given return period and duration. The traditional way of doing this is to use the Rational Method. This uses the simple formula:

$$Q = 0.00278CAi \tag{19.3}$$

where

Q is the peak discharge (m^3/s)
A is the total area (ha)
C is a runoff coefficient
i is the rainfall intensity corresponding to a given return period and duration (mm/hr)

Normally the return period is prescribed, and the duration is the same as the time of concentration. This is the time taken for water particles to travel from the furthest point of the catchment to the location in the pipe or channel network where the peak discharge is being calculated. Typically, the time of concentration is equal to a 'time of entry' for water to enter the pipe or channel over the ground surface plus the longest time of travel along the network to the location. The time of entry is typically 2 minutes for small contributing areas, but can be longer for larger areas. The time of travel along the pipes or channels is determined by the length divided by the pipe full velocity or the flow velocity at some given depth in the channel.

The runoff coefficient can be determined as described above. Note that for the UK a (percentage) runoff coefficient which takes into account a design value of the antecedent wetness of the catchment is used.

When pipes or channels are being sized it is important to ensure that each pipe or channel upstream of a given pipe or channel has been sized so that the time of concentration can be properly defined. The convenient way of doing this is through ordering the sizing of the pipes or channels in 'end order'. This requires starting with the pipe furthest away (in terms of number of pipes) from the outfall. The ordering continues with the pipes downstream of the current pipe, unless a new branch of pipes joins the current pipe at its downstream end. In this case the next pipe is the pipe on the new branch furthest away from the junction. This procedure is followed strictly until the outfall of the system is reached; See Figure 19.23.

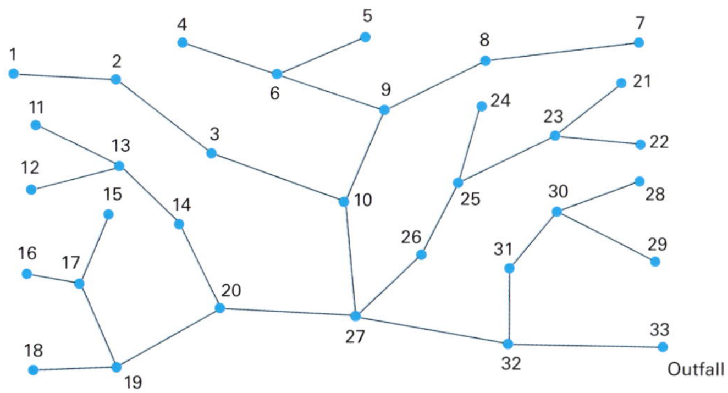

Figure 19.23 Pipes in a dendritic network with the nodes identified in end order

19.11 USING DESIGN RAINFALL EVENTS FOR DESIGN OF PIPE NETWORKS

The use of design rainfall events in modelling is appreciated because of the identification of the return period of flow with the return period of the rainfall depth, and the economy of running the simulation model. Strictly however, there is no guarantee that the return period of the flow is the same as the return period of the rainfall. What is more, it can be difficult to be confident that the return period of the rainfall is well defined for return periods less than a year, especially in areas with a low annual rainfall. For these reasons, modellers tend to make increasing a use of time series rainfalls when they can accommodate the lengthy computer run-times involved.

Design rainfall events are adopted for prescriptive design of pipes and channels in dendritic networks. In particular, the Wallingford Procedure advocates the use of rainfall events (depth, duration, profile and antecedent wetness) for sizing pipes. The concept is that a pipe network has all of its pipes sized for each run of a sequence of rainfall events with the given design return period, 50 percentile profile, a design Urban Catchment Wetness Index (UCWI) of 120, and with increasing duration (e.g., 5 min, 10 min, 20 min, 30 min, 60 min, 120 min, etc). Every pipe is sized to convey the peak discharge from the pipes immediately above it unless the pipe is already large enough. In this way the pipes can be designed consistently.

19.12 MODELLING TREATMENT WORKS

A fuller description of the treatment works simulation model is given in Price and Vojinović (2011). The important point made here is that in the majority of cases the outfall of the collection network discharges sewage to the treatment works via a pump, which is often an Archimedes screw pump. Upstream of the pump will be an overflow to divert excess flows to the adjacent receiving waters. Combined with the overflow is some form of storage, which is designed to capture the first foul flush or its equivalent, and to delay as much polluted wastewater for treatment over a longer period than the duration of the storm.

19.13 MODELLING RECEIVING WATERS

Modelling rivers, lakes, watercourses, and coastal influences in urban areas can be undertaken by either 1D, 2D or even 3D models. Rivers and watercourses are typically modelled with the use of 1D models and the representation of floodplains with either 1D or 2D models. The open channel geometry is represented as a series of irregularly spaced cross-sections perpendicular to the 1D flow. Boundary conditions typically contain an inflow hydrograph at the upstream cross section and a stage hydrograph at the downstream end. If necessary, rivers and watercourses can be modelled in a common framework together with the pipe network models. This is usually done in cases where there is a potential for greater influence of rivers/watercourse on hydraulic conditions at stormwater pipe network outlets. Also, the same approach can be applied in cases where the impacts of sewer networks are evaluated.

In some cases, the elevation of the groundwater table during wet weather can cause direct (above ground surcharge) or indirect (infiltration and overload of the pipe network system) flooding in urban areas and as such it might be of importance for modelling purposes. Depending on the complexity of issues associated with groundwater influences different modelling approaches may be chosen (e.g., conceptual, detailed, etc.).

Coastal models (2D or even 3D) can be also used in direct connection with pipe (or open channel) network models to assess drainage and flooding problems. However, in many cases, the use of historical tidal variation water level data as a downstream boundary condition for pipe (or open channel) network models is adequate. Certainly, if there is a risk of coastal inundation and more substantial influence on a drainage pipe (or open channel) network then the use of a fully-fledged coastal and/or estuarial model might be required.

The discharges diverted by the overflows and the effluent from the treatment works enter the corresponding receiving waters, whether a lake, urban stream or river, or coastal waters. The pollutants in the discharges will adversely affect the receiving waters, and the associated damage should strictly be taken into account when considering the performance of the collection system. Modelling is used to quantify the impact of the pollution on the receiving waters, and thereby to deduce the economic consequences of the damage incurred. Again, a fuller description is given in Price and Vojinović (2011), where the impact of sewage from CSOs and from treatment works on the stream or coastal waters is discussed.

19.14 INSTANTIATING AN URBAN DRAINAGE SIMULATION MODEL

The previous sections have described the concepts and assumptions that are made in isolating the essential features of urban water collection systems for modelling purposes. These features are included generically in a number of simulation modelling software products such that the application of the software to a particular collection system requires the addition of specific data describing the geometry of the model space (sub-catchments, pipes, channels, manholes, ancillary structures, urban surface topography, receiving water channels or 2D coastal bed topography, etc) and the appropriate boundary inputs (rainfall, wastewater inflows, groundwater levels, river discharges and water levels, ancillary structure settings, astronomic tides, etc). Considerable expertise is needed to instantiate such a model, that is, to construct and confirm the integrity of the structural data for the model space, and to calibrate and verify the model parameters. The model is a significant part of the virtual world which parallels particular aspects of the real world. As such it needs to be 'fit for purpose', that is, it should generate information that is safe and reliable to use in formulating interventions in the real world, whether directly affecting the structural aspects of the collection network or the non-structural consequences for the community benefiting from the network. For this reason, experience gained in instantiating models for collection networks is accumulated in engineering procedures such as the Wallingford Procedure (Wallingford 1981a), the WaPUG Code of Practice for the Hydraulic Modelling of Sewers (WaPUG 2002), the UK Sewerage Rehabilitation Manual (WRc 1994) and the UK Urban Pollution Management Procedure (FWR 1998); see Chapter 6. In addition, many water companies have produced their own in-house manuals on model building requirements, consistent with internal QA and standards. *HarmoniQuA* is an EU project that has focussed on a general procedure for computational model instantiation. In that there are hundreds of urban drainage networks world-wide, such procedures are intended to provide quality assurance guidelines for model instantiation and application. These procedures are part of the procedural world in which professionals operate on behalf of society. The key elements of an instantiation procedure for a collection model are given in Price and Vojinović (2011).

19.15 MODEL APPLICATION

A number of applications of simulation modelling in urban drainage can be identified:

- Conduit design
- Hydraulic analysis (including flooding)
- Infiltration and inflow analysis.
- CSO analysis
- Storage facility analysis
- Real time control and warning

- Rehabilitation of sewerage systems
- Urban pollution management

For the above applications, either event-based or time-series simulations are used. A general procedure for applying hydraulic and water quality models within the context of these applications is given in Figure 19.24. Each kind of application is now considered in turn.

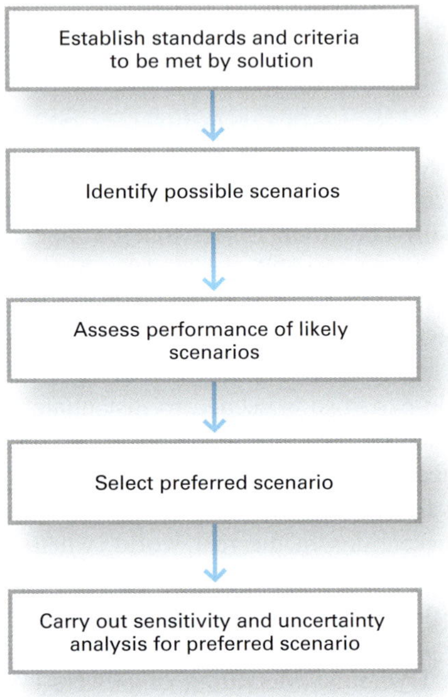

Figure 19.24 Procedure for application of models

19.16 EVENT-BASED AND LONG-TERM SIMULATIONS

There are two fundamental types of model simulations which can be applied in the assessment of urban drainage system's performance. They are event-based and time series simulations.

Event-based simulations are used when an impact is required to meet a prescribed frequency-based standard. Flood levels are a typical example where a 1 in 100 year event might be defined as the acceptable house floor level for protection against flooding. These events are typically of short duration and can be analysed either on the basis of a frequency analysis of measured data over a long period, or by applying rainfall patterns which have been developed after analysis of long-term rainfall records. The major assumption in the event-based approach is normally that the effect (i.e., the flood level) is assumed to have the same frequency of occurrence as the cause (the rainfall). This approach is applicable when the frequency of the effect is closely linked to the frequency of the cause, as is the case in urban flooding.

Time-series simulations require the application of a long series of measured rainfalls to develop a long time series of effects. This approach is used when analysing systems where the frequency of the effect is not directly related to the frequency of the cause, or when the effect is more related to the accumulated effects of catchment behaviour, such as the ecological health of receiving waters. A time-series is appropriate when the time scale of the effect is long in comparison to the time scale of the cause.

The event-based approach employs a short-term design rainfall applied to the stormwater catchment model. As discussed earlier, a design rainfall relates to a specified frequency (AEP) and has a design total depth and a design hyetograph (profile in time). Some design rainfalls represent a single duration and since different sections of a drainage system will respond with different response times, a range of durations must be investigated in order to ensure that the critical duration is covered for all parts of the system. Other design rainfalls inherently encompass a range of durations within a single design rainstorm therefore removing the necessity of examining all durations.

The 1 in 1, 1 in 2, 1 in 5, 1 in 10, 1 in 25, 1 in 50 and 1 in 100 year AEP design events are usually used as criteria for standard levels of flood protection, and as such they are commonly used for flood model simulations.

The time-series approach is normally used for the analysing the combined sewer system. The approach follows long-term (typically, 10-year plus 1 year for initial conditions) model simulations using historical or synthetic rainfall and evaporation time series as input in addition to the wastewater dry weather flow. The long-term time series are simulated using the rainfall-run-off model (calibrated or built by using the regionalised parameters) and the overflow performance is then statistically evaluated. In the analysis of the combined sewer system performance, the long-term statistical analysis is considered a vast improvement on the traditional steady-state approach. It provides a better understanding of the system and leads to more appropriate and cost-effective solutions.

Under the long-term statistical analysis approach, the input requirements are rainfall, potential evapotranspiration and dry weather flow applied to a rainfall-runoff (i.e., hydrological) model and routed through the system by a hydraulic (pipe or open channel flow) model. Inputs are:

- Long-term Historical Rainfall: Long-term historical (e.g., 10-year, plus 1 year for initial conditions) rainfall time-series, taken at time intervals applicable to the study area from the nearest available rain-gauge data, are normally used. If the study area is large and the standard deviation of spatial variation (i.e., storm event-based variation) within the study area is more than $+/-10\%$, long-term time series data of rainfall from several rain-gauge stations can be applied.
- Long-term Synthetic Rainfall: Upon completion of the historical rainfall analysis, system performance is normally analysed using synthetic stochastic rainfall to enable long-term system performance, the assessment of long-term loading to the receiving environments, and to examine a range of storm events not captured in the historical analysis.
- Long-term Historical Potential Evapo-transpiration: Potential evapo-transpiration data (daily/monthly) applicable to the study area (e.g., from Meteorological Service Office for the same period as the rainfall time-series data) is normally used in combination with the long-term historical rainfall. The same historical potential evapo-transpiration timeseries can be applied across the entire catchment.
- Long-term Synthetic Potential Evapo-transpiration: Average monthly potential evapo-transpiration applicable to the urban catchment area (e.g., from Meteorological Service Office) is normally derived and used in combination with the long-term synthetic rainfall. The same synthetic potential evapo-transpiration time-series can be applied across the entire catchment.
- Dry Weather Flow: Where flow gauging has identified the average dry weather flow and the daily profile then these are normally used for that subcatchment. Where no such flows are available then the regionalised values are used.

19.17 DESIGN OF SYSTEMS

The sizing of pipes in reticulated sewerage systems has depended largely on the use of the Rational Method with local IDF curves. The basic notion is that the pipe layout, seen as a dendritic collection network, is given, but with the pipes largely following the road network for ease of construction and maintenance. The pipes are laid such that their gradients induce flow under gravity down the main (collecting) branches of the pipe network. A proper engineering practice is that the sizes of the pipes increase downstream. The outfall of the network discharges to a treatment works or directly to receiving waters. During the sizing of the pipes adjustments may be made to the gradients of the pipes so that self cleansing velocities for reasonable flows are maintained to limit sediment deposition.

With the use of the Rational Method the pipes can be sized in such way that they convey the Rational design discharges flowing just full (or at a certain proportion of pipe full). For details of the design process see Butler and Davies (2000). See also the Wallingford Procedure (Wallingford 1981a, b).

The shortcoming of this approach to designing the pipes is that it says nothing about the frequency of surface flooding, which the pipe network is designed to limit. However, because the pipes are laid at a depth such that the cover over the pipe is typically more than a metre, then the frequency of surface flooding might be considerably less than the design frequency for the pipes. The frequency of significant surface flooding at nodes throughout the network defines the level of service provided by the network in the drainage catchment, and this will vary from place to place depending on the topography of the catchment and the pipe layout. Obviously, the actual levels of service in the catchment are an important factor in justifying the design, but to obtain these a simulation of the designed system for events rarer than the design frequency is needed. For this reason the Wallingford Procedure for the design of urban storm drainage was developed in the late 1970s as a computer-based method which gives the options of using the Rational method or a hydrograph method for the design of the pipe network followed by the assessment of the performance of the designed network using a simulation method; see Wallingford (1981). Where the simulation model gives depths of flooding at nodes in the network, there is the potential to estimate the damage costs as a function of the flooding depth. Given a designed pipe network and the cumulative damage costs associated with a set of rainfall events rarer than the design event, it is possible to trade off the cumulative damage cost against the construction cost of the pipe network. That is, the pipe network can be redesigned for a larger design rainfall, and the corresponding cumulative damage cost, which should be lower than the damage cost for the first design, can again be compared with the new construction cost. The damage costs should decrease as the design rainfall for the sizing of the pipe network is increased, while the construction cost of the pipe network increases. This gives a standard problem in economics that leads to an optimal size of the pipes: see Figure 19.25.

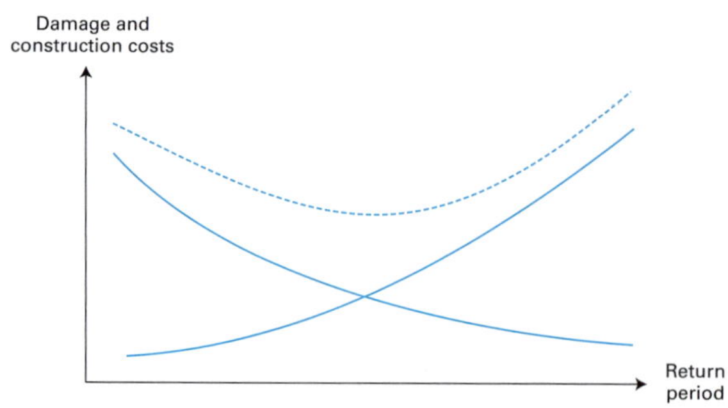

Figure 19.25 Deriving the minimum cost

The Wallingford Procedure includes a number of radical innovations in modelling rainfall-runoff:

- A (UK) rainfall depth for a prescribed duration and a design frequency (or return period) specifically for the geographic location of the catchment
- A (UK) specific symmetric rainfall profile (a summer 50 percentile profile)
- The (UK) volume of runoff from a catchment given by a regression equation based on the percentage paved area, the soil type and the antecedent catchment wetness
- A (UK) calibrated design value for the antecedent catchment wetness
- A rate of runoff calculated through one or two calibrated linear reservoirs (in series)
- A sequence of design rainfall events with increasing duration
- A simulation model consistent with the hydrograph design method
- (Actually, the simulation model is limited and was subsequently replaced by a more complete solution of the full Saint Venant equations)

19.18 DESIGNING FOR EXCEEDANCE

Since it is inevitable that in certain situations due to extreme rainfall the capacities of sewers, covered watercourses and other drainage systems will be exceeded (i.e., residual risk). Such situations will occur when the rate of surface runoff exceeds the drainage system inlet capacity, when the pipe system becomes overloaded, or when the outfall becomes restricted due to flood levels in the receiving water.

The excess flow that occurs on the surface is referred to as the *exceedance flow*. Such flow is conveyed on the ground by surface flood pathways such as roads, paths or depressions in the surface (Figure 19.26). If such pathways have not been specifically designed they are referred to as default pathways, otherwise they are referred to as designed pathways. CIRIA C635 (2006) discusses these issues and emphasises the need for engineers to track overland flows. This document suggests that surface pathways should be linked together in the same way as conventional drainage networks, so as to provide a system of conduits that effectively conveys the exceedance flows off the developed site. Flows should be prevented from accumulating at low spots except where temporary surface storage is incorporated into the design strategy.

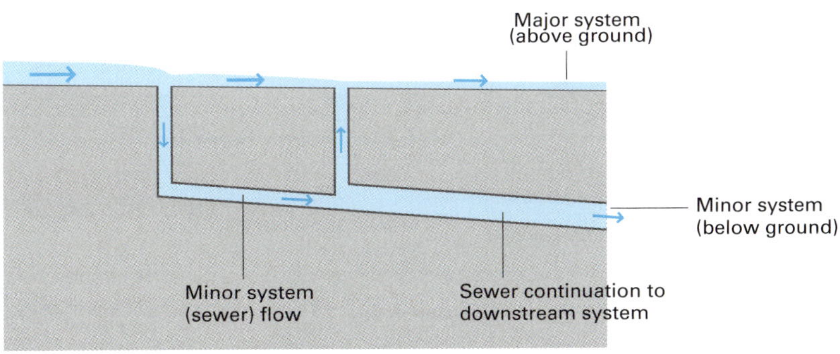

Figure 19.26 Designing for exceedance (After: CIRIA C635, 2006)

19.19 HYDRAULIC ANALYSIS

Once simulation modelling of sewerage networks became feasible, its potential to analyse the performance of a system led to an explosion in methods of hydraulic analysis. Firstly, simulation modelling is used to assess the performance of a designed system for events rarer than the design event. This leads to the

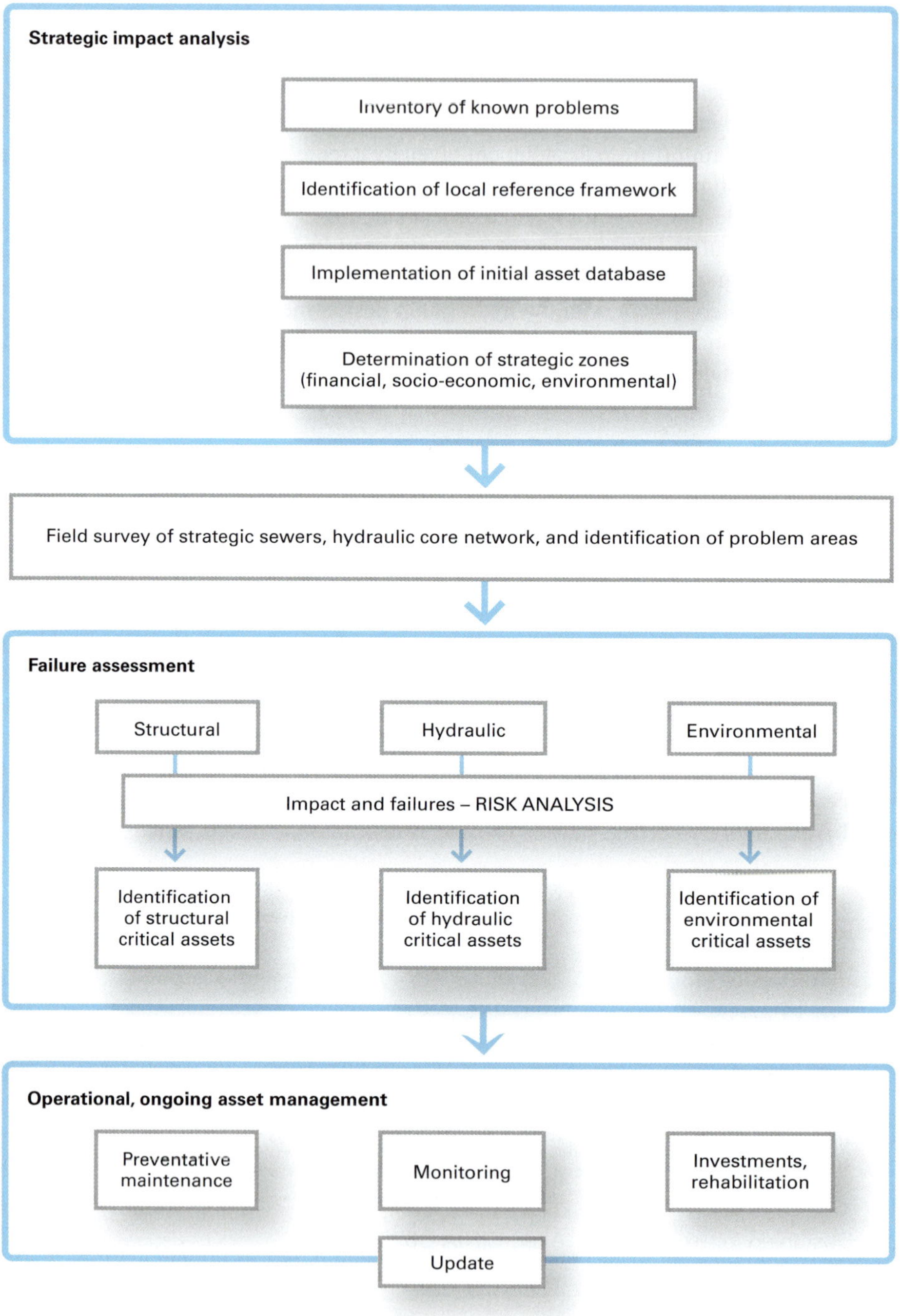

Figure 19.27 Hydroplan – a procedure for defining optimal cost-efficient investment and operation and maintenance scenario (Vojinović *et al.* 2005)

determination of the frequency of surcharging in each pipe and surface flooding at all nodes in the network, and therefore the level of service (in terms of flooding) throughout the catchment. For a combined sewerage system the frequency of CSO operation can be determined by running the model for, say, an annual time series of rainfall, or a series of rainfall events with higher frequency than the design event for the CSO. By running the model for the complete range of higher- and lower-frequency events it is possible to study the frequency of filling the tanks, operating the pumps at maximum flow, and so on. Hydraulic analysis therefore, is directed towards the assessment of the performance of sewerage or drainage networks in terms of the service provided. The aim is to formulate upgrading scenarios

When carrying out a hydraulic analysis of a drainage network, the main problems that arise include such items as:

- Shortage of network data
- Unknown dry weather flows
- Unknown infiltration
- Uncertainty about pipe blockages, sediment deposits etc
- Shortage of temporal and spatial rainfall data
- Shortage of measured flows in the network
- Unknown head-losses at structures

Figure 19.27 shows the procedure that can be followed when applying a model to hydraulic analysis (see also Price & Vojinović 2011).

19.20 INFILTRATION AND INFLOW ANALYSIS FOR WASTEWATER SYSTEMS

Actual infiltration and inflow in wastewater systems were not intended in the original design of the networks but occur because of deficiencies in the structural integrity of the pipes or because of illegal or wrongly connected storm flows from streams, gullies draining yards, roof downpipes or manhole covers. Infiltration can arise because of cracks and fissures in pipes and manholes or inadequate joints between pipes. There may be many reasons why the integrity of the elements of the network are compromised, including poor construction, adverse surface loadings due to passing traffic, aging of pipes, thinning of pipe walls due to high shear velocities, high groundwater levels, aggressive chemicals in the soil, frost damage, ground subsidence, penetration by tree roots, human interference, and so on. On the other hand, changes on inflow are usually due to some form of human intervention. Both infiltration and inflow can lead to a sewer capacity that is inadequate to convey the designed wastewater flow, leading to surcharging and possibly flooding, overloading of pumping stations and treatment works, higher frequency of operation of sewer overflows installed to protect the treatment works, and possible higher sediment loads due to the ingress of soil leading to a greater risk of blockages. There are also the likely requirements of increased maintenance as well as possible ground subsidence and the associated risk of collapse of the affected sewers.

There is then a value in using simulation modelling for separate wastewater systems. Here the dry weather flow should strictly follow a slowly time-varying profile dependent on the largely diurnal cycle of the domestic use of potable water. But it is well established that separate wastewater systems deteriorate over time, especially from illegal stormwater connections and infiltration from groundwater. In the latter case, wastewater systems can in some cases act as land drains. Consequently there can be a considerable stormwater component in the flow, so much so that overflows have to be constructed to drain-off excess flows that the wastewater treatment works cannot cope with. This is harmful for the environment and therefore expensive for the community. This leads to rehabilitation techniques applied to the wastewater network, involving the identification of stormwater inflows and the sealing of pipes, sometimes with relining, to reduce the infiltration to a tolerable level. It may be difficult to completely eliminate infiltration in house connection pipes.

Simulation modelling is frequently used to investigate the performance of a separate wastewater collection system and to identify the presence, flow rate, and type of infiltration/inflow conditions which exist in the sewer system. The investigation focuses on describing the existing wastewater collection system using estimates of average residential, industrial, commercial and institution wastewater flows. Continuous flow monitoring, and in some cases flow isolation monitoring, is used to determine infiltration and inflow rates, and to reproduce these in the model. On the basis of the results for the existing situation, a detailed survey of the system may be needed. Obviously, prior to any modelling or survey, all available data and information on the performance of the system will have been collected and archived.

19.21 CSO ANALYSIS

As described earlier, combined (sanitary or separate) sewer overflows are included in a drainage system to divert excess flows directly to receiving waters in order to limit the overloading of a treatment works or to provide a relief outside of the system in case of overloading. Their performance is usually evaluated by statistical analyses which are used to:

- Assess the frequency of occurrence of surcharging and percentage of deficiency with respect to existing capacity,
- Assess the flow, duration and volume of overflow frequencies,
- Assess the average annual overflow volumes,
- Perform various statistical tests and produce the charts to assist in conceptualisation of remedial/mitigation options,
- Tabulate combined sewer overflow (CSO) spill volumes, together with peak and average daily dry weather flows,
- Summarise system performance based on the above analyses.

Unfortunately the CSOs (or SSOs) discharge may dilute sewage that can be heavily polluted. Many of the problems of pollution from sewerage systems are due to inadequate overflows that do not meet current design guidelines. This causes them to spill too much and too frequently and to have poor retention of gross solids and of finer polluting solids. Drainage area plans will therefore frequently involve the re-construction and rationalisation of overflows to reduce these problems. However these changes can have further side effects on the operation of the rest of the system.

Some systems exhibit what is called a first foul flush. This is a phenomenon that occurs early on in a storm, when washoff of deposited solids from the catchment, the flushing of gully pots, heavily polluted wastewater and near-bed organic solids in pipes, together with the erosion of deposited solids also in pipes, produce a high concentration of pollutants at the beginning of the runoff hydrograph for the event. A combined sewer overflow should be designed in conjunction with storage to minimise the discharge of any highly polluted first foul flush and to retain the polluted sewage until the pollutant concentration is acceptable.

An effective CSO should provide adequate hydraulic relief from the sewerage system so as to meet the targets for flood control in the catchment, and control the continuing flow to ensure there is no overloading of the downstream sewerage system and wastewater treatment works. The regulatory requirements for the intermittent discharges to the receiving water must be met, while minimising operational requirements and whole-life costs.

Reconstruction of overflows to spill less volume and less frequently may require the construction of new sewers downstream of the overflow to take the increased continuation flow. However, even in this case, the increase in flows through the treatment works may reduce the quality of the effluent.

Rationalisation of overflows to replace several old overflows with one new one can have the side effect of causing a larger impact on the receiving water at the new discharge point than at the old discharge points by discharging the entire spill into one reach of river. The use of a river model will identify this problem.

New overflows based on proven designs, such as stilling pond, high side weir and vortex overflow, exhibit a much better retention of gross solids within the sewer than old hole-in-wall or low side-weir overflows. As the emission of gross solids is one of the most obvious forms of pollution caused by overflows, this is a significant benefit. However, even these more recent overflows are not very effective at retaining the fine sediment particles which carry much of the polluting matter. They are of course also unable to retain dissolved pollutants from the spill. The concentration of fine sediment in the spill should be assumed to be equal to that in the continuation flow.

Before 1970 in the UK it was customary to define the CSO setting as 6 times the dry weather flow. Subsequently the UK Technical Committee on Storm Overflows and the Disposal of Storm Sewage (HMSO 1970) reported that it would be better to make the setting more dependent on the population and its consumption of water and industrial use of water. They proposed 'Formula A':

$$Q_0 = DWF + 1360P + 2E \tag{19.4}$$

where

Q_0 is the overflow setting (litres/day)
$DWF = PG + I + E$
P is the population
G is the water consumption per person (litres/day)
I is the pipe infiltration rate (litres/day)
E is the average industrial effluent (litres/day)

Eq. (19.4) does not include any allowance for the flow in the receiving stream. The Scottish Development Department recommended in a report from the Working Party on Storm Sewage (Scotland), Storm Sewage: Separation and Disposal (HMSO 1977) that, for a dilution of more than 7 to 1, Formula A is satisfactory, but for a dilution of 6 to 1 or less then Q_0, this should be increased by $455P$ and storage should be introduced with the CSO.

In 1993 the National Rivers Authority in the UK introduced the AMP2 Guidelines (NRA 1993); see Table 19.2. These were designed to classify existing CSOs in terms of their significance, and subsequently to identify CSOs that were not performing well and needed to be improved. For bathing beaches, the guideline was that CSOs should not spill more than 3 times per bathing season.

Table 19.2 AMP2 guidelines.

Significance	Dilution	Interactions	Population equivalent	Fisheries	Assessment criteria
Low	>8:1	None	–	–	Emission control (Formula A)
Medium	<8:1	Limited	>2000	Cyprimid	Simple models (QUALSOC, CARP + sewer hydraulic model)
High	<2:1	Significant	>10,000	Cyprimid or salmonid	Complex models (sewer and river quality model)

Further to these approaches, the Urban Pollution Management Manual (FWR 1998) has been introduced. The Manual provides a procedure for assessing not only the pollution runoff from a sewerage system, but also the impact of CSOs and effluent form treatment works on the receiving waters. In addition, the Manual provides quantitative standards for intermittent discharges; see Table 19.3.

Table 19.3 Intermittent standards for dissolved oxygen and ammonia concentration/ duration thresholds for sustaining cyprinid fisheries (after FWR (1998)).

Return period (months)	**DO concentration (mg/l)***		
	1 h	6 h	24 h
1	4.0	5.0	5.5
3	3.5	4.5	5.0
12	3.0	4.0	4.5
	NH_3-N concentration (mg/l)**		
1	0.150	0.075	0.030
3	0.225	0.125	0.050
12	0.250	0.150	0.065

*Applicable when NH_3-N < 0.02 mg/l
**Applicable when DO > 5 mg/l, pH > 7 and T > 5 °C

There are several aspects of the CSO performance that should be examined during an upgrading of the structure. One is that care should be taken over the design of the hydraulic storage volume associated with a storm overflow. In order to correctly deal with the volumes as well as the peak flows during storm events the analysis of storage is normally carried out using a rainfall time series approach, such as produced by StormPac^{TM} or the synthetic design storm time series developed at HR Wallingford. The storage tank is normally designed to provide some settling of sediment such that a greater concentration of sediment goes on to treatment than is conveyed in the overflow discharge. A scumboard will in this case be used to deflect gross floating solids from the overflow.

Another aspect is the ability of the CSO to remove gross solids from the effluent. There are two standards: the first is the separation from the effluent of a ‘significant quantity’ of solids with a size of 6 mm in any two dimensions, and a screen that removes 10 mm solids.

19.22 PERFORMANCE ANALYSIS OF PIPES AND CHANNELS

The pipe (or channel) capacities (i.e., performances) in relation to different design events can be established and presented for the current and future development scenarios in colour-coded GIS maps: see Figure 19.28.

Furthermore, in case of the sediment transport analysis, calculated maximum bed shear stresses in the drainage network system can be displayed in a similar manner.

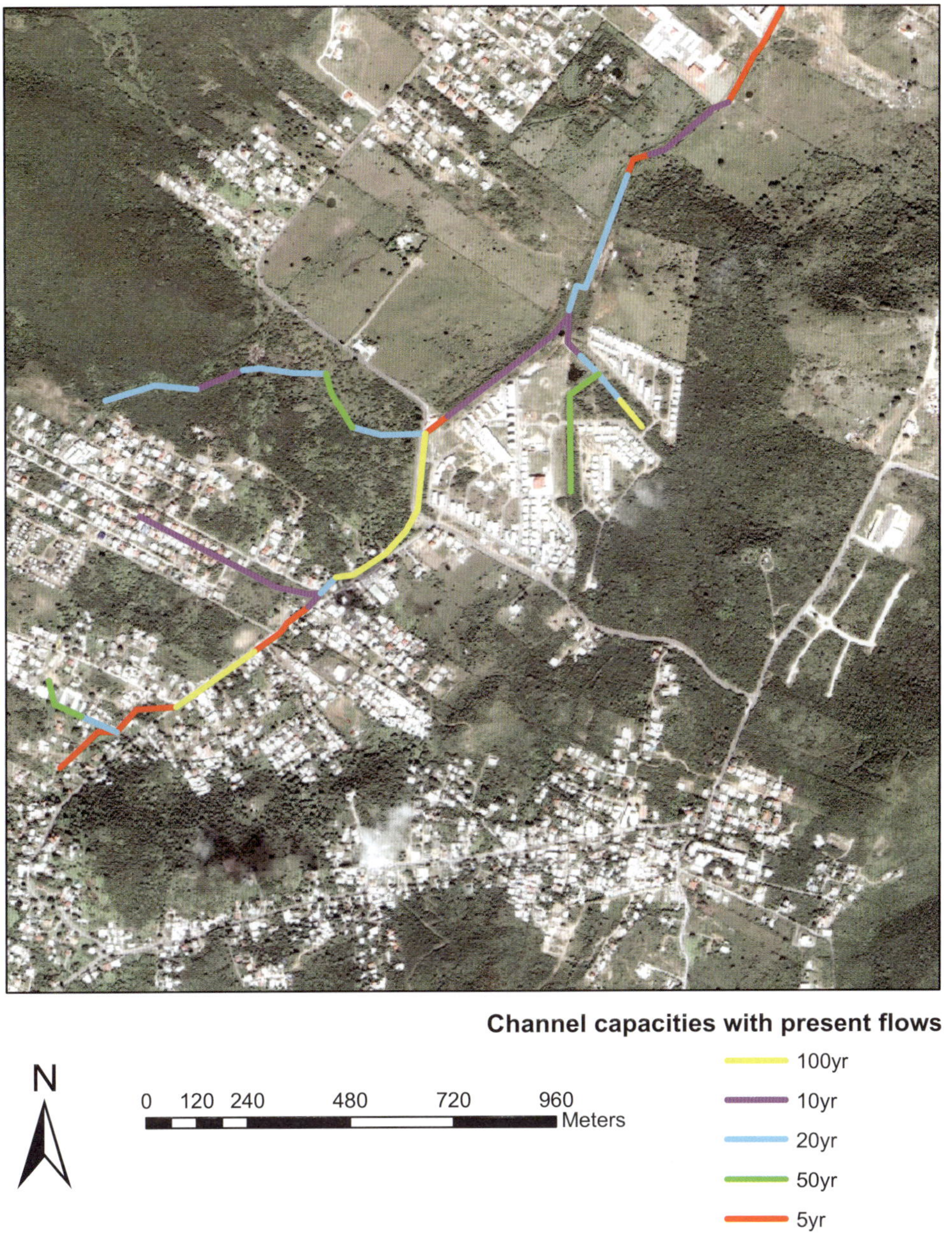

Figure 19.28 Example of a GIS map displaying channel capacities against design events

19.23 STORAGE FACILITY ANALYSIS

A detention tank can be used to limit the peak discharge from a catchment. It does this by storing excess flow in an on- or off-line storage tank or pond until the storm has passed, and then the stored water is released from the tank to the pipes downstream or directly to the sewage treatment works. The volume continuing on downstream can be reduced through source control techniques (as introduced through the UK Sustainable Urban Drainage Systems (SUDS) initiative; see Woods-Ballard *et al.* 2007) or through diverting excess flows through an overflow to the nearest receiving waters. In both cases the sizes of the pipes downstream of the tank or the overflow can be reduced. The introduction of storage is therefore a means of upgrading existing sewerage systems and of reducing the pollution impact on receiving waters downstream of new sewerage systems.

A storage tank is also needed to deal with a possibly increased first foul flush of pollutants at the start of storms as described above. Sediment will be transported into a tank where fine sediment will tend to be deposited as the tank drains down at the end of a storm and flushed out at the start of the next storm. If there is another overflow further downstream it may spill a more polluted flow, or the treatment works may be unable to cope with the pollutant loading especially the final clarifiers of activated sludge plants. The long detention times in the tank can also lead to the generation of methane and hydrogen sulphide in septic sewage causing both pollution and operational problems. The use of a sewerage water quality model in conjunction with a treatment works model can help with the investigation of these problems and with the design of the tank.

The sizing of a tank is based on standard storms of the desired return period, whereas the verification of the model is done with long events, typically those that include runoff from permeable areas, (such as are generated during the winter in the northern hemisphere): see Figure 19.29. Accordingly, adjustments may be needed to the designed tank size.

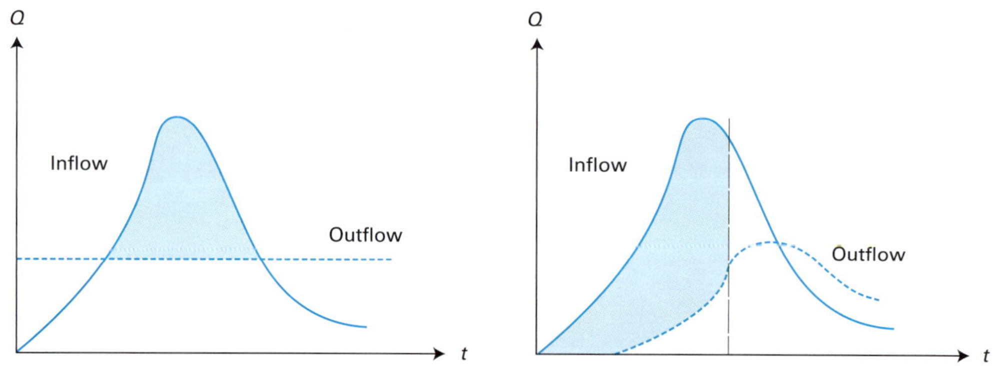

Figure 19.29 Use of storage

19.23.1 On-line tanks

On-line tanks form an integral part of the sewer system, consisting of enlarged flow sections that fill when the inflow exceeds the required continuation flow to the downstream sewer. The flow into the downstream sewer is usually controlled using an orifice plate, penstock, throttle pipe or vortex flow regulator. Pumps are rarely used in on-line tanks.

19.23.2 Off-line tanks

Off-line tanks are physically separate from the sewer system, with the flow entering the tank via an overflow device incorporated into a diversion structure. The flow to the downstream sewer is again controlled in a similar way to that of the on-line tank. The stored flows are returned to the system either by gravity, where the site allows, or by pumping.

19.23.3 Pumped storage tanks

Pumped storage tanks may be used to prevent flooding, to balance hydraulic loading, to reduce the frequency of spill, to reduce pollution spilled from the sewer system to the receiving watercourse, and to allow adequate response times in the event of pump or power failure at pumping stations.

The UPM manual states that "detention tanks should be designed to be, as far as possible, self cleansing whilst in operation". In pumped storage tanks, which are designed to meet such criteria, it is preferable that the weir is located at the upstream end of the tank so that the operation of the tank results in the retention of the heavily polluted first flush in the tank downstream of the weir. This means that the spilled flow to the watercourse is comprised of the more dilute stormwater that enters the tank during the later stages of the storm event.

In contrast, pumped storage tanks which are designed to settle out the sediments rely on the creation of a tranquil flow pattern within the tank that is conducive to the settlement of the gross and finely suspended solids. In this case the tank should be so dimensioned as to create conditions of ideal settlement with all sediments deposited at the bed of the tank as they are discharged through the tank. Hence, in this type of tank, the position of the overflow weir should be located at the downstream end of the tank.

The retention time of effluents in the pumped storage tank is a particularly important parameter in respect of the deterioration of the quality of the effluents retained in the tank. Long retention times, in excess of 12 hours, may lead to problems of septicity and odour. Microbiological activity in the effluent and in the sediment deposits on the bed of the tank results in an uptake of oxygen from the effluent in the tank with a corresponding reduction in the concentration of dissolved oxygen. The rate of oxygen uptake is a function of many parameters which include temperature, the initial concentration of DO, the constituents and pollutant strength of the effluent, the characteristics of the bed sediment deposits and the re-entrainment of oxygen via the water surface. Equations have been developed by Pomeroy and Parkhurst (1972) to predict the uptake of DO and the potential for the generation of hydrogen sulphide in pumped rising main systems. Their applicability to large pumped storage tank systems needs further research and developments.

19.23.4 Designing for sediments

Flow velocities are greatly reduced on entry to a storage structure, and the hydraulic conditions favour sediment deposition. Deposits of sediments in storage chambers may however, lead to blockages, surcharging, flooding and premature spill operation. In addition, the ageing of deposited sediments, and their subsequent re-suspension in later storms, can lead to poor quality in the continuation flow and in the overflow spilled discharges, and may also lead to further odour and septicity problems. It is therefore important that sediments are not allowed to build up in storage tanks. Saul and Ellis (1992) recommended criteria for the design of on-line tanks, taking sediment into account.

19.23.5 Levels of service for receiving waters

Where a first flush occurs, pumped storage tanks can be designed to capture the flush, thus allowing the storm flow with lower pollutant concentration to enter the river during the later stages of the storm event. Hedley and King (1971) observed that the major part of the pollution load in storm sewage was carried either in solution, or as a very fine suspension, and could not, therefore, be retained within the sewerage system by the settling processes occurring at a conventional overflow structure. They suggested therefore that the storage volume should be equivalent to the excess volume of the rising limb of a design storm hydrograph above that of the continuation flow. They recommended that the return frequency of the design storm is a function of the size and quality of the receiving stream, ranging from 6 months for a small stream to an overflow frequency of 10 times per year for large rivers.

Where a storage tank is required because river amenity is of special concern, the Scottish Development Department, (HMSO 1977), suggested a simple method to estimate the desired storage capacity based on available dilution in the receiving watercourse. The guidelines are presented in Table 19.4 below. The dilution is based on the dry weather flow of sewage and on the minimum flow in the stream at the discharge point, that is, the mean daily discharge which is exceeded in the long term for 95% of the time. The UK Environment Agency normally asks for 1 to 2 hours storage at 3 DWF at pumping stations before spills can be made to a water course.

Table 19.4 SDD (HMSO 1977) Design guidelines for storm overflows.

Dilution*	Percentage pollution**	Type or storm overflow
8	100	Formula A, no tank.
6	75	Formula A+455P, no tank, if sewer capacity available. Alternatively adopt next recommendation.
4	50	Formula A+tank capacity 40 l/hd.
2	25	Formula A+tank of capacity 80 1/hd.
1	12	Formula A+tank of capacity 120 1/hd.

*Minimum dilution in terms of 95% exceedence mean daily discharge in receiving watercourse and sewage DWF.
**Approximate percentage of total polluting load of storm sewage spilled from sewer that may be discharged to stream (%).

19.23.6 Sizing of tanks: European practice

Historically, in Europe, it has been common practice to design the size of the tank based on the storage of a rainfall amount falling on the impervious area of the catchment. Typically 1.5 mm to 4.0 mm may be used for design purposes, which gives a volume of storage equivalent to 15–40 m^3/hectare of impervious area. This design philosophy was based on the recommendations of Krauth which have subsequently been written into the (German) ATV A128 design guidelines for storage tanks (ATV 1992).

19.24 MODELLING REAL TIME CONTROL OPTIONS

An urban drainage system is controlled in "real time" if flows within the system are governed by moveable regulators which operate based on information of the behaviour of the system at one or more sites at each moment in time. The reason for using real time control (RTC) is to make optimum use of the sewers and facilities, especially the storage available in them. This allows operating and maintenance costs to be minimised, flooding and environmental impacts to be reduced and capital expenditure to be deferred or limited. In contrast to passive systems, RTC enables the control structures to manage the flows actively. The degree to which flows can be managed is clearly related to the extent and amount that available storage can be utilised.

With RTC, operators have a degree of freedom that does not exist in passive systems. Control rules are adopted that improve the performance of a system, and these rules are normally the subject of an agreement with the Regulator such that he can grant a consent for the structure. Then it is important that monitoring procedures are in place to log the RTC actions and demonstrate that the control rules are implemented in the event of a failure in the system performance. Such a failure may be due to a malfunction of the monitoring system that is providing the information for the RTC decisions. Safeguards are therefore needed for these situations to ensure that the system does not perform worse than an equivalent passive system.

It is also important to ensure that the effects of any RTC action on the whole drainage system (collection network, treatment works and receiving waters) are properly understood, such that the effects of any action can be predicted and can be shown to be of benefit. Modelling therefore becomes strategically important in RTC in producing safe and efficient rules for controlling the flows, and in training operators to run the system reliably.

The essential feature of RTC is that, during a rainfall event, parts of the sewerage system may be overloaded while other parts are under utilised, and the parts affected change dynamically during the event. The flow control structures are operated such that the stress on the overloaded parts of the network is alleviated by storing sewage or moving it to the under utilised parts. In this way RTC can benefit the following:

- Deciding when to divide flows
- Accounting for spatial rainfall
- Flushing sediment
- Controlling diurnal dry weather flow
- Minimising pumping station costs
- Controlling spill to rivers
- Minimising the transient impact on receiving waters
- Providing improved flood protection

An RTC system consists of regulators, sensors and controllers; see Table 19.5. Rules are used to operate the regulators controlling the flows and these rules depend on the inflow conditions. The controllers receive information from sensors and this information is used to control the regulators in order to produce the optimum performance of the network under various conditions.

Table 19.5 The elements of RTC.

Item	Description	Examples
Regulators	Physical structures such as auto penstocks that move or store flow.	Auto penstocks Variable level weir crests Pumps
Sensors	Equipment such as level sensors that record aspects of system behaviour.	Flow monitor Depth monitor Velocity probe Raingauge Timer/clock pH meter Water quality meters
Controllers	Decision making units which communicate the required actions to the regulators	Manual – An operative observes the current state of the system and physically alters the setting of a regulator(s) based on this information. Remote – An operative observes the current state of the system and directs the setting on a regulator remotely via a telemetry link. Automatic – Direct communication between sensors and regulators. Supervisory – The system normally operates under automatic control but can be over-ridden by an operative.

It is generally acknowledged that there are three levels of RTC, namely, local, area and global: see Table 19.6. Obviously, the complexity increases according to the area affected by the RTC. The strategy adopted in each case depends on the peculiarities of the sewerage system. If rainfall is uniform across a catchment then the best strategy to minimise the volume and frequency of spills is to employ a uniform filling of the available storage elements, even though this does not consider the quality of the discharges or the sensitivity of the receiving watercourses. In theory, all storage should be utilised before any spills from CSOs start to occur, and this condition has the advantage of meeting the current requirements of overflow consents more closely.

Table 19.6 The levels of RTC.

Level	Description
Local	One regulator controlled with information from one sensor
Area	Parts of a system are controlled in a coherent manner
Global	Multiple regulators across the network are controlled with information from multiple sensors

There are two basic ways in which the rules of an RTC system can be developed. The first way is to deduce rules for the regulators and then to demonstrate their effectiveness by analysing historical events. This is off-line control. Alternatively, on-line control is about continuously revising and updating the rules based a prediction of future system behaviour, and finding the best way of minimising the failure or damage. For guidelines about implementing RTC, see CIWEM (1996).

One of the principles of RTC is that pass-forward rates from overflows can be varied. In fact, under certain circumstances auto-penstocks can be fully closed, with all the flow being retained in a certain part of the catchment. Sensible design criteria should be developed in consultation with the appropriate regulatory body.

In developing scenarios for RTC, care should be taken over the selection of the appropriate rainfall:

- Synthetic (design) rainfall data
- Annual time series rainfall data
- Stochastic rainfall generator (such as StormPac™)
- Local historical time series data

Design storms may be of value for certain types of analysis, but a series of events should normally be used.

It is important to ensure that during the failure of any element of RTC, the sewerage system does not perform any worse than it would have done if it was operated under passive control. It is therefore necessary to identify the 'safe' condition that the system should revert to in the event of a breakdown of the RTC system.

19.25 SYSTEM REHABILITATION

Rehabilitation of an urban water utility network is a major aspect of its management. Any sewerage network deteriorates in time for a variety of reasons: pipes degenerate with age, structural stresses and strains lead to pipe fracture and collapses, there is unacceptable infiltration, the levels of service against flooding in the catchment become unacceptable in time due to increased loading from rainfall-runoff, overflows operate with increased frequency, and so on. In view of the large number of sewerage networks around the world, considerable effort has been put into deriving an appropriate procedural approach to sewerage rehabilitation (see also Part IV).

Note that the word 'rehabilitation' is not strictly correct for the task in hand. 'Rehabilitation' conveys the notion of restoring the system to its *original* performance. But this is not what the sewerage authority is normally attempting to do. Its concern is to improve the performance of a system that has inevitably changed its structure and loading. The population being served by the system may well have changed significantly, and the drainage network may have been considerably extended. Also, community expectations may have been raised such that they seek less frequent flooding in a combined system. In other words, levels of service may have increased and therefore standards have been raised. Consequently the sewerage manager is not necessarily seeking to restore the performance of the system to its original level of service, but to improve the service provided.

A number of procedures for sewerage rehabilitation have been put forward. The original version in the UK was the Sewerage Rehabilitation Manual promoted by WRc in 1983. It is currently in its 4th edition (WRc 1994). Subsequently, extensions of this procedure have been advocated by the Guide for the Rehabilitation of Sewerage Systems and the HydroPlan procedure (Vojinović *et al.* 2005), among others, see also Price and Vojinović (2011).

Stage 1 (Initial Planning) of the Guide procedure concerns the definition of the extent of the rehabilitation project; see Figure 19.30. It may be initiated by changes in legislation, poor performance of the network, evidence of a physical deterioration of the system, or the threat of additional loading on the network. Tangible and intangible indicators are used to determine the performance of the system. A 'trigger' is defined for each indicator as a criterion for giving attention to the particular aspect of performance. A 'target' criterion is then defined as an objective for the performance as measured by the indicator. In both cases, trigger and target levels will normally be specified by local, regional or national government. Criteria for the added value of the improvements in system performance are then specified, along with criteria that take into account the intangible benefits.

The performance requirements for a rehabilitated system are similar to those for a new system. BSI (2008) EN752-2 is entitled "Performance Requirements", and in Clause 6, it lists the following "Basic performance requirements":

(a) The pipework operates without blocking.
(b) The flooding frequencies shall be limited to prescribed values.
(c) Public health and life shall be safeguarded.
(d) The sewer surcharge frequencies shall be limited to prescribed values.
(e) The health and safety of operator personnel shall be safeguarded.
(f) Receiving waters shall be protected from pollution within prescribed limits.
(g) Drains and sewers shall not endanger existing adjacent structures and utility services.
(h) The required design life and structural integrity shall be achieved.
(i) Drains and sewers shall be watertight in accordance with testing requirements.
(j) Odour nuisance and toxicity do not arise.
(k) Appropriate access shall be provided for maintenance.

Indicators therefore relate to hydraulic performance, structural condition, environmental impact and social response, and are selected to minimize overlap. (Note that the social response has been introduced here in recognition of the importance of consulting stakeholders; it is not in the original Guide). Performance criteria for the system must be selected, along with objectives. For example, the number of spills for overflows, the life expectancy of the rehabilitated system, and the flooding frequency

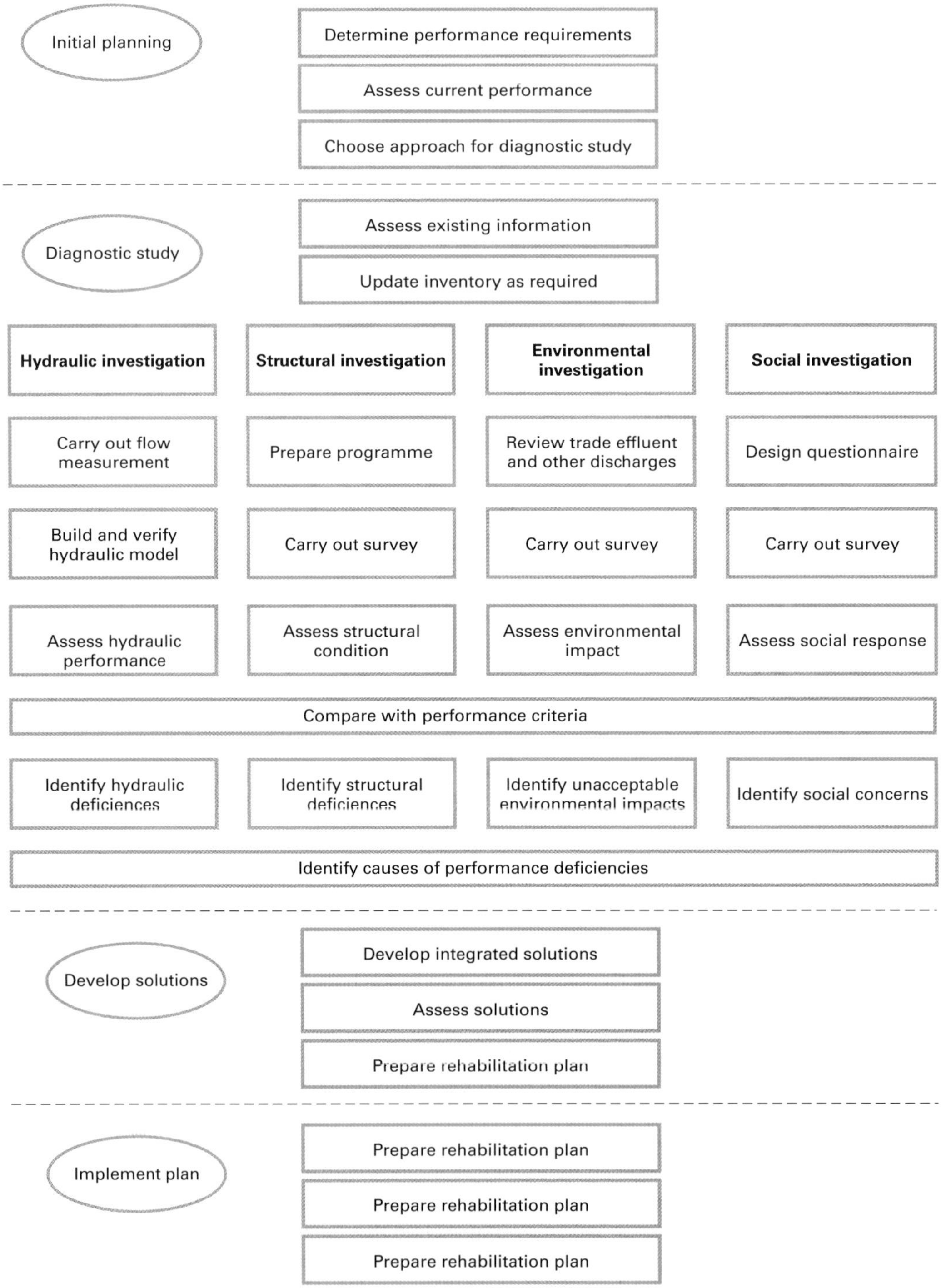

Figure 19.30 Sewerage rehabilitation (after WRc (1994))

at specified locations. Basic hydraulic, structural and environmental performance requirements are specified in Part 2 of EN 752 (BSI 2008), whilst recommendations for design criteria and procedures are given in Part 4; basic design recommendations for structural performance are given in EN 1295 (BSI 1998).

Given that the trigger levels indicate that a full diagnostic study is required then the objective is to assess the hydraulic, structural, environmental and social performance of the sewerage system against defined performance criteria, and to identify the causes of performance deficiencies. For this a comprehensive sewer system inventory should be developed and maintained using a GIS.

Given a large sewerage network it is likely that 80% of the problems are caused by 20% of the network. Therefore there is clearly a need where possible to focus the rehabilitation on a limited set of pipes, and to identify three categories of pipe: critical, strategic and normal. Every pipe is normal (and acceptable in its current state) or critical or strategic. A strategic pipe is one that is key to the performance of the system whereas a critical pipe is one that might be difficult and expensive to rehabilitate, such as a pipe under a main railway line or major road; see HydroPlan (Vojinović *et al.* 2005) below.

A hydraulic simulation model is an essential tool for the assessment of both the hydraulic and environmental performance; see Table 19.7. The integrity of the structural data for the model should be confirmed together with any calibration and verification of model parameters, including parameters relevant to pollutant loading and transport. The hydraulic model may then be used to assess the service levels of flooding throughout the catchment and the pollutant concentrations reaching the treatment works or being discharged through the overflows. If necessary, the modelling can be extended to include the impact on the receiving waters (lake, river, estuary or coastal waters) and the interaction with groundwater; see also Price and Vojinović (2011).

Table 19.7 Impacts.

State	Environmental impact
Hydraulic	Sewage spills to rivers
	Aesthetic (litter)
	Dissolved and organic solids
	STW discharges
	Stormwater discharges
	Flooding
Structural	Groundwater pollution
Other	Odour
	Noise
	Operational and maintenance activities
	Illicit connections to surface water systems
	Industrial effluent
	Domestic effluent

In order for the pipe network to fulfil its purpose the pipes must not collapse or leak, thereby contaminating the environment, and infiltration should be minimal. Excessive infiltration reduces the hydraulic performance of the network and the efficiency of the treatment works, and can create voids in the backfill and soil around the pipes which can put unacceptable stresses on the pipe structures leading to their collapse. A structural study using man-entry or CCTV should therefore be carried out to ensure that the integrity of the pipe system is maintained. A scoring system may be used to highlight those pipes that are failing structurally, or could fail in the future.

A cost-benefit approach can be applied to the development and selection of appropriate rehabilitation proposals in order to improve the performance of a sewer network. Account is taken of the costs of phased solutions, maintenance and operating costs and also the estimated costs of the consequences of failure. Other intangible costs such as public perception of the service provision may also be included. In parallel with costing, a model of the network should be used to evaluate the projected level of service of the future, preferably taking into account the deterioration of the sewer fabric, the operation and maintenance proposed, future developments and other influences on the system performance. The model should be used to ensure trigger levels are complied with for the duration of the planning period. The main limitations of this approach are in determining the deterioration rate of the sewer and assessing the risks and consequent costs of failures. The full analytical process involved in assessing the best option for rehabilitation can be summarised as follows:

(a) Understand the "Trigger" failures of the current system
(b) Establish client criteria "Targets" for performance indicators
(c) Develop Performance scoring criteria for assessment
(d) Develop intangible indicators for Value Engineering
(e) Score the baseline network for comparative purposes
(f) Develop engineering solutions for rehabilitation that generally meet Client Target levels for indicators and ensure all Trigger failures are removed
(g) Cost the engineering solutions
(h) Score the engineering solutions to determine Performance scores;
(i) Score Value Engineering indicators
(j) For each engineering solution, evaluate costs against the system performance score and the Value Engineering scores and compare with the baseline network performance score
(k) Eliminate high cost, low value options
(l) Further develop high value options
(m) Choose the Best Engineering solution.

The sewerage rehabilitation strategy to find a solution can consider the following mix of options:

(a) Rehabilitating individual sewers
(b) Introducing additional storage or sewerage ancillary structures
(c) Improving performance of CSOs or other existing structures
(d) Introducing RTC to make better use of existing (or new) storage

The choice on how to rehabilitate an individual sewer will depend on a variety of *local* factors. However, solutions cannot just be local as every change to the system will have an impact downstream and possibly a short distance upstream as well. This means that there are also important *global* considerations for nearly all solutions. Therefore, the design of a rehabilitation scheme to meet certain levels of service in terms of frequency of (significant) surface flooding and pollution impact must consider the whole system. In addition, it should be understood that a system is not 'frozen' at the time the scheme is implemented. In fact, the system continues to change: pipes age, the population changes and development takes place within the catchment. The design life of the system is therefore also important. Not only might it be better to have a shorter design life, but it might also be better to phase the rehabilitation into the future. We have therefore a multi-criteria optimisation problem.

It should also be noted that predicted failure of a sewer poses a probablistic problem due to hydraulic failure being related to the frequency of rainfall events. It is impossible to take all issues into account because the future, by definition, is uncertain, but these issues illustrate the importance of deriving solutions which take into account whole-life costs, the future of the system and the risks and consequences of delaying rehabilitation activities.

A catchment that has a range of problems that need to be addressed over a period of time will require a range of different rehabilitation options. Due to the interactive nature of networks, each solution may affect the performance of the system at other locations as well. This will influence the requirements for other rehabilitation solutions. It is therefore clear that some proposals, particularly for complex systems, will benefit from comparing the rehabilitation costs with an objective evaluation of a measurement of the improvement in system performance. Other proposed solutions may have very little influence on other parts of the network and the choice of option is dictated by the cost of the proposals.

The rehabilitation plan should be part of a continuous procedure of monitoring the system performance, taking account of changes in the required levels of service and minimising operating and maintenance costs.

The performance score enables a continuous evaluation of the system performance, drawing attention to any reduction in performance over time and also flagging up "Trigger" failures.

One of the problems faced by operators is that of treating such an exercise as an end in itself and not keeping an adequate record of the changes made: a rolling programme of phased rehabilitation changes requires a continuously updated model and the related GIS database needed to provide all the relevant data.

Stage 4 of the Guide provides guidance on the activities relating to the needs during the construction process, as one which will ensure a satisfactory outcome for the construction, and also on the various testing procedures available to confirm that the renovated sewers meet the specified performance criteria.

19.26 URBAN POLLUTION MANAGEMENT

The discharges from combined sewer overflows will normally contain significant concentrations of sewage pollutants. Even the effluent from wastewater treatment works will include residual quantities of different pollutants. These pollutants have an adverse effect on the receiving waters, whether a river or stream or coastal waters. In each case the receiving waters have a self-treatment capacity, but in terms of normal runoff from urban areas this capacity is limited and the impact of the pollutants can have serious short-term and long-term consequences for the water quality of the receiving waters. The impact on different rivers and coastal waters has become so significant, with serious consequences for environmental and public health that many governments have had to bring in legislation to control urban discharges to receiving waters through the setting of environmental quality standards (EQOs) to meet agreed environmental quality objectives (EQOs) for the various target uses of the receiving waters. So, for example, in 1994 the UK advocated an Urban Pollution Management (UPM) programme which adopts such an approach to enable the control of CSO discharges such that levels of impact are compatible with the desired uses of the receiving water. Water quality criteria are used to define pollutant concentration profiles in terms of magnitude (pollutant concentration), duration, and frequency, against which output from a river model can be evaluated in order to assess the performance of the sewerage system in respect of the EQO's.

The UPM approach is to use a variety of models in an integrated methodology for management of pollution on the river catchment scale. The tools include water quality models to simulate the behaviour of pollutants and sediments in sewer systems, sewage treatment works, and rivers. The UPM procedure is not one specific methodology, but a set of methods selected on the basis of appropriateness. In general, the method uses detailed water quality models (including river, sewage treatment works and sewers) though simplified approaches are also recommended.

The UPM procedure has four key stages

(1) Initial planning
(2) Data collection and modelling
(3) Development of solutions and testing for compliance
(4) Obtaining approval and detailed design

These are discussed in detail in Price and Vojinović (2011).

The WaPUG guide to Quality Modelling of Sewerage Systems (WaPUG 2006) outlines a procedure for building, calibrating and operating a model of water quality in a sewerage system. In summary, the procedure is as follows:

- First, define the purpose of the model, such as to determine intermittent discharges to receiving waters from CSOs or from storm tanks at sewage treatment works, loads discharged to sewage treatment works, and sediment transport and deposition in sewer systems for planning maintenance activities.
- Select the right type of model to use for the purpose defined. For example, a hydraulic model without an explicit water quality component may be used if the only objective is to determine the frequency of operation of CSOs. Also, the UPM manual recommends the use of simple models such as SIMPOL, which is a tank-based model that is calibrated against the full physically-based model. Otherwise a full physically-based water quality model will be needed, superimposed upon a formally calibrated hydraulic model, which gives a particular attention to the dry and wet weather components.
- Seek any regulatory approval that may be required.
- Build and test the model such that the hydraulic model is verified and initial values for each determining feature in dry weather flow and for each parameter in wet weather processes are defined. There will be a focus on checks such as mass balances, ranges of determining features and unexpected accumulations of sediment. Values of the quality parameters can be obtained from standard publications.
- Collect site specific data, particularly from flow and sediment surveys for calibration purposes.
- Calibrate model parameters within valid ranges to make the model fit the data for both dry and wet weather flows. Note however that force-fitting should be avoided at all costs. Additionally the values of the parameters should be confirmed by checking the model against observed events not used in the calibration.
- When running the calibrated model it is necessary to define properly the antecedent dry weather period before the rainfall event. This affects the build up of sediment on the surface and in the pipes. Alternatively a continuous rainfall record including dry periods can be run with the software. Dry and wet weather flows should be calibrated separately.
- Report on model building and calibration, including sections on project definition, model building, model calibration, and model testing.

19.27 RIVER MODELLING

Typically, there are different modelling approaches that can be applied to river modelling and they range from simplified conceptual models to the more complex 1D-2D models. The digital solution to the 1D

Saint Venant equations is also relevant to the 1D flow in river channels. Unlike the pipe situation separate cross-sections are defined at each node where h is calculated. Cross-sections are typically introduced at 100 to 500 meter intervals, whereas on steep rivers this distance may need to be decreased, Figure 19.31. However, if sections are too close together models may suffer from stability problems. The cross-sections may include flood plains, though where there are embankments along the river channel, or the flow is strictly two-dimensional on the flood plain, then a 2D flow model is needed in interaction with the 1D model for the river channel. Alternatively, a full 2D model can be used for both the channel and the flood plain. For the channel, A-h and B-h relationships are established for each model cross section, and a rating curve is more likely to be used as the downstream boundary condition.

Figure 19.31 Illustration of a detailed 1D model of river (cross-sections and branches)

In terms of river hydraulics, features that can affect conveyance as the water level change are very important (e.g., locations with variable cross sections, large changes in gradient, large changes in depth, etc.). Weirs and level control structures are particularly important and they require careful incorporation into the river models. Once a model has been instantiated a number of tests must be undertaken to verify that the model runs in a stable manner.

19.28 COASTAL SYSTEMS MODELLING

Navier-Stokes equations are the governing equations for modelling hydrodynamic processes of coastal systems, Figure 19.32. Elevations are typically obtained from the solutions of the depth-integrated continuity equation and the velocity is obtained from the solution of the momentum equations. These equations are often solved using either the finite difference or the finite volume method. The boundary conditions and conditions within the domain refer to the following forces: wind, wave and tides. The wind action is normally considered as a constant or as a variable action in space and time. Wind action determines the movement of the superficial layer of the water and the rise of the sea level due to this phenomenon. Waves need to be taken into account in order to determine the currents as well as the rise of the sea level close to the chore due to the breaking of the waves.

With respect to the tides, it is necessary to distinguish between the astronomical and the meteorological tides. The astronomical tide description can be either a boundary condition or a domain condition. The meteorological tide is caused by the wind action and by the pressure variation that has to be introduced in the model.

Depending on the size of the study area, the Coriolis effect may also need to be taken into consideration. Furthermore, such factors as evaporation, precipitation, the friction with the bottom, salinity, temperature conditions and the sources and sinks of water flows are also important to be incorporated into coastal models.

In terms of the interaction between coastal and inland floods, the main influence of coastal systems can arise due to reduction of drainage capacity and backing up of water. This situation can occur due to combined effects of rainfall events, sea level rise as well as fluvial discharges. Therefore, such interaction needs to be carefully considered and appropriately represented in coastal models.

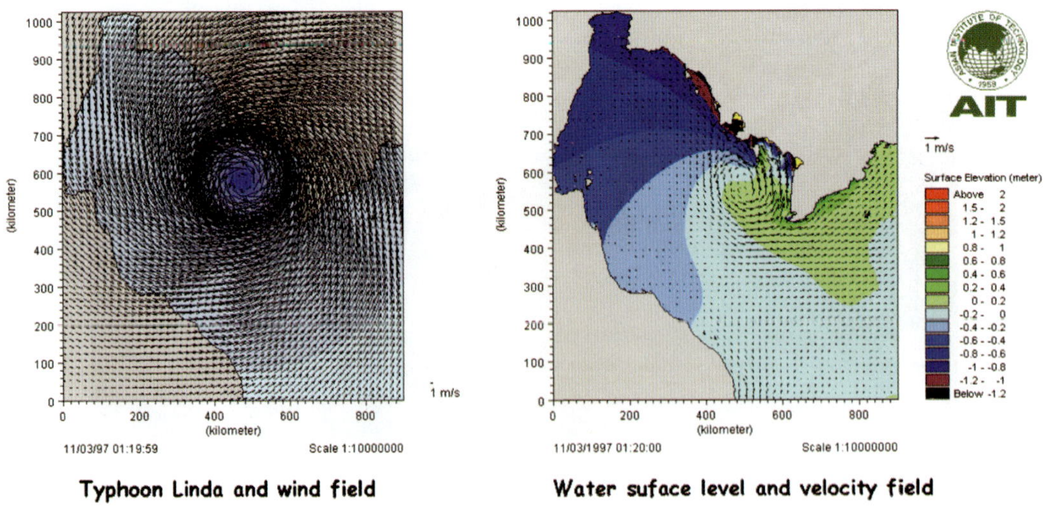

Figure 19.32 Modelling of Tropical Storm Linda which occurred in 1997; This system was the worst typhoon in southern Vietnam in at least 100 years, killing thousands of people and leaving extensive damage. (Source: Khaing 2010)

19.29 GROUNDWATER MODELLING

As mentioned earlier, groundwater models aim to describe groundwater flow and fate and transport processes by solving Richards equation and they are based on several simplifying assumptions. These assumptions typically involve the direction of flow, geometry of the aquifer, the heterogeneity or anisotropy of sediments or bedrock within the aquifer, the contaminant transport mechanisms and chemical reactions. Execution of groundwater studies requires the following information:

- Hydraulic properties of the aquifers and confining units.
- Areal extent and thickness of aquifers and confining units;
- Boundary conditions which control the rate and direction of movement of groundwater.
- Horizontal and vertical distribution of hydraulic head throughout the modeled area for beginning (initial conditions), equilibrium (steady-state conditions) and transitional conditions when hydraulic head may vary with time (transient conditions).
- Details of the groundwater recharge/discharge, pumping or injection, leakage to or from surface-water bodies, sources or sinks (i.e., stresses). Sources or sinks may be constant (i.e., unvarying with time) or may change with time (i.e., transient).

The typical outputs from the model simulations are the hydraulic heads and flow rates. Figure 19.33 shows simulated groundwater flow vectors and hydraulic head.

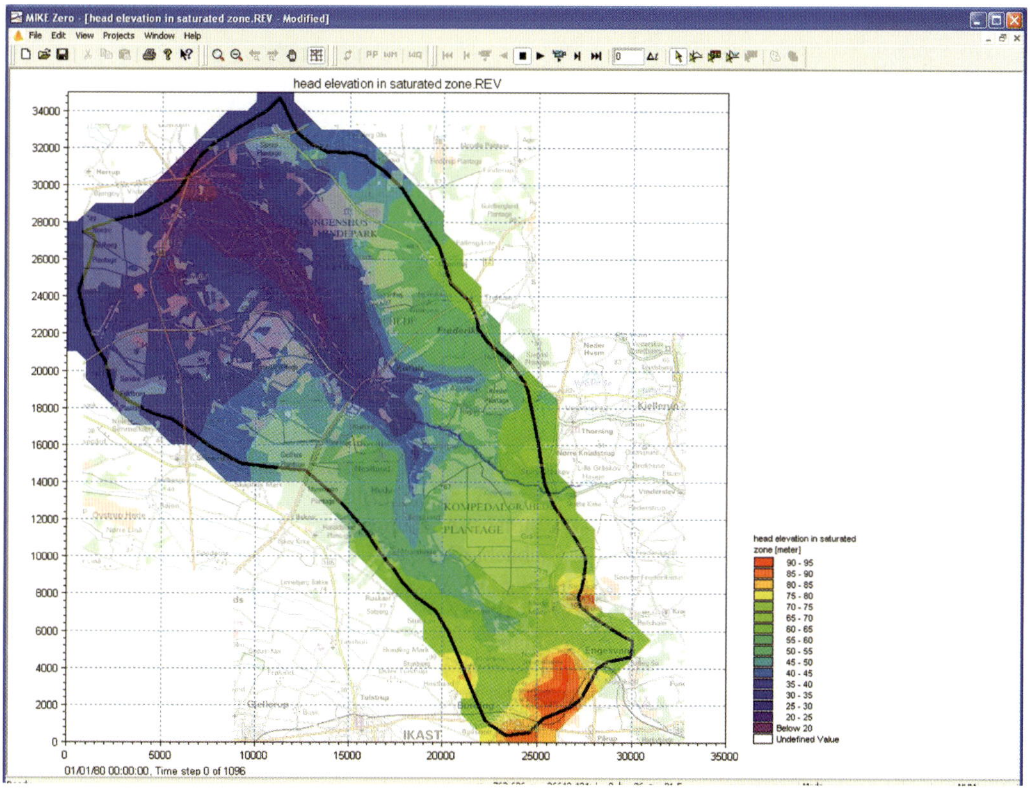

Figure 19.33 Map view of groundwater head elevations (typical MIKE SHE model output)

Once the groundwater model is calibrated and verified it can be then used to simulate possible future changes to hydraulic head or groundwater flow rates as a result of future changes in stresses on the aquifer system. In recent time, such models are coupled with drainage models to simulate groundwater infiltration into the pipe network systems.

19.30 INTEGRATED MODELLING

Integrated modelling refers to modelling more than one system and its interactions with other systems. It aims to improve understanding of different processes and their interactions and it can involve different scales of modelling complexity (see also Price and Vojinović 2011). In the past, different systems have been modelled in a sequential way without the feedback between different models; see Figure 19.34. As an improvement to this approach, efforts have been made to incorporate feedback mechanism and to enable models to exchange information at each time step (this is also commonly referred to as simultaneous or parallel modelling); see Figure 19.35.

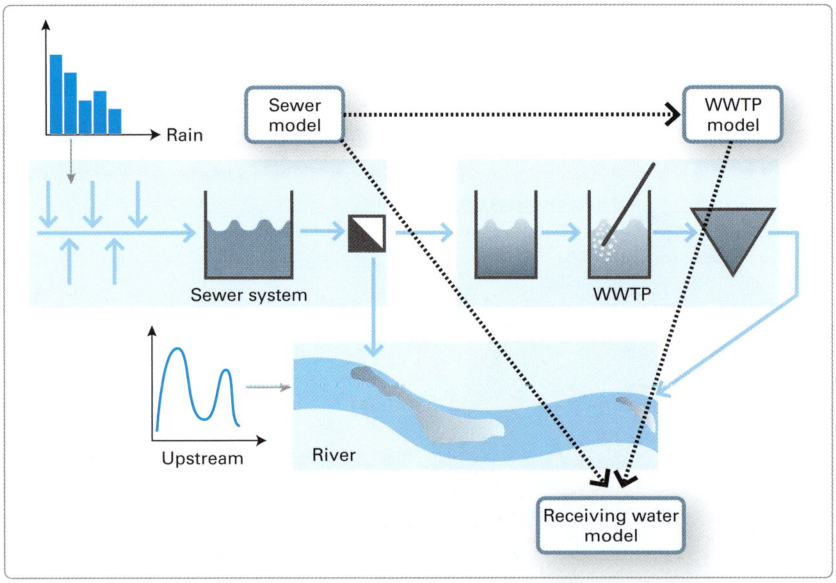

Figure 19.34 Example of a sequential modelling framework

Typical examples of the more common interactions are:

- Sewer system, wastewater treatment plant (WWTP) and rivers;
- Surface runoff and sewer system (this depends largely on the number of catchpits or gullies and their hydraulic characteristics);
- Minor and major systems;
- Minor and coastal systems;
- Minor and groundwater systems;

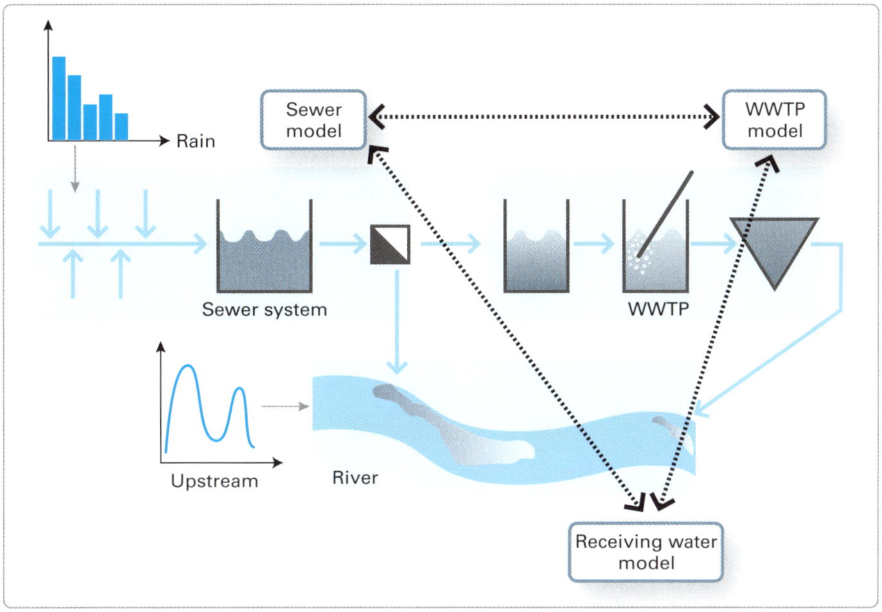

Figure 19.35 Example of a simultaneous (or parallel) modelling framework (i.e., modelling with feedback)

The following Table 19.8 provide an overview and application of different modelling systems.

Table 19.8 Model application areas.

Model category	Application areas
Compartmentalised (0D)	Rainfall-runoff, reservoirs, flood routing
1D hydrostatic, sub- or super-critical flow, free surface or pressurised 1D Boussinesq (first order non-hydrostatic) 1D compressible	Rivers, flood plains, estuaries, irrigation, water distribution, wastewater and stormwater collection, treatment works, water hammer
2D hydrostatic, vertical (depth – layers, horizontally averaged) 2D hydrostatic, horizontal (spatial, depth averaged, turbulence, wind (for shallow depths) pressure, Coriolis density variations, wave stress) 2D Boussinesq (first order non-hydrostatic) 2D wave refraction, reflection, diffraction	Lakes, wide river channels, flood plains, estuaries, coastal waters, groundwater
3D hydrostatic (turbulence, wind, pressure, Coriolis, thermoclines, density variations) 3D non-hydrostatic, (Navier Stokes plus turbulence)	Lakes, river channels, channel-flood plains, coastal waters, estuaries, treatment processes almost anything, river channel-flood plain interaction, structures

All sewer (minor system) modelling is with 1D models. Most of the river (major system) modelling is also with 1D models. Major system overland flow can either be modelled withy 1D or 2D models. Traditionally, the floodplains alongside a river (or a watercourse) have been modelled with 1D models where the wider river channel cross-section (including parts of the floodplain) are introduced. However, it is important to note that this practice is dependent on extrapolation of computed channel water levels to fit the terrain and if there are only few cross sections then this procedure may result in unrealistically high/or low water depths. This is further discussed and illustrated in Part IV.

Part IV

Practical Aspects of Flood Risk Assessment and Mitigation

Chapter 20

Flood risk assessment

20.1 OVERVIEW OF PART IV

The purpose of flood risk assessment and mitigation is to identify areas at risk, assess the severity of risk, investigate how the community perceives such risk, define what would be an appropriate mitigation measure (or a set of measures) that can minimise (if not completely eliminate) its negative impacts, produce a plan for implementation of mitigation measures, implement the measures accordingly and monitor their effectiveness (see Figure 20.9). One of the important aspects of this work is to introduce and maintain *an holistic perspective* throughout the entire process. This is one which will enable us to look at risk as a whole, as this emerges through its parts and their interactions (see also Section 2.1, Part I). Certainly, the use of physically-based computational modelling coupled with GIS mapping is invaluable for this purpose. With instantiated models and specialist GIS flood mapping techniques it is possible to explore the generation of floods and evaluate effects of different measures in response to any extreme event. This process may then proceed in a *hydroinformatics graphical environment* by simulating a number of alternative scenarios in ways that are not only suited to the immediate needs of the stakeholders but will lead them to deepen their understandings and awaken their own individual and collective imaginations, thereby activating their own creative imaginations of the phenomenon (i.e., flood risk). Such environment would then represent a central instrument for facilitating and conducting qualitative and quantitative risk assessment. Information sourced from a number of related (suitably detailed, coloured and dynamic) pictorial situation-orienting scenarios can enable stakeholders to gain better understanding of risk and to contribute actively to building more resilient societies.

This part of the book addresses some of the practical aspects of flood risk assessment and mitigation work. It starts with an overview of some of the current practices and it describes how the qualitative and quantitative data and methods can be used in the related work. An approach based on the holistic perspective which takes into account not only technical but also social and ethical considerations is presented and discussed. A brief overview of some of the typical flood mitigation measures concerning different types of floods is also provided. Finally, this part of the book ends with discussion concerning production of flood-related plans and presentation of four case studies where different aspects of flood mitigation practice are described and illustrated.

20.2 CURRENT PRACTICES

20.2.1 US practice

The Federal Emergency Management Agency (FEMA) has undertaken several risk analysis programs that assess the impact of natural hazards that lead to effective strategies for reducing risk for the entire US (see also http://www.fema.gov/hazard/flood/index.shtm). These programs support the Department of Homeland Security objective to "strengthen nationwide preparedness and mitigation against natural disasters". FEMA aims to foster informed risk management decisions and actions to mitigate risk through a consistent risk-based approach to assessing potential vulnerability and losses, and tools to communicate the related messages. By analyzing and delineating risk in flood zone areas, communities in the US can better understand their risk and make more informed decisions to reduce vulnerability. Flood zones are geographic areas that the FEMA has defined according to varying levels of flood risk. Each zone reflects the severity and/or type of flooding in the area taking into account the presence and status of Federal flood control systems.

The Department of Homeland Security (DHS) has developed a national infrastructure plan (NIPP) which is supported by 18 sector-specific plans with the purpose of increasing the protection and resilience of their critical infrastructure and key resources (CIKR). The NIPP risk management framework is illustrated in Figure 20.1, see also (http://www.dhs.gov).

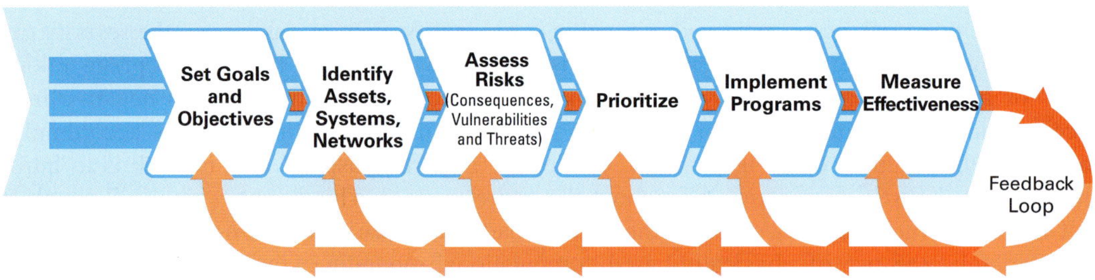

Figure 20.1 'The NIPP risk management framework', US practice (Source: DHS 2009)

The NIPP framework calls for assessment of risk from any scenario as a function of consequence, vulnerability, and threat:

$R = f(C, V, T)$

where:

C – Consequence (social, economic and environmental effects of an event);
V – Vulnerability (a degree of susceptibility to disruption);
T – Threat (something with the potential to cause the harm).

The risk assessment process within the NIPP framework combines the following aspects:

- characterisation of the system, including its objectives and the services it provides;
- identification and prioritisation of adverse consequences that have to be avoided;
- determination of critical assets that might be subject to malevolent acts that could result in undesired consequences;
- assessment of the likelihood (qualitative probability) of such malevolent acts by adversaries;
- evaluation of existing countermeasures;
- analysis of current risk and development of a prioritised plan for risk reduction.

20.2.2 EU practice

The EU Floods Directive 2007/60/EC outlines the requirements for flood risk assessment, mapping and management. In this document, a considerable emphasis is placed on the role of floodplains and sustainable land-use practices. The document requires Member States to undertake a national preliminary flood risk assessment by 2011 to identify areas where significant flood risk exists or might be likely. It also requires the preparation of catchment-based flood risk management plans by 2015 by setting out flood risk management objectives, actions and measures for those areas. As such, this directive is driving advances in flood risk assessment and hazard mapping across Europe and so far the majority of Western European countries have produced the related flood maps.

A handbook which facilitates the exchange of information, knowledge and experience in the field of flood risk mapping was produced by the European Exchange Circle on Flood Mapping. It provides an overview of current practice across Europe (EXCIMAP 2007). The flood mapping procedures outlined in this document focus on river and coastal flooding. Also, the same document addresses "torrent flows" in mountainous areas on the basis of the methodology developed in Austria. Issues concerning groundwater flooding refer to the research carried out by Defra, the Environment Agency and the British Geological Survey (BGS). Some aspects concerning flash flooding are also discussed in this document.

20.2.3 UK practice

The UK government has prepared a National Risk Register (NRR) and guidance on risk assessment and the preparation of Community Risk Registers (CRR). Many of these documents are now publicly available. The risk assessment process is outlined in the HM Government document (HM 2005) and illustrated in Figure 20.2.

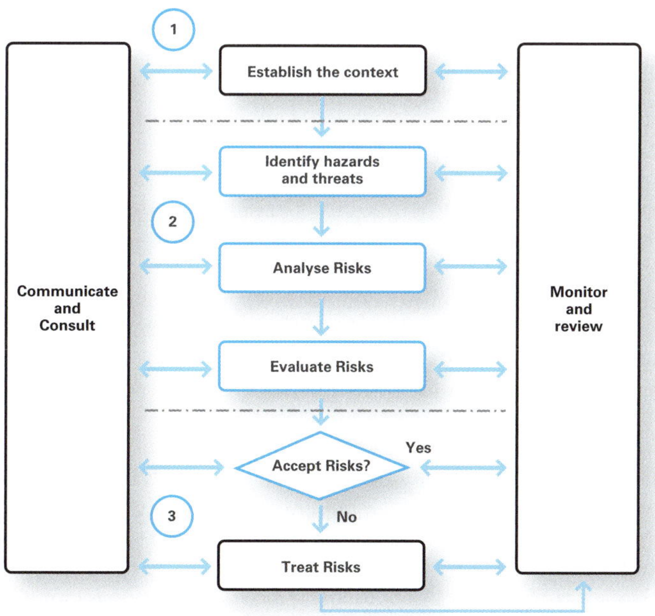

Figure 20.2 'The risk management process within emergency preparedness', UK practice (HM 2005)

Since the assessment of flood risk in the UK has been in a focus for many researchers and practitioners over a number of years, all parts of the UK have prepared hazard maps that show areas that are liable to flood from rivers and the sea. The Environment Agency's flood map, which is accessible through their website (http://www.environment-agency.gov.uk/subjects/flood/floodwarning/), illustrates the areas at risk from a one per cent (1 in 100) annual probability flood (inland), a 0.5 per cent (1 in 200) annual probability flood (in tidal areas) and the 0.1 per cent (1 in 1000) annual probability flood across England and Wales. The map is limited to the hazards associated with rivers and sea. It does not take account of the presence of flood defences, nor of potential climate change effects. The map is kept alive, being continuously updated to reflect the outputs of the Agency's ongoing detailed local hydrological and hydraulic modelling and mapping studies. The Environment Agency also has a historic flood map, which is updated as floods occur.

In Scotland, the Scottish Environment Protection Agency (SEPA) has developed a river and coastal flood map which indicates the areas likely to flood from a 0.5 per cent (1 in 200) annual probability event under existing climatic conditions. In Northern Ireland, the Northern Ireland Rivers Agency's (NIRA) have produced a strategic flood map which highlights areas that have flooded historically, areas likely to flood under current climatic conditions (one per cent (1 in 100) annual probability events for rivers, 0.5 per cent (1 in 200) annual probability for the sea and for land areas potentially affected by climate change as estimated up to 2030.

The British Geological Survey (BGS) has been involved in flood research and data provision for the UK. They have produced maps showing susceptibility to groundwater flood risk and the Environment Agency has recently produced pluvial flood risk susceptibility maps for England.

Table 20.1 Levels of flood risk assessment – FRA (Source: Lancaster *et al.* 2004).

FRA level	Description
Level 1	Screening study to identify whether there are any flooding or surface water management issues related to a development site that may warrant further consideration. This should be based on available existing information, including national flood maps, advice from the Environment Agency, SEPA or rivers agency.
Level 2	Scoping study to be undertaken if the Level 1 FRA indicates that the site may lie within an area that is at risk of flooding or that the site may increase flood risk due to increased runoff. This study should confirm the sources of flooding that may affect the site and should include: – an appraisal of the availability and adequacy of existing information – a qualitative appraisal of the flood risk posed to the site, and potential impact on the development on flood risk elsewhere – an appraisal of the scope of possible measures to reduce the flood risk to acceptable levels. The scoping study may show that sufficient quantitative information is already available to complete a flood risk assessment appropriate to the scale and nature of the development.
Level 3	Detailed study to be undertaken if the Level 2 FRA concludes that further quantitative analysis is required to assess flood risk issues related to the development site. The study should include: – quantitative appraisal of the potential flood risk to the development – quantitative appraisal of the potential impact of development site on flood risk elsewhere – quantitative demonstration of the effectiveness of any proposed mitigation measures.

In terms of new development activities, flood risk assessments are required in support of the majority of applications. For example, Planning Policy Statement (PPS) 25, (Scottish Planning Policy) SPP7, Technical Advice Note (TAN) 15 and PPS15 provide guidance to planning bodies on how they should take account of flood risk in their spatial planning activities. In England, regional planning bodies are responsible for the preparation of regional flood risk assessments and local authorities have to prepare strategic flood risk assessments. Both levels of study require assessment of flood risk from all sources. CIRIA C624 (Lancaster *et al.* 2004) and the practice guide to PPS25 (CLG 2010) refer to three levels of site-specific flood risk assessments: see Table 20.1.

20.2.4 Australian and New Zealand practice

In Australia and New Zealand, the risk management standard AS/NZS 4360:2004 provides a basic framework for managing and assessing the risk posed by any kinds of natural hazards. The steps involved in this process are: establish the context, identify the risks, analyse the risks, evaluate the risks and treat the risks (Figure 20.2). The standard requires continuous communication with stakeholders throughout each step of the process. It also requires that this process is monitored and reviewed. The steps in the shaded subsection in Figure 20.3 (i.e., identify risks, analyse risks and evaluate risks) form the risk assessment process. As AS/NZS 4360:2004 is generic, professionals from different disciplines have tailored individual sections to suit their areas of expertise and responsibility. For example, Emergency Management Australia (EMA) has developed a detailed risk management process for emergency management (EMA 2004), and the Australian Geomechanics Society (AGS 2007) has developed guidelines for landslide risk management. The reader is referred to AS/NZS 4360:2004 and its companion guide HB 436:2004 for further information on the steps in the related risk management process.

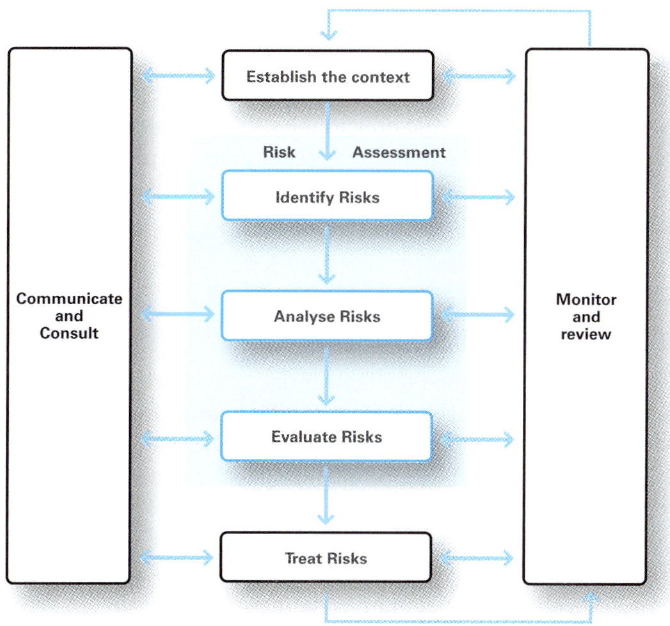

Figure 20.3 'The risk management and assessment process', Australian and New Zealand practice; Source: AS/NZS 4360:2004

As the Australian and New Zealand population has in many cases settled very close to rivers for the supply of water and transport, riverine (or fluvial) flood hazard assessments have been undertaken regularly. Middelmann (2007) states that the extension of flood hazard assessment to a risk analysis that looks more objectively at likelihood and consequence is, however, much less widespread, and there are heavily populated urban areas for which no such analysis of flood risk has been made. This deficit is exacerbated by the fact that urban centers are more affected by flash flooding. In many instances, assessments have led to structural and non-structural mitigation measures. Such measures are usually framed around a 'flood standard' expressed in terms of the annual exceedance probability (AEP) or its inverse, the return period. As mentioned earlier, AEP represents a statistical benchmark used for flood comparison, defined as the probability of a flood event of a given magnitude being equalled or exceeded in any one year (see also Pilgrim 2001).

The flood risk management process is described in SCARM (2000) and other documents (e.g., DIPNR 2005). The approach is not prescriptive and requires identifying and managing the full range of flood risks in partnership with the stakeholders who are at risk of flooding, and taking account of local issues.

The 1% AEP flood or the largest known historical flood (plus an appropriate safety/error margin) is typically used in this flood risk assessment process and for determining the design standard for structural flood mitigation works. However, more extreme floods (e.g., 0.2% AEP) or the probable maximum flood (PMF) are used as the development controls for the building of critical facilities such as hospitals, emergency services, police facilities, power stations and water treatment plants (e.g., Queensland Government 2003).

In terms of the risk analysis process, the following key stages have been identified in the Australian and New Zealand practice:

- assessing the likelihood of flooding by developing an understanding of flood behaviour through modelling studies using floods of different frequencies; and
- calculation of consequences of the full range of flood risks by assessing the exposure and/or vulnerability of the elements at risk of inundation, such as people, buildings and infrastructure. It also examines how risk can most effectively be managed to enable informed decision making.

These stages require making important decisions that relate to the choice and scope of the modelling work. The selection of models would typically depend on variables such as catchment characteristics, the purpose of the modelling, budget, availability of data and time constraints. Typical outputs from the related work would include GIS maps showing the extent of flooding and, the variations in water depth and flow velocity across the floodplains.

20.3 OVERVIEW OF DATA AND METHODS FOR FLOOD RISK ASSESSMENT

20.3.1 Classification of data and methods

Risk assessment, at its simplest, is any method that categorises, or attempts to categorise, risk. The data and methods that are used in the risk assessment work are typically divided into qualitative and quantitative data and methods. Qualitative data and methods aim to gather an in-depth understanding of aspects that cannot be easily (or not at all) expressed in numerical and/or monetary terms and which are based on human values, judgments, feelings and perceptions. As distinct from the qualitative data and methods, quantitative data and methods are used for systematic empirical and analytical investigation of quantitative properties of a process or phenomena and its relationship with other processes or other phenomena. As described earlier, this is also known as *positivism.* The objective of quantitative methods is to develop and employ the forms of calculations, computations, mathematical models, theories and/or hypotheses pertaining to the phenomena. It is important to note that the process of measurement is central to quantitative methods because it provides the fundamental connection between empirical observation and mathematical expression of quantitative relationships. Therefore, we could say that the basis for quantitative methods is scientific (i.e., value-free, or value-neutral, or 'objective') assessment of the process or phenomena under consideration, whereas the basis for qualitative methods is, in the modern-scientific sense, a 'non-scientific' (i.e., intrinsic-value based or 'subjective'). In standard risk assessment practice, qualitative methods are often used for enquiries into the phenomena and the hypothesis generating activities, whereas quantitative methods are used for hypothesis testing. In addition to the straight qualitative and quantitative flood risk assessments there can also be a combined form of these two types of assessments. That form of assessment is typically referred to as semi-quantitative flood risk assessment as it is largely based on quantitative analysis with some minor qualitative aspects appended.

In general terms, the data needed for flood risk assessment work can be grouped into the following four types:

- Social data;
- Economic data;
- Technical data;
- Other data.

20.3.2 Social data requirements for flood risk assessment

Social data refers to various kinds of data and information that can reflect the characteristics of social conditions (or social functioning and well-being) with respect to flood risk over time and across different geographical areas. Such data is typically divided into:

- Direct data – data that can be directly related to a particular aspect of flood risk (e.g., policies, legislations, standards, institutional and organisational roles and responsibilities, etc.)
- Indirect data – data about some other variable that allows an inference to be drawn about flood risk (e.g., timely and accurate provision of information, educational characteristics across the city, etc.)
- Subjective data – data that reflects values, attitudes, beliefs, feelings and perceptions of individuals (e.g., knowledge about flood risk, cultural ideology, ethnic and religious beliefs, risk perception, experiences, etc.);
- Objective data – data that represents counts of behaviour or conditions (e.g., housing conditions, number of people living on the floodplain, age and income, possession of insurance covers, etc.).

20.3.3 Economic data requirements for flood risk assessment

Economic data that is needed for flood risk assessment relates mainly to the following data sets:

- Capital investment and operational & maintenance costs of various flood mitigation (and generally, stormwater or drainage management) measures;
- Economic indicators and cost indices,
- Flood damage data;
- Value of land and properties;
- Etc.

20.3.4 Technical data requirements for flood risk assessment

Technical data refers to different elements of urban drainage systems, flood protection structures, floodplain (coastal zone or river basin) characteristics, and so on (see also Chapter 14, Part III). For example, an urban drainage system database represents a collection of spatial, temporal and other data (e.g., pipe network data, water levels, flow rates, pumping station data, rainfall measurements, etc.). Various technologies can be utilised for collecting technical data. In addition to the traditional ground survey technologies, the use of remote sensing technologies also plays an important role in the collection of technical data. This is true whether for a fluvial system, coastal zone or a drainage network.

In terms of the temporal data, for example, drainage networks engineers have developed the strategy of carrying out short term flow surveys in which as many as 50 flow meters may be installed in a (large) drainage network to collect flow data over a period of several weeks. Such data is also used for calibrating and verifying a physically-based model and for improving the understanding of the structure and the performance of the network.

There are several aspects that need careful consideration and planning in order to achieve successful collection campaign of technical data. Obviously, the first aspect is to clearly define the needs and objectives. Together with this, consideration needs to be given to the minimum degree of accuracy in order to select the right equipment, measurement techniques and calibration of instruments. It is very important that the survey contractor is in direct contact with the person who has made the plan for the campaign because the position of survey sites is likely to be a compromise between where the data is required and where it can be effectively and safely collected. This is equally important for both spatial and temporal data surveys.

Since the data collection is a very costly and time consuming task, it is important to have a systematic and coordinated approach during the implementation of all related activities. Once the necessary data is collected it will then need to go through the process of validating and processing before it can be stored and used. This is usually done within a data management framework, which links various types of databases with GIS technologies.

20.3.5 Other data

In addition to the above mentioned data, we may also need to collect other kinds of data such as:

- Meteorological data;
- Environmental data;
- Geological and soil data;
- Customer-related data;
- Property and infrastructure data;
- Insurance data;
- Actual flood damage data from the past;
- Data concerning other infrastructural services and sectors (e.g., electricity, water distribution, gas, etc.);
- Etc.

20.3.6 Qualitative flood risk assessment

As mentioned earlier, qualitative methods have a descriptive or categorical nature as they are regarded not to be directly tied to a more precisely quantified (or numerical) measure of risk. Therefore, typical sources of information used in qualitative risk assessment are: signs, drawings, statements (or qualitative expressions), experiences (e.g., narrative stories), evidence (which may be sourced from interviews, consultations, questionnaires, etc.) and other forms of subjective expressions. These can be obtained from publicly available documents, site visits, expert opinions, discussions, consultations and workshops with flood authorities as well as community representatives and stakeholders. Typically, the interviewees (or participants) are asked to describe their view of potential impacts from different flood events and situations. They are then asked to give their estimate of frequencies (or likelihoods) of such events and to define scenarios under which these events are likely to occur. In addition to the consultation with community representatives, consultation with directly responsible authorities is usually undertaken in order to gain an understanding of current and future planning policies and their potential implications for those sites that are exposed to flood risk. Furthermore, consultation with other water bodies such as stormwater and wastewater utilities, together with coastal and river basin commissions provides a source of information of any existing relevant national or local technical guidance, existing information of flood risk, known or potential stormwater, wastewater, coastal and river related problems and issues, any proposed plans, and advice on potential interventions that may be relevant for the mitigation of flood risk.

The key methods used to process such information are judgments and reasoning, and the conclusions that are drawn are typically derived in relation to some qualitative criteria (e.g., values, beliefs, feelings, perceptions, etc.), Figure 20.4.

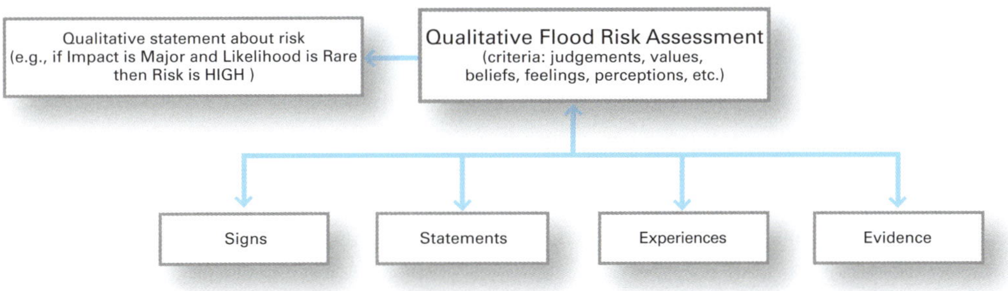

Figure 20.4 Qualitative flood risk assessment

In traditional risk assessment practice, qualitative assessments aim to arrive at some conclusion about the probabilities of outcomes for a baseline risk and/or any mitigation strategies that have been proposed. Interviewees are asked to describe the likelihood and impact of events, selecting from a predetermined set of phrases, such as rare, possible, likely and almost certain, with a range of numerical values assigned to each phrase: see Table 20.2. These expressions are considered subjective as they are defined on the basis of participants' experiences rather than on the basis of statistical methods. Impacts are also defined in a similar way: see Table 20.3. The results from such work are then presented in a form of a decision tree, a risk ranking matrix, a table or any other similar means. Table 20.4 illustrates an example of a risk matrix which categorises different levels of risk.

Table 20.2 An example of likelihood categories.

Category	Likelihood (per year)
Almost Certain	>20
Likely	10
Possible	5
Rare	<2

Table 20.3 An example of impact categories. A similar table can be produced for other risk related aspects (e.g., exposure, etc.).

Category	**Impact on Community life (per event)**
Chatastrophic	Complete collapse
Major	Wider disruption
Moderate	Local disruption
Insignificant	Nuisance

Table 20.4 An example of a risk matrix table expressing different levels of risk.

Likelihood	**Almost Certain**	Moderate	Very High	Very High	Extreme
	Likely	Moderate	High	Very High	Extreme
	Possible	Low	Moderate	High	Very High
	Rare	Low	Moderate	High	Very High
		Insignificant	**Moderate**	**Major**	**Chatastrophic**
			Impact		

The key output from the qualitative risk assessment is a categorisation of risk and a recommendation on the way forward, which may, in case of a relatively simple problem, be a recommendation for a small scale intervention, or in a case of a more complex problem the scope of a more detailed study and requirements for sophisticated topographic and hydrographic surveys as well as the use of physically-based models.

20.3.7 Quantitative flood risk assessment

Quantitative risk analysis is regarded as a more precisely quantified measure of risk and its aim is to provide an expression of risk which involves a (modern) scientific undertaking. Typical sources of data and information used in quantitative risk assessments are: measurements and experimental data, calculations, models and economic analysis: see Figure 20.5. Examples of quantitative methods for risk assessment are: probability-based methods, first- and second-order Monte Carlo simulations, Latin Hypercubes, sensitivity analyses, Bayesian techniques, multi-criteria analysis, etc.

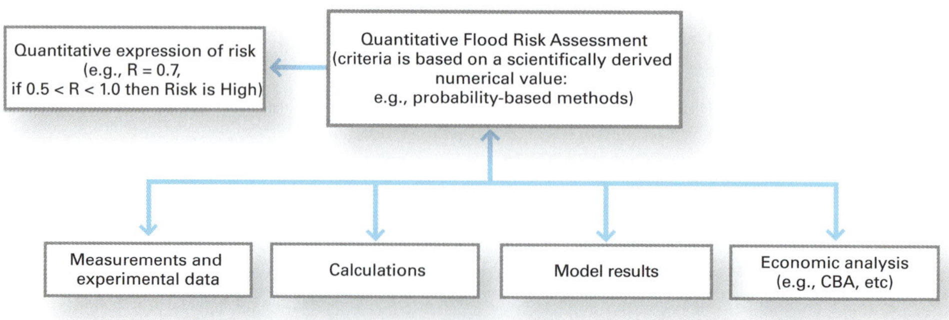

Figure 20.5 Quantitative flood risk assessment

The following Table 20.5 provides an example of an output from a numerical modelling study which can be used as a basis for quantitative risk assessment. In addition to that, other forms of outputs such as charts, tables, GIS maps, 1D, 2D and 3D animations of model results are also used to communicate the results from this work. These outputs are typically produced for a range of different scenarios in order to address different causes and their potential impacts.

Table 20.5 An example of a table that is often used in quantitative risk assessment.

ARI (years)	AEP (%)	Peak depth (m)	Peak velocity (m/s)	Duration (hr)	Impact (m$)			
					Buildings and other private belongings	Infrastructure	Services	Environment
1	63.2	<0.1	<0.5	<1	0.6	0.1	0.1	0.1
2	39.3	0.1–0.3	0.5–1	1–3	1.2	0.4	0.2	0.3
5	18.1	0.3–0.5	1–2	3–5	3	0.8	0.4	0.4
10	9.5	0.5–1	2–3	5–10	4	1.5	1	0.5
20	4.9	1–1.3	3–4	10–15	5	2	1.2	0.6
50	2.0	1.3–1.5	4–5	15–24	6	3	1.5	0.7
100	1.0	>1.5	>5	>24	8	4	2	0.8

Cost benefit analysis represents a standard supplement to the quantitative risk assessment work and it aims to assess economic aspects between different mitigation measures. In this context, such analysis is aimed at determining the economically optimum intervention i.e. maximising net benefit at the level of a measure and achieving the point of allocative efficiency at the programme level (the point where the programme marginal costs equal the marginal benefits). CBA is commonly used to determine the economic level of intervention and, thereby, the economically efficient level of risk (for further discussion of economic analysis for flood risk mitigation see Section 21.12, Chapter 21, Part IV).

Hazards are typically estimated for different magnitudes of events and expressed in either graphical GIS form or through a frequency — magnitude relationship, Figure 20.6.

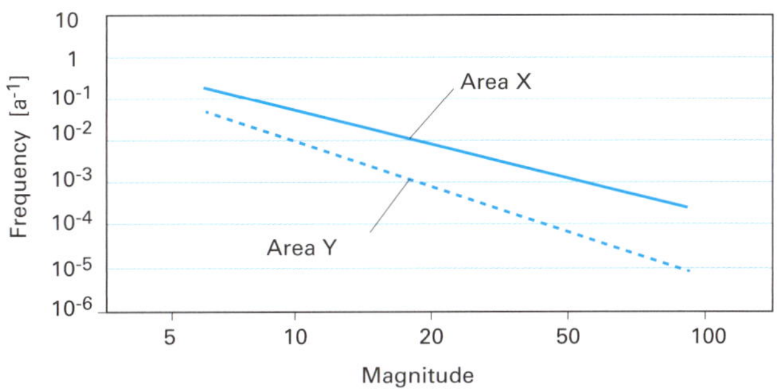

Figure 20.6 Quantification of hazards through a frequency — magnitude relationship

Similarly, different types of vulnerabilities are also assessed for different events (or scenarios) and resulting impacts (which are often related to damages) are presented in GIS and/or plotted as damage – magnitude relationship, Figure 20.7.

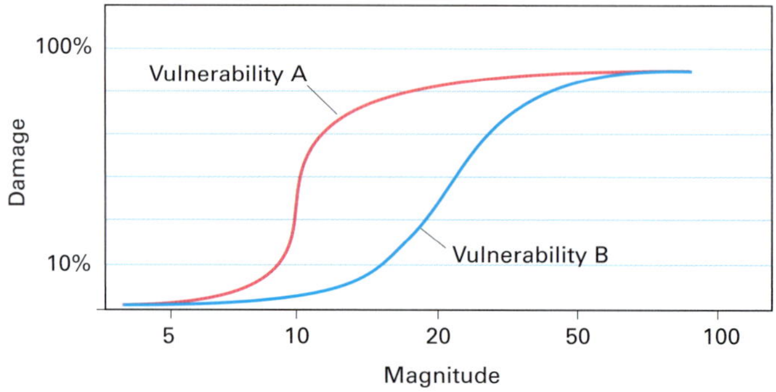

Figure 20.7 Quantification of vulnerability through a frequency – magnitude relationship

Finally, risk is then calculated on the basis of hazards and vulnerabilities and expressed through a frequency–damage relationship and/or a GIS map, Figure 20.8. It is important to note that quantitative risk assessment aims to categorise risk in different categories using a range of numerical values assigned to each category.

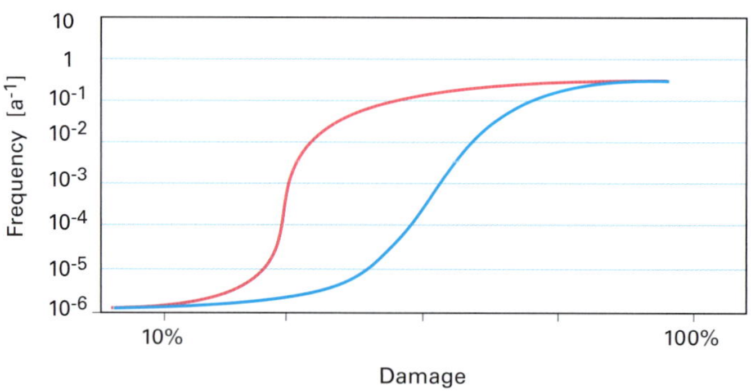

Figure 20.8 Quantification of risk from hazards and vulnerability

The results from quantitative risk assessment are normally synthesised in a report which includes the following aspects:

- Description of the study area;
- Description of the existing and future risks;
- Quantification of risks (on the basis of hazards, vulnerabilities, exposure, coping capacity, resilience, etc.) associated with different types of floods;
- Description of supporting calculations and assumptions;
- Description of sources, together with the amount and quality of data used in the work;

- Description of the modelling work including: technical details of the model(s) used, data sources, calibration/validation results, description of modelled scenarios (existing and future land-use, climate change, etc.), etc.;
- Quantification of uncertainties and description of methodologies for dealing with them;
- Description of mitigation measures (e.g., type of mitigation measures, flood defence levels, recommended operation and maintenance programmes, estimated lifecycle costs, etc.) and details of their evaluation;

Further to all of the above, we can conclude that the straight quantitative risk analysis approach is very much a consequentialist approach which does not take into account ethical and other important qualitative aspects of risk.

20.3.8 Strengths and weaknesses of current methods

Table 20.6 provides an overview of the commonly perceived strengths and weaknesses of qualitative and quantitative risk assessment methods.

Table 20.6 Strengths and weakness of qualitative and quantitative risk assessment methods.

	Qualitative Risk Assessment	Quantitative Risk Assessment
Strengths	• Allows encounter of personal experiences; • Can be applied in cases where there is no expertise or sufficient amount of data for quantitative assessment; • Easier to understand by stakeholders;	• Convenient for comparisons; • Compatible with standard statistical and probability-based methods; • Easier to incorporate probability-based uncertainty analyses;
Weaknesses	• Subject to ambiguity; • Subject to biases; • Subject to complexity; • Incompatible with standard statistical and probability-based methods;	• Does not allow reasoning and judgment; • Subject to expressing everything in numbers; • Can create the feeling of objectivity even though it may be far from it; • Subject to complexity; • Not easy to understand by stakeholders;

From the social justice perspective, neither of the two traditional methods can produce fruitful results on its own. Whether the challenge is with the protection of an individual, a community, or the catchment area as a whole, both types of methods must be carefully formulated and applied to attain to socially just decisions. The following sections describe the details of such combined approach.

20.4 TOWARDS SOCIALLY-JUST FLOOD RISK ASSESSMENT AND MITIGATION

20.4.1 The overall framework

As introduced earlier, flood risk assessment and mitigation can be regarded as a process of identifying and assessing (or attempting to assess) risk – and selecting and implementing strategies that can minimise, if not eliminate, that risk. The key steps that can be identified in this process are correspondingly illustrated in Figure 20.9. In principle, this process is equally applicable to different spatial and temporal scales. As such, it can be carried out locally or across a larger area (e.g., a city, a suburb, a basin, a coastal zone or a region). It should be also noted that this is a recurrent process and not a one-off activity. With ever increasing complexities, rates of urbanisation and the amount of available knowledge and data, this process is repetitive as it aims to produce a better *praxis* which supersedes, and necessarily transcends, the previous *praxis* (see also Section 7.2, Chapter 7, Part II).

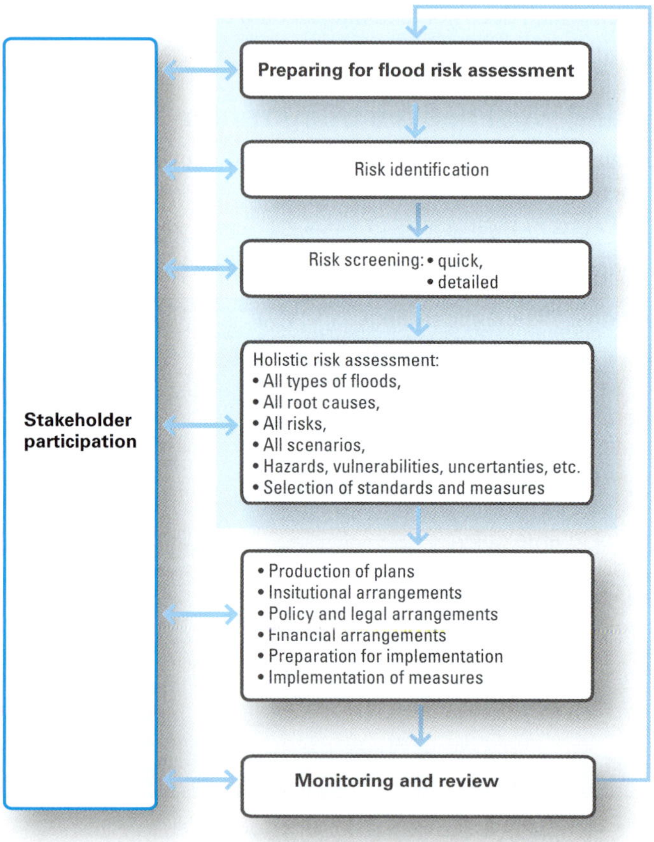

Figure 20.9 The overall framework; the area shaded in blue refers to the flood risk assessment activities

As illustrated with an overview of current practices, there is no generic framework that is universally acceptable and equally good for each and every situation. Therefore, our intention here is to discuss some of the critical aspects of this framework and to provide readers with ideas of how to develop their own framework for each occasion as it presents itself to them.

In what follows we shall go through some of the key aspects of the framework that can be used to support the kinds of critical thinking, reflective decision making and effective action and which can work in the interest of social justice.

20.4.2 Preparing for flood risk assessment

As stated earlier, risk assessment, at its simplest, is any method (or a combination of methods) that categorises, or attempts to categorise, risk. Such categorisation is necessary in order to identify and implement appropriate flood mitigation measures. Therefore, once the risk areas from all potential types of floods are identified, a risk assessment framework that can suit particular occasions and purposes, while also taking into account local historical, cultural, human and social contexts, would need to be formulated. Some of the preparatory activities for flood risk assessment work are:

- Development of appropriate mechanisms that can enable effective participation of stakeholders;
- Sourcing preliminary information that can be used in differentiating between 'socially acceptable' and 'socially unacceptable' levels of risk;
- Development of criteria for identification, screening and prioritisation of risk areas;
- Data collection and development of an approach that combines qualitative (i.e., perceptive, sensuous-intuitive, phenomenological, etc.) and quantitative (i.e., analytical) methods for assessment of all possible categories of risks (i.e., primary, residual, technological, legal, institutional, etc.);
- Seeking for synergies between activities of other water-related planning authorities and sectors;

It is important to emphasize that the flood risk assessment process is a process which requires *holistic* view and application of qualitative and quantitative approaches in addressing the needs and issues of present and future generations by considering different *scenarios* of population growth, urbanisation and climate change (see also Section 2.1, Chapter 2, Part I). The same process should also ensure that each and every important aspect of the work has been communicated to stakeholders on time and that the necessary feedback has been obtained and incorporated into the recommendations.

20.4.3 Identification of flood risk areas

Following clear definition of work objectives one of the first steps in the overall risk assessment process is to identify all areas that are potentially at risk. This is initially done through community consultations and through interviews with flood authority managers and other experts who may have the necessary experience and knowledge. In this process, each risk (or flood prone) area is analysed in terms of all possible types of floods (e.g., pluvial, fluvial, coastal, etc.) and different categories of risk (i.e., primary, residual, technological, legal and other risks): see Section 12.2 (Chapter 12, Part III) and Section 3.4 (Chapter 3, Part I). For example, to analyse how reliable a particular detention storage facility is, not only is it necessary to analyse its storage capacity limits for a particular location, but also all its possible types of failure need to be taken into consideration (e.g., structural, mechanical, electrical, human, etc.). Similarly, if we are dealing with new developments then the risks associated with such developments need to be identified (e.g., Does the proposed development lie within an area that is at risk of flooding? Does the design of the proposed development take into account all potential impacts of flood waters? Will the proposed development increase flood risk in upstream and/or downstream areas?).

Indeed, a complete list of all risks is hardly ever attainable, but nevertheless even an incomplete list can serve to focus attention on the missing identifications of risks. Also, whether a risk falls into a particular category or not is not all that important: what is important at this point is that the risk gets identified.

20.4.4 Screening of flood risk areas

The main purpose of *screening* of flood risk areas is to develop the list of all areas that are exposed to risk and to prioritise them. This is often needed due to the exigencies of financial and other constraints. Screening can be undertaken with different methods and they range from a *quick scan analysis* to more detailed model-based studies. It is not unusual that this process initially starts by conducting a quick scan analysis first and then gradually grows into a more detailed risk assessment which involves fully-fledged numerical modelling studies.

Quick scans (or rapid appraisals) are a common practice for problem investigation and formulation. They represent means for problem specification and structuring, and they are concerned with assessing the essential characteristics of the problem area and its immediate surroundings. Their purpose is also to recognise constraints and opportunities for problem solving. Their main advantages are that they do not require a considerable amount of time and effort and they are reasonably transparent. Since they are used to assess and pre-select promising alternatives they are regarded as systematic and semi-structured activities which have the potential to generate new information in a short period of time. Thus, quick-scan methodologies may involve the following aspects:

- Collection of best available data, information and knowledge;
- Communication and exchange of information and knowledge between experts and stakeholders;
- Quick qualitative (i.e., perceptive, sensuous-intuitive, phenomenological, etc.) assessment of risk;
- Quick quantitative (i.e., analytical) assessment of risk;
- Prioritisation of at risk areas;

Final screening and prioritisation is carried out in accordance with the criteria that are developed through the stakeholder participation process. At least, it should reflect severity of risk, importance of at-risk areas and different kinds of vulnerability. Furthermore, flood-risk areas can also be screened in terms of drainage assets and flood defence structures that present a risk to providing a desired level of service. Generally, areas and assets at very low risk might not be considered initially, whereas those areas and assets that appear to have a higher risk may be subjected to a more detailed risk assessment. Certainly, in doing quick scans, there needs to be a 'sense check' to ensure that some high risk areas and assets have not been omitted, or that the assigned risk values in some other areas are not overestimated.

A more detailed risk-screening (or assessment) can be carried out by embarking on a more complex analytical study which may employ various kinds of computations and calculations as well as the use of physically-based models.

20.4.5 Combining qualitative and quantitative data and methods

Further to the strengths and weaknesses associated with different flood risk assessment methods (see Table 20.6), we shall now discuss a combined approach that can be potentially placed in the context of social justice and which can build upon the strengths of these two methods. An illustration of such an approach is given in Figure 20.10. This approach is essentially *holistic and qualitative* in its nature as it is not based on the mere mathematical and probabilistic expression of risk (which is the essence of quantitative risk assessment). However, it also differs from the standard qualitative risk assessment approach in a way that it is not limited to a certain type of data and it does not aim to arrive at a conclusion about the probabilities or likelihoods of outcomes. Thus, the qualitative side of this approach is rather a subject of *phenomenological enquiry*[1] into flood risk (i.e., analysis of its root causes and their

---

[1]As introduced in Section 2.1 (Chapter 2, Part I), *phenomenology* is a form of philosophy which offers the means to take into account perception of risk and other such qualities of human experience. Phenomenology literally means the study of phenomena and it serves as the rationale behind efforts to understand individuals by entering into their field of perception in order to the see the phenomenon as these individuals see it. There is not one phenomenological methodology but rather a variety of methods that can be employed to gain the necessary understanding of subjective experience. The most suitable phenomenological methodologies for a phenomenon under consideration would be those methodologies that allow human experience to arise in a rich, unstructured, multidimensional and

interactions, as a phenomenon in the exact Husserlian sense, as defined in Section 2.1, Chapter 2, Part I) and it should represent the starting point in the overall risk assessment process. Once the initial qualitative phenomenological assessment is undertaken one may then proceed to do quantitative assessment and return back again to another round of qualitative assessment and do such iterations as many times as it is necessary in order to allow new understanding and new experience to emerge (a process similar to the process given in Figure 7.2, Chapter 7, Part II). This should be undertaken in a *hydroinformatics graphical environment* by presenting a number of scenarios graphically (e.g., GIS maps, animations, multimedia presentations, etc.) and examining people's perception while encountering the phenomenon (i.e., flood risk). The important point is that the results obtained from such assessments will need to provide sufficient information so that social justice principles and ethical reasoning can be finally applied in a manner that is more responsive to stakeholder defined values, beliefs, feelings and perceptions, while remaining technologically and economically efficient.

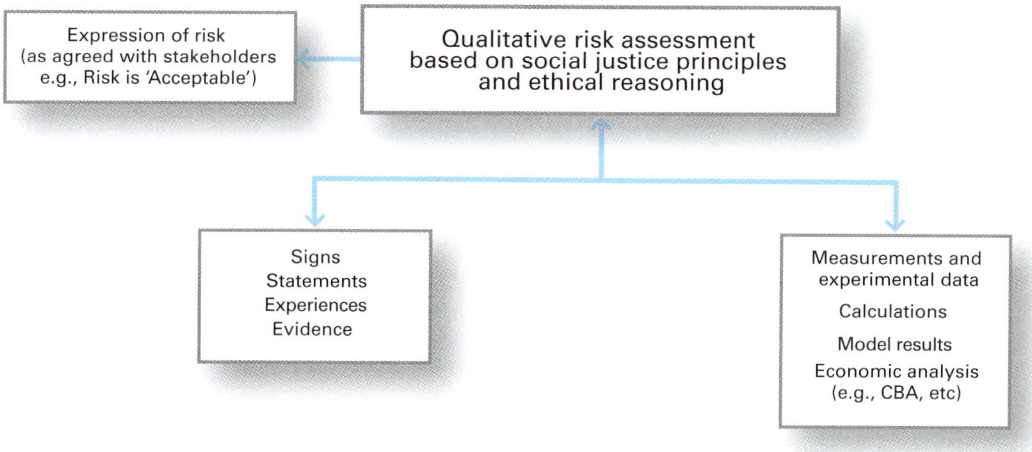

Figure 20.10 An approach that combines qualitative and quantitative data and methods

The above described approach, if conducted properly, has the potential to become much more suitable for addressing the question of what is the socially acceptable level of risk than any other traditional approach as it does not strive to simply place the risk in one of the predetermined categories (e.g., low, moderate, high, etc.) which are often found to be meaningless and impractical for decision making purposes. As stressed earlier, to assess risk from a social justice perspective, one must consider not only its materialistic consequences and compare them according to some sensible criteria, but, also, one must evaluate risk in a sensuous-intuitive and ethically defensible way. This implies that amongst other issues we must also consider its impact on inner experience of people where the means of phenomenological investigation

---

holistic way. In practical terms, this normally translates into gathering 'deep' information and perceptions through inductive, qualitative methods such as interviews, discussions and participant observation, and representing it from the perspective of the research participant(s). The philosopher and mathematician Edmund Husserl (1859–1938) is founder of the 20th century philosophical school of phenomenology. Others who have also used and substantially contributed to phenomenological ideas include Heidegger, Levinas, Sartre, Merleau-Ponty, Schütz, Derrida and many other philosophers and thinkers. For an overview of phenomenology, see for example Moran (2000), Farber (1943) and Spiegelberg (1982), and for discussion of Husserl, see for example Zahavi (2003).

can play an important role (practicalities concerning phenomenological investigation in flood risk assessment work are discussed further in the following section).

Once the information obtained from qualitative and quantitative assessments is sourced and processed, we can then go on to address acceptance of risk from an ethical point of view. Typically, people do not accept or reject risks in isolation. Instead, they make choices among courses of actions and the corresponding risks. If people would accept a course of action, like deciding to live at risk of floods for the sake of other benefits that they may weigh more important to their life, then those risks might be considered acceptable in the context of the other consequences of that action. They may choose such riskier course of action (e.g., living on a floodplain), if it can bring a compensating benefit in one way or the other. Also, they might choose a less risky course of action (e.g., postponing such decision until more reliable flood protection measures are implemented), if that could be done at a reasonable benefit. This implies that a level of risk that is acceptable for one activity might be unacceptable in other contexts. In some situations, if a particular level of flood risk is acceptable to most of the members of society at risk, then such risk might be considered satisfactory to the society as a whole. In some other situations, defining what is acceptable and what is not acceptable is part of a qualitative discourse that needs to be applied on a case by case basis. Thus, in general terms, ethical judgment of risk may involve the following two-step approach:

(1) developing a set of criteria for defining socially acceptable levels of risk; and
(2) applying such criteria to all risk locations, thereby determining their individual perception and acceptance.

A variety of aspects must be considered and carefully formulated in a set of criteria that will be used to define socially acceptable levels of risk. In doing so, one must also consider what kinds of tradeoffs are acceptable. Such activity primarily requires consideration of social and ethical aspects and assessment of community's general attitudes towards risk and risk-benefit tradeoffs. A straightforward approach of determining the acceptability of a particular level of flood risk might be simply to ask all those at risk whether they are satisfied with how it affects them personally. Thus, acceptable tradeoffs might be ones that the affected community endorses in principle.

Adopting the social justice context corresponds to the practical application of social justice principles and concerns that have been discussed at length in Part II of this book. Since these principles cannot have a single interpretation that can suit each and every problem at hand, we can only try to illustrate how these principles can be related to flood mitigation practice in the attempt to provide readers with the idea of how to develop their own specialised frameworks for each occasion as it presents itself to them.

The following four key social justice principles were identified earlier:

- *Maximising welfare* – This principle can be interpreted as a principle of maximising the benefits of flood mitigation measures (in accordance to the stakeholder defined values), or maximising the use of public funds (in which case the CBA analysis or some other economic evaluation method may be applied, which again needs to be agreed upon by stakeholders);
- *Fairness* – This can be interpreted as a principle which concerns fair use of public resources. This could mean favouring the most vulnerable while addressing the needs of present and future generations (i.e., intergenerational efficiency) as well as efficiency in terms of allocation of resources between different sectors and different spatial scales (national, regional, local);
- *Respecting individual rights and equality* – This can be interpreted as a principle that combines both a restricted liberty principle and a reciprocity principle (see also Section 7.3, Chapter 7, Part II). These principles can be formulated as:
    - *Restricted liberty principle*: every person has the right to be safeguarded from the negative impacts of floods that are caused by another person's actions;

  - *Reciprocity principle*: all flood mitigation activities must be given permission by those who are at risk through their informed consent. Those who engage in flood-related activities without consent from those at risk can be held to a full and unconditional liability for the negative impacts that these activities may cause to those at risk who did not give their consent.

  These two principles are in direct relation to the recognition of all parties including those exposed to risk, minorities and other disadvantaged groups in all aspects of decision making process.
- *Promoting virtue* – This can be interpreted as a principle that promotes active stakeholder participation, encourages sharing of risks and enables the establishment of ethical environments through carefully formulated policies, standards and legislations;

Whilst these principles may be considered in many cases as welcome, the practical reality of implementing them through policy, guidance, strategies and decision-tools at regional and local levels will undoubtedly prove to be more difficult and challenging. At the first place, the embodiment of these principles will require an enquiry into questions that concern definitions of what is just, fair and equitable. It will also require development of appropriate institutional and organisational arrangements and the ability of current practices and procedures to account fully for these principles. Whatever the case might be, our aim must be to maintain critical thinking and to augment traditional engineering (or technocratic) practice that can eventually lead towards states of social justice. As discussed earlier, engaging in social justice work is a matter of learning to read qualities and it is an ongoing process which strives for repeated processes of transcendence in which the means of transforming technocratic and materialistic cultures into the sociotechnical culture where the primary values and social concerns are favoured and continuously sought.

20.4.6 Holistic risk assessment: addressing different root causes

As introduced earlier, an approach that combines qualitative and quantitative data and methods must be holistic in its nature (i.e., it should reflect perceptive and analytical modes of consciousness) and responsive to social and ethical considerations, while remaining technologically and economically efficient. Also, as we recall from Part II of this book, social justice needs to get the moral direction of its work orientated in the direction of 'absolute truth', and not just 'correctly'. Placing this in the context of flood risk mitigation this implies that we would need to maintain a sharp analysis while dealing with all flood related issues and concerns. Central to this analysis is the process of critical thinking and posing and investigating questions that arise not only from analytical and technical analyses but also from morally deep and caring perspectives, ascending to a level that we can only intimate while referring to Bortoft (1996) and the precious few of his illustrious predecessors – Rudolf Steiner's *Riddles of Philosophy*, of 1973; Ernst Cassirer's *The Problem of Knowledge*, of 1974; Hans Georg Gadamer's *Truth and Method* of 1960/75 – just to mention some of the earliest. It is these who are showing us how we can attain to a far more holistic view of our world, and we have shown earlier that to a large extent this can be achieved by following the path of active stakeholder participation within effective hydroinformatics environments.

Indeed, to return for a moment to Bortoft (1996, p. 94): "Seeing in the holistic perspective is more a matter of learning to read qualities". This is to say that the holistic perspective cannot be understood in the mere analytical mode of consciousness. For example, when the illustration given in Figure 2.1 (Section 2.1, Chapter 2, Part I) is seen analytically from the perspective of the intellectual mind it will appear numerically as the "unity in multiplicity" between different levels of root causes. However, if it is seen holistically from the perspective of the sensuous-intuitive mind it will appear nonnumerically as the "multiplicity in unity". Such way of seeing is based on the notion that the properties of almost any one part of risk are determined by all other parts as every part is a reflection of all others in its own way. As introduced earlier, the qualitative enquiry through phenomenology views the phenomena holistically and as such it provides the necessary means to address risk and its parts in this way.

Over the past thirty years, there has been a growing number of authors that discuss the relation of phenomenology to the scholarly and professional worlds in general terms (Burch 1989, 1990, 1991; Embree 1996; Stewart & Mukunis 1990) and to specific disciplines – for example, anthropology (Jackson 1996; Weiner 1991); art (Berleant 1991; Davis 1989; Jones 1989); education (Van Manen 1990); environmental design (Berleant 1992; Condon 1991; Corner 1990; Dovey 1993; Mugerauer 1994, 1995; Vesely 1988); geography (Cloke *et al.* 1991, chap. 3; Relph 1989a, b, 1990; Seamon 1997); psychology (Moustakis 1994; Valle 1998); philosophy (Casey 1993, 1997); and natural science (Bortoft 1996; Heelan 1983; Riegner 1993; Seamon and Zajonc 1998). Even though phenomenology has a long history and a growing number of authors who regard phenomenology as an important qualitative method there is no universal step-by-step procedure that can be applied to each and every situation. Since the ultimate aim of qualitative enquiry through phenomenology is to discover commonalities that mark the essential core of the phenomenon, the key activities of such enquiry that can be placed in the context of flood risk assessment work may include the following:

(1) Gathering participants (e.g., stakeholders, representatives of those at risk, etc.);
(2) Identifying and intuiting the phenomenon (i.e., physical and emotional risk identification) from individual and shared experiences;
(3) Formulating questions and employing graphical means that can reflect the phenomenon (i.e., flood risk);
(4) Collecting descriptive accounts;
(5) Studying the respondents' accounts of risk with the aim of identifying underlying commonalities and patterns that mark the general perception of risk;
(6) Identifying what was experienced and how it was experienced; It is important to note that this does not refer to the process of counting different experiences but rather to the process of searching for meanings and essences of such experiences;
(7) Organising, analysing, interpreting, validating and presenting the respondents' accounts of risk;

Since the holistic thinking and questioning is central to the phenomenological enquiry of a particular phenomenon the following discussion aims to list some of the questions that are important for addressing different root causes of flood risk (i.e., different parts of the whole). Correspondingly, this should lead to a better formulation of strategies for flood risk mitigation, Figure 20.11.

20.4.7 Addressing the natural root causes

The assessment of natural root causes should be addressed from a holistic perspective in relation to all other root causes. This is necessary in order to identify those human activities that can be adjusted to minimise flood risk. In this regard, examples of questions are:

- *Is there anything that a city (or a region, or a nation) can do about warming temperatures, rising sea levels, and more frequent storms?*
- *What are the measures that a city can do to minimise surface runoff and its side effects?*

Indeed, we cannot stop the rain or prevent a storm surge, but we can certainly adjust our activities to minimise their impacts on climate change (e.g., carbon emissions, etc.) and the related sea level rise or more intense precipitation. In this respect, there are numerous reports which suggest that green house gas

(GHG) emissions have increased 70 percent over the last thirty years. As a result of this, cities are experiencing the consequences of increased levels of heat trapped in the atmosphere that modify weather patterns. While changes in average conditions can have serious consequences by themselves, the main impacts of climate change will be felt due to weather extremes and the consequent risk of natural disasters. In many places, this means more frequent and intense floods, storm surges, and winds. Reducing the production of GHGs would result in improvements to the regional environment and could contribute to better well-being of urban communities. Reduction of GHG emissions would over time reduce the extent of increase in mean global temperature. In this respect, a range of mechanisms (e.g., regulations and standards, taxes and charges, tradable permits, financial incentives, voluntary agreements, research and development, etc.) have been identified as mitigating actions. Their effectiveness depends on national and sectoral policies and practices. Such actions are most efficient if they are suitably integrated with the corresponding programs of all those who live and work in a city.

In terms of a better control of surface runoff and other natural processes, Sustainable Urban Drainage Systems (SUDS) have proved to be useful for this purpose. The idea behind SUDS is to promote the use of natural systems for a better management of surface water run-off at the source level. Apart from this benefit, other benefits of SUDS solutions are: easy operation, little or no requirements for energy input and the promotion of developments of an environmental/aesthetic nature. Examples of SUDS solutions are described and illustrated later in this part of the book.

Furthermore, we can also decide to redirect urbanisation activities in those areas where the natural root causes are minimal.

Figure 20.11 Addressing different root causes

20.4.8 Addressing the technical root causes

The assessment of technical root causes should be also addressed in relation to all other root causes and simultaneous perception of the whole. As introduced earlier, technical issues are often associated with poor drainage systems (e.g., lack of the system's capacity, poor design, bad construction practice, lack of operation and maintenance, poor asset management practice, etc.), inadequate flood defences and local obstructions across overland flowpaths, and as such they are directly attributable to the actions of human beings. In addressing such causes of flood risk, the following examples of site-specific questions can be posed:

- What is the evidence and memory of historical flood events?
- What are the eye witness accounts of such events?
- How, where and when did these events occur?
- Which standards were followed in the design of drainage systems and flood protection structures in this area? Do they reflect community's needs and concerns and do they address present and future land use characteristics?
- Were the nature-related issues taken into account in the design activities?
- Which kinds of operational and maintenance regimes and programmes were followed?

Similarly, examples of hazard-related questions are:

- What are the dominant types of flooding (pluvial, fluvial, coastal, groundwater, etc.)?
- What are the system-related causes of flooding (e.g., inadequate system capacity, inadequate flood defence structures, poor maintenance, poor asset management practice, etc.)?
- What are the scenarios that can lead to flood events?
- What are the frequencies of flood events?
- What is the severity (i.e., magnitudes and extents) of flood events?

As part of the same analysis, aspects such as asset characteristics, local vulnerability of structures and interactions with other high risk infrastructure (e.g., water supply, power, communications, gas, etc.) would also need to be considered.

20.4.9 Addressing the social root causes

When speaking about flood risk as a function of social processes, we refer at the first place to those institutions and organisations whose activities can be related to the formation and/or amplification of flood risk at both planning and operational levels. We also refer to different sectors of society, their functioning and interrelation. The social functioning also addresses issues of resources sharing, financial support and other practical measures. Therefore, a social framework should address the following issues:

- Establishment of procedures that encourage ethical behavior;
- Establishment of the necessary level of coordination and cooperation between various organisations, institutions, sectors and stakeholders in order to achieve sustainable and socially acceptable (or socially just) flood risk mitigation;
- Development and implementation of policies for timely sharing of data and information between different organisations, institutions, sectors and stakeholders for informed decision making; and
- Creation of enabling environments for all stakeholders to participate and make collective decisions;

Certainly, laws and regulations are vehicles for implementing the necessary procedures and processes. They obviously play an important role in the flood risk mitigation process. It should be noted that a number of ethical policy–related questions must be addressed before the law or regulation can be formulated. Examples of such questions which deserve special consideration at the governmental level are:

- Which resources (and their amounts) are needed to achieve sustainable flood protection for a socially acceptable (or socially just) level of risk?

- Which resources (and their amounts) can be obtained for sustainable and socially acceptable (or socially just) flood risk mitigation from the private sector and which resources can be obtained from the government?

Since there are many cities around the world which do not have a defined flood protection policy or standards we shall now list several questions which are important to be addressed from the social justice perspective. Examples of such questions are:

- What would be socially and ethically legitimate considerations in judging the acceptability of primary, residual, technological and other risks?
- Who should be involved in setting up acceptable standards for flood protection?
- Is there in place an environment that can promote and enable effective participation by all stakeholders?
- Should the variable or unique standards (or risks) across the city be considered?
- What is a fair and just distribution of risks and allocation of flood protection funds?
- Apart from standard CBA analysis, how should other considerations (e.g., non-monetary) be introduced in the decision–making framework?
- How do we measure community's well-being?
- How should the community assess and perceive different levels of risk and how should the victims be compensated in case of flood-related disasters?
- If it is not possible to provide equal level of protection to all communities, how should the priorities be defined?

In a similar way we can come up with a list of questions that concern areas where the flood protection standards have already been defined:

- What are the current flood protection standards?
- Do they take into consideration all types of floods and all causes of floods?
- How are they defined and by whom?
- Are they fair and adequate?
- Do they give advantages to those that are most vulnerable?
- Do they reflect community's well-being?
- Which procedures (or methodologies) were used for the selection of standards, the distribution of risks and the allocation of flood protection funds?
- Were the risk-exposed persons involved in the decision-making process?
- Did the risk-exposed give their voluntary consents for risk-related decisions and, if yes, under which conditions?
- Are the risk-exposed the same people as the beneficiaries of the flood protection measures?
- Are the risk-exposed dependents on those who were decision-makers?
- Did the risk-exposed have access to all relevant information?
- Did decision-makers benefit from other people's exposure?
- Were all community members (including minorities) involved in the procedure of setting up the standards?
- Do the current standards reflect the needs and concerns of all community members, and if not why not?
- Do they vary or are they unique across the city area?
- Which parts of the community are more protected and which parts are less protected and why?

- What criteria and procedures were used to define priorities for the implementation of mitigation works?
- Which methods and techniques were used in communicating different aspects of risk to stakeholders?
- Did all stakeholders have equal opportunities to express their needs, concerns and choices?
- How does the community perceive risks and how are victims compensated in case of flood-related disasters?

These are just some examples of questions that can be posed in the process of attaining to a socially-just flood risk mitigation practice. There could be also many other important questions which are specific to a particular situation and location.

20.4.10 Addressing the root causes related to decision makers

In Section 2.5 (Chapter 2, Part I) we have explained how the root causes related to the human mind can lead towards wrong choices and wrong decisions. We have also grouped the reasons for making such choices and decisions into unintentional or intentional. Examples of intentional reasons for making wrong choices and decisions are: selfishness, ignorance, greed, wrong values, ideology of money, the desire for power, corruption, self-centredness and other such failures of human character. These are obviously examples of failed (or wrong) values and beliefs. Since values and ethics are some of the central aspects within a socially-just flood risk mitigation practice we shall now focus on aspects that are relevant to those individuals who are involved in decision–making process (i.e., strategic leaders, public officials, stakeholder representatives, managers and other individuals who enjoy greater influence and power within the society). We shall now discuss the distinctive nature of ethics for these individuals together with the forces which influence their ethical behavior, and we shall also explore the actions that these individuals can take to create ethical environments.

As introduced in Chapter 2, values have been defined as broad preferences concerning someone's courses of action or outcomes. That someone can be either an individual, an organisation or a society. They are considered subjective as they can vary across individuals and cultures and they are in many ways aligned with belief and belief systems (see e.g., definition given in Wikipedia). One typical place where the right values are important is in relation to stakeholder participation process. When the right values are shared by all stakeholders (or decision-makers), they are extraordinarily important tools for making judgments, assessing possible outcomes of potential actions, and choosing among different alternatives. Correspondingly, they put those involved in the decision making process on an equal level with regard to what all members may consider as being important and acceptable.

It is a well known fact that an individual's ethical development occurs much earlier in life, before he/she enters an organisation or a group of stakeholders. In most cases, the influence of family, school, church and society determines individual values. Therefore, an organisation or a group of stakeholders is, to a large extent, composed of individuals whose basic values have already been determined.

Values provide the basis for judgments about what is important in a particular situation and how to deal with the different issues arising in each such situation. They reflect either an actual or an idealised set of criteria for determining what is the 'right' thing to do, based on individual experiences and a corresponding set of goals.

To behave ethically is to behave in a manner that is considered to be right, or moral. In stakeholder participation, ethical behavior is the essence of mutual trust. So, what does "considered to be right, or moral" mean? That is a crucial question, and the first place to look in determining what is right or wrong in dealing with flood related problems is within the potentially afflicted society. As discussed earlier, in all cultures a determination of what is right or moral is usually related to religious structures.

Since values are the essence of what the stakeholders are striving for, a problem may arise when the members of a particular stakeholder group do not share the same values. Obviously, this situation can create dysfunction amongst the group and a difficulty for determining what is "acceptable."

Public officials, and more generally, stakeholders, are given the trust of the public to develop and carry out policies that are in the community's best interest. Therefore, they carry greater responsibility to follow the basic moral codes and laws as they have more power than the ordinary member of the society, and hence more opportunity to abuse those codes or laws. Furthermore, conflict of interest is an important issue as it deals with the conflict between advancing the community's interest and advancing one's self-interest. The important point here is to stress the need for maintaining high ethical standards by establishing procedures and policies that will ensure that the community's interest comes first, and that one's self-interest does not advance at the expense of the community's interest.

However, there are numerous reasons why public officials and other decision makers have shown unethical behaviour. As discussed earlier, the most common reason for unethical behaviour appears to be greed and an associated tendency to gain a competitive advantage in the race for position or power. Willbern (1984) uses the case of embezzlement of public funds, bribery, corruption and contract kickbacks to demonstrate examples of how individuals pursue their personal interests at the expense of those of the public. There is often an issue of conflict of interest as these individuals struggle to separate personal and public interests. The areas where government officials are involved in various dealings with private organisations, issuing development permissions, managing contracts and acquisition activities are typical areas where the possibilities of conflicts of interest have been the most common. Again, this brings us back to more or less the same issues: greed and pursuing of personal and political interests, conflicting loyalties, inability to grasp complexity of issues, negative attitudes towards opposing positions, etc. It should be observed in passing that these features are by no means restricted to so-called 'third world' countries, but are all too common in societies in the 'first world' also.

Certainly, measures such as implementing appropriate and transparent procedures are important measures and they may prevent such issues to a certain extent, but above all the character and personal values are important characteristics that should be sought in individuals that need to fulfill these positions, and especially when forming stakeholder groups.

In addition to the question concerning character, there are also other qualities that individuals must possess in order to be able to exercise ethical reasoning. Some of these are: ability to recognise ethical issues and to reason through ethical consequences of decisions, understanding of complexities and ability to look at the problem holistically and from different points of view.

To conclude, the quest for transcendence to pursue *higher levels of consciousness* and to strive for social justice must begin with a personal commitment within each individual. Therefore, no other transformation is possible without personal transformation and self-transcendence: see also Chapter 2.

In what follows we shall discuss different aspects concerning development of scenarios for risk assessment.

20.5 SCENARIO DEVELOPMENT FOR RISK ASSESSMENT

The risk assessment process is typically undertaken in relation to various what-if scenarios in order to address the question of whether the resulting risks are acceptable or not. Development of such scenarios needs to be undertaken through a process whereby a number of situations which have been perceived as critical are defined by stakeholders and expressed in a form that can be used in the subsequent analyses. These scenarios are based on projections made at different levels of root causes (see Section 2.1, Chapter

2, Part I) that can describe risk as a whole and which can be used for assessment of potential impacts. There could be a range of scenario requirements for a particular flood risk assessment work. They can range from those that address social and policy related issues (i.e., social root causes) to those that are necessary for developing operational and maintenance plans and procedures (i.e., technical root causes). The approach to scenario development is important for the holistic assessment and it acts as an important bridge from the analysis of causes to the estimation of impact, adaptation and vulnerability assessment. In this respect, the use of agent-based and cellular automata models situated within *hydroinformatics environments* for modelling land use change has proved to be beneficial.

Since there is a considerable degree of uncertainty in predicting future changes (even though some uncertainties are likely to be narrowed with time) a number of scenarios are often needed. If numerical models are used for the evaluation of such scenarios then the available computer power and model complexity may represent an obstacle that can limit the number of scenarios that can be evaluated. Consequently, there could be the need for prioritising the most critical scenarios according to those risks that pose the greatest threat and to evaluate them accordingly.

20.5.1 Modelling land-use change

As introduced earlier, land-use change is a complex process that links together natural and human systems. As such it has direct impacts on flooding and flood related problems. Modelling of land-use change can make possible a gain in the understanding of the processes concerning urbanisation and, as such, is has been used to develop different scenarios. An extensive discussion of the most common land-use change models and some potential future research directions can be found in Briassoulis (2000) and Verburg *et al.* (2004).

Land-use change models are typically divided into static models (i.e., models which can calculate the situation at a given point in time) and dynamic models (i.e., models which deal with intermediate time-steps). They are also differentiated in terms of transformation and allocation. Transformation models are based on transformation probability and the status of surrounding locations. Allocation models are based on allocating a certain type of land-use to a location based on its characteristics. Furthermore, land-use change models are also divided into deterministic (based on cause and effect relations) and probabilistic (based on the probability of land-use changes).

The process of land use change, and particularly in urban areas, is now being looked at in new ways. A city can be viewed as an open and complex self-organising system that is far from being in equilibrium. The structure of this system emerges from local actions where uncoordinated local decision making may give rise to coordinated global patterns. Land use change modelling for urban areas is then considered as a spatially dynamic process, exhibiting some fundamental features of a self-organising system (Benenson and Torrens 2004; Batty 1997; White & Engelen 1994; Liu 2009).

Agent-based and cellular automata models have been tested for this purpose and proved to provide appealing and effective tools for analysing different population migration and land-use change scenarios (see e.g., Sanchez *et al.* 2011). With agent-based models (ABM), actors have a strategy that makes them react to their environment and the actions of other actors. The cellular automata (CA) models operate on the basis of decision rules. Every cell has a certain state (or function) and its characteristics are dependent on surrounding cells as well as the characteristics of the cell itself: see also Wolfram (1984).

De Leon (2007) demonstrated how various socio-economic scenarios of low-income residents can be analysed by an agent-based model in order to come up with different design interventions for the case of Tijuana city in Mexico: see Figure 20.12. Such a model allows the input of migration rates and border crossing rates relative to employment and service centres in order to define population types (e.g., migrants or full-time residents). Population densities relative to block sizes can be adjusted via the interface in order to steer the simulation toward specific land-uses such as urban centres or peripheral (rural) development. The rate of residential building can be also be adjusted, based on the relative community size, land value (approximate), required (per-capita) capital and the carrying capacity of (potentially) existing infrastructure. The simulation allows inspecting specific information about agents such as current income, savings, living expenses, job, etc. Based on this set of indicators a group of agents can settle and establish a Colony that with time can develop their own infrastructure, to be later attached to the city.

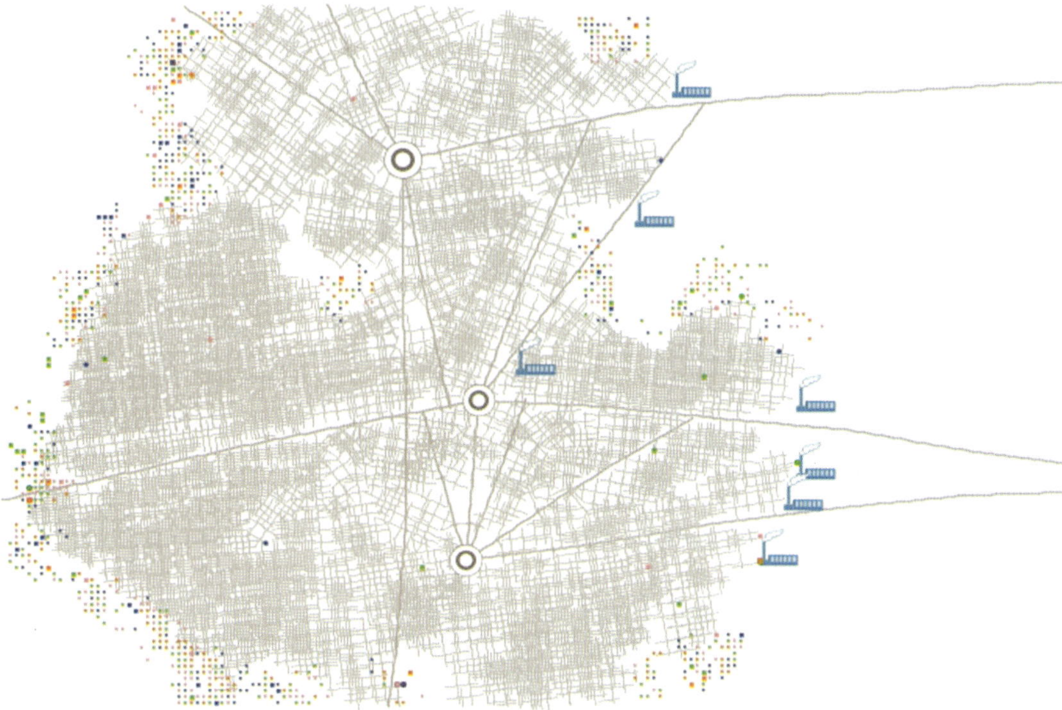

Figure 20.12 Interface of Tijuana's border town model

In the work of Lechner *et al.* (2004), a 'Cities Model' has been developed which can allow a user to create a terrain and environment in which a set of builders (agents with different roles) can create a city, Figure 20.13. This model represents a good example of how an agent-based system can be used to analyse different scenarios of urban dynamics and complexity.

The role of the agents is defined for each type of development or class: residential, commercial, industrial, and park. The agent starts to move through the terrain, grouping patches into parcels and then attempting to "develop" them by putting a new building upon the parcel, or increasing the size of the current building. There is another set of agents with the role of road developers. The road builders move through the terrain building roads between areas and increasing their value.

Figure 20.13 Interface of the Cities model

For many years the use of urban growth models was mostly restricted to academia. The improvement in spatial data availability, computing technology and geographic information systems have enabling more research, providing a foundation for the development of new applications and case studies to test the different theories in reality. Land use changes as mentioned above are complex systems with several components, objects, factors, variables and agents from natural as well as human systems. Therefore a holistic or integrated approach is more appropriate for model design. Like any model it is essential to identify the key variables and the relations that better represent the urban dynamics and reality.

Due to the increase in popularity nowadays there are several models that have been developed and tested with real data. The results of those models have successfully probe that the application of ABM and CA for land use changes modelling is possible and that the outcome of the models are similar of what is observed in reality. Moreover, the Moland-Metronamica framework, developed by RIKS (Research Institute for Knowledge Systems, Ltd, www.riks.nl) in the Netherlands, has been successfully applied and calibrated to real life cases in several European cities, as presented by Barredo *et al.* (2003), Engelen *et al.* (2007), van Vliet and van Delden (2008), among others. The Moland-Metronamica modelling framework takes different types of spatially referenced digital data as inputs:

- *Land-use maps*, showing the distribution of land-use types in the area of interest.
- Suitability maps, showing the inherent suitability of the area of interest for different land-use types. These maps are created using an overlay analysis of maps of various physical, environmental and institutional factors.
- *Zoning maps*, showing the zoning status (i.e., legal constraints) for various land-uses of the area of interest. These maps are derived from existing planning maps (e.g., master plans, zoning plans, designated areas, protected areas, historic sites, natural reserves, land ownership).
- *Accessibility maps*, showing the accessibility to transportation networks for the area of interest, and they are based on the importance of access to transport networks for the various land-uses.

- *Socio-economic data* for the main administrative regions of the area of interest, comprising demographic statistics (i.e., population and income), and data on production and employment for the four main economic sectors (i.e., agriculture, industry, commerce, and services).

The main components of the MoLAND system are:

- A two-dimensional grid or cell space, each grid-cell having its own unique set of attributes (i.e., land-use, suitability, zoning, accessibility, socio-economic);
- A cell neighborhood, consisting of a circular area of radius eight pixels around each cell;
- A set of discrete cell states (i.e., 24 MoLAND land-use classes or the Corine land-use classification);
- Transition rules (describe the effect of the neighbouring cells on the central cell).

The socio-economic information is used to build a regional model that accounts for the demand of the different land-uses in the future or modelled time frame. Thus, the regional model constrains the local model or cellular automata by assigning a number of cells according to the demand. Once all the cells for a particular land-use are assigned no other cell can change to that land-use even if it has the highest potential, instead it will get the immediate land-use class in the highest potential list. So far, MoLAND has been applied in several European cities to study different aspects concerning urban dynamics and patterns (e.g., spatial planning and hazard mitigation, strategic environmental assessment, development of sectoral policies and assessment of their spatial impacts, etc.).

Despite these successful examples the identification of the rules and their domain (spatial, temporal, physical) that drives dynamics of urban growth and their incorporation into the rules of ABM and CA models is still a challenging task. The development of such models required larger datasets and knowledge and it is essentially an art. Innovative models that can allow integration of different information and knowledge are very much needed. The use of land use change models in the framework of scenario-based simulations is necessary to identify trends or patterns that provide useful information required for land use planning and regulation, as well as flood impact assessment.

Scenario planning is a systemic method which has been tested in several fields as a tool to generate scenarios that can deal with complexity and uncertainty (see e.g., Petterson *et al.* 2003; Kok and van Delden 2007). A scenario describes possible future, or futures that could be rather than will be (see also Peterson *et al.* 2003). In this sense scenarios describe a vision of possible future pathways considering critical uncertainties. Scenarios can be defined as storylines that can be told in words and numbers which offer a description of how events can evolve over time, like a movie (see also Kok and Van Delden 2007).

Kahn and Weiner (1967) were pioneers in developing scenarios for strategic planning applications. Scenarios were refined at Royal Dutch/Shell by Wack in the 1970s and 1980s, and Shell became a leader in the scenario approach to business planning (see also Peterson *et al.* 2003; Kok and van Delden 2007). Nowadays, scenario development is used in a variety of different contexts ranging from political decision making, to business planning. Increasingly, decision makers and other stakeholders are being involved in the process in different ways (see Van Asselt and Rijkens-Klomp 2002). The combination of quantitative models and more qualitative participatory methods to develop scenarios is currently being advocated as a promising way forward (Alcamo 2001; Kok *et al.* 2007).

One of the disadvantages of scenario planning is that without a clear direction the numbers of possible scenarios are almost endless and the key factors and description are difficult to limit (Mietzner *et al.* 2004). Also, the practice of scenario planning and development can be very time-consuming and demanding in resources (see also Barredo *et al.* 2005; Kok and Van Delde 2007).

Dynamic population models have been developed also within the computational hydraulics community (e.g., Abbott and Warren 1974). There was also a paper, by the present second author and his colleagues, on modelling transit flows of the population of the city of Miami in responses to a range of specific scenarios associated with a proposed nuclear power plant. The very fast implicit-algorithmic approach that was employed could well be used to estimate the population distribution at any specific time and date in the year in the case of unforeseen inundations.

20.5.2 Addressing climate change scenarios

Currently available approaches for addressing different climate change scenarios range from using design event uplifts to applying more sophisticated climate – impact models. Some researchers and practitioners find the uncertainties in the uplift values to be too great to enable the use of single values directly and prefer to design adaptable solutions that can cope with a range of possible impacts. Even though this approach may sound reasonable, there is still a requirement to define the range of possible impacts. In this respect, the use of synthetic 'future' time series rainfall data has been used in a number of studies despite the difficulties associated with convective rainfall representation.

Typically, for less detailed analysis simple uplifts ratios are used to make a baseline assessment of the potential impact of climate change on an area (or an asset). A time series rainfall approach has been used for more detailed analysis as it allows the modelling of the antecedent catchment conditions. Furthermore, a hybrid approach has been used for cases where the likely investment is significant, or the uncertainty needs to be reduced before making investment decisions.

20.6 USE OF MODELS AND SPATIAL VISUALISATION TECHNOLOGIES FOR FLOOD RISK ASSESSMENT WITHIN HYDROINFORMATICS ENVIRONMENTS

The construction of *hydroinformatics environments*, as information providers of virtual worlds using all the resources of the current and future information and communication technologies and practises, is an important aspect in the risk assessment and mitigation process. Such means have the capacity to change the ways in which individual stakeholders think and feel about the world in which they live and function, even as the stakeholder group as a whole then mutates correspondingly so as to provide one collective voice, as a voice that is experienced as one of social justice – just so long as this process is "conducted correctly". Graphical presentation of model results, visual images and media are particularly important in communicating various aspects of risk to the general public. They can have major influence on sensuous-intuitive perception of risk by shaping the way how people perceive the potential reality (for definition of *reality* see Section 4.3.3, Chapter 4, Part II).

As stated earlier, physically-based computational modelling has been invaluable for the purpose of evaluating different flood risk scenarios and assessing their potential impacts. With instantiated models, it is possible to explore the generation of flooding and to simulate the consequent effects in response to any control actions. Geo-referenced results from 1D or 1D-2D coupled models can readily be used to evaluate the potential scenario, communicate the risk of flooding and to gain insights into the nature of floods and their impacts on communities. The knowledge gained can then be transformed into a set of effective and acceptable actions to be taken by all who are affected. In this context, flood maps and spatial visualisation technologies play an essential role to facilitate an exchange of information and views, Figures 20.14, 20.15, 20.16, 20.17 and 20.18. Typically, when speaking about flood maps we refer to flood depth, flood hazard and vulnerability, as well as flood risk maps. Such maps cover the geographical areas which could be flooded according to different scenarios (see also EXCIMAP 2007). These maps may also include flood damage maps, groundwater aquifer maps, trans-boundary maps, emergency maps, insurance maps, etc. The flood mapping task is usually carried out for the entire catchment, coastal zone or river basin with the following objectives:

- To identify the flood extent, flood depths and quantify the flood hazard,
- To provide input into the risk assessment procedure.

Typically, the flood extent maps are developed for:

- all relevant AEP and storm surge events,
- existing and future development scenarios,
- all overland flow paths in which the depth and/or velocity pose a possible safety hazard to pedestrians and vehicles.

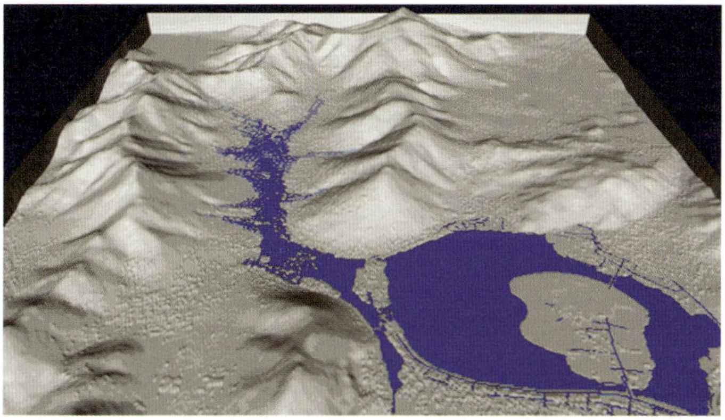

Figure 20.14 3-dimensional view of the water surface simulated using a 1D-2D coupled model (St Maarten, N.A.). The coupling in this example involves channel flows generated by a 1D model superimposed on the flood plain identified by a Digital Terrain Model (DTM) of the 2D ground surface which contains building and roads

Figure 20.15 Example of a Depth map

Figure 20.16 Example of a Velocity map

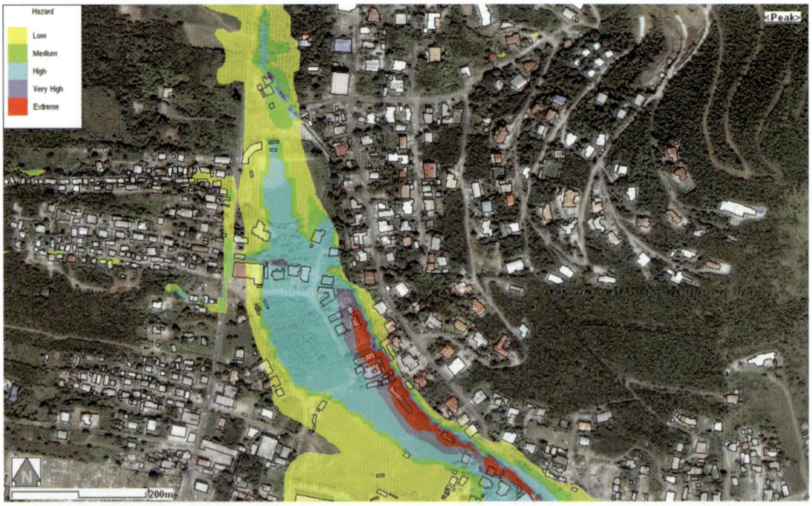

Figure 20.17 Example of a Hazard map

Effective communication of information and knowledge gained from flood modelling work is the key to ensuring that all those concerned have a common understanding and can jointly implement each phase of the flood risk mitigation work. It follows that effective planning of mitigation works depends on a correspondingly effective communication process. For such communication to take place between all stakeholders, it is important that the data and technical information are presented in a form that can be assimilated by everyone. For example, information from a model becomes knowledge–generating only in so far as the end user interprets that information correctly. Therefore, the provision of appropriate means to enable such communication process to take place is a challenge that faces all those involved in this process.

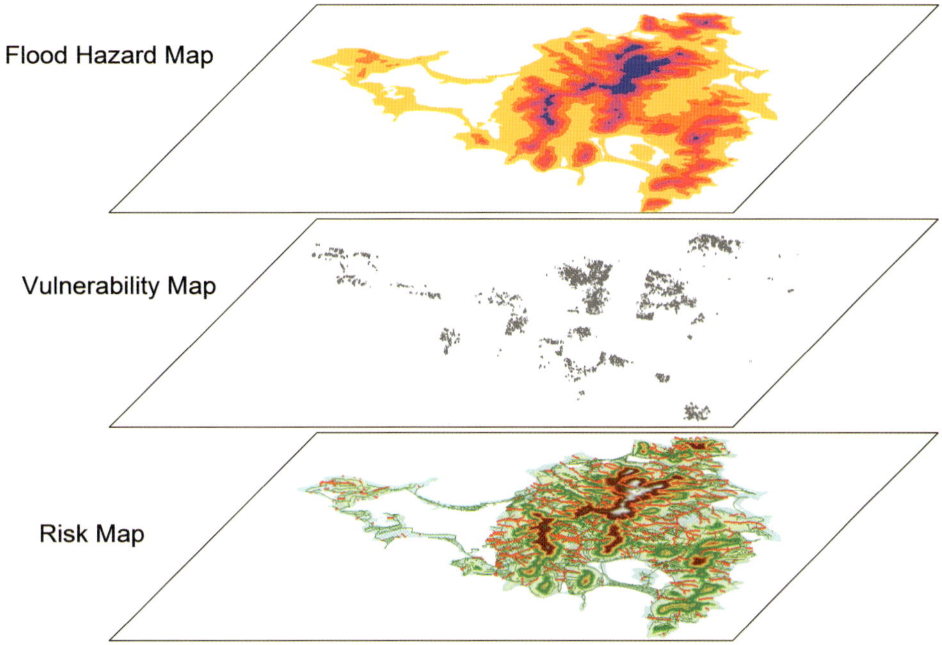

Figure 20.18 Example of a risk map which combines the flood hazard map and the vulnerability map

The important role of modelling is in complementing the acquisition of data to improve the information and understanding about the performance of a given drainage network (river channel or a coastal flood protection structure), taking into account the associated urban terrain (Abbott and Verhoog 1968). Considerable attention needs to be given to the acquisition of good geometric and topographical data at adequate resolution in order to describe the primary features of the flow paths through the urban area. Given such instantiated flood models, it is then possible to begin optimising the performance of the drainage network in terms of rehabilitating the drainage network itself and even introducing necessary modifications to the urban topography that are consistent with the appropriate planning criteria (see, e.g., Vojinović *et al.* 2006b). Obviously, there needs to be a close cooperation between the drainage and planning authorities in order to develop an effective and efficient drainage system.

The monitoring system and its associated database, coupled with an effective modelling system can form the basis of a reliable decision support system, which has at least four major functions. First, through scenario building and evaluation it can inform stakeholders, engineers and planners on better ways of determining risks associated with potential disasters, and formulating and implementing appropriate mitigation works, whether structural or non-structural. Second, the conjunction of data monitoring, historical databases and modelling systems can enable reliable forecasts to be used as the basis of a warning system for the authorities, emergency services and public. Third, both the decision support system for disaster mitigation and warning can involve all the relevant stakeholders in the decision making process, ensuring that all who are potentially affected by a disaster have a say in what is done in response to it. Fourth, the decision support system can also be adapted as an educational tool by raising public awareness of potential disasters through use in educational programmes, games and role-plays. In this respect, GIS and other visualisation technologies enable a wide range of presentational material to be easily generated and tailored for the target audiences, and in the first place for the entire variety of active stakeholders. Such material is invaluable for demonstrating and evaluating the probable impact of an impending flood to residents more generally.

20.7 FLOOD HAZARD ASSESSMENT

20.7.1 Quantifying and mapping hazards from numerical model results

Specialist GIS mapping techniques need to be applied in order to map the results of 1D and 2D models onto a 2D map. In this respect, 2D models (especially if they use a structured grid network) are much easier to handle since they have their own spatial framework which can be easily translated onto the 2D map, whereas 1D models do not have a 2D spatial framework and the development of such a framework needs to be created. The following text describes one of the possible procedures for translating 1D model results onto the 2D spatial map.

As described in Part II of the present book, the spot height data provided by the ALS survey will typically include tens of millions to even a hundred million points which are usually provided in a single long file or a series of files on a DVD. The format is typically an ASCII xyz file or it may have been converted to a GIS (e.g., MapInfo, ArcGIS) layer. Whatever the format, the file or files are usually very large and require careful consideration of storage, processing and usage issues. Unless specifically captured for floodplain management, the ALS coverage area is typically a large rectangular region that extends beyond the limits of the floodplain. Post-processing of the dataset to reduce the number of points is necessary to ensure some further efficiency in its use. The first step in this process is to clip the full dataset to the extent of a floodplain using a polygon derived on the basis of the knowledge about the flood-prone area. An example of such an activity is given in Figure 20.19.

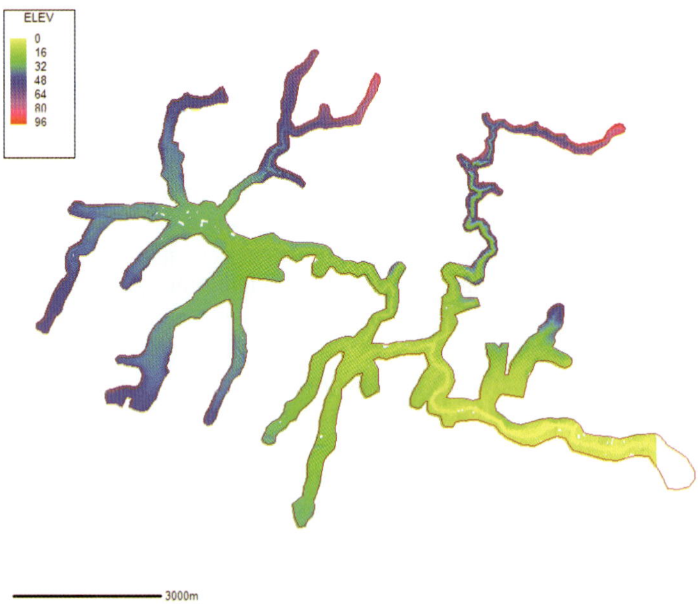

Figure 20.19 Clipping of full ALS data sets to the extent of a floodplain (ALS points outside the floodplain have been removed). Full ALS sets can be 10's to 100's of millions of points

The second stage is to thin the remaining points by neighbour tolerance thinning (removing neighbours that are within a defined elevation tolerance) or volume thinning (removing points that affect the volume by less than a defined amount). For example, for an area that contains a football field the number of survey points across the field (since it is entirely flat) does not need to be covered to a great level of details: only few points are needed to describe a flat topography!

The final step in processing the ALS data is to convert the points into a triangulated irregular network (TIN) and to create a seamless faceted surface, see Figure 14.13 in Part III of this book.

The benefit of a TINed ALS dataset is its ability to allow the use of variable levels of detail according to changes in the terrain. Figure 14.13 shows the ground elevations thematically mapped with the ALS TIN superimposed. In this example the triangles have a nominal size varying from 7 to 15 metres.

An alternative to triangulating the points is to create a raster structured grid, Figure 20.20. However, this has the disadvantage of having to be a rectangular coverage and not providing a continuous surface directly. Each rectangular grid cell has a constant elevation rather than the sloping surface provided directly for each triangle in a TIN, as one that can be so conveniently interpolated at any point.

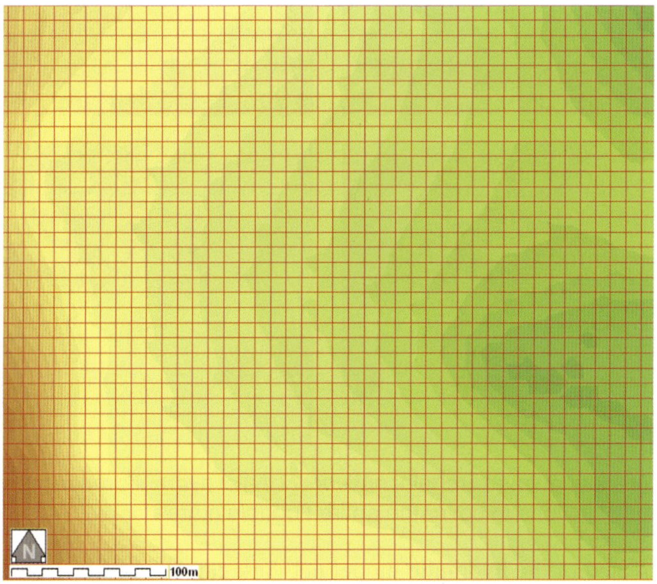

Figure 20.20 Representation of the terrain shape by structured grids

The results from 1D hydraulic models contain water level/pressure and velocity data computed in time increments for a simulated rainfall event. For an open channel network, the water levels and velocities are determined only at the cross-sections and along the connecting branches, respectively, of a branched model network and at the nodes and links, respectively, of a linked node model network. Similarly, for a pipe network model, the water levels are determined at each node (or manhole). As such the model data is discrete. The data can be converted to a spatial framework by extending the cross-section to cover the flood plain and then by creating a TIN of the model network. Essentially, this ensures that all model results are linked by triangles and an interpolated value of water level or velocity can be determined at any point by planar geometry; see Figure 20.21.

Figure 20.21 Creation of a triangulated irregular network (TIN) of hydraulic model network data. It is important to enlarge the model cross sections to cover the entire floodplain area

The TIN structure created of the hydraulic model network can be used as a basis for processing model results at each time step. The model results obtained using the TIN are on a scale dictated by the model network and are much coarser than an ALS spatial framework. Because water level, and velocity to a lesser extent, usually represent gradually–varying surfaces, estimates of the variation of these fundamental model results can be interpolated within this coarse framework. The model results from the TIN are mapped onto the much finer spatial framework of the ALS data by taking every node of the fine–meshed ALS framework and determining the model results for water levels and velocities within the triangles of the model results of the TIN. This translation is repeated for every time set of model results. The translation means that at every node or point of the ALS the ground elevation as well as the interpolated values of the time–varying water levels and velocities are stored. This can be conceptualised as a "cube of data" where the first slice is the model result set from the first time step interpolated to the scale of the ALS spatial framework; see Figure 20.22. Following slices represent the spatial frameworks of subsequent time steps.

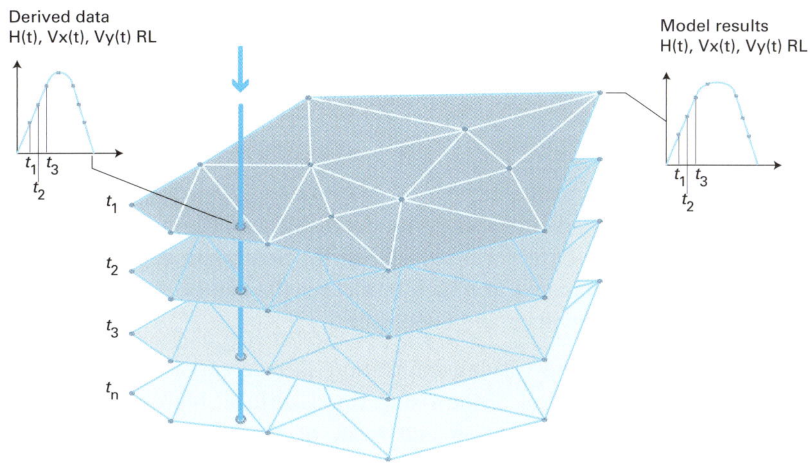

Figure 20.22 Mapping model results for each time step into a number of surfaces

The cube of data is a homogeneous dataset from which the derived hydraulic data can be easily determined and mapped. The time–varying water depth is usually determined by subtracting the ground elevation from the water level stored at each node, in each slice of the cube. Also, the flood hazard is usually determined by multiplying the depth with the velocity at each node of the ALS framework.

The derived variables are calculated for every node at every time step in the cube, and they can be animated as a time–varying thematic map. A derived data set which is of particular value to floodplain management is the peak data set, that is, the peak value of any variable at each node, regardless of time. This is easily determined by searching through the slices of the cube, for each node, and selecting the peak value and preparing a thematic map of the peak value of all nodes.

The confidence of hydraulic models is invariably measured against how well the model can replicate an evolving flood inundation surface. The appreciation of flood behaviour embodies observations of where the channel banks (or manholes) first overtop, the direction of movement of early flood waters over the land, where roads are first cut, which parts of an urban area floods first and the direction of flow across different parts of the floodplain. All these observations reflect the rising flood waters and spreading across a flood plain to occupy low–lying gullies and depressions. If a model can be shown to replicate these observations, it will gain in credibility with the community. Most hydraulic models simulate the longitudinal water surface gradient well. When model results are mapped onto the fine–scale spatial framework of an ALS, the time–varying flood inundation surface is determined by the intersection of the water surface with the DTM. This process is directly analogous to what happens in nature and, because the ALS provides accurate, fine–scale ground elevations, the simulated water surface can replicate most community observations. Hence the availability of ALS data also provides flood models with more public credibility.

Having the model results mapped to the spatial framework of an ALS allows a flood to be more realistically simulated against an aerial image. A topographically accurate animation correctly portrays the risk associated with rapid inundation of low-lying land and the possible cutting of evacuation routes, leading to potentially dangerous isolations of communities. A clear appreciation of flood hazards is important when considering the risks associated with floods larger than those of living memory — a very common situation in and around urban communities.

Having model results converted into a spatial framework provides substantial analysis capabilities such as:

- The primary model variables of water level and velocity are freely available in time and space throughout a simulated flood;
- Derivatives of these primary variables, such as depth or hazard (which is usually a product of velocity and depth) are similarly freely available;
- The spatial framework of primary and derived model results can be integrated with other spatial data. These capabilities provide the basis for a very powerful floodplain decision support system. Flood models are given a planning context in which model results can be shown interactively with aerial photographs, cadastres, property data and GIS layers.

Figure 9 depicts a snapshot of the time varying inundation surface spreading over an aerial image. The animation can mimic the real progression of the rising flood surface, highlighting which areas inundate first, which roads get cut in which succession and areas of the 'floodplain' susceptible to isolation. Similarly, Figure 20.17 displays a flood hazard map.

Depending on the particular type of flood as well as the local conditions, there can be different numerical values that can be applied to reflect different hazard categories. The values used to produce a flood hazard map in Figure 20.23 are:

- Low hazard – depth < 0.3 m, and velocity < 0.5 m/s which can be **still just** suitable for cars;
- Medium hazard – depth < 0.8 m, velocity < 2 m/s, and velocity times depth < 0.5 which can be **still just** suitable for heavy vehicles and wading by able bodied adults;
- High hazard – depth < 1.8 m, velocity < 3 m/s, and velocity times depth < 1.5 which can be **still just** suitable for light construction *(timber frame, brick veneer, etc.)*;
- Very high hazard – velocity > 0.5 m/s and < 4 m/s, and velocity times depth < 2.5 which can be **still just** suitable for heavy construction *(steel frame, concrete, etc.)* but not acceptable for places of assembly or critical institutions;
- Extreme hazard – greater than very high significant flowpath where any development, may be considered unsuitable and likely to adversely impact flood levels. It can be certainly considered unsuitable for light construction and for places of assembly or critical institutions.

Figure 20.23 shows one of the possible relationships between velocities and depths that can be used to derive a range of flood hazard categories.

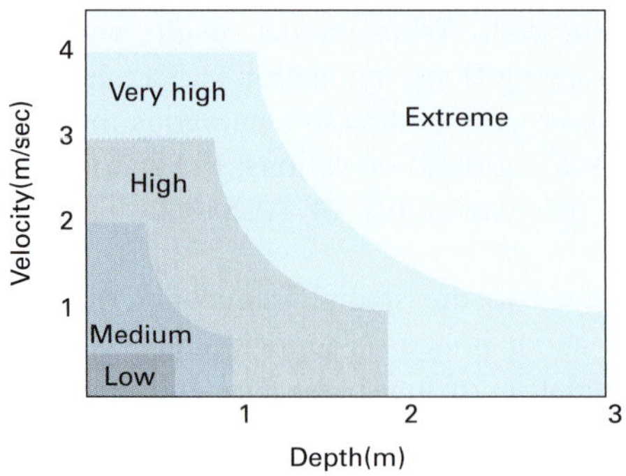

Figure 20.23 Flood hazard categories

It is important to note that the procedure described here is dependent on extrapolations of computed channel water levels to fit the terrain (which is a standard procedure when used for a 1D modelling approach) and if there are only a few cross sections then this procedure may result in unrealistically high or low water depths: see Figure 20.24. Such a problem can be particularly evident in a steep and irregular terrain topography for which the water level extrapolation along the floodplain can produce misleading results. This can be also observed from Figure 20.25 and Figure 20.26.

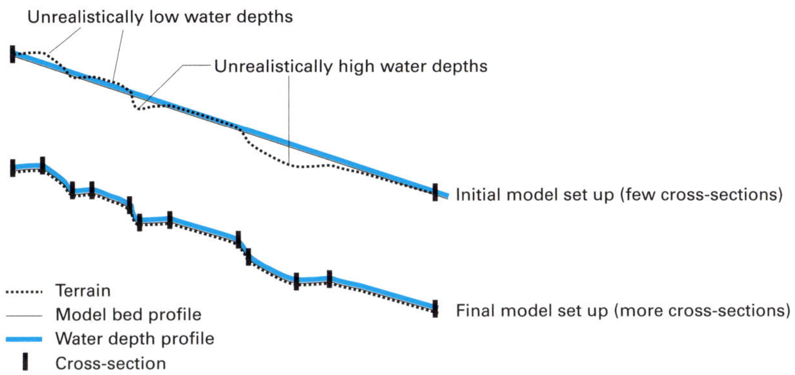

Figure 20.24 Potential problems of mapping 1D model results onto the 2D spatial map (longitudinal profile view)

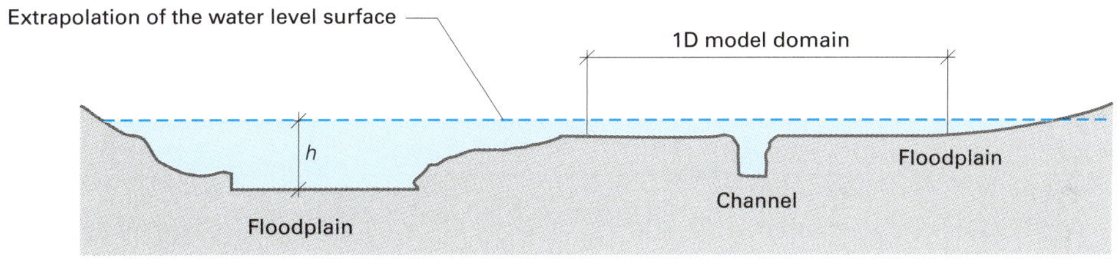

Figure 20.25 Potential problems of mapping 1D model results onto the 2D spatial map (the case where the floodplain level is below the channel cross-sectional view)

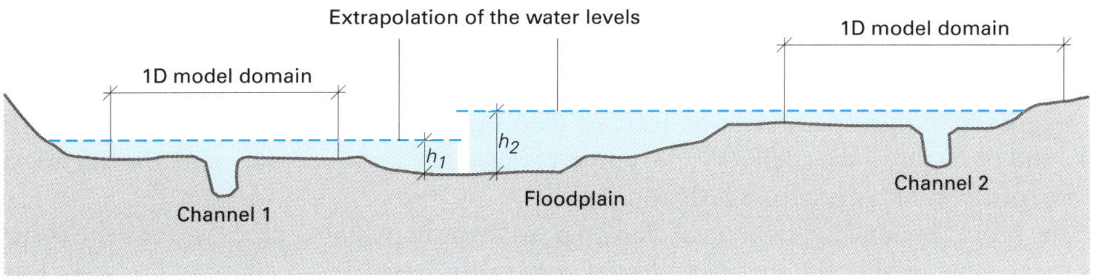

Figure 20.26 Potential problems of mapping 1D model results onto the 2D spatial map (the case of two channels and extrapolation of the flood depths across the same floodplain cross-sectional view)

Since the velocities of a 1D model across the cross-section are uniform, the 1D model results have to be adjusted in order to represent the more realistic flow conditions across the floodplain and to finally produce the flood hazard map. This can be done in the following way. After intersecting with the terrain (as appropriate), the cross-sections can be divided into a number of panels and velocity values can recomputed and redistributed across the flood plain using the following procedure; see Figure 20.27.

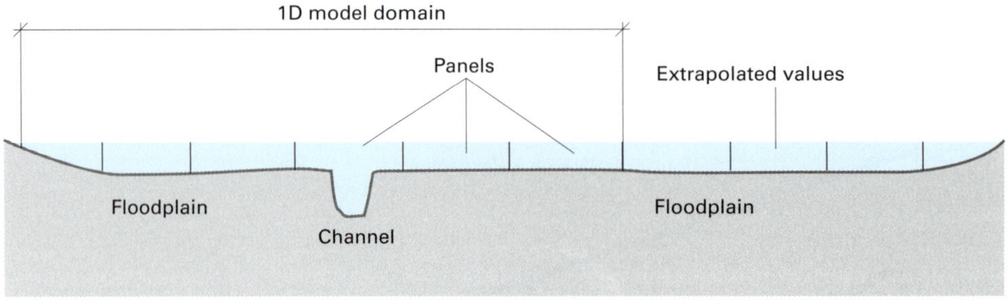

Figure 20.27 Terrain topography and redistribution of velocities for each panel

From 1D model results, velocity values can be extracted at each cross-section and applied in Manning's equation as follows:

$$Q = V \times A = \frac{1}{n} A \times R^{\frac{2}{3}} \times \sqrt{S} \tag{13.1}$$

The distribution of flows in each panel can be calculated as:

$$q = v \times a = \frac{1}{n} a \times r^{\frac{2}{3}} \times \sqrt{S} \tag{13.2}$$

from the ratio of flows:

$$\frac{q}{Q} = \frac{(v \times a)}{(V \times A)} \Rightarrow v = \frac{(q \times V \times A)}{(Q \times a)} = \frac{\left(a \times r^{\frac{2}{3}} \times \sqrt{S} \times V \times A \times n\right)}{\left(n \times A \times R^{\frac{2}{3}} \times \sqrt{S} \times a\right)} \tag{13.3}$$

where, Q, V, A, R are the values associated with the modelled cross-sections, whereas, q, v, a, r are the values specific for each panel.

With the assumption that n and S are constant (i.e., the same for both the cross-section and the panels), the velocity in each panel was calculated as:

$$v = \frac{V \times r^{\frac{2}{3}}}{R^{\frac{2}{3}}} \tag{13.4}$$

The above described procedure can be used to map the hydraulic model results onto an ALS spatial framework and transform the delivery of property specific flood advice and the appreciation and quantification of flood hazards across a floodplain.

Finally, the proper use and application of flood hazard maps in planning processes and the corresponding awareness campaigns require the incorporation of some very basic information on the map. Examples of such information are:

- Title of the map: making clear reference to the map content such as:
  - Flood parameter: Flood extent, depth, flow velocity, duration;
  - Hazard legend: defining more precisely the meanings of low, medium, high, very high and extreme hazards;

- Geographic location of the map as part of the catchment or region: provision of a small inset map
- North arrow and scale bar;
- Responsible authority or institution with logo, address, website (and/or telephone number);
- Base date for the data and date of publication;
- If necessary, a disclaimer, including remarks on the quality of information can also be added.

20.7.2 Quantifying and mapping hazards from satellite images

Over the past two decades a considerable number of publications was produced, which show actual and potential application of remote sensing information to hydrology and overall water management (Østrem *et al.* 1981; Rango & Ritchie 1996; Meijerink 1996, Schultz 1996; Menenti 1996). This information may be used as survey of areas covered by snow, detection of areas inundated by floods, precipitation measurements, water quality monitoring etc. It is obvious that the use of remote sensing data for flood management is beyond doubt and some of the techniques developed are already operational, while others are still in the research stage.

In March 2002, the European Space Agency (ESA) launched *Envisat*, an advanced polar-orbiting Earth observation satellite which provides measurements of the atmosphere, ocean, land, and ice. The Envisat satellite has as its objective to enable the continuity of the data measurements of the ESA ERS satellites. Envisat data supports earth science research and allows monitoring of the evolution of environmental and climatic changes. An active remote sensor called Advanced Synthetic Aperture Radar (ASAR) was launched on the Envisat mission to study features of the Earth such as sea surface winds, topography of the Earth and natural hazards. It uses such radar instruments that can work during the night time and also has a higher resolution compared with other sources of satellite observations. Moreover, the radar system used is able to penetrate clouds, whereby such observations are weatherproof. This has the advantage that the satellite images can be received at different times and from different satellite orbits. The following describes an approach that can be used to delineate flood hazard areas from ASAR or other satellite images.

What is displayed on the ASAR image is the backscatter coefficient of the targets on the ground; see Figure 20.28. As is usually the case with all satellite image data, the data obtained from ASAR also does not have embedded projections and as such it needs to be geographically positioned; see Figure 20.28. This can be easily done by the use of any of the commercial GIS or remote sensing packages.

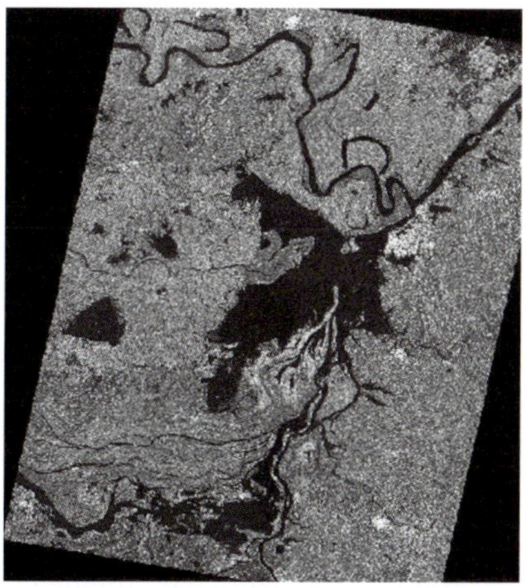

Figure 20.28 Example of ASAR image of Dongting lake in China (Chen 2008)

Once the image is placed in the right projections, the next step would be to distinguish flooded from non-flooded areas. This can be done by using a suitable classification algorithm. After the classification of the water cells, the terrain elevation values along the water boundary need to be extracted. This is normally done by cross-referencing the water boundary cells against the digital terrain data.

Once the terrain elevation values along the water boundary are known, one of the suitable interpolation algorithms can be used to produce the potential water surface across the floodplain. The water surface can be then subtracted form the terrain elevation data in order to obtain potential flood depths. What is important to note is that although there are many interpolation methods embedded in all of the commercial GIS packages, most of them are non-linear and as such they are limited for interpolating water surfaces, Figure 20.29. Therefore, a particular attention should be given to the selection of an interpolation method that can be used for the estimation of flood depths: Figure 20.30.

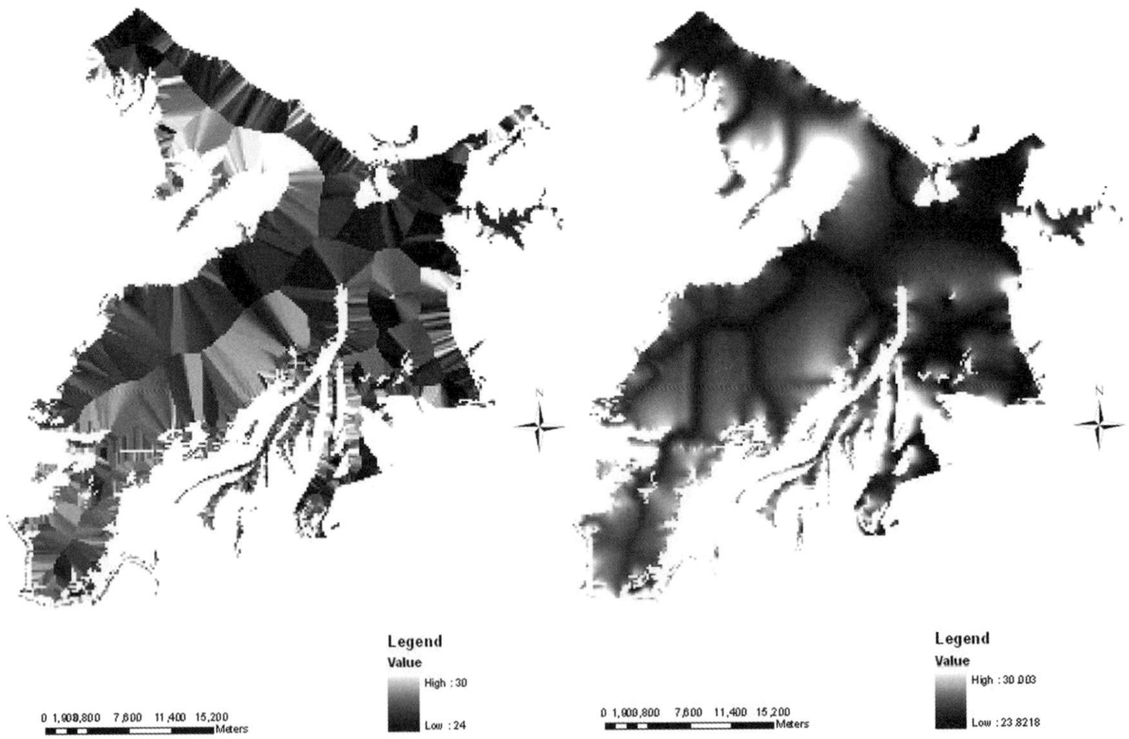

Figure 20.29 Interpolation of a water surface with IDW (left) and the 'Topo to Raster' method (right): as implemented in the ArcGIS package

Certainly, in the case that there are ground flood depth measurements available, such information should be used to improve this process.

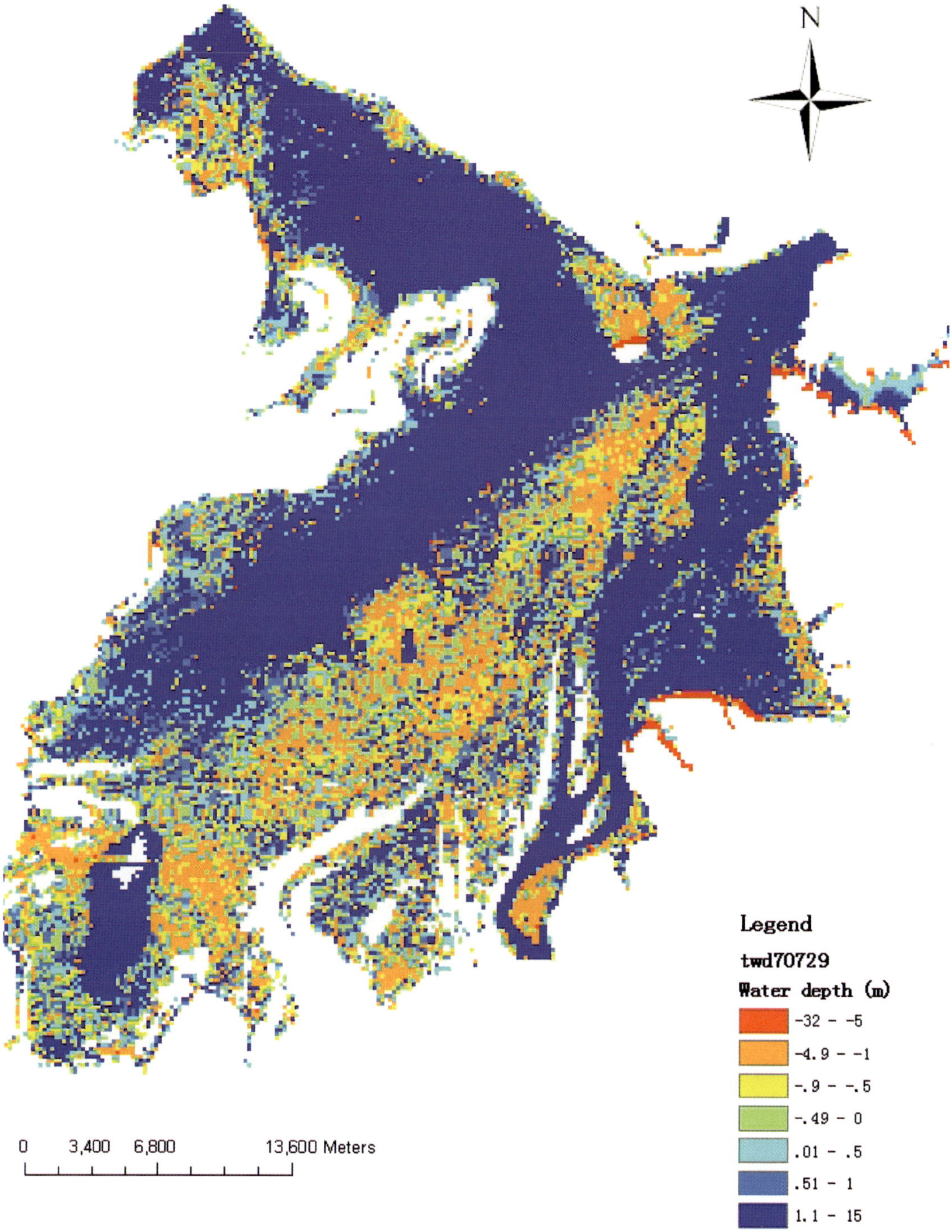

Figure 20.30 Example of a water depth map produced from a satellite image (ASAR)

Further to the production of a flood depth map, the next step would be to produce the flood hazard map. In addition to the use of flood depth data, information related to the frequency of each cell being flooded can be used as another parameter for the derivation of flood hazards. This implies that several images (taken during different flood events) should be obtained. An example of a map showing the flood frequency obtained from several images is given in Figure 20.31.

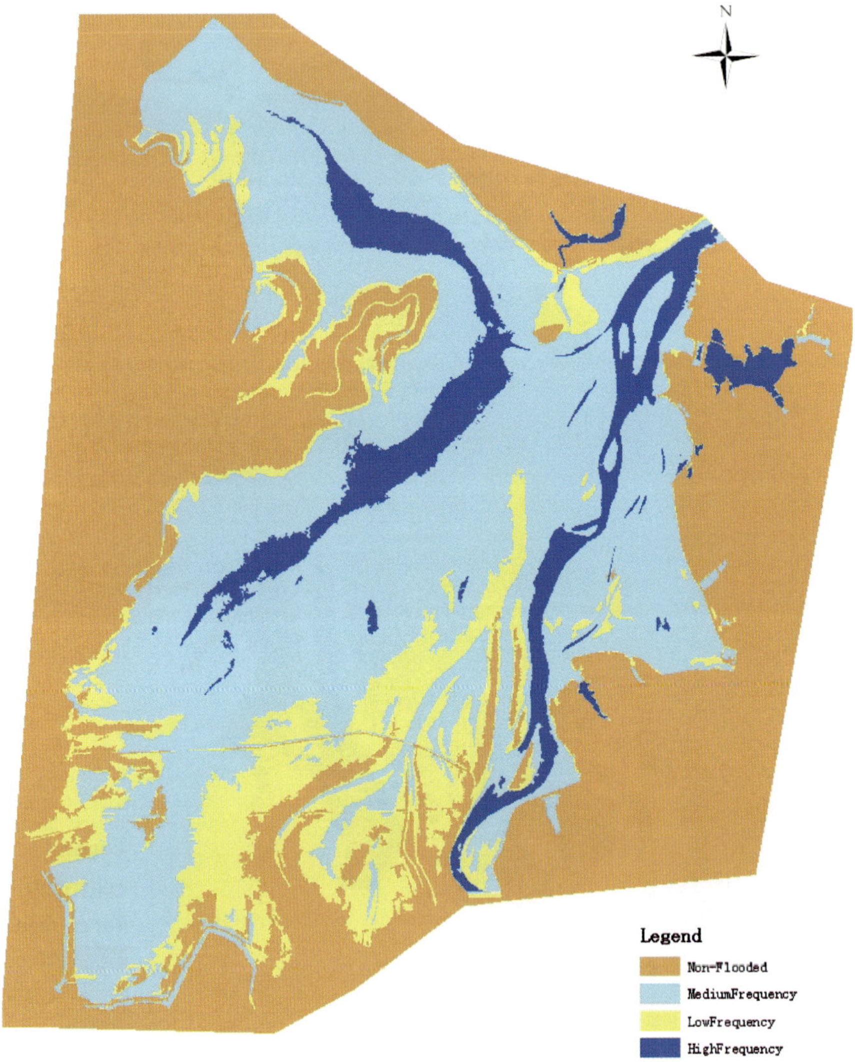

Figure 20.31 A flood frequency map

Finally, an example of a flood hazard map produced by combining the previous maps is given in Figure 20.32.

The methodology used in this example is outlined in a flow chart in Figure 20.33.

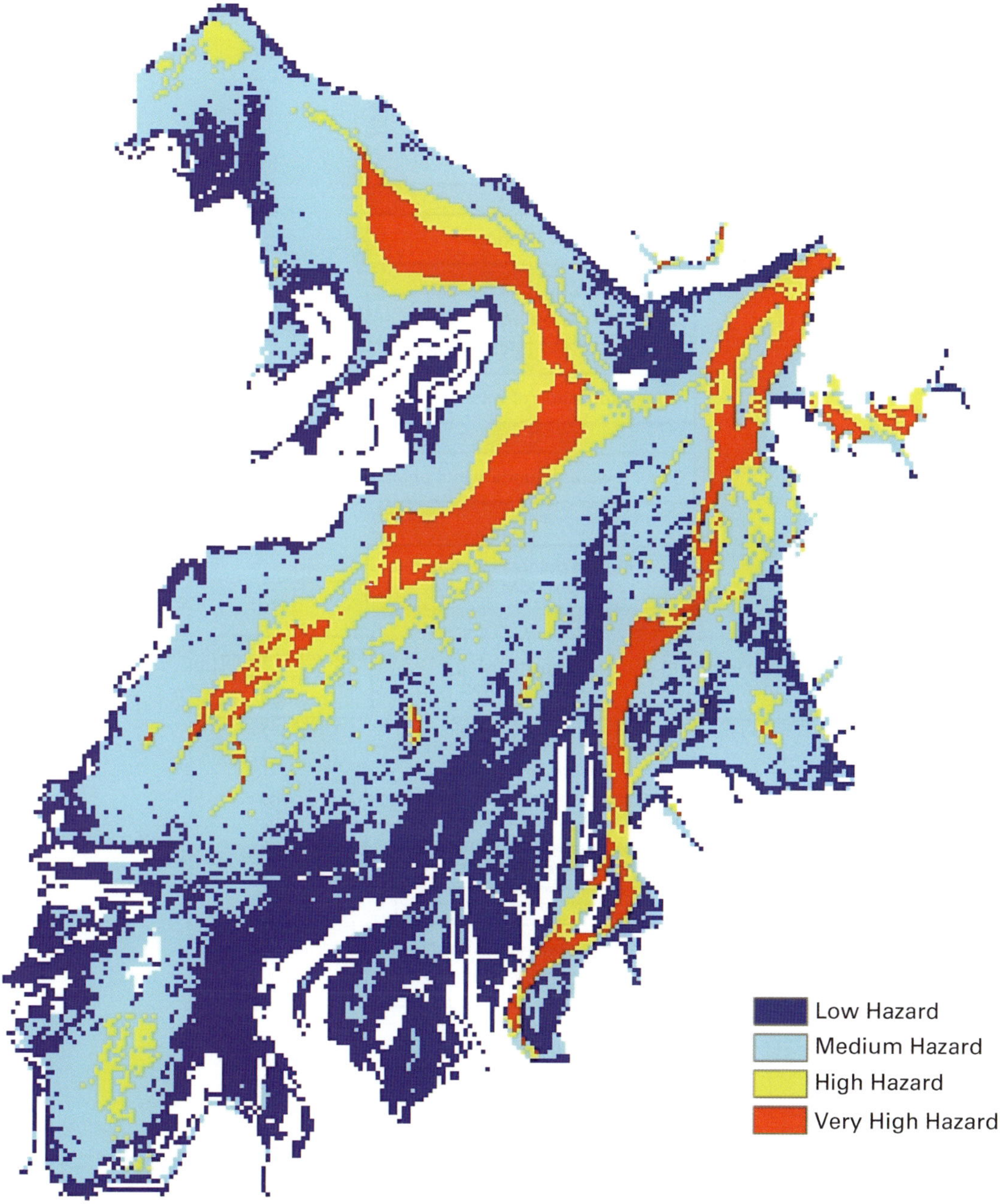

Figure 20.32 Example of a flood hazard map produced from 6 Satellite Images (the Case of Dongting Lake in China, Chen 2008)

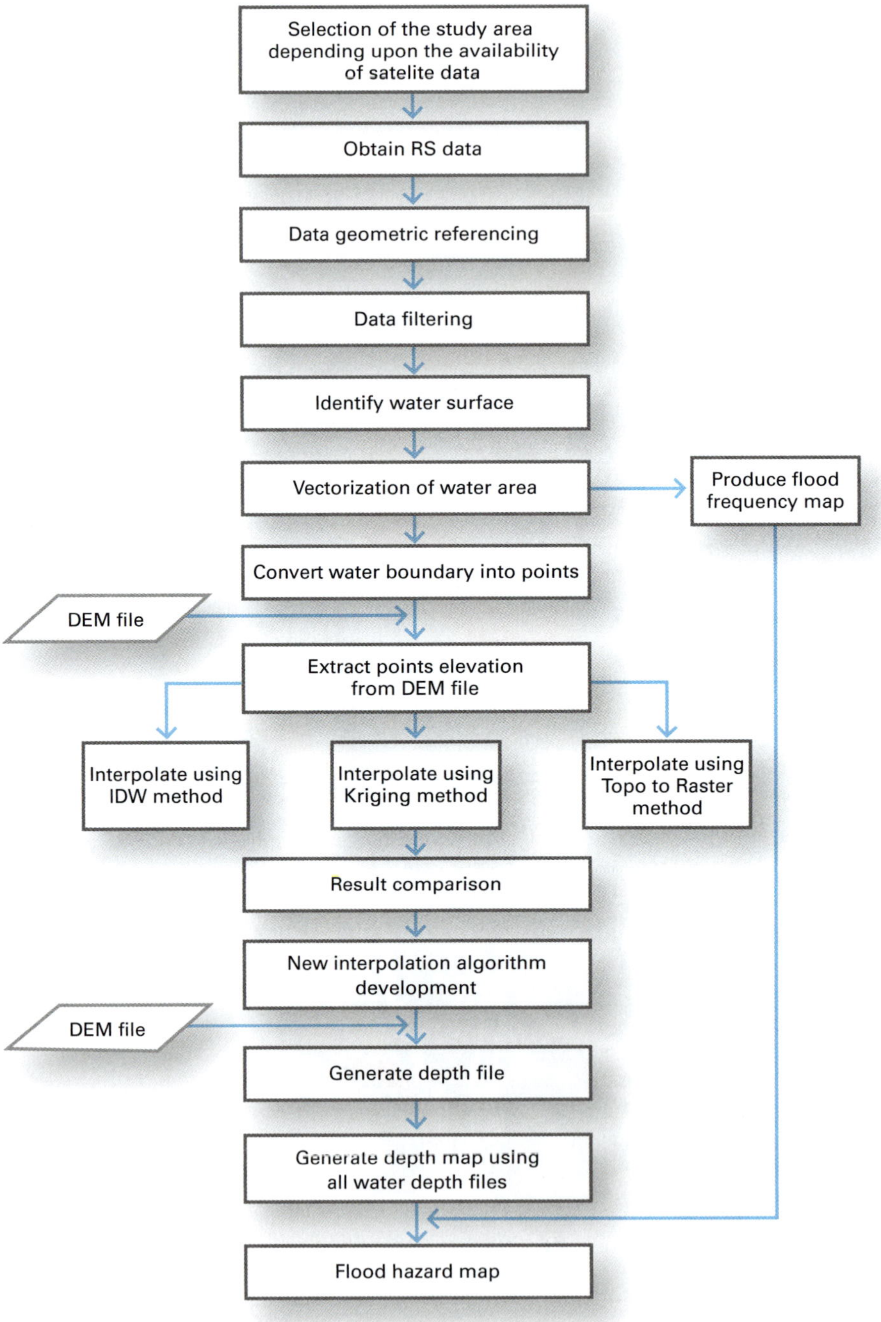

Figure 20.33 A methodology for the production of flood extent and flood hazard maps obtained from satellite images (Chen 2008)

20.8 FLOOD VULNERABILITY ASSESSMENT

The aim of flood vulnerability assessment is to evaluate how vulnerable are urban communities to flood related hazards that can be incurred by all kinds of floods under present circumstances, and if possible, in the future. Such work typically addresses five different dimensions of vulnerability: social, physical, economic, institutional and environmental. It involves the application of different types of techniques and they range from ground surveys and questionnaire-based interviews to the use of numerical models and remote sensing. There are several frameworks which have been developed to support vulnerability assessment (e.g., BBC framework developed by Bogardi/Birkman (2004) and Cardona (1999/2001), Turner's framework (Turner *et al.* 2003), Bohle's framework (Bohle 2001), etc.): see Section 2.8, Chapter 2, Part I of this book.

Analysis of the vulnerability in a particular area should address why and to what extent different elements of the society and environment are affected. A demographic analysis based on surveys is normally required for this purpose. Furthermore, a close involvement of the communities in these assessments along with the experts would yield more reliable results. Estimation of damages in monetary and non monetary terms would also be required as these form the basis for the economic analysis of various options and the deciding of priorities.

Outcomes of these analyses should also be translated into understandable presentations for the active stakeholders and the public in general to move towards a common understanding.

Some of the typical questions that are addressed in flood vulnerability assessments are:

- What and where are the flood defence structures in the affected area and how were they designed and constructed?
- What are the structural characteristics of buildings and infrastructure in the affected area?
- Which parts of the community are at risk?
- What are the demographic characteristics of that area?
- Which institutions are affected?
- Which economic sectors are affected?
- Which ecological sites are affected?

Currently, there is no consensus on how to approach vulnerability assessment in flood risk studies and the following text describes one of the possible procedures that can be used for this purpose.

20.8.1 Physical vulnerability assessment

As discussed earlier, the physical dimension of vulnerability usually refers to the capacity of the built environment (e.g., buildings, structures, etc.) to cope with floods and flood-related disasters and it encompasses the characteristics of objects to sustain or resist potential physical (i.e., structural) damage.

Once the necessary model simulations are carried out, the model results can be processed and incorporated into the GIS maps following the procedure described in the previous section.

Estimation of physical vulnerability usually refer to the estimation *tangible direct* damages. Those damages are damages that can be estimated and expressed directly in monetary terms (e.g., damages to properties, infrastructure, etc.); see for example Penning-Rowsell and Chatterton (1977), Kanchanarat (1989), FloodSITE documents (see www.floodsite.net), Price and Vojinović (2011). As described in Price and Vojinović (2011) flood damages can be broadly divided into *tangible* and *intangible* damages, Figure 20.34.

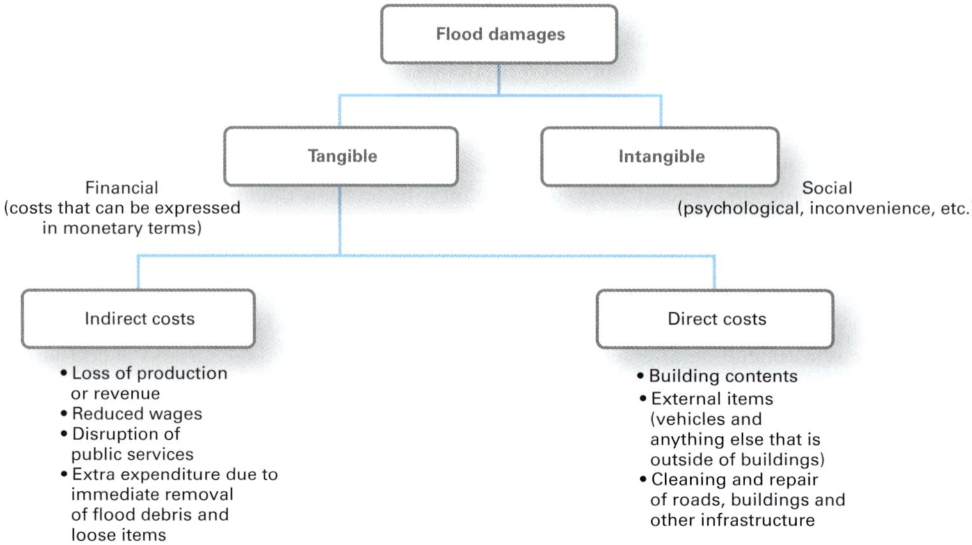

Figure 20.34 Classification of different flood damages (After Price and Vojinović 2011)

Tose damages that can be estimated and expressed directly in monetary terms are referred to as *tangible* damages (e.g., damages to properties, infrastructure, etc.); see Penning-Rowsell and Chatterton (1977), Kanchanarat (1989). Damages that are difficult to identify in monetary terms are referred to as *intangible* damages (e.g., loss of social values, loss of life, anxiety, etc.); see Lekuthai and Vongvisessomjai (2001). Tangible damages may be classified further into *direct* and *indirect* damages. *Direct* damages are those that have occurred from a direct interaction with the flood (e.g., property and infrastructure damages). The extent and magnitude of direct damages that can be incurred in a flood event greatly depend on the *vulnerability* of the objects and structures exposed (e.g., depth damage curves for different buildings, etc.). Vulnerability here refers to the characteristics that limit a household, a society, an economy, or even a whole ecosystems' capacity to anticipate, manage, resist or recover from the impact of a flood event.

Indirect damages include reduced economic activity and individual financial hardship, as well as adverse impacts on the social comfort of a community. Indirect damages also take into account disruptive impacts such as loss of trading time and loss of market demand for products. The methodologies for estimating these categories of flood damages may differ from country to country, but essentially they all tend to address and quantify various aspects of possible damages; see for example Tang *et al.* (1992), Nascimento *et al.* (2007).

The extent and magnitude of direct damages that can be incurred in a flood event greatly depend on the *structural characteristics* of objects and the level of impact to which they are exposed (e.g., water depth, velocity, duration, etc.). Such a type of vulnerability here refers to the characteristics that limit a household's or physical infrastructure's capacity to sustain the impact of a flood event.

Therefore, the requirements for the estimation of tangible direct flood damages are:

- flood depth or flood water-surface maps for the specified AEPs;
- a digital elevation model (DEM);
- a georeferenced property database with appropriate attributes including land-use type;
- flood depth vs. damage relationships (i.e., depth-damage curves) for all objects that are spread over different land-use types, Figure 20.35.

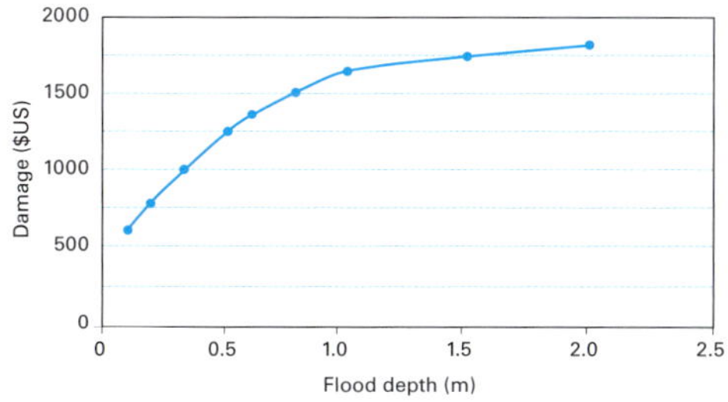

Figure 20.35 Example of a depth-damage relationship for a certain type of residential buildings

The procedure for costing the expected annual cost of flooding for the existing situation or any particular remedial works options generally goes through the following steps. For each AEP, the appropriate maximum water-level flood map is imported and subtracted from the DEM to give a flood–depth surface. Alternatively, flood–depth maps may be used directly. A flood depth is related to each property through the geo-referenced property database. Direct damage is estimated for each property using the appropriate flood–depth vs. damage relationship: see Figure 20.35. In many cases, indirect damages are estimated based on a multiplier of the direct damage. Total direct and indirect damages are summed over all properties giving the total cost of flooding for that AEP. Once a range of AEPs has been costed, the expected annual damage is estimated by integrating the cost vs. AEP relationship. This procedure gives the total expected annual costs due to flooding for that particular case. The total cost of flooding in present worth is estimated using a carefully selected inflation rate. In addition to the existing case a number of other cases, e.g., remedial works options can also be costed. The benefit gained is the difference in the total present-worth cost of flooding between a proposed remedial works options and the existing case.

This procedure for assessing the value of tangible direct damages is normally carried out for the existing and future development scenarios. Finally, a set of GIS maps is produced, see Figure 20.36.

Figure 20.36 Illustration of a GIS presentation of flood damages for critical objects under existing (left) and future (right) scenarios

20.8.2 Economic vulnerability assessment

Typically, the assessment of the economic dimension of a vulnerability would focus on determining the vulnerabilities of the centres of economic activity, and the largest employers within the community. The first step in this process would be to identify the major sectors of the economy and mapping their primary centres of activity in those sectors. These are the areas where hazard risks could have major impacts on the local economy and therefore they represent potential locations for certain hazard–mitigation strategies. This can be done by overlaying the economic centre maps with hazard maps.

Priority areas for mitigation measures can be assessed by identifying the community's most critical organisations (e.g., the largest employers) and those economic sectors that are most vital for community life. In this respect, economic census data can help to identify employment numbers for each economic sector and organisation.

20.8.3 Institutional vulnerability assessment

This analysis focuses on determining the vulnerability of key institutions and their facilities (or resources). Since it is very difficult and time consuming to conduct a detailed analysis for every structure in a community, the focus should be placed on identifying the categories of structures that can be considered as *critical objects* or *critical facilities*. They may include the following:

- Shelters;
- Schools;
- Rest homes;
- Hospitals and nursing homes;
- Fire and Rescue;
- Police;
- Utilities;
- Communications;
- Transportation;
- Government;
- Etc.

The next step would be to establish a critical facilities database and to collect the most relevant data. Such data may include the following:

- Facility type and name;
- Location (street address, etc.);
- Contact name and telephone number;
- Etc.

The vulnerability assessment of critical facilities should be carried out in such way that each critical facility is evaluated in relation to its location with respect to the hazard risk areas and its potential vulnerability to the hazard impacts (i.e., structural vulnerability and operational vulnerability).

20.8.4 Environmental vulnerability assessment

Typically, an environmental vulnerability assessment would attempt to identify environmentally sensitive areas that may be impacted by flood hazards. From this assessment, environmental areas are then analysed in relation to potentially hazardous sites (e.g., wastewater treatment plants, industrial toxic facilities, etc.) and such an analysis is used to identify potential hazard–mitigation activities. Examples of such environmental areas are:

- Recreational water bodies;
- Wetlands;
- Estuaries, endangered or threatened species habitats;
- Water resources and other protected areas;
- Etc.

The first step in this process would be to identify locations of potential hazardous sites and overlay them with the map of environmental sites. The next step would be to overlay the environmental sites with flood hazard maps. Once this is done it would be then be necessary to prioritise each and every site in terms of its sensitivity and potential for hazard mitigation.

20.8.5 Social vulnerability assessment

Social vulnerability usually refers to the inability of people, organizations, and societies to withstand adverse impacts from flood hazards to which they are exposed. On one hand, such impacts are due to characteristics that are inherent in social conditions, arrangements, institutions, systems and cultural values and on the other hand they are closely related to human characteristics. Most of the work to date concerning social vulnerability is limited to the following indicators: age, gender, income, employment, education, residence type, household type, tenure type, health insurance, house insurance, car ownership, possession of insurance, disability, language skills, debts/savings, minority population data, single parent families and so on. All these aspects are certainly relevant as they do reflect some characteristics of a society. However, they can only produce a partial picture of society's well being and there is much more to be gained by addressing the physical and psychological characteristics of people. In this respect, there is a dimension of vulnerability that relates to people's physical abilities and as such it can restrict their potential to react in a particular situation. Another dimension is their capacity for analytical reasoning which in turn calls for logic to assess hazards and to react accordingly. Also, there is a third dimension which is based on experiential perception of risk (or perceptive mode of consciousness) and it involves feelings, intuitions, emotions and affect (see also Chapter 3). Such aspects differ from one person to another to the extent that almost any form of generalisation can be found misleading and dangerous. This in turn makes social vulnerability to be the most difficult aspect to assess and measure from the scientific point of view and as such it calls for the use of qualitative form of investigation (e.g., *phenomenological* investigation).

As discussed in Part II of this book, science has been very successful in addressing the material world, but when it comes to the inner world of the mind (i.e., our thoughts, feelings, sensations, intuitions, beliefs and so on) science has very little to say. There is very little in modern science that can explain and predict *perceptive consciousness*. If we recall from our discussion, we are using the word *consciousness* to describe the capacity of experiencing. Such experience comes from our personal view of reality, sense perceptions and believing the pictures and thoughts that our mind had constructed. Therefore, if our means of addressing social vulnerability are restricted to those that have become so characteristic of modern science generally, namely those of *ordering, numbering, counting and computing*, then our thinking will be potentially distorted and our actions misguided. By considering perceptive consciousness as an important quality of reality and by introducing the means that can incorporate various forms of graphics and colour, which we refer here to as hydroinformatics' environments, we can increase our capabilities in addressing social vulnerability. All this necessitates much deeper qualitative considerations than those that are used in the current practice.

One of the potentially fruitful approaches to social vulnerability assessment would be to develop such hydroinformatics' environments that will contain a variety of graphical images and multimedia technologies and which can activate such mode of consciousness to facilitate phenomenological enquiry about the phenomenon. The aim of such facilitated phenomenological investigation process would be to extend the feelings, whereby the community representatives will experience a deeper, more direct contact with the phenomenon imagined (i.e., a particular flood scenario). In this way, a more complete and comprehensive encounter with the phenomenon could be accomplished, and aspects that are otherwise not easily noticeable would be able to come to light. Moreover, the people would feel themselves to be more in harmony with the phenomenon, as if they themselves were participating in it. Such phenomenological approach would allow things to manifest as they show themselves without forcing our own categories on them. All this could then lead towards higher states of consciousness (i.e., capacity for experiencing) and a more profound encounter of people's perceptions of flood risk.

20.9 EXPOSURE

As discussed earlier, exposure is often characterised by the number of objects, structures, population, etc. which are exposed to risk. Typically, questions that are addressed for assessment of exposure are:

- What is the number of people at risk?
- What is the number of buildings at risk?
- What is the number of buildings that are insured for such events?
- What is the number of institutions at risk?
- What is the number of economic sectors at risk?

Addressing the above–mentioned questions requires a knowledge of the existing land-use and the type of activities that are undertaken in the affected areas.

Exposure can be easily determined in a GIS environment by overlying hazard maps with various demographic datasets.

20.10 RESILIENCE AND COPING CAPACITY

As introduced earlier, the term *resilience* refers to the capacity to adjust to threats and to mitigate or preferably to avoid them altogether. Along the same lines, the term *coping capacity* refers to the combination of all strengths and resources available within a community or organisation that can reduce the level of risk, or the effects of a disaster. Typical resilience– and coping–capacity related questions are:

- What is the physical ability of the population to cope with flood events?
- What is the financial ability of the population to cope with and recover from flood events?
- What is the structural ability of buildings and infrastructure to cope with and recover from flood events?
- What is the ability of different institutions to cope with and recover from flood events?
- What is the ability of different economic sectors to cope with and recover from flood events?
- What is the ability of ecological sites to cope with and recover from flood events?

Similar to the social vulnerability assessment, assessment of aspects such as community's resilience and coping capacity is a very difficult, if not impossible, task. Certainly, resilience of structures and other material objects can be assessed in many ways and presented in a form of GIS maps. But crucially, to assess the resilience of humans and our fellow creatures such appraisal would necessitate a different worldview and much more complex and deeper qualitative considerations than those that are used in the current practice. Some of these important considerations are addressed in the first two parts of the book.

20.11 UNCERTAINTY IN FLOOD RISK ASSESSMENT

20.11.1 Treatment of uncertainly

It is important to note that there will be always a degree of uncertainly due to incompleteness in the data, information and methods used throughout the entire process of flood risk assessment. In the context of the present book, we shall define uncertainty as a degree of incompleteness of knowledge about the process under consideration and our inability to define certain aspects of it. Incompleteness of the data and information can be overcome by acquiring more information and data, but where this applies to historic data the only possible way left open is to use a simulation model to replicate what happened historically, provided the relevant input data can be found. The treatment of uncertainty in the data and information is another issue affecting this process. In fact, there is uncertainty not only in the data and information but in the tools used to analyse and make use of the data as well as the related methodologies. Therefore, the treatment of uncertainties aims to identify potential sources of uncertainty and to develop strategies to eliminate them or minimise their propagation. There could then be different types of uncertainties between different risk assessment methods.

20.11.2 Uncertainty associated with qualitative risk assessment

Typically, in a qualitative risk assessment, we are dealing with three types of uncertainties: *linguistic* uncertainty, uncertainty due to *variability* and uncertainty due to the *method*. Linguistic uncertainty occurs due to the language used to describe different aspects of the phenomena or the process under consideration. Such uncertainty can arise from ambiguity, complexity and vagueness. Ambiguity arises of course when words have more than one meaning and when it is not clear which meaning is applicable in a particular context. Ambiguity can be minimised by introducing more clear definitions of terms. Complexity may arise from an inability to comprehend the context in which a particular term is to be used. Similar to the way of dealing with ambiguity, uncertainty due to complexity can be dealt with by gaining a deeper understanding of the context and by being more specific. Vagueness arises when there is a potential borderline overlap between the words (e.g., high, low, severe, acceptable, widespread, etc.). The usual solution here is to better–define the related terms.

One of the methods which has been used to deal with linguistic uncertainty due to vagueness is fuzzy set theory (a case of 'possibility theory'), in which the terms (or linguistic variables) are represented by a membership function. The membership function is the representation of the degree of belief in the statement (for further descriptions of fuzzy set theory, see e.g., Kecman 2001).

In terms of the practical application of phenomenological methods, one of the main concerns has been the question of trustworthiness, or in other words, the appropriateness of criteria that is used to establish the reliability of phenomenological descriptions and interpretations. Furthermore, there has been also a concern with the possibility of arriving at different interpretations and meanings. In spite of these concerns, there have been growing efforts by phenomenologists to establish qualitative criteria that can help to assess the validity of phenomenological interpretation (see e.g., Van Manen 1990; Polkinghorne 1983). Polkinghorne (1983, p. 46), for example, presented how four qualities can help to assess the trustworthiness of phenomenological interpretation: vividness, accuracy, richness, and elegance. Ultimately, the most appropriate phenomenological work is the work that can move people from their usual recognitions and transcend them along new paths of understanding. Furthermore, given that the methodological activities in conducting phenomenological investigation are those that allow human experience to arise in rich, unstructured and multidimensional way, then the researcher himself (or herself) must be open to respondents and adapt such activities throughout the entire investigation process. This implies that phenomenological method itself needs to incorporate a degree of uncertainty and spontaneity which must be then accepted and transformed into possibility and pattern.

20.11.3 Uncertainty associated with quantitative risk assessment

With respect to the quantitative risk assessment, we typically distinguish between *inherent* or *aleatory uncertainty*, which consists of random (stochastic) fluctuations that are intrinsic to the problem being studied, and *epistemic uncertainty*, which results from a limited knowledge of fundamental phenomena; see Korving *et al.* (2002). An example of aleatory uncertainty is the temporal and spatial variability of rainfall over an urban area. Epistemic uncertainty can be classified as *model uncertainty* (due to a lack of understanding of the physics) and as *statistical uncertainty* (due to a lack of sufficient data), where the latter can be reduced when the relevant knowledge becomes available. Traditionally, probability and possibility theories have been used to address these types of uncertainty. Probability depends on the frequency of occurrence of an event, while possibility focuses on the meaning of an event. It in no way implies that a high degree of possibility leads to a high degree of probability. If, however, an event is not possible, it is also improbable. Therefore, possibility can be viewed as an upper bound for probability. This weak connection between probability and possibility is defined by Zadeh (1999) as the *possibility/probability consistency principle*.

In probability theory all uncertainties are assumed to be random and therefore they can be represented by probability distributions of events. Uncertainty in possibility theory however, is quantified in terms of imprecision and vagueness, and is represented by a possibility distribution. Possibility is a *fuzzy measure*, which is represented by a function taking a value between 0 and 1. This indicates the *degree of evidence* or *belief* that a certain element x belongs to a set: see again Zadeh (1999) and Dubois and Prade (1988).

Quantitative uncertainty is expressed using numerical values. Typically, these numerical values have a meaning from a probabilistic point of view. For example, "*the probability for the water level at the road junction to exceed a value of one meter during a 10 year rainfall event is 4%*". In the quantitative approach, the laws that describe the physical phenomena are assumed to be known. However, the value of some of the parameters, for example, surface–runoff coefficients or pipe–roughness, may not be known precisely. The uncertain parameters and variables are treated as random variables, that is, variables which have known probabilities to exceed (or not) certain values. The random variables are characterised by a probability density function (pdf) and its moments (mean, variance, etc.). The two most commonly used methods for estimating uncertainty are Monte Carlo Simulation (MCS) and the First Order Second Moment (FOSM) method, both of which are based on probability theory. For further discussion of these methods see Price and Vojinović (2011).

Furthermore, expert elicitation (i.e., the synthesis of opinions of experts) which is also known as a scientific consensus methodology is also used to address issues in relation to uncertainty, accepting of course that when we speak of 'scientific' we continue to imply 'modern scientific'.

20.11.4 Uncertainty associated with numerical modelling work

The sources of uncertainty start from the variability of the phenomena that we are capturing with our measuring devices (e.g., due to an insufficient number of rain gauges to adequately capture spatial variability of rainfall, or difficult conditions at the measuring location in a pipe, or poor calibration of instruments, etc.). Such uncertainties may propagate further through the process of acquiring the required knowledge (e.g., using the model to predict/extrapolate outside of the range of rainfall events which are used in the calibration phase) and may continue further to the decision-making process (e.g., sensitivity/robustness of the decision with respect to the choice of a particular flood mitigation measure). It is then obviously beneficial to make the necessary efforts in quantifying the significance of uncertainty within the entire flood risk assessment process. Certainly, the potential for quantifying uncertainty in often limited as it depends on many things (e.g., the level of complexity of the modelling system used, the amount and quality of data, experience of the modeller, etc.). In summary, sources of uncertainty can be grouped in the following classes:

(1) Different issues related to various domains of data;
(2) Equations for conservation laws and other relationships describing water flow in different systems;
(3) Numerical solutions and specifics of their implementation;
(4) Modelling practice;
(5) Calibration procedures;
(6) Vulnerability and hazard assessment approach;
(7) Spatial mapping techniques;

With regards to the spatial data collection, even though the availability of exceptionally detailed and accurate terrain and land-use data in urban surfaces has risen dramatically in the past few years, there is still a challenge to capture and process all features that are necessary for flood modelling work at a sufficient level of detail. A similar comment applies to the collection of coastal bathymetric data. Furthermore, the capturing of rainfall data at very fine urban local scales with weather radars and rain gauges is still a challenging task. The collection of flow–gauging data is also subject to the following uncertainties:

- Sampling errors: an example of this is sampling at a location in the network that does not give relevant values of flow.
- Gauge errors: gauge measurements are often biased: they can give measured values consistently below or above the true value.
- Measurement errors of model input: gauge rainfall not being measured accurately or the gauge is far from the centroid of the local catchment where flow rates are measured.
- Measurement density: the number of monitoring points.

A primary source of uncertainty in numerical solutions in coupled 1D/1D and 1D/2D urban flood models comes from the simulation of pressurised flow in pipe networks. Most sewer models deal with this using the Preissmann open-slot approach, whereby pressurised pipe flow is simulated by solving free-surface flow equations while introducing an imaginary narrow open slot on the top of it but with a short transition region beginning with the width of the top and narrowing rapidly in the vertical. This conveniently enables the use of a single solution procedure for both free-surface flow and pressurised flow, however some problems are caused. During the transition from free-surface flow to pressurised flow i.e. at sudden transition from the free-surface flow with relatively wide cross-section to the flow with a very narrow fictitious water table width, the wave celerity suddenly drops for two to three orders of magnitude, and vice versa. For very narrow slots that replicate theoretical wave celerity at pressurised flow, this typically leads to numerical instabilities even with the most robust schemes. Hence, slot widths used in practice are inevitably a trade-off between accuracy and robustness. Furthermore, a theoretical wave celerity in pressurised pipes is in practice reduced by the pressure relaxation effect of lateral connections further complicated by the fact (Watanabe and Kurihara 1993). Secondly, pipe area fictitiously extended by the slot is incorrect, which effectively increases the volume of the sewer system. One way of reducing thus introduced uncertainty is by reducing the effective volume of manholes (Ukon *et al.* 2008). Finally, models based on the open-slot approach assume a one-to-one relationship between water level in the pipe (including the slot) and the pressure; consequently such models cannot handle negative pressure (in the pressurised flow) nor the movement of trapped air. The solution to this problem has been suggested by Vasconcelos *et al.* (2006) who developed a so called two-component pressure approach.

The assumptions introduced in numerical solution procedure are also a source of uncertainty in the simulation of supercritical and transcritical flows in 1D models. A very common approach is the implementation of two approximations (Havnø *et al.* 1985):

- gradual reduction of the inertia term in the momentum equation as the flow approaches the supercritical regime, and
- the use of an algorithmic structure for the boundary conditions inherent to subcritical flow to all regimes.

When the former assumption is applied at a relatively small Froude number, it typically results in an unrealistically diffused hydraulic jump. On the other hand, if it is applied only at Froude numbers close to one, the jump is less diffused, however persistent non-amplifying water level oscillations downstream from the hydraulic jump are caused due to its jerky shifting from one computational distance step to another (Djordjević *et al.* 2004). The latter approximation enforces an inadequate representation of internal boundary conditions which is a further source of uncertainty, but this approach remains widely used because it enables the application of the efficient general solution algorithm for finite difference problems to both sub-critical and super-critical flow regimes.

With respect to the flood modelling practice, the physical characteristics of pipe–network or open–channel system (effective diameters, cross-sections, slopes, ancillaries, connectivity, structural condition, etc.) are only partially known. Also, the instantiation of urban drainage models often require a simplification of system details in order to allow numerical computations to be undertaken within practical timeframes. Such work may impose limits on how well the model can explain the phenomena. Furthermore, the size of the area under investigation will dictate how computationally demanding the simulation may become and consequently if a particular type of uncertainty analysis is feasible at the current level of technology or not. One approach is to conduct uncertainty analysis based on a large number of runs using a series of storms, with or without varying catchment and drainage system parameters. However, when it comes to the application of this and any other formal method of analysis the greatest challenge in uncertainty analysis of urban flood modelling is dependent on how computationally demanding the models are. It might be expected that the computational expense will gradually become less of an issue with the ever increasing computer speed, particularly with progress in the domain of parallel computing and cloud computing. In addition, the split between the hydrological and hydraulic phases of urban flood modelling, whereby the former only produces input to the latter, also represents a source of uncertainties.

When combined *pluvial/fluvial* or *pluvial/fluvial/coastal* flooding is the problem at hand, urban flood models need to take river flow or the sea water level variations into account either as a downstream boundary condition or as a third sub-system that should be simulated simultaneously with the two dual drainage sub-systems. Uncertainties related to these interactions have not yet been studied in detail and as such they continue to represent a challenge. Furthermore, undertaking large numbers of simulations with such models (e.g., 1D-2D models) can only be efficient by assuming a rather coarse spatial resolution i.e. with a simplified pipe (or open channel) network and/or a very low resolution of the surface flow domain or with approximate urban flood models in which the computational treatment of flow either in the minor or in the major system is significantly simplified or not even considered. The alternative approach is to run simulations only for a limited number of rainfall events using high resolution, higher dimensionality fully-dynamic coupled models. This implies limitations in terms of what kind of analysis can be done because formal probability theory is out of question here and the application of formal statistical methods is rather limited. However, with this approach even some simple forms of sensitivity analysis may enable issues specific for urban flood risk assessment to be quantified.

Model calibration represents yet another challenge. In addition to the errors associated with the input to the model there are many other forms that modelling uncertainty can take. Some of them are: the model physics is too simple to describe the phenomena, there are suboptimal values of model parameters, the calibration data does not cover the full range of response values the model needs to predict and therefore the model fit is poor in the range of values not represented in the calibration, etc. The process of adjusting model parameters may range from a simple trial-and-error procedure to the most sophisticated mathematical optimisation approaches under conditions of uncertainty. Since A multi-criteria calibration procedure based upon a global optimisation algorithm by which different objective functions are satisfied simultaneously (e.g., Khu *et al.* 2006) is preferred to the manual calibration process. Essentially, this is a trail-and-error process of minimising the differences between the computed and measured values. Whatever calibration procedure is used, the reults might be deemed meaningless if the geometrical features of the domain were not represented accurately or if calibration parameters are allowed to vary outside boundaries that physically make sense. For example – if Manning roughness coefficient is used to describe friction losses for all flow regimes – it must not be allowed to have values smaller than those that correspond to smooth turbulent flow in a pipe. Therefore, a modeller should first remove the uncertainty associated with input data as best as possible and adopt sensible range of parameter values before proceeding with model calibration. Once the model calibration has been accomplished, the model validation exercise should be carried out on a set of observed data different from the calibration data set, in order to analyse and explain the difference between model results and measurements, and if necessary, further adjust model parameters. High quality field data for urban flood model calibration/validation is usually modest.

In addition to the challenges associated with the data collection and numerical modelling work, we should also bear in mind that flood mapping is not an exact science. Therefore, there is also a level of uncertainty for any work concerning the mapping of flood risks. This uncertainty dictates the accuracy of mapping products, and needs to be understood in order to know how closely the mapping represents what users would see as an accurate representation. The European project Floodsite, more precisely its Taskgroup 3, has undertaken a study about the Flood hazard mapping. One of their activities were to study the question of the uncertainty in flood mapping and to test and suggest appropriate techniques to handle this (see www.floodsite.net).

20.11.5 Communication of uncertainties

When disseminating the flood risk assessment work to the public it is important to provide a description about uncertainties concerning map production and presentation of results within the related report. Description of the data collection and the modelling work could be very helpful, as well as a disclaimer. Particular considerations should be paid to avoid misunderstandings or to producing something that is counterproductive to public flood risk awareness.

For dissemination of the same work to *the professional users*, maps and explanatory texts (metadata) which indicate methodological uncertainties and uncertainties in data collection, modelling and mapping in a direct way, should be provided.

20.12 ADDRESSING THE ATTRIBUTION OF FLOOD RISK FROM DIFFERENT SOURCES

20.12.1 The holistic risk appraisal

As introduced earlier, flood risk assessment and mitigation work is concerned with finding the effective mitigation measures from all types of floods within ever changing social, economic and environmental contexts. This approach is very difficult to implement from the practical point of view as the ownership and responsibility for different water bodies and related infrastructure lie in the hands of a variety of public and private organisations. Such challenge has already been recognised in the USA (Rangarajan 2005) and elsewhere (Andjelkovic 2001). For this to be achieved, a new approach where different stakeholders can act together, communicate and exchange information is necessary. Therefore, the new holistic approach should not concern only a design–standard analysis but also situations that exceed that standard (i.e., residual risk analysis) and situations where the potential solutions can fail (i.e., technological risk or some other forms of risk analysis). What is then needed is a holistic risk analysis approach where risks from different sources, as they are often managed by different organisations, can be addressed and compared on a common basis. The intention would be to develop the means to *disaggregate* the holistic risk and *attribute* it to different components in the system and/or agents with responsibilities for risk reduction. This approach would aim to minimise obstacles due to organisational issues and as such it would enable a better synergy of data collection and modelling activities as well as the allocation of resources for different mitigation measures.

20.12.2 Risk attribution

Risk attribution has been defined as the process of calculating the relative contribution towards risk from different flooding sources and components of flooding pathways, including infrastructure components, see Dawson *et al.* (2008). As described in Dawson *et al.* (2008), the concept of risk attribution is important for a number of flood–risk management purposes:

(1) Risk ownership (i.e., a portion of risk for which an organisation is responsible);
(2) Estimation of the capacity to reduce risk (i.e., the capacity to change the characteristics of the flooding system by improving the infrastructure or modifying the surface flow paths);
(3) Asset management practice (i.e., a practice that concerns flood defence or the asset management of the drainage infrastructure).
(4) Data acquisition strategies (i.e., data–acquisition plans and programmes concerning different organisations).

A number of possible methods for risk attribution have been developed. Some of these methods are: source attribution, design standards attribution and sensitivity-based risk attribution (see: Djordjevic *et al.* 2011).

Chapter 21

Flood mitigation measures

21.1 TYPES OF MEASURES

Within the broader flood risk mitigation framework, three distinctive modes of mitigation can be identified: pre-flood, during flood and post-flood modes. With the knowledge of the inherent risks for the various flood hazards there is a range of possible mitigation measures that can be applied to each of the three modes. The pre-flood mode aims to reduce the flood risk in the long term, whereas, the second and third modes aim to enable timely issuing of flood warnings and the undertaking of emergency (or disaster) response and recovery activities.

Flood mitigation measures are often divided into *structural* and *non-structural*. Another classification that can be found is into *conventional* (i.e., *piped*) and *natural* (i.e., *sustainable or ecological*). It is important to note that a single category of measures is not sufficient on its own: what is needed is an integrated approach that designs and balances different measures in relation to stakeholder needs and concerns.

Structural measures can be defined as physical interventions in the physical conditions of a drainage (river or coastal zone) system. They involve building facilities such as channel/pipe networks, flood embankments, retaining walls or underground or above ground storages or areas (such as parks), or designing interconnected, designated flow paths over the urban surface (usually along roads and between certain structures following the natural ground topography). Such measures are normally constructed according to a certain level of protection (i.e., primary risk), usually based on national standards for flood protection (e.g., 10-year ARI event) and the optimum level of costs and benefits. For example, a useful source of information on the design and construction of urban drainage facilities is CIRIA which provides access to a wide and rich variety of information (http://www.ciria.org); see also Balmforth *et al.* (2006). These are certainly important considerations, but we should not forget that the need for flood mitigation arises from the community and the resulting measures should be constructed in accordance with social needs and concerns rather than some predetermined standards which may not necessarily be reflective of such needs.

Moreover, it is obvious that all structural elements of drainage systems have their own hydraulic capacity limits. At some point in their operational life time they will reach their limits and eventually cause flooding. This is due to the fact that they are designed to handle flows/volumes up to a certain standard (e.g., 10 year or even 100 year return period) and storms beyond that will exceed their limits (i.e., residual risk). It is therefore important to accept that flooding is inevitable in the urban environment. To compensate for the limitations associated with structural measures it is important to consider the use of appropriate non-structural measures in order to reduce residual and other risks. Non-structural measures can be defined as interventions which are based on mechanisms that influence human behaviour (e.g., regulatory, enforcement of standards and codes, financial, legal, etc.). Examples of such measures are: introduction of a legislation that can enforce planning criteria, changing land-use, especially restricting construction on natural flood plains (or in flood affected areas) introducing flood insurance schemes where the national economy permits, and the implementation of real-time forecasting and warning systems.

There are also so-called natural (or sustainable, ecological) measures which are often referred to as 'best management practices' (BMPs in US), 'sustainable drainage systems' (SUDS in UK) or 'water sensitive urban design' (Australia). The main aim of these measures is to enable better management of runoff (in terms of both quantity and quality) at source by preserving natural drainage patterns and emulating the natural hydrological cycle. Examples of these measures are: runoff reuse, soakaways, infiltration trenches, swales, green roofs, permeable paving, ponds, wetlands, and so on.

Irrespective to the type and nature of measures, the process of evaluating and selecting preferred measures should take into account the following aspects:

- Development of criteria for selection of measures that reflects social needs and concerns;
- Evaluation of measures against that criteria while taking into account primary, residual, technological and other risks; Measures should be also evaluated in terms of their flexibility and potential for future expansion/modification;
- Potential for synergy with measures concerning other types of disasters as well as future planning initiatives;
- Practicality for implementation and monitoring;

Once the measures are identified and agreed upon with the stakeholders, they should be carefully described in one or more flood related plans (e.g., integrated water resources management plans, river basin management plans, catchment flood management plans, coastal management plans, disaster management plans and asset management plans) and these plans should become public documents.

21.1.1 Fluvial flood protection measures

Fluvial flood protection measures involve a variety of the above described measures. Structural river flood protection measures refer to those measures that alter the river system and prevent flood water overflowing into the floodplain by means of structures along the floodplains (a.k.a. as extensive structural measures), or in the river (a.k.a. as intensive structural measures). Extensive structural measures are those measures that act in the floodplain, aiming to change the relationships between precipitation and flow, such as altering the vegetation cover of the soil, which reduces and delays the flood peaks and controls erosion. Intensive structural measures are those measures that act in the river and may be of three types (see also Simons *et al.* 1977):

(1) those that accelerate flow and regulate water levels: levees (or levées), dikes (or dykes), Figure 21.1, embankments and floodbanks (or stopbanks), polders and cutting of meanders; Levees, dikes, embankments and floodbanks measures contain an elongate naturally occurring ridge or artificially constructed fill or wall, which regulates water levels. These are normally positioned parallel to the course of a river in its floodplain or along low-lying coastlines.
(2) those that delay flow: flood control dams, attenuation reservoirs and basins; The purpose of these structural measures is to store a portion of the flood waters during flood events and then to release the water slowly.
(3) those that divert flow: structures such as diversion canals.

It should be noted that even though all of these measures can yield beneficial results in terms of flood alleviation they may also produce negative effects on the river as a habitat.

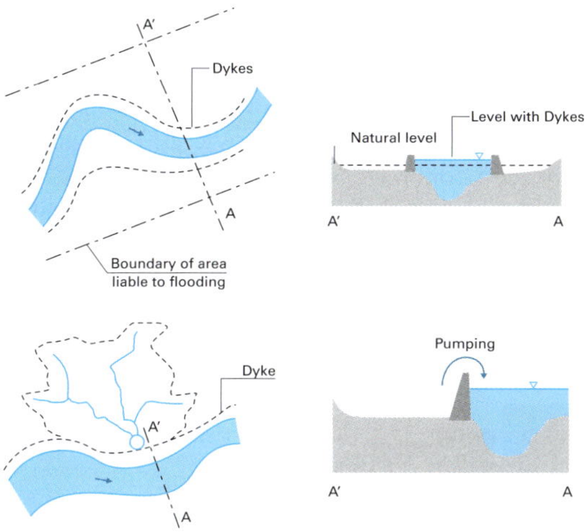

Figure 21.1 Illustration of a dyke

Non-structural flood protection measures (a.k.a. as preventive measures) involve measures such as zoning of risk areas (Figure 21.2), flood warnings, flood insurance, flood proofing, and so on.. River flood control aims to implement such measures that can enable the riverside population to minimise its losses and continue to live in a harmony with the river. These include engineering and social, economic and administrative measures. Planning of protection against flooding and its effects involves the search for the ideal combination of these measures.

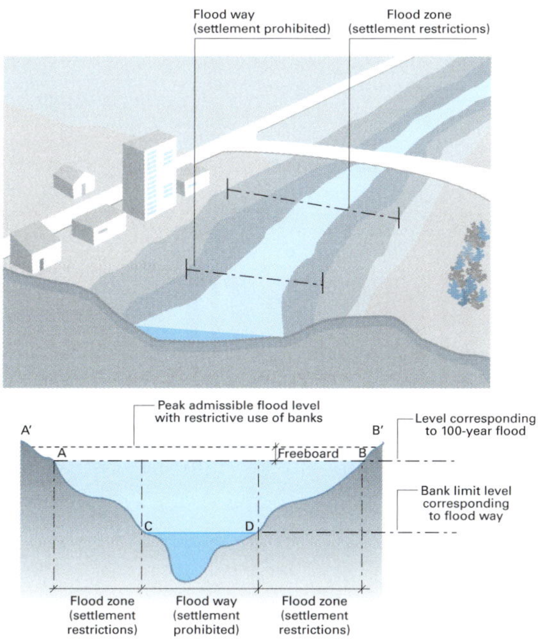

Figure 21.2 Flood zoning

21.1.2 Coastal flood protection measures

Typical structural coastal flood protection measures are sea walls, tidal barrages, barrier islands and breakwaters. A sea wall (Figure 21.3) is a structure which is used as a protection from the action of tides and waves (see also Kamphius 2010; Shipman & Stojanovic 2007). Such structures are constructed from a variety of materials (e.g., reinforced concrete, boulders, steel, gabions, rocks, etc.). In some parts of the world, a sea wall refers to an earthen bank used to create a polder, or a dike.

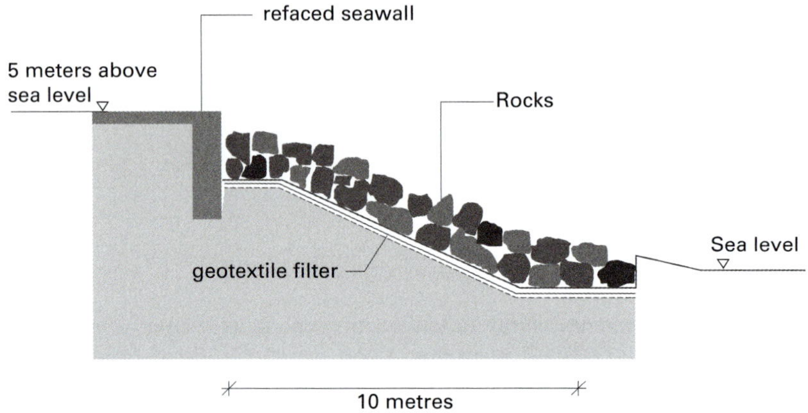

Figure 21.3 Illustration of the sea wall

A tidal barrage is a dam-like structure which allows water moving in and out of a bay (also a river) due to tidal forces, Figure 21.4. This is done by controlling the sluice gates at different phases of the tidal cycle. Turbines are also often introduced and they may be placed at the sluices so as to capture the energy as the water flows in and out.

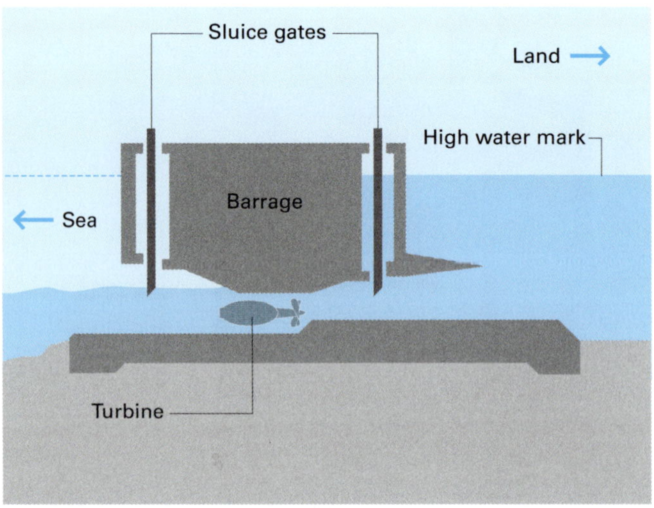

Figure 21.4 Illustration of a tidal barrage (Source: Strathclyde University, New South Wales)

Barrier islands are narrow strips of sand parallel to the mainland coast, Figure 21.5. They usually occur in chains consisting of a number of islands. The length and width of barriers and overall morphology of barrier coasts are determined on the basis of parameters such as tidal range, wave energy, sediment supply, sea-level trends, and so on.

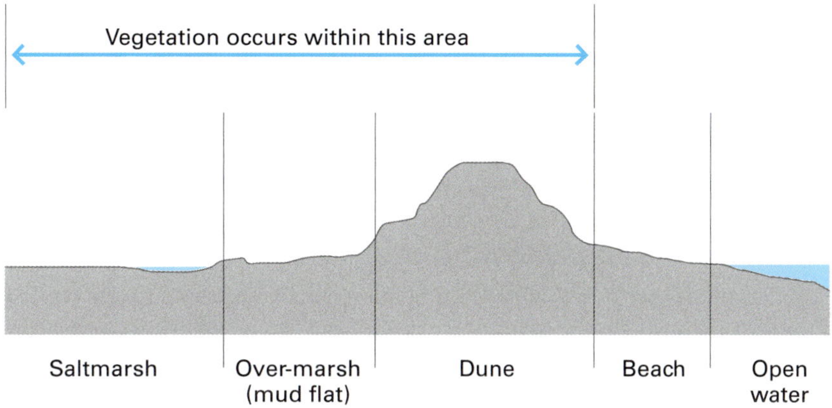

Figure 21.5 Illustration of a typical barrier island

Offshore breakwaters (or bulkheads) are structures which are used to reduce the wave energy in inshore waters and thereby reduce flooding and coastal erosion, Figure 21.6. They are usually constructed parallel or perpendicular to the coast and placed away from the coast or built with one end linked to the coast.

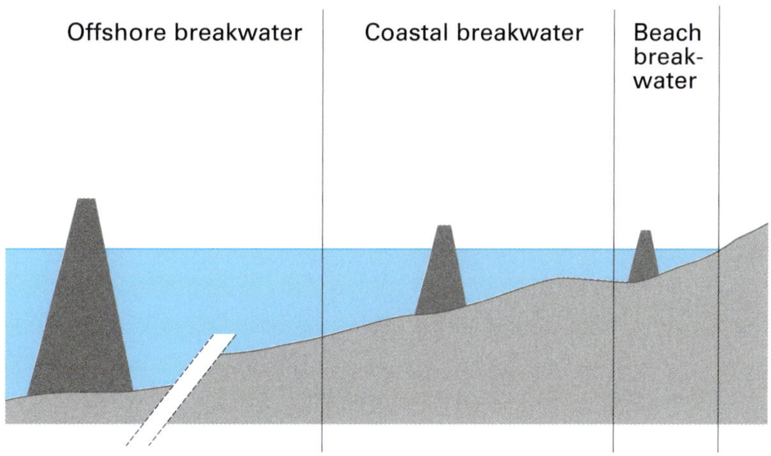

Figure 21.6 Illustration of breakwaters

In terms of the non-structural measures for coastal flood management, typical measures are flood warnings, development controls, erosion controls, waterproofing, flood insurance, and so on.

21.1.3 Flash flood protection measures

As introduced in Part III of this book, flash floods are floods which are caused by the rapid response of ephemeral streams to heavy rainfall, as related, *inter alia*, to steep slopes. Since they are characterised by exceptionally the high velocities of the flood wave, the main purpose of structural measures for flash flood mitigation is to reduce the spread of the floods and to reduce the flood–affected area by constructing various kinds of control structures along the rivers and streams (see fluvial flood protection measures), retention or detention reservoirs (or ponds) and other features that can lead to a reduction in the flood culmination (e.g., reshaping the land, construction of various kinds of walls, reinforcing steep slopes, stabilising the bottoms and sides of drainage ditches, etc.).

In terms of non-structural measures, similarly to the previously–mentioned non-structural measures, typical instruments would be flood warnings, development controls, erosion and landslide controls, waterproofing, flood insurance, etc.

21.1.4 Groundwater flood protection measures

As discussed earlier, groundwater floods are caused by the groundwater table rising: see Figure 21.7. Even though groundwater flooding has a shallower depth and a lower flow velocity, many of the previously-mentioned fluvial flood mitigation measures are also applicable to this type of flooding. The main measures for protection from groundwater flooding are: dewatering of the groundwater aquifer (i.e., groundwater pumping), construction of diversion works, improvement of the local drainage system and various property–improvement measures (e.g., removal of all sensitive goods and materials from basements and other flood-prone areas, flood proofing, floor raising, etc.).

In terms of the non-structural measures for groundwater flood management, similar to the previously mentioned non-structural measures, such measures would be flood warnings, development controls, flood insurance, and so on.

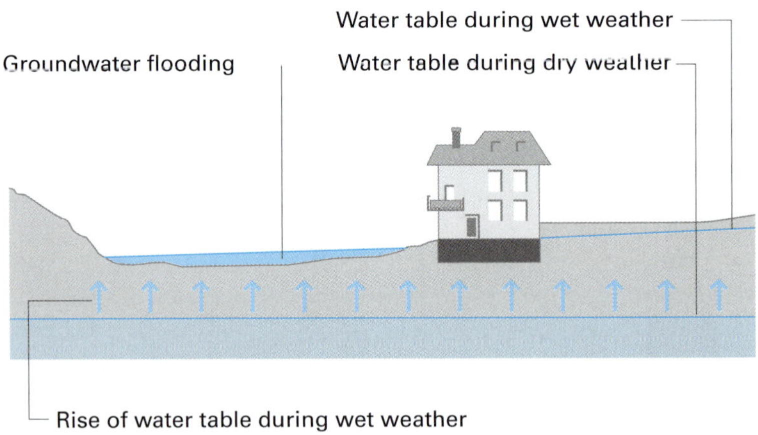

Figure 21.7 Groundwater flooding

21.1.5 Pluvial flood protection measures

As introduced earlier, pluvial floods are typically caused by a heavy rainfall that occurs directly on the urban area and where the runoff exceeds the capacity of the drainage systems. Therefore, such floods can occur due to two types of processes: overland flows which cannot enter the natural or a man-made drainage system (Figure 21.8a) and a rainfall event that exceeds the drainage system's capacity (Figure 21.8b). Typical pluvial structural flood protection measures can be grouped into the following:

- runoff control measures (e.g., control over housing developments, infiltration and percolation into the ground, parking places and other public surfaces);
- drainage control measures (construction of new or enlargement of existing drainage channels and pipes, above-ground and below-ground storages, construction of new or enlargement of existing dykes, installation of pumping stations, etc.).

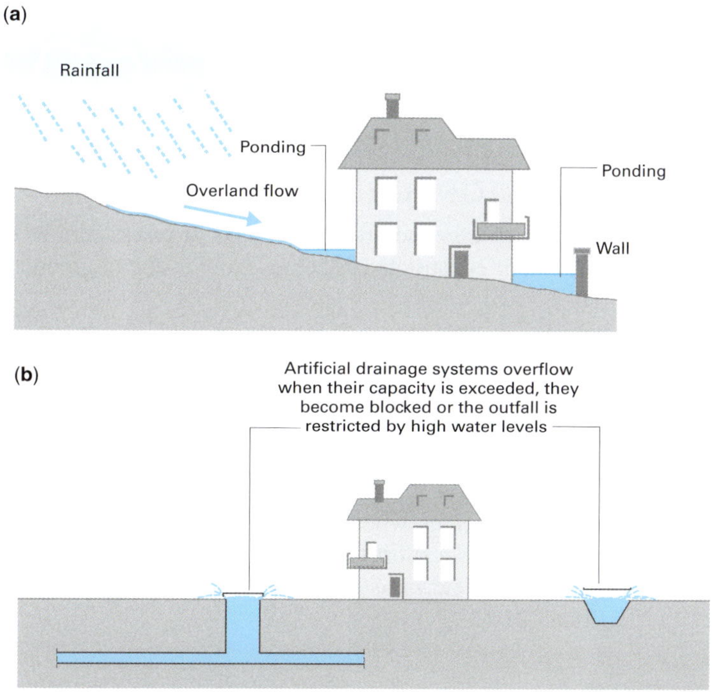

Figure 21.8 Overland flow flooding (a) and flooding from a drainage system (b) (After: CIRIA, 2004)

Again, similar to the previously mentioned non-structural measures, typical non-structural measures for pluvial flooding would be flood warnings, development controls, flood insurances, and so on.

21.2 UPGRADING OF CHANNELS AND PIPES

The upgrade of an open channel network may include: lining (in the case that the channel bed is natural), widening of sections (Figure 21.9), fitting of energy dissipation devices (steps, baffles, checkwalls, etc.) and enlargements of culverts and construction of diversions. In the case of a pipe network drainage system, typical upgrading work may include: replacing existing pipes with larger diameter pipes, construction of new parallel pipes and diversions, construction of larger and/or longer outlets (e.g., deep sea outfalls, etc.), re-grading the slopes of existing pipes and construction of catchpits and larger manholes/chambers.

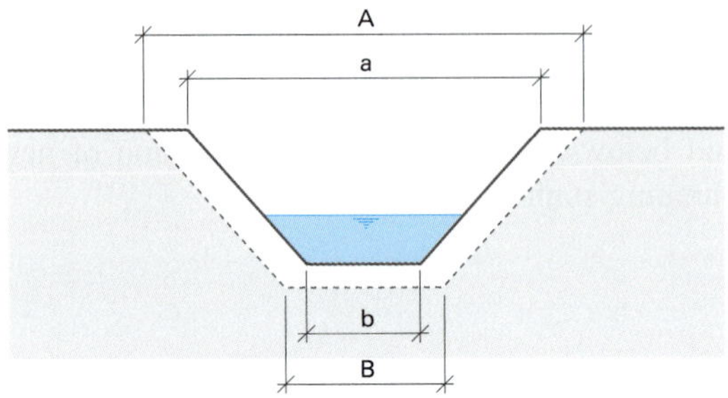

Figure 21.9 Widening of open channel's cross-sections

21.3 CONSTRUCTION OF STORAGE FACILITIES

As introduced in Part III of this book, storage facilities (e.g., tanks or ponds) can be used to limit the peak discharge from a catchment. They can be constructed as below-ground, above-ground, open, closed, on-line and off-line.

Detention ponds (or *dry ponds*) are a very common measure for flood mitigation as they are designed to detain the excessive runoff and then drain it completely after the end of the rainfall event (typically, over a period of up to 2 days). Basically, they are dry during the dry weather periods and wet during the wet weather periods. They are normally placed upstream of the problem areas and as such they can be either on-line or off-line. Their purpose is to detain excess flows which are exceeding the channel capacity (e.g., utilising parks or any another suitable area within the catchment). They can be constructed to act as a long-term as well as a short-term measure. In situations where such a measure is not desirable or feasible in the longer term, it might be then considered as a short term measure until some other measure (e.g., upgrade of channels/pipes, stormwater diversion, etc.) has been constructed to accommodate the design flows: see Figures 21.10, 21.11, 21.12 and 21.13.

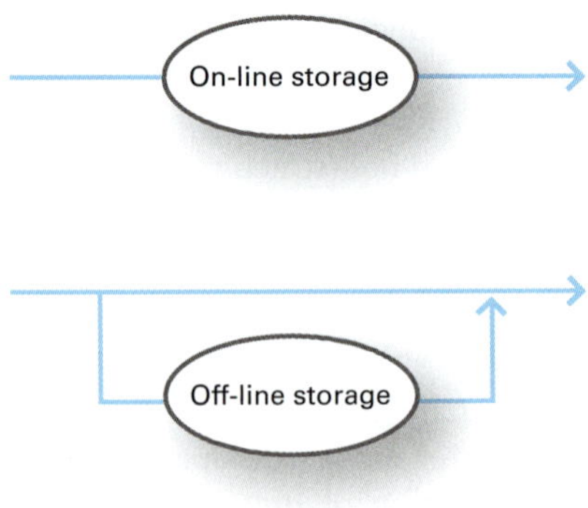

Figure 21.10 Schematics of on-line and off-line storages

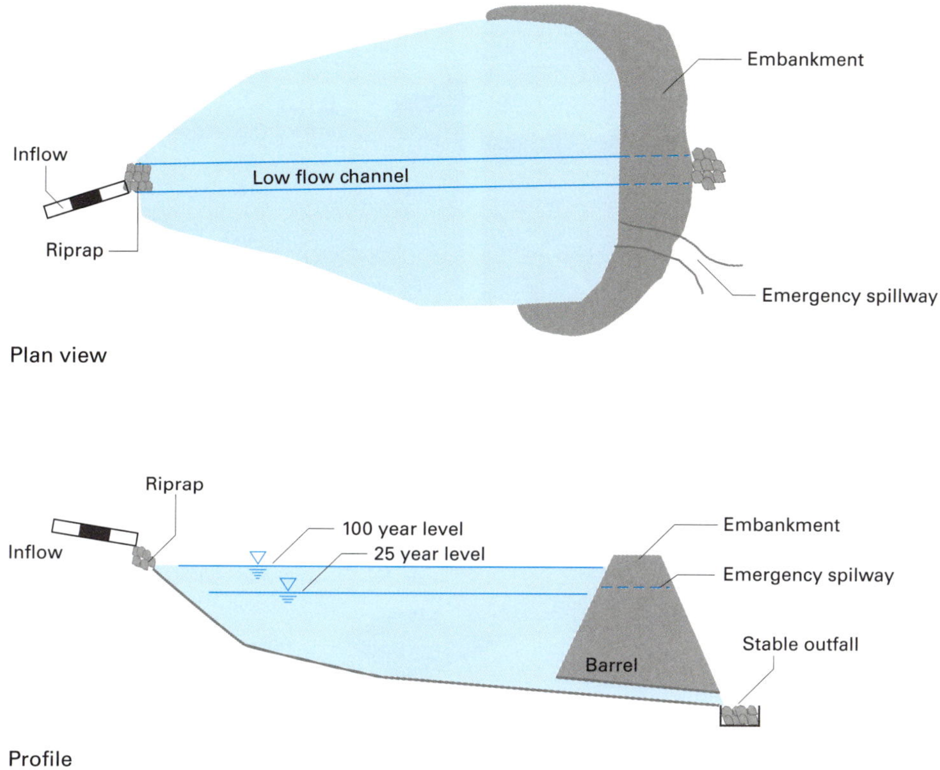

Figure 21.11 Schematics of an open stormwater detention pond

Figure 21.12 A sport field designed to act as a stormwater detention pond in Greve, Denmark (Photograph: Ole Mark)

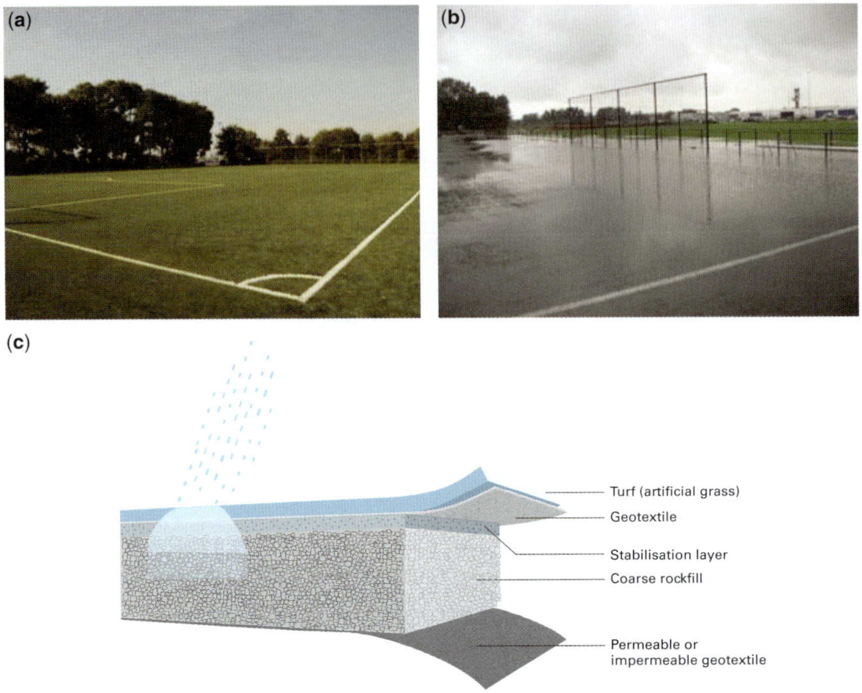

Figure 21.13 Underground storage under the sport field in Rotterdam, The Netherlands: (a) Dry weather conditions; (b) Wet weather conditions; (c) Typical cross-sectional view of the storage; Contractor: GKB Barendrecht, (Photograph: E. van der Burg)

In addition to detention ponds, retention ponds (or *wet ponds*) are also commonly used for flood mitigation. As differentiated from detention ponds, retention ponds are designed to retain the water permanently. Excessive runoff is captured by these ponds and some portion is drained out after the end of the rainfall event. The water in these ponds is displaced and replaced in part by the stormwater. Similar to detention ponds, they are also placed upstream of problem areas; see Figure 21.14.

Figure 21.14 A stormwater retention pond in Belo Horizonte, Brazil (Photograph: Z. Vojinović)

21.4 ON-SITE-DETENTION AND RAINWATER REUSE

On-Site-Detention and rainwater reuse (or rainwater harvesting) is usually done by installing rainwater tanks or On-Site-Detention (OSD) facilities. Such facilities are constructed at individual households in order to provide temporary storage of stormwater runoff. The stored rainwater is either discharged to the drainage system after the rainfall event or it is used as a source for toilet flushing, garden watering and car washing.

21.5 SUSTAINABLE URBAN DRAINAGE SYSTEMS (SUDS)

As introduced earlier, surface water drainage systems developed as a more natural means for achieving sustainable development are often referred to as Sustainable Urban Drainage Systems (SUDS). The aim of SUDS is to mitigate many of the adverse effects of urban stormwater runoff as close to its source through the following key processes:

- reducing runoff rates;
- reducing the runoff volumes;
- conveying the runoff;
- encouraging natural groundwater recharge;
- reducing pollutant concentrations in stormwater before it enters receiving waters;
- enhancing natural and aesthetic areas; and
- providing habitats for wildlife and biodiversity enhancement.

There are many cases where these measures have proved to be effective in reducing pollution, flow rates and volumes of runoff. With successful implementation of SUDS (which are often implemented in a series of individual elements and referred to as a "management train") the need for large flow attenuation and flow control structures can be reasonably reduced. The range of SUDS elements and their design variations allows designers and planners to take into consideration many important issues (e.g., existing drainage system, land-use, future planning scenarios, community needs, etc.) when undertaking design of measures for flood mitigation.

Typical examples of SUDS measures are: soakways, green roofs, permeable paving, swales, detention ponds (or basins), wetlands, etc.

Soakways, pits and other infiltration devices (e.g., basins) are normally constructed to allow the water to soak away into the ground, Figure 21.15.

Figure 21.15 Infiltration basin (Angmering, England; Photograph: Alison Duffy)

Green roofs are vegetated layers that sit on top of the conventional roof surfaces of a building, Figure 21.16. They are distinguished between 'extensive' and 'intensive'. These terms refer to the degree of maintenance the roofs require. Their main purpose is to reduce and slow down the runoff.

Figure 21.16 Green roofs (Malmo, Sweden; Photograph: Alison Duffy)

Permeable paving aims to attenuate the surface runoff by using the permeable paving materials, Figure 21.17. They may or may not allow infiltration to the ground.

Figure 21.17 Permeable paving (Dundee, Scotland; Photograph: Alison Duffy)

Swales are natural open channels which may be designed to have multiple purposes: overland flow paths, storage/attenuation, reduction of runoff velocity and infiltration, Figure 21.18.

Figure 21.18 Swale (Dundee, Scotland; Photograph: Alison Duffy)

Constructed wetlands are also used for the control of urban runoff as well as for water quality improvements. There are different types of such facilities in terms of their hydraulic, physical and biological characteristics. Their surface can be planted with different vegetation to improve pollutant–removal mechanisms through different processes (filtration, infiltration, biosorption, etc.). They are typically designed to have a short residence time in order to prevent mosquito breeding.

21.6 LITTER MANAGEMENT

Urban litter (alternatively called waste or trash, debris, flotsam, jetsam, floatables, gross pollutants, rubbish or solid waste) can be a major problem in urban areas, Figure 21.19. It accumulates in public areas until either someone removes it or it is transported by the wind and/or stormwater runoff into the drainage system. Once in the drainage system, the litter is potentially able to travel via the stormwater pipes or channel, either until it eventually reaches the receiving water or creates an obstruction and creates blockages along the way. Besides the educational, public awareness and enforcement activities, there are also other measures that can be put in place in order to control urban litter. These are mainly policy related measures and they are directed towards the restriction of the litter-generating activities to areas where their impact can be controlled and reduced most effectively. These include:

- Introduction of legislation to enforce planning criteria (i.e., flood zoning), changing land-use, especially restricting construction on natural flood plains (or in flood affected areas) and introducing flood insurance schemes where the national economy permits.
- Adoption of proactive management practices such as improved maintenance programmes (e.g., more frequent street sweeping, gully pot cleaning, etc.), prevention of inappropriate practices (e.g., illegal wastewater connections into separate stormwater systems, pesticide management, etc.) and the use of flood forecasting/warning and real-time control systems.
- Capacity building and professional training: As one of the critical steps in addressing urban flood management issues in many cities around the world (and particularly in developing countries) it is

necessary to strengthen the knowledge and capacity building processes, and then not only in terms of general flood management principles but also in terms of the instantiation and operation of numerical models.

Figure 21.19 Catchpit (or gullypot) maintenance in Auckland, New Zealand (Photograph: John Tetterroo)

21.7 FLOOD FORECASTING AND WARNING SYSTEMS

The purpose of real-time forecasting and warning systems is to provide for better preparations activities in advance of the flood event (e.g., lowering the water level in a reservoir or a pond, activation of pumps, evacuation of residents, traffic regulation, road closure, etc.) and enhancing the preparatory lead time. This measure aims to reduce the loss of life and property in vulnerable areas, especially in the inner floodplains. Typically, a combination of real-time measurement/communication technologies and modelling tools are used to forecast rainfall and other hydrometeorological factors. A well–proven modelling system is required to provide reliable forecasts of flows and flood levels along the floodplain. In order to make the warnings effective for wide dissemination, the forecasts have to be translated into readily understandable forms, and related to the population and infrastructure by linking model outputs to the GIS flood maps. Conceptually, the construction of such a system consists of three distinctive units: *data acquisition and processing*, *model simulations and results–mapping*, and *information dissemination*.

In the first part, telemetry (i.e., data acquisition) and pre-processing are concerned with the use of meteorological data and simulations using a hydraulic model to predict an upcoming flood, given the required lead time. With the development of such weather radar as Next Generation Radar (NEXRAD) and Local Area Weather Radar (LAWR), it is possible to make very accurate forecasts of the upcoming rainfall events with a lead time of up to 3 hours. In certain cases (e.g., in the case of flash floods), the short response time of flooding between peak rainfall and runoff makes the introduction of warning systems in urban areas more difficult. This may necessitate the forecasting of rainfall some hours or even days ahead. Many countries are nowadays making increasing use of ensemble rainfall forecasts produced by models of global circulation of the atmosphere and associated local area models. These can provide forecasts up to 10 days ahead. Remote sensing data or weather radar can supplement such forecasts by generating reliable information up to 3 hours ahead.

In the second part, either the model results are obtained or the pre-prepared library of model results is searched for appropriate information (e.g., risk maps, safe areas notification, evacuation routes, traffic control etc.).

Finally, the third part is concerned with information dissemination. At this stage, all thw necessary real-time information, such as is applicable not only to flood managers but also to the public in general and the active stakeholders in particular, is crucial for the success of such a system. Technologies from the fields of wireless inter-linkage and spatial IT such as GPS, GSM, UMTS, 3G, WAP and MGS are opening–up new possibilities for flood management and dissemination of warnings using real-time data services and high-speed data capture. Furthermore, XML-based Common Alerting Protocol (CAP) data–format technologies have been developed for the purpose of exchanging public warnings and emergencies between alerting technologies. Such technologies based on mobile communications, spatial systems, mobile positioning, mobile terminals, embedded software, mobile database, and application–server architectures are aimed at establishing an efficient and dynamic information service to reach individual users at different times and at different places. Further to the warnings issued by the flood warning system, a set of specific actions (e.g., control and diversion of traffic by the police or fire brigade, evacuation of children from schools, etc.), as defined within the flood hazard emergency response management plan, would be triggered.

Schematisation of such a real-time flood warning system is given in Figure 21.20. The system schematised in Figure 59 would need to be able to release warnings in advance of the storms by the use of different communication means, and primarily it would include very valuable real-time-modelled flood risk forecasts. Such information can provide substantial enhancements and make possible informed decision making in adopting proper measures towards disaster preparedness, mitigation, control, planning and operations.

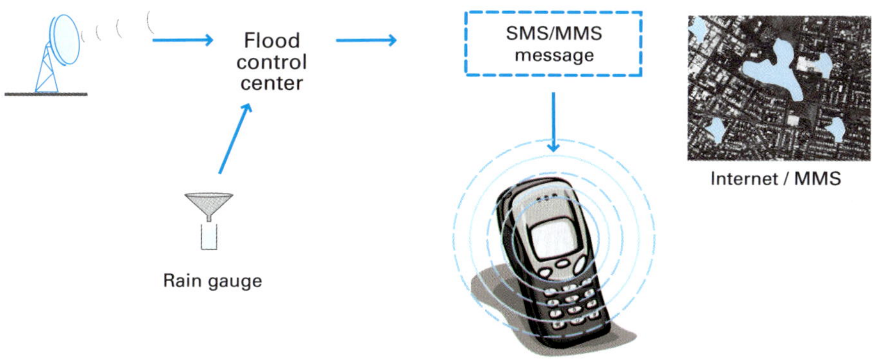

Figure 21.20 Schematisation of a real-time flood control centre: Alerts are automatically sent in case of flood risks

21.8 REAL-TIME CONTROL SYSTEMS

Traditional solutions to the urban drainage problems are based on the concept of providing sufficient capacity for a given rainfall event (i.e., design event). With this approach, the dynamics of the system and the lack of flexibility in operating the system are often neglected. For example, a pumping facility is activated on the basis of a predefined water level at a wet well, independently of the situation elsewhere in the system. This could lead to the situation where the capacity of the system is exceeded locally, whereas elsewhere in the system some storage capacity is still available. Therefore, the system will only be fully utilised when it is loaded with the design load, while for all other loadings the system will perform sub-optimally, in which case a certain amount of storage capacity will never be utilised. In such situations, it is always beneficial to investigate the benefits of applying real-time control facilities before upgrading the system.

A structure of real-time control systems that aims to achieve a higher efficiency in the existing drainage system at relatively low implementation/operational costs can have a different level of complexity. Generally, there are three different levels of control: *local*, *regional* (or *process*) and *central* (or *systems*) control. Any control system is comprised of the following elements: the *sensor* (its purpose is to monitor the systems state and variables), the *flow regulator* (its purpose is to control or maintain the process; examples are: pumps, gates, sluices, movable weirs, valves, orifices), the *controller* (its purpose is to control the flow regulator; examples are: two point controller – on/off, the Proportional Integral Differential – PID controller, predictive controllers, etc.), and the *communications system* (its purpose being to link the sensors, controllers and regulators together in the system): see Figure 21.21. It is often used together with the SCADA system which is positioned on top of a real-time control system in order to control a process that is external to the SCADA system. Furthermore, in order to send and receive the data remotely, the entire system can be telemetered. Using the resulting telemetry, the systems engineer or operator can obtain the sensor measurement information remotely in real time.

An efficient control strategy can reduce (or optimise) substantially the amount and frequency of CSO spills, flooding from a critical manhole or pump running time. In addition, it may lead to substantial cost savings as the number and storage capacities of the reservoirs required to keep spilling (from either an overflow structure or a manhole) below a certain limit depends on the efficiency of the applied control strategy.

For aspects concerning modelling real time control options see Chapter19, Part III.

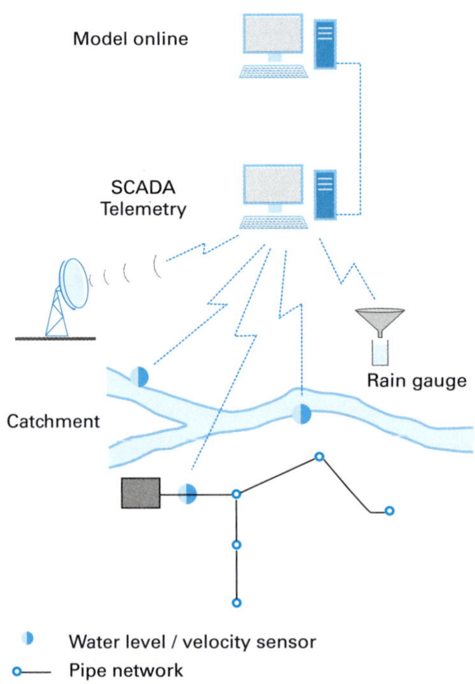

Figure 21.21 Schematics of a real-time control system for a drainage network

21.9 DISASTER PREPAREDNESS AND RESPONSE

In general terms, a natural disaster can be defined as an event that overwhelms the capacity of a household, community, city, business or a utility to resist or recover from the impacts without external assistance. Within the same context, and as described in the opening section of this chapter, the hazard is defined as a natural or other phenomenon with the potential to bring harm. Usually, there are three types of events that can cause a disaster:

- Low frequency – high magnitude events (usually 'declared disasters') with long return period and large losses;
- High frequency – low magnitude events (often considered 'normal'/'routine' threats) that may occur daily/weekly/monthly but with relatively small impacts;
- High frequency – high magnitude events.

Disaster management usually refers to a *suite of measures* intended to strengthen the management of risks and reduce the consequences of disasters. These measures are broadly divided into:

- *Prevention*; This refers to measures designed to provide permanent protection or reduce the intensity of a hazardous event to such a level that it does not become a disaster (e.g., relocation of households from a flood-prone area, etc.).
- *Mitigation*; This refers to the measures taken well in advance of a hazard alert to minimise the vulnerability of communities/households to a known/expected threat (e.g., construction of a detention pond, etc.).
- *Preparedness*;
- This refers to advance measures taken to predict, respond to and manage a hazard event, or in other words, measures that prepare people to react appropriately before, during and after this event (e.g., dissemination of early flood warning information, etc.).
- *Response/relief*; This refers to measures taken to alleviate immediate hardship and meet basic needs for shelter, water, sanitation and health care, as well as aiding search, rescue and protection operations directed to those affected
- *Recovery*; This refers to processes undertaken by a disaster-affected community to fully restore itself to its pre-disaster level of functioning and which enables it to become even more disaster-resilient (e.g., flood-proofing essential community buildings, schools and clinics, etc.).

When speaking within the context of disaster risk mitigation, disaster risk refers to the probability that a household, community, city, province or country is unable to anticipate, resist or recover from the losses sustained from a hazard or other threat without external assistance. Within the same context, disaster risk mitigation refers to the increase in the likelihood that a household, community or a city will be able to anticipate, resist or recover from the losses sustained from a hazard or other threat without external assistance.

Risk mitigation (or reduction) is a fundamental activity that leads to the evaluation of schemes for reducing but not necessarily eliminating the overall risk, as in many cases the risk cannot be entirely eliminated. It includes assessing the potential for a hazard to occur and a vulnerability analysis to provide an understanding of the consequences should an event of a certain magnitude and frequency occur. Based on such work, various mitigation measures can be evaluated to assess their ability for reducing risk exposure. Once the preferred measures are identified, risk assessment of those measures needs to be carried out in order to develop effective disaster management plans and specific actions. In all these activities, of course, it is possible, and even increasingly likely, that the management will be in the hands of the active stakeholders, despite the tension that may then arise with the so-called "professional managers".

As described earlier, risk identification is comprised of three complementary processes:

- *Hazard assessment;*
- *Vulnerability assessment;*
- *Coping capacity (or resilience);*

It is important to mention here that the measures for flood related disasters should be placed in the context of other disasters (e.g., chemical emergencies, dam failures, earthquakes, fires (including wildfires), hazardous materials, heat, hurricanes/typhoons, landslides, nuclear power plant emergencies, terrorism, thunderstorms, tornados, tsunamis, volcanic eruptions and winter storms) and wherever possible joint solutions should be sought.

21.10 FLOOD PROOFING OF STRUCTURES

In principle, all structures that are located on the floodplain should be flood-proofed to reduce future damages. In this respect, the development of building codes can ensure that beneficial uses of buildings are located above the design flood elevation. New buildings can be constructed above the design flood level by placement of fill, stilts or piles. Ground floors can be constructed in such way that little flood damage can occur. Such measures are much easier to implement for new buildings, whereas they may not be so easy to implement for the existing buildings. Some of the typical measures concerning existing structures are:

- raising of structures;
- relocation;
- change of use;
- installation of protective walls and waterproof closures, Figure 21.22.

Figure 21.22 Door gates made of steel designed to deal with flash floods: Barreiro, Belo Horizonte, Brazil (Photograph: Cristiane Parisi Jonov)

In many cases, buyout and relocation of vulnerable buildings represents one of the measures concerning flood proofing initiatives. Typically, such measures would appear favorable if it is more economical to buy out and relocate an existing building than to protect it.

It is often the case that a number of critical services, such as water mains, gas and power supplies, telephone services, cross the flood plain. Such services are usually protected against the flooding through placement at deeper levels or elevating above design flood levels. Water supply reservoirs, treatment plants and other critical industrial premises should be protected against extreme events and designed to prevent cross contamination from floodwaters or sewers.

21.11 LAND-USE PLANNING REGULATIONS AND ENFORCEMENT OF STANDARDS AND CODES

Land-use planning represents an important non-structural measure when dealing with flood related problems. Therefore, it is vital to integrate the land-use planning for flood-prone lands into the related flood management plans. The land along a river or a coastline is often highly desirable for various recreational uses, as well as for ecological reserves. Since these areas are prone to floods, their supportive facilities should be flood-proofed.

Furthermore, the enforcement of standards and codes represents yet another important non-structural measure. Introducing requirements, such as for a development impact analysis, surveyor certificates and inspector reports to check whether flood-proofing measures for flood prone areas have been designed and implemented, would be very beneficial.

21.12 ECONOMIC ANALYSIS FOR FLOOD RISK MITIGATION

The economic analysis of flood mitigation measures is concerned with the study of how to obtain maximum benefits from different measures in economic terms. It is important to emphasise that although economic efficiency is certainly an important aspect, it is not the only aspect in the decision making process. It involves the application of analytic frameworks which make it possible to compare different alternatives. Therefore, the purpose of economic analysis in the decision-making process is to provide an evaluation of the economic benefits and costs concerning various flood mitigation measures (or projects) so as to enable their comparison and judge their economic viability, see Figure 21.23, as provided by WMO (2007).

There are various economic techniques that provide means for assessing a range of direct and indirect benefits and costs of alternatives. Cost-benefit analysis (CBA) is the most common approach used in economic evaluation of alternatives. Basically, CBA aims to assess whether the total benefits of a particular alternative, as evaluated in monetary terms, exceeds the costs of utilising the resources necessary for its implementation and operation. As discussed earlier, being a monetary-based technique, CBA does not take into account any ethical (and other qualitative) considerations and therefore it should be used with caution and it should be treated as only one criteria amongst others.

In summary, the key aspects of CBA are:

- Identifying all benefits and costs that can be expressed in monetary terms for different measures;
- Computing the whole life cycle costs (i.e., investment, operational and other costs for the whole duration); Adjusting the future costs and benefits to present values (e.g., Net Present Value analysis);
- Comparing different measures according to the following indicators: net present value (NPV), benefit cost (BC) ratio or internal rate of return (IRR);

The full lifecycle cost estimates must consider all relevant costs, from initial planning and design through to disposal. Different costs can be identified within a lifecycle of different assets and they can be broadly divided into capital, operation and maintenance costs. Capital investment costs are those costs incurred during the planning, design, construction or acquisition phase of a particular asset and they generally include all advance expenditure on research, feasibility studies, program planning, land acquisition and improvement, building and site facilities, machinery and equipment, management services, quality control and commissioning, duties and taxes, consulting service fees, cost of raising funds (financial charges), interest charges during the construction period and the construction labour costs.

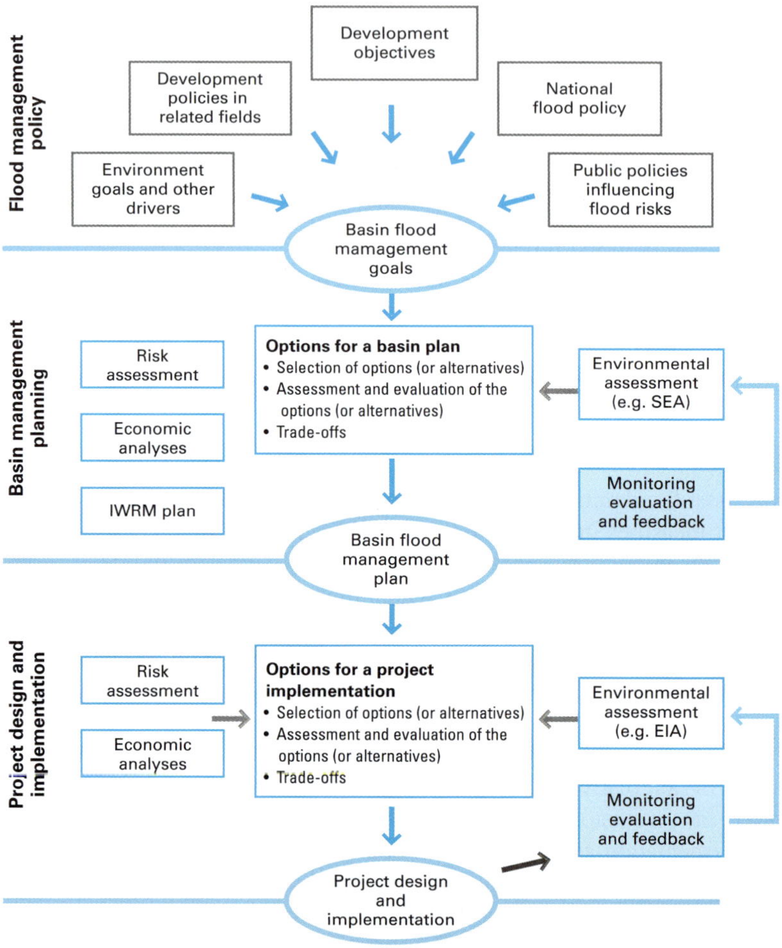

Figure 21.23 Example of a flood management decision-making framework (Source: WMO, 2007)

Operational costs include costs for operations personnel, materials, fuel, chemicals and energy consumption, etc. The operation costs can be greatly affected by factors such as age and condition of the assets, external environment (e.g., during wet season stormwater discharge may require additional pumping), or maintenance related issues which affect the efficiency of the system.

Maintenance costs include costs of maintenance personnel, plant and equipment, spares required during emergency (unplanned) breakdown, schedulable corrective/predictive planned maintenance or the intermittent maintenance for major mid-life refurbishment, alterations etc. The objective of effective maintenance management is, therefore, to have an optimal maintenance regime where the overall maintenance costs would be minimum whilst providing the required level of service or output (e.g., certain reliability or availability). This requires an appropriate blend of planned and unplanned maintenance. Figure 21.24 illustrates the typical profile of costs throughout the life of an asset.

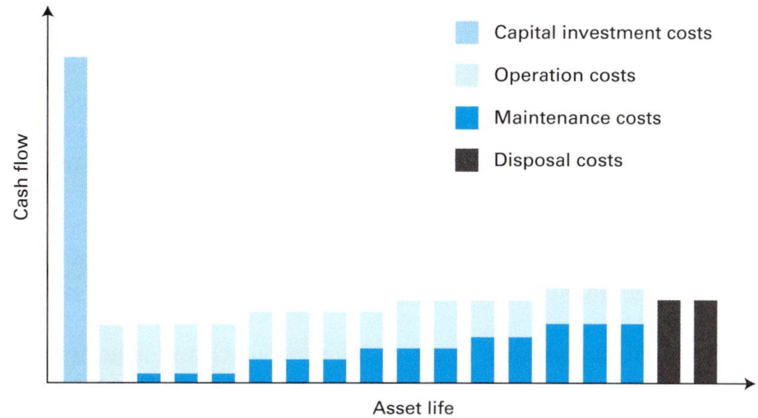

Figure 21.24 Lifecycle costs include capital investment costs, operational and maintenance costs as well as disposal costs.

As illustrated in Figure 21.24, in an asset lifecycle there are also other costs apart from the capital investment costs and they certainly need to be applied in the process of selecting an alternative that would yield the same level of performance or benefits. Life cycle cost assessment enables such selection not to be based solely on the lowest initial costs (capital costs), but also considers all the future costs (appropriately discounted) over the project's usable life. Also, it is common that if benefits vary among the design alternatives (e.g., they would accommodate surface-flooding standards), then the alternatives clearly cannot be compared solely on the basis of cost.

Once the performance and costs associated with each alternative have been established, they must be compared over the asset period. This is typically done by calculating net present value (NPV) for each rehabilitation alternative. NPV is determined by discounting all alternatives costs to the base, or present, year (usually the present year, year of construction or year of authorisation). Thus the entire cost can be expressed as a single base year, or present year, cost. Alternatives can be then compared by comparing these base-year costs. NPV is usually calculated using Eq. 1:

$$NPV = Capital\_Costs + \sum_{k=1}^{N} O\&M\_Costs_k \left[\frac{1}{(1+i)^{n_k}} \right] \tag{21.1}$$

where:

i is the discount rate
n is the year of expenditure
$\left[\frac{1}{(1+i)^{n_k}} \right]$ is the Present Value (PV) factor

WMO (2007) divides the benefits into two parts: primary and secondary. Primary benefits are of two types – direct and indirect. In the case of a flood management project, direct benefits include the reduction in physical damage. Thus, flood damage to crops, cattle, houses, commercial and industrial buildings, infrastructure etc. can be avoided because of the construction of embankments, flood retention reservoirs or flood warning systems provide direct benefits. Indirect benefits may occur by avoiding disruptions to business, transport networks and public services and by avoiding the costs of emergency response and recovery. The direct benefits from flood management projects accrue from the prevention of flood damage.

There are different methodologies for flood damage estimation (e.g., unit area approach, percentage of property area approach, synthetic approaches, etc.). The unit area approach to flood damage estimation is based on individual property data and an assessment of damage per square meter of floor space. The "percentage of property–value" approach is based on the market value of the property concerned, and then preferably just for the building rather than the land the building occupies, and it expresses flood damage potential as percentages of that value. Synthetic approaches are based on a synthesis of all available data, from both secondary sources and from the real experience of floods. Example of a methodology for computing flood damages from a damage-frequency curve can be found in Penning-Rowsell (2006).

Further to the primary effects, whether direct or indirect, which originate from physical aspects or technological linkages of a particular measure, secondary effects include those activities which originate from economic linkages through the market mechanism. Therefore, the secondary effects are then induced by the resulting output (see WMO 2007).

Similar to the benefits, the costs of a particular measure can also be defined as primary and secondary. In this respect, construction quantities and their latest market prices are at the core of cost analysis.

Multi-criteria analysis (MCA) is often used as a complementary technique to CBA which is based on a two-step procedure. In the first step, a set of goals or objectives is established and in the second step various weights or (scores) are assigned to these objectives.

Although the above mentioned techniques have proved to be useful for evaluating certain economic aspects they have not been very successful in taking into consideration the non-numerable (or qualitative) social and environmental aspects. It is important to note that any flood risk assessment and mitigation practice that omits a social and ethical component will be severely restricted in its scope and its reach.

21.13 IMPLEMENTATION, MONITORING AND REVIEW OF MITIGATION MEASURES

In many cases there will be several requirements for the implementation of flood mitigation measures (e.g., legal, institutional, etc.). Legal and institutional arrangements are required to ensure that roles and responsibilities are clearly assigned and that the stakeholder participatory process is established and maintained. Also, some of the institutions may not have the adequate capacity to undertake the responsibilities assigned to them during the implementation work as well as in the post implementation (or operation and maintenance) phase.

After implementation of measures, their monitoring and review is a very important aspect in order to follow if the measures that have been put in place are efficient in meeting previously defined levels of performance and whether the assumptions used in the scenario analysis provide a reasonable reflection of reality. Typically, various performance indicators (and indices) are used to monitor technical effectiveness of the measures and to decide whether any modification and upgrade work (or corrective action) is necessary. In case that adverse affects beyond the specified threshold values are observed, modification and adjustment of the works may need to be undertaken.

Chapter 22

Production of plans

22.1 THE NEED FOR FLOOD MANAGEMENT PLANS

The need for producing flood-related management plans can be traced to several sources. Since the flood risk mitigation process is a stakeholder-driven process, some of the sources arise from the needs of the community in the case of adverse consequences from a flood event (or a series of flood events) while other reasons may be due to regulatory and legal requirements. These plans are also needed for interactions with the general public. Whatever the case may be, the successful production of these plans requires inputs from all active stakeholders, the involvement of experts, institutional and organisational support and adequate financial resources. Different plans would discuss different aspects of flood related issues and they are typically divided into: integrated water resources management plans, river basin management plans, catchment flood management plans, coastal management plans, disaster management plans and asset management plans. Since these plans provide an overview of the framework for evaluating, developing, and implementing various strategies, programs and activities they are also sometimes referred to as *master plans* (or *strategic plans*). They are the result of the periodic need to update programs and activities to satisfy current needs and interests, to communicate flood-related issues and strategies, to accommodate changing trends, to take account of philosophies, regulations and standards, to ensure maximum effectiveness and cost efficiency, and to keep up with continuously evolving community goals and objectives. Therefore, productions of these plans is a recurring process, and indeed a transcendental process, and not a one-off activity (a process similar to the process given in Figure 7.2, Chapter 7, Part II). Some of these plans may need to be repeated every so many years (e.g., every five to ten years) whereas some other plans may need to be produced on an annual basis (e.g., asset management plans).

In short, the need for producing flood-related plans can be summarised as follows:

- To raise awareness and understanding of current and future flood-related issues (in relation to ever-continuing urbanisation, population growth, land-use and climate change issues);
- To prevent inappropriate developments in flood risk areas;
- To outline and document the details of the scope and implementation of short- medium- and long-term measures;

This chapter aims to provide an overview of current practice and the practicalities of producing various flood-related plans that address flood management issues from different perspectives.

22.2 DATA AND INFORMATION

A wide range of data and information is needed in order to produce the above mentioned plans:

- Drainage data;
- River data;
- Coastal data;
- Hydrological data;
- Environmental data;
- Land-use data;
- Census data;
- Social data;
- Economic data;
- Future planning data;
- Flood related data (e.g., information from flood studies, flood incidents, damages, etc.);
- Insurance data;
- Other data;

Some of the above mentioned data may not be readily available and may require special efforts and, in many cases, collections of all necessary data may not be feasible due to budgetary and time-related constraints. Therefore, it is often the case that the best available data is used initially, and later-on the remaining data gaps are filled-in. This implies that some of the early plans will have a somewhat greater uncertainty due to the lack of adequate data and this uncertainty will need to be addressed together with the production of subsequent plans.

22.3 SOME COMMON STEPS IN THE PRODUCTION OF FLOOD RELATED PLANS

Some of the common steps used in the production of plans are:

- Initiation of a plan;
- Initiation of the stakeholder participation process;
- Agreement on the scope and communication/reporting means (e.g., hard and soft copies of reports, use of web-based technologies, etc.);
- Formation of the working group;
- Collection of necessary data and information (e.g., standards and policies, stakeholder needs and concerns, present and future issues, details concerning selection and implementation of measures, etc.);
- Report production and dissemination;

Production of a plan should obviously start with a valid statement of the reasons for its initiation and this should be agreed in consultations with all stakeholders. The analyses of present and future issues may require execution of a fully-fledged numerical modelling study that will take into account projected conditions of land-use change, future developments (e.g., urbanisation, infrastructure development, etc.) and potential climate change issues. Such work would typically conclude with the production of flood maps which will then need to be incorporated into the related plan of action. In the case of a disaster management plan, such maps should also include supplementary information concerning evacuation routes, use of flood warning systems, instructions on how to address different phases of any disasters (pre-disaster, during-disaster and post-disaster), etc.

22.4 RISK COMMUNICATION

As practised today, risk communication is in some cases taken seriously but in some cases it is reduced to an ad hoc activity. Even if taken seriously, it is still a challenging task to define what needs to be communicated and what are the procedures to follow that can ensure the credibility of information that has to be communicated. As used here, communication refers to the process and means of delivering and exchanging information to and amongst institutions, stakeholders and the general public. Since communication represents a vital aspect within the entire flood risk mitigation process, the success of the work of the hydroinformatician often depends upon how good is his or her communication strategy and how well this is implemented. Therefore, an effective communication strategy should focus on the issues that those who are receiving information need to understand in order to make informed and independent judgements. If the communication process fails to include critical information then it subverts the most obvious responsibilities of those who are in charge of such activities. In some cases, failure to supply critical information may create serious communication problems, whereas in some other cases it may subvert the empowerment of those dealing with the risk – and so it may amplify consequent disasters.

Once the critical information is determined, it should be then communicated in a way to ensure that it is understood as intended. Effective communication requires credible and trustworthy sources. If those who are communicating information are perceived as having vested interests, or conflicts of interest, then those who are receiving such information may not know what to believe. This situation can make risk decisions seem intractable and it may erode relations between the stakeholders (see also Fischhoff 1995, 2005, and 2007). Therefore, serious efforts should be made to develop the appropriate communication strategy and the necessary means to implement it. Some of these efforts are imbedded in the production and formulation of flood related plans.

22.5 INTEGRATED WATER RESOURCES MANAGEMENT PLANS

The purpose of producing integrated water resources management (IWRM) plans is to maintain a systematic process for the sustainable management of all the water resources concerned. These plans are normally produced while taking into consideration not only the water-related issues and concerns but also social, economic and environmental objectives. As described in CAPNET (2005) and CAPNET (2008), integrated water resources management refers to the process whereby all the different uses of water resources are considered together and it refers to ensuring coherent policy-making relative to all sectors concerned. Integrated Flood Management (IFM) aims to harmonise human activities and flood risks through appropriate interventions to modify the water regime and the adaptation of human behaviour, thereby reducing such risks while maintaining the beneficial aspects of floods. The planning cycle of IWRM as described in CAPNET document is given in Figure 22.1.

The key phases of the IWRM planning cycle are given in Figure 22.1. Since it is very much a multidisciplinary undertaking it requires a great deal of effort to facilitate and coordinate the related activities among larger teams of experts and stakeholders. The resulting IWRM plan would need to address a range of issues which may differ from country to country in light of the national strategies and longer-term objectives. At a minimum, an IWRM plan should include the following aspects:

- Details of the water management approach (including flood-related issues) which has been used to date;

- Discussion of the present and future water resources situation (including individual aspects of different water resources, their technical, hydrological, environmental, social, institutional, political, legal and other aspects and constraints, land-use plans, etc.);
- List of all stakeholders, their roles and mechanisms for participation and interaction;
- Description of the strategic goals and objectives which are intended to be achieved with the IWRM plan;
- Description of strategies and activities that are needed to be put in place in order to achieve strategic goals and objectives (e.g., action plans and programmes, priorities, etc.);
- Monitoring strategy;
- Communication strategy;

Once the IWRM plan is complete it needs to be accepted and approved by the body of stakeholders, preferably unanimously.

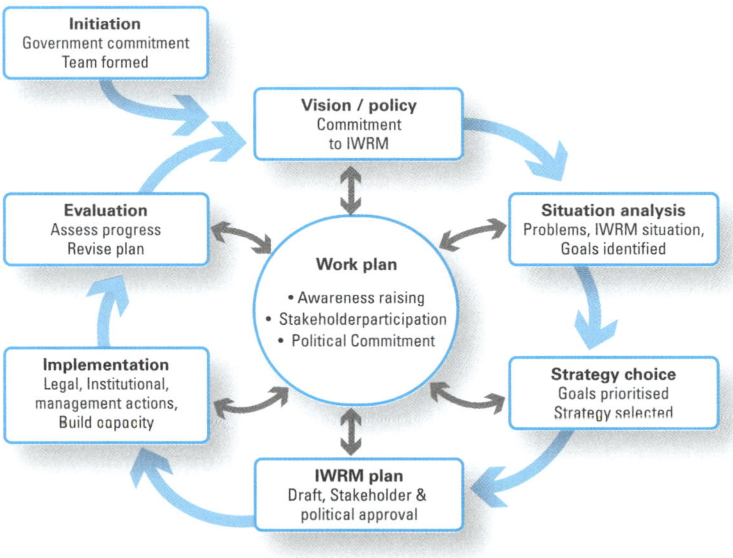

Figure 22.1 The cycle for developing and adjusting an IWRM plan (Source: CAPNET 2005)

22.6 RIVER BASIN MANAGEMENT PLANS

River basin management plans are plans for protecting and improving the quality of life in relation to the water environment within the river basin. They form an integral part of IWRM planning and as such they should be formulated in an atmosphere of close coordination. In dealing with flood related issues, the overall objective of such plans is to prevent hazards turning into disasters even while maintaining the benefits of floods. An overview of the basic steps involved in the production of river basin management plans, as defined by (WMO 2007), is given in Figure 22.2.

Similar to IWRM plans, river basin management plans also need to be produced in consultation with various organisations and individuals. Their initiation would also require a clear listing of objectives and communications with stakeholders.

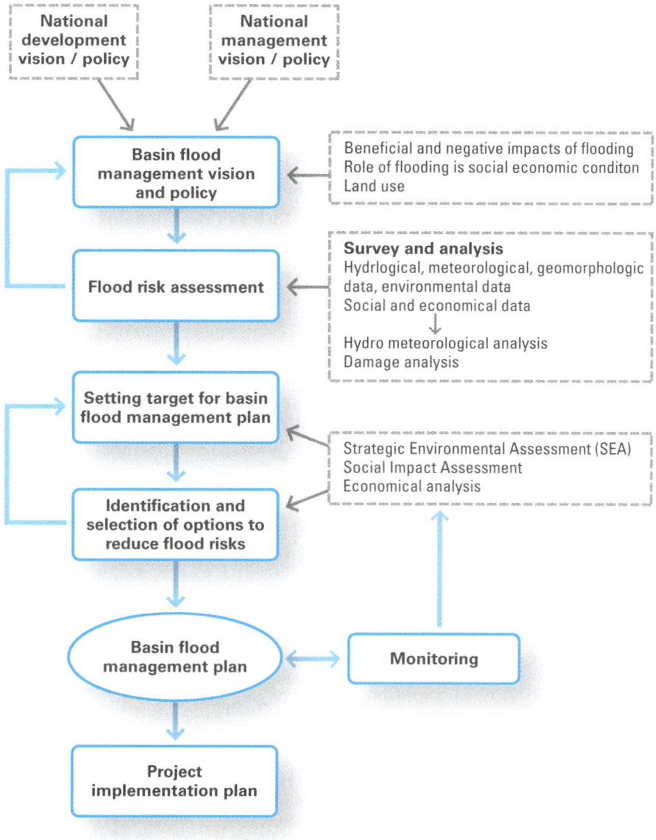

Figure 22.2 Production of river basin flood management plans (Source: WMO 2007)

At a minimum, a river basin management plan should include the following aspects:

- Details of the present river management approach;
- Discussion of the present and future situation concerning the river and its entire basin (including technical, hydrological, environmental, social, institutional, political, legal and other aspects and constraints, land-use plans, etc.);
- List of all stakeholders, their roles and mechanisms for participation and interaction;
- Description of the strategic goals and objectives which are intended to be achieved with the plan;
- Description of strategies and activities that are needed to be put in place in order to achieve strategic goals and objectives (e.g., implementation details, action plans and programmes, priorities, etc.); It also needs to have a clear description of how the proposed strategies will contribute towards the objectives of the IWRM plan;
- Monitoring strategy;
- Communication strategy;

Once the river basin management plan is complete it needs to be accepted and approved by the concerned active stakeholders.

22.7 CATCHMENT FLOOD MANAGEMENT PLANS

Catchment flood management plans aim to address all types of inland flooding which may result from rivers, groundwater, surface water, drainage infrastructure and tidal flooding, but not necessarily coastal flooding. Coastal flooding is usually covered in separate Coastal Management Plans. Catchment management plans would also aim to take into account the likely impacts of climate change, the land-use management in force at present and as envisioned for the future. These plans need to define those objectives that can be identified at the catchment scale and that are complementary to the wider objectives of the river basin management and integrated water resources management plans. An illustration of interactions and links between different flood management planning activities is given in Defra (2004): see Figure 22.3.

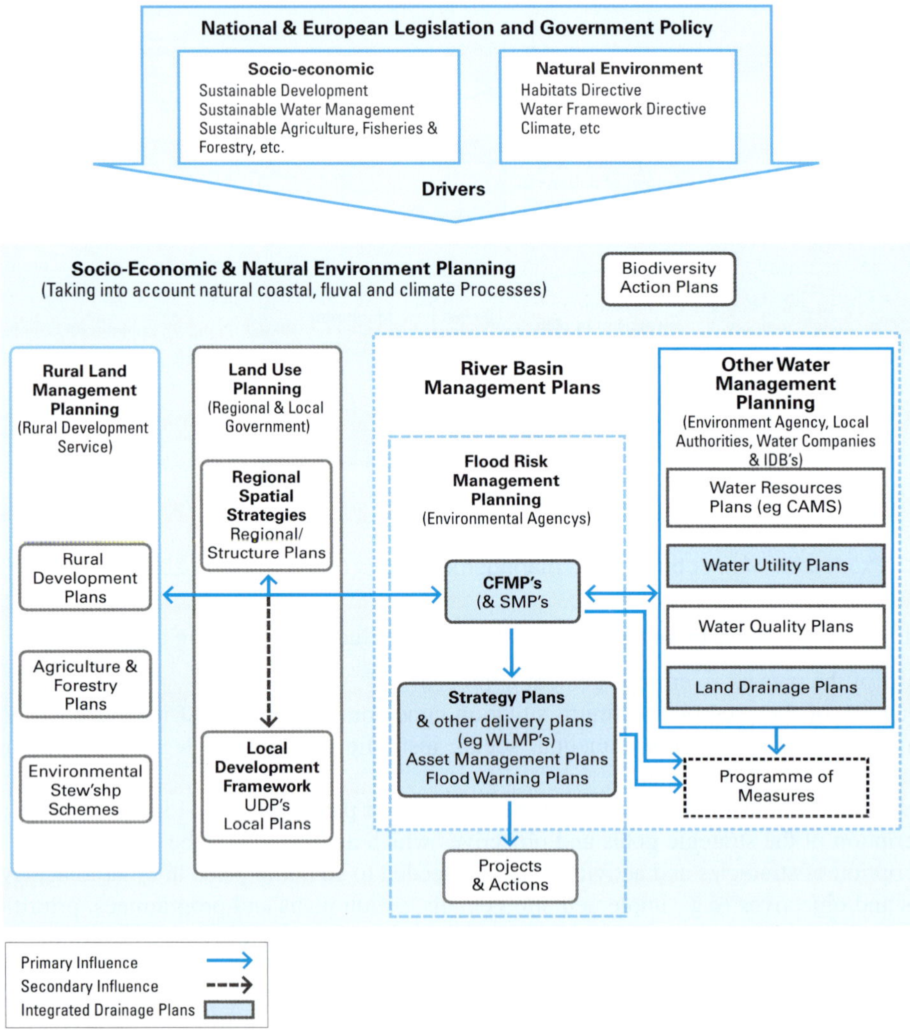

Figure 22.3 Illustration of links between catchment flood management plans and the wider planning framework (Source: Defra 2004)

Similar to the previously-discussed plans, a catchment flood management plan should include the following aspects:

- Details of the present catchment flood management approach;
- Discussion of the present and future situation concerning drainage systems and catchment characteristics (including technical, hydrological, environmental, social, institutional, political, legal and other aspects and constraints, land-use plans, etc.);
- List of all stakeholders, their roles and mechanisms for participation and interaction;
- Description of the strategic goals and objectives which are intended to be attained with the plan;
- Description of strategies and activities that need to be put in place in order to achieve strategic goals and objectives (e.g., implementation details, action plans and programmes, priorities, etc.); It also needs to provide a clear description of how the proposed strategies will contribute towards the wider objectives of the river basin management and integrated water resources management plans;
- Monitoring strategy;
- Communication strategy;

These plans also need to be approved and accepted by all stakeholders.

22.8 COASTAL MANAGEMENT PLANS

The aim of coastal management plans is to provide guidance on how to achieve sustainable and liveable environments along the coastal zone. They should normally take into consideration the risks from coastal flooding and erosion that can be experienced in specific areas and to identify those assets that are exposed to risks so as to come up with effective strategies that can minimise such risks. Low-lying coastal areas are particularly vulnerable to coastal flooding and will certainly experience a more significant level of flooding in case of rising sea levels. Erosion is another process that occurs in coastal areas. Such issues, together with some other issues again need to be addressed and discussed in a coastal management plan. Usually, apart from the more general group of stakeholders, asset managers, engineers, coastal geomorphologists and sea-level-rise experts are also involved in the development of coastal management plans.

Similar to the previously discussed plans, these plans should include the following aspects:

- Details of the present coastal management approach;
- Discussion of the present and future issues concerning coastal zone (including technical, geomorphological, environmental, social, institutional, political, legal and other aspects and constraints, land-use plans, etc.);
- List of all stakeholders, their roles and mechanisms for participation and interaction;
- Description of the strategic goals and objectives which are intended to be achieved with the plan;
- Description of strategies and activities that are needed to be put in place in order to achieve strategic goals and objectives (e.g., implementation details, action plans and programmes, priorities, etc.);
- A monitoring strategy;
- A communication strategy;

Coastal management plans also need to be formulated within the stakeholder community until this community is able to accept and approve them.

22.9 DISASTER MANAGEMENT PLANS

22.9.1 The aim of disaster management plans

The aim of disaster management plans is to provide a framework on how to prepare and react in case of disasters. The most critical elements in the suite of activities associated with disaster management are emergency preparedness and response activities. Since the response to a natural disaster warning must be immediate, comprehensive, and demonstrate very clear lines of command, the hierarchy in the chain of command needs to be established for all responsible personnel. Further to the specification of roles and responsibilities for those involved in disaster management activities, the main tasks responding to different disaster scenarios need to be identified. Figure 22.4 displays a summary of disaster management activities that can be grouped according to the following three major stages:

(1) Pre disaster (or preparation phase);
(2) During disaster (or supportive phase);
(3) Post disaster (or restoration phase);

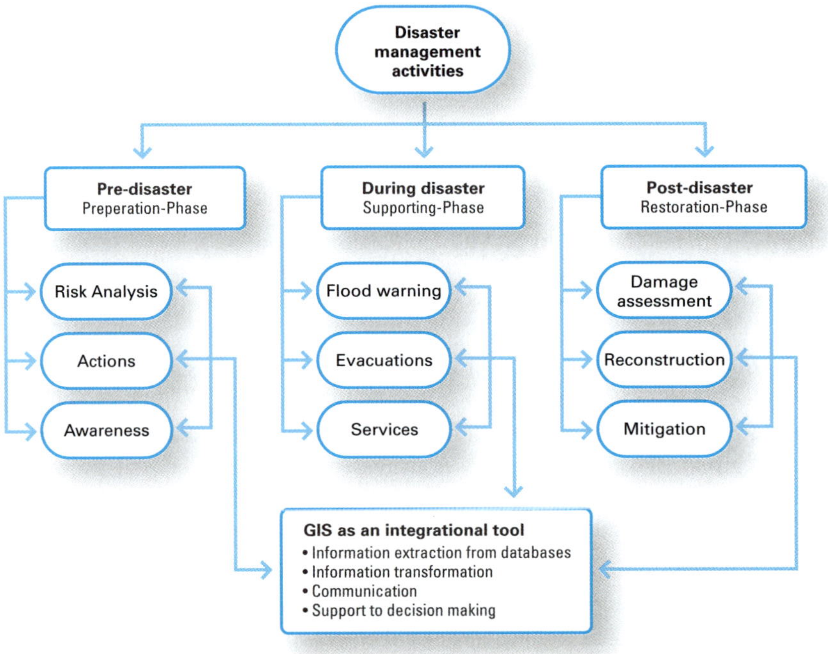

Figure 22.4 Disaster management activities

For all phases detailed response actions need to be formulated, outlined in a disaster management plan and agreed upon by all key stakeholders. The disaster management team needs to be well-resourced and trained in advance, and individual skills should be continually upgraded.

Flood-related disaster management plans are typically produced in conjunction with other disaster management plans. Sometimes a single disaster management plan which contains actions concerning all potential types of disasters is produced. In such a case, different types of disasters are identified and ranked according to their severity and scale of impact. This is done by analysing historical events and considering aspects such as: lives lost, economic damages, adverse impact on most people (e.g., an earthquake can cause 10,000 deaths in one place, and adversely impact the lives of 100,000 people over a wider area, but a volcano can disrupt the lives of 100,000,000 people – as was the case with Island's volcano Eyjafjallajokull in April 2010, causing long term damage which takes the longest time not only for rebuilding but for otherwise recovering), psychological impacts and others. Such analyses should then seek for joint solutions wherever they are possible.

22.9.2 Pre-disaster phase

One of the first activities of the pre-disaster phase is to evaluate the hazards and quantify the extent of potential flood impacts (e.g., estimation of damages, etc.). Once the hazard areas are identified for each flood scenario and the corresponding flood impacts assessed, the next step would be to evaluate the risks associated with the road infrastructure; see examples given in Figures 22.5 and 22.6.

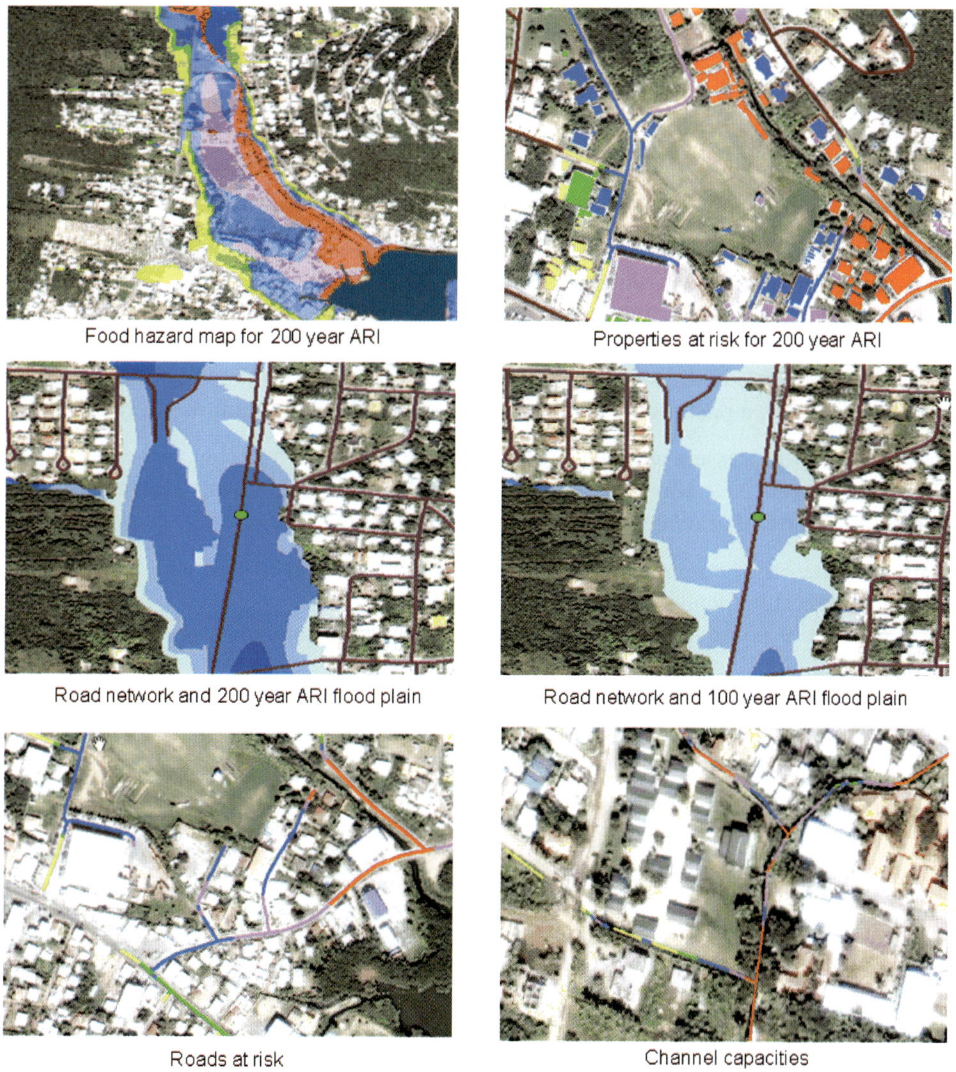

Figure 22.5 One example of an analysis of roads for disaster management purposes

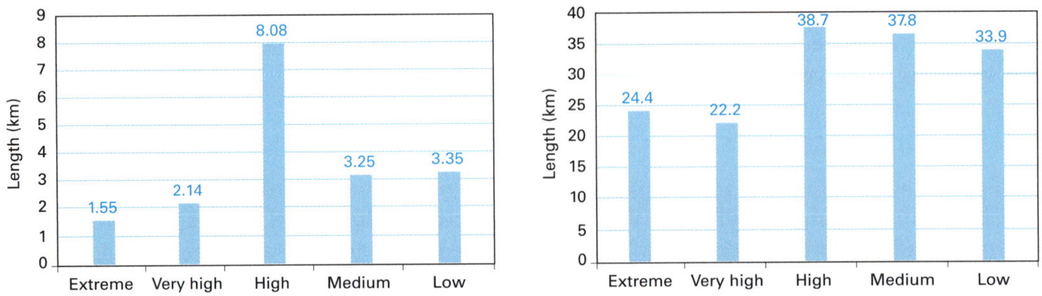

Figure 22.6 Another example of an analysis of roads for disaster management purposes

GIS technology continues to provide an invaluable tool in the work of integrating different sources of data, together with the results from the numerical models, and extracting and displaying various components of the information so provided for communication to the various groups which are required to respond to a flood emergency. Furthermore, as shown earlier, a GIS platform can be used to interpret the basic hydraulic and hydrologic inputs and other spatial data to provide flood intelligence, including:

- Flood affected properties in terms of the degree of damage associated with different rainfall events,
- Stream capacities in relation to different rainfall events,
- Evacuation and exit routes, including the nearest locations for evacuation and predicted flood depths and relative timing along evacuation routes to and from evacuation zones,
- Affected road/bridge crossings in terms of the predicted impact on road crossings and bridges,
- Flood damage, using a post–flood spatial assessment of the distribution of likely damages.

With the model results processed within the GIS package, the most relevant information needs to be synthesised and displayed in a way that is best suited to the disaster management planning process. Among of the other actions that relate to the pre-disaster phase are:

- establishing emergency communications procedures for different scenarios;
- developing and testing plans;
- training relevant response–personnel on operational implementation;
- considering and planning the finances and capital required;
- implementing the real-time flood warning system,
- establishing or refining procedures regarding the real-time flood warning system (once it is implemented and becomes functional), and
- rapid public consultation;

22.9.3 During the disaster phase

This is the phase of combating the impacts of the flood disaster and providing immediate assistance to affected persons and sections of the community. A number of scenarios need to be identified and all necessary measures and actions need to be worked out. Figure 22.8 depicts an example of one of the decision trees used for the development of a disaster management plan.

Activities such as notification and warning, evacuation of residents from affected areas, traffic control, stockpiling of sandbags, distribution of emergency food and water supplies, and the evacuation of high value goods from flood prone areas are typically undertaken, Figure 22.7.

Figure 22.7 Sandbags in the close vicinity of Chao Praya River, Bangkok, Thailand (Photograph: Zoran Vojinović)

Once the flood-related emergency is identified, the first task for a disaster coordinator (during working hours) or for an on-call duty officer (outside working hours) is to assess the situation and inform people on a need-to-know basis. The coordinator (or an officer) then verifies and associates the actual situation with one of the scenarios described in the disaster management plan, and advises other people on appropriate immediate actions by referring to the disaster actions of the selected scenario as set out in the plan (Figure 22.8). The plan differentiates actions on the basis of different flood scenarios (including those with a potential for spreading waterborne diseases due to the mixing of stormwater and wastewater and a further drinking water contamination due to the infiltration of stormwater/wastewater within the pipe network system). Each set of actions concerning the different scenarios needs to be derived according to the potential distribution of flood hazards based on model simulations for a range of storms.

When a flood-related disaster is declared, the disaster management team's office is the focal point for event-related information, for implementing disaster management priorities and for coordinating response and recovery after the disaster. It maintains regular liaison and communications with community leaders and with other authorities.

22.9.4 Post-disaster phase

A guide on how to assess the causes and effects of the flood need to be identified for the post-disaster (or restoration) phase and discussed in a disaster management plan. Once the event has passed, the emergency response does not end, but continues through cleanup, restoration and resettlement activities. This is described as the post-disaster phase. As part of the post-disaster phase, it is beneficial to conduct an assessment of the causes and effects of the flood and to come up with recommendations that would better prepare for the next event and so reduce future losses.

Typically, these activities lead towards recommendations that improve preparedness for the next event and reduce future flood losses. The main objective of the reconstruction phase is to facilitate

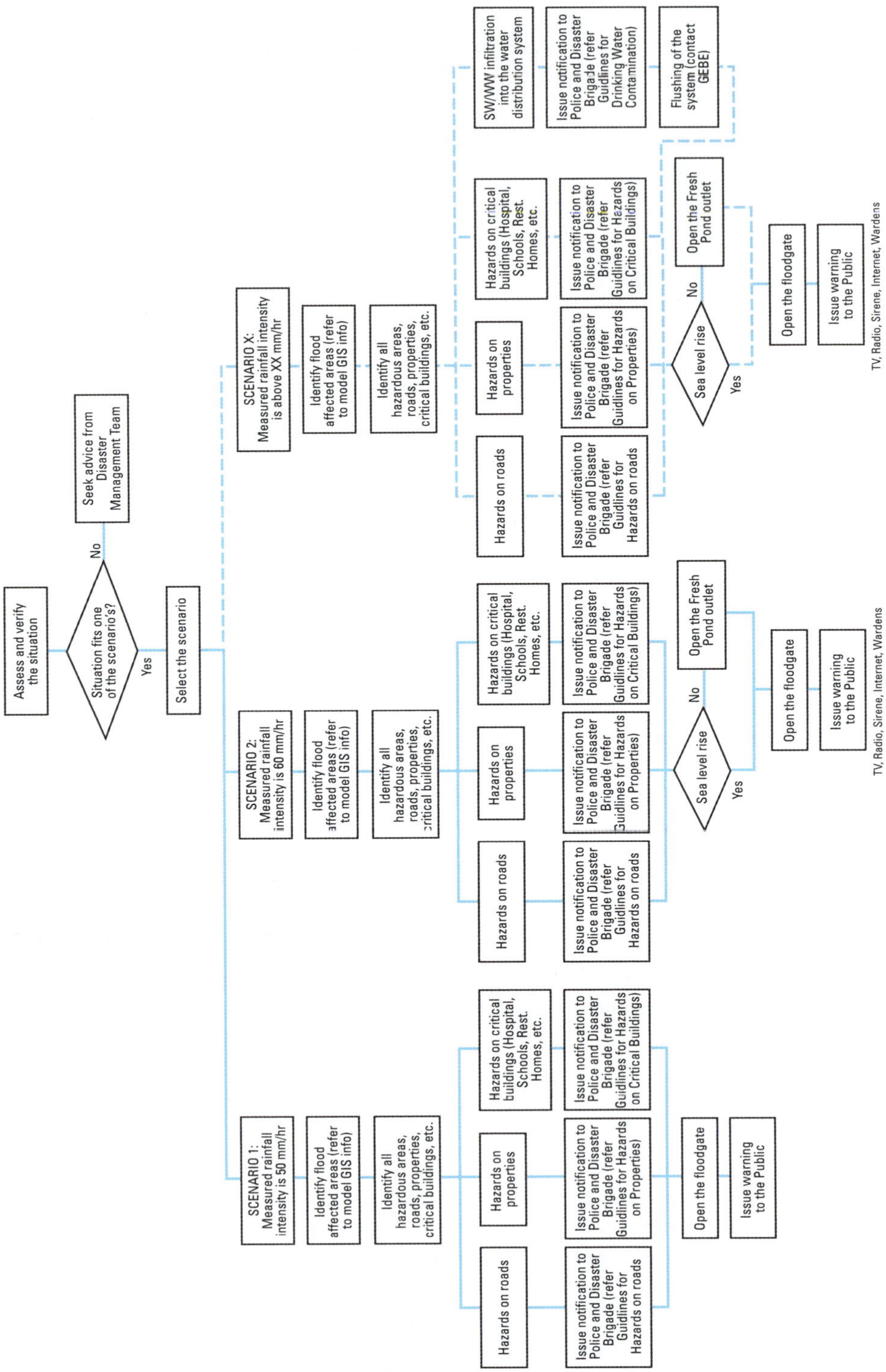

Figure 22.8 Example of a decision tree for the development of disaster management plan for St Maarten, N.A

flood-affected communities, properties and services to improve upon their pre-disaster condition. Rehabilitation includes repairing and reconstructing houses, commercial establishments, public buildings, and infrastructure, restoring and coordinating vital community services, and so on. Recovery can take a few weeks or several years, depending upon the magnitude of the disaster and the availability of reconstruction resources. The important recommendation for reconstruction and rehabilitation projects is that they should reduce future vulnerability and avoid recreating prior vulnerable conditions.

The flood visualisation component of the disaster management framework needs to be produced in such way so that the disaster management team can become familiar with the potential behaviour of flooding including rates of rise, evolution of flood extents, areas of high flood hazard and lead times prior to roads being cut. Further refinements to the built environment and associated demographic changes also need to be considered in the plan, because they are likely to be initiated after any significant flood event. What the flow visualisation does in this situation therefore is to act as an awareness-raising tool and as a simulator within *hydroinformatics environments* described earlier.

Similar to the previously discussed plans, these plans also need to be communicated and approved by all stakeholders.

22.10 ASSET MANAGEMENT PLANS

22.10.1 The need for asset management

Drainage systems together with various flood defence structures represent a vast investment which is made over many generations by both public and private sector organizations in order to support the life of urban communities. These assets are strictly non-stationary systems, but for maintenance purposes they are often regarded as being stationary in that they are maintained to provide certain levels of service (or standards) by the continuing replacement and refurbishment of their components. In this context, the term 'asset' refers to any physical object which enables the provision of related services (e.g., pipe, manhole, valve, pump, weir, dyke, floodwall, reservoir, detention storage, channel, etc.). One of the main concerns of drainage management utilities and various flood defence authorities is to secure and manage their asset investments in such a way that they meet appropriate standards and levels of service with an acceptable number and severity of failures while ensuring maximum return on their investments regardless of whether they are publicly or privately owned. All of these issues highlight the need for a structured and full life-cycle management approach to the management of these assets, starting from planning and design, through construction and operation to dismantling and disposal. A formal approach to the asset management procedure necessitates a number of activities which can be grouped into the following three groups: creation of assets (planning, prioritising and acquiring), operation of assets (including their maintenance and rehabilitation) and asset rationalisation (including their potential reuse, decommissioning and disposal). All these activities are typically described and discussed in an organisational asset management plan.

The overall concept of asset management has been described by a number of authors (see e.g., ALGENZ 1998; Marlow *et al.* 2007; Price and Vojinović 2011). This section aims to provide an overview of the basic principles behind current asset management practices and the related details that are necessary for the production of asset management plans.

22.10.2 Asset management cycle

As described in Price and Vojinović (2011) asset management may be defined as a comprehensive and structured approach to the long-term management of assets to serve the needs of urban communities at the lowest possible cost. Asset management has been also viewed as a *business discipline* as its success is measured using business and financial indicators. As illustrated in Figure 22.9, the span of application of asset management extends from the definition of desired standards, through the risk assessment to the asset valuation, monitoring and review.

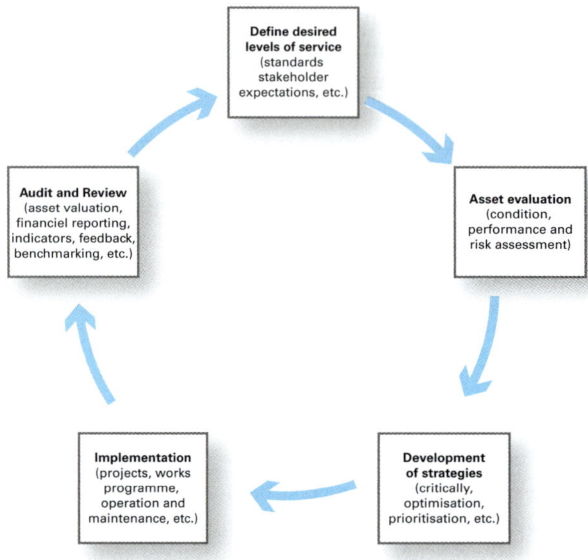

Figure 22.9 Asset management process (After Price and Vojinović 2011)

The term *levels of service* refers to the selection of a certain quality of a particular service (e.g., frequency of and extent of flooding, etc.) against which the performance of an asset can be measured. It is directly related to reliability, quality, quantity, and safety (or risk). Levels of service for drainage management and flood defence assets are normally defined according to the standards and policies/regulations as well as stakeholder expectations and requirements. Typically, the key performance indicators or quality assurance issues in relation to level of service are various national industry-related performance, structural and environmental indicators and standards. In that respect, the elements of asset management are focused around the structural integrity of assets, their hydraulic performance and environmental impacts.

The term *asset evaluation* usually refers to the condition, performance and risk aspects. Condition relates to the physical state of assets, whereas the performance is closely aligned to the level of service which can be measured in terms of capacity, reliability, availability, etc. Within this context, *risk* relates to the process of risk identification and the assessment and development of strategies that can minimise that risk.

The development of strategies refers to the process whereby the information gathered from the asset evaluation phase is used to optimise and prioritise asset management strategies that relate to the operation and maintenance of existing assets, their rehabilitation (renewal and replacement), the acquisition of new assets, and the dismantling and disposal of surplus assets. In addition to the condition and performance data, it is common to include criticality as yet another aspect in the prioritisation and decision-making process. Criticality in this instance refers to the importance of a particular asset with respect to financial, business and service level terms. Typically, assets for which the financial, business or service level consequences of failure are significant are often regarded as being more critical (e.g., a sewer main that runs under the road of a central business district area), whereas those assets for which such consequences are insignificant are regarded as less critical (e.g., a sewer pipe that conveys the wastewater to a toilet in a peripheral park which is only used on rare occasions).

The implementation phase refers to the realisation of the necessary strategies which are needed to satisfy or improve a given level of service. This is typically done by undertaking various operational and maintenance activities (i.e., the day-to-day running and upkeep of assets), projects, works programme, etc. Also, periodic asset management audit and review are usually undertaken in order to evaluate the utility's performance in achieving its asset management objectives, so as to establish a continuous asset management improvement cycle and to maintain best practice. Usually, this is done by assessing various aspects against a set of indicators, with corresponding benchmarking and customer feedback. Benchmarking refers to the analysis and comparison of asset management processes by gathering and exchanging the data with other utilities nationally and/or worldwide. Typically, the asset evaluation process ends up with indicating the service quality gaps and observing how well a particular utility performs in meeting the desired levels of service. In this respect, poor service relates to an inability to provide a particular service to the customers (e.g., an inability to evacuate wastewater from housing areas and industry, flooding due to excess rainfall-runoff which is unable to enter the sewerage network, etc.).

A critical part of the asset management process is to create an environment for a continuous and meaningful engagement of stakeholders. The general purpose of stakeholder participation here, as elsewhere, is to induce a change in the management of assets in a way that aligns with a positive change in the social environment (Abbott 2007). Therefore, establishing an effective information management and communication framework is a very challenging task as it needs to provide a balance between technical aspects and those values that are important to stakeholders. In the case of such framework it is important to translate the data and technical information into those forms of knowledge and understanding that can be assimilated by the non-technical population so that this population may intervene meaningfully and responsibly in the decision-making process. The aim of such an asset management framework (as illustrated in Figure 22.10) is to integrate asset performance, condition and criticality with legislative requirements and other constraints defined by regulations, standards and stakeholder interests. The outputs from this framework are the policies, procedures and actions that need to be adopted in order to achieve cost-effective creation, acquisition, maintenance, rehabilitation and disposal of assets that will meet the required level of service and predicted future demands.

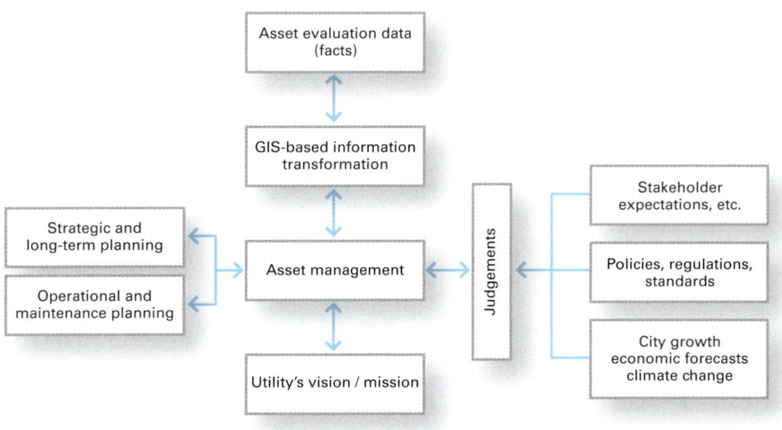

Figure 22.10 A sociotechnical asset management framework (After Price and Vojinović 2011)

22.10.3 Asset management practice

In the past, many assets were utilised until they failed, after which they were simply repaired or replaced as quickly as possible. This is referred to as reactive asset management. However, following developments in asset management theory, data collection and processing capabilities, the focus has been increasingly placed on a more proactive and structured condition- and performance-based asset management, which requires anticipating failures and planning ahead of them – and even more recently there has been a further shift towards service level and risk-based asset management practices.

With respect to the condition- and performance-based asset management practice, decision making is based on condition and performance assessments for either all assets or for a selected group of assets (e.g., critical assets). In this respect, various approaches on how to identify and rank all assets with respect to their condition and performance characteristics have been developed. Some of the important asset condition and performance aspects are: age, condition, performance (with respect to technical, environmental, public health, community standards), maintenance history, etc.

Asset condition is usually determined by applying various visual inspection techniques which range from direct visual inspections, acoustic and electrical/electromagnetic devices to portable Closed Circuit Television (CCTV) techniques. Some of the more recent technologies are laser profiling, gamma-gamma logging, ground penetrating radar, infrared thermography, micro-defloction and Impact Echo/Spectral Analysis of Surface Waves (SASW): see for example Feeney *et al.* (2009). Asset condition assessment aims to detect the presence of defects (e.g., cracks, corrosion, displacement of joints, root intrusion, etc.) and their severity, as collected over a period of time, can assist in developing typical asset deterioration curves. Furthermore, it is also necessary to analyse such defects and to establish the level of their severity. Various decay (or deterioration) curves for different groups of assets have been developed with the purpose of determining their remaining life and probability of failure, even as the deterioration process is rather uncertain and influenced by random events; see Figure 22.11.

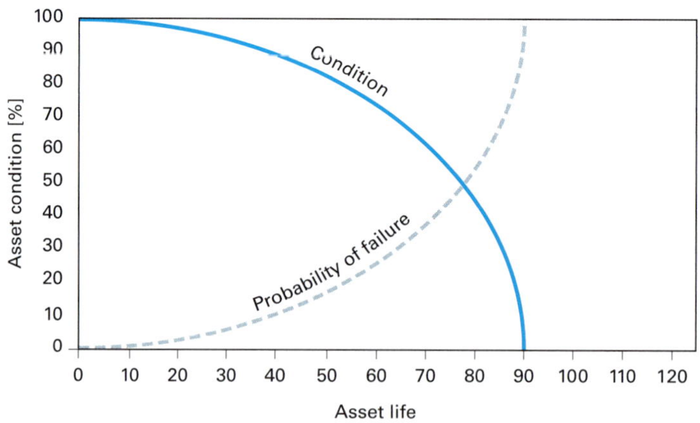

Figure 22.11 The relationship between asset condition and probability of failure (Marlow *et al.* 2007)

It should be noted that the rate of deterioration of assets is very much asset- and site-specific. The same type of asset within a system may demonstrate different lives due to different service conditions, such as design standards, quality of construction, material, operational stresses, maintenance history/practice, asset working environment, external stresses, etc.

Some examples of different assets and their typical manufacturer's claimed life durations are given in Table 22.1. It should be notes that the effective (or economic) life of an asset is not necessarily the same as the manufacturer's claimed life. The effective (or economic) life of an asset is the period from its acquisition to the time when it has become unserviceable or obsolete or when the present value of operating and maintenance costs is greater than the cost of replacement. Typically, the effective lives of assets can be estimated from physical inspection, examination of asset service records, discussion with technical staff and benchmarking against known retirement practices.

Table. 22.1 Example of typical manufacturer's claimed life durations for different assets.

Asset	Average manufacturer's claimed life (years)
Concrete pipe	100
PVC pipe	100
Vitrified Clay pipe	100
HDPE pipe	70+
CMP pipe	30 to 80 depends on soil resistance
Ductile Iron pipe	80
Cast Iron pipe	80
Kerb and channel	65
Catchpit (or road gully)	65
Concrete culvert	65
Pump	40
Pump chamber	60
Pump mechanical and electrical	30

As documented from various sources (see e.g., WRc (1994)), the process of pipe collapse can be divided into three stages: initial defect (cracks, leaking joints, etc.), deterioration (erosion of the pipe material under different circumstances) and actual collapse (which can be related to the magnitude of deterioration and/or to some random event). Some of these examples of defects are given in Figures 22.12 and 22.13.

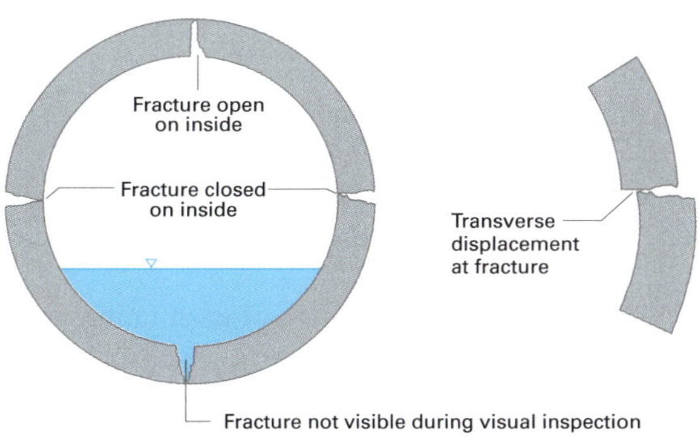

Figure 22.12 Examples of longitudinal fractures

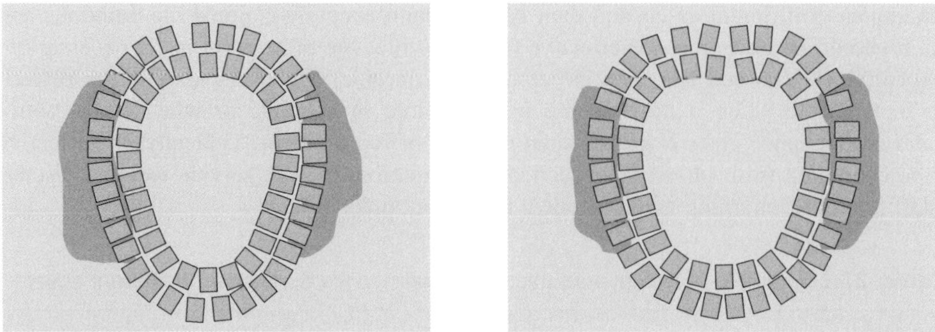

Figure 22.13 Examples of pipe defects: infiltration/exfiltration and the formation of fractures at the crown (Adopted from Sewerage Rehabilitation Manual (WRc 1994))

Since the condition surveys can be very expensive and time consuming undertakings they are often carried out on one portion of the assets only. In such cases, the process of selecting assets for such a survey is done by applying one of the sampling techniques (e.g., stratified sampling, random sampling, systematic sampling, etc.). This is typically done for those assets that are assessed to be at a higher risk level or those assets that are defined as being critical. The most common technique is the stratified sampling technique in which all assets are initially divided into groups to estimate their proportions and then sampling programmes are designed according to their proportional value and variability.

Performance assessment is based on various indicators and standards (customer-related, environmental, legislative, design, etc.). Certainly, the choice for indicators should represent what all stakeholders regard as important and necessary to meet the required levels of service.

In more recent times, risk-based asset management has gained a greater importance than the pure condition- and performance-based asset management practice. Such practice seeks to accomplish its aims through the consideration of the quality of service delivery and associated risks. In standard asset management applications, risk has been defined as a probability of a given number of occurrences in excess of a given magnitude for more than a given duration. It is then further determined by taking into account both the probability and the consequence of asset failure. For example, a risk associated with the provision of surface water (or stormwater) drainage can be formulated as: *probability of flooding occurring more than once in five years in excess of a depth of 30 cm for a duration of more than 30 minutes (and thus relating to a particular damage cost).*

As introduced in Price and Vojinović (2011) and now taken further, typical risk assessment methodology should contain the following steps:

- Agree expected services and consequences of failure;
- Developing scenarios;
- Condition and performance assessments (i.e., asset vulnerability and exposure assessments);
- Identifying probabilities, frequencies and durations of events and scenarios leading to a failure of the service;
- Determining the probability, frequency and duration of each failure for each event and scenario;
- Determining the corresponding consequences;
- Qualitative and quantitative risk assessment;
- Determining an optimal system management strategy to minimise damage (fail-safe) through communication with stakeholders;

- Determining a business risk for an optimal system management strategy;
- Identifying the list of actions and programmes in support of the new strategy;

An example of a quantitative risk assessment is given in Figure 22.14. Findings from the risk assessment work concerning each asset (or asset group) can be plotted on the matrix and the relative position of the asset will determine the priorities for a more detailed risk assessment and other actions.

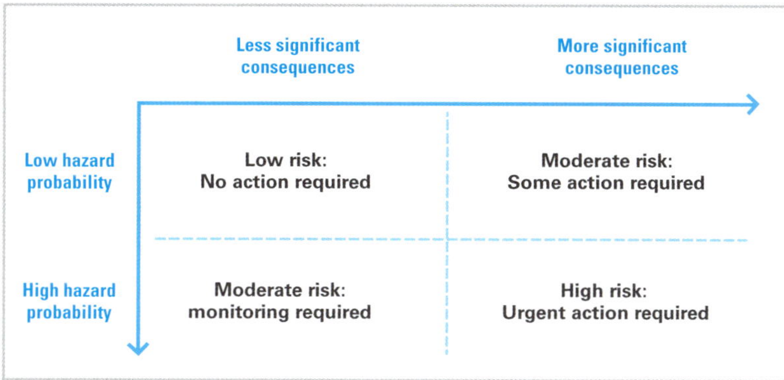

Figure 22.14 Example of a quantitative risk assessment

22.10.4 Asset rehabilitation and optimised decision making

An evaluation of different asset rehabilitation measures can be undertaken either manually or using automatised optimisation approaches. Due to the limitations of the manual trail-and-error processes and following the increase in computer power, the research into automatic calibration procedures has led to the development of different automatic optimisation approaches that can be used for the evaluation of different rehabilitation alternatives. These approaches are based on a general task of optimising (minimising or maximising, as appropriate) the value of a suitable objective function, which is a numerical measure expressing the effectiveness of different rehabilitation options against the *do nothing* scenario. Some examples of such approaches which combine numerical models and optimisation algorithms (i.e., genetic algorithms) are described in Barreto *et al.* (2010), Price and Vojinović (2011), Vojinović *et al.* (2006) and Vojinović and Sanchez (2008).

22.10.5 Asset management decision support systems

Over the past years, a number of asset management decision-support tools, which vary in their scope and focus, have been developed. An overview of some of these tools and the stages they cover within the infrastructure asset management is given in Figure 22.15; see also Ana and Bauwens (2007). In Figure 22.15, the tools are grouped according to their functionalities and capabilities. The left-hand side of the figure shows the main aspects of a particular asset management system.

Group one consists only of performance analysis tools (tools that deal mainly with performance modelling), group two consists of performance and decision analysis tools (they cover the first three asset management stages – from data collection, to performance modelling and decision analysis) and group three consists of the complete set of asset management tools (they cover all aspects of the asset management stages). The group one tools are: the Baik model (Baik, Jeong *et al.* 2006), Bengassem and Bennis model (Bengassem and Bennis 2000) and Hasegawa *et al.* (1999) model. The group two tools are: APOGEE system (Rommel *et al.* 1989), AQUA-WertMin (Baur and Herz 2001) system, Edmonton models (Ruwanpura *et al.* 2004), KureCAD tool (Stone *et al.* 2002) and PRISM model (Ariaratnam and MacLeod 2002). The group three tools are: CARE-S system (Saegrov 2006) and Hydroplan model (Vojinović *et al.* 2005).

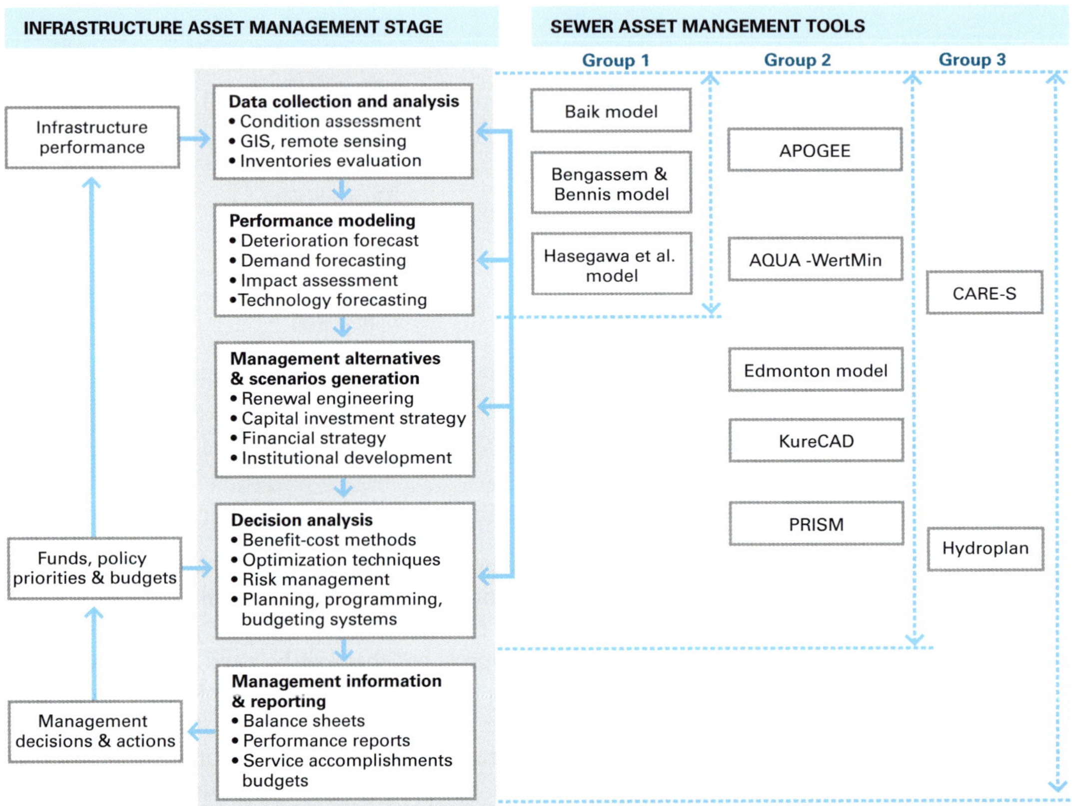

Figure 22.15 The generic infrastructure asset management system with the corresponding sewer asset management tools applicable at different stages (After: Ana and Bauwens 2007)

22.10.6 Outline of an asset management plan

In general terms, an asset management plan should include the following aspects:

- Description of the asset management approach;
- Discussion of the present and future issues concerning different levels of service;
- List of all stakeholders, their roles and mechanisms for participation and interaction;
- Description of an asset management system;
- A whole lifecycle asset overview with respect to condition, performance, criticality, risk, etc.;
- Financial summary (cash flow projections, tariff structures, asset value, sources of funding, etc.);

- Description of the strategic goals and objectives which are intended to be achieved with the plan;
- Description of strategies and activities that are needed to be put in place in order to achieve strategic goals and objectives (e.g., implementation details, action plans and programmes, priorities, operation and maintenance issues, etc.);
- Monitoring strategy;
- Communication strategy;

Asset management plans also need acceptance and approval by the relevant stakeholders. Furthermore, they are also subject to formal reviews and monitorings.

Chapter 23

Case studies

23.1 CASE STUDY 1: FLOOD VULNERABILITY ASSESSMENT, BELO HORIZONTE (BRAZIL)

23.1.1 Introduction

Risk is the result of exposure of people to a natural event (hazard), leading to their vulnerability (Green, Pareker *et al.* 2000), (Bollin, Cardenas *et al.* 2003). During a flood event, risk is identified by the probability of its impacts times their consequent damages. Vulnerability is the relationship between the system being considered and its environment, which varies over time (Green 2004), and is measured by the impacts of the natural hazard on the environment in affecting the goals of the system.

In this case study the system is the urban household. It is assumed the main objective is maximizing the wellbeing of the household. Adopting a simplified economic perspective, welfare can be measured by the wealth provided through access to income and goods (Wisner 2004); the more the household welfare is affected by floods, the greater is its vulnerability. This involves a combination of factors that determine the levels of risk to life, livelihood conditions, property and other assets due to an event or sequence of events (Green 2004).

In order to capture the various dimensions of flood risk, we adopt a set of vulnerability indicators referring to flood hazard, exposure and resilience. These indicators are applied and tested in a prototype built as a simplified model of significant elements of the intra-urban environment and simulated as agent networks.

23.1.2 Hypotheses and research methodology

Indicators are conceived as in Eq. (13.1) and the Figure 23.1. They are derived from hypotheses about risk factors leading to household damages, using literature and empirical experiences (e.g., Penning-Rowsell and Chatterton (1977)). Hydrologic, hydraulic and socioeconomic variables are standardized on a 0.0 to 1.0 scale, allowing comparisons and associations between them. Higher values for the indicators indicate a greater impact of flooding, with the exception of the resilience indicator, for which higher values indicate greater resilience.

$$Risk = hazard \times exposure \times vulnerability \tag{23.1}$$

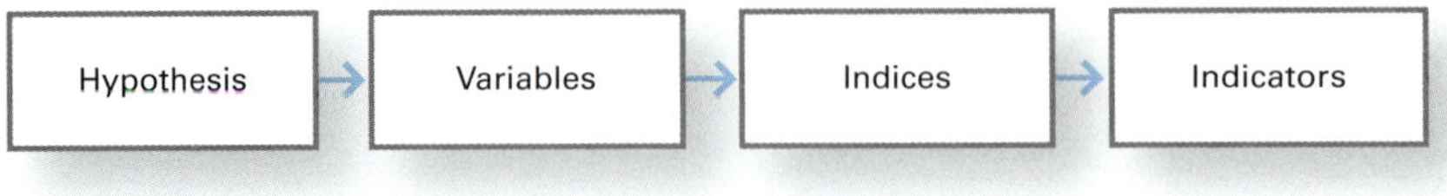

Figure 23.1 Analytical scheme for indicator development

23.1.3 Application example

The urban area studied has about 179,000 households and 24,000 firms spread over an area of 332 km^2, in which there is a watercourse and a flood prone area. Flood maps (area, depth, flow velocity) were obtained for 5, 25 and 100-year return periods (Figure 23.2) using hydrologic and hydraulic modelling based on HEC-RAS (USACE) and ArcGIS (ESRI).

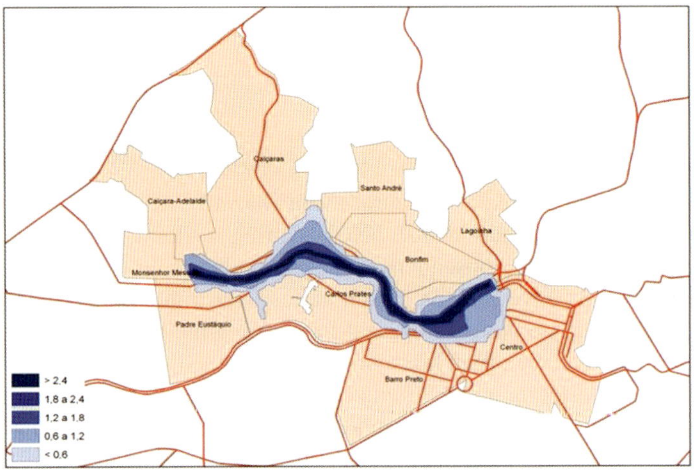

Figure 23.2 Prototype: distribution of flood depths for T = 100

Households were individually classified according to seven socio-economic classes (A1, A2, B1, B2, C, D and E), A1 being the richest, establishing for each of them their assets and consumption habits. Firms were defined by the number of employees, type of goods offered, wages paid to employees and maximum revenues. The relationships between households and firms were established through consumption and labour markets according to census data from Belo Horizonte, Brazil, and assumptions about consumer behaviour (e.g., allocations of expenditures, consumption and work places). *NetLogo 4.0.4*, a platform for agent-based modelling, was employed to simulate this complex economic network.

Flood hazard index

Hazard indexes (HI) have five components: material (buildings and contents), human, access to units of consumption and income generation, and flooding occurrence. HI_{M1} and HI_{M2} refer to flow characteristics (depth and velocity) capable of producing physical damage to a building and its content (RESCDAM 2000). The human index (hazard to residents), HI_H, highlights the relationship between the flow characteristics and direct harm to people (i.e., injury, death); it is based on the flow depth-velocity factor and the individuals' height and mass (RESCDAM 2000); (Zonensein 2007). HI_A aims at capturing the consequences of the flood hazard to a household's access to consumption and income generation; three household inundation depth intervals guide residents on the decision to leave home for shopping, to go to work, or to abandon the house ($d < 0.15$, $0.15 \leq d \leq 0.45$ d, $d > 0.45$ m). HI_F refers only to the occurrence of the flood in the local area.

Exposure index

Buildings, contents and inhabitants are the components of the household exposure index. The first two components are the material exposure indices (EI_{M1} and EI_{M2}) and the third (EI_H), the human exposure index. The reference variable for EI_{M1} and EI_{M2} is the residential area; the bigger the area, the greater the exposure of buildings and contents. EI_H is defined by the number of household residents, and indicates individual exposure to the physical damage due to floods (i.e., injuries, drowning, etc.). A location index, EI_L, refers to the household exposure to floods (i.e., location in the flooding area, household floor level).

Vulnerability and susceptibility indices

There are six indices for assessing population vulnerability and material susceptibility to flooding. Dwelling construction quality and inventory (content) quality, VI_{M1} and VI_{M2} respectively, are the main factors influencing flooding susceptibility. It is assumed that construction and content quality are associated to socioeconomic classes. VI_{H1}, the first index of human vulnerability, indicates the possibility of death, disease and helplessness due to flooding. Eleven census variables are used in its definition (e.g., age, sight, mental health, etc.). These attributes are weighted by an analytical hierarchical process according to four criteria: mobility, disease propensity, perception of risk, and level of dependence on other household residents (Jonkman and Kelman 2005). The second index, VI_{H2}, is based on the relationship between a family typology, defined by eleven models (e.g., elderly couples, young couples without children, one-parent families, couples with young children, etc.) and the vulnerability to poverty (Montali 2000); (Golgher 2004). VI_{H2} is defined through weighting family types by four analytical criteria: dependence, mutual aid, responsibility for the household vulnerability, and unequal access to income opportunities (Montali 2000); (Golgher 2004); (Green 2004); (Emergency Management Australia, 2004), (Wisner 2004). VI_{A1} and VI_{A2} reflect restrictions to access to consumption and work places due to flooding respectively. VI_{A1} is estimated by a weighting process taking into account the characteristics of goods (essentiality, substitutability and offer spatial distribution). VI_{A2} is calculated as the ratio between the rate of production of household income in the flood prone area and that produced outside this area.

Resilience indices

Resilience refers to the ability of a system to adjust to hazards and to mitigate or avoid damage (Pelling 2003). The shorter the time required for the system to return to the conditions prior to the impact, the greater its resilience. In this study, two factors of resilience were considered: access to information (S1I) and to income (S2I). S1I assumes that information provides an improvement in risk perception and also in strategies for dealing with floods. The education level of those responsible for the household is measured by S1I. S2I measures the household financial capacity for mitigation and reconstruction based on socioeconomic census classes. Flood damage has less impact on the family welfare according to its preparedness (mitigation) and speed in repairing damage.

Risk indicators

The fifteen indexes are grouped into eight indicators (Table 23.1) using Eq. (23.1), adopting equal weighting factors for hazard and consequences (exposure and vulnerability).

Table 23.1 Risk and resilience indicators composition.

Indicators	Components
Damage to construction (RI_{M1})	HI_{M1}, EI_{M1}, VI_{M1}
Damage to contents (RI_{M2})	HI_{M2}, EI_{M2}, VI_{M2}
Direct damage to residents (RI_{H1})	HI_{H}, EI_{H}, VI_{H1}
Damage to family (RI_{H2})	HI_{A}, EI_{L}, VI_{H2}
Restriction on access to consumption unities (RI_{A1})	HI_{A}, EI_{L}, VI_{A1}
Restriction on access to income generation (RI_{A2})	HI_{A}, EI_{L}, VI_{A2}
Access to information (RI_{S1})	HI_{F}, S1I
Financial ability (RI_{S2})	HI_{F}, S2I

Risk spatial representation

A spatial representation of the risk to flooding provides useful information for flood fighting, land-use regulation, adoption of flood proofing and other urban policies related to flood risk. Indicators were therefore grouped spatially by cluster analysis. Several cluster centres were initially created in the flood prone area (using the parallel reference method), with Euclidian distance as a criterion of similarity. Four risk maps were developed according to material and human, access to consumption together with income and resilience indicators.

23.1.4 Results

Table 23.2 contains the distribution of 3,780 flooded households according to the eight risk indicators for the 100-year return period scenario, with indicator ranges established according to the hierarchical approach proposed by Saaty (1977).

Table 23.2 Distribution of flood households according to risk and resilience indicators, 100-year return period event (%).

Indicators range	RI_{M1}	RI_{M2}	RI_{H1}	RI_{H2}	RI_{A1}	RI_{A2}	RI_{S1}	RI_{S2}
Indicator < 0.33	50.3	63.9	32.3	8.1	11.6	86.7	6.2	76.7
$0.33 \leq$ Indicator < 0.56	42.0	14.6	28.5	37.3	45.7	4.6	59.8	5.9
$0.56 \leq$ Indicator < 0.78	7.6	21.4	21.8	39.8	32.0	2.1	23.3	15.5
Index ≥ 0.78	0.0	0.0	17.5	14.7	10.6	6.6	10.7	1.9

A relatively low indicator of material risk (<0.33) prevails in this case study. Human risk tends to be higher, especially RI_{H2} (family typology), illustrating that there is not always a strong relationship between human and material vulnerability. This difference may also be due to methodological reasons, since there is more subjectivity in the definition of human than material vulnerability. RI_{A2} suggests that access to income is not highly affected, in spite of high RI_{A2} figures affecting 9% of the households; this is because either low flood depths did not disturb access to work, or most workplaces were not flooded. Consumption access was impacted more severely, possibly due to the spatial distribution of consumption points, and characteristics of goods and flood disturbances on consumers' behaviour. Indicators of family type associated risks deserve more attention. Also, resilience indicators suggest little recovery capacity in the area.

Through cluster analysis, exposed households are grouped according to similarity of risk, as illustrated in Figure 23.3 for the material risk indicator.

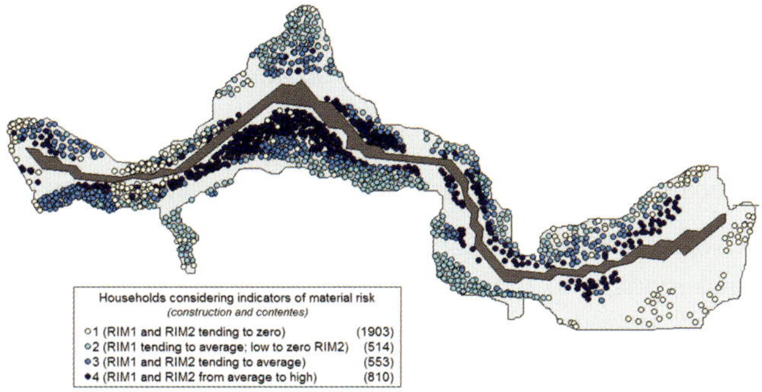

Figure 23.3 Material risk indicator

Each point of Figure 23.3 represents a household, with households in black being the most affected. These results show the relevance of analyzing the various dimensions of risk according to their particularities and of mapping them for flood protection policy formulation and implementation.

23.1.5 Acknowledgment

This case study was prepared by Vanessa Cançado, Nilo Nascimento, and Ricardo Ruiz (Federal University of Minas Gerais, Belo Horizonte, Brazil) and Sylvie Barraud (INSA-Lyon, France). It is based on research supported by the European Project SWITCH (switchurbanwater.eu) and the Brazilian agencies CAPES, CNPq and FAPEMIG.

23.2 CASE STUDY 2: FLOOD CONTROL CENTRE, BANGKOK (THAILAND)

23.2.1 Introduction

Bangkok, the capital of Thailand, is the "Venice of the East" with more than 1,000 canals and a population of about 10 million. It is situated in the floodplain of the Chao Phraya River. The monsoon climate brings heavy rainfall every year with 1,200–1,400 mm annual rainfall. Floods have occurred more frequently in the recent decades with severe flooding in 1978, 1980, 1983, 1995 and 2006. In particular, the Chao Phraya River, which drains one-third of the country, overflows regularly each year due to high discharges. At the same time, overland flow from the east of Bangkok brings in more flood water, intensifying the degree of flooding in the city. In addition, high tides in the Gulf of Thailand during November can exacerbate flooding in low lying areas of Bangkok. A number of master plans for flood protection and drainage improvement have been studied since 1960. In 1996, the Department of Drainage and Sewerage started to implement the latest master plan, with structural measures being implemented for flood protection. However, the high investment required to improve the drainage network caused delays in completing a planned drainage system covering 2,250 sq. km. The system is designed for rainfall with only a 5 years return period, so floods will still occur frequently. Therefore, non-structural measures were reconsidered. With new information and communication technologies (ICT) readily available, information about floods and flooding can be disseminated to the population in simple language so that they can take action to minimise the consequences for them of any flooding.

23.2.2 Flood Control Centre

The Flood Control Centre (FCC) in Bangkok was established in 1977 under the Department of Drainage and Sewerage (DDS). The main tasks of the centre are to co-ordinate and ensure collaboration between DDS departments involved in operations in the field and to monitor rainfall and water levels in the drainage system during urban flooding. The hydrological and hydraulic data are important for real time decision making and operational control in the field, especially in the city where many of the canals are linked to the primary drainage networks. New communication systems have been gradually introduced, and a number of systems have been installed and upgraded in the centre. The present system consists of:

(a) SCADA (Supervisory Control and Data Acquisition): This is a system that collects data from remote stations and sends it to the central computer where it can be managed and controlled. There are 75 network stations linked to the main station at the FCC. Each station collects real time data and sends it once every fifteen minutes using a radio frequency with a remote terminal unit (RTU). The present telecommunication connection is gradually being changed to fiber-optic in order to increase the stability of the system and to avoid data jamming caused by transferring too much data during the same time window. The following parameters are measured: rainfall (tipping-bucket type), water levels both inside and outside the regulator (floating type), the number of pumps operating at each station, and the openings of gates. There are back up central processor units and hard disks at each station, storing data for 7 days so that non-transferred data can be retrieved later. All data is collected and presented graphically at the FCC. Figure 23.4 shows an example of a graphic presentation at E01 station. Figure 23.5 shows a plot of rainfall over a 24 hour period at all stations from E01 to E39 covering East Bangkok and W01 to W20. These data are useful, not only for monitoring the operations of pumping stations and regulators, but also in identifying the location of heavy rainfall where flooding may occur.

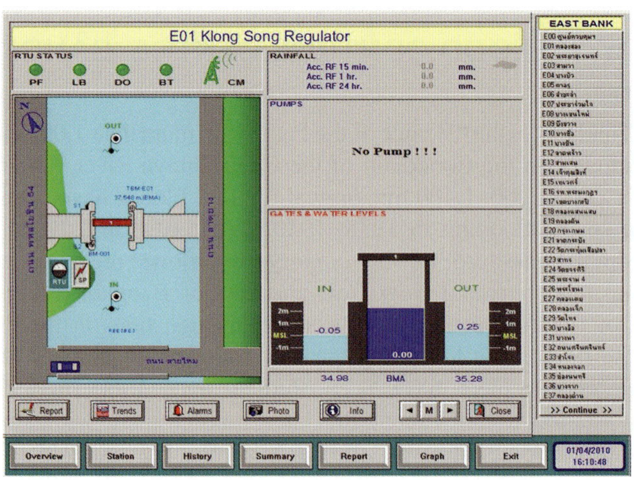

Figure 23.4 Example of graphic presentation at E01station

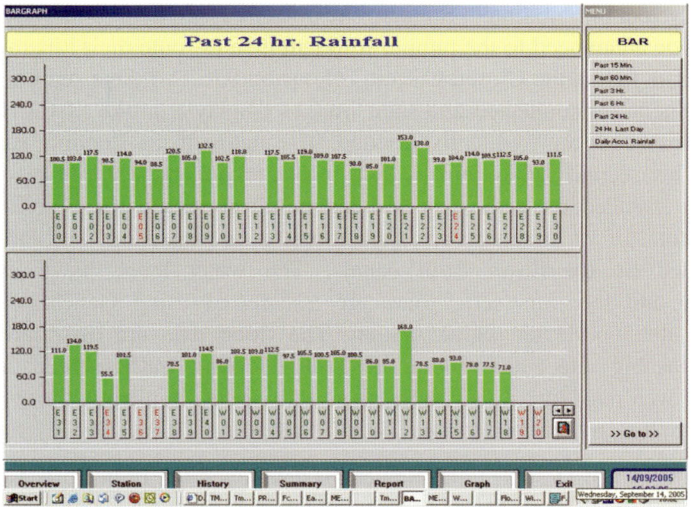

Figure 23.5 Plot of rainfall in 24 hour at all stations

(b) Weather radar: AC band radar was installed west of Bangkok in 2005. It can operate in 3 ranges namely 30, 60, and 120 km. The images and binary files are stored every 10 minutes and published on the web site http://dds.bma.go.th. The time series of radar images provides the direction, magnitude and extent of the observed rainfall, which provides useful data to forecast the near–future rainfall distribution. It can be used for warning and encouraging the preparedness of the local people. Figure 23.6 shows examples of radar images at 60 km radius.

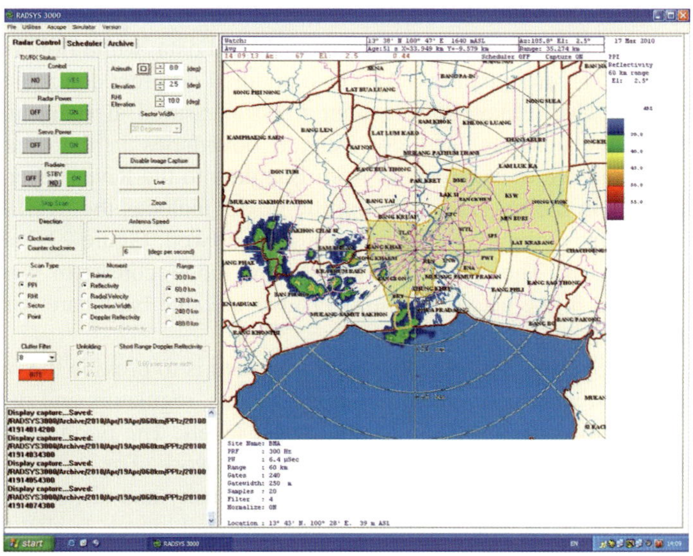

Figure 23.6 Example of radar images at 60 km radius

(c) Monitoring System of Flooding in Streets: The objective is to determine which streets will flood so that they can be announced to the public, who can then re-route their itinerary. It consists of three subsystems. The first is the Closed Circuit Television (CCTV). Selected CCTV cameras from the traffic department are connected and their images displayed at the FCC. There are a total of 16 locations where roads are frequently flooded. The status of the flooding can be directly seen on television. The second subsystem is the measurement of water levels in the streets. The equipment is installed on walkways and the monitors record the time series of water levels. There are a total of 71 streets where such monitors are installed. The last subsystem concerns streets flooded by overflows from canals. The water levels in 6 canals are monitored automatically and the time series data are transferred using General Packet Radio Service (GPRS) and the mobile telephone system. The flood data are disseminated through various media, such as big screen televisions at street junctions, traffic screens at bus stations, radio channels, the web page http://dds.bangkok.go.th, wireless application protocol (wap) http://dds.bangkok.go.th/lfcc_mobile and twitter: http://twitter.com/BKK_BEST.

In 2005, the Department of Drainage and Sewerage developed the flood forecasting and management system for East Bangkok in order to provide systematic operation and coordination. The system consists of 4 subsystems as follows:

(1) Rainfall Forecasting System: The SCOUT system provides the rainfall forecast using the recorded binary file in polar PPI type from the weather C band radar described above. The forecast time is 3 hours ahead. The computed results from the numerical weather prediction (NWP) system at the Thai Meteorological Department are used to forecast rainfall 6 hours ahead. The back up system utilizes the Artificial Neural Network (ANN) model proposed by Hung *et al.* (2008). The rainfall forecast system is updated every 15 minutes with the required accuracy of 65% of the accumulated rainfall over 3 hours. The output data is the series of forecast rainfall on grid system which provides the input to a hydraulic flooding model.

(2) Flood Modelling System: MIKE 11 is used to compute surface flow in the urban drainage network. The study area is divided into 31 catchments. The unit hydrograph method is used to model rainfall-runoff after calibration and verification with field data, while flow routing in the canals is done using MIKE 11 Hydrodynamic (HD). Two hundred different control structures in canals are used in the model, which takes into account flows measured at Bangsai and the tide at Fort Chula, which are the upstream and downstream boundaries in the Chao Phraya River: see Figure 23.7. The drainage of storm water from Bangkok to the River depends on the water level and flow conditions in the river. In addition, the model includes the overland flow from East of Bangkok, which drains to the River through Bangkok as well. Figure 23.7 shows the model setup for the East of Bangkok. The results provided by the model consist of time series for the water levels and discharges in the network. The overflows at canal banks are computed and shown in the Geographic Information System (GIS).

(3) Flood Management System: The objective of this system is to provide the operation strategies for the control structures, which consist of pumping stations, regulators and detention storage ponds under different scenarios of rainfall and initial water levels. The flood modelling system is used to simulate the hydraulic conditions in the drainage network. The scenarios can be set up for a combination of factors such as variations in rainfall intensity and distribution which are analysed from past records with 3 hours duration, and planned water levels for different forecast conditions. The optimum operation can be achieved by comparing the model results with the agreed flood criteria at particular locations in the network. Otherwise, the operation of control

structures must be modified, and the model must be rerun until the criteria are met. Figure 23.8 shows the steps in the formulation of a flood management plan. Management plans are stored in a database ready for use when a real flood condition occurs, when the plan that most closely matches one of the scenarios can be retrieved.

(4) Decision Support System (DSS): The software "Floodwatch" is used to integrate the subsystems and provide communication with the real time database of the Flood Control Centre (FCC). All information is gathered and audited prior to incorporation in the flood management system. The system chooses the appropriate flood management plan and provides the forecast results. The functions of database management, data communication and flood warning and dissemination are included in this subsystem. The media used for dissemination include Short Message Service (SMS) and Multimedia Messaging Service (MMS). The system provides a choice for operators to choose different management plans and to obtain the corresponding forecast results.

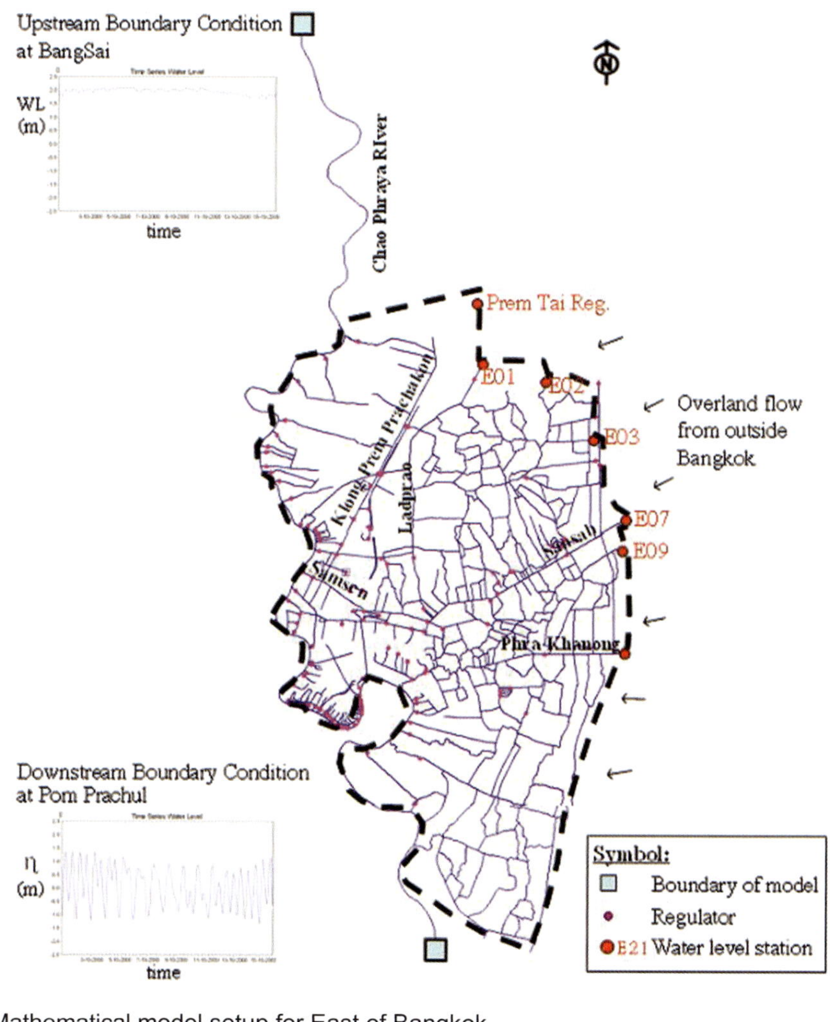

Figure 23.7 Mathematical model setup for East of Bangkok

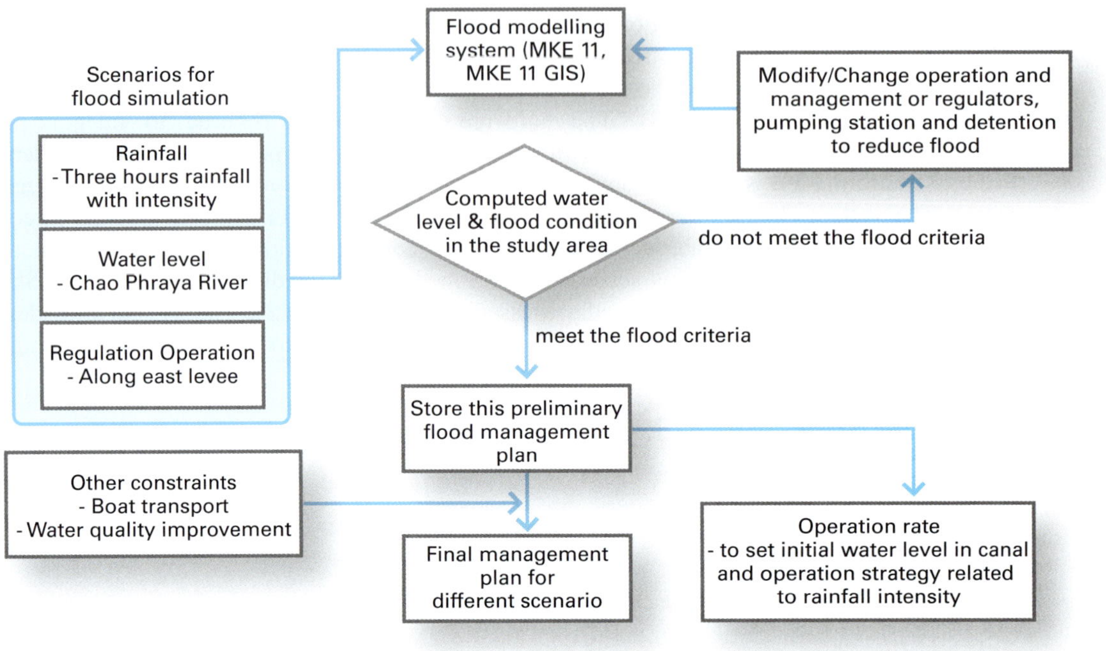

Figure 23.8 The steps in formulation of flood management plan

The flood forecasting and management system for Bangkok is operational and is expected to include more details of the urban pipe drainage network in the future. This essentially hydroinformatics system showcases different techniques in information technology to reduce losses due to flooding in the urban area.

23.2.3 Acknowledgment

This case study was prepared by S. Weesakul (Water Engineering and Management Program, School of Engineering and Technology, Asian Institute of Technology, P.O. Box 4, Klong Luang, Pathumthani, 12120, Thailand), P Supataratan (Asdecon Corporation Co., Ltd, 2782 Ladprao Soi 130, Khlongchan, Bangkapi, Bangkok 10240, Thailand) and O. Mark (DHI–Institute for Water and Environment, Agern Allé5 DK-2970, Horsholm, Denmark).

23.3 CASE STUDY 3: *FLOOD FORECASTING – FROM RAINFALL TO EMERGENCY RESPONSE* (AUSTRALIA)

23.3.1 Introduction

The objective of this case study is to demonstrate a procedure that has been implemented to serve the needs of flood emergency managers by providing timely information. The process capitalises on real time field data and design flood modelling results to provide flood extent mapping and associated intelligence linking the flood surfaces with GIS data covering affected properties, evacuation routes, critical infrastructure, and so on.

Effective flood forecasting and warning is an important part of the floodplain management process. If managed well, it can greatly reduce the risk to life and, to a lesser extent, the risk to property during a flood event. In Australia, the responsibility for flood forecasting rests with the Bureau of Meteorology (BoM) which provides forecasts for catchments with a response time generally in excess of 24 hours. There are many smaller catchments with insufficient response time for a BoM forecast, but sufficient time to implement effective flood emergency management procedures. The candidate catchment in this study covers an area of 700 km^2 with an average rainfall of 1100 mm and a response time of roughly 18 hours for large floods.

The procedure follows a discrete number of steps as outlined in Figure 23.9.

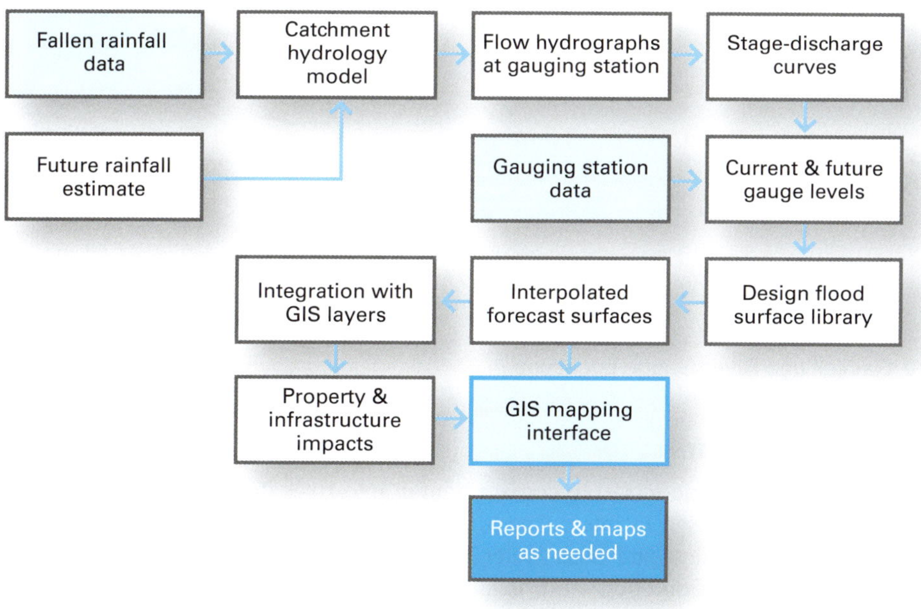

Figure 23.9 Flood Forecasting Concept

23.3.2 Input data

An extensive network of 38 rainfall and 6 river telemetered gauges are read and their content exported every 5 minutes. Figure 23.10 shows the location of the rainfall and river gauges. Hydraulic modelling for a range of design floods (1 in 10, 20, 50, 100, 500, 1000, 10000 year ARI and PMF) had been carried out offline. The modelling consisted of 9 separate models from 5 different modelling packages (MIKE11, MIKE21, TUFLOW, SOBEK and HEC-RAS). A Digital Elevation Model (DEM) derived from LiDAR and 2 m contour data was available. A range of GIS data was available, including high resolution aerial photography, cadastral and road network layers and building footprint and property data layers including surveyed floor levels of each flood-affected property.

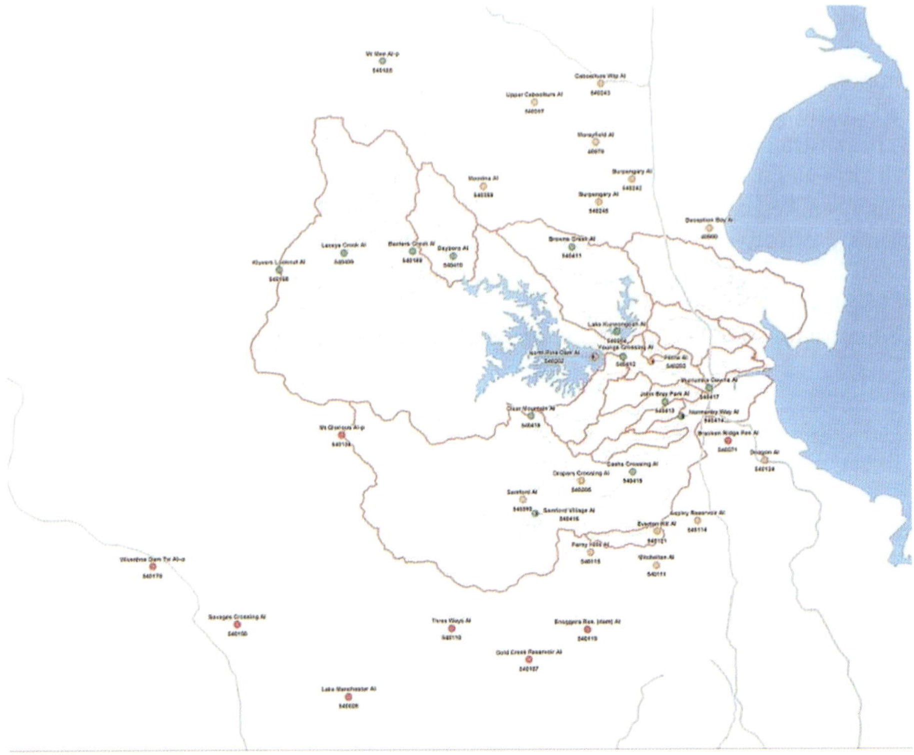

Figure 23.10 Rainfall and River Gauge Locations

23.3.3 Hydrologic model

The process includes a hydrologic model to convert rainfall into river flows corresponding to the stream gauges, which in turn can be used to interpolate a forecast time varying flood surface. The natural lag between rainfall in the catchment and peak water levels in the river allows the hydrologic model to provide an indication of likely flood behaviour in sufficient time for emergency management procedures to be initiated. Incorporating forecast rainfall into the hydrologic model extends the effective warning time facilitating "what if" analyses and leading to improved decisions. The hydrologic model consists of 674 sub-catchments. The necessary input files can be created and can be run automatically, executing typically in less than a minute. By referencing the recorded level data with values predicted by the model, a degree of automated calibration can be applied for the loss and lag parameters in the model.

23.3.4 Flood surface determination

Peak and time varying design flood surfaces represent the dynamics of flow across the floodplain. When a library of such surfaces for a range of probabilities are available for a catchment, predictions of water levels at various locations (*gauges*) can be used to interpolate a likely flood surface. For the simple case of a single gauge, a predicted level can be used to identify the two surfaces from the library that bracket the predicted level. The ratio between the predicted level and each reference surface can then be used to interpolate a likely flood surface across the entire floodplain, Figure 23.11 and Figure 23.12.

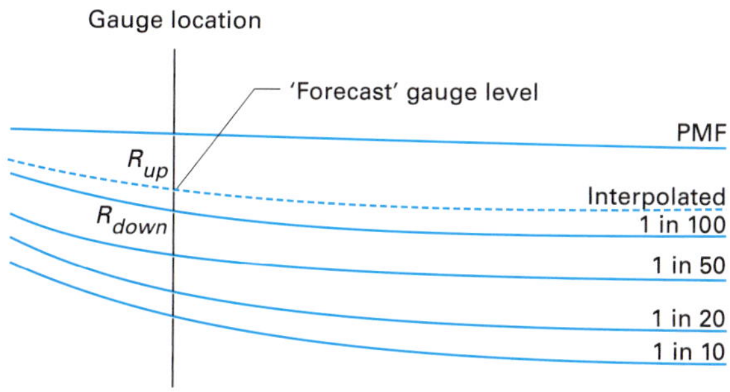

Figure 23.11 Preservation of water surface slope in interpolated surface

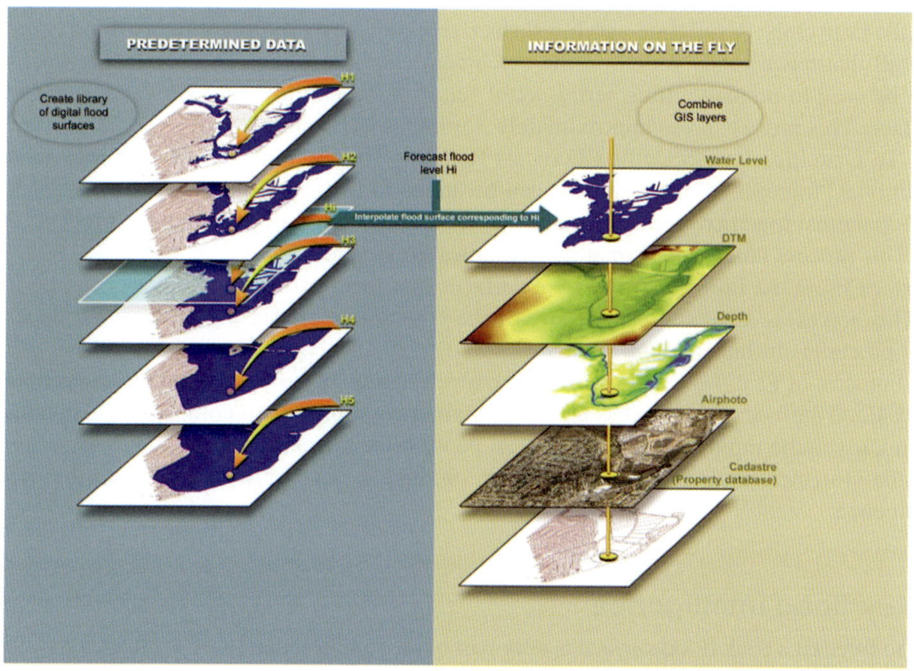

Figure 23.12 Interpolation of flood surface from a "library of design flood surfaces"

In the study catchment, the spatial variation in rainfall and resulting flooding is accounted for through the use of multiple gauges (*the six river gauges*) covering the different tributaries. In such a case the interpolated surface can be warped to account for the varying magnitudes of flooding in different parts of the catchment. Such warping is determined by developing a "zone of influence" for each gauge. Inside each zone, the magnitude of flooding is governed by the gauge within that zone. Outside of these zones, the magnitude of flooding is influenced by surrounding gauges and the relative distance to them. The forecasting engine simply requires a library of design flood surfaces covering the catchment, and a continuous, 2D flood surface covering the extent of all 9 models for each flood probability was created by mapping the variety of model results to a common spatial framework.

23.3.5 Operation

The process can operate on two levels. Firstly, the hydrologic model and the generation of the forecast flood surfaces can be run continuously and independently with a preset interval on a server. Secondly, at any time, the forecast surface can be interrogated and integrated with the property and infrastructure GIS datasets to generate the desired flood intelligence which can then be accessed by all users. The use of a simple and intuitive interface with preset views of the datasets minimises the training requirements and facilitates ready access to the flood intelligence or information inherent in the data. Typical hydraulic data that is useful for interrogation and use in deriving flood intelligence includes:

Current River Levels :A static surface showing the current situation can be predicted by reading the levels at each of the 6 river gauges and interpolating a continuous surface. This surface can then be mapped to the DEM on the fly to provide likely flood depths and extents. This process typically takes less than 1 minute to complete.

Rainfall To Date Only: A full time varying surface showing the evolution of likely flood behaviour based on the rainfall that has actually fallen on the catchment can be created. The forecast hydrographs at the gauges can be used to prepare a time series set of future flood surfaces in a similar manner to the single current surface. This process takes approximately 3 minutes to complete.

Rainfall To Date Plus User-Specified Forecast Rainfall: This is similar to the previous process, except that an additional "forecast" rainfall amount can be added to the actual rainfall. Allowances for temporal and spatial patterns together with a sensitivity assessment can be applied to bracket the forecast outcomes. This process takes approximately 5 minutes to complete.

Any Surface From the Library: Any of the surfaces that form part of the "library" can be immediately accessed. This serves as a backup should any of the continuous processing routines produce unforeseen errors. This process is instantaneous.

Manual Forecast of Peak Levels: Forecast peak levels at any of the gauges can be entered to interpolate a static peak flood surface. This process takes approximately 1 minute to complete. This feature serves as a backup and, given the rapid execution time, can be used for manual "what if" analyses.

Any of the aforementioned modes of operation produce a flood surface which can be used to derive flood intelligence.

23.3.6 Key flood intelligence

The ability to forecast a peak flood surface is only part of the function of a flood forecasting system. The key to an effective system is its ability to translate the likely flood behaviour into consequences of flooding (i.e., "*what will the flood impact upon and who will it affect?*"). The availability of the likely time-varying flood surface allows a range of flood intelligence to be derived through the integration of the surface with key GIS datasets.

23.3.7 Predicted flood surface interrogation

As the forecast flood surface is mapped to the finer scale DEM, the likely peak flood extent and flood depth can be displayed at any location on the floodplain: see Figure 23.13.

Such information makes it clear to the emergency manager what areas are likely to be affected by flooding and to what degree. The use of GIS tools such as zooming, panning and spot polling allow the emergency manager to "drill down" to any level of detail.

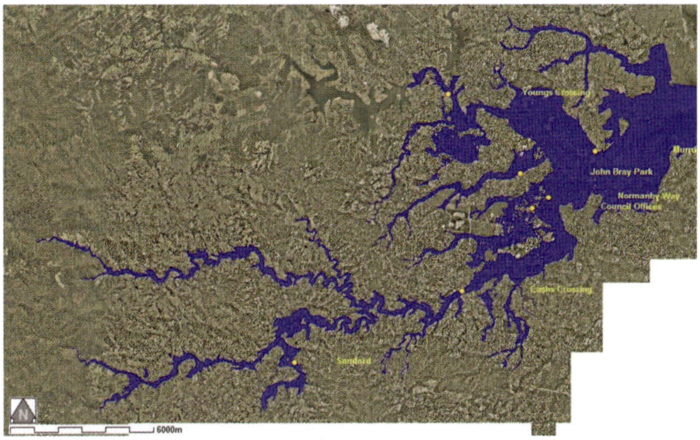

Figure 23.13 Likely Peak Flood Extent

23.3.8 Forward looking surface

An indication of the likely flood outcome is invaluable for a flood emergency manager. However, it is of potentially greater value to know when flooding will occur and how a flood is likely to develop over time prior to the peak when evacuation needs have to be assessed and prioritised. The time based forecast flood surface can be used to derive a "forward looking surface" which encapsulates the likely time-based flood behaviour in a single view, Figure 23.14.

This surface is created by superimposing the following surfaces on the DEM:

- Current flood extent,
- Flood extent in 2.5 hrs time, and
- Peak flood extent.

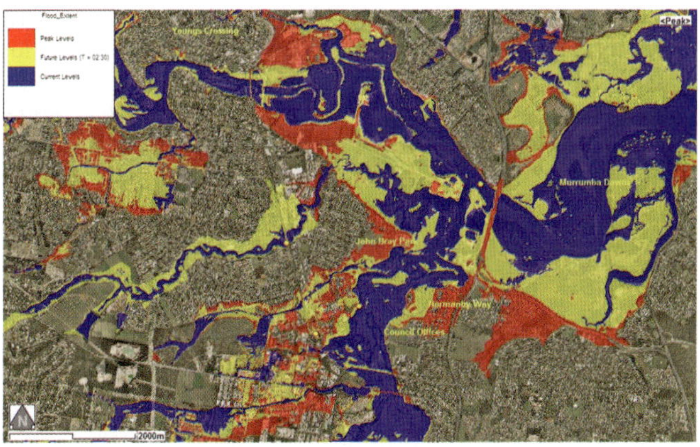

Figure 23.14 Forward Looking Flood Surface- (*blue = current, yellow = 2.5hr time, red = peak*)

The "forward looking surface" provides a readily understandable snapshot of current and future flood behaviour. It is derived from a combination of the actual river levels at the gauges, and the forecast levels from the hydrologic modelling.

23.3.9 Flood affected properties

Where the local authority has a property GIS layer containing, amongst other information, building footprints and floor levels the predicted flood surface can be integrated with this layer to determine properties likely to be affected. Key property specific flood intelligence that can be derived may include:

- Likely depth over floor/ground,
- When a property is likely to be inundated,
- A list of flood affected properties,
- Likely flood damages

Figure 23.15 shows the depth over floor at the peak of the flood for a selected part of the catchment.

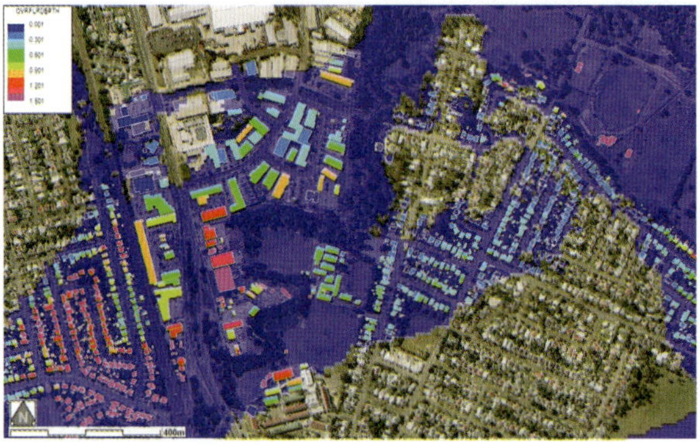

Figure 23.15 Depth over floor at the flood peak.

23.3.10 Flood affected critical infrastructure

Critical infrastructure vital to the community and flood emergency management procedures may be at risk during a flood event. Such infrastructure may include: telecommunications, electricity, refuge centres, hospitals, etc. Determining if these facilities are likely to be operational during the flood (*and if not, when they are inoperable*) can be critical to smooth emergency management.

In a similar process to the determination of flood affected properties, a depth above (*or freeboard to*) a critical level for various infrastructure items can be determined by integrating the critical infrastructure GIS layer with the predicted flood surface. The time at which the item of infrastructure is inoperable can also be determined, thus assisting in devising/revising emergency management procedures as required.

23.3.11 Evacuation routes

Of key importance to flood emergency managers is the status of evacuation routes (i.e., *are they open or closed*). Having an evacuation route layer showing the low points along key evacuation routes allows integration with the time varying forecast flood surface to determine when the flood reaches the specified level for each low point. Figure 1 shows the time until the flood reaches the low point (i.e., *the time available for evacuation*) for a selection of low points. Note that a value of "9999.99" indicates that the flood does not overtop the evacuation route. Note that a value of "9999.99" indicates that the flood does not overtop the evacuation route. A time-series plot of water level at each location is automatically generated by the system, see Figure 23.16, which can be useful in examining "rates of rise" of floodwaters in relation to the evacuation route. The process can also determine the freeboard to an evacuation route low point if it is not inundated.

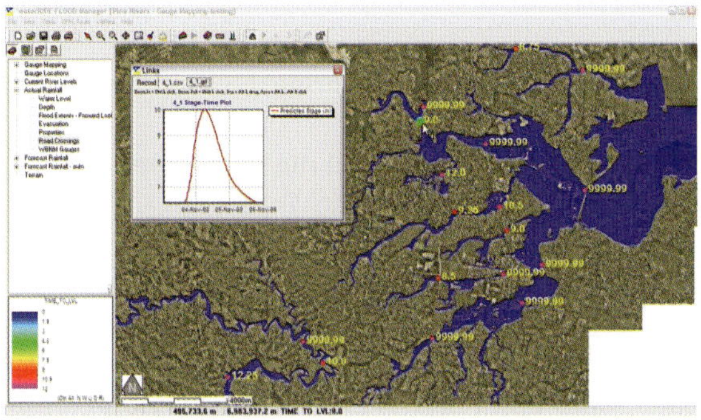

Figure 23.16 Time To Inundation of Key Evacuation Route Low Points

Such information, in conjunction with the "properties affected" intelligence can be used by flood emergency managers to prioritise areas for evacuation. This information may also be useful during planning and training operations in dry weather.

23.3.12 Sensitivity to forecast rainfall

There is a degree of "guesswork" associated with estimating what rainfall is likely over the ensuing duration of a storm. To assist flood emergency managers in planning for variances in such estimates, the process can generate a "sensitivity surface" showing the difference in likely peak flood extent for a preset variation in forecast rainfall about the estimate (e.g., +/−20%).

An example surface is shown in Figure 23.17. On this figure, the green surface shows the peak flood extent for the forecast rainfall amount. The red and pink surfaces show the peak flood extents for the higher and lower variations in estimated rainfall, respectively. This figure implies that, in this particular region of the catchment, the peak flood extent is more sensitive to an under estimation of the rainfall. Such information can be valuable for the flood emergency manager when adding contingencies to the forecast rainfall.

Figure 23.17 Automatic Sensitivity Analysis on Forecast Rainfall Estimates. (*red = forecast rainfall + 20 %, green = forecast rainfall, pink = forecast rainfall – 20%)*

23.3.13 Checking and validation

Any process that is predominantly automated (*or indeed, partially automated*) should include a degree of redundancy to ensure that it can always be hand-operated to some degree. The process includes a number of such features:

- *Extensive error reporting* – errors encountered during the continuous operation are written to a log file and a status alert is issued.
- *Hydrologic Result Logs* – each hydrologic model run saves summary results including the automated calibration which can be inspected at any time.
- *Forecast vs. Actual Plots* – at each river gauge, the process exports the actual, recorded water level time series as well as the forecast time series: see Figure 23.18.
- *Manual Data Entry* – At any time, forecast/actual gauge levels can be used to derive a likely flood surface. This process is infallible as it can operate independently on a laptop as a fallback position in the event of catastrophic issues (e.g., *loss of power/gauge readings*). Multiple laptops can also be used by field staff, with updated forecast information delivered via phone and entered directly into the system in the field.

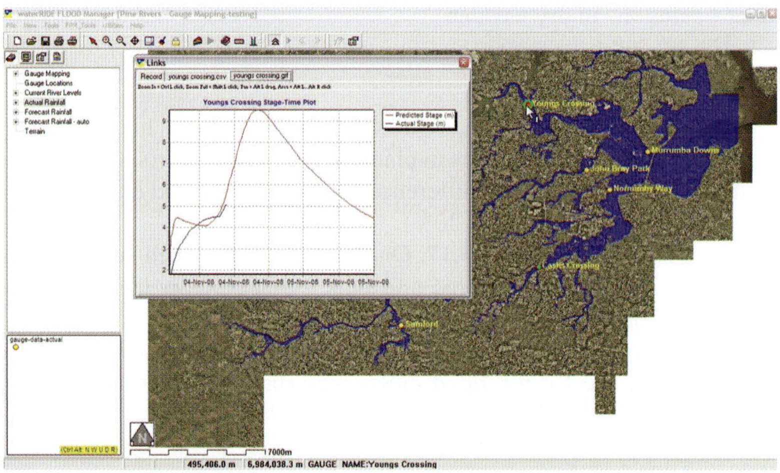

Figure 23.18 Actual Water Level Time Series vs. Forecast Time Series at a gauge

23.3.14 Future direction

The data processing for this study was undertaken with the waterRIDE™ spatial application development environment. It can be readily enhanced to suit the evolving needs of the flood emergency response community and could include a number of future enhancements to the system and delivery of flood intelligence including:

- Web-delivery expansion
- Email/SMS delivery of flood triggers
- Email/SMS delivery of derived flood intelligence

23.3.15 Acknowledgment

This case study was prepared by David McConnell, Advanced Technologies Manager, WorleyParsons, Australia.

23.4 CASE STUDY 4: *THE SMART TUNNEL, KUALA LUMPUR* (MALAYSIA)

23.4.1 Introduction

The Klang River basin is the most densely populated area of Malaysia with an estimated population of over 3.6 million and growing at almost 5% per year. Although major flood mitigation works within Kuala Lumpur city have, to a large extent, been implemented, flooding in the city is still frequent and severe. This is causing disruption to various socio-economic activities in the city as well as extensive flood damages.

Some severe floods have been recorded in the city of Kuala Lumpur and these have prompted the Government through the Department of Drainage and Irrigation (DID) to review, reanalysis and supplement the on-going flood mitigation projects that have been implemented for the city. A thorough and comprehensive review was done in the Klang River Basin Environmental Improvement and Flood Mitigation Project, carried out by Dr. Nik & Associates Sdn. Bhd (DNASB) in association with the Snowy Mountain Engineering Corporation (SMEC) and the Danish Hydraulic Institution (DHI), and completed in 2002. During the project period, options were submitted to the Government to solve the frequent flooding in the city of Kuala Lumpur following a severe flood event on 26th April 2001.

Among the options presented to solve the frequent flooding in the city of Kuala Lumpur is the Storm water Management and Road Tunnel (SMART) Project. The project involves the construction of a bypass tunnel to divert a large portion of the flood discharge from the Sungai Klang, just downstream of the confluence with Sungai Ampang to the Taman Desa pond before releasing it back into Sungai Klang through Sungai Kerayong.

23.4.2 Project description

The SMART project focuses on diverting flows to solve the flooding problems in the Kuala Lumpur city centre along the Sungai Klang from upstream of Masjid Jamek to the confluence with Sungai Gombak. It forms part of the integrated flood mitigation project for the Klang river basin. By passing flood flows through the proposed tunnel and detention ponds prior to returning the flow to Sungai Klang downstream via Sungai Kerayong, flood peaks at the city centre can be substantially reduced. It has been found that this project will mitigate the flooding problems and improve the flood protection level up to 100-year ARI for the city centre if it is implemented along with other flood mitigation projects as proposed in the Klang River Basin Environmental and Flood Mitigation Project.

In that the tunnel will be used infrequently for conveying flood water, part of the tunnel is also designed for tolled motorway usage. The dual functions of the tunnel will also alleviate traffic congestion problems at the Kuala Lumpur's Southern Gateway.

23.4.3 Project corridor

The SMART project area is largely located in the Federal Territory of Kuala Lumpur except for a very small area at the upstream intake point which is located in the District of Ulu Langat in the State of Selangor. The SMART project alignment commences just downstream of the confluence of Sungai Klang and Sungai Ampang. This is designed to divert the maximum amount of flood flows from both the Sg. Ampamg and upper Sg. Klang catchments. A holding basin with an offtake structure and diversion weir will be constructed at Kampung Berembang, the starting point of the SMART system, with facilities for the trapping debris and sediment brought down by the flood discharges. The SMART project has a tunnel length of approximately 9.7 km which traverses in a southerly direction under the existing TNB reserve before skirting to the west towards the Kampung Pandan roundabout. The alignment then continues to traverse in the southerly direction under the existing roads, viz. Jalan Desa Pandan, Jalan Tun Razak, Jalan Chan Sow Lin and the KL-Seremban Highway alongside the Sungai Besi Airfield. The SMART project alignment then crosses the Federal Highway near the Jalan Salak interchange and finally terminates at the existing ex-mining pond at Taman Desa. This existing pond has an important role in the SMART system as it acts as a storage reservoir to detain the flood discharges before releasing them back into the river system at Sungai Kerayong, about 1.8 km upstream from the confluence of Sungai Klang and Sungai Kerayong. Figure 23.19 shows the SMART layout.

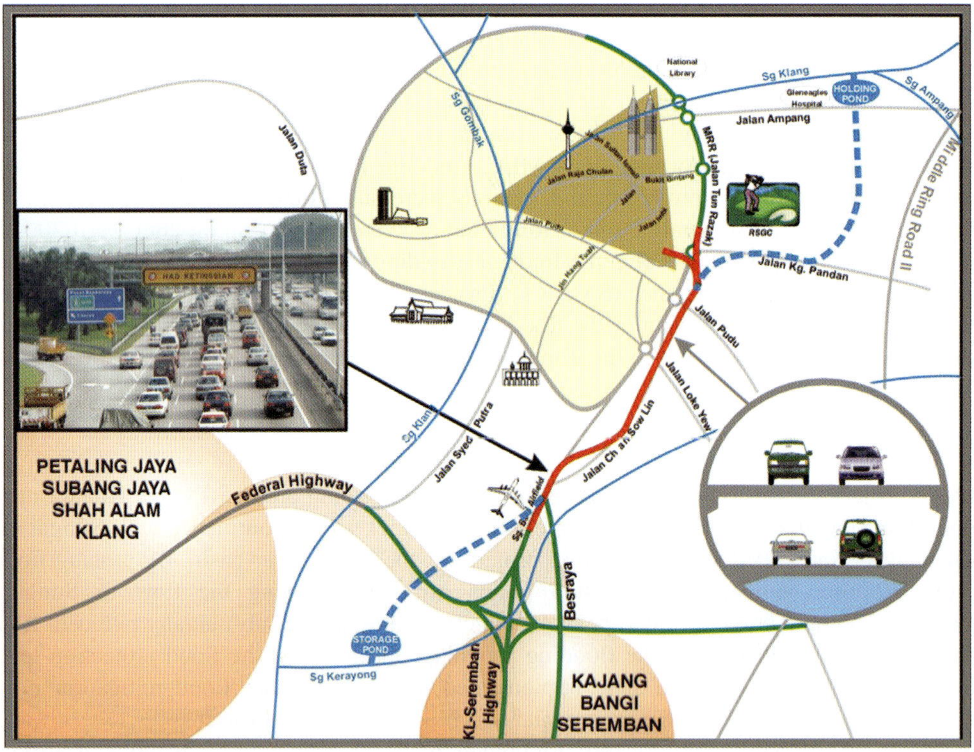

Figure 23.19 Alignment Plan of the SMART Project

23.4.4 Traffic study

The SMART project through the innovation of the Company has been engineered to perform the dual functions of management for flood discharges as well as a motorway tunnel. While the storm water bypass tunnel is essential in mitigating the flooding problems, it will only be used on a few occasions per year. As such, to make full economic use of the storm water bypass tunnel and given its strategic location adjacent to the Kuala Lumpur–Seremban Highway, the motorway tunnel thus performs another essential and important role of easing the traffic congestion along Kuala Lumpur's Southern Gateway.

Congestion along the Kuala Lumpur–Seremban Highway is a daily and familiar sight as traffic travels towards the city centre. The daily traffic queues have been observed to stretch some 6 km back from the Jalan Loke Yew to the Desa Petaling area. Based on historical data, there is a rapid increase in the traffic volume along the Kuala Lumpur–Seremban Highway where it is over 14% per annum. However, the recent growth has been curtailed by the constrained capacity. In essence, the traffic congestion in this area results from a relatively short constraint in the road network, as traffic from the Kuala Lumpur–Seremban Highway, Besraya Highway and Jalan Istana/Federal Route 2 converge to the MRR1 and create a "pinch point" condition. This condition continues through Jalan Loke Yew, Jalan Pudu and the Kpg. Pandan roundabout.

While the main purpose of the tunnel is to bypass the flood flows, its dual purposes including the motorway tunnel offers an alternative road route to relieve congestion by providing much needed additional capacity for traffic entering the city centre efficiently through bypassing this "pinch point".

23.4.5 Major components of SMART project

The major components of the SMART Project include:-

- A storm water bypass tunnel is about 9.7 km in length and 11.8 m internal diameter, providing an efficient diversion of 280 m^3/s of flood discharge and with a storage capacity of about 1,000,000 m^3;
- A holding basin having a maximum storage capacity of 600,000 m^3;
- Diversion structures to divert flood discharge into the holding basin from the river which comprise two strings of floating booms, a 300 m^3/s capacity multiple-bay gated diversion weir at Sungai Klang and a 280 m^3/s capacity multiple–bay gated off-take structure located between the river and the holding pond;
- A multiple-bay bell mouth intake structure with capacity of 280 m^3/s to divert the flood discharge from the holding basin into the bypass tunnel;
- Construction of a gated outfall structure from the bypass tunnel to the ex-mining pond at Taman Desa;
- Slope protection works for the Taman Desa storage reservoir which will have a flood storage capacity of about 1,400,000 m^3;
- A gated intake structure and an approximately 500 m long twin-cell box culvert complete with an outfall structure to Sungai Kerayong for controlling flood discharge released from the storage reservoir to Sungai Kerayong;
- An operation control centre complete with administrative, storm water management and traffic management facilities;
- A double deck motorway within the bypass tunnel for a length of approximately 3 km that starts near the Kpg. Pandan roundabout and ends near the Jalan Istana Interchange at the Kuala Lumpur–Seremban Highway. The motorway will be equipped with an efficient ventilation system and emergency escape provisions;
- The motorway will be connected to existing road networks by ingress and egress connection links to Jalan Tun Razak, Jalan Sultan Ismail and the Kuala Lumpur–Seremban Highway;
- Flood gates located at the upstream and downstream ends of the motorway tunnel as well as within the motorway at the ingress and egress connection links.

The general layout of all the major components of the tunnel with its motorway is given in Figure 23.20 and Figure 23.21.

Figure 23.20 Major components of the SMART project

Figure 23.21 Major components of the SMART project

23.4.6 Operational procedure

In the SMART Project the operations of the gated diversion structure, the off-take structure at the river and the holding pond as well as flood gates in the tunnel, are generally configured into three principle modes as illustrated in Figure 23.22 and in Table 23.3.

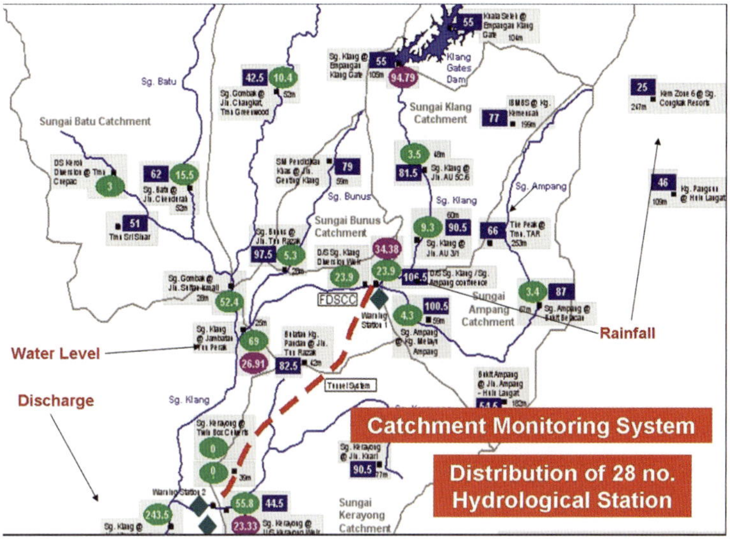

Figure 23.22 Flood Detection System (FDS)

Table 23.3 Operation modes of the tunnel.

Mode	Discharge (Q) at Tun Perak	1.1 SMART Operation Procedure	
		1.2 Stormwater Management	Traffic Management
I	Q < 70 cumec	All storm water flow through city centre. No diversion of water into tunnel.	Traffic uninterrupted
II	150 > Q > 70 cumec	Storm water will be diverted into the holding pond and flow through the tunnel below the lower deck without opening the floodgates in the tunnel. The diverted flood volume will be stored in the holding pond, the tunnel (minus volume in the motorway decks) and the Taman Desa storage reservoir before releasing slowly into Sg. Kerayong.	Traffic uninterrupted
III	Q > 150 cumec	Flood flow will be diverted into the holding pond. The traffic in the tunnel (motorway decks) will be evacuated before the floodgates are opened and the flood flow will be allowed to flow through the whole tunnel section. The diverted flood volume will be stored in the holding pond, the whole tunnel section and the Taman Desa storage reservoir before releasing slowly into Sg. Kerayong.	Closed to Traffic

The tunnel begins at Kg. Berembang where the Diversion Weir at the confluence of Sg. Klang/Sg. Ampang and Off-take Structure divert flood water to the Holding Pond. The water then flows through the Intake Bell Mouth structure into the Tunnel and discharges at the Attenuation Pond at Taman Desa. Flood water is stored in this pond before being released to Sg. Kerayong through the Twin Box Culvert.

The SMART Control Centre is in operation 24 hours a day. Once high flow is measured at the confluence, the SMART Control Centre will send a signal to the Motorway Control Centre to evacuate the traffic from thc tunnel. After ensuring the road tunnel is cleared, the traffic compartment will be flooded. Once the flood has receded, the tunnel will be cleaned and opened to the public.

A sophisticated and modern flood Detection System (FDS), as shown in Figure 23.23 provides real-time flood forecasting information. This enables the efficient and safe management at the operation of the tunnel. For the safety of motorists, the motorway traffic is unidirectional and safety features such as ventilation shafts, cross passages, vms, fire extinguisher and air ventilators have been incorporated in the design in compliance with international and Malaysian Standards. Some features in SMART are shown in Figure 23.24. This project has been a great challenge for the local engineers involved in the management and construction of the SMART Project as it runs below congested roads, near sensitive structures and through varied geological ground conditions.

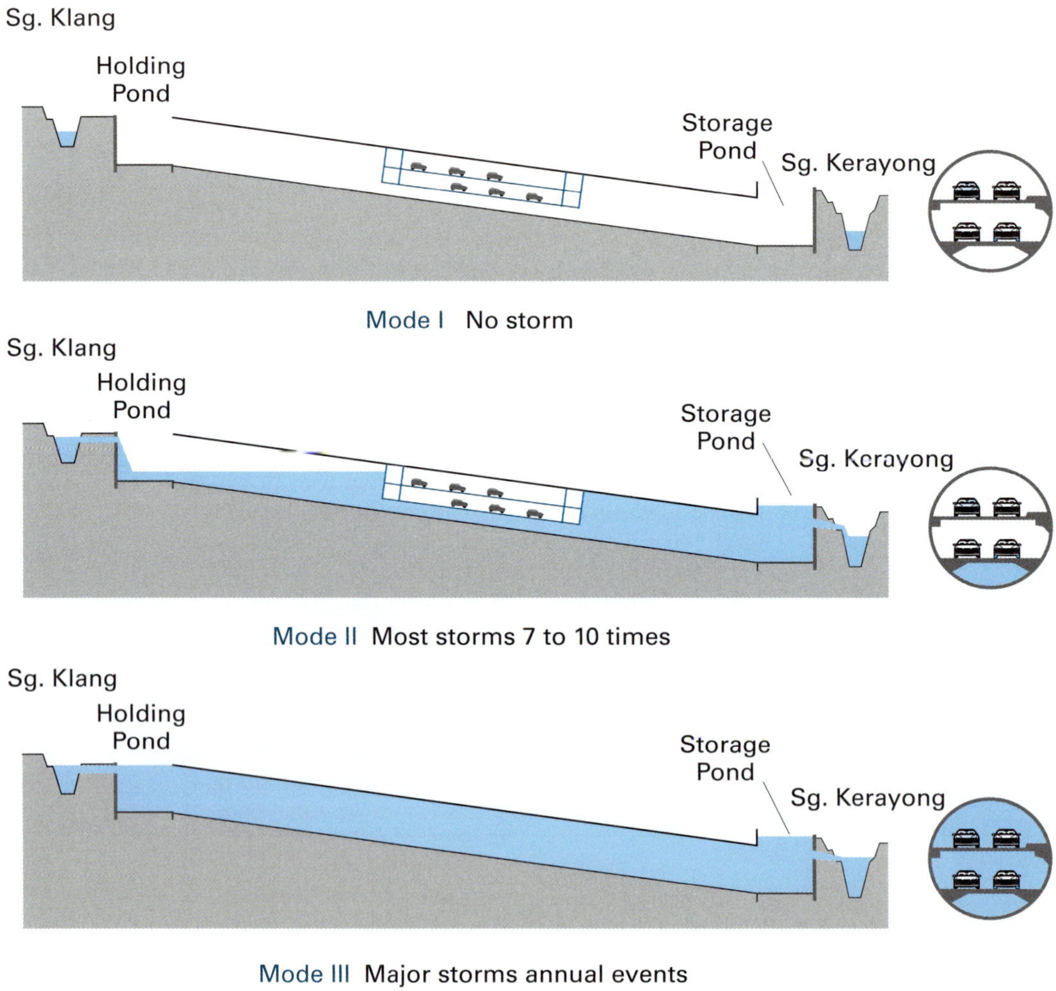

Figure 23.23 Three operation modes for the SMART project

Figure 23.24 Features in SMART: real-time CCTV cameras

23.4.7 Acknowledgment

This case study was prepared by Ahmad Fikri Abdullah with permission from Drainage and Irrigation Department of Malaysia.

Afterword

As the title of the epilogue to his best-known and most popular book, *Enten-Eller//Either-Or*, Søren Aabye Kierkegaard, as 'the father of existentialism', used the word *Ultimatum*.

This word is coming back with vengeance today as the warnings of the destruction of the Earth's ecosystem and the menaces of global warming and climate change pile up alongside the deprivation of ever more persons, deprivations that are now penetrating into the lives of an increasing number of people of European descent who not long ago could identify themselves with 'The First World'. These warnings that have now become realities in an increasing number of minds are associated with an ever-more cogent number of writings that are essentially just that: *Ultimatums*.

In a lecture delivered by HRH The Prince of Wales on 8 July 2010, and reproduced in the *Temenos Academy Review* of 2010 (No. 13, pp. 9–22), its most cogent observation (p. 12) was posed as follows:

> That earlier shift, away from seeing ourselves within Nature to us standing apart from it, gradually undermined what I have always felt, deep down, to be the true situation – that if we wish to maintain our civilizations then we must look after the Earth and actively maintain its many intricate states of balance so that it achieves the necessary active state of harmony which is the prerequisite for the health of everything in creation. In other words, that which sustains us must be sustained, and I am afraid that I have come to the unavoidable conclusion that we are failing to do that. We are not keeping to our side of the bargain and, consequently, the sustainability of the entire harmonious system is collapsing – in failing the Earth we are failing Humanity …

As our world moves to its forecast of 80 percent of our populations living in cities – and even facing the arrogance of those who can so-accurately predict an '81 percent' – the consequences of an ecosystem that is moving ever more rapidly to a state of total disintegration in so much of our world appear as too dreadful to contemplate. The risks of flooding of our cites is a direct function of the states of our ecosystems, and these risks are becoming incalculable. And so we come to an *ultimatum* (ibid, p. 22) which we may perhaps be allowed to share (with our italics added):

> We are, Ladies and Gentlemen, as I said at the beginning, at an historic moment – because we face a future where there is a real prospect that if we fail the Earth, we fail Humanity. To avoid such an outcome, which will comprehensibly destroy our children's future, we must urgently confront the issues and make the choices which carry monumental implications. In this, we are the masters of our fate ….
>
> On the one hand, we have every good reason to believe that carrying on as we are will lead to a depleted and divided planet, incapable of meeting the needs of its nine billion citizens, let alone sustaining its other life

> forms. On the other hand, we can adapt our technologies, our lifestyles, and, crucially, *our way of thinking and perceiving the world* so as to transform our relationship with the Earth that sustains us. The choice is certainly clear to me.

The thrust of the present book has been that this tectonic shift – of *adapting our way of thinking and perceiving the world* – will not happen as and by itself, but has to be explained, nurtured and introduced in such a way that the decision–making processes, and especially in the field of urban flooding, have to be placed in the hands of those who care most directly, and indeed in many cases care passionately, about the consequences of interventions in their local, and often even more global, living spaces and ecosystems – and the consequent influences on their qualities of life. These persons must become the new decision makers in our societies, even if in some cases they may appear to some as 'decision makers of last resort'. We have tried to explain how the hydroinformatician has a vital role to play in bringing this mode of decision-making to presence, and indeed how this process provides a paradigm, or at the very least a provisional template, of the tasks of the hydroinformatician. The construction of hydroinformatics' environments, as information providers of virtual worlds using all the resources of the current and future information and communication technologies and practises, is the new task of such a hydroinformatician. These concepts are illustrated with colour and dynamics in the web site that is dedicated to the contents of this book: www.urbanhydroinformatics.com. This web site includes a reference to several of our fellow creatures here on Earth and what we owe to them, even as we so often misuse them.

It would be naive not to see that the means and the processes that are being promulgated in this book must arouse opposition in many quarters, where they will be seen as subversive in the eyes of many present-day engineers and scientists, who prefer Panglossian recommendations and simple solutions whereby "all is for the best in the best of all possible worlds". For the most part, albeit with a few honourable exceptions, these experts prefer to ignore how Voltaire demonstrated the disenchantments that the young Candide was obliged to suffer by holding to such illusions, as they were propagated by his mentor, Monsieur Pangloss. We should have no illusions concerning the opposition that this book will encounter in some circles, even as our experience to date has been that the new (for most readers) paradigm will be taken up by those in the hydroinformatics community, and many more in other fields besides, who will see in this the only way forward if we are to maintain any kind of social justice in our lives and our communities, and thus play our own part in saving our own and our children's lives and 'our' planet.

As explained at various points in the book, the kinds of processes that occur within a group of active stakeholders working within a hydroinformatics environment provide the members of the group with the means to attain to higher levels of consciousness, even as these in turn mutate into a higher levels of conscientiousness during the process. Thus these participants in the joint effort come to *transcend themselves* through this process in which they are active participants, and then, when we observe this as we do, we must return to one of the aphorisms given in the body of the book itself, where we refer to:

> … H. H. John Paul II himself, where (1994, p. 104) he echoed a theme that also passed from Kierkegaard to Jaspers, and further again to the great Jewish thinker, Emmanuel Levinas (e.g. 1995/2006), namely that "Man, perhaps subconsciously, waits for such a challenge, indeed, man feels the inner need to transcend himself. Only by transcending himself does man become fully human (cf. Blaise Pascal, *Pensées*, Ed. Brunschvicg, (1670/1953) p. 434, "apprenez que l'homme passe infinement l'homme" [// "understand that the man infinitely surpasses the man"])".

It is on the authority of these witnesses and for the purposes of the present book that *social justice is defined as that which provides all its citizens with the possibility to transcend the limits of their earlier selves*, a definition that is prepared by references to several other sources in the body of the book itself. Writing this book has provided it's authors with something of this transcendence – and we hope that it will also do so for such persons as those who Kierkegaard always addressed most reverently as "My Reader".

Zoran Vojinović
Michael B. Abbott

References

PART I

Abbott M. B. (1991). Hydroinformatics: Information Technology and the Aquatic Environment. Avebury Technical, Aldershot, UK and Gower Publishing, Brookfield, USA.

Abbott M. B. (2004). European Industrial Leadership. Knowledge Engineering, Brussels.

Abbott C. J. and Cheung C. H. (2005). From a Knowledge Society to an Understanding Society. Knowledge Engineering and European Institute for Industrial Leadership (EIIL), Brussels.

Alexander D. (1993). Natural disasters. Chapman and Hall, New York.

Alterman E. (2005). Found in the flood. *The Nation*, **281**(9), 33.

Blaikie P., Cannon T., Davis I. and Wisner B. (1994). At risk – natural hazards, people's vulnerability and disasters. Routledge, London.

Bateson G. (1972). Steps to an Ecology of Mind: Collected Essays in Anthropology. Psychiatry, Evolution and Epistemology, University Of Chicago Press.

Birkmann J. (ed.) (2006). Measuring Vulnerability to Natural Hazards. Towards Disaster Resilient Societies. United Nations Publications, New York, ISBN: 978 92 808 1135 3.

Bohle H. (2001). 'Vulnerability and Criticality: Perspectives from Social Geography'. Newsletter, International Human Dimensions Programme on Global Environmental Change, **2**(1), 3–5.

Bogardi J. J. and Birkman J. (2004). Vulnerability assessment: the first step towards sustainable risk reduction. In: Disasters and Society – From Hazard Assessment to Risk Reduction, D. Malzahn and T. Plapp (eds), Logos Verlag Berlin, Berlin, pp. 75–82.

Bollin C., Cárdenas C., Hahn H. and Vatsa K. S. (2003). Natural Disaster Network; Disaster Risk Management by Communities and Local Governments. Inter-American Development Bank, Washington, D.C. http://www.iadb.org/sds/doc/GTZ%2DStudyFinal.pdf.

Bortoft H. (1996, Fourth edition, 2010). The Wholeness of Nature: Goethe's Way of Science. Lindisfarne Books and Floris Books, Edinburgh.

Burton I. and Kates R. W. (1964). The perception of natural hazards in resource management. *Natural Resources Journal*, **3**(3), 412–441; also in H. Kunreuther and A. Z. Rose (eds) (2004). The Economics of Natural Hazards. Edward Elgar, Cheltenham, UK, Vol. 1, pp. 53–82.

Cardona O. D. (1999). Environmental management and disaster prevention: Holistic risk assessment and management. In: Natural Disaster Management, I. Ingleton (ed.), Tudor Rose, London.

Cardona O. D. (2001a). "Estimación Holística del Riesgo Sísmico utilizando Sistemas Dinámicos Complejos" Universidad Politécnica de Cataluña, Barcelona. http://www.desenredando.org/public/varios/2001/ehrisusd/index.html.

Crompton T. and Kasser T. (2009). Meeting Environmental Challenges: The Role of Human Identity, WWF-UK, Panda House, Godalming, Surrey, UK.

Cutter S. L. (1996). Vulnerability to environmental hazards. *Progress in Human Geography*, **20**(4), 529–539.

Davidson R. A. (1997). An Urban Earthquake Disaster Risk Index, The John A. Blume Earthquake Engineering Center, Stanford University, Report no. 121, Blume Center, Stanford, California.

DETR, Environment Agency, Institute for Environment and Health (2000). Guidelines for Environmental Risk Assessment and Management, The Stationary Office, London.

Durning D. (1999). The transition from traditional to postpositivist policy analysis: a role for Qmethodology. *J. Pol. Anal. and Manag.*, **18**(3), 389–410.

Dupuy J.-P. (2002). Pour un catastrophisme éclairé. Quand l'impossible est certain, Le Seuil, Paris.

Farber M. (1943). The Foundation of Pehnomenology: Edmund Husserl and the Quest for a Rigorous Science of Philosophy. Harvard University Press, Cambridge, MA.

Feather N. T. (1992). Values, valences, expectations, and actions. *Journal of Social Issues*, **48**, 109–124, Wiley-Blackwell, US.

Flew A. (1993). Social Justice Isn't Any Kind of Justice. Libertarian Alliance, London.

Good J. (2007). Shop'til we drop? Television, materialism and attitudes about the natural environment. *Mass Communication and Society*, **10**, 365–383, Routledge, Taylor and Francis, US.

Hall J. W., Meadowcroft I. C., Sayers P. B. and Bramley M. E. (2003). Integrated flood risk management in England and Wales. *Natural Hazards Review*, **4**(3), 126–135.

Hardoy J. E., Mitlin D. and Satterthwaite D. (2001). Environmental Problems in an Urbanizing World: Finding Solutions for Cities in Africa, Asia and Latin America. Earthscan Publications, London, 470 p.

Heyman B. N., Davis C. and Krumpe P. F. (1991). An assessment of world wide disaster vulnerability. *Disaster Management*, **4**, 3–26.

Hoppe R. (1999). Policy analysis, science and politics: from 'speaking truth to power' to 'making sense together'. *J. Science and Public Policy*, **26**(3), 201–210.

Husserl E. (1900, 1901/1913). Logische Untersuchungen. Niemeyer, Halle. Available in English as Findlay J. N. (transl.) (1970) Logical Investigations. Routledge, London.

ICE (2001). Learning to live with rivers. Final report of the Presidential Commission to review the technical aspects of flood risk management in England and Wales, November 2001.

IFRC/RC (1999). World Disasters Report 1999, International Federation of the Red Cross and Red Crescent, Edigroup, Switzerland.

Kaplan S. and Garrick B. J. (1981). On the Quantitative Definition of Risk. *Risk Analysis*, **1**, 11–27.

Kelly J. S. (1976). Arrow Impossibility Theorems. Academic, New York.

Latour B. (1991). Nous n'avons jamais été modernes. La Découverte, Paris. Available in English as Porter C. (transl.) (1993) We Have Never Been Modern. Harvard University Press, Harvard and Harvester, London.

Miller D. (1999). Principles of Social Justice. Harvard University, Cambridge, Mass.

Moran D. (2000). Introduction to Phenomenology. Routledge, London.

Munich Re (2002). "Topics, Annual review: natural catastrophes 2002", Munich Re Group, http://www.munichre.com/

National Research Council (1994). Committee on Risk Assessment of Hazardous Air Pollutants, Science and Judgment in Risk Assessment. National Academy Press, Washington, DC.

NEF (2006). The Happy Planet Index: An Index of Human Well-being and Environmental Impact. New Economics Foundation, London. Available at: www.happyplanetindex.org.

Pelling M. (2003). The Vulnerability of Cities: Natural Disaster and Social Resilience. Earthscan, London.

Rashid S. F. (2000). The Urban Poor in Dhaka City: Their Struggles and Coping Strategies during the Floods of 1998, Disasters, Volume 24, Number 3, September 2000. Blackwell Publishing, Oxford, pp. 240–253(14).

Rawls J. (1971). A Theory of Justice. Harvard University, Harvard, MA.

Rokeach M. (1973). The Nature of Human Values. Free Press, New York, US.

Saunders S. and Munro D. (2000). The construction and validation of a consumer orientation questionnaire (SCOI) designed to measure Fromm's (1955) 'marketing character' in Australia. *Social Behaviour and Personality*, **28**, 219–240, Psychology Press, Taylor and Francis Group, US.

Sayers P. B., Hall J. W. and Meadowcroft I. C. (2002). Towards risk-based flood hazard management in the UK. *Proceedings of ICE*, May 2002, pp. 36–42.

Schultz P. W., Gouveia V. V., Cameron L. D., Tankha G., Schmuck P. and Franek M. (2005). Values and their relationship to environmental concern and conservation behaviour. *Journal of Cross-cultural Psychology*, **36**, 457–475, SAGE Publications, US.

Schwartz S. H. (1992). Universals in the content and structure of values: Theory and empirical tests in 20 countries. In: Advances in Experimental Social Psychology (Vol. 25). M. P. Zanna (ed.), Academic Press, New York, US, pp.1–65.

Spiegelberg H. (1982). 3rd edn, The Phenomenological Movement: A Historical Introduction (Vols 1 & 2). Martinus Nijhoff Publishers, The Hague, Netherlands.

Talberth J., Cobb C. and Slattery N. (2006). The Genuine Progress Indicator 2006: A Tool for Sustainable Development. Redefining Progress, Oakland, US. Available at: www.rprogress.org.

Turner II B. L., Kasperson R. E., Matson P. A., McCarthy J. J., Corell R. W., Christensen L., Eckley N., Kasperson J. X., Luers A., Martello M. L., Polsky C., Pulsipher A. and Schiller A. (2003). A framework for vulnerability analysis in sustainability science. *Proceedings of the National Academy of Sciences of the United States of America*, **100**, 8074–8079.

Vis M., Klijn F., De Bruijn K. M. and van Buuren M. (2003). Resilience strategies for flood risk management in the Netherlands. *Intl. J. River Basin Management*, **1**(1), 33–40.

Wats M. J. and Bohle H. G. (1993). The space of vulnerability: the casual structure of hunger and famine. *Progress in Human Geography*, **17**, 43–67.

Wiest R. E., Mocellin J. S. P. and Motsisi D. T. (1994). The Needs of Women in Disasters and Emergencies, Report prepared for the Disaster Management Programme of the UNDP and Office of the UN Disaster Relief Coordinator, Winnipeg, Manitoba. http://www.apu.ac.uk/geography/gdn/resources/papers.html.

Wisner B., Blaikie P., Cannon T. and Davis I. (2004). 2nd edn, At risk – natural hazards, people's vulnerability and disasters. Taylor and Francis Group.

WMO (2004). Integrated Flood Management Concept Paper, APFM Technical Document No. 1. Second Edition. http://www.apfm.info/pdf/concept_paper_e.pdf

WMO (2008). Urban Flood Risk Management, APFM Technical Document No. 1. Second Edition. http://www.apfm.info/pdf/ifm_tools/Tools_Urban_Flood_Risk_Management.pdf

WMO (2009). Integrated Flood Management Concept Paper. http://www.apfm.info/pdf/concept_ paper_e.pdf

PART II

Abbott M. B. (1991). Hydroinformatics: Information Technology and the Aquatic Environment. Avebury, Aldershot, UK and Brookfield, USA.

Abbott M. B. (1999). Forchheimer and schoklitsch: a postmodern retrospection. In: *Memorial Symposium P. Forchheimer and A. Schoklitsch, Proc. 28th IAHR Congress, Graz*, W. H. Graf (ed.), Also available at: http://www.knowledge-engineering.org.

Abbott M. B. (2000). The gender issue in hydroinformatics, or Orpheus in the Underworld. *J. Hydroinf.*, **3**(2), 87–104. Also available in a slightly extended form at: http://www.knowledge-engineering.org.

Abbott M. B. and Jonoski A. (2001). The democratisation of decision-making processes in the water sector, I and II. *Journal of Hydroinformatics, 2001*, **3**(1), 22–48.

Abbott M. B. (2002). On definitions. *J. Hydroinf.*, **4**(2), electronic version only. Also available at: http://www.knowledge-engineering.org.

Abbott M. B. (2004). European Industrial Leadership. European Institute for Industrial Leadership and Knowledge Engineering, Brussels.

Abbott M. B. (2007). Stakeholder participation in creating infrastructure. *New Civil Engineer*, **160**(1), 26–32, [Awarded the Telford Gold Medal of the Institution of Civil Engineers in 2008].

Abbott M. B. (2008a). Towards a hydroinformatics for Asia (I). In: *Proc. First International Conference on Human Resources and the Country's Structure*, Islamic Azad University, Esfahan, Iran.

Abbott M. B. (2008b). Knowledge transfer and sharing at UNESCO-IHE: the next 25 years. In: Water for a Changing World: Enhancing Local Knowledge and Capacity, N. Dickinson and G. Alaerts (eds), UNESCO-IHE Institute for Water Education, Delft, pp. 287–296.

Abbott M. B. (2009). Towards a Hydroinformatics for Asia II. *Proc. World City Water Forum 2009*, Incheon, Korea.

Abbott M. B. and Basco D. R. (1989). Computational Fluid Dynamics: An Introduction for Engineers. Longman, London and Wiley, New York.

Abbott M. B., Chambers J. and Hing C. C. (eds) (2005). From a Knowledge Society to an Understanding Society. European Institute for Industrial Leadership and Knowledge Engineering, Brussels.

Abbott M. B. and Jonoski A. (2001). The democratisation of decision-making processes in the water sector, I and II. *J. Hydroinf.*, **3**(1), 22–48.

Abbott M. B., Larsen J. and Tao J. (1985). Modelling circulation in two-dimensional flows; Part 1: the accumulation of the evidence. Part 2: a reconciliation. *J. Hydraul. Res.*, **23**(309–326), 397–420.

Abbott M. B. and Minns A. W. (1998). 2nd edn, Computational Hydraulics. Ashgate, Aldershot, UK, and Brookfield, USA.

Abbott M. B. and Vojinović Z. (2009). Applications of numerical modelling in hydroinformatics. *Journal of Hydroinformatics*, **11**(3–4), 308–319.

Abbott M. B. and Vojinović Z. (2012). The Meaning of Colour in Web-based Hydroinformatics Environments. Hamburg Hydroinformatics Conference.

Abbott M. B. and Vojinović Z. (2012). The Hydroinformatics of the Three Economies. Hamburg Hydroinformatics Conference.

Abelson R. (1967). The Responsibility of Intellectuals: A Special Supplement from the February 23, 1967 issue of the New York Review of Books.

Alexander D. (1993). Natural Disasters. Chapman and Hall, New York.

Ababarnel H. D. I. (1996). Analysis of Observed Chaotic Data. Springer, New York.

Alterman E. (2005). Found in the flood. *The Nation*, **281**(9), 33.

Anderson E. (1988). Values, risks and market norms. *Philosophy & Public Affairs*, **17**(1), 54–65.

Aristotle (1980[350 BC]). Nicomachean Ethics (translated by Ross W. D.). Clarendon Press, Oxford.

Augustine (398//e.g. 1961). (trans. Pine-Coffin R. S.) Confessions, Penguin, Harmondsworth.

Bakunin M. (1980). Bakunin on Anarchism. In: Sam Dolgoff (ed.), 2nd edn, Black Rose Books, Montréal.

Barth K. (1938–1950). Die Kirchliche, Evangelischer, Zollikon-Zurich. Bromiley, G. W. and Torrance, T. F. Church Dogmatics (transl.)(1961), Clark, Edinburgh.

Barry B. (1989). The Ethics of Resource Depletion. Democracy, Power and Justice: Essays in Political Theory. Clarendon Press, Oxford, pp. 259–273.

Barth K. (1938–1950). Die Kirchliche Dogmatik. Evangelischer, Zollikon-Zurich. Available in English as Bromiley G. W. and Torrance T. F. (transl.) (1961) Church Dogmatics Clark, Edinburgh.

Beauchamp T. L. and Childress J. F. (2001[1994]). 5th edn, Principles of Biomedical Ethics. Oxford University Press, New York.

Bentham J. (1948[1789]). An Introduction to the Principles of Moral and Legislation. Hafner Press, Oxford.

Birkmann J. (ed.) (2006). Measuring Vulnerability to Natural Hazards. Towards Disaster Resilient Societies, United Nations Publications, New York.

Blaikie P., Cannon T., Davis I. and Wisner B. (1994). At risk: Natural Hazards, People's Vulnerability and Disasters. Routledge, London.

Bogardi J. J. and Birkman J. (2004). Vulnerability assessment: the first step towards sustainable risk reduction. In: Disasters and Society – From Hazard Assessment to Risk Reduction, D. Malzahn and T. Plapp (eds), Logos Verlag Berlin, Berlin, pp. 75–82.

Böhner P. and Gilson E. (1954). 3rd edn, Christliche Philosphie von ihren Anfängen bis Nikolaus von Cues. Schöningh, Paderborn.

Bohle H. (2001). Vulnerability and criticality: Perspectives from social geography. *Newsletter, International Human Dimensions Programme on Global Environmental Change*, **02**(1), 3–5.

Bollin C., Cárdenas C., Hahn H. and Vatsa K. S. (2003). Natural Disaster Network; Disaster Risk Management by Communities and Local Governments. Inter-American Development Bank, Washington, DC, http://www.iadb.org/sds/doc/GTZ%2DStudyFinal.pdf

Bookchin M. (1980). Toward an Ecological Society. Black Rose Books, Montreal.

Bortoft H. (1997, 4th edition, 2010). The Wholeness of Nature: Goethe's Way of Science. Lindisfarne Books and Floris Books, Edinburgh.

Brentano F. V. (1930). Wahrheit und Evidenz: Erkenntnistheoretische Abhandlungen und Briefe, ausgewählt erläutert und eingeleitet von Oskar Kraus. Meiner, Leipzig.

Brillouin L. (1959). La science et la théorie de l'information. Masson, Paris. Second edition in English: Science and Information Theory (2004). Dover Publications, Mineola, NY.

Bullard R. (2000). 3rd edn, Dumping in Dixie. Westview press, Boulder, CO.

Bullard R. (ed.) (2005). The Quest for Environmental Justice: Human Rights and the Politics of pollution. Sierra Club Books, San Francisco, CA.

Burton I. and Kates R. W. (1964). The Perception of Natural Hazards in Resource Management. *Natural Resources Journal*, **3**(3), 412–441; also in The Economics of Natural Hazards, H. Kunreuther and A. Z. Rose (eds.) (2004). Edward Elgar, Cheltenham, UK, **1**, 53–82.

Cantor G. (1895, 1897). Math. Annalen xivi, 481–512; xiix, 207–246. Available in English as Jourdain P. E. B. (transl.) (1915). Contributions to the Founding of the Theory of Transfinite Numbers. Dover, New York.

Carey J. (2002). Etymology and Time. *Temenos Academy Review*, **5**, 117–136, London.

Carey J. (2007). Editorial. *Temenos Academy Review*, **10**, 7–13.

Cardona O. D. (1999). Environmental management and disaster prevention: Holistic risk assessment and management. In: Natural Disaster Management, I. Ingleton (ed.), Tudor Rose, London.

Cardona O. D. (2001a). Estimación Holística del Riesgo Sísmico utilizando Sistemas Dinámicos Complejos. Universidad Politécnica de Cataluña, Barcelona, http://www.desenredando.org/public/varios/2001/ehrisusd/index.html

Chan G. K. Y. (2008). The relevance and value of confucianism in contemporary business ethics. *Journal of Business Ethics*, **77**(3), 347–360.

Chavis Rev and Benjamin F. Jr. (1993). 'Forward'. In: Confronting environmental racism: Voices from the grassroots. D. Robert and Bullard (ed.), South End Press, Boston, MA.

Cheetham T. (2002). Consuming Passions: The Feast, the Stars and the Science of the Balance. *Temenos Academy Review, 5*. Ashford, Kent.

Cheetham T. (2005). The Prophetic Tradition and the Battle for the Soul of the World. *Temenos Academy Review 8*, Ashford, Kent.

Chisholm R. (1954). Brentano and Meinong Studies. Rodopi, Amsterdam.

Chisholm R. (1966/1977/1989). Theory of knowledge. Prentice Hall, Englewood Cliffs, NJ.

Corbin H. (1958, 1964, 1974, 1986). L'Imagination créatrice dans le soufisme d'Ibn 'Arabî. Flammarion, Paris. Entre Lacs, Paris (2006). Creative Imagination in the Sutism of Ibn ' Arabî, Bloom L. (ed.) (1969), Princeton University Press, Princeton, New Jersey.

Copleston F. (1946, 1950, 1953). A History of Philosophy. *Image Books*, Doubleday, Garden city, NY.

Cranor C. F. (1990). Some moral issues in risk assessment. *Ethics*, **101**(1), 123–143. Princeton University Press, Princeton, NJ. Bollinger series XCI (1997), Princeton University Press, Princeton, NJ.

Cutter S. L. (1996). Vulnerability to environmental hazards. *Progress in Human Geography*, **20**(4), 529–539.

Davidson R. A. (1997). An Urban Earthquake Disaster Risk Index, The John A. Blume Earthquake Engineering Center, Stanford University, Report no. 121, Blume Center, Stanford, California.

Dean R. G. (2002). Beach Nourishment: Theory and Practice. World Scientific, New Jersey.

Depelsenaire Y. (2004). Une analyse avec Dieu. La Lettre Volée, Brussels.

Derrida J. (1967). La voix et le phénomène. Presses Universitaires de France, Paris. Available in English as Allison D. B. and Garver N. (transl.) (1973) 5th printing 1989 Speech and Phenomena and Other Essays on Husserl's Theory of Signs.

Dilthey W. (1976). Selected Writings. H. P. Rickman (ed. and transl.), Cambridge University Press, Cambridge.

Dixon N. F. (1976). The Psychology of Military Incompetence. Jonathan Cape, London.

Donington R. (1963). Wagner's Ring and Its Symbols. Faber, London.

Drury M. O. C. (1981). Some notes on conversations with Wittgenstein. In: Ludwig Wittgenstein, Personal Recollections R. Rhees (ed.), Blackwell, Oxford.

Duhem P. and Brenner A. (1997). L'aube du savoir: Épitomé du Système du monde, Paris, Hermann.

Durning D. (1999). The transition from traditional to postpositivist policy analysis: a role for Q methodology. *J. Pol. Anal. Mngmnt.*, **18**(3), 389–410.

Dworkin R. M. (2000). Sovereign Virtue: Equality in Theory and Practice. Cambridge, Harvard University Press.

Evers M. (2008). Decision Support Systems for Integrated River Basin Management: Requirements for Appropriate Tools and Structures for a Comprehensive Planning Approach. Shaker, Aachen.

FAO (2003). State of the World's Forests. Food and Agriculture Organization of the United Nations, Rome.

FAO and CIFOR (2005). Forests and floods – Drowning in fiction or thriving on facts RAP Publication 2005/03 and Forest Perspectives 2. Food and Agriculture Organization of the United Nations, Bangkok, and Center for International Forestry Research, Bogor.

Findley J. N. (1963/1995). 2nd edn, Meinong's Theory of Objects and Values. Gregg Revivals and Ashgate, Aldershot, UK, and Brookfield, USA.

Findlay J. N. (1977). Foreword to Hegel, G.W.F. Phenomenology of Spirit. (transl. Miller A. V.). Oxford University Press, Oxford.

Flew A. (1993). Social Justice Isn't Any Kind of Justice. Libertarian Alliance, London.

Foltz R. C. (2002). Iran's water crisis: Cultural, political, and ethical dimensions. *Journal of Agricultural & Environmental Ethics*, **15**(4), 357–380.

Foucault M. (1966). Les Mots et les Choses. Gallimard, Paris. The Order of Things (1970). Tavistock, London (1989) Routledge, London.

Foucault M. (1972–1977/1980). Power/Knowledge: Selected Interviews and Other Writings. In: C. Gordon (ed.), Pantheon, New York.

Fraser J. T., Laurence N. and Park D. (eds) (1978). The Study of Time, III. Springer, Heidelberg.

Freyd P. J. and Scedrov A. (1990). Categories, Allegories. North Holland, Amsterdam.

Gilson E. (1955/1980, 1982, 1985). History of Christian Philosophy in the Middle Ages. Sheed and Ward, London.

Granger G.-G. (1994). Sur l'idée de concept mathématique 'naturel'. In: Formes, Opérations, Objets, G.-G. Granger. (ed.) Vrin, Paris.

Hall J. W., Meadowcroft I. C., Sayers P. B. and Bramley M. E. (2003). Integrated Flood Risk Management in England and Wales. *Natural Hazards Review*, **4**(3), 126–135.

Hamed S. E. D. (1993). Seeing the environment through Islamic eyes: Application of Shariah to natural-resources planning and management. *Journal of Agricultural & Environmental Ethics*, **6**(2), 145–164.

Hansson S. O. (2003). Ethical criteria of risk acceptance. *Erkenntnis*, **59**(3), 291–309.

Hansson S. O. (2007). Philosophical Problems in Cost-Benefit Analysis. *Economics and Philosophy*, **23**(2), 163–183.

Hardoy J. E., Mitlin D. and Satterthwaite D. (2001). Environmental Problems in an Urbanizing World: Finding Solutions for Cities in Africa, Asia and Latin America. Earthscan Publications, London.

Hart L. (1970, 8th printing 1983). Liddell Hart's History of the second World War. Cassell, London.

Harvey H. (2009). The hydroinformatics platform. *Proc. World City Water Forum 2009*, Incheon, Korea.

Hegel G. W. F. (1807). Phänomenolgie des Geistes. Goebhardt, Bamberg and Würzburg. 1970 with an afterword by Lukács, G. Ulstein, Frankfurt am Main, Berlin and Vienna. / 1977, Hegel's Phenomenology of the Spirit, (transl. Miller A. V. with a text analysis foreword by Findlay J. N.). Oxford University Press, Oxford.

Heidegger M. (1927). Sein und Zeit. Niemeyer, Tübingen. Available in English as Macquarrie J. and Robinson E. (transl.) (1962) Being and Time. Blackwell, London.

Heidegger M. (1963). Die Technik und Die Kehre, Niemyer, Tübingen (Tenth reprint: Cotta, 2002). Available in English as Lovitt W. (transl.) (1977) The Question Concerning Technology and Other Essays. Harper, New York.

Heidegger M. (1967). Der Europäische Nihilismus. Günther Neske, Pfulligen.

Hermans H. J. M. (1996). Voicing the self: from information processing to dialogical interchange. *Psych. Bull.*, **199**(1), 31–50.

Heyman B. N. Davis C. and Krumpe P. F. (1991). An assessment of world wide disaster vulnerability. *Disaster Management*, **4**, 3–26.

Hoppe R. (1999). Policy analysis, science and politics: from 'speaking truth to power' to 'making sense together'. *J. Science and Public Policy*, **26**(3), 201–210.

Huntley H. E. (1970). The Divine Proportion: A Study in Mathematical Beauty. Dover, New York.

Husserl E. (1945). Erfahrung und Urteil: Untersuchung zur Geneologie der Logik. Claessen and Govaerts, Hamburg. Available in English as Churchill J. S. and Ameriks K. (transl.) (1973) Experience and Judgement: Investigations in a Geneology of Logic. Routledge, London.

Husserl E. (1945). Erfahrung und Urteil: Untersuchung zur Geneologie der Logik. Claessen and Govaerts, Hamburg. In: Churchill J. S. and Ameriks K. (trans.) (1973), Experience and Judgement: Investigations in a Geneology of Logic. Routledge, London.

Hume D. (1739–1740). A Treatise of Human Nature. D. F. Norton and M. J. Norton (ed.) (2000), Oxford University Press, Oxford.

Hume D. (1748). An Enquiry Concerning Human Understanding. T. L. Beauchamp. (ed.) (1999) Oxford University Press, Oxford.

Ibn 'Arabi 1194/590H //1997. Divine Governance of the Human Kingdom, Interpreted by Tosun Bayrak. Fons Vitæ, Louisville.

ICE (2001). Learning to live with rivers. Final report of the Presidential Commission to review the technical aspects of flood risk management in England and Wales, November 2001.

IFRC/RC (1999). World Disasters Report 1999. International Federation of the Red Cross and Red Crescent. Edigroup, Switzerland.

James W. (1890/1902). The Principles of Psychology. Vol. 1. Macmillan, London.

James W. (1976). Essays in Radical Empiricism (collection of articles and correspondence from 1889 to 1909. edited by F. Bowers). Harvard University Press, Cambridge, MA.

Jaspers K. and Bultmann R. (1954). Die Frage der Entmythologisierung. Piper, Munich.

John Paul H. H. II. (1994). Crossing the Threshold of Hope. Jonathan Cape, London.

John Paul H. H. II. (1998). Fides et Ratio (Faith and Reason). Libreria Editrice Vaticana, Vatican City. Veritas, Dublin.

Jonoski A. (2002). Hydroinformatics as Sociotechnology: Promoting Individual Stakeholder Participation by Using Network-Distributed Decision Support Systems. Swets and Zeitlinger, Lisse, UNESCO-IHE Institute for Water Education, Delft/Balkema, Rotterdam.

Jung C. G. (1944). Psychologie und Alchemie. Rascher. Zurich // 1952 Psychology and Alchemy. Routledge, London.

Kant I. (1786). Groundwork for the metaphisics of morals (original German title: Grundlegung zur Metaphysik der Sitten). Oxford, Oxford University Press.

Kant I. (1787/1924/1979). Kritik der reinen Vernunft. Facsimile reproduction of the 1924 Kehrbachsche Ausgabe, Reclam, Leipzig // 1855 Critique of Pure Reason. (trans. Meiklejohn J. M. D.) /1934, (A. D. Lindsay(ed.)) Everyman, London / see also, as a more precise but less mellifluous translation, 1929/1987, Immanuel Kant's Critique of Pure Reason (transl. Kemp Smith, N.), Macmillan, London.

Keeble B. (2004). An interview with Kathleen Raine. *Temenos Academy Review*, **7**, 27–58.

Kelly J. S. (1976). Arrow Impossibility Theorems. Academic, New York.

Kierkegaard S. Aa. (1843/1976). Frygt og Bæven. Gyldendal, Copenhagen (reprinted from Danish 5th edition of the Collected Works, Gyldendal, Copenhagen) // 1983 Fear and Trembling (ed. and transl. H. V. Hong and E. H. Hong). Princeton University Press, Princeton. NJ / but see also, as a valuable alternative translation, 1985 Fear and Trembling (ed. and transl. Hannay, A.). Penguin, Harmondsworth.

Kierkegaard S. Aa. (1844/185I/1960). Begrebet Angest. (reprinted from the Danish 5th edition of the Collected Works, Introduction and Notes by Sørensen, V., Gyldendal, Copenhagen // 1980. The Concept of Anxiety, (transl. Thomte R. and Anderson B.). Princeton University Press, Princeton, NJ.

Kierkegaard S. Aa. (1849/1920–31//1989/1965). Sygdommen til Døden, Reitzel, Copenhagen / 1920–31, Samlede Værker. (ed. A. B. Drachman, J. L. Heiberg, H. O. Lange Gyldendal Copenhagen) / 1965, Sygdommen til Døden, Gyldendal, Copenhagen // 1989, The Sickness unto Death (ed. and tranl. Hong, H.V., and Hong, E.H.),

Princeton Univ., Princeton, NJ; (but see also, for comparison, 1985, The Sickness unto Death, transl. Hannay, A., Penguin, London).
Kirkman R. (2010). The Ethics of Metropolitan Growth: The future of the Built Environment. Continuum, London.
Klinkenberg J.-M. (1996). Précis de sémiotique générale. *De Boeck et Larcier*, Seuil, Paris.
Koehn D. (1999). What can Eastern philosophy teach us about business ethics? *Journal of Business Ethics*, **19**(1), 71–79.
Kolmogorov A. N. and Fomin S. V. (1953). Elements of the Theory of Functions and Functional Analysis vol. 1: Metric and Normed Spaces; 1961 vol. 2: Measure, the Lebesgue Integral, Hilbert Space. Graylock, Albany and Academic, New York.
Kuhn T. S. (1970). 2nd edn, The Structure of Scientific Revolutions. University of Chicago Press, Chicago.
Kuusela O. (2011). Key Terms in Ethics. Continuum, London.
Lacan J. (1964/1973). Les quatre concepts fondamentaux de la psychanalyse. In: J. A. Miller (ed.), Seuil, Paris. Available in English as Sheridan A. (transl.) (1977/2004) The Four Fundamental Concepts of Psycho-Analysis. Hogarth, London.
Lacan J. (1959–1960/1986/1992/2008). Le Séminaire, Livre VII. de Seuil, Paris // The Ethics of Psychoanalysis: The Seminar of Jacques Lacan: Book VII. Norton, New York (1992) and Routledge, London (2008).
Latour B. (1991). Nous n'avons jamais été modernes. La Découverte, Paris. Available in English as Porter, C. (transl.) (1993) We Have Never Been Modern. Harvard University Press, Harvard and Harvester, London.
Lave L. B. (1984). Eight frameworks for regulation. Technological Risk Assessment. Ricci P. F. Sagan L. A. and Whipple C. G. The Hague, Martinus Nijhoff.
Law J. (ed.) (1991). Power, Action and Belief. A New Sociology of Knowledge, Routledge, London.
Levinas E. (1995/2006). Altérité et transcendence. Fata Morgana, Paris.
Lévi-Strauss C. (1958). Anthroplogie structurale. Plon, Paris.
Locke J. (1960[1690]). Two Treatises of Government. Dutton, New York.
Lovelock J. E. and Margulis L. (1974). Atmospheric homeostasis by and for the biosphere – The Gaia hypothesis. *Tellus*, **26**(1), 2–10.
Machiavelli N. (1516+/−/1929). Tutte le opere storiche e letterarie di Niccolò Machievelli. In: Mazzoni G. and Casella M. (ed.), Barbèra, Florence. Available in English as Bull G. Prelude by Mussolini B (transl.) (1970) The Prince, London.
MacIntyre A. (1984). After virtue. Notre Dame University Press, Notre Dame.
Manin Yu. I. (1977). A Course in Mathematical Logic, (transl. Koblitz N.). Springer, Heidelberg.
Marin L. (1997). Pascal et Port-Royal. Collège International de Philosophie, Presses Universitaires de France, Paris.
Martin M. W. and Schinzinger R. (2005). Ethics in Engineering. McGraw-Hill, New York.
Marx C. (1872). Capital, Vol. I, Preface to the French Edition, March 18, 1872, IP Ed.
Meinong A. V. (1913). Abhandlungen zur Erkentnistheorie und Gegenstandstheorie. Barth, Leipzig.
Michel S. and Beuret P. (2008). La Chinafrique. Grasset. Paris.
Mill J. S. (1859). On Liberty. John W. Parker and Son, London.
Mill J. S. (1979[1863]). Utilitarianism. Collins, London.
Miller D. (1999). Principles of Social Justice. Harvard University Press, Cambridge, MA.
Monot J. (1970). Le hasard et la nécessité. Essai sur la philosophie naturelle de la biologie moderne, éditions du Seuil, Paris.
Nozick R. (1974). Anarchy, State, and Utopia. Basic Books, New York.
Nussbaum M. (1995). Poetic Justice: The Literary Imagination and Public Life. Beacon Press, Boston.
Pascal B. (1670/1953). Les pensées. In: L. Brunschvicg (ed.) Vrin, Paris.
Pascal B. (1670/1962/2003). Les pensées. In: L. Lafuma (ed.) de Seuil, Paris.
Pascal B. (1670/1977/2004). Les pensées. in: M. Le Guern (ed.) Gallimard, Paris.
Pareto V. (1966[1909]). 4th edn, Manuel d'économie politique. (4th edn), Geneva, Droz.
Proudhon J. (1994) "What is Property", Cambridge University Press, Cambridge, England.
Rashid S. F. (2000). The Urban Poor in Dhaka City: Their Struggles and Coping Strategies during the Floods of 1998. *Disasters*, **24**(3), 240–253(14), Blackwell Publishing.
Rawls J. (1971). A Theory of Justice. Harvard University Press, Harvard, MA.

Rawls J. (1999 [1971]). A Theory of Justice. The Belknap Press of Harvard University Press, Cambridge, MA.

Rawls J. (2001). Justice as Fairness: A Restatement. The Belknap Press of Harvard University Press, Cambridge, MA.

Read H. (1949). Existentialism. Marxism and Anarchism, London.

Regan T. (1983). The case for animal rights. University of California Press, Berkeley, CA.

Rhees R. (1982). Wittgenstein on language and ritual. In: Wittgenstein and His Times, McGuinness B. (ed.), Blackwell, Oxford.

Riley D. (2008). Engineering and Social Justice. Morgan & Claypool, San Rafael, CA.

Roy A. (1999). The Cost of Living. Flamingo/Harper Collins, London.

Sagoff M. (1988). Some problems with environmental economics. *Environmental Ethics*, **10**, 55–74.

Salter J. and Howsam P. (2002). The Precautionary Principle and the law on risk. http://www.cranfield.ac.uk/cuplu/papers.htm.

Sands P. (2000). Principles of International Environmental Law. Vol. I. Manchester University Press, Manchester.

Sartre J.-P. (1943). L'Être et le néant. Gallimard, Paris. Available in English as Barnes H. E. (transl.) (1958), Being and Nothingness, Methuen, London.

Sayers P. B., Hall J. W. and Meadowcroft I. C. (2002). Towards risk-based flood hazard management in the UK. *Proceedings of ICE*, May 2002.

Shrader-Frechette K. S. (1991). Risk and Rationality: Philosophical Foundations for Populist Reform. University of California Press, Berkeley.

Schopenhauer A. (1813//1974). The Fourfold Root of the Principle of Sufficient Reason. (trans. E. F. J. Payne), Open Court, LaSalle, Illinois.

Simons P. (1996). Meinong's theory of sense and reference. In: Meinong und die Gegenstandstheorie: Meinong and the Theory of Objects, R. Haller (ed.), Rodopi, Amsterdam.

Shiva V. (1988). Staying Alive: Women, Ecology and Development. Kali for Women: New Delhi.

The Newcomen Society for the Study of the History of Engineering and Technology.

Thorkilsen M. and Dynesen C. (2001). An owner's view of hydroinformatics: its role in realising the bridge and tunnel connection between Denmark and Sweden. *J. Hydroinf.*, **3**(2), 105–136.

Tremmel J. C. (ed.) (2006). Handbook of Intergenerational Justice. Edward Elgar Publishing, Cheltenham.

Troisfontaines R. (1953/1968). De l'éxistance à l'étre: La Philosophie de Gabriel Marcel. Nauwelaerts, Louvain.

Tumwesigye E. (2009). The Water Webframe Modelling Framework and Business Context in the Development towards the Fifth Generation of Hydroinformatics. Doctoral Thesis, National University of Ireland: University College, Cork.

Turner B. L. II., Kasperson R. E., Matson P. A., McCarthy J. J., Corell R. W., Christensen L., Eckley N., Kasperson J. X., Luers A., Martello M. L., Polsky C., Pulsipher A. and Schiller A. (2003). A framework for vulnerability analysis in sustainability science. *Proceedings of the National Academy of Sciences of the United States of America*, **100**, 8074–8079.

Van de Poel I. R. and Royakkers L. M. M. (2007). The ethical cycle. *Journal of Business Ethics*, **71**(1), 1–13.

Van de Poel I. R. and Royakkers L. M. M. (2011). Ethics, Technology, and Engineering: An Introduction. West-Sussex, Wiley-Blackwell.

van Velsen J. F. C. (2000). Relativity, universality and peaceful coexistence. *Archiv für Rechts-und Sozialphilosophie*, **86(1)**, 88–108.

Vis M., Klijn F. De Bruijn K. M. and van Buuren M. (2003). Resilience strategies for flood risk management in the Netherlands. *Intl. J. River Basin Management*, **1**(1), 33–40.

Vojinović Z. and Abbott M. B. (2012). Ethics as a means to improve flood risk analysis. *Hamburg Hydroinformatics Conference*.

Wahl J. (1959). Les philosophies de l'existence, Armand Colin, Paris. Available in English as Lory, F. M. (transl.) (1969), The Philosophies of Existence. Routledge, London.

Wats M. J. and Bohle H. G. (1993). The space of vulnerability: the casual structure of hunger and famine. *Progress in Human Geography*, **17**, 43–67.

Warren K. (2000). Ecofeminist philosophy: A Western perspective on what it is and why it matters. Rowman & Littlefield, Lanham, MD.

WCED (1987). Our Common Future (Brundtland report). Oxford University Press, Oxford.

Wersal L. (1995). Islam and environmental ethics: Tradition responds to contemporary challenges. *Zygon*, **30**(3), 451–459.

Whitbeck C. (1998). Ethics in Engineering Practice and Research. Cambridge University Press, Cambridge.

Wiest R. E., Mocellin J. S. P. and Motsisi D. T. (1994). The Needs of Women in Disasters and Emergencies. *Report prepared for the Disaster Management Programme of the UNDP and Office of the UN Disaster Relief Coordinator*. Winnipeg, Manitoba. http://www.apu.ac.uk/geography/gdn/resources/papers.html.

Williams R. (1978/1990/2002). The Wound of Knowledge. Darton-Longman-Todd, London.

Winch P. (1967). The Encyclopaedia of Philosophy. P. Edwards (ed.), **6**, 44–46, Collier-Macmillan, London.

Wisner B., Blaikie P., Cannon T. and Davis I. (2004). 2nd edn, At Risk – Natural Hazards, People's Vulnerability and Disasters. Taylor and Francis Group, Routledge, London.

Witkovsky M. S. (ed.) (2011). Avant Guarde Art in Everyday Life: Early twentieth-century European modernism. Art University of Chicago and Yale University Press.

Wittgenstein L. (1953). 3rd edn, Philosophical Investigations, trans. Anscombe G. E. M., Rees R. and von Wright G. H., 1958, Oxford University Press.

WMO (2004). Integrated Flood Management Concept Paper. *APFM Technical Document No. 1, Second Edition*. http://www.apfm.info/pdf/concept_paper_e.pdf

WMO (2008). Urban Flood Risk Management. *APFM Technical Document No. 1, Second Edition*. http://www.apfm.info/pdf/ifm_tools/Tools_Urban_Flood_Risk_Management.pdf

WMO (2009a). Integrated Flood Management Concept Paper. Available at: http://www.apfm.info/pdf/concept_paper_e.pdf

WMO (2009b). Risk Sharing in Flood Management, A Tool for Integrated Flood Management, Associated Programme on Flood Management. Available at: http://www.apfm.info/pdf/ifm_tools/Tools_Risk_Sharing_in_FM.pdf

Wolf C. (2003). Intergenerational justice. In: A Companion to Applied Ethics, Frey R. G. and Wellman C. H., Blackwell Publishing, Malden.

Wong P. H. (forthcoming). Dao, harmony and personhood: Towards a confucian ethics of technology. Philosophy & Technology.

Woodcock G. (1974). Anarchism, Encylopædia Britannica. Vol. 1, 808–813.

Woodcock G. (1967). Anarchism, Encyclopedia of Philosophy. 111–115.

Yao Bin (2008). Transparency, It's the Law. *Beijing Review*, **51**(10), 28–29.

Zandvoort H. (2008). Risk zoning and risk decision making. *International Journal of Risk Assessment and Management* **8**(1–2), 3–18.

PART III

Abarbanel H., Brown R., Sidorowich J. and Tsimring L. (1993). The analysis of observed chaotic data in physical systems. *Reviews of Modern Physics*, **65**(4), 1331–1392.

Abbott M. B. (1966). An Introduction to the Method of Characteristics. Thames & Hudson, London and American Elsevier, New York.

Abbott M. B. and Ionescu F. (1967). On the numerical computation of nearly horizontal flows. *Journal of Hydraulic Research*, **5**(2), 97–117.

Abbott M. B. (1980). Computational Hydraulics - Elements of the Theory of Free Surface Flows. Pitman Publishing Ltd, London, UK.

Abbott M. B., Bathurst J. C., Cunge J. A., O'Connell P. E. and Rasmussen J. (1986). An introduction to the European hydrological system - Systeme Hydrologique Europeen, "SHE", 1: History and philosophy of a physically-based, distributed modelling system. *Journal of Hydrology*, **87**, 45–49.

Abbott M. B. and Minns A. (1998). Computational Hydraulics. 1st edn, Pitman, London/Longman, London, and Wiley, New York/2nd edition, Ashgate, Aldershot, UK, and Brookfield, USA.

Abdullah A. F., Vojinović Z. and Price R. K. (2011a). Improved Methodology for Processing Raw LIDAR Data to Support Urban Flood Modelling - Accounting for Elevated Roads and Bridges. *Journal of Hydroinformatics, IWA Publishing 2011*, doi:10.2166/hydro.2011.009.

Abdullah A. F., Vojinović Z. and Price R. K. (2011b). A Methodology for Processing Raw LIDAR Data to Support 1D/2D Urban Flood Modelling Framework. *Journal of Hydroinformatics, IWA Publishing 2011*, doi:10.2166/hydro.2011.089.

Ackers P. (1993). Sediment transport in open channels: Ackers and White update. *Proceedings of the Institution of Civil Engineers: Water, Maritime and Energy*, **101**(4), 247–249.

Ackers P. and White R. W. (1973). Sediment transport: new approach and analysis. *J. Hyd. Div. ASCE*, **99**(HY11), 204–254.

AIAA (1995). Assessment of Wind Tunnel Data Uncertainty. American Institute of Aeronautics and Astronautics, Reston, USA. S-071.

Alcrudo F. (2004) Mathematical modelling techniques for flood propagation in urban areas. *Proceedings Ist IMPACT Project* WorkshopProject, Wallingford, UK, p. 22.

Allitt R., Foreman H., Djordjević S. and Chen A. S. (2009). WM/17 – UKWIR/FRMRC Integrated Urban Drainage Demonstration Projects – Cowes, Isle of Wight, Phase 3.

Andréassian V., Oddos A., Michel C., Anctil F., Perrin C. and Loumagne C. (2004). Impact of spatial aggregation of inputs and parameters on the efficiency of rainfall-runoff models: A theoretical study using chimera watersheds. *Water Resour. Res.*, **40**, W05209, p. 9, doi:10.1029/2003WR002854.

ANSI/ASME (1985). Measurement Uncertainty: Part 1, Instruments and Apparatus. American National Standards Institute (ANSI)/American Society of Mechanical Engineers. Washington, DC. PTC 19.1-1998.

ASME (1998). Test Uncertainty: Instruments and Apparatus. American Society of Mechanical Engineers, New York. PTC 19.1-1998.

ATV (1992). Standards for the Dimensioning and Design of Stormwater Structures in Combined Sewers. ATV-A 128E. Sankt Augustin Gesellschaft zur Forderung der Abwassertechnik e.V. (GFA).

Bardossy A. and Duckstein L. (1995). Fuzzy rule-based modelling with applications to geophysical, biological, and engineering systems. CRC.

Batchelor G. (2000). An introduction to fluid dynamics. Cambridge University Press. In: Bernardo J. and Smith A. (2001). Bayesian theory. *Measurement Science and Technology*, **12**, 221–222.

Berggren K., Olofsson M., Viklander M. and Svensson G. (2007). Tools for measuring climate change impacts on urban drainage systems. In: *Proceedings of the 6th NOVATECH International Conference: Sustainable Techniques and Strategies in Urban Water Management*, Lyon, France, June, pp. 239–246.

Bhattacharya B. and Solomatine D. P. (2005). Neural networks and M5 model trees in modelling water level-discharge relationship. *Neurocomputing*, **63**, 381–396.

Bhave P. (1988). Calibrating water distribution network models. *Journal of Environmental Engineering*, **114**, 120.

Bilham E. G. (1936). Classification of heavy falls in short periods. *British Rainfall*, **1935**, HMSO.

Box G., Jenkins G. and Reinsel G. (1976). Time Series Analysis: Forecasting and Control. Holden-day, San Francisco.

Boonya-Aroonnet S. (2008). Applications of the Innovative Modelling of Urban Surface Flooding in The UK Case Studies. *11th International Conference on Urban Drainage*, Edinburgh, UK, 1–10.

Broomhead D. and Lowe D. (1988). Radial Basis Functions, Multi-Variable Functional Interpolation and Adaptive Networks. Royal Signals and Radar Establishment, Malvern, England.

Brown D. J., Spencer T. and Moeller I. (2007). Modelling storm surge flooding of an urban area with particular reference to modelling uncertainties: A case study of Canvey Island, UK. *Water Resources Research*, **43**, W06402.

BSI (1998). Structural Design of Buried Pipelines Under Various Conditions of Loading, General Requirements. BS EN 1295-1:1998. British Standards Institute, London.

BSI. (2008). Drain and Sewer Systems Outside Buildings, Performance Requirements. BS EN 752-2. British Standards Institute, London.

Butler D. and Davies J. (2000). Urban Drainage. Spon Press, London.

CEC (2000). Directive Concerning Water Framework - Integrated River Basin Management for Europe. 2000/60/EEC.

CIRIA C624 (2004). Development and Flood Risk: Guidance for the Construction Industry. Construction Industry, Research & Information Association (CIRIA), Published by CIRIA, Classic House, 174–180 Old Street, London EC1V 9BP, UK.

CIRIA C635 (2006). Designing for Exceedance in Urban Drainage–good practice, CIRIA Research Project 699. Published by CIRIA, Classic House, 174–180 Old Street, London EC1V 9BP, UK.

CIWEM (1996). Evaluating real time control for waste water systems, A managerial and technical guide. Report of RTC Steering Group. CIWEM, London.

Chaudhry M. (1987). 2nd edn, Applied Hydraulic Transients. Van Nostrand Reinhold Co., New York.

Chen A. S., Hsu M. H., Teng W. H., Huang C. J., Yeh S. H. and Lien W. Y. (2005). Establishing the Database of Inundation Potential in Taiwan, Natural Hazards, Springer.

Choi J. Y., Engel B. A. and Farnsworth R. L. (2005). Web-based GIS and spatial decision support system for watershed management. *Journal of Hydroinformatics*, **7**(3), 165–174.

Corucci L. and Masini A. (2011). A Study on Bathymetry Estimation From High Resolution Multi-Spectral Satellite Images. http://ieee.uniparthenope.it/chapter/_private/proc10/15.pdf. (accessed on 06 September 2011).

Cowpertwait P. (1991). Further developments of the Neyman-Scott clustered point process for modelling rainfall. *Water Resources Research*, **27**(7), 1431–1438.

Crabtree R. (2007). Sediments in sewers. *Water and Environment Journal*, **3**(6), 569–578.

Crank J. and Nicolson P. (1996). A practical method for numerical evaluation of solutions of partial differential equations of the heat-conduction type. *Advances in Computational Mathematics*, **6**(1), 207–226.

Crank J. and Nicolson P. (1947). A practical method for numerical evaluation of solutions of partial differential equations of the heat conduction type. *Proc. Camb. Phil. Soc.*, **43**, 50–67.

Cross H. (1936). Analysis of Flow in Networks of Conduits or Conductors. Bulletin No. 286, 1-29, University of Illinois Engg. Experiment station, III. University of Illinois, Urbana.

Crossley A. J. (1999). Accurate and efficient numerical solutions for the Saint Venant equations of open channel flow. PhD thesis, University of Nottingham.

Cunge J., Holly F. and Verwey A. (1980). Practical aspects of computational river hydraulics. Pitman Publishing Ltd., London, UK.

Cunge J. and Wegner M. (1964). Intégration numérique des équations d'écoulement de Barré de Saint Venant par un schéma implicite de différences finies. *La Houille Blanche*, **1**, 33–39.

Dee D. P. (1994). Guidelines for documenting the validity of computational modelling software. Special publication, International Association for Hydraulic Research.

Defra (2005a). Joint probability: Dependence mapping and best practice: Technical report on dependence mapping. Defra/Environment Agency R&D Technical Report FD2308/TR1 (also referenced as HR Wallingford Report SR 623).

Defra (2005b). Use of joint probability methods in flood management: A guide to best practice. Defra/Environment Agency R&D Technical Report FD2308/TR2 (also referenced as HR Wallingford Report SR 653).

Djordjevic S., Prodanovic D., Maksimovic C., Ivetic M. and Savic D. (2005). SIPSON - Simulation of Interaction between Pipe flow and Surface Overland flow in Networks. *Water Science and Technology*, **52**(5), 275–283.

Einfalt T., Denoeux T. and Jacquet G. (1990). A radar rainfall forecasting method designed for hydrological purposes. *Journal of Hydrology*, **114**, 229–244.

Entekhabi D. I., Rodriguez-Iturbe I. and Eagleson P. S. (1989). Probabilistic representation of the temporal rainfall process by a modified Neyman-Scott rectangular pulses model: Parameter estimation and validation. *Water Resources Research*, **25**(2), 295–302.

Environment Agency/Heriot-Watt University (2010). The DEM with 0.5m resolution Digital Terrain Model (no vegetation or buildings) created from LiDAR data collected on 13th August 2009 and provided by the Environment Agency (http://www.geomatics-group.co.uk).

Evans B., Chen A. S., Djordjević S. and Savić D. A. (2009). A Cellular Automata based approach to Generalising Digital Terrain Models for 2D flood modelling, 8th International Conference on Urban Drainage Modelling, Tokyo, Japan.

Evans B. (2010). A Multilayered Approach to Two-Dimensional Urban Flood Modelling. PhD thesis, University of Exeter.

Farmer J. and Sidorowich J. (1987). Predicting chaotic time series. *Physical Review Letters*, **59**(8), 845–848.

Fewtrell T. J., Bates P. D., Horitt M. and Hunter N. M. (2008). Evaluating the effect of scale in flood inundation modelling in urban environments. *Hydrological Processes*, **22**, 5107–5118.

Fleming G. (2001). Learning to Live with Rivers. Final Report of the Institution of Civil Engineers' Presidential Commission to Review the Technical Aspects of Flood Risk Management in England and Wales. Institution of Civil Engineers, London.

Fraser A. and Swinney H. (1986). Independent coordinates for strange attractors from mutual information. *Physical Review A*, **33**(2), 1134–1140.

FWR (1998). Urban Pollution Management Manual. Foundation for Water Resesarch, Marlow.

Garcia-Navarro P. and Saviron J. (1992). McCormack's method for the numerical simulation of one-dimensional discontinuous unsteady open channel flow. *Journal of Hydraulic Research*, **30**(1), 95–105.

Garcia-Navarro P. and Burguete J. (2006). Numerical methods for the shallow water equations: 1D approach, Chapter 19. In: River basin modelling for flood risk mitigation, D. W. Knight and A. Y. Shamseldin (eds) Taylor & Francis Group plc, London, UK, 387–408.

Girossi F. and Poggio T. (1989). Representation qualities of neural networks: Kolmogorov's theorem is irrelevant. *Neural Computation*, **1**(4), 465–469.

Goldberg D. (1989). Genetic Algorithms in Search and Optimization. Addison-Wesley.

Grum M., Jørgensen A., Johansen R. M. and Linde J. J. (2006). The effect of climate change on urban drainage: an evaluation based on regional climate model simulation. *Water Science and Technology: A Journal of the International Association on Water Pollution Research*, **54**(6–7), 9.

Guinot V. and Soares-Frazão S. (2006). Flux and source term discretization in two-dimensional shallow water models with porosity on unstructured grids. *International Journal for Numerical Methods in Fluids*, **50**(3), 303–341.

Harms R. W. and Verworn H. R. (1984). HYSTEM - A hydrological urban drainage model; Part 1. *Model description., Korrespondenz Abwasser*, **31**, 1112–1117.

Hartnack J. M., Enggrob H. G. and Rungø M. (2009). 2D overland flow modelling using fine scale DEM with manageable runtimes. Flood Risk Management: Research and Practice – Samuels *et al.* (eds), 119–124.

Hassoun M. H. (1995). Fundamentals of Artificial Neural Network. MIT Press, Cambridge.

Havnø K., Brorsen M. and Refsgaard J. C. (1985). Generalised mathematical modelling system for flood analysis and flood control design, *Proceedings of 2nd International Conference on Hydraulics of Floods and Flood Control.* Cambridge, BHRA, Cranfield, 301–312.

Hedley G. and King M. (1971). Suggested Correlation Between Storm Sewage Characteristics and Storm Outflow Performance. Thomas Telford–ICE Virtual library, London.

Hjelmfelt M. (1990). Numerical study of the influence of environmental conditions on lake-effect snowstorms over Lake Michigan. *Mon. Wea. Rev*, **118**, 138–150.

HMSO (1970). Technical Committee on Storm Overflows and The Disposal of Storm Sewage. Final Report, Ministry of Housing and Local Government, London.

HMSO (1977). Storm Seweage: Separation and Disposal. Report of the Working Party on storm sewage (Scotland), Scottish Development Departement, Edinburgh.

Horton R. (1933). The role of infiltration in the hydrologic cycle. *Trans. Am. Geophys. Union*, **14**, 446–460.

Hsu M. H., Chen S. H. and Chang T. J. (2000). Inundation simulation for urban drainage basin with storm sewer system. *Journal of Hydrology*, **234**, 21–37.

Hughes T., Franca L. and Hulbert G. (1989). A new finite element formulation for computational fluid dynamics: VIII. The galerkin/least-squares method for advective-diffusive equations* 1. *Computer Methods in Applied Mechanics and Engineering*, **73**(2), 173–189.

Ilachinski A. (2001). Cellular Automata: Discrete Universe. World Scientific Publishing Company.

IPCC (2000). IPCC Special Report, Emission Scenarios, International Panel on Climate Change. Contribution of Working Group III to the IPCC.

IPCC (2001). Climate Change 2001, The Scientific Basis, International Panel on Climate Change. Contribution of Working Group I to the Third Assessment Report of the IPCC.

ISO (1993). Guide to the Expression of Uncertainty in Measurement. International Organisation for Standardisation, Geneva.

Karney B. and Ghidaoui M. (1997). Flexible discretization algorithm for fixed-grid MOC in pipelines. *Journal of Hydraulic Engineering*, **123**(11), 1004–1011.

Kecman V. (2001). Learning and Soft Computing: Support Vector Machines, Neural Networks, and Fuzzy Logic Models. MIT Press, Cambridge, MA, 121–191.

Kecman V. and Pfeiffer B. (1994). Learning Fuzzy Rules Equals to Radial Basis Function Neural Network Training.

Kellagher R. (1993). Sewer Modelling for Rehabilitation. EU SPRINT Project reports, HR Wallingford, Wallingford, UK.

Kennel M., Brown R. and Abarbanel H. (1992). Determining embedding dimension for phase-space reconstruction using a geometrical construction. *Physical Review A*, **45**(6), 3403–3411.

Khaing M. M. (2010). Simulation of Cyclone-Induced Flooding in Myanmar, MSc thesis, AIT, Bangkok, Thailand.

Koza J. and Rice J. (1992). Genetic Programming, Springer.

Kutija V. (1993). On the numerical modelling of supercritical flow. *Journal of Hydraulic Research*, **31**, 841–858.

Li H. (2009). Spatial pattern dynamics in aquatic ecosystem modelling. *Delft, UNESCO-IHE. PhD*, 181.

Liou C. (1998). Limitations and proper use of the Hazen-Williams equation. *Journal of Hydraulic Engineering*, **124**(9), 951–954.

Lobbrecht A. and Solomatine D. P. (1999). Control of water levels in polder areas using neural networks and fuzzy adaptive systems. *Water Industry Systems: Modelling and Optimization Applications*, **1**, 509–518.

Lorenz E. (1963). Deterministic Nonperiodic Flow. *Journal of the Atmos. Sci.*, **20**, 130–141.

Lotka A. (1925). Elements of Physical Biology. Williams & Wilkins company.

Lyzenga D. R. (1978). Passive remote sensing techniques for mapping water depth and bottom features. *Applied Optics* **17**, 379–383.

Maksimović Č., Prodanović D., Boonya-aroonnet S., Leitão J., Djordjević S. and Allitt R. (2009). Overland flow and pathway analysis for modelling of urban pluvial flooding. *Journal of Hydraulic Research*, **47**(4), 512–523.

Mark O., Weesakul S., Apirumanekul C., Aroonnet S. B. and Djordjevic S. (2004). Potential and limitations of 1D modelling of urban flooding. *Journal of Hydrology*, **299**, 284–299.

Marquez L., Hill T., Worthley R. and Remus W. (1991). Neural network models as an alternative to regression. System Sciences, *Proc. of the 24th Annual Hawaii International Conference*, **4**, 129–135

Mathieu J. and Scott J. (2000). An Introduction to Turbulent Flow. Cambridge University Press, Cambridge, UK.

May W., Christensen B. O. and Christensen J. H. (2005). Potential Future Changes in Heavy Rainfall Events in Europe Simulated by HIRHAM Regional Climate Model. 10th International Conference on Urban Drainage. Copenhagen, Denmark.

McCarthy G. T. (1938). The Unit Hydrograph and Flood Routing. *Conf. North Atlantic Div.*, U.S. Corps of Engineers, New London, Conn.

Mees A. I., Rapp P. E. and Jennings L. S. (1987). Singular-value decomposition and embedding dimension. *Physical Review A*, **36**(1), 340.

Mignot E., Paquier A. and Haider S. (2006). Modelling floods in a dense urban area using 2D shallow water equations. *Journal of Hydrology*, **327**(1–2), 219–225.

Miles J. (1985). The representation of flows to partially penetrating rivers using groundwater flow models. *Journal of Hydrology*, **82**(3), 341–355.

Morton K. and Mayers D. (2005). Numerical Solution of Partial Differential Equations: An Introduction. Cambridge University Press.

Moulin L., Gaume E. and Obled C. (2009). Uncertainties on mean areal precipitation: assessment and impact on streamflow simulations. *Hydrol. Earth Syst. Sci.*, **13**, 99–114.

Nash J. (1957). The form of the instantaneous unit hydrograph. *General Assembly of Toronto*, **101**, 114–121.

Neal J. C., Bates P. D., Fewtrell T. J., Hunter N. M., Wilson M. D. and Horritt M. S. (2009). Distributed whole city water level measurements from the Carlisle 2005 urban flood event and comparison with hydraulic model simulations. *Journal of Hydrology*, **368**(1–4), 42–55.

NERC (1975). Flood Studies Report (in 5 volumes). Natural Environment Research Council, London.

NRA (1993). Guidelines for AMP (2). Periodic Review, Volume 2, Water Resources. National Rivers Authority, London.

Oreskes N., Shrader-Frechette K. and Belitz K. (1994). Verification, validation, and confirmation of numerical models in the earth sciences. *Science*, **263**(5147), 641.

Osuna E., Freund R. and Girosi F. (1997). An improved training algorithm for support vector machines. *Proc. of IEEE Neural Networks Signal Proceedings*, Amelia Island, FL, USA.

Packard N., Crutchfield J., Farmer J. and Shaw R. (1980). Geometry from a time series. *Physical Review Letters*, **45**(9), 712–716.

Pomeroy R. and Parkhurst J. (1972). Self Purification in Sewers. *6th International Water Pollution Research Conference.*

Price R. K. (2009). Volume-Conservative Nonlinear Flood Routing. *Journal of Hydraulic Engineering*, **135**, 838.

Price R. K. and Vojinović Z. (2011). In Urban Hydroinformatics Series: Data, Models and Decision Support for Integrated Urban Water Systems. IWA Publishing, London, UK.

Priestley A. (1989). Roe-type schemes for super-critical flows in rivers. *Numerical Analysis Report*, **13**, 89.

Quinlan J. (1992). Learning with Continuous Classes. 5th Australian Joint Conference on Artificial Intelligence.

Roe P. L. (1981). Approximate Riemann solvers, parameter vectors, and difference schemes. *Journal of Computational Physics*, **43**, 357–372.

Roesner L., Shubinski R. and Aldrich J. (1983). Stormwater Management Model User's Manual Version III: Addendum I, EXTRAN. EPA Report in press, Cincinnati, OH.

Rossman L. (2000). EPANET 2 users manual. US Environmental Protection Agency, Cincinnati, Ohio.

Ruin I., Creutin J.-D., Anquetin S. and Lutoff C. (2008). Human exposure to flash floods - Relation between flood parameters and human vulnerability during a storm of September 2002 in Southern France. *Journal of Hydrology*, **361**, 199–213.

Salum M. H. (2010). Towards The Methodology For Capturing Small-scale Urban Features into The Coarse Resolution 2D Models. MSc thesis (WSE-HI.10–17), UNESCO-IHE.

Sanders B. F., Schubert J. E. and Gallegos H. A. (2008). Integral formulation of shallow-water equations with anisotropic porosity for urban flood modelling. *Journal of Hydrology*, **362**, 19–38.

Sangati M. and Borga M. (2009). Influence of rainfall spatial resolution on flash flood modelling. *Nat. Hazards Earth Syst. Sci.*, **9**, 575–584.

Savic D., Walters G. and Davidson J. (1999). A genetic programming approach to rainfall-runoff modelling. *Water Resources Management*, **13**(3), 219–231.

Saul A. and Ellis D. (1992). Sediment deposition in storage tanks. *Water Sci. Technol.* WSTED 4, **25**(8), 189–198.

Saulnier G. M. and Le Lay M. (2009). Sensitivity of flash-flood simulations on the volume, the intensity, and the localization of rainfall in the Cevennes-Vivarais region. *Water Resource Res.*, **45**, W10425, p. 9, doi:10.1029/2008WR006906.

Seyoum S. D., Vojinović Z. and Price R. K. (2009). Integrated Modelling of Urban Wastewater Systems. *8th International Conference on Urban Drainage Modelling*, Tokyo, Japan.

Seyoum S. D., Vojinović Z. and Price R. K. (2010). Urban Pluvial Flood Modelling: Development and Application. *9th International Conference on Hydroinformatics, HIC 2010*, Tianjin, China.

Seyoum S. D., Vojinović Z. and Price R. K. (2012). Generalisation of topographical resolutions for two-dimensional urban flood modelling. *ASCE Journal of Hydraulic Engineering*, ASCE (Accepted, In Press).

Shakhnarovish D., Darrel T. and Indyk P. (2005). Nearest-Neighbor Methods in Learning and Vision: Theory and Practice. The MIT Press.

Sherman L. (1932). Streamflow from rainfall by the unit-graph method. *Engineering News Record.*, **108**, 501–505.

Sithole G. and Vosselman G. (2004). Experimental Comparison of Filter Algorithm for Bare Earth Extraction from Airbone Laser Scanning Point Clouds. *ISPRS Journal of Photogrammetry & Remote Sensing*, **56**, 85–10.

Soares-Frazão S., Lhomme J., Guinot V. and Zech Y. (2008). Two-dimensional shallow water model with porosity for urban flood modelling. *Journal of Hydraulic Research*, **46**(1), 45–64.

Solomatine D. P. and Dulal K. N. (2003). Model trees as an alternative to neural networks in rainfall–runoff modelling. *Hydrological Sciences Journal*, **48**(3), 399–411.

Solomatine D. P. and Siek M. B. (2006). Modular learning models in forecasting natural phenomena. *Neural Networks*, **19**(2), 215–224.

Solomatine D. P. and Xue Y. (2004). M5 Model Trees and Neural Networks: Application Flood Forecasting in the Upper Reach the Huai River in China. *Journal of Hydrologic Engineering*, **9**(6), 491, doi:10.1061/(ASCE) 1084-0699(2004)9.

Speziale C. (1991). Analytical methods for the development of Reynolds-stress closures in turbulence. *Annual Review of Fluid Mechanics*, **23**(1), 107–157.

Stravs L. and Brilly M. (2007). Development of a low-flow forecasting model using the M5 machine learning method. *Hydrological Sciences Journal*, **52**(3), 466–477.

Stumpf R. P., Holderied K. and Sinclair M. (2003). Determination of water depth with high-resolution satellite imagery over variable bottom types. *Limnology and Oceanography*, **48**, 547–556.

Sturm T. W. (2001). Open channel hydraulics. McGraw-Hill, New York, USA.

Swamee P. and Jain A. (1976). Explicit equations for pipe-flow problems. *Journal of the Hydraulics Division*, **102**(5), 657–664.

Syme W. J. (2008). Flooding in Urban Areas – 2D Modelling Approaches for Buildings and Fences. *Engineers Australia, 9th National Conference on Hydraulic in Water Engineering*, Darwin Convention Centre, Australia.

Takens F. (1981). Detecting strange attractors in turbulence. *Lecture Notes in Mathematics*, **898**(1), 366–381.

Temam R. (2001). Navier-Stokes equations: Theory and Numerical Analysis. American Mathematical Society.

Tjalling J. (1995). Historical development of the Newton–Raphson method. *Society for Industrial and Applied Mathematics*, **37**(4), 531–551.

Toth E., Brath A. and Montanari A. (2000). Comparison of short-term rainfall prediction models for real-time flood forecasting. *Journal of Hydrology*, **239**(1–4), 132–147.

UNESCO-IHE (2011). Bathymetry Estimation from Satellite Imagery and Sonar Measurements: A case of St Maarten. Project Report.

USDA NRCS (1986). Urban Hydrology for Small Watersheds, Technical Release No. 55. United States Department of Agriculture Natural Resource Conservation Service (USDA NRCS), Washington, DC.

USEPA (2008). Guidance on the Development, Evaluation and Application of Environmental Models. Council for Regulatory Environmental Modelling, USEPA, Washington, DC.

Verwey A., Muttil N., Liong S. Y. and He S. (2008). Implementing an urban rainfall-runoff concept in Sobek for a catchment in Singapore, *Proceedings 31st Hydrology and Water Resources Symposium (Water Down Under 2008)*, ISBN 0-858-25735-1.

Vojinović Z., Bonillo B., Chitranjan K. and Price R. K. (2006). Modelling flow transitions at street junctions with 1D and 2D models. *7th International Conference on Hydroinformatics*, Acropolis - Nice, France.

Vojinović Z., Price R. K. and Den Broek W. V. (2005). HYDROPLAN-EU Knowledge Management Framework for Urban Water Asset Management. *10th International Conference on Urban Drainage*, Copenhagen/ Denmark.

Vojinović Z., Seyoum S. D., Mwalwaka J. M. and Price R. K. (2011). Effects of Model Schematization, Geometry and Parameter Values on Urban Flood Modelling. *Water Science and Technology*, **63**(3), 462–467.

Vojinović Z., Seyoum S., Salum M. H., Price R. K., Fikri A. F. and Abebe Y. (2012). Modelling floods in urban areas and representation of buildings with a method based on adjusted conveyance and storage characteristics. *Journal of Hydroinformatics*, IWA (Accepted, In Press).

Vojinović Z. and Tutulic D. (2009). On the use of 1D and coupled 1D–2D approaches for assessment of flood damages in urban areas. *Urban Water Journal*, **6**(3), 183–199.

Volterra V. (1926). Fluctuations in the abundance of a species considered mathematically. *Nature*, **118**(2972), 558–560.

Wallingford H. R. (1981a). Design and Analysis of Urban Storm Drainage - The Wallingford Procedure. Department of Envrionment, National Water Council, Standing Technical Committee Reports, Wallingford, Hydraulics Research Limited.

Wallingford H. R. (1981b). The Wallingford Procedure for The Design of Urban Storm Drainage, HR Wallingford.

Wang Y., Zhang P., Dong W. and Zhang Y. (2007). Study on Remote Sensing of Water Depths Based on BP Artificial Neural Network. *Marine Science Bulletin*, **9**, 26–35.

WaPUG (2002). Code of Practice for the Hydraulic Modelling of Sewers. CIWEM, London.

WaPUG (2006). Guide to the Quality Modelling of Sewer Systems. CIWEM, London.

Watanabe M., Etoh T. and Murota A. (1990). Runoff simulation model of sewer pipe systems considering pressure-relaxation effect by lateral pipes. *Fifth Int. Conference on Urban Storm Drainage*, Iwasa Y. and Sueishi T. (ed.) Vol. 1, 275–280. Suita, Osaka,

Weigend A. and Gershenfeld N. (1994). Time Series Prediction: Forecasting the Future and Understanding the Past. *Proc of the NATO Advanced Research Workshop on Comparative Time Series Analysis*, Santa Fe, New Mexico.

Witten I. H. and Frank E. (2000). Data mining: Practical Machine Learning Tools with JAVA Implementations. Morgan Kaufmaan, San Francisco, CA, USA.

Wolfram S. (1986). Theory and Applications of Cellular Automata. World Scientific, Singapore.

Woods-Ballard B., Kellagher R., Martin P., Jefferies C., Bray R. and Shaffer P. (2007). The SUDS Manual CIRIA C697. Construction Industry Research and Information Association, London.

World Health Organisation (2002). Flooding: Health Effects and Preventative Measures. Copenhagen and Rome; Europe; September 2002; http://www.who.dk/document/mediacentre/fs0502e.pdf (accesed June 2010).

WRc (1994). The Sewerage Rehabilitation Manual. WRc Engineering, Swindon, UK.

Wylie E. B. and Streeter V. L. (1978). Fluid Transients. McGraw-Hill International Book Co, New York, USA.

Yen B. (2004). Hydraulics of sewer systems, Chapter 4. In: Urban Stormwater Management Tools, L. W. Mays McGraw-Hill Engineering Reference.

Yevjevich V. (1972). Probability and Statistics in Hydrology. Water Resources Publications, Littleton, Colorado, USA.

Yu D. and Lane S. N. (2006). Urban fluvial flood modelling using a two-dimensional diffusion-wave treatment, part 1: mesh resolution effects. *Hydrological Processes*, **20**, 1541–1565.

Zadeh L. A. (1965). Fuzzy sets. *Information and Control*, **8**(3), 338–353.

Zijderveld A. (2003). Neural Network Design Strategies and Modelling in Hydroinformatics. Technical University of Delft, Delft.

PART IV

Abbott M. B. and Warren R. (1974). A Dynamic Population Model. *J. Environmental Management*, **2**, 281–297.

Abbott M. B. and Verhoog F. H. (1968). Étude numérique des ondes de crue dans un Fleuve avec Plaines d'Inondation. *Dixième Journées de L'Hydraulique,* Question II, Rapport II, Paris.

Abbott M. B. (2007). Stakeholder participation in creating infrastructure. *New Civil Engineer*, **160**(1), 26–32, [Awarded the Telford Gold Medal of the Institution of Civil Engineers in 2008].

AGS (2007). Guideline for landslide susceptibility, hazard and risk zoning for land use planning. *Australian Geomechanics*, **42**(1), 13–36.

ALGENZ (1998). Infrastructure Asset Management Manual, National Asset Management Steering Group. Association of Local Government Engineers, New Zealand.

Alcamo J. (2001). Scenarios as Tools for International Assessments'. Prospects and Scenarios Number 5, European Environment Agency, Copenhagen. http://www.usf.uos.de/IA/Scenarios_Report.pdf

Ana E. Jr. and Bauwens W. (2007). Sewer Network Asset Management Decision-Support Tools: A Review. International Symposium on New Directions in Urban Water Management, UNESCO, Paris.

Andjelkovic I. (2001). Guidelines on Non-Structural Measures in Urban Flood Management. IHP-V Technical Documents in Hydrology 50, UNESCO, Paris.

Ariaratnam S. and MacLeod C. (2002). Financial outlay modeling for a local sewer rehabilitation strategy. *Journal of Construction Engineering and Management*, **128**(6), 486–495.

AS/NZS 4360 (2004). Risk Management: AS/NZS 4360:2004. Standards Australia/Standards New Zealand, Sydney and Wellington.

Baik H.-S., Jeong H. S. and Abraham D. M. (2006). Estimating transition probabilities in Markov chain-based deterioration models for management of wastewater systems. *Journal of Water Resources Planning and Management*, **132**(1), 15.

Balmforth D., Digman C., Kellagher R. and Butler D. (2006). Designing for exceedance in urban drainage - good practice. CIRIA C635, London.

Barredo J., Kasanko M., McCormick N. and Lavalle C. (2003). "Modeling dynamic spatial processes: simulation of urban future scenarios through cellular automata." *Journal of Landscape and Urban Planning*, **64**, 145–160. Elsevier Science.

Barredo J. I., Petrov L., Sagris V., Lavalle C. and Genovese E. (2005). Towards an integrated scenario approach for spatial planning and natural hazards mitigation. European Commission, DG-Joint Research Centre, Italy.

Barreto W., Vojinović Z., Price R. K. and Solomatine D. P. (2010). A multi-objective evolutionary approach for rehabilitation of urban drainage systems. *ASCE Journal of Water Resources Planning and Management, ASCE*, **136**(5), 547–554.

Batty M. (1997). Cellular automata and urban form: a primer. *Journal of the American Planning Association*, **63**, 266–274.

Baur R. and Herz R. (2001). Selective Inspection Planning with Aging Forecast for Sewer Types. *Proceedings IWA 2nd World Water Congress: Efficient Water Management - Making it Happen*, Berlin, Germany.

Benenson I. and Torrens P. M. (2004). Geosimulation: Automata-Based Modeling of Urban Phenomena. John Wiley, London.

Bengassem J. and Bennis S. (2000). Fuzzy Expert System for Sewer Networks Diagnosis. *Proceedings International Conference on Decision Making in Urban and Civil Engineering*, Lyon, France.

Berardi L., Giustolisi O., Kapelan Z. and Savić D. (2008). Development of pipe deterioration models for water distribution systems using EPR. *Journal of Hydroinformatics*, **10**(2), 113–126.

Berleant A. (1991). Art and Engagement. Temple University Press, Philadelphia, PA.

Berleant A. (1992). The Aesthetics of Environment. Temple University Press, Philadelphia, PA.

Bogardi J. and Birkman J. (2004). Vulnerability assessment: the first step towards sustainable risk reduction. In: Disasters and Society – From Hazard Assessment to Risk Reduction, D. Malzahn and T. Plapp (eds) Logos Verlag Berlin, Berlin.

Bohle H.-G. (2001). Vulnerability and Criticality: Perspectives from Social Geography. *IHDP Update 2/01, Newsletter of the International Human Dimensions Programme on Global Environmental Change*, 1–7.

Bollin C., Cardenas C., Hahn H. and Vatsa K. S. (2003). Red de Desastres Naturales: gestión de riesgo de desastres por comunidades y gobiernos Locales. Inter-American Development Bank, Washington, D.C.

Bortoft H. (1996). The Wholeness of Nature: Goethe's Way Toward a Science of Conscious Participation in Nature. Lindesfarne Press, Hudson, New York.

Briarssoulis H. (2000). Analysis of land use change: theoretical and modelling approaches. In: The web book of regional science, W. R. Jackson (ed.), Regional Research Institute, West VIrginia University, USA.

Burch R. (1989). On phenomenology and its practices [part I]. *Phenomenology+pedagogy*, **7**, 187–217.

Burch R. (1990). On phenomenology and its practices [part II]. *Phenomenology+pedagogy*, **8**, 130–160.

Burch R. (1991). On phenomenology and its practices [part III]. *Phenomenology+pedagogy*, **9**, 167–193.

CAP-NET (2005). Integrated Water Resources Management Plans. Training Manual and Operational Guide. Pretoria, Cap-Net. http://www.cap-net.org/sites/cap-net.org/files/Manual_english.pdf [Accessed: 21.07.2010].

CAP-NET (2008). Integrated Water Resource Management for River Basin Organisations. Training Manual. Pretoria, Cap-Net. http://www.cap-net.org/sites/cap-net.org/files/RBO%20Manual.pdf [Accessed: 21.07.2010].

Cardona O. D. (1999). Environmental management and disaster prevention: Holistic risk assessment and management. In: Natural Disaster Management, J. Ingleton (ed.), Tudor Rose, London.

Cardona O. D. (2001). Estimación Holística del Riesgo Sísmico utilizando Sistemas Dinámicos Complejos. Universidad Politécnica de Cataluña, Barcelona. http://www.desenredando.org/public/varios/2001/ehrisusd/index.html.

Casey E. S. (1993). Getting Back into Place. Indiana University Press, Bloomington.

Casey E. S. (1997). The Fate of Place: A Philosophical History. University of California Press, Berkeley.

Chen P. (2008). Flood Hazard Analysis using GIS and Remote Sensing Data. MSc thesis (WSE-HI. 08-05), UNESCO-IHE.

CLG (2010). Planning Policy Statement 25: Development and Flood Risk. HMSO, London.

Cloke P., Philo C. and Sadler D. (1991). Approaching human geography: An introduction to contemporary theoretical debates. Guildford Press, New York.

Condon P. M. (1991). Radical romanticism. *Landscape Journal*, **10**, 3–8.

Corner J. (1990). A discourse on theory I: "Sounding the depths–origins, theory, and representation. *Landscape Journal*, **9**, 61–78.

Davis T. (1989). Photography and landscape studies. *Landscape Journal*, **8**, 1–12.

Dawson R. J., Speight L., Hall J. W., Djordjević S., Savić D. and Leandro J. (2008). Attribution of flood risk in urban areas. *J. Hydroinformatics*, **10**, 275–288.

Defra (2004). Catchment flood management plans. Volume I - Policy guidance, Environment Agency, July 2004.

De Leon F. D., Felsen M. and Wilensky U. (2007). NetLogo Urban Suite - Tijuana Bordertowns model. Center for Connected Learning and Computer-Based Modeling, Northwestern University, Evanston, IL. http://ccl.northwestern.edu/netlogo/models/UrbanSuite-TijuanaBordertowns.

Department of Drainage and Sewerage Bangkok Metropolitan Administration (1996). "Master Plan for Basic Infrastructure Systems and Preliminary Design for Flood Protection and Drainage Systems in Eastern Sub-Urban Bangkok".

Department of Drainage and Sewerage Bangkok Metropolitan Administration (2005). "Flood Forecasting and Management System" Main report.

DHS (2009). National infrastructure protection plan, Partnering to enhance protection and resiliency. Report: Department of Homeland and Security, Science and Technology, Washington DC. http://www.dhs.gov/xlibrary/assets/NIPP_Plan.pdf accessed June 2011.

DIPNR (2005). Floodplain Development Manual. The Management of Flood Liable Land. New South Wales Department of Infrastructure, Planning and Natural Resources, Sydney.

Djordjević S., Prodanović D. and Walters G. A. (2004). Simulation of transcritical flow in pipe/channel networks. *J. Hydraul. Eng.*, **130**(12), 1167–1178.

Djordjević S., Vojinović Z., Dawson R. and Savic D. (2011). Chapter 10: Flood modelling in urban areas. In: Applied Uncertainty Analysis for Flood Risk Management, K. Beven and J. Hall (eds), (In press). Imperial College Press/World Scientic.

Dovey K. (1993). Putting geometry in its place: Toward a phenomenology of the design process. In: Dwelling, Seeing and Designing: Toward a Phenomenological Ecology, D. Seamon (ed.), SUNY Press, Albany, New York. pp. 247–269.

Dubois D. and Prade H. (1988). Possibility Theory: An Approach to Computerized Processing of Uncertainty. Plenum Press, New York.

EMA (2004). Emergency Risk Management Applications Guide: Manual 5. Emergency Management Australia, Canberra. www.ema.gov.au.

Embree L. (1996). The Encyclopedia of Phenomenology. Kluwer, Dordrecht, The Netherlands.

Emergency Management Australia (2004). Guidelines for Psychological Services: Emergency Managers Guide. Australian Emergency Manual Series, Guide 25. www.ema.gov.au/www/emaweb/emaweb.nsf/Page/Publications_AustralianEmergencyManualSeries_AustralianEmergencyManualSeries

Engelen G., Lavalle C., Barred J. I., van der Meulen M. and White R. (2007). The Moland modeling framework for urban and regional land-use dynamics. In: Modeling Land Use Change: Progress and Applications, E. Koomen, J. Stillwell, A. Bakema and H. J. Scholten (eds) Springer, The Netherlands.

EXCIMAP (2007). Handbook on good practices for flood mapping in Europe. *European Exchange Circle on Flood Mapping*.

Feeney C. S., Thayer S., Bonomo M. and Martel K. (2009). Condition Assessment of Wastewater Collection Systems. State of Technology Review Report, EPA/600/R-09/049, May 2009.

Fischer H. B., List E. J., Koh R. C. Y., Imberger J. and Brooks N. H. (1979). Mixing in Inland and Coastal Waters. Academic Press, New York.

Fischhoff B. (2009). Risk perception and communication. In: Oxford Textbook of Public Health, 5th edn, R. Detels, R. Beaglehole, M. A. Lansang and M. Gulliford (eds), Oxford, Oxford University Press. pp. 940–952. Reprinted in N. K. Chater (ed.), Judgement and Decision Making. London, Sage.

Fischhoff B. (2007). Non-Persuasive Communication about Matters of Greatest Urgency: Climate Change. *Environmental Science & Technology*, **41**, 7204–7208.

Fischhoff B. (2005). The psychological perception of risk. In: McGraw-Hill Handbook of Terrorism and Counter-terrorism, D Kamien (ed.), McGraw-Hill, New York, pp. 463–492.

Fischhoff B. (1995). Risk perception and communication unplugged: Twenty years of process. *Risk Analysis*, **15**, 137–145.

Farber M. (1943). The Foundation of Pehnomenology: Edmund Husserl and The Quest for A Rigorous Science of Philosophy. Harvard University Press, Cambridge, MA.

Friazinov I. V. (1970). Solution algorithm for finite difference problems on directed graphs. *J. Comp. Math. and Math. Phys.*, **10**, 474–477 (in Russian).

Giustolisi O. and Savic D. (2006). A symbolic data-driven technique based on evolutionary polynomial regression. *Journal of Hydroinformatics*, **8**(3), 207–222.

Golgher A. (2004). Família. In: BH Século XXI, M. Lemos, C. Dini, J. A. Carvalho and F. Santos (Orgs), Belo Horizonte, CEDEPLAR, UFMG. http://www.cedeplar.ufmg.br/pesquisas/projetos-concluidos

Goulter I. C., Davidson J. and Jacobs P. (1993). Predicting water-main breakage rates. *J. Water Res. Plan. Manage.*, **119**(4), 419–436.

Green C. (2004). The evaluation of vulnerability to flooding. *Disaster Prevention and Management*, **13**(4), 323–329.

Green C. H., Pareker D. J. and Tunstall S. M. (2000). Assessment of Flood Control and Management Options. World Commission on Dams (WCD), Cape Town, 183p, http://www.dams.org/docs/kbase/thematic/drafts/tr44/finaldraft.pdf

Hall J. W., Boyce S., Dawson R. J. and Tarantola S. (2009). Sensitivity analysis for hydraulic models. *J. Hydraul. Eng.*, **135**(11), 959–969.

Harada H. (1988). Statistics of the cold wave in the temperate region and prediction of the number of damaged service pipes. *J. AWWA*, **57**(8), 12–15.

Hasegawa K., Wada Y. and Miura H. (1999). New Assessment System for Premeditated Management and Maintenance of Sewer Pipe Network. *Proceedings 8th International Conference on Urban Storm Drainage*, Sydney, Australia, 586–593.

Havnø K., Brorsen M. and Refsgaard J. C. (1985). Generalized mathematical modeling system for flood analysis and flood control design. *2nd Int. Conf. on the Hydraulics of Floods and Flood Control*, Cambridge, 301–312.

HB 436. (2004). Risk Management Guidelines Companion to AS/NZS 4360:2004, HB 436:2004. Standards Australia/Standards New Zealand, Sydney and Wellington.

Heelan P. A. (1983). Space-Perception and The Philosophy of Science. University of California Press, Berkeley.

HM (2005). Emergency preparedness, Ref: 267619/1105/D40. Library and Information Services, Cabinet Office, York. http://www.cabinetoffice.gov.uk/media/131903/emergprepfinal.pdf

HM Government (2005). Risk, Good Practice in Government.

Homma T. and Saltelli A. (1996). Importance measures in global sensitivity analysis of nonlinear models. *Reliability Engineering and System Safety*, **52**, 1–17.

Hung N. Q., Babel M. S., Weesakul S. and Tripathi N. K. (2008). "An Artificial Neural Network Model for rainfall Forecasting in Bangkok, Thailand". *Hydrol. Earth Syst. Sci.*, **5**, 183–218.

Jackson M. (1996). Things as They Are: New Directions in Phenomenological Anthropology. University of Indiana Press, Bloomington.

Jones E. (1989). Reading the Book of Nature: A Phenomenological Study of Creative Expression in Science and Painting. Ohio University Press, Athens, OH.

Jonkman S. and Kelman I. (2005). An analysis of the causes and circumstances of flood disaster deaths. *Disasters, Oxford e Malden*, **29**(1), 75–97.

Kamphuis W. J. (2010). Introduction to Coastal Engineering and Management. World Scientific Publishing Co Ltd., Singapore.

Kanchanarat S. (1989). Estimation of Flood Damage Function for Bangkok and Vicinity. MSc thesis, Asian Institute of Technology, Bangkok, Thailand.

Kahn H. and Weiner A. (1967). The Year 2000: A Framework for Speculation on the Next Thirty-three Years. Macmillan, New York.

Kecman V. (2001). Learning and Soft Computing: Support Vector Machines. Neural Networks and Fuzzy Logic Models, MIT Press, Cambridge, MA, 121–191.

Kettler A. J. and Goulter I. C. (1985). An analysis of pipe breakage in urban water distribution networks. *Can. J. Civil Eng.*, **12**(2), 286–293.

Kok K. and Van Delden H. (2007). Combining Two Approaches of Integrated Scenario Development to Combat Desertification in the Guadalentin watershed, Spain. Environment and Planning B: Planning and Design. Vol. 34.

Kok K., Verburg P. H. and Veldkamp A. (2007). "Integrated assessment of the land system: the future of land use". *Land Use Policy*, **24**, 517–520.

Khu S.-T., di-Pierro F., Savić D., Djordjević S. and Walters G. (2006). Incorporating spatial and temporal information for urban drainage model calibration: An approach using Preference ordering genetic algorithm. *Adv. in Wat. Res.*, **29**, 1168–1181.

Korving H., Van Gelder P., Van Noortwijk J. and Clemens F. (2002). Influence of Model Parameter Uncertainties on Decision-Making for Sewer System Management. Citeseer.

Lancaster J. W., Preene M. and Marshall C. T. (2004). CIRIA C624: Development and Flood Risk – Guidance for The Construction Industry. London, UK.

Lechner T., Watson B. A., Tisue S., Wilensky U. and Felsen M. (2004). Procedural Modeling of Land Use in Cities. Technical Report, NWU-CS-04-38.

Lekuthai A. and Vongvisessomjai S. (2001). Intangible Flood Damage Quantification. *Water Resources Management*, **15**, 343–362.

Liu Yan (2009). Modeling Urban Development with Geographical Information Systems and Cellular Automata. CRC Press, Boca Raton, FL.

Marlow D., Heart S., Burn S., Urquhart A., Gould S., Anderson M., Cook S., Ambrose M., Madin B. and Fitzgerald A. (2007). Condition Assessment Strategies and Protocols for Water and Wastewater Utility Assets. Water Environment Research Foundation (WERF): Alexandria, VA and IWA Publishing, London, UK.

Meijerink A. M. J. (1996). Remote sensing applications to hydrology: groundwater. *Hydrol. Sci. J.*, **41**(4), 549–561.

Menenti M., Azzali S. and d'Urso G. (1996). Remote sensing, GIS and hydrological modelling for irrigation management. In: Sustainability of Irrigated Agriculture, L. S. Pereira, R. A. Feddes, J. R. Gilley and B. Lesaffre (eds), Kluwer Academic Publishers, The Netherlands, pp. 453–472.

Middelmann M. H. (ed.) (2007). Natural Hazards in Australia: Identifying Risk Analysis Requirements. Geoscience Australia, Canberra.

Mietzner D. and Reger G. (2004). "Scenario Approaches – History, Differences, Advantages and Disadvantages". EU-US Seminar, New Technology Foresight, Forecasting & Assessment Methods, Seville.

Montali L. (2000). Família e Trabalho na Reestruturação Produtiva: ausência de políticas de emprego e deterioração das condições de vida. *Revista Brasileira de Ciências Sociais*, **15**(42), 55–71.

Moran D. (2000). Introduction to Phenomenology. London, Routledge.

Moustakas C. (1994). Phenomenological Research Methods. Newbury Park, CA, Sage.

Mugerauer R. (1994). Interpretations on Behalf of Place: Environmental Displacements and Laternative Responses. SUNY Press, Albany, New York.

Mugerauer Re (1995). Interpreting environments: Tradition, Deconstruction, Hermeneutics. University of Texas Press, Austin.

Munich Re (2002). Topics, Annual Review: Natural Catastrophes 2002. Munich Re Group. http://www.munichre.com/

Nascimento N., Machado M. L., Baptista M. and de Silva A. (2007). The assessment of damages caused by floods in The Brazilian context. *Urban Water Journal*, **4**(3), 195–210.

Østrem G., Anderson T. and Odegaard H. (1981). Operational use of satellite data for snow inventory and runoff forecast. *Satellite Hydrology*, 230–234. American Water Resources Association, Minneapolis.

Pelling M. (2003). The Vulnerability of Cities: Natural Disasters and Social Resilience. Earthscan, London, 212p.

Penning-Rowsell E. C. and Chatterton J. B. (1977). The Benefits of Flood Alleviation: A Manual of Assessment Techniques. Gower Publishing Company Limited, Aldershot, England.

Penning-Rowsell E. C., Tapsell S., Johnson C. and Wilson T. (2006). Development of Economic Appraisal Methods for Flood Management and Coastal Erosion Protection, Objective 14: Strategies for The Future Updating of Flood

Damage Tables and Related Indices. DEFRA R&D Technical Report FD2014/TR2, Joint DEFRA/EA Flood and Coastal Erosion Risk, Management R&D Programme.

Peterson G. D., Cumming G. S. and Carpenter S. R. (2003). Scenario planning: a tool for conservation in an uncertain world. *Conservation Biology*, **17**, 358–366.

Pilgrim D. (2001). Australian Rainfall and Runoff (Volume 1). Engineers Australia, Canberra.

Polkinghorne D. (1983). Methodology for the Human Sciences. SUNY Press, Albany.

Price R. K. and Vojinović Z. (2011). In Urban Hydroinformatics Series: Data, Models and Decision Support for Integrated Urban Water Systems. IWA Publishing, London, UK.

Queensland Government (2003). State Planning Policy 1/03 Guideline: Mitigating the Adverse Impacts of Flood, Bushfire and Landslide. Queensland Department of Local Government and Planning and Queensland Department of Emergency Services, Brisbane.

Rangarajan S. (2005). Challenges and solutions in urban drainage planning and management. *Geophysical Research Abstracts*, **7**, 1288.

Rango A. and Ritchie J. C. (eds) (1996). Remote sensing applications to hydrology. *Special Issue, Hydrol. Sci. J.*, **41**(4), 477–494.

RESCDAM (2000). The use of physical models in dam-break flood analysis. Final Report, Helsinki University of Technology, Helsinki, Finland.

Relph E. (1989a). A curiously unbalanced condition of the powers of the mind: Realism and the ecology of environmental experience. In: The Behavioral Environment: Essays in Reflection, Application, and Re-evaluation, F. W. Boal and D. N. Livingston (eds), Routledge, London.

Relph E. (1989b). Responsive methods, geographical imagination and the study of landscapes. In: Remaking Human Geography, A. Kobayashi and S. MacKenzie (eds), Unwin Hyman, Boston, pp. 149–163.

Relph E. (1990). Geographical imagination. *National Geographical Journal of India*, **36**, 1–9.

Riegner M. (1993). Toward a holistic understanding of place: Reading a landscape through its flora and fauna. In: Dwelling, Seeing, and Designing: Toward a Phenomeno-Logical Ecology, D. Seamon (ed.), SUNY Press, Albany, New York, pp. 181–215.

Rommel Y., Bouyssou D. and MacGilchrist R. (1989). Planning Rehabilitation Works on a Sewer Network. Computing in Civil Engineering: Computers in Engineering, 108–115.

Ruwanpura J., Ariaratnam S. T. and El-assaly A. (2004). Prediction models for sewer infrastructure utilizing rule-based simulation. *Civil Engineering and Environmental Systems*, **21**(3), 169–185.

Saaty T. L. (1977). A scaling method for priorities in hierarchical structures. *Journal of Mathematical Psychology*, **15**, 234–281.

Saegrov S. (2006). Care-s: Computer Aided Rehabilitation of Sewer and Storm Water Networks. *Intl Water Assn.*

Saltelli A., Chan K. and Scott M. (2000). Sensitivity Analysis. Wiley, New York.

Sanchez Torres A., Vojinović Z. and Price R. (2011). Determining the Route for A Water Main in A New Urbanising Area. *11th International Conference on Computing and Control for the Water Industry: "Urban Water Management: Challenges and Opportunities"*. CCWI 2011. Exeter, UK.

SCARM (2000). Floodplain Management in Australia - Best Practice Principles and Guidelines. Agriculture and Resource Management Council of Australia and New Zealand, Standing Committee on Agriculture and Resource Management, Report No. 73, CSIRO Publishing, Collingwood, Victoria.

Schultz G. A. (1994). Remote sensing for forecasting of floods (Chapter 21) and remote sensing for control of floods (Chapter 36). In: Coping with Floods, G. Rossi, N. Harmancioglu and V. Yevjevich (eds), Kluwer Academic Publishers, The Netherlands, pp. 459–472 and 697–703.

Seamon D. (1997). Phenomenology and behavioral geography. In: Encyclopedia of Phenomenology, L. Embree (ed.), Kluwer, Dordrecht, The Netherlands, pp. 53–56.

Seamon D. and Zajonc A. (1998). Goethe's way of science: Toward a phenomenology of nature. SUNY Press, Albany, NY.

Shamir U. and Howard C. D. D. (1979). An analytical approach to scheduling pipe replacement. *J. AWWA*, **71**, 248–258.

Shipman B. and Stojanovic T. (2007). Facts, fictions, and failures of integrated coastal zone management in Europe. *Coastal Management*, **35**(2), 375–398.

Simons B. D., Ponce V. M., Li R. M., Chen Y. H., Gessler J., Ward T. J. and Duong N. (1977). Flood Flows, Stages and Damages. Colorado State University, Fort Collins.

Sobol I. (1993). Sensitivity analysis for non-linear mathematical models. *Mathematical Modelling & Computational Experiment*, **1**, 407–414.

Spiegelberg H. (1982). The phenomenological movement: A historical introduction (Vols 1 & 2) 3rd edn. Martinus Nijhoff Publishers, The Hague, Netherlands.

Stone S., Dzuray E. J., Meisegeier D., Dahlborg A. S. and Erickson M. (2002). Decision-Support Tools for Predicting the Performance of Water Distribution and Wastewater Collection Systems, US EPA.

Stewart D. and Mukunas A. (1990). Exploring Phenomenology: A Guide to The Field and Its Literature. 2nd edn, Ohio University Press, Athens, Ohio.

Su Y.-C., Mays L. W., Duan N. and Lansey K. E. (1987). Reliability-based optimization model for water distribution systems. *J. Hydraul. Eng, ASCE*, **113**(12), 1539–1556.

Tabesh M., Soltani J., Farmani R. and Savic D. (2009). Assessing pipe failure rate and mechanical reliability of water distribution networks using data-driven modeling. *Journal of Hydroinformatics*, **11**(1), 1–17.

Tang J. C. S., Vongvisessomjai S. and Sahasakmontri K. (1992). Estimation of flood damage cost for Bangkok. *Water Resources Management*, **6**, 47–56, Kluwer Academic Publishers, The Netherlands.

Turner II B. L., Kasperson R. E., Matson P. A., McCarthy J. J., Corell R. W., Christensen L., Eckley N., Kasperson J. X., Luers A., Martello M. L., Polsky C. and Pulsipher A. and Schiller A. (2003). A framework for vulnerability analysis in sustainability science. *Proceedings of the National Academy of Sciences of the United States of America*, **100**, 8074–8079.

Ukon T., Shigeta N., Watanabe M., Shiraishi H. and Uotani M. (2008). Correction methods for dropping of simulated water level utilizing Preissmann and MOUSE slot models. *11th Int. Conf., on Urban Drainage*, Edinburgh, CD-ROM.

Valle R. (1998). Phenomenological Inquiry in Psychology. Plenum, New York.

Van Asselt M. B. A. and Rijkens-Klomp N. (2002). A look in the mirror: reflection on participation in integrated assessment from a methodological perspective. *Global Environmental Change*, **12**, 167–184.

Van Delden H., Luja P. and Engelen W. (2007). Integration of multi-scale dynamic spatial models of socio-economic and physical processes for river basin management. *Journal Environmental Modeling and Software*, **22**, 223–238.

Van Manen M. (1990). Researching Lived Experience. SUNY Press, Albany, New York.

Van Vliet J. and Van Delden H. (2008). An activity based cellular automaton model to simulate land use changes. *International Congress on Environmental Modelling and Software*. (iEMSs). Barcelona, Spain.

Vasconcelos J. G., Wright S. J. and Roe P. L. (2006). Improved simulation of flow regime transition in sewers: Two-component pressure approach. *J. Hydraul. Eng.*, **132**, 553–562.

Verburg P. H., Schot P. P., Dijst M. J. and Veldkamp A. (2004). Land use change modelling: current practice and research priorities. *GeoJournal*, **61**, 309–324.

Vesely D. (1988). On the relevance of phenomenology. In: Form; being; absence: Pratt journal of architecture, S. Perrella (ed.), **2**, pp. 54–60. Rizzoli, New York.

Vojinović Z., Bonillo B., Chitranjan K. and Price R. (2006). Modelling Flow Transitions at Street Junctions with 1D and 2D Models. *7th International Conference on Hydroinformatics*. Acropolis - Nice, France.

Vojinović Z., Price R. K. and van der Broek W. (2005). Hydroplan-EU Knowledge Management Framework for Urban Water Asset Management. *International Conference on Urban Drainage*. Copenhagen, Denmark.

Vojinović Z. and Sanchez A. (2008). Optimising Sewer System Rehabilitation Strategies between Flooding, Overflow Emissions and Investment Costs. *11th International Conference on Urban Drainage*, Edinburgh, Scotland.

Vojinović Z., Solomatine D. P. and Price R. K. (2006). Dynamic Least-Cost Optimisation of Wastewater System Remedial Works Requirements. *Water Science and Technology*, **54**(6-7), 467–475, IWA Publishing.

Watanabe K. and Kurihara T. (1993). Practical Simulation Method of Surcharged Flow Using Pressure-Relaxation Effect in Manhole. In: J. Marsalek and H. C. Torno (ed.), *6th Int. Conf. on Urban Storm Drainage,* Vol. I, Niagara Falls, pp. 128–133.

Weiner J. F. (1991). The Empty Place: Poetry, Space and Being Among the Foi of Papua, New Guinea. Indiana University Press, Bloomington, IN.

White R. and Engelen G. (1994). Cellular dynamics and GIS: modelling spatial complexity. *Geographical Systems*, **1**, 237–253.

Willbern Y. (1984). Types and Levels of Public Morality, Public Administration Review, Vol. 44, No. 2 (Mar. – Apr., 1984), pp. 102–108, Blackwell Publishing.

Wisner B., Blaikie P., Cannon T. and Davis I. (2003). 2nd edn, At Risk: Natural Hazards, People's Vulnerability and Disasters. Routledge Taylor & Francis Group, London, New York.

Wisner B., Blaikie P., Cannon T. and Davis I. 2003. 2nd edn, At Risk: Natural Hazards, People's Vulnerability and Disasters. Routledge Taylor & Francis Group, London, New York, 471p.

WMO (2007). Economic Aspects of Integrated Flood Management. APFM Technical Document No. 5, Flood Management Policy Series, WMO, Geneva, Switzerland. http://www.wmo.int/

WMO (2007). Formulating basin flood management plan, a tool for integrated flood management. APFM technical document No.6, Flood management tools series, The associated programme on flood management.

Wolfram S. (1984). Cellular automata as models of complexity, *Nature*, **311**, 419–424.

WRc (1994). The Sewerage Rehabilitation Manual. WRc Engineering, Swindon.

Zadeh L. (1999). Fuzzy sets as a basis for a theory of possibility. *Fuzzy Sets and Systems*, **100**(1), 9–34.

Zahavi D. (2003). Husserl's Phenomenology. Stanford University Press, Stanford, CA.

Zonensein J. (2007). Índice de Risco de Cheia como Ferramenta de Gestão de Enchentes. Master of Science in Civil Engineering, COOPE, UFRJ, Rio de Janeiro.

Index

Note: t = Table, f = Figure

Ø

P

Q

R

S